The Ecology of Sulawesi

The Ecology of Indonesia Series
Volume IV

THE ECOLOGY OF INDONESIA SERIES

Volume IV: The Ecology of Sulawesi

Other titles in the Series
Volume I: The Ecology of Sumatra
Volume II: The Ecology of Java and Bali
Volume III: The Ecology of Kalimantan
Volume V: The Ecology of Nusa Tenggara and Maluku
Volume VI: The Ecology of Irian Jaya
Volume VII: The Ecology of the Indonesian Seas

Produced by
Environmental Management Development in
Indonesia Project, a cooperative project of the
Indonesian Ministry of the Environment
and
Dalhousie University, Halifax, Nova Scotia
under the sponsorship of the
Canadian International Development Agency

The Ecology of Sulawesi

Tony Whitten
Muslimin Mustafa
Gregory S. Henderson

PERIPLUS

First Edition © 1987 Gadjah Mada University Press
First Periplus Edition © 2002

Published by Periplus Editions (HK) Ltd.

ISBN 962-593-075-2

Publisher: Eric Oey
Typesetting and graphics: JWD Communications Ltd.
Copyediting: Marylouise Wiack

Distributors:

Indonesia:
PT Java Books Indonesia
Jl. Kelapa Gading Kirana
Blok A14 No. 17, Jakarta 14240

Japan
Tuttle Publishing
RK Bldg. 2nd Floor 2-13-10 Shimo-Meguro
Meguro-ku Tokyo 153 0064

Asia Pacific
Berkeley Books Pte Ltd
130 Joo Seng Road, #06-01/03
Olivine Building
Singapore 536983

USA
Tuttle Publishing
Distribution Center, Airport Industrial Park
364 Innovation Drive, North Clarendon, VT 05759-9436

Printed in Singapore

Table of Contents

Foreword to
the first edition (1987)

Indonesia, with its large population and vast and varied natural resources, must strive for economic development at the same time as protecting and enhancing the environment. Development must involve a harmonious relationship between Man and God, Man and his fellow man, and Man and Nature.

This book has been written as one important element in the incorporation of environmental thinking into development activities on Sulawesi. It provides important ecological information which will assist government planning agencies and project developments in including ecological considerations in development activities. *The Ecology of Sulawesi* is part of a series of books on the ecology of Indonesia. The first book in the series, *The Ecology of Sumatra*, was published in 1984 and is now in its third printing. A continuing demand for books in the series is evident.

The Ecology of Sulawesi has been written with the involvement of scientists from the Centre for Resource and Environmental Studies at Hasanuddin University, Ujung Pandang. The books in this *Ecology of* series serve not only to provide basic information on the environment. The process of preparation of the books also allows for training and technology transfer in environmental research between the authors and the junior scientists at the university environmental study centres.

We hope that scientists will be encouraged by this book to pursue further research on the ecological conditions in Sulawesi, and to apply the results of this continuing research to solutions to the environmental challenges posed by development in Sulawesi.

Emil Salim
Minister of State for Population and Environment
Republic of Indonesia

Acknowledgements
to the first edition (1987)

This book was produced within the Environmental Management Development in Indonesia (EMDI) Project, implemented by the Indonesian Ministry of State for Population and Environment (KLH) and the School for Resource and Environmental Studies (SRES) at Dalhousie University, Halifax, Nova Scotia, Canada. The project is funded by the Canadian International Development Agency (CIDA). The commitment to the publication of *The Ecology of Sulawesi* of those people involved in the administration of the project is much appreciated and special thanks are due to Koesnadi Hardjasoemantri, Sjafran Sjamsuddin (KLH, Jakarta), Arthur Hanson, Geoffrey Hainsworth, and George Greene (SRES, Halifax). The book has been born out of the experience of producing *The Ecology of Sumatra* which was financed by the United Nations Development Programme (UNDP) between 1982 and 1984.

The Rectors of the four state universities on Sulawesi namely Fachruddin, (Hasanuddin University, Ujung Pandang), Mattulada (Tadulako University, Palu), W.J. Waworoento (Sam Ratulangi University, Manado) and Eddy Agussalim Mokodompit (Haluoleo University, Kendari) and their staffs have cooperated by supplying information and have supported the effort to further the ecological understanding of Sulawesi.

The entire English text has been critically reviewed and examined by Jane Whitten (Bogor), Anoma Santiapillai (Bogor) and Gembong Tjitrosoepomo (Yogyakarta), the third of whom made the translation for the Indonesian version, *Ekologi Sulawesi*. Major portions were reviewed by Peter Bellwood (Canberra), Chris Bennett (Manado), David Bulbeck (Ujung Pandang), S.C. Chin (Kuala Lumpur), James Davie (Rockhampton), David Dudgeon (Hong Kong), Ian Glover (London), Atmadja Hardjamulia (Bogor), Duncan Parish (Kuala Lumpur), Nicholas Polunin (Port Moresby), and Nengah Wirawan (Ujung Pandang). Great thanks are due to these people for their constructive criticisms, but it must be stressed that the mistakes remaining in the text are entirely the responsibility of the authors.

The Executive Director of Gadjah Mada University Press, H.J. Koesoemanto†, has always been willing to advise, listen, and cooperate and we owe him a considerable debt of gratitude.

We also express gratitude to the Royal Entomological Society of London for permission to visit their Project Wallace research site in the Bogani Nani Wartabone (formerly Dumoga-Bone) National Park, Bolaang Mongondow, in 1985.

In addition, many people have provided considerable assistance (often more than they realize) by sending reports, papers, unpublished information and other forms of information, by advising, cajoling, identifying specimens, by making helpful suggestions, or by helping the authors in the field. They are: Abdul Rachman Abudi, Amran Achmad (Ujung Pandang), Mohamad Amir (Bogor), Dick R. Askew (Manchester), Michael Audley-Charles (London), Andy Austin (Adelaide), Max van Balgooy (Leiden), Henry S. Barlow (Kuala Lumpur), Wim Bergmans (Amsterdam), David Bishop (Chesterhill and NSW), Roger Blackith (Dublin), Peter Bloks (Leiden), Boeadi (Bogor), Hans A.J. in den Bosch (Leiden), W. Boudewijn (Ujung Pandang), Martin Brendell (London), Francois Brouquisse (Tolouse), Sean Brown (London), Arie Budiman (Bogor), Elisa Bung'alo (Lumuk), Burhan (Ujung, Pandang), Roger Butlin (Norwich), Diane Calabrese (Carlisle, Penn.), Ray Catchpole (Kendari), Ailsa Clark (London), Lynn Clayton (Oxford), Nigel Collar (Cambridge), Mark Collins (Cambridge), David Coyle (Watampone), Wempy Dahong (Ujung Pandang), Rokhmin Dahuri (Bogor), Sengli Damanik (Medan), Anthony Davis (Nagercoil), Louis Deharveng (Tolouse), Rene Dekker (Dumoga), Peter Dinwiddie (Swindon), Henry Disney (Cambridge), Machfudz Djajasasmita (Bogor), John Dransfield (Kew), Julian Dring (London), Jans Duffels (Amsterdam), Lance Durden (Nashville), Rusly Durio (Ujung Pandang), Siegfied Eck (Dresden), Chris Escott (Saskatoon), C.H. Fernando, (Waterloo), Theodore Flemming (Coral Gables, FL), Ben Gaskell (London), S.S. Gasong (Palu), E. Gittenberger (Leiden), Emily Glover (London), Michael Green (Cambridge), Penny Greenslade (Canberra), Steven Greenwood (Oxford), Colin Groves (Canberra), James Guiry (Soroako), Surastopo Hadisumarno (Yogyakarta), Gavin Hainsworth (Vancouver), Tony Harman (Canterbury), Hanna Hardjono (Ujung Pandang), A.M. Hashi (Palopo), Loky Herlambang and the staff of the Nusantara Diving Centre (Manado), John E. Hill (London), Bert Hoeksema (Leiden), Ian Hodkinson (Liverpool), Jeremy Hollowax (London), Derek Holmes (Jakarta), L.B. Holthuis (Leiden), Marinus Hoogmoed (Leiden), Geoff Hope (Canberra), Hans Huijbregts (Leiden), Jaffre (Palopo), Paula Jenkins (London), Clive Jermy (London), Ahdul Rachman Kadir (Ujung Pandang), John Katili (Jakarta), Peter Kevan (Guelph), Ashley Kirk-Spriggs (Cardiff), David Kistner (Chico), Roger Kitching (Armidale), Robby V.T. Ko (Bogor), Jan de Korte (Amsterdam), Maurice Kottelat (Courrendlin), Jan Krikken (Leiden), Jaroslav Klapste (Melbourne), Bill Knight (London), Andrew Lack (Swansea), David de Lauberfels (Syracuse), Philippe Le Clerc (Tolouse), Cecile Lomer (Bogor), C.H.C. Lyal (Auckland), Colin McCarthy (London), Ron and Patty McCullogh (Palu), Odilia Maessen (Halifax, N.S.), Ayub Mahmud (Malili), Syafii Manan (Bogor), Adrian, G. Marshall (Aberdeen). Joe Marshall (Washington), G.A. Matthews (Ascot), John Miksic (Yogakarta), Andrew Mitchell (London), Willem Moka (Ujung

Pandang), Robert Molenaar (Manado), Hans Moll (Leiden), Kate Monk (Safat), Barry P. Moore (Canberra), Evelyn Mundy (Bogor), Guy Musser (New York), Safiruddin Natsir (Ujung Pandang), Baharuddin Nurkin (Ujung Pandang), Grace O'Donovan (Dublin), Sharifuddin Andi Omar (Ujung Pandang), H.S. Padeato (Manado), S.W. Padiato (Luwuk), Raymondus Palete (Dumoga), J.L. Panelewan (Manado), P.E.A. Pangalila (Manado), Duncan Parish (Kuala Lumpur), Michael Pearce (London), Totok Prawitosari (Ujung Pandang), T. Racheli (Rome), Musaka Rachmat (Ujung Pandang), Anthony Reid (Canberra), Christopher Rees (York), Mien Rifai (Bogor), Willem Rodenburg (Koudekerk a/d Rijn), Frank G. Rozendaal (Bilthoven), Didi Rukmana (Ujung Pandang), A. Sabani (Palopo), Peter Sane (Kendari), Sjahril T. Selamat (Ujung Pandang), Victoria Selmier (San Francisco), Dan Sembel (Manado), Siahaan (Luwuk), Alison Skene (Bogor), the late C.G.G.J. van Steenis (Leiden), Nigel Stork (London), J.S.W.D. Subroto (Manado), N. Sulaiman (Jakarta), Mulyadi Susanto (Kendari), Stephen Sutton (Leeds), Antius Tolesa (Luwuk), Michael Tremble (Albuquerque), John Uttley (Durham), Charlotte Vermeulen (Dumoga), Ed de Vogel (Leiden), Michael Wade (Bogor), B. Wahyu (Soroako), Sarah Warren (Yale), Dick Watling (Suva), Chris Watts (Adelaide), Alice Wells (Adelaide), Chris Wemmer (Washington), Michael Walters (Tring), Alwyne Wheeler (London), Tim Whitmore (Oxford), Willem de Wilde (Leiden), Johanna Wilson (Bogor), Mike Wilson (London) and John Winter (Townsville).

The principal typist has been Lisbet[†] who has worked with great diligence, care and commitment. The majority of the figures were drawn by Syamsul, to whom our considerable thanks are due.

Finally, and mainly, we thank Almighty God for His grace and caring love shown throughout the preparation of this book, without which it would never have been completed.

[†]Deceased

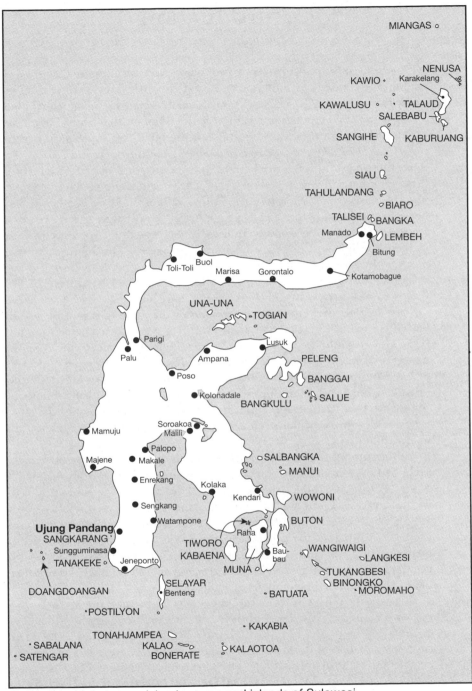

Figure I.1. Main towns, island groups and islands of Sulawesi.

Introduction

We arrived in Sulawesi in 1991, our bags overstuffed with the gear of ecologists—binoculars, notebooks, two dictionaries and one extremely heavy reference book, *The Ecology of Sulawesi*. *The Ecology of Sulawesi* provided a summary of the current knowledge of the island at that time. There was no argument about the thoroughness of the book; if a topic wasn't mentioned in *The Ecology of Sulawesi*, it probably hadn't been studied. At the time, there wasn't even a field guide to the birds, and our first identifications of endemic mynas relied on a small, black and white sketch in chapter one. Because few other references were available, *The Ecology of Sulawesi* became our bible. Over a four-year period of constant use by visitors, students, and biologists—not the least ourselves—our copy developed a broken binding and dog-eared, underlined pages—the highest compliment to the authors.

Much has changed in the past 13 years since the first edition of *The Ecology of Sulawesi* was published. There have been discoveries that we never imagined, and losses beyond what only the worst of pessimists would have dreamed in 1987. Unfortunately, funding is not yet available for a full revision that would incorporate all these changes into a new edition of *The Ecology of Sulawesi*. This second edition of *The Ecology of Sulawesi*, however, will still be invaluable in any library. The updated bibliography should provide readers with numerous sources for more recent information about the island. Our first edition copy (now rebound) never stays on the shelf long and remains a primary reference when we, our staff and students are writing manuscripts.

Surely the most dramatic discovery of the last 13 years was the completely unexpected find of a coelocanth fish in a Manado market (Erdmann et al. 1998). Indeed, it was heralded by some as 'the zoological find of the century'. Only one other population of this 'living fossil' (Forey 1998), located 1,000 kms off the coast of east Africa, has ever been discovered. Unfortunately, the excitement of discovery was marred by politics of taxonomy and currently controversy rages over whether or not the Manado population is a separate species. But the arguments certainly do not take away from the sheer excitement of such a discovery.

Perhaps the most satisfying find, especially for the authors of *The Ecology of Sulawesi*, was the rediscovery of the endemic Cerulean paradise flycatcher on Sulawesi's satellite island of Sangihe (Wardill and Riley 1999). The caption of plate 4 of the first edition states that the Cerulean paradise flycatcher bird is probably extinct. The flycatcher became become a poster bird for extinction when plate 4 appeared on the cover of the first issue of the journal for *Conservation Biology* and the species was declared 'almost

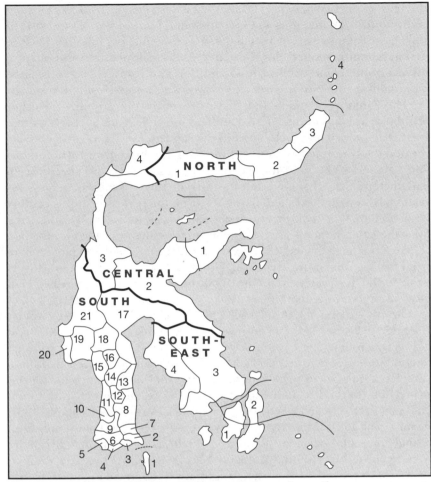

Figure I.2. Division of Sulawesi into provinces and counties.

NORTH
1 - Gorontalo
2 - Bolaang Mongondow
3 - Minahasa
4 - Sangihe-Talaud

CENTRAL
1 - Luwuk-Banggai
2 - Poso
3 - Donggala
4 - Toli-Toli

SOUTH
1 - Salayar
2 - Bulukumba
3 - Bontaeng
4 - Jeneponto
5 - Takalar
6 - Gowa
7 - Sinjai
8 - Bone
9 - Maros
10 - Pangkajene Kep.
11 - Barru
12 - Soppeng
13 - Wajo
14 - Sinderng Rappang

15 - Pinrang
16 - Enrekang
17 - Luwu
18 - Tana Toraja
19 - Polewati Mamasa
20 - Majene
21 - Mamuju

SOUTHEAST
1 - Buton
2 - Muna
3 - Kendari
4 - Kolaka

certainly' extinct (Whitten et al. 1987). First described in 1873, the bird was known only from one museum specimen and was last sighted in 1981. In 1998 however, members of a small non-governmental organization (NGO), Action Sampiri, spotted the flycatcher while conducting ornithological surveys of the island. Although uncommon and restricted to an isolated mountaintop, the bird survives and provides a rare spark of optimism.

The Sangihe-Talaud Islands have proven fertile grounds for other ornithological discoveries. Frank Lambert and colleagues uncovered three species new to science, including two rails and an owl (Lambert 1998a, b), verified the continued existence of the Sangihe shrike-thrush (Rozendaal and Lambert 1999), and uncovered two distinct scops owl species from Sangihe and Siau (Lambert and Rasmussen 1998). Similarly, surveys for mammals on Sangihe-Talaud have produced the first records this century of an endemic rat, a distinct form of bear cuscus, and a the Talaud flying fox—the latter found in a small restaurant, presumably headed for the pot.

Back on the mainland, biologists and other modern-day explorers have been making similar discoveries, including taxa as diverse as fungi (Rogers et al. 1987), freshwater crabs (Ng 1993), turtles (McCord et al. 1995; Platt et al. in press), rodents (Musser 1991; Musser and Holden 1991), snakes and lizards (Iskandar 1999a; Lazell 1987; Bosch and Ineich 1994), insects (Conde 1992a, b, 1994; van Tol 1987, 1994; Monk and Butlin 1990) and plants (Dransfield 1989, 1992) as well as co-evolved relationships between ants and trees (Maschwitz and Fiala 1995). Parks and protected areas have revealed a number of secrets, underscoring their importance for species protection. In the late 1980s and early 1990s, expeditions to Lore Lindu National Park revealed two new tarsier species (Musser and Dagosto 1987; Nemitz et al. 1991). More recently, a *Ninox* owl, previously known only from a single museum specimen (Rasmussen 1999), and the Matinan flycatcher were recorded in Gunung Ambang Nature Reserve (Lee and Riley in press). Most recently, a Heinrich's nightjar was observed in Panua Nature Reserve and unexpectedly large populations of anoa and babirusa have been found in Bogani Nani Wartabone (formerly known as Dumoga-Bone) National Park (Lee pers. comm.). The Sulawesi palm civet, thought to be tumbling towards the vortex of extinction, was photographed by a team from The Nature Conservancy in the forests of Lori Lindu National Park (*Indonesian Observer* 2000).

In addition to species discoveries and rediscoveries, the 1990s experienced an explosion of medium- and long-term research projects. These projects focused primarily on large vertebrates, especially Sulawesi endemics. Sulawesi's seven species of endemic macaques received the lion's share of attention with research covering taxonomy (Bynum 1999; Evans et al. 1999; Watanabe et al. 1991), conservation status (Bynum 1999; O'Brien and Kinnaird 1996), ecology (Lee 1997; Matsumura 1993, 1996, 1998; O'Brien and Kinnaird 1997; Rosenbaum et al. 1998), and the dramatic scale of hunting (Alvard 2000; Lee 2000a, b; O'Brien and Kinnaird 1996, 2000). The previously unknown natural history of the rare babirusa was unfolded by Clayton (1996) and Clayton and MacDonald (1999) during her intensive studies at a salt lick in Paguyaman forest, North Sulawesi. First-ever studies were published on the ecology of bear cuscus (Dwiyaherni et al. 1999), endemic forest kingfishers (Sunarto 1999), and hornbills (O'Brien 1996; Kinnaird and O'Brien 1993, 1999a; Kinnaird et al. 1996; Suryadi et al. 1994, 1996). Dekker and colleagues (Argeloo 1994; Dekker 1990; Dekker et al. in press) provided the first detailed information on the status and distribution of Sulawesi's oddest avian endemic—the Maleo. Gursky (1994, 1998) has greatly improved our knowledge of tarsier behavior and ecology, Alvard and Winarni (1999) assessed bird communities in relation to habitat disturbance in Morowali Nature Reserve, and Bynum (1999) developed new habitat monitoring techniques for use in Lore Lindu National Park.

Sadly, all this ground-breaking research and discovery is occurring against a background of destruction. Since the publication of the first edition of *The Ecology of Sulawesi*, deforestation has continued unchecked. In spite of the efforts of international donor agencies (e.g., United States Agency for International Development [USAID], the World Bank, United Nations Development Program [UNDP], Asian Development Bank [ADB]) and the dedication of many NGOs, habitat loss on Sulawesi has actually accelerated. In a shocking report by Derek Holmes (2000) for the World Bank and the Ministry of Forestry and Estate Crops, Holmes shows that between 1985 and 1997 Sulawesi lost 20% of its natural forest cover. This figure does not account for forest quality, and land classified as forest could contain logged or burned forest, areas of reduced value for biodiversity conservation. Most startling, Holmes states that lowland dry forest, the most valuable type of Indonesian forest for logging and biodiversity conservation is 'essentially defunct as a viable resource in Sulawesi'. This should have been a wake-up call to the Indonesian government but instead the list of causes of forest destruction grew longer and more complex.

The leading cause of deforestation has been large-scale logging by a few conglomerates and the use of timber concessions for political patronage—huge tracts of state-owned forests were held in the 1980s and 1990s by family and business associates of former president Suharto (Richardson 2000). Ironically, central government policy has encouraged deforestation. Logging concessions, sugarcane plantations, and increasingly, oil palm plantations are not regulated and as Walton and Holmes (2000) state, 'perverse incentives exist that make it more lucrative to clear forested land for plantations than to plant open and unproductive land'. As timber exports shifted from logs to plywood and then to pulp and paper, more log processing plants were established until processing capacity exceeded sustainable yields. Finally, poor enforcement has allowed excesses to go unchecked. Illegal logging has become rampant, even in national parks, on a scale that exceeds the volume of legal logging. Authorities look the other way while the government loses tax revenue at the rate of roughly $500 million each year (Walton and Holmes 2000).

An increasingly important agent of deforestation on Sulawesi is the influx of new migrants into forested areas. In the late 1990s, as Indonesia plunged into economic crisis, unemployed urbanites returned to the countryside in need of land. Long-term residents also cleared new land to expand production of low-value vegetable crops and high-value cash crops (e.g., coffee, cocoa). As we write, new immigrants are arriving from areas of ethnic strife in the Moluccas. Thousands of displaced and land-hungry refugees are arriving daily; many observers report that there are now more than 200,000 refugees in North Sulawesi alone. Unless the problem is resolved soon, these people will resettle and require more land to support themselves and their families.

Forests that are not being cut are not necessarily safe. One of the most alarming discoveries of the 1990s is the degree to which people of Sulawesi are destroying their wildlife heritage (Lee 1997; O'Brien and Kinnaird 1996, 2000). Researchers showed that a number of endemic mammals, including the babirusa, anoa, and crested black macaque were on the edge of extinction as a result of over-hunting for subsistence and commercial bushmeat trade. Further evidence came from modeling exercises by Clayton and colleagues (1997) who showed that under current practices, babirusa populations will be reduced to very small and probably nonviable population numbers. Lee (1997, 2000a, b, c) conducted an extensive analysis of commercial and subsistence hunting and concluded that hunting for markets was having the greatest impact on local populations of wildlife, and Alvard (2000) showed a lesser, but still considerable, impact on wildlife through subsistence hunting by the Wana people of Morowali Nature Reserve. Overcollecting of sulphur-crested cockatoos for the pet trade has almost eliminated the species from Sulawesi except for a few remnant birds near Palu (PHPA/LIPI/BirdLife International-IP 1998). Finally, the breakdown of traditional customs and harvest regimes has played a major role in the on-going decline of the Maleo. Although more breeding sites have been discovered since the classic work of Dekker (1990), most are in danger from over-collecting and many sites are no longer active.

Sulawesi has a diverse system of parks and nature reserves that were set up to provide sanctuary for the island's biodiversity, free from hunting, and habitat destruction. The island boasts three national parks, including Lore Lindu National Park, a World Heritage Site, two marine parks, nineteen nature reserves, and an assortment of tourist parks and wildlife refuges. These areas cover most major habitats, contain populations of most endemic species, and several are large—all good qualities for reserves design. But in today's climate, the subject of reserve design is largely one of academic debate. Sulawesi's parks are not safe and although most of the protected areas have staff, the areas function as little more than paper parks. The question of what percentage of species will become extinct in 50, 500 or 5,000 years' time is not of great relevance when it is by no means certain how much of its reserves will be intact in even 25 years time. In Sulawesi today the major priority is simply to maintain the integrity of reserves against legal and illegal forms of habitat disturbance and resource exploitation.

Hunting and habitat destruction continue unchecked throughout Sulawesi's protected area system. Development projects such as road building, new transmigration schemes and dam development are being permitted in the parks, and in most cases supported by local government officials. Mining, both illegal and legal, continue to plague areas such as Bogani Nani Wartabone National Park. As we write, more than 4,000 illegal miners scour the hillsides of the park in search of gold, poisoning themselves and the environment with the mercury they use to separate the precious metal (*Sydney Morning Herald* 2000). An additional 5,000 illegal miners—supported by the military, local police and government officials—have invaded the Perth-based Aurora Gold's claim on the outskirts of Manado, similarly poisoning themselves and the surrounding waters of Bunaken Marine National Park (*Sydney Morning Herald* 2000). Logging operations encroach on reserve boundaries in many areas. Fires set by plantation companies and local people ravaged reserve land in 1994-95 and again in 1997-98. Illegal fishing nets, set in protected waters of Tangkoko drowned thousands of marine mammals and sea turtles, and killed manta rays and other pelagic fish. In spite of legal protection, hunters, rattan collectors and other forest product collectors roam freely throughout many reserves in North and Central Sulawesi, undaunted by apathetic, or in some cases, participating enforcement officials.

All of the abuses described have been exacerbated by the prolonged economic crisis, the collapse of central authority and the proposal to decentralize natural resource management. In the confusion and uncertainty, local natural resource managers are more reluctant than ever to take action. As local people move into protected areas and even logging concessions, they find that there is little resistance by the government. In addition, military and land speculators are getting into the act on the belief that if the land is cleared, they will be able to keep it. In Lore Lindu National Park, park management turned over approximately 2,000 ha of park land to local people for illegal coffee and cacao plantations, rather than confront the issue (*Indonesian Observer* 2000). As management breaks down, guards feel less responsible, are demoralized and stop working. In Tangkoko Nature Reserve in 1997, guards watched, but failed to report the forest fires that damaged more than half of the reserve.

Although the problems of Sulawesi's forests, parks and protected areas are severe, they are not insurmountable. Today, a growing number of organizations and individuals are committed to the conservation and rational management of Sulawesi's natural resources. The Natural Resource Management and Coastal Resource Management Programs of USAID have been working with local stakeholders including the government, local NGOs, and communities to develop policy and implement plans for sustainable use of marine and terrestrial resources. They were responsible for developing the management plan and basic infrastructure for Bunaken Marine National Park, one of Indonesia's most exquisite and popular marine sanctuaries. Conservation International and Biological Conservation Network have been working on the Togian Islands developing community-based management of coral reef tourism and protection of local forests. Sahabat Morowali, an Indonesian NGO, focuses on the conservation issues of Morowali National Park. They are especially concerned with involving the resident Wana people in management decisions concerning the park and have developed ecotourism ventures. The Nature Conservancy concentrates its efforts in Lore Lindu National Park. They have provided resources and training to the Directorate of Nature Protection and Conservation (PKA) staff, and are trying to develop alternative, environmentally friendly income sources aimed at reducing pressure on the park. Group Sampiri, a local NGO based on the work of Action Sampiri, is carrying out a rural awareness project on Sangihe-Talaud. Our organization, the Wildlife Conservation Society, has been active in Sulawesi since 1991 carrying out wildlife research and training of conservationists in protected areas such as Tangkoko, Gunung Ambang, and Bogani Nani Wartabone in North Sulawesi. The Wildlife Conservation Society conducts in-depth ecological research and baseline surveys on a variety of endangered species, conducts conservation education campaigns in North Sulawesi and trains park guards and provincial officials in conservation enforcement. At present, we are carrying out an island-wide field survey of protected areas and wildlife, and combining these surveys with training and management assistance.

In the final analysis, all the hard work of foreign and domestic conservation organizations will pale unless Indonesia as a nation—including its government, military, police forces, local NGOs, universities and communities—begin to show a serious commitment to conservation on Sulawesi. Unless the talk stops and the action begins, we will lose everything that made Sulawesi a special place in Alfred Russell Wallace's heart, as well as our own. And the next edition of *The Ecology of Sulawesi* could very easily become a history book.

M.F. Kinnaird and T.G. O'Brien
September, 2000

REFERENCES

Alvard, M.S. 2000. The impact of traditional subsistence hunting and trapping on prey populations: Data from the Wana of upland Central Sulawesi, Indonesia. In *Hunting for Sustainability in Tropical Forests*, eds. J.G. Robinson and E.L. Bennett, 214–230. New York: Columbia University Press.

Alvard, M.S. and Winarni, N.L. 1999. Avian biodiversity in Morowali Nature Reserve, Central Sulawesi, Indonesia and the impact of human subsistence activities. *Tropical Biodiversity* 6(1&2): 59–74.

Andrew, P. 1992. *The Birds of Indonesia: A checklist (Peter's Sequence)*. Indonesian Ornithological Society: Jakarta.

Argeloo, M. 1994. The maleo *Macrocephalon maleo*: New information on the distribution and status of Sulawesi's endemic megapode. *Bird Conserv. Int.* 4: 383–393.

Baker, G. 1997a. Birds of the Spice Islands. Mollucan Megapode conservation project. An expedition from the University of Sussex in association with the Indonesian Department of Forestry (PHPA). November 1996–January 1997. Final Report: 1–43. Rep., Brighton: University of Sussex.

Baker, G. 1997b. Birds of the Spice Islands: Moluccan megapodes in Halmahera. *Megapode Newsl.* 11(1): 3–4.

Baker, G. 1997c. The Moluccan megapode: Bird of the Spice Islands. *Wld Pheasant Assoc. J.* 53: 21–24.

Baker, G.C. 1999. Temporal and spatial patterns of laying in the Moluccan Megapode *Eulipoa wallacei* (G.R. Gray). In *Proceedings of the Third International Megapode Symposium, Nhill, Australia, December 1997*, eds. R.W.R.J. Dekker, D.N. Jones and J. Benshemesh, 327: 53–59. Leiden: Zoologische Verhandelingen.

Baker, G.C. and Butchart, S.H.M., in press. (August/September 2000) Threats to the maleo and solutions for its conservation. *Oryx*.

Baker, G.C. and Dekker, R.W.R.J. 2000. Lunar synchrony in the reproduction on the Mollucan Megapode *Megapodius wallacei*. *Ibis* 142: 382–388.

Bergmans, W. and Rozendaal, F.G. 1988. Notes on collections of fruit bats from Sulawesi, Indonesia, And some off-lying islands (Mammalia: Megachiroptera). *Zool.*

Verh. 248: 1–74.

Bororing, R.F., Hunowu, I., Hunowu, Y., Maneasa, E., Mole, J., Nusalawo, M.H., Talangamin, F.S. and Wangko, M.F. 2000. Birds of the Manembonembo Nature Reserve, North Sulawesi, Indonesia. *Kukila* 11: 58–72.

Bosch, H.A.J. in den 1985. *Snakes of Sulawesi: checklist, key and additional biogeographical remarks*. Zoologische Verhandelingen no. 217. Leiden: Rijksmuseum van Natuurlijke Historie.

Bosch, H.A.J. in den and Ineich, I. 1994. The Typhlopidae of Sulawesi: A review with the description of a new genus and a new species. *J. Herpetol.* 28: 206–217.

Breed, W.G. and Musser, G.G. 1991. *Sulawesi and Philippine rodents (Muridae): A survey of spermatozoal morphology and its significance for phylogenetic inference*. American Museum novitates no. 3003. New York: American Museum of Natural History.

Bridle, J., Garn, A-K, Monk, K.A., and Butlin, R.K. Submitted. Speciation in Chitaura grasshoppers on the island of Sulawesi: Colour patterns, morphology, and population history.

Brown, R.M. and Iskandar, D.T. 2000. Nest site selection, larval hatching and advertisement calls of *Rana arathooni* from Southwestern Sulawesi (Celebes) Island, Indonesia. *J. Herpetol.* 34(3): 403–413.

Brown, R.M., Supriatna, J. and Ota, H. 2000. A new *Luperosaurus* from Sulawesi, with notes on the status of *Luperosaurus serraticaudus*. *Copeia*

Burger, J. and Gochfeld, M. 1997. Heavy metal and selenium concentrations in feathers of egrets from Bali and Sulawesi, Indonesia. *Archives of Environmental Contamination and Toxicology*. 32: 217–221.

Butchart, S.H.M. and Baker, G.C. 2000. Priority sites for conservation of maleos *Macrocephalon maleo* in Central Sulawesi. *Biological Conservation* 94: 79–91.

Butlin, R.K. and Monk, K.A. 1990. Catantopine grasshoppers in Sulawesi. In *Insects and the rain forests of South East Asia (Wallacea)*, eds. J.D. Holloway and W.J. Knight, 89–93. London: The Royal Entomological Society of London.

Butlin, R.K. Walton, C., Monk, K.A. and Bridle, J.R. 1998. Biogeography of Sulawesi grasshoppers, genus Chitaura, using DNA sequence data. In *Biogeography and Geological Evolution of S.E. Asia*, eds. R. Hall and J.D. Holloway, 355–359. Leiden: Backhuys Publishers.

Bynum, D.Z. 1999. Assessment and Monitoring of Anthropogenic Disturbance in Lore Lindu National Park, Central Sulawesi, Indonesia. *Tropical Biodiversity* 6(1&2): 43–58.

Bynum, E.L. 1999. Biogeography and evolution of Sulawesi macaques. *Tropical Biodiversity* 6(1&2): 19–36.

Bynum, E.L., Kohlhaas, A.K. and Pramono, A.H. 1999. Conservation status of Sulawesi macaques. *Tropical Biodiversity* 6(1&2): 123–144.

Casson, D. and Hodkinson, I.D. 1991. The Hemiptera (Insecta) communities of tropical rain forest in Sulawesi, Indonesia. *Zool. J. Linn. Soc.* 102: 253–276.

Chen, P. and Nieser, N. 1992. Gerridae mainly from Sulawesi and Pulau Buton, Indonesia. *Tijdschr. Entomol.* 135: 145–162.

Clayton, L. 1996. Conservation Biology of the Babirusa, Babyrousa babyrussa, in Sulawesi, Indonesia. Unpubl. Ph.D. dissertation, Wolfson College, University of Oxford.

Clayton, L. and MacDonald, D. 1999. Social organization of the babirusa (*Babyrousa babyrussa*) and their use of salt licks in Sulawesi, Indonesia. *Journal of Mammalogy* 80(4): 1147–1157.

Clayton, L., Keeling, M. and Milner-Gulland, E.J. 1997. Bringing home the bacon: A spatial model of wild pig hunting in Sulawesi, Indonesia. *Ecol. Applications* 7: 642–652.

Coates, B.J., Bishop, K.D. and Gardner, D. 1997. *A guide to the birds of Wallacea: Sulawesi, the Moluccas, and Lesser Sunda Islands, Indonesia.* Alderley: Dove Publications.

Collette, B.B. 1995. *Tondanichthys kottelati*, a new genus and species of freshwater halfbeak (Teleostei: Hemiramphidae) from Sulawesi. *Ichthyol. Explor. Freshwaters* 6: 171–174.

Conde, B. 1992a. Campodeidae from caves in Celebes (Insecta: Diplura). *Mem. Biospeleol.* 19: 155–158.

Conde, B. 1992b. Endogean and cave-dwelling palpigrades from Thailand and Sulawesi. 1. *Rev. Suisse Zool.* 99: 655–672.

Conde, B. 1994. Endogean and cave dwelling palpigrades from Thailand and Sulawesi. 2. *Rev. Suisse Zool.* 101: 233–263.

Coode, M.J.E. 1996. Elaeocarpus for Flora Malesiana - *E. kraengensis* and ten new

species from Sulawesi. *Kew Bull.* 50: 267–294.

Corbett, G.B. and Hill, J.E. 1992. *The mammals of the Indomalayan Region: A systematic review.* Oxford: Oxford University Press.

Daws, G. and Fujita, M. 1999. *Archipelago: The Islands of Indonesia.* Berkeley, CA: University of California Press.

Deharveng, L. and Leclerc, P. 1989. Studies on the caverniculous fauna of Southeast Asia. *Mem. Biospeleol.* 16: 91–110.

Dekker, R.W.R.J. 1990. The distribution and status of nesting grounds of the maleo *Macrocephalon maleo* in Sulawesi, Indonesia. *Biol. Conserv.* 51: 139–150.

Dekker, R.W.R.J., Fuller, R.A. and Baker, G.C., eds., in press. Megapodes: Status survey and conservation action plan 2000-2004. Gland, Switzerland and Cambridge: IUCN.

Deuve, T. 1990. *Mateuis troglobioticus* n. g., n. sp. a caverniculous beetle from the Sulawesi karsts (Harpalidae: Pterostichinae: Abacetini). *Rev. Fr. Entomol. (Nouv. Ser.)* 12: 95–99.

Disney, R.H.L. 1987. A most remarkable new genus of scuttle fly (Diptera: Phoridae) from Sulawesi, Indonesia. *Syst. Entomol.* 12: 29–32.

Disney, R.H.L. 1990. A striking new species of *Megasella* (Diptera: Phoridae) from Sulawesi with re-evaluation of related genera. *Entomol. Fenn.* 1: 25–31.

Dransfield, S. 1989. *Sphaerobambos*: New genus of bamboo (Graminae: Bambusoideae) from Malesia. *Kew Bull.* 44: 425–434.

Dransfield, S. 1992. A new species of *Racemobambos* (Graminae: Bambusoideae) from Sulawesi with notes on generic delimitation. *Kew Bull.* 47: 707–711.

Dudgeon, D. 1999a. The future now: Prospects for the conservation of riverine biodiversity in Asia. *Aquatic Conservation* 9: 497–501.

Dudgeon, D. 1999b. *Tropical Asian streams: Zoobenthos, ecology and conservation.* Hong Kong University Press. 830 pp.

Dudgeon, D. 2000. Riverine biodiversity in Asia: A challenge for conservation biology. *Hydrobiologia* 418: 1–13.

Dudgeon, D. 2000. Riverine wetlands and biodiversity conservation in tropical Asia. In *Biodiversity in Wetlands: Assessment, function and conservation*, eds. B. Gopal, W.J. Junk and J.A. Davis, 35–60. Leiden: Backhuys Publishers.

Duffels, J.P. 1990. *Dilobopyga janstocki* n. sp., a

cicada endemic to Sulawesi (Homoptera: Cicadiade). *Bijdr. Dierk.* 60: 323–326.

Durden, L.A. 1986. The reinfestation of Forest Rats (*Maxomys musschenbroekii*). *J. Trop. Ecol.* 2(3): 283–286.

Durden, L. and Page, B.F. 1991. Ectoparasites of commensal rodents in Sulawesi Utara, Indonesia, with notes on species of medical importance. *Med. Vet. Entomol.* 5: 1–8.

Durden, L. and Musser, G.G. 1991. A new species of sucking louse (Insecta: Anoplura) from a montane forest rat in central Sulawesi and a preliminary interpretation of the sucking louse fauna of Sulawesi. *Am. Mus. Novitates* 3008: 1–10.

Dwiyahreni, A.A., Kinnaird, M.F., O'Brien, T.G., Supriatna, J. and Andayani, N. 1999. Diet and activity of the Bear Cuscus, *Ailurops ursinus*, in North Sulawesi, Indonesia. *J. of Mammalogy* 80: 905–912.

Erdmann, A. 1999a. An assesment of the conservation.status of the Indonesian coelacanth, *Latimera menadoensis. Wallacean Conservation Bulletin* 1: 6–8.

Erdmann, A. 1999b. Coelacanth: 400 million years and counting. *Asian Diver* 8: 29–32.

Erdmann, M.V., Caldwell, R.L. and Moosa, M.K.K. 1998. Indonesian 'King of the Sea' discovered. *Nature* 395: 335.

Erdmann, M.V., Caldwell, R.L., Jewett, S.L. and Tjakrawidjaja, A. 1999. The second recorded living coelacanth from North Sulawesi. *Env. Biol. Fishes* 54: 445–451.

Erftemeijer, P.L.A. 1993. *Factors limiting growth and production of tropical seagrass: Nutrient dynamics in Indonesian seagrass beds.* Katholieke Universiteit Nijmegen.

Erftemeijer, P.L.A. and Middelburg, J.J. 1995. Mass balance constraints on nutrient cycling in tropical seagrass beds. *Aquat. Bot.* 50: 21–36.

Erftemeijer, P.L.A., Djunarlin, and Moka, W. 1993. Stomach content analysis of a dugong *Dugong dugon* from South Sulawesi, Indonesia. *Aust. J. Mar. Freshwater Res.* 44: 229–233.

Evans, B.J., Morales, J.C., Supriatna J. and Melnick, D.J. 1999. Origin of the Sulawesi macaques (Cercopithecidae: Macaca) as suggested by mitochondrial DNA phylogeny. *Biol. J. Linn. Soc.* 66(4): 539–560.

Forey, P. 1998. A home away from home for coelacanths. *Nature* 395: 319.

Fujita, K. and Watanabe, K. 1995. Visual preference for closely related species by Sulawesi macaques. *Am. J. Primatol.* 37(3): 253–261.

Fujita, K., Watanabe, K., Widarto, T.H. and

Suryobroto, B. 1997. Discrimination of macaques by macaques: the case of Sulawesi species. *Primates* 38(3): 233–245.

Gay, H. 1991. Ant-houses in the fern genus *Lecanopteris*: The rhizome morphology and architecture of *Lecanopteris sarcopus* and *Lecanopteris darnaedi. Bot. J. Linn. Soc.* 106: 199–208.

Gay, H., Hennipman, E., Huxley, C.R. and Parrott, F.J.E. 1993. The taxonomy, distribution and ecology of the epiphytic Malesian ant-fern *Lecanopteris* (Polypodiaceae). *Grdns' Bull. S'pore* 45: 293–335.

Goff, M.L. and Durden, L.A. 1987. A new species and new records of chiggers (Acari: Trombiculidae) from North Sulawesi, Indonesia. *Int. J. Acarol.* 13(3): 209–211.

Goff, M.L., Durden, L.A. and Whitaker, Jr., J.O. 1986. A new species of *Schoengastia* (Acari: Trombiculidae) from mammals collected in Sulawesi, Indonesia. *Int. J. Acarol.* 12(2): 91–93.

Gursky, S.L. 1994. Infant care in the spectral tarsier (*Tarsius spectrum*) in Sulawesi, Indonesia. *Int. J. Primatol.* 15(6): 843–853.

Gursky, S.L. 1998. Conservation status of the spectral tarsier *Tarsius spectrum*: Population density and home range size. *Folia Primatol.* 69: 191–203.

Hamada, Y., Watanabe, T., Takenaka, O., Suryobroto, B. and Kawamoto, Y. 1988. Morphological studies on the Sulawesi macaques. 1. Phyletic analysis of body color. *Primates* 29: 65–80.

Hasegawa, H. and Mangali, A. 1996. Two new nematode species of *Bunomystrongylus* n. gen. (Trichostrongylina: Heligmonellidae) collected from *Bunomys* spp. (Rodentia: Muridae) of Sulawesi, Indonesia. *J. Parasitol.* 82(6): 998–1004.

Heppner, J.B. 1989. Lepidoptera diversity in North Sulawesi, Indonesia. *Orient. Insects* 23: 349–364.

Hill, J.E. 1988. A record of Rhinolophus arcuatus (Peters, 1871) (Chiroptera: Rhinolophidae) from Sulawesi. *Mammalia* 52(4): 588–589.

Hill, J.E. 1991. Bats (Mammalia: Chiroptera) from the Togian Islands, Sulawesi, Indonesia. *Bull. Am. Mus. Nat. Hist.* 206: 168–175.

Hill, J.E., Robinson, M.F. and Boeadi 1990. Records of Theobald's tomb bat, *Taphozous theobaldi* Dobson, 1872 (Chiroptera: Emballonuridae) from Borneo and Sulawesi. *Mammalia* 54(2): 314–315.

Hill, N.P., and Bishop, K.D. 1999. Possible winter quarters of the Aleutian tern? *Wilson*

Bull. 111(4): 559–560.

Hodkinson, I. and Casson, D.S. 1987. A survey of food plant utilization by Hemiptera in the understorey of primary lowland rain forest in Sulawesi, Indonesia. *J. Trop. Ecol.* 3: 75–85.

Hodkinson, I.D. and Casson, D. 1991. A lesser predilection for bugs (Hemiptera: Insecta) diversity in tropical rain forests. *Biol. J. Linn. Soc.* 43: 101–110.

Holder, M.T., Erdmann, M.V., Wilcox, T.P., Caldwell, R.L. and Hillis, D.M. 1999. Two living species of coelacanths? *Proc. Natl. Acad. Sci.* 96: 12616–12620.

Holmes, D. and Phillipps, K. 1996. *The Birds of Sulawesi.* Kuala Lumpur: Oxford University Press.

Holmes, D. 2000. *Deforestation in Indonesia: A review of the situation in 1999.* Jakarta: The World Bank.

Indonesian Observer 2000. *Rare Brown Civet Rediscovered*, 7 January, 2000.

Indonesian Observer 2000. *Locals Take Over Most of Lore Lindu National Park*, 12 January, 2000.

Iskandar, D.T. 1999a. The biogeography of Cylindrophis (Cylindrophidae, Ophidia) in the Wallacean region. In *Proceedings of the Second International Conference on Eastern Indonesian-Australian Vertebrate Fauna*, eds. D.M. Prawiladilaga, M. Amir, and J. Sugardjito, 31–40.

Iskandar, D.T. 1999b. The relationships and biogeography of the giant frogs in Eastern Indonesia (Limnonectes, Amphibia, Anura, Ranidae). In *Proceedings of the Second International Conference on Eastern Indonesian-Australian Vertebrate Fauna*, eds. D.M. Prawiladilaga, M. Amir, and J. Sugardjito, 41–47.

Iskandar, D.T. and Tjan, K.N. 1996. The amphibians and reptiles of Sulawesi, with notes on the distribution and chromosomal number of frogs. In *Proceedings of the First International Conference on Eastern Indonesian-Australian Vertebrate Fauna*, eds. D.L. Kitchener and A. Suyanto, 39–46.

Janvier, P. 1999. Coelacanth à la Marseillaise. *Nature* 401: 854–856.

Jarvie, J.K. and Hardiono, M. 2000. The vegetation types of Lore Lindu National Park. Final report to The Nature Conservancy - Indonesia.

Jepson, P. and Ounsted, R. 1997. *A bird watcher's guide to the world's largest archipelago.* Singapore: Periplus Edition.

Kevan, P.G. and Gaskell, B.H. 1986. The awkward seeds of Gonystylus macrophyllus (Thymelaeaceae) and their dispersal by the bat *Rousettus celebensis* in Sulawesi, Indonesia. *Biotropica* 18(1): 76–78.

Kinnaird, M.F. 1996a. Indonesia's hornbill heaven. *Nat. Hist.* 105: 24–29.

Kinnaird, M.F. 1996b. *North Sulawesi: A natural history guide.* Yayasan Pengembangan Wallacea, Jakarta, Indonesia. (Indonesian Edition 1997: Sulawesi Utara: Sebuah Panduan Sejarah Alam).

Kinnaird, M.F. 1998. Evidence for effective seed dispersal by the Sulawesi red-knobbed hornbill, Aceros cassidix. *Biotropica* 30: 50–55.

Kinnaird, M.F. 2000. Big on fig. *International Wildlife* 30: 12–21.

Kinnaird, M.F. and O'Brien, T.G. 1993. Preliminary observations on the breeding of the endemic Sulawesi red-knobbed hornbill (*Rhyticeros cassidix*). *Tropical Biodiversity* 1(2): 107–112.

Kinnaird, M.F. and O'Brien, T.G. 1996. Ecotourism in the Tangkoko-Dua Saudara nature Reserve: opening a Pandora's box? *Oryx* 30: 65–73.

Kinnaird, M.F. and O'Brien, T.G. 1999a. Breeding ecology of the Sulawesi red-knobbed hornbill, Aceros cassidix. *Ibis* 141: 60–69.

Kinnaird, M.F. and O'Brien, T.G. 1999b. Contextual analysis of the loud call of the Sulawesi Crested Black Macaque. *Tropical Biodiversity* 6(1&2): 37–42.

Kinnaird, M.F., O'Brien, T.G. and Suryadi, S. 1996. Population fluctuation in Sulawesi red-knobbed hornbills: Tracking figs in space and time. *Auk* 113: 431–440.

Kitchener, D.J., Schmitt, L.H. and Maharadatunkamsi 1994. Morphological and genetic variation in *Suncus murinus* (Soricidae: Crocidurinae) from Java, Lesser Sunda Islands, Maluku and Sulawesi, Indonesia. *Mammalia* 58(3): 433–451.

Kitching, R.L. 1987a. A preliminary account of the metazoan food webs in phytotelmata from Sulawesi, Indonesia. *Malay. Nat. J.* 41: 1–12.

Kitching, R.L. 1987b. Aspects of the natural history of the lycaenid butterfly *Allotinus major*, Sulawesi. *J. Nat. Hist.* 21: 535–544.

Kostermans, A.G.J.H. 1992. Two remarkable Lindera species (Lauraceae) probably representing an undescribed genus. *Reinwardtia* 11: 23–26.

Kottelat, M. 1989a. Der Matano-See. *Aquar.-Terrar-Zeitschr.* 42: 616–618.

Kottelat, M. 1989b. Der Towuti-See. *Aquar.-Terrar-Zeitschr.* 42: 681–684.

Kottelat, M. 1989c. Die Suesswasser-Fauna von Sulawesi. *Aquar.- Terrar-Zeitschr.* 42: 555–558.

Kottelat, M. 1990a. Der Mahalona-See. *Aquar.- Terrar-Zeitschr.* 43: 485–488.

Kottelat, M. 1990b. Der Wawontoa-See. *Aquar.- Terrar-Zeitschr.* 43: 618–619.

Kottelat, M. 1990c. Sailfin silversides (Pisces: Telmatherinidae) of lakes Towuti, Mahalona and Wawontoa (Sulawesi, Indonesia) with descriptions of two new genera and two new species. *Ichthyol. Explor. Freshwaters* 1: 227–246.

Kottelat, M. 1990d. Synopsis of the endangered Buntingi (Osteichthyes: Adrianichthyidae and Oryziidae) of Lake Poso, Central Sulawesi, Indonesia, with a new reproductive guild and description of three new species. *Ichthyol. Explor. Freshwaters* 1: 49–67.

Kottelat, M. 1990e. The ricefishes (Oryziidae) of the Malili Lakes, Sulawesi, Indonesia, with a description of a new species. *Ichthyol. Explor. Freshwaters* 1: 151–166.

Kottelat, M. 1991. Sailfin silversides (Pisces: Telmatherinidae) of Lake Matano, Sulawesi, Indonesia, with descriptions of six new species. *Ichthyol. Explor. Freshwaters* 1: 321–344.

Kottelat, M., Whitten, A.J., Wiryoatmodjo, S. and Kartikasari, S.N. 1993. *Freshwater fishes of Western Indonesia and Sulawesi.* Reprinted in 1996 with a separate summary of new species, name changes, etc. Singapore: Periplus Edition.

Kottelat, M. and Whitten, A.J. 1996. *Freshwater Fishes of Western Indonesia and Sulawesi: Additions and corrections.* Singapore: Periplus Edition.

Kunz, T.H., Fujita, M., Brooke, A.P. and McCracken, G.F. 1994. Convergence in tent architecture and tent-making behavior among neotropical and paleotropical bats. *J. Mammal. Evol.* 2: 57–78.

Lambert, F.R. 1998a. A new species of *Amaurornis* from the Talaud Islands, Indonesia, and a review of the taxonomy of bush hens occurring from the Philippines to Australasia. *Bull. B.O.C.* 118: 67–82.

Lambert, F.R. 1998b. A new species of Gymnocrex rail from the Talaud Islands, Indonesia. *Forktail* 13: 1–6.

Lambert, F.R. 1999. The conservation status of the red-and-blue Lory *Eos histrio* and other threatened birds of the Sangihe and Talaud islands. *Tropical Biodiversity* 6(1&2): 99–112.

Lambert, F.R. and Rasmussen, P.C. 1998. A new Scops Owl from Sangihe Island, Indonesia. *Bull. B.O.C.* 118: 204–217.

Larson, H.K. and Kottelat, M. 1992. A new species of *Mugiligobius* (Pisces: Gobiidae) from Lake Matano, Central Sulawesi, Indonesia. *Ichthyol. Explor. Freshwaters* 3: 225–234.

Lazell, J.D. 1987. A new flying lizard from the Sangihe Archipelago, Indonesia. *Breviora* 488: 1–9.

Lee, R.J. 1997. Impact of hunting and habitat disturbance on the population dynamics and behavioral ecology of the crested black macaque (*Macaca nigra*). Unpubl. Ph.D. dissertation, University of Oregon, Eugene, OR.

Lee, R.J. 1999. Market hunting pressures in North Sulawesi, Indonesia. *Tropical Biodiversity* 6(1&2): 124–144.

Lee, R.J. 2000a. A taste of Sulawesi. *BBC Wildlife* February: 53–60.

Lee, R.J. 2000b. Impact of subsistence hunting in North Sulawesi, Indonesia, and conservation options. In *Hunting for Sustainability in Tropical Forests,* eds. J.G. Robinson and E.L. Bennett, 455–472. New York: Columbia University Press.

Lee, R.J. 2000c. No lions, tigers and bears here, oh my! *Asian Geographic* 3: 20-39.

Lee, R.J. and J. Riley, in press. *Notes on the newly discovered Cinnabar Hawk-owl Ninox ios from North Sulawesi, Indonesia.* Coopers.

Maassen, W.J.M. and Kittel, K. 1996. Notes on terrestrial molluscs of the island of Sulawesi. 1. The Pupinidae (Gastropoda: Prosobranchia: Pupinidae). *Basteria* 60: 171–176.

Macdonald, A.A. and Fadrich, J. 1991. Les suides: que sont-ils? (Pigs and peccaries: What are they?) In *Biology of Suidae , Biologie des Suides,* eds. Barrett R.H. and F. Spitz, 7–19. IRGM. France: Imprimerie des Escartons; Briancon.

Macdonald, A.A., Leus, K., Florence, A., Clare, J. and Patry, M. 1996. Notes on the behaviour of Sulawesi warty pigs (*Sus celebensis*) in North Sulawesi, Indonesia. *Malay. Nat. J.* 50: 47–53.

MacKinnon, J. 1997. *Protected Areas Systems Review of the Indo-Malayan Realm.* Cambridge: World Conservation Monitoring Centre.

Maschwitz, U. and Fiala, B. 1995. Investigations on ant plant associations in the southeast Asian genus *Neonauclea* Merr. (Rubiaceae). *Acta Oecologia* 16: 3–18.

Mason, I.J.and Forrester, R.I. 1996. Geographical differentiation in the channel-billed cuckoo Scythrops novaehollandiae Latham, with description of two new subspecies from Sulawesi and the Bismark Archipelago. *Emu* 96(4): 217–233.

Matsumura, S. 1993. Female reproductive cycles and the sexual behavior of Moor macaques *Macaca maura* in their natural habitat, South Sulawesi, Indonesia. *Primates* 34: 99–103.

Matsumura, S. 1996. Postconflict affiliative contacts between former opponents among wild Moor macaques (*Macaca maurus*). *Amer. J. Primatol.* 38(3): 211.

Matsumura, S. 1998. Relaxed dominance relations among female moor macaques (*Macaca maurus*) in their natural habitat, south Sulawesi, Indonesia. *Folia Primatol.* 69(6): 346–356.

McCord, W.P., Iverson, J.B. and Boeadi 1995. A new batagurid turtle from northern Sulawesi, Indonesia. *Chelonian Conserv. Biol.* 1: 311–316.

Melisch, R. 1994. Observations of swimming babirusa in Lake Poso, Central Sulawesi, Indonesia. *Malay. Nature J.* 47: 431–432.

Meyburg, B.-U. and van Balen, B. 1994. Raptors on Sulawesi (Indonesia): The influence of rain forest destruction and human density on their populations. In *Raptor conservation today*, eds. B.-U. Meyburg and R.D. Chancellor, 269–276. WWGBP/The Pica Press.

Miksic, J., ed. 1996. *Ancient History. Indonesian Heritage Encyclopaedia Vol. 1.* Singapore: Archipelago.

Milliken, T., ed. 1990. Aspects of sea turtle exploitation in Indonesia. Tokyo, TRAFFIC Japan, 57pp.

Mol, A.W.M. 1989. *Echinobaetis phagus* n. g., n. sp. a new mayfly from Sulawesi, Indonesia (Ephermeroptera: Baetidae). *Zool. Meded.* 63: 61–72.

Monk, K.A. and Butlin, R.K. 1990. A biogeographic account of the grasshoppers (Orthoptera: Acridoidea) of Sulawesi, Indonesia. *Tijdschr. Entomol.* 133: 31–38.

Morales J.C and Melnick, D.J. 1998. Phylogenetic relationships of the macaques ercopithecidae: Macaca, as revealed by high resolution restriction site mapping of mitochondrial ribosomal genes. *J. Hum Evol* 34(1): 1–23.

Muroyama, Y. and Thierry, B. 1998. Species differences of male loud calls and their perception in Sulawesi macaques. *Primates* 39(2): 115–126.

Musser, G.G. 1990. Sulawesi rodents: Species traits and chromosomes of *Haeromys minahassae* and *Echiothrix leucura* (Muridae: Murinae). *Am. Mus. Novitates* 2989: 1–18.

Musser, G.G. 1991. Sulawesi rodents: Descriptions of new species of *Bunomys* and *Maxomys* (Muridae: Murinae). *Am. Mus. Novitates* 3001: 1–42.

Musser, G.G. and Breed, G. 1991. Sulawesi and Philippine rodents (Muridae): A survey of spermatozoal morphology and its significance for phylogenetic inference. *Am. Mus. Novitates* 3003: 1–15.

Musser, G.G. and Dagosto, M. 1987. The identity of *Tarsius pumilus*, a pygmy species endemic to the montane mossy forests of Central Sulawesi. *Am. Mus. Novitates* 2867: 1–53.

Musser, G.G. and Holden, M.L. 1991. Sulawesi rodents (Muridae, Murinae): Morphological and geographical boundaries of species in the *Rattus hoffmanni* group with a new species from Pulau Peleng. *Bull. Am. Mus. Nat. Hist.* 206: 1–164.

Musser, G.G. and Mary, E.H. 1991. Sulawesi rodents (Muridae: Murinae): Morphological and geographical boundaries of species in the *Rattus hoffmanni* group and a new species from Pulau Peleng. *Bull. Am. Mus. Nat. Hist.* 206: 322–413.

Neboiss, A. 1994. A review of the genus *Paranyctiophylax* Tsuda from Sulawesi, Papua New Guinea and northern Australia (Trichoptera: Polycentropodidae). *Mem. Mus. Victoria* 54: 191–205.

Ng, P.K.L. 1991. *Cancrocaeca xenomorpha*, n. g., n. sp., a blind troglobitic freshwater hymenosomatoid (Crustacea: Decapoda: Brachyura) from Sulawesi, Indonesia. *Raffles Bull. Zool.* 39: 59–74.

Ng, P.K.L. 1993. On *Parathelphusa ceophallus* n. sp. (Crustacea: Decapoda: Brachyura: Parathelphusidae) from Pulau Buton, Sulawesi. *Zool. Meded.* 67: 1–26.

Niemitz, C., Nietsch, A., Wartyer, S. and Rumpler, Y. 1991. Tarsius dianae, a new primate species from Central Sulawesi, Indonesia. *Folia primatol.* 56: 105–116.

Nieser, N. and Chen, P. 1992. Revision of Limnometra Mayr Gerridae in the Malay Archipelago. Notes on Malesian aquatic and semiaquatic bugs (Heteroptera) II. *Tijdschr. Entomol.* 135: 11–26.

Nietsch, A. and Kopp, M-L. 1998. Role of vocalization in species differentiation of Sulawesi tarsiers. *Folia Primatol.* 69(Suppl. 1): 371–378.

Nooteboom, H.P. 1988. A new species of *Symplocos* Symplocaceae from Sulawesi. *Blumea* 33: 263–264.

O'Brien, T.G. 1996. Behavioural ecology of the North Sulawesi tarictic hornbill

Penelopides exarhatus exarhatus during the breeding season. *Ibis* 139: 97–101.

O'Brien, T.G., ed. 1999. *Sulawesi Special Issue.* Tropical Biodiversity. 6(1&2): 1–3.

O'Brien, T.G. and Kinnaird, M.F. 1996. Changing populations of birds and mammals in North Sulawesi. *Oryx* 30: 150–156.

O'Brien, T.G. and Kinnaird, M.F. 1996. Effect of harvest on leaf development of the Asian palm *Livistona rotundifolia*. *Conserv. Biol.* 10: 53–58.

O'Brien, T.G. and Kinnaird, M.F. 1994. Notes on the density and distribution of the endemic Sulawesi Tarictic hornbill (*Penelopides exarhatus exarhatus*) in the Tangkoko-Dua Saudara Nature Reserve, North Sulawesi. *Tropical Biodiversity* 2: 252–260.

O'Brien, T.G. and Kinnaird, M.F. 1997. Behavior, diet and movement patterns of the Sulawesi crested black macaque, *Macaca nigra. Intl. Journal of Primatology* 18: 321–351.

O'Brien, T.G. and Kinnaird, M.F. 2000. Differential vulnerability of large birds and mammals to hunting in North Sulawesi, Indonesia and the outlook for the future. In *Hunting for Sustainability in Tropical Forests*, eds. J.G. Robinson and E.L. Bennett, 199–213. New York: Columbia University Press.

O'Brien, T.G., Kinnaird, M.F., Dierenfeld, E.S., Conklin-Brittain, N.L., Wrangham, R. and Silver, S. 1998. What's so special about figs? *Nature* 392: 668.

O'Byrne, P. 1996a. A look at *Dendrobium* section Aporum, Part 1. *Malayan Orchid Review* 30: 19–24, 53–59.

O'Byrne, P. 1996b. Three new orchid species from Indonesia. *Malayan Orchid Review* 30: 67–75.

O'Byrne, P. 1998. New and unusual Aeridinae from Borneo and Sulawesi. *Malayan Orchid Review* 32: 47–54, 65–70.

O'Byrne, P. 1999. New orchid species from Sulawesi. *Malayan Orchid Review* 33: 43–47, 93–96.

O'Donoghue, P.J., Watts, C.H.S. and Dixon, B.R. 1987. Ultrastructure of Sarcocystis spp. (Protozoa: Apicomplexa) in rodents from North Sulawesi and West Java, Indonesia. *J. Wildl. Dis.* 23(2): 225–232.

Oliver, W.L.R., ed. 1993. Pigs, peccaries, and hippos: Status survey and conservation action plan. IUCN/SSC Pigs and Peccaries Specialist Group and IUCN/SSC Hippo Specialist Group, 202p+XIII.

Paarmann, W. and Stork, N.E. 1987. Canopy fogging: A method of collecting living insects for investigations of life history strategies. *J. Nat. Hist.* 21: 563–566.

Patry, M.K.L. and Macdonald, A.A. 1995. Group structure and behaviour of babirusa (*Babyrousa babyrussa*) in northern Sulawesi. *Aust. J. Zool.* 43: 643–655.

PHPA/LIPI/BirdLife International-IP 1998. *Yellow-crested cockatoo recovery plan.* PHPA/LIPI/BirdLife International-Indonesia Programme, Bogor, Indonesia.

Platt, S.G., Lee, R.J. and Klemens, M.W., in press. Notes on the distribution, life history, and exploitation of turtles in Sulawesi, Indonesia. Chelonian Biology.

Prabowo, A. 1989. Evaluation of the mineral status of grazing ruminants in South Sulawesi, Indonesia. Ph.D. dissertation, Univ. Fla.

Prawiradilaga, D.M. 1997. The maleo Macrocephalon maleo on Buton. *Bull. Br. Ornithol. Club* 117(3): 237.

Proctor, J. 1992. The vegetation over ultramafic rocks in the tropical Far East. In *The ecology of areas with serpentized rocks: A world view*, eds. B.A. Roberts and J. Proctor, 249–270. Kluwer, Dordrecht.

Rasmussen, P.C. 1999. A new species of hawk-owl Ninox from north Sulawesi, Indonesia. *Wilson Bull.* 111(4): 457–464.

RePPProT 1988. Sulawesi: Regional physical planning programme for transmigration, Natural Resources Institute and Direktorat Bina Program, Direktorat Jenderal Penyiapan Pemukiman, Departemen Transmigrasi, Jakarta.

Richardson, M. 2000. *Indonesia Faces Forest Dilemma.* International Herald Tribune. Tuesday, February 1, 2000.

Riggs, J., ed. 1996. *Human Environment. Indonesian Heritage Encyclopaedia Vol. 2.* Archipelago, Singapore.

Riley, J. 1997. The birds of Sangihe and Talaud Islands, North Sulawesi. *Kukila* 9: 3–36.

Riley, J., Hunowu, I., Wangko, M.F. and Wardi, J.C. 1999. The status, ecology and behavior of the elegant sunbird *Aethopyga duyvenbodei* on Sangihe, Indonesia. *Tropical Biodiversity* 6 (1&2): 113–123.

Rogers, J.D., Callan, B.E., and Samuels, G.J. 1987. The Xylariaceae of the rain forests of North Sulawesi, Indonesia. *Mycotaxon* 29: 113–172.

Rosenbaum, B., O'Brien, T.G., and Kinnaird, M.F. 1998. Population densities of Sulawesi crested black macaques (*Macaca nigra*) on Bacan and Sulawesi, Indonesia: effects of habitat disturbance and hunting. *Am. J. Primatol.* 44: 89–106.

Rozendaal, F.G. and Lambert, F.R. 1999. The taxonomic and conservation status of *Pinarolestes sanghirensis* Oustalet 1881. *Forktail* 15: 1–13.

Ruedi, M. 1995. Taxonomic revision of shrews of the genus Crocidura from the Sunda Shelf and Sulawesi with description of two new species (Mammalia: Soricidae). *Zool. J. Linn. Soc.* 115(3): 211–265.

Russell-Smith, A. and Stork, N.E. 1994. Abundance and diversity of spiders from the canopy of tropical rainforests with particular reference to Sulawesi. *J. Trop. Ecol.* 10: 545–558.

Saeed, B. and Ivantsoff, W. 1991. *Kalyptatherina*, the first telmatherinid genus known outside of Sulawesi. *Ichthyological Expl. Freshwaters* 2(3): 227–238.

Sakagami, S.F. and Inoue, T. 1989. Stingless bees of the genus *Trigona* (*Geniotrigona*) (Hymenoptera: Apidae) with description of *Triogona incisa* n. sp. from Sulawesi, Indonesia. *Jpn. J. Entomol.* 57: 605–620.

Schileyko, A.A. 1996. *Guamampa* n. g. (Gastropoda: Pulmonata), a bradybaenid land snail with monadeniid chararcters. *Bull. Mus. Natn. d'Hist. Nat. Sec. A* 18: 4.

Schoorl, J.W. 1987. Notes on the birds of Buton (Indonesia, Southeast Sulawesi). *Bull. Br. Ornithol. Club* 107(4): 165–168.

Schreiber A., Seibold, I., Nötzold, G. and Wink, M. 1999. Cytochrome *b* gene haplotypes characterize chromosomal lineages of Anoa, the Sulawesi dwarf buffalo (Bovidae: *Bubalus* sp.). *J. Heredity* 90(1): 165–176.

Severns, M. 1995. *Sulawesi Seas: Indonesia's majestic underwater realm.* Honolulu: University of Hawaii Press.

Sidiyasa, K. 1986. Tree flora on the ridges and upper slopes of dry climate area at Paboya Nature Reserve, Central Sulwesi. *Bul. Pen. Hutan* 485: 31–38.

Sidiyasa, K. 1989. Some aspects of the ecology of *Diospyros celebica* at Sausu and adjacent area, Central Sulawesi, Indonesia. *Bul. Pen. Hutan* 508: 15–26.

Sidiyasa, K., Whitmore, T.C., Tantra, I.G.M. and Sutisna, U. 1989. *Tree flora of Indonesia: Check list for Sulawesi.* Bogor: Ministry of Forestry Agency for Forestry Research and Development Forest Research and Development Centre.

Siebert, S.F. 1997. Economically important rattans of Central Sulawesi, Indonesia. *Principes* 41: 42–46.

Siluba, M. 1987. Species abundance or *Corbicula possoensis* and *Melanoides granifera* in Poso Lake, Central Sulawesi. *Berita Biol.* 3: 317-320.

Sinclair, J.R., O'Brien, T.G., and Kinnaird, M.F. 1999. Observations on the breeding biology of the Philippine Megapode (*Megapodius cumingii*) in North Sulawesi, Indonesia. *Tropical Biodiversity* 6(1&2): 76–86.

Smith, R.M. 1991. New species of *Alpinia* from Sulawesi and Sarawak. *Edinb. J. Bot.* 48: 347–352.

Soeyatman, H.C. and Sutisna, U. 1988. Tree species composition analysis of the tropical lowland rain forest at Masaba, South Sulawesi. *Bul. Pen. Hutan* 497: 7–20.

Sota, T. and Mogi, M. 1996. Species richness and altitudinal variation in the aquatic metazoan community in bamboo phytotelmata from North Sulawesi. *Res. Pop. Ecol. (Kyoto)* 38: 275–281.

Stork. N.E. and Paarmann, W. 1992. Reproductive seasonality of the ground and tiger beetle (Coleoptera: Carabidae: Cicindelidae) fauna in North Sulawesi, Indonesia. *Stud. Neotrop. Fauna Environ.* 27: 101–115.

Sugardjito, J., Southwick, C.H., Supriatna, J., Kohlhaas, A., Baker S., Erwin, J., Froehlick, J., Lerche, N. 1989. Population survey of macaques in northern Sulawesi. *Am. J. Primatol.* 18(4): 285–301.

Sujatnika, Jepson, P., Suhartono, T.R., Crosby, M.J., and Mardiastuti, A. 1995. *Melestarikan keanekaragaman Hayati di Indonesia: Pendekatan daerah burung endemik (Conserving Indonesian biodiversity: The endemic bird area approach).* Jakarta: PHPA/BirdLife International-Indonesia Programme.

Sunarto 1999. Resource partitioning among kingfishers in Tangkoko-Duasudara Nature Reserve, North Sulawesi. *Tropical Biodiversity* 6(1&2): 75–86.

Supriatna, J., Froehlich, J.W., Erwin, J.M. and Southwick, C.H. 1992. Population, habitat and conservation status of *Macaca maurus*, *Macaca tonkeana* and their putative hybrids. *Tropical Biodiversity* 1(1): 31–48.

Suryadi, S., Kinnaird, M.F., and O'Brien, T.G. 1998. Home ranges and daily movements of the Sulawesi red-knobbed hornbill during the non-breeding season. In *Proceedings of the Second Asian Hornbill Workshop, Bangkok, BRTP, Bangkok*, ed. P. Poonswad, 159–170.

Suryadi, S., Kinnaird, M.F., O'Brien, T.G., and Supriatna, J. 1996. Time budget of the

Sulawesi red-knobbed hornbill during the non-breeding season at Tangkoko-Dua Sudara Nature Reserve, North Sulawesi. In *Proceedings of the First International Conference on Eastern Indonesian-Australian Vertebrate Fauna*, eds. D.L. Kitchener and A. Suyanto, 123–126.

Suryadi, S., Kinnaird, M.F., O'Brien, T.G., Supriatna, J. and Somadikarta, S. 1994. Food preferences of the red-knobbed hornbill during the non-breeding season. *Trop. Biodiv.* 2: 377–384.

Suryobroto, B. 1992. Estimation of the biologial affinities of seven species of Sulawesi macaques based on multivariate analysis of dermatoglyphic pattern types. *Primates* 33: 429-449.

Sydney Morning Herald 2000. *The Rush is on*, 1 July, 2000.

Taiti, S., Ferrara, F., and Kwon D.H. 1992. Terrestrial Isopoda (Crustacea) from the Togian Islands, Sulawesi, Indonesia. *Invertebr. Taxon.* 6(3): 787–842.

Tong, X. and Dudgeon, D. 1999. Two new species of Platybaetis (Ephemeroptera: Baetidae) from Sulawesi, Indonesia. *Entomological News* 110: 290–296.

Tremble, M., Muskita, Y. and Supriatna, J. 1993. Field observations of *Tarsius dianae* at Lore Lindu National Park, Central Sulawesi, Indonesia. *Tropical Biodiversity* 1(2): 67–76.

Van Den Berg, A.B. and Bosman, C.A.W. 1986. Supplementary notes on some birds of Lore Lindu Reserve, Central Sulawesi. *Forktail* 1: 7–13.

van Tol, J. 1987. The Odonata of Sulawesi and adjacent islands. 1 and 2. *Zool. Meded.* 61: 155–176.

van Tol, J. 1994. The Odonata of Sulawesi and adjacent islands. 3. *Tijdschr. Entomol.* 137: 87–94.

Vermeulen, J.J. 1992. Two species of the genus Enteroplax (Gastropoda: Pulmonata: Strobilopsidae) from Sulawesi, Indonesia. *Basteria* 546: 77–81.

Vermeulen, J.J. and Whitten, T. 1999. *Biodiversity and cultural property in the management of limestone resources: Lessons from East Asia.* Washington, D.C.: The World Bank.

Walton, T. and Holmes, D. 2000. *Indonesia's forests are vanishing faster than ever.* Opinion/Commentary, International Herald Tribune. Tuesday, January 25, 2000.

Walton, C., Butlin, R.K. and Monk, K.A. 1997. A phylogeny for grasshoppers of the genus *Chitaura* (Orthoptera: Acrididae) from Sulawesi, Indonesia, based on mitochondrial DNA sequence data. *Biol. J. Linnean Soc.* 62: 365–382.

Wardill, J.C. and Hunowu, I. 1998. First observations of the endemic subspecies of black-fronted white-eye on Sangihe, North Sulawesi. *Bull. Oriental Bird. Cl.* 27: 48–49.

Wardill, J.C. and Riley, J. 1999. Sangihe and Talaud Islands, Indonesia. *Bull. Orient. Bird Cl.* 29: 30–35.

Wardill, J.C., Fox, P.S., Hoare, D.J., Marthy, W. and Anggraini, K. 1999. Birds of Rawa Aopa Watumohai National Park, South-east Sulawesi. *Kukila* 10: 91–114.

Watanabe, K. and Matsumura, S. 1991. The borderlands and possible hybrids between three species of macaques *Macaca nigra*, *M. nigrescens*, and *M. hecki* in the northern peninsula of Sulawesi. *Primates* 32: 365–370.

Watanabe, K., Lapasere, H. and Tantu, R. 1991. External characteristics and associated developmental changes in two species of Sulawesi macaques *Macaca tonkeana* and *Macaca hecki* with special reference to hybrids and the borderland between the species. *Primates* 32: 61–76.

Watanabe, K., Matsumura, S., and Watanabe, T. 1991. Distribution and possible intergradation between *Macaca tonkeana* and *M. ochreata* at the borderland of the species in Sulawesi. *Primates* 32: 385–390.

Watanabe, K., Matsumura, S., Watanabe, T., and Hamada, Y. 1991. Distribution and possible intergradation between *Macaca tonkeana* and *M. ochreata* at the borderland of the species in Sulawesi. *Primates* 32(3): 385–389.

Weinberg, S. 1999. *A Fish Caught in Time.* N.Y.: Fourth Estate.

Wells, M., Khan, A., and Jepson, P. 1998. *Investing in Biodiversity: Integrated Conservation Development Projects in Indonesia.* Washington, D.C.: The World Bank.

Wemmer, C. and Watling, D. 1986. Ecology and status of the Sulawesi Palm Civet *Macrogalidia musschenbroekii* Schlegel. *Biol. Conserv.* 35(1): 1–17.

Whitmore, T.C. 1984. *Tropical Rain Forests of the Far East.* Oxford: Clarendon.

Whitmore, T.C. 1998. *Introduction to tropical rain forests. 2nd ed.* Oxford: Oxford University Press.

Whitmore, T.C. and Tantra, I.G.M. 1986. *Tree Flora of Indonesia - Sulawesi.* Bogor: Forest Research and Development Centre.

Whitten, T. and Whitten, J. 1992. *Wild Indonesia.* London: New Holland.

Whitten, T. and Whitten, J. 1996a. *Plants. Indonesian Heritage Encyclopaedia Vol. 4.*

Archipelago, Singapore.

Whitten, T. and Whitten, J. 1996b. *Wildlife: Indonesian Heritage Encyclopaedia Vol. 5.* Archipelago, Singapore.

Whitten, T., Nash, S.D., and Bishop, K.D. 1987. One or more extinctions from Sulawesi? *Conserv. Biol.* 1: 42–48.

Whitten, T., Soeriaatmadja, R.E., and Afiff, S. 1996. *The Ecology of Java and Bali.* Singapore: Periplus.

Whitten, A., Whitten, J., Mittermeier, C.G., Supriatna, J., and Mittermeier, R.A. 1998. Indonesia. In *Megadiversity: Earth's Biologically Wealthiest Nations,* eds. R.A. Mittermeier, P.R. Gill, and C.G. Mittermeier, 74–97. Cemex, Prado Norte.

Whitten, A., Whitten, J., Mittermeier, C.G., Supriatna, J. and Mittermeier, R.A. 1999. Wallacea. In *Hotspots: Earth's Biologically Wealthiest Places,* eds. R. Mittermeier, P.R. Gill, and C.G. Mittermeier, 296–304. Cemex, Prado Norte.

Wiles, G. J. and Masala, Y. 1987. Collapse of a nest tree used by Finch-billed Mynas Scissirostrum dubium in North Sulawesi. *Forktail* 3: 67–68.

Williams, A.K., Ashley, M.V. and Melnick, D.J. 1989. Evolutionary relationships among Sulawesi macaques as revealed by mitochondrial DNA analysis. *Am. J. Phys. Anthropol.* 78: 324.

Wuester, W. 1996. The status of the cobras of the genus Naja Laurenti, 1768 (Reptilia: Serpentes: Elapidae) on the island of Sulawesi. *Snake* 27(2): 85–90.

Zwahlen, R. 1992. The ecology of Rawa Aopa: A peat swamp in Sulawesi, Indonesia. *Environ. Conserv.* 19(3): 226–234.

Chapter One

Physical, Biological and Human Background

GEOLOGY

Geological History

The geology of an area and its geological history are the major determinants of the soils, plants and animals that occur there. For this reason a brief account is given below concerning the physical conditions and history of Sulawesi.

About 250 Ma[1] ago the earth comprised of two great continents: Laurasi–comprising present-day North America, Europe and much of Asia and Gondwanaland–comprising present-day South America, Africa, India, Australia, Antarctica and the remainder of Asia. Until the last few years the once widely accepted view of the geological history of Indonesia and surrounding regions was that the western half (Malay Peninsula, Sumatra, Java, Borneo and western Sulawesi) had been part of Laurasia, separated until recently from the eastern half (eastern Sulawesi, Timor, Seram, Buru, etc.), which had been part of Gondwanaland, by the broad Tethys Ocean.

This picture has had to change in the light of recent geological and palaeontological evidence. The current view, not without its critics however, is that southern Tibet, Burma, Thailand, Peninsular Malaysia and Sumatra were once part of Gondwanaland and rifted from the northern Australia-New Guinea continental margin some 200 Ma ago. This continental fragment then formed a dissected land connection between Australia and Asia and may have carried with it an evolving higher-plant flora (Audley-Charles 1987). Western Sulawesi together with Sumatra, Borneo, and land that would later form the islands of the Banda Arc[2] are considered to have separated from Gondwanaland in the middle Jurassic (Audley-Charles 1983). Australia broke away from Antarctica much later, perhaps in the early Cretaceous (90 Ma), and Australia, New Guinea, and east Sulawesi proceeded to travel northwards at about 10 cm per year. At least part of eastern Sulawesi probably separated from New Guinea before its mid-Miocene collision with western Sulawesi after which

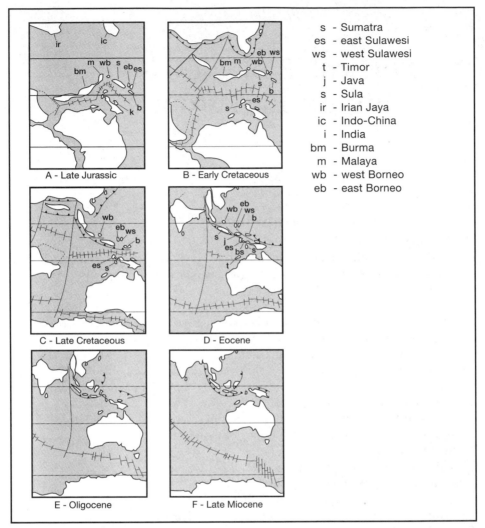

s - Sumatra
es - east Sulawesi
ws - west Sulawesi
t - Timor
j - Java
s - Sula
ir - Irian Jaya
ic - Indo-China
i - India
bm - Burma
m - Malaya
wb - west Borneo
eb - east Borneo

A - Late Jurassic

B - Early Cretaceous

C - Late Cretaceous

D - Eocene

E - Oligocene

F - Late Miocene

Figure 1.1. Changing locations of components of Southeast Asia since the first rifting from east Gondwanaland, showing schematic ocean spreading ridges, subduction trenches (triangles) and continental margins (dotted lines). Present coastlines shown for reference only. Horizontal lines: 0°, 30° and 60° South.

After Audley-Charles 1987

the eastern half began to emerge as an island (Audley-Charles 1987) (fig. 1.1; table 1.1).

The most dramatic event in Indonesian geological history occurred in the Miocene when the northward-drifting Australian Plate caused the

Table 1.1. Summary of geological timescale and events relevant to Sulawesi over the last 350 million years.

Era	Period	Epoch	Began (millions of years ago)	Geological events	Biological events
Cenozoic	Quaternary	Holocene	0.01		Modern man
		Pleistocene	1		Man's earliest ancestors
	Tertiary	Pliocene	10	Possible connections with Borneo either via Doang-doang shoals or a reduced Makassar Straits	Large carnivores
		Miocene	25	Sula/Banggai together with east Sulawesi collide with west Sulawesi; northern peninsula starts rotating; eastern and western Sulawesi begin to fuse; widespread volcanism in west Sulawesi	Abundant grazing animals
		Oligocene	40	Western Indonesia and western Sulawesi in more or less present positions	Large running animals
		Eocene	60	Australia breaks away from Antarctica, volcanism in western Sulawesi begins	Many modern types of mammals evolve
		Palaeocene	70		First placental mammals
Mesozoic	Cretaceous		145		First flowering plants and extinction of dinosaurs and ammonites at end of period
	Jurassic		215	Western Indonesia with Tibet, Burma, Thailand, Malaysia and western Sulawesi break away from Gondwanaland	First birds and mammals, dinosaurs and ammonites abundant
	Triassic		250	Pangea rifts into two: Laurasia and Gondwanaland; insular and some mainland parts of Southeast Asia part of eastern Gondwanaland	First dinosaurs, abundant cycads and conifers
Palaeozoic	Permian		280	All land together as one continent, Pangea	Extinction of many forms of marine animals including trilobites
	Carboniferous		350		Great coal-forming conifer forests, first reptiles, sharks and amphibians abundant

bending to the west of the eastern part of the Banda Arc. This westward movement, coupled with the westward thrust along the east-west Sorong fault system from western Irian Jaya, modified the two major landmasses that would form the peculiar shape of Sulawesi we recognize today. It has been proposed that this collision occurred 19-13 Ma ago (Sasajima et al. 1980; Audley-Charles 1987). The Banggai-Sula Islands formed the continental platform section of the east Sulawesi fragment. The Talaud Islands and the small islands of Mayu and Tifore between North Sulawesi and Halmahera are probably also part of the collision suture that formed between Sundaland and Gondwanaland (Audley-Charles 1987).

Thus eastern Sulawesi was like a spearhead that hit western Sulawesi and caused the southwest peninsula to rotate anticlockwise by about 35°, thereby opening the Gulf of Bone (Haile 1978) and causing the northern peninsula to pivot around its northern end, rotating clockwise through nearly 90°. This would have caused subduction[3] along the North Sulawesi Trench in Gorontalo Bay (Otofuji et al. 1981) and obduction[4] of the ultrabasic rocks of east and Southeast Sulawesi over the erosion debris or molasse deposits of younger rocks.

The physical history of eastern Indonesia has made it one of the most geologically complex regions in the world (Audley-Charles 1981). It is this complexity and the strange shape of Sulawesi, described variously as an orchid, a demented spider and a wobbly 'K', which have long attracted the interest of geologists and others (Davis 1976; Otofuji et al. 1981).

Sulawesi comprises three distinct geological 'provinces' brought together by movements of the earth's crust as described above. These are West and East Sulawesi (divided by the north-northwest fault between Palu and the Gulf of Bone–the Palu-Koro fault), and the Banggai-Sula province comprising the Tokala region behind Luwuk on the northeast peninsula, the Banggai Islands, Butung Island and the Sula Islands (actually part of the political province of Maluku) (fig. 1.2).

West Sulawesi is underlain in the south by a basement of schists (metamorphic rocks of continental origin that split easily along their mineral plates) and ultrabasic rocks (derived from the mantle), and in the north by schists and gneiss (banded, coarse, metamorphic rocks that do not split easily). These are overlain by marine sediments including limestone (primarily calcium carbonate from animal shells), sandstones (consolidated sand), cherts (a compact flint-like variety of silica) and shales (thin layers of consolidated mud, clay and silt). Volcanism began in the Eocene but became widespread in the Miocene, and the volcanic arc that ran from the south to the north deformed the existing sedimentary rocks. The ash and dust from the volcanoes mixed with eroded sedimentary rock derived from the uplifting of the eastern part of the Sulawesi collision zone, to form the Celebes molasse, a generally poorly-consolidated conglomerate rock of gravels, sands, silts and muds formed in terrestrial or shallow-

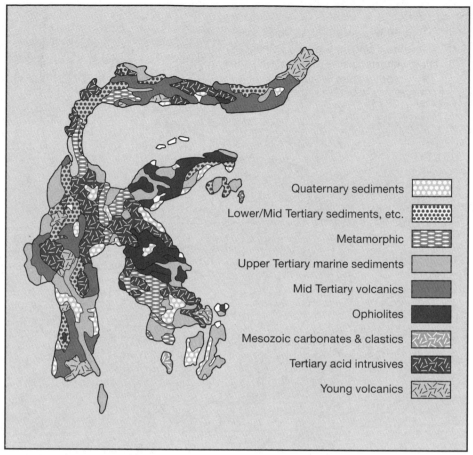

Figure 1.2. Geological map.
After Katili 1978

water environments. Molten igneous rocks such as granite and diorite have forced their way by intrusion into Miocene and older rocks at a number of locations (Sukamto 1975a, b; Otofuji et al. 1981). East Sulawesi consists mainly of basic and ultrabasic igneous rocks associated with schists in the west and with Mesozoic limestones in the east and the south.

The Banggai-Sula province comprises a basement of Palaeozoic metamorphic rocks intruded by granites. Triassic and Permian effusive rocks were deposited locally on the basement. These rocks are overlain by Mesozoic shale, sandstone, conglomerate and marl (consolidated mud and

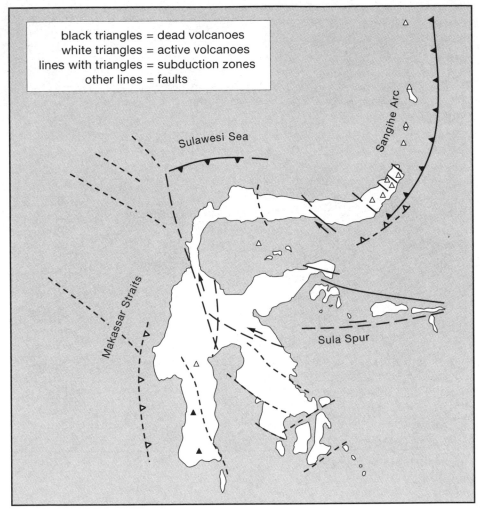

black triangles = dead volcanoes
white triangles = active volcanoes
lines with triangles = subduction zones
other lines = faults

Sangihe Arc

Sulawesi Sea

Makassar Straits

Sula Spur

Figure 1.3. Tectonic features of Sulawesi.
After Katili 1978

calcium carbonate) deposited in both continental and marine shelf environments. Butung Island is included in this province by virtue of distinctive Jurassic shales which it shares with Banggai-Sula but which do not occur in East Sulawesi (Sukamto 1975b).

It has been suggested that western Sulawesi collided with eastern Borneo in the late Pliocene (3 Ma ago) thereby closing the Makassar Straits which opened again only during the Quaternary (Katili 1978), although there is no great weight of data to support this. Thick sediments in the Straits from before the Miocene indicate that Borneo and Sulawesi have been separated for at least 25 Ma. During periods of low sea-level (p. 16) it is quite likely that islands would have existed particularly in the area west of Majene and around the Doangdoangan shoals (Audley-Charles 1981). In the latter area a drop in sea-level of 100 m would have exposed an expanse of almost continuous land between southeast Borneo and southwest Sulawesi. One interesting observation, however, is that along the northern (and deeper) section of the Makassar Straits the 1,000 m submarine contour of eastern Borneo more or less exactly matches that of western Sulawesi (Katili 1978) so it is possible that the Straits was once narrower.

As can be deduced from the active lateral movement along the Gorontalo, Palu-Koro, Matano and Sorong faults (fig. 1.3), the island of Sulawesi is at present undergoing a process of fragmentation. The end result could be a cluster of islands separated by narrow straits resembling the complex pattern of the Philippine Archipelago in which the original double island-arc structure is no longer recognizable (Katili 1978).

Volcanoes

The most devastating eruption in Sulawesi in recent times was that of Colo volcano on Una-una Island in Tomini Bay. As far as is known this had erupted only once before, in 1898, when a large eruption spewed out 2.2 km^3 of ash which was deposited over 303,000 km^2, reaching nearly as far as the border between Sarawak and East Kalimantan 800 km away (Umbgrove 1930). For some days after 14 July 1983 a large number of earthquakes—up to 100 per day—shook Una-una and on 18 July a large 'phreatic' eruption occurred; that is, water caught below hot volcanic material was turned into high pressure steam and exploded, generating a column of steam and debris some 500 m high. The first magmatic eruption occurred in the morning of 23 July, and this sent a plume of ash and other material 1,500 m into the air. The violent climax of the activity came that afternoon when most of the island blew apart and most of the vegetation was destroyed by a 'nuée ardente' (Katili and Sudrajat 1984),[5] a glowing mass of turbulent, superheated gases and incandescent solid particles (Francis 1978). Ash from the 15,000 m-high cloud reached Pulau Laut, an island 900 km away off southeast Kalimantan, and covered 90% of the remains of Una-una. All the island's inhabitants were evacuated in time, a wonderful achievement, but all houses, crops, animals, coral and inshore fish were destroyed except along a narrow strip on the east of the island (Katili and Sudrajat 1984).

Colo volcano is just one of a number of volcanoes on or near Sulawesi which have greater and lesser effects on the surrounding human population. By far the most destructive of human life has been Mt. Awu on

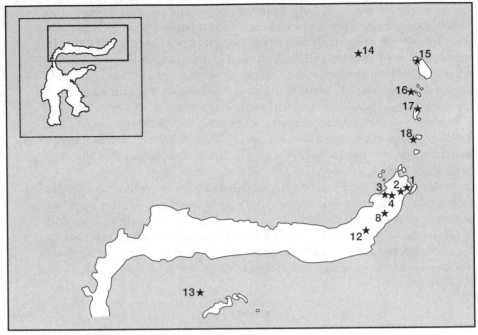

Figure 1.4. Location of active volcanoes. Numbers refer to table 1.2.

Sangihe Island (fig. 1.4; table 1.2). Apart from the obvious damage to the surrounding land, ash clouds can have serious effects on aircraft. For example, a Qantas jet had to be grounded for major repairs in 1985 after it had flown into an ash cloud from Soputan volcano.

Sulawesi has 11 active volcanoes, compared with 17 on Java, 10 on Sumatra and 6 on Halmahera, and numerous old cones, the most beautiful of which is probably Manado Tua in Minahasa. The volcanoes are associated with the subduction zones north of Toli-Toli (in the case of Colo on Una-una), and east of Minahasa and Sangihe (in the case of the remainder). These regions are often shaken by earthquakes (McCaffrey and Sutardjo 1982). The epicenters or positions on the earth's surface where these originated are frequently cited, but less attention is paid to the depths at which the earthquakes originate. For example, the floor of the Sulawesi Sea is moving southwards but instead of buckling and piling up it is forced down at an angle of about 60° under the northern peninsula. The

Table 1.2. Active volcanoes and solfatara- and fumarole-fields with structure, dates of known eruptions and known casualties caused by the active volcanoes.

	Type	Volcano description	Years of known activity	Known deaths
1 Tangkoko-Batu Angus	A	Strato	1680, 1683, 1694, 1801, 1821, 1843-45, 1980.	0
2 Klabat	B		-	-
3 Lokon-Empung	A		1350, ?1750, 1829, 1893, 1930, 1942, 1949, 1952, 1953, 1958, 1959, 1961, 1969, 1970, 1986.	0
4 Mahawu/Rumengan	A		1789, 1846, 1904, 1958.	1
5 Tampusu	C		-	0
6 Lahendong	C		-	0
7 Sarongsong	C		-	0
8 Soputan-Aeseput	A		1785 or 86, 1819, 1833 or 38, 1845, 1890, 1901, 1906, 1907, 1908-09, 1919, 1911-12, 1913, 1915, 1917, 1923-24, 1925, 1947, 1966, 1984, 1985.	0
9 Sempu (Kawah Masem)	A		-	-
10 Batu Kolok	C		-	-
11 Tempang (Tempaso)	C		-	0
12 Ambang	C		-	0
13 Una-una	A	Strato + crater lake	1898, 1983.	0
14 Submarine volcano with no name	A		1922.	0
15 Awu	A	Strato + crater lake	1640, 1641, ?1870, 1711, 1812, 1856, 1875, 1885, 1892, 1893, 1913, 1921, 1922, 1931, 1952, 1966.	7,300
16 Bonua Wuhu			1835, 1889, 1895, 1904, 1918, 1919, ?1930.	0
17 Api Siau	A	Strato, lava rich	1675, 1712, 1825, 1864, 1883, 1886, 1887, 1892, 1899, 1900, 1905, 1921, 1922, 1924, 1926, 1930, 1935, 1940, 1941, 1947, 1948, 1949, 1953, 1961, 1974, 1983, 1985.	1
18 Ruang (Duang)	A	Strato with lava dome	1808, 1810, 1840, 1856, 1866, 1871, 1874, 1889, 1904-05, 1914, 1946, 1949.	350

A - volcano with eruptions in historical time
B - volcano in solfatara and fumarole stage
C - solfatara and fumarole field

After van Bemmelen 1970; Anon. 1979

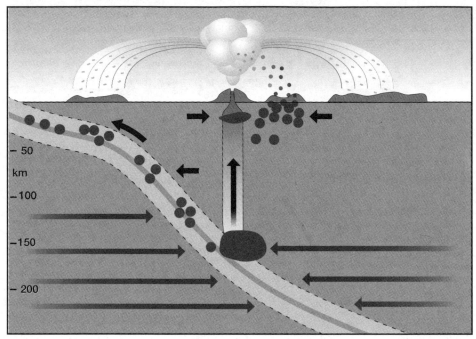

Figure 1.5. Vertical section through North and Central Sulawesi showing the Benioff Zone below Una-una.
Based on Katili and Sudrajat 1984

enormous forces and friction involved generate both earthquakes and heat, and the zone where these occur is referred to as the *Benioff Zone*. The heat can be so intense that the rocks of the descending plate melt, and the molten material or magma forces its way upwards. In most cases this magma never reaches the surface, hundreds of kilometers above, but cools down in pockets in the earth's crust. The magma which does reach the surface, however, is ejected in volcanic eruptions as lava or pyroclastic deposits such as ash and larger rock particles (fig. 1.5).

Minerals

Petroleum. Petroleum deposits are the remains of microscopic plants and animals which were buried in mud and sand of shallow prehistoric seas, underwent slow decomposition by bacteria and left a residue of hydrocarbon compounds which, under conditions of high temperature and pressure, were converted into oil and gas. Petroleum in Indonesia is usually found in thick Tertiary deposits (fig. 1.6) but there are leakages of natural gas in Tanjung Api Nature Reserve on the northern coast of the northeast peninsula of Sulawesi, and of oil at various inland and coastal locations in the northeast arm. The largest is in a mangrove area in Kolo Bay on the southern shore where the people have used the viscous oil for corking boats. The first oil well on Sulawesi was drilled in 1902 over one such deposit near the mouth of the Lariang River in northwest South Sulawesi. A gas field to the northeast of Lake Tempe has recently been found by British Petroleum and it is possible that an ammonia or urea plant might one day be built nearby. Development has been postponed, however, due to the low international price of liquified natural gas (Anon. 1986). The first commercially-viable source of oil in the Sulawesi region was found in 1985 by Union Texas 25 km south of Kolo Bay on the southern coast of the northeast Peninsula, and more test wells are planned.

Asphalt. Asphalt is a black, sticky mixture of bitumen or tarry hydrocarbons and mineral matter. About 70,000 ha of limestone in the southeast of Butung is impregnated with asphalt to the extent of 10%-40% by weight (fig. 1.7). These hydrocarbon mixtures have migrated upwards along faults above deep deposits into recent, relatively porous rocks and the lighter fractions have evaporated, leaving the viscous asphalt behind. The asphalt deposits with concentrations of 20%-30% were first exploited in the 1920s, primarily for tarring roads, and remain the only source of natural asphalt in Southeast Asia (van Bemmelen 1970; Anon. 1985a). About 500,000 tons of asphaltic rock are processed each year (Anon. 1984a).

Coal. Coal, the fossilized remains of plants, has never been mined in Sulawesi and occurs only around Pangkajene, Enrekang, Makale and the Karama River, all in the southwest peninsula. None of these young Tertiary deposits is economically viable (van Bemmelen 1970).

Limestone. Just north of Maros at the edge of the karst hills of elevated Miocene coral reefs (p. 468), are the P.T. Semen Tonasa limestone quarry and cement factories from which about 400,000 tons of cement is produced each year, which supplies the entire needs of eastern Indonesia (Anon. 1984a).

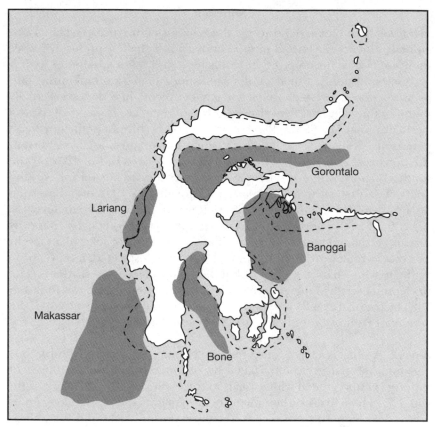

Figure 1.6. Distribution of Tertiary deposits (dark shade) and the location of past and present petroleum exploration activities (dashed lines).
After Anon. 1984a

Copper. Copper ores are found primarily in the northern arm and near the nickel areas of Soroako. None has yet been mined but there were plans to do so just east of Gorontalo[6] (Lowder and Dow 1978), and possibly in the Latimojong Mountains of the southwest peninsula. An analysis of leaves from herbs collected on Salayar Island, south of the southwest peninsula, revealed high concentrations of copper (80-600 mg/g compared with normal concentrations of <50 mg/g), indicating that rich deposits might also be present there[7] (Brooks et al. 1978).

Gold. Gold is generally associated with copper deposits, and the exploration for copper in North Sulawesi has revealed locations of potential

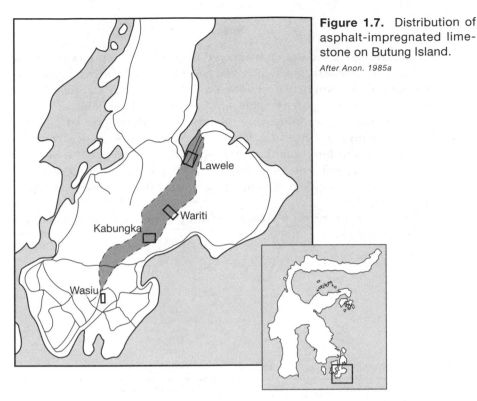

Figure 1.7. Distribution of asphalt-impregnated limestone on Butung Island.

After Anon. 1985a

large, commercial gold mines. Four mines that were worked early this century have been exhausted for commercial exploitation. New exploration licenses, however, have recently been issued.

Nickel. Nickel ores are derived from the weathering of ultrabasic rocks and are also found in the molasse deposits derived from these rocks. The ore deposits around Soroako and Lake Matano are mined by P.T. Inco and those around Pomala'a, south of Kolaka, are mined by P.T. Aneka Tambang.

Minor Products. Deposits of sulphur are known only from Soputan and Mahawu volcanoes in Minahasa and these have been exploited since the 1920s, although production is currently not very active. Kaolin or white clay is produced on a small scale in North Sulawesi for the ceramics industry. Salt is produced in coastal salt pans by small-scale operators in South Sulawesi. Quartz sands and silica were mined in South Sulawesi up to 1977 since when it has not been economically viable to do so.

SOILS

The description of soil types is hampered because there is no system of soil classification that has gained universal acceptance in terms of definitions or names. From experience and observation it has been said that "scientists who are otherwise reasonable and unemotional are liable to behave quite differently when discussing this topic" (Mulcahy and Humphries 1967). In many cases, hybrid systems have evolved using names from various systems with the result that those with little knowledge can become extremely confused. For the purposes of this section the FAO system is used because it has wide recognition and because the names are easily pronounceable.

The major soil types on Sulawesi and their approximate distributions are shown below (table 1.3; fig. 1.8).

Table 1.3. Major soils of Sulawesi and their characteristics.

	Alternative	Characteristics
Histosol	Organosol	Peaty soils.
Fluvisol	Alluvium	Soils transported by rivers; any horizons* generally relate to the history of deposition rather than to processes of soil development.
Regosol	Entisol	Weakly developed on sands.
Rendzina	Mollisol	Shallow calcareous soils over limestone.
Vertisol	Grumusol	Dark, cracking clays.
Andosol	Inceptisol	Soils on recent volcanic material.
Luvisol	Mediterranean	Soils with clay particles deposited in the B horizon (argillated), and a base saturation >50%.
Ferric luvisol	Ferrugineous, Red-yellow podsolic	As luvisol but with red mottles or concretions, apparent cation exchange capacity of clay fraction <24 meq./100 g.
Ferralsol	Latosol, oxisol	Highly weathered soils with few or no weatherable minerals, clay minerals dominantly kaolinite and capacity of clay fraction <16 meq./100 g.
Podzol	Spodosol	Soils with a bleached A horizon and a B horizon rich in secondary humus or iron.

* Horizons are layers of soil roughly parallel to the surface which have fairly distinctive characteristics.

After Young 1976

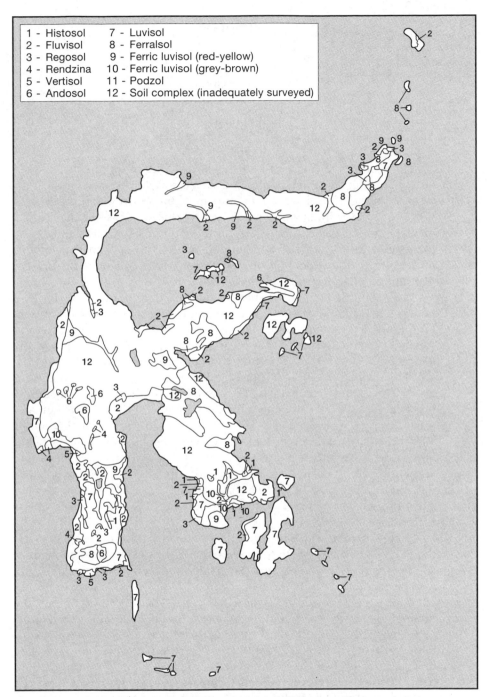

1 - Histosol	7 - Luvisol
2 - Fluvisol	8 - Ferralsol
3 - Regosol	9 - Ferric luvisol (red-yellow)
4 - Rendzina	10 - Ferric luvisol (grey-brown)
5 - Vertisol	11 - Podzol
6 - Andosol	12 - Soil complex (inadequately surveyed)

Figure 1.8. Approximate distribution of major soil types on Sulawesi.

After Anon. 1976

CLIMATE

Palaeoclimate

The palaeoclimate of Sulawesi does not appear to have been studied specifically, but analyses and summaries of palaeoclimate in Southeast Asia are available (Verstappen 1980; Flenley 1980; Whitmore 1981; Walker 1982; Morley and Flenley in press).

The climates of Sulawesi and the rest of Indonesia today, are quite unlike the climates which dominated the region during most of the Quaternary and before. As was shown on page 2, the world's landmasses have moved around, joining and separating, and this has led to changes in climatic regimes. Tropical and subtropical conditions, and the animals and plants associated with them, extended further away from the equator during the Tertiary than they do now and this has influenced the present distribution of animals and plants (p. 63).

During the latter part of the Quaternary, temperatures in the temperate areas of the world repeatedly rose and fell above and below present temperatures. In the cooler periods the ice sheets of the Arctic and Antarctic extended and this took great quantities of water out of the hydrological cycle (fig. 1.9). This in turn caused sea-levels around the world to fall. The maximum fall in Southeast Asia was about 150 m below present levels. This exposed large areas of dry land beyond the present coastlines–indeed, it uncovered three times the present area of the Sunda Shelf and twice the present area of the Sahul Shelf[8] When the sea-level was only 40 m below present levels it would have been possible to walk in a straight line between Banjarmasin and Surabaya, Saigon and Kuching, Singapore and Pontianak, and Merauke and Darwin. The effective area of Sulawesi was also increased but to a much lesser extent (fig. 1.10). Examination of bathymetric contours reveals features that may have been river valleys when the sea-level was lower (fig. 1.11).

Ocean currents which now enter the Indonesian Archipelago through the Torres Straits, the South China Sea, the narrow straits between many of the Lesser Sunda Islands and the straits between Mindanao and the Sangihe/Talaud Islands would have been blocked and their buffering effect on climate would have been lost. The Sulawesi Sea (between the Philippines and Sulawesi) and the Makassar Straits would have been much more enclosed but the currents entering the Molucca Sea (between Halmahera and Sulawesi) towards and from eastern Sulawesi would have been only marginally obstructed. The main Sunda and Sahul landmasses would have experienced a more continental climate (greater diurnal temperature range, lower rainfall and humidity) but this would have been rather less pronounced on Sulawesi. The lowering of sea and land temperatures

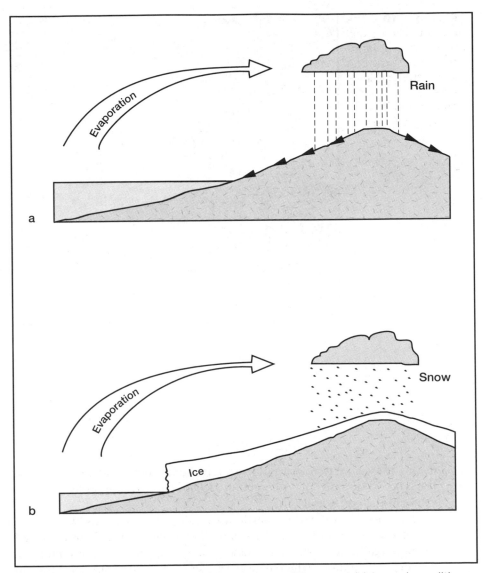

Figure 1.9. Hydrological cycles (a) in warm conditions and (b) in cool conditions. Note the fall in sea-level because water in the form of ice is unable to flow to the sea.

Figure 1.10. The area of Sulawesi and neighbouring land-masses exposed when the sea-level was 100 m below present levels.

After van Balgooy in press

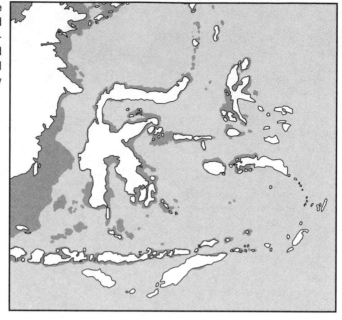

during these cool periods would also have reduced rainfall and humidity. It has been estimated that the rainfall 11,000 years ago was 30% of present values in the equatorial zone.

The cooling of the whole earth during the glacial periods lowered the lowest altitudes at which ice remained all year and snow fell, and lowered the upper altitude at which montane trees grew. The maximum temperature depression during the Quaternary occurred 18,000 years ago when, in New Guinea, temperatures at 2,500 m were 10°C lower, but at sea-level only 2° or 3°C lower, than at present. The tree line (the altitude at which trees are no longer able to grow) was lowered about 1,500 m in New Guinea but only about 350-500 m in Sumatra. The sea-level changes also had considerable impact on corals (p. 215).

The zone where the curtains of rising air of the northern and southern hemisphere meet is called the Intertropical Convergence Zone (ITCZ), and in the cool, dry glacial periods this was probably south of its present position (that is roughly over the equator). It is a zone of frequent, showery

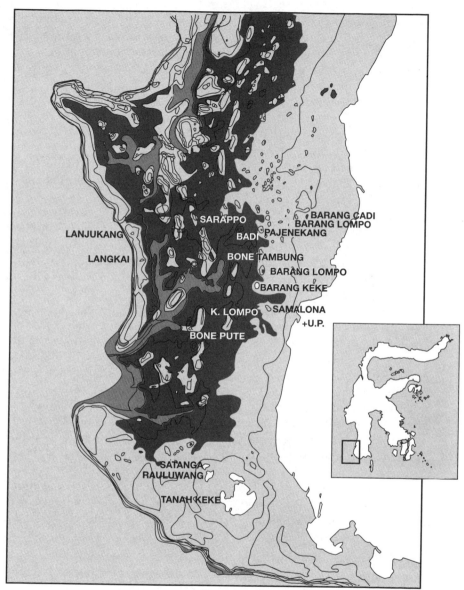

Figure 1.11. Bathymetric map of the Sangkarang Archipelago to show the deep channels that are probably drowned riverbeds.

rain, to the north of which is a high pressure belt with relatively low rainfall. Thus, when the ITCZ lay south of the equator, parts of northern Indonesia (including Sulawesi) would have experienced drier, more seasonal climate with lower rainfall and humidity and greater seasonal change in mean daily temperature (Verstappen 1980).

Temperatures during the warmest parts of the Quaternary were only 1° or 2°C higher than now (at sea-level) and at this time the water released from the polar ice caps caused sea-levels to rise. There is no unequivocal evidence that shorelines have been more than 6 m higher than at present during the warm periods of the Holocene, but sea-levels could have reached up to 25 m above present levels during the Pleistocene. The most recent sea-level maxima detected off the southwest peninsula were 4,500 and 1,600 years ago when sea-level was 5 m and 2.5 m higher respectively (fig. 1.12) (de Klerk 1983). This agrees closely with evidence from elsewhere in Southeast Asia (Tjia 1980; Tjia et al. 1984). Whereas this rise had a marked effect on the long shorelines of low-lying parts of eastern Sumatra and southern Borneo, the most marked effect on Sulawesi would have been the separation of the blocks of land either side of the Tempe depression. Evidence for this has been found in the vegetation record (p. 29) and there are even stories among local people of a time when travellers did not have to sail around the southern tip of South Sulawesi but could instead sail from the Gulf of Bone, through Lake Tempe and emerge in the Makassar Straits (Sartono 1982). With the exception of the Tempe depression and a few other flat plains (such as Malengke), most of Sulawesi's coastline slopes quite sharply and minor rises in sea-level would not have had significant effects. During this period seasonality in rainfall would have been less, rainfall would have been similar to or even greater than now, and mountain zones of climate, vegetation and fauna would have been raised. However, as stated above, this period occupied only a small fraction of the Quaternary. The majority of the period was characterized by lower rainfall and humidity, greater diurnal and seasonal variations, and by more marked rain shadows. Thus, the seasonal areas of Sulawesi and elsewhere would have been more extensive and, conversely the areas subject to more stable, wetter climates would have been reduced in area.

Present Climate

The climate of Sulawesi is best described with reference to rainfall since temperature is relatively constant, and other climatic variables such as wind velocity, evaporation and humidity change within even small areas. Between September and March, cool northwesterly winds pick up moisture while crossing the South China Sea (between East and West Malaysia, Philippines and Vietnam) and arrive in North Sulawesi via the Sulawesi Sea in about November, and in the west coast of South Sulawesi via the

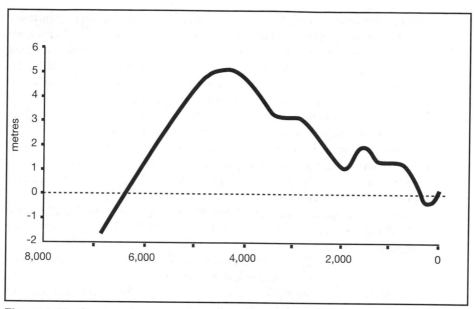

Figure 1.12. Changes in sea-level over the last 7,000 years determined from a study in the southwest peninsula.

After de Klerk 1983

Java Sea in late November or early December. The west coast of the central part of Sulawesi is sheltered from the effects of these winds by Borneo.

After this period, variable, humid, southeasterly winds blow towards eastern Sulawesi and rainfall peaks on the southeast coast occur between April and June, and on the northeast coast somewhat later. The southeasterly winds from the dry and wintery Australian landmass become stronger and these dry winds have a significant influence on the southern tips of the southwest and southeast peninsulas. Manado experiences a short dry season from August to October, but Jeneponto in the south of the southwest peninsula is subject to a long dry season between April and November.

Areas on the west coast of Sulawesi therefore tend to have their highest rainfall in December whereas those on the east coast have their wettest month around May. One might expect to find intermediate areas with two dry seasons (a bimodal distribution) and this is indeed the case; Pendolo and Pinrang in the middle of the southwest peninsula are examples.

Where the orientation of a range of mountains is more or less at right angles to the prevailing winds, the rainfall is higher on the windward side because the water in the air rises and cools as it climbs over the mountains, and this moisture is released as rain. Thus Maros receives over 500 mm per month between December and February but towns on the leeward side of the peninsula receive little rain. Valleys orientated in a north-south direction are in a rain shadow for virtually the whole year and the sheltered nature of the central western coast results in the Palu valley being one of the driest areas of Indonesia with less than 100 mm of rain, on average, falling in each month and an annual total of less than 600 mm.

Various authors have mapped the climatic zones of Sulawesi. The map which corresponds closest to the distribution of vegetation is that which uses the ratio between dry and wet periods (fig. 1.13) (Schmidt and Ferguson 1951; Whitmore 1984a, b).

A second map based on suitability criteria for growing rice has also been devised (fig. 1.14) (Oldemann and Darmiyati 1977), in which five major zones are recognized.

Zone A - an area with ten to twelve consecutive wet months and two or less consecutive dry months;

Zone B - an area with seven to nine consecutive wet months and three or less consecutive dry months;

Zone C - an area with five or six consecutive wet months and three or less consecutive dry months;

Zone D - an area with three or four consecutive wet months and two to six consecutive dry months;

Zone E - an area with zero to two consecutive wet months and up to six consecutive dry months.

'Wet' and 'dry' are defined as more than 200 mm and less than 100 mm of rain per month respectively.

Sulawesi has a greater percentage of its area in agroclimatic Zone E than have the islands around it, but more in Zones B and C than Borneo (most of which is in Zone A), Nusa Tenggara and Bali, or the Moluccas (table 1.4).

The most recent and complex climatic map of Sulawesi concerns bioclimate (Fontanel and Chantefort 1978) and recognizes many zones determined by three criteria: mean temperature of the coldest month, mean annual rainfall (fig. 1.15), and number of dry months[9] (fig. 1.16). One of the major differences between this map and that on agroclimatology is that it takes into account altitude effects, although not in an absolute sense since the degree of exposure to prevailing winds has a significant effect (p. 21).

The maps discussed above are based on long-term averages but uncommon climatic events, particularly periodic drought, can be extremely important in determining the distribution of certain animals and plants. Most crops, for example, experience stress after only about four days without rain. The variation in annual rainfall is quite considerable (fig.

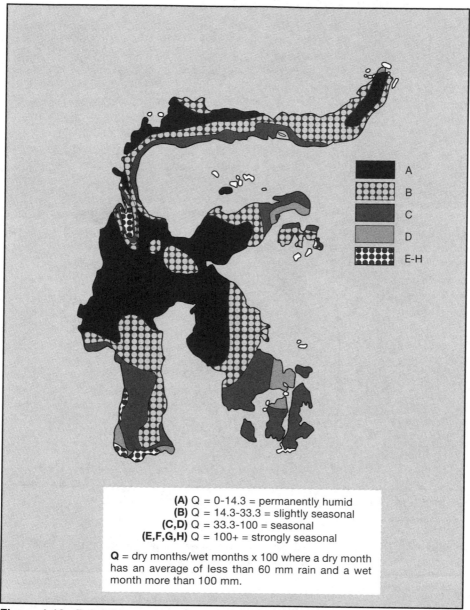

Figure 1.13. Rainfall types based on dry/wet period ratios.

After Schmidt and Ferguson 1951; Whitmore 1984a, b

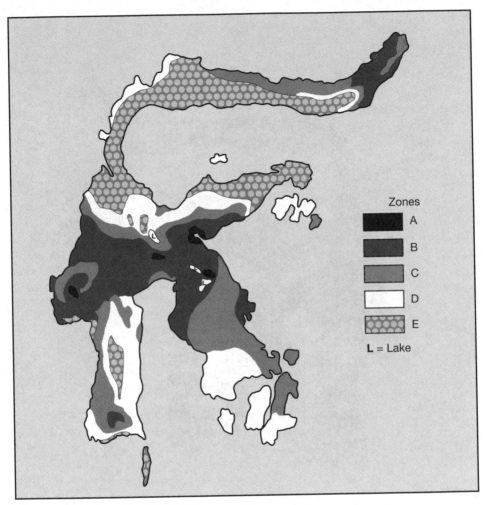

Figure 1.14. Agroclimatic zones.

After Oldemann and Darmiyati 1977

Table 1.4. The percentage of land area of Sulawesi and neighbouring regions of Indonesia falling within five agroclimatic zones.

| | Agroclimatic Zones | | | | |
	A	B	C	D	E
Kalimantan	43	32	15	4	6
Java	4	23	39	25	9
Sulawesi	1	25	25	22	27
Lesser Sundas + Bali	< 1	< 1	3	69	26
Moluccas*	5	11	40	12	20
Irian Jaya	48	23	17	8	4

* The figures in the text do not add up to 100%.

After Oldemann and Darmiyati 1977; Oldemann et al. 1980

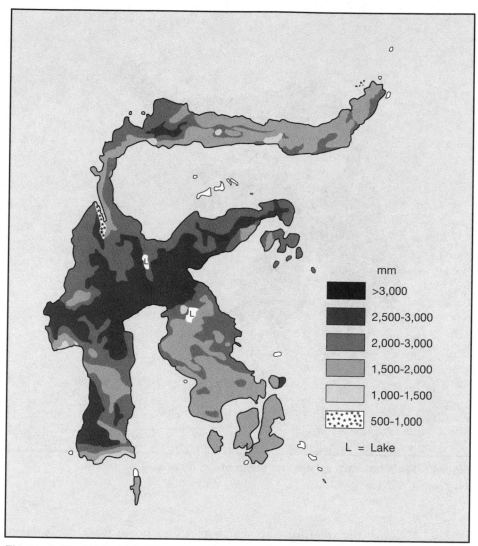

mm

■ >3,000

■ 2,500-3,000

■ 2,000-3,000

□ 1,500-2,000

□ 1,000-1,500

✴ 500-1,000

L = Lake

Figure 1.15. Areas with different mean annual rainfall.

After Fontanel and Chantefort 1978

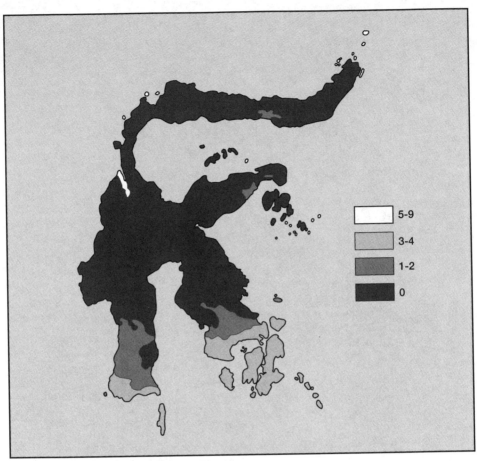

Figure 1.16. Areas with different numbers of dry months.
After Fontanel and Chantefort 1978

1.17) with the total for one year sometimes being twice the total of another. Meteorological records normally extract maximum rainfall figures but these are not particularly meaningful ecologically because above a certain quantity, rain will simply run off saturated soils to rivers. Lack of water, with its associated cloudlessness, high temperatures and low humidity, is a much more potent factor and an examination of rainfall minima and their distributions reveal that these too are extremely variable (fig. 1.18). Dry seasons can in fact only be defined by probability since, for the stations examined, at least six different months were recorded as 'driest months' in at least one year. A single location (e.g., Mapanget or Watampone) can have minimum monthly rainfalls between years ranging from 0 to 100+ mm.

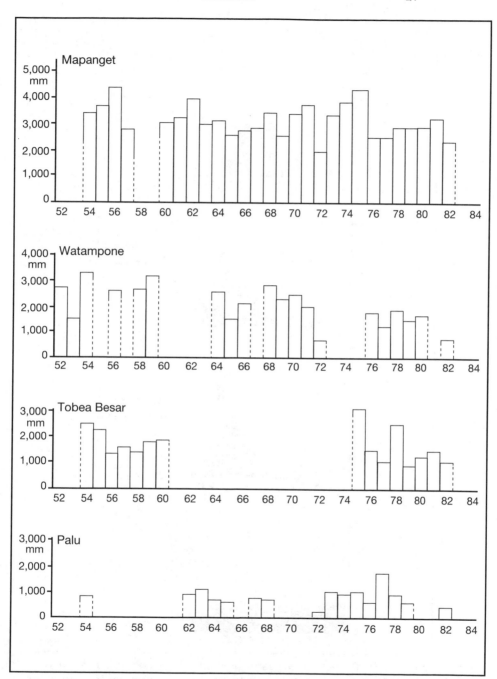

Figure 1.17. Annual rainfall at Mapanget (Manado airport), Watampone, Tobea Besar (an island between Butung and the mainland) and Palu.

Data obtained from the Directorate of Meteorology, Jakarta

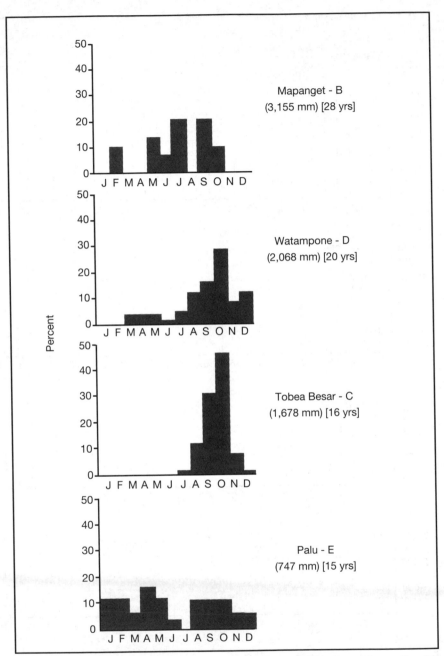

Figure 1.18. Percentage of years' driest months occurring in each month with agroclimatic zone, average annual rainfall in parentheses, and years of complete data in brackets.

From data obtained from the Directorate of the Meteorology, Jakarta

Droughts in Sulawesi seem to occur about every 20-30 years but information on their ecological effects have not been found. The grassy Sook Plain in Sabah, however, was formed when fire burned rainforest and the peaty topsoil on which it grew after an exceptionally dry period in 1915 (Cockburn 1974). In 1972-73 a prolonged dry spell caused changes on Mt. Kinabalu, Sabah, which it has been estimated will take a century to be reversed (Whitmore 1984a). In most areas of the world, but particularly around the Pacific region, 1982-83 was exceptionally warm and dry, and weather patterns were most unusual. For example, Watampone had 5, and Tobea Besar 6, consecutive rainless months at the end of 1982. This was related to the anomalous sea surface temperatures which in parts of the Pacific rose by as much as 8°C above normal. Around Sulawesi, however, the sea was only about 0.2°C warmer than normal. The sea surface temperature anomaly first appeared in the region of the Sulawesi Sea during the middle of 1982 but its major effects were not felt until December 1982-February 1983 (Barber and Chavez 1983; Cane 1983; Gill and Rasmusson 1983; Rasmusson and Wallace 1983; Chavez et al. 1984; Whitmore 1984a). Although 1982 was generally a very dry year on Sulawesi, 1972 seems to have been even drier.

Vegetation

Palaeovegetation

Impressions of leaves of grassy plants and rattans have been found in rocks in Minahasa (Koorders 1895) but the vast majority of information regarding palaeovegetation is derived from the careful analysis of pollen remains in deep organic sediments from the bottom of swamps or lakes. The oldest-known Sulawesi pollen is of a relative of *Sonneratia* mangrove trees from Tertiary rocks in north Central Sulawesi (Sohma 1973).

Attention of palaeobotanists on Sulawesi has been concentrated on two areas: Lake Tempe and in topographic depressions in ultrabasic rocks around Lake Matano. One particularly long core of possibly considerable age has been collected recently from one of the latter sites but results will not be available for some while (Hope 1986).

In the Lake Tempe region cores were taken from the swampy Lake Rawa Lampulung about 4 km east of Sengkang and the pollen shows that at least part of the surrounding area was covered by mangrove vegetation, that is, it was inundated by the sea, from about 7,100 to 2,600 years ago. The rise in sea-level needed to produce this effect is about 5 m which matches the palaeo-climate information (p. 21). A core from the edge of

Lake Tempe, 5 km northwest of Sengkang, reveals that at least from 4,400 years ago the area was dominated by freshwater vegetation and that the sea was probably prevented from reaching the lake by a squat molasse ridge to the east.

The history of the vegetation during the Quaternary and late Tertiary is very closely linked with the climatic changes. During the drier periods of the Pleistocene the area of seasonal forest would have extended while the area of rain forest became less. Thus the populations of rain forest species would have been reduced but isolated populations of certain species probably maintained some level of contact or gene flow along riverine areas or wetter soils. In the driest parts of Sulawesi it is quite likely that monsoon or even savannah forest predominated.

Many mountain plants are unable to live successfully in the lowlands, but the lowering of zones on mountains during the cool periods gave opportunities for these plants to spread because the available stepping-stones of suitable habitat increased in area and number.

Present Vegetation

Fewer botanical specimens have been collected on Sulawesi than on any other major island/region in Indonesia. To date only about 23 specimens per 100 km^2 have been placed in herbaria whereas over 200 per 100 km^2 are known from Java. A density of 100 specimens per 100 km^2 would represent an adequately-known flora. Allowing for this, the number of higher plant species[10] may be about 5,000. Only seven genera are known to be endemic compared with 17 in Sumatra, 59 in Borneo and 124 in New Guinea (E. de Vogel pers. comm.). In addition, some species are barely known: a forest shrub *Thottea celebica* (Aris.), for example, has been collected only once, from Lambarese, northeast of Palopo (fig. 1.19) (Ding Hou 1984).

The natural vegetation[11] growing in a particular area is dependent on various factors such as soil chemistry, soil water, climate, altitude, distance from the sea, and distance from areas of similar conditions (table 1.5; fig. 1.20).

The coasts of Sulawesi are fringed by coral reefs, mudflats, mangrove forests and rocky or sandy beaches (chapters 2 and 3). The freshwater habitats are generally rather nutrient-poor (although there are striking exceptions) and as a result freshwater vegetation is not well developed (chapter 4). The lowland and hill forests of Sulawesi (chapter 5) have the most tree species of all forest types (table 1.5) but have only seven species of dipterocarp trees, the mainstay of forestry operations in Borneo and Sumatra where there are 267 and 106 species respectively (Ashton 1982). The major trees of commerce are the tall *Agathis* (Arau.)[12] trees with broad, flat leaves; the magnificent yellow-flowered legume *Pterocarpus indicus* (Legu.) (huge pollarded specimens of which are commonly seen

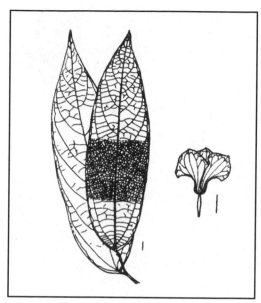

Figure 1.19. *Thottea celebica* leaves and flower; one of a number of endemic plants known from just one specimen. Scale bars indicate 1 cm.

After Ding Hou 1984

Table 1.5. The natural vegetation types on Sulawesi. A more detailed classification has been constructed by Kartawinata (1980).

Climate	Soil water	Location	Soils	Altitude	Vegetation type
Ever-wet	Dryland	Inland	Zonal	Lowlands, <1,000 m	Lowland forest
				Montane, 1,000-2,100 m	Lower montane forest
				Montane, 2,100-3,250 m	Upper montane forest
				Montane, 3,250-3,450 m	Subalpine forest
			Ultrabasic	Mainly lowlands	Forest on ultrabasics
			Limestone	Mainly lowlands	Forest on limestone
		Coastal			Beach vegetation
	High water-table (at least periodically)	Freshwater	Peat		Peatswamp forest
			Muck		Freshwater swamp forest
		Brackish	Saline clays		Brackish water forest
		Salt water	Saline clays		Mangrove forest
Clear seasonal shortage of rain			Zonal	Lowlands	Lowland monsoon forests
			Ultrabasic	Mainly lowlands	Monsoon forest on ultrabasic
			Limestone	Mainly lowlands	Monsoon forest on limestone

Based on van Steenis 1950; Whitmore 1984a

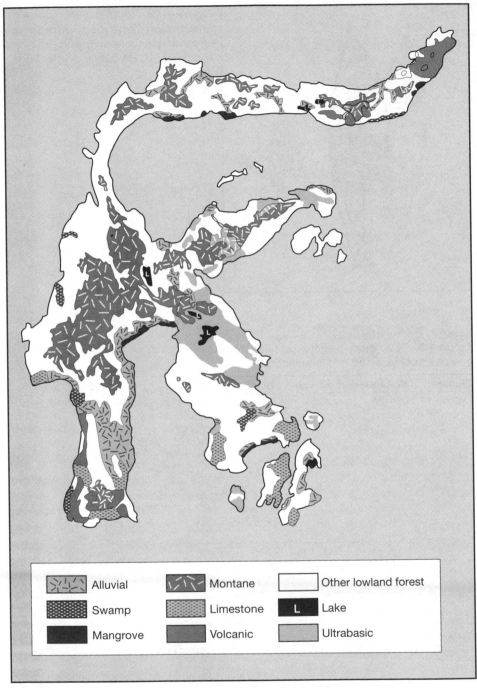

Figure 1.20. Major natural vegetation types on Sulawesi.

Based on Anon 1982a

around town squares such as Ujung Pandang) which is deciduous and found most commonly in the more seasonal areas; the gum tree *Eucalyptus deglupta* (Myrt.), usually found wild in riverine habitats and which is extensively used in reforestation projects; beremban *Duabanga mollucana* (Sonn.), and gutta percha *Palaquium* spp. (Sapo.). The hemi-parasitic[13] sandalwood *Santalum album* (Sant.), from the heartwood of which sandal wood oil is extracted, used to be found in the dry Palu valley (chapter 6) but even in the Poboya reserve set up to protect one of the last stands of this tree, only 62 small individuals exist (Sidiyasa and Tantra 1984).

Sulawesi has both freshwater and peatswamp forests, though neither is particularly extensive, as well as forests growing on volcanic, limestone and ultrabasic soils (chapter 6). The mountain forests have tree species absent from or rarely found in the lowlands, and the vegetation of the mountain summits includes many brightly-coloured shrubs and herbs (chapter 7).

A large number of areas, particularly in the southwest peninsula, have lost their original vegetation because of the activities of man, and are now made up largely of agricultural landscapes with areas of dry, barren land used to some extent for cattle grazing. The driest areas, in the Palu valley and in the south of the southwest arm, have a generally open appearance with a scrub of prickly pear cactus and other drought-tolerant species. Many grasslands are dominated by alang-alang grass *Imperata cylindrica* (Gram.) (Steup 1939), but it is not correct to label all grasslands as alang-alang wastelands; some grasslands comprise communities of numerous

Table 1.6. Approximate number of species of trees over 15 cm diameter at breast height to be found in 0.5 ha of various habitats in Sulawesi and adjacent regions.

	Kalimantan	Sulawesi	Moluccas	Irian Jaya
Upper Montane Forest	22	10	11	15
Fertile Lower Montane Forest	61	49	32	62
Infertile Lower Montane Forest	43	23	25	41
Fertile Hill Forest	64	45	35	63
Infertile Hill Forest	44	34	19	40
Fertile Lowland Forest	89	58	37	75
Infertile Lowland Forest	50	34	28	44
Monsoon Forest	-	19	30	32
Beach Forest	25	18	13	20
Volcanic Scrub	-	10	11	17
Complex Mangrove	15	7	8	13
Simple Mangrove	6	4	4	6
Swamp Forest	45	25	29	46

Based on Anon. 1982a

grasses (not including alang-alang) and small legumes (Steup 1939). Sword grass occurs predominantly along roadsides and around villages. Grassy savannas often have scattered trees of a variety of species such as *Morinda tinctoria* (Rubi.), *Albizia procera* (Legu.) and, where burning has been frequent, *Fagraea fragrans* (Loga.). There are also areas where teak *Tectona grandis* (Verb.) is common but these are usually the remnants of old plantations (Steup 1939). Teak was introduced to Java and South Sulawesi many centuries ago from India and Burma (Altona 1922; Carthaus 1909).

FAUNA

Palaeofauna

The palaeofauna of Sulawesi is known from just two sets of sites: river sediments near Sompoh, Beru and Celeko in Soppeng district about 100 km northeast of Ujung Pandang, and various limestone caves near Maros. The animals found in the first of these have been called the Cabenge fauna and probably date from the Late Pliocene (more than 1 Ma ago) (Sartono 1979; Hooijer 1982). The animals in the cave sites are known as the Toalian fauna and are of relatively very recent origin, dating from perhaps 30,000 years ago (Hooijer 1950) (table 1.7).

The stegodonts would have looked similar to modern elephants except that the males had huge curving tusks that were so close together that the trunk must have been draped over the sides of the tusks. The pygmy elephant (fig. 1.21), was descended from the prehistoric African elephant *Elephas ekorensis* (the extant African elephant is *Loxodonta africana*) and probably left the lineage of the modern Asian elephant *E. maximus* about 3 million years ago.

The extinct giant pig had peculiarly large tusks in the upper jaw, nearly triangular in cross section, which pointed sideways and extended beyond the lower tusks. The fossil babirusa teeth are larger than those of their living relatives, and the giant tortoise had a carapace nearly two metres long (Hooijer 1948b, 1982), larger than those confined today to the Galapagos Islands in the eastern Pacific and small islands in the western Indian Ocean. The Pleistocene anoa was a similar or possibly slightly smaller size than modern specimens. The large relative size of many Pleistocene fossils is a commonly observed phenomenon, possibly a result of the cooler temperatures then prevailing. Larger animals have a lower ratio of skin area to body volume and a consequently lower rate of heat loss. As a result, larger animals are better able to survive at low temperatures (Edwards 1967). Dwarfing of animals on islands generally occurs in species that are relatively large, such as elephants and anoas, whereas gigantism on islands generally occurs in small species such as rats.

The Pleistocene sharks are all still found around Sulawesi and occasionally in rivers, but *Hemipristis* is now represented only by a single rare species in the Red Sea and Zanzibar (Hooijer 1958). The stingray is also common in Indo-Pacific seas and rivers.

The animal remains found to date are much poorer in numbers and species than those from the Trinil fauna deposits in East Java. Whereas the Trinil deposits may be used to some extent to illustrate the fauna present

Table 1.7. Prehistoric fauna of Sulawesi known from two sites. Human artefacts found with the Cabenge fauna date from a much later period (Sartono 1979; Bartstra 1977, 1978).

Cabenge fauna		Toalian fauna	
Pygmy stegodont	*Stegodon sompoensis**	Bear cuscus	*Ailurops ursinus*
Java stegodont	*S.* cf *trigonocephalus**	Dwarf cuscus	*A. celebensis*
Sulawesi elephant	*Elephas celebensis* *	House shrew	*Suncus murinus*
Giant pig	*Celebochoerus heekereni* *	Moor macaque	*Macaca maura*
Sulawesi pig	*Sus celebensis*	Human	*Homo sapiens*
Babirusa	*Babyrousa babyrussa* ?	Giant rat	*Lenomys meyeri*
Lowland anoa	*Bubalus depressicornis*	Common forest rat	*Paruromys dominator*
Giant tortoise	*Geochelone atlas* *	Gray's rat	*Taeromys celebensis*
Soft-shelled turtle	*Chitra ?indica*	Pinadapa rat	*Rattus punicans*
Celebes crocodile	*Crocodylus* sp.*	Yellow rat	*R. xanthurus*
Tiger shark	*Galeacerdo cuvier*	Roof rat	*R. rattus*
Serra shark	*Hemipristis* cf *serra* *	Hoffman's rat	*R. hoffmani*
Ground shark	*Carcharhinus* cf *gangeticus*	Rat	*R.* sp.
Whale shark	*C.* cf *brachyurus*	Spiny rat	*Maxomys*
Sand shark	*Caricharias* cf *cuspidatus*		*musschenbroeckii*
Mackerel shark	*Isurus glaucus*	Sulawesi civet	*Macrogalidia*
Giant stingray	*Dasyatis* sp.*		*musschenbroeckii*
		Sulawesi pig	*Sus celebensis*
		Babirusa	*Babyrousa babyrussa*
		Lowland anoa	*Bubalus depressicornis*
		Mountain anoa	*B. quarlesi*
		Rice-field tortoise	*Cuora ?amboinensis*
		Snake	?
		Fish	?
		River molluscs	*Brotia perfecta*
			Melanoides crenulata
			M. cf *granifera*
			Thiara scabra
			Viviparidae

* extinct forms

After Hooijer 1948a, b, 1949, 1950, 1954, 1958, 1964, 1967, 1969, 1972; Clason 1976; 1982; Musser 1984

Figure 1.21. Relative sizes of the giant tortoise, 1.65 m man and a pygmy elephant.

over some of Sundaland, it is not possible to use the Trinil finds as a basis for discussion of the Sulawesi palaeofauna. Two of the four elephants but none of the three pigs in the Trinil deposits are known from Sulawesi (Hooijer 1974), and the pygmy stegodont on Sulawesi is now also recognized as the same species as the fossil stegodont found on Timor and Flores. The Plio-Pleistocene fauna of Sulawesi is thus not as distinctive as was once thought and the only species not known from outside Sulawesi is the giant pig.

The ancestors of the elephant and stegodonts could have reached Sulawesi via Taiwan and the Philippines (stegodont remains have been found on both islands). The 3,000-m deep sea between Java, Flores, Alor, Timor and Sulawesi might be thought to be too great for even water-loving elephants to cross by swimming (Hooijer 1967, 1970, 1972). If it is assumed that the presence of similar pygmy stegodonts on all these islands must be a result of migrations across land and only short water gaps, it is necessary to postulate massive downfaulting of at least 3,000 m in the intervening straits and seas. Before the downfaulting occurred there would have been routes from Flores to Sulawesi along the Kaloatoa ridge and then via Tanahjampea and Salayar, and from Java to Sulawesi via

Madura, the Kangean Islands and the Doangdoangan Islands (Audley-Charles and Hooijer 1973) or perhaps more easily from the Kangean Islands via the Sabalana and Postilyon Islands. The routes from Java may in fact have been viable even without downward faulting during the glacial periods of the Pleistocene during which sea-level dropped 100-130 m below present levels (p. 16).

A recent examination of the swimming powers of elephants has suggested, however, that such massive downfaulting need not have occurred. Elephants have been recorded swimming across lakes and seas to islands in Africa and Asia and the distance record is held by an unfortunate elephant that swam ashore after being washed overboard from a boat 48 km off the South Carolina coast in 1856. Aspects of elephant morphology such as the spongy skull bones and absence of a pleural cavity, the need to bathe regularly and their ancestry (dugongs, or sea cows, are among their closest relatives), suggest that their ancestors may have lived in a semi-aquatic habitat. Some of the distances prehistoric elephants would have had to have swim are in the upper limits of known ability and so the crossing of elephants to and from Sulawesi and other islands does not necessarily require that the sea bed was much higher, and exposed land more extensive, in the recent past (Johnson 1980).

Giant tortoises are said to be able to float and survive long periods in seawater. Thus *Geochelone atlas* could have drifted to Sulawesi from another landmass (Hooijer 1982), and in fact remains of the same species are found in deposits from Java, Timor and India.

Present Fauna

The fauna of Sulawesi is one of the most distinctive in all Indonesia particularly among the mammals (table 1.8). Of the 127 indigenous mammal species,[14] 79 (62%) are endemic[15] and the percentage rises to 98% if the bats are excluded (table 1.9). The mammal fauna is also characterized by its relatively primitive characters (Musser in press). New species of mammals continue to be found (e.g., Musser 1981, 1982; Bergmans and Rozendaal 1982; Hill 1983; Boeadi pers. comm.) and revisions of old and new specimens combined with fieldwork help to clarify the exact number and identity of species (e.g., Musser 1971a, b, 1973, 1977; Musser et al. 1982; Groves 1980a, b; Takenaka 1982).

One of the earliest descriptions by a European of a Sulawesi animal was of the curly-tusked babirusa *Babyrousa babyrussa* by Piso in 1658. It was kept and even bred by early rulers, perhaps as a gift for visiting diplomats, and it is conceivable that some were taken to Bali by Buginese seafarers. It is possible that knowledge of the babirusa influenced the way in which 'Raksasa', a Balinese demonic man-beast, was first drawn, for it has a curved tusk piercing each cheek (Groves 1980a).

The babirusa are enigmatic animals as their name, which means 'pig-deer', implies. They are variable in hair covering (from nearly naked to densely hairy), hair colour (from off-white to brown), and eye colour (from whitish, through grey to brown) (Wemmer and Watling 1982) and none of these characteristics appear to be related to sex. There may be geographical trends in these characters, however, but too few specimens with detailed location data are available. The babirusa is generally grouped with pigs but they have had no common ancestor since the Oligocene (about 30 Ma), and they are barely more similar to pigs than they are to hippos, remains of primitive forms of which have been found in Java (C. Groves pers. comm.).

Two dwarf buffaloes, or anoa, *Bubalus depressicornis* and *B. quarlesi* are said to be found in the lowlands and mountains respectively (Groves 1969) but it has been suggested that the differences in horn-shape, a major means of distinguishing the species,[16] may simply be a function of age (Wind and Amir 1978). Also the 'mountain' anoa is sometimes found at sea-level and the 'lowland' anoa is sometimes found on high mountains (Thornback 1983). The diurnal bear cuscus *Ailurops ursinus* is quite a large animal, with a head and body 45 cm long and a tail 55 cm long, and is quite frequently seen in lowland forests. It is very distinct morphologically from other cuscus and is in its own genus in recognition of its primitive characters (C. Groves pers. comm.). The smallest cuscus of all, *Strigocuscus celebensis* (head and body only 34 cm), is nocturnal, and consequently rarely seen.

The composition of the Sulawesi mammal fauna is very different from that of Borneo or Irian Jaya with many fewer families represented (fig. 1.22). The rats and bats are major constituents of the fauna as they are in the insular regions of the Moluccas and Lesser Sundas.

Table 1.8. Percentage of species of indigenous mammals (including bats), birds and reptiles endemic to the seven main areas of Indonesia.

	Mammals	Birds	Reptiles
Sumatra	10	2	11
Java	12	7	8
Borneo	18	6	24
Sulawesi	61	34	26
Lesser Sundas	12	30	22
Maluku	17	33	18
Irian Jaya	58	52	35

Based on Anon 1982b

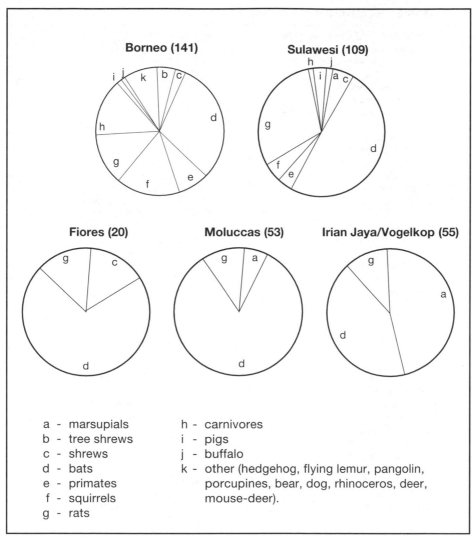

Borneo (141)

Sulawesi (109)

Flores (20)

Moluccas (53)

Irian Jaya/Vogelkop (55)

a - marsupials
b - tree shrews
c - shrews
d - bats
e - primates
f - squirrels
g - rats

h - carnivores
i - pigs
j - buffalo
k - other (hedgehog, flying lemur, pangolin,
   porcupines, bear, dog, rhinoceros, deer,
   mouse-deer).

Figure 1.22. The faunas of Borneo, Sulawesi, Flores, Maluku and Irian Jaya (Vogelkop) compared.

There are 332 species of birds known from Sulawesi of which 92 (27%) are endemic, and 81 (25%) are migratory (p. 145; table 1.10) (White 1974, 1976, 1977; White and Bruce 1986). New records of species not previously known from Sulawesi are still being made (Escott and Holmes

Table 1.9. Mammals of Sulawesi.

PHALANGERS Phalangeridae	Sulawesi naked-backed bat
Peleng cuscus	*Dobsonia exoleta*
Phalanger pelengensis (p)	Greenish naked-backed bat
Dwarf cuscus	*D. crenulata*
Strigocuscus celebensis	Toraja naked-backed bat
Spotted cuscus	*D.* sp. (n)
Spilocuscus maculatus (i,s)	Lesser dog-faced fruit bat
Bear cuscus	*Cynopterus brachyotis*
Ailurops ursinus	Sulawesi dog-faced fruit bat
SHREWS Soricidae	*C. **minor***
Long-tailed shrew	Black-capped fruit bat
*Crocidura **elongata***	*Chironax melanocephalus*
Black shrew	Swift fruit bat
*C. **lea***	*Thoopterus nigrescens*
Brown shrew	Sulawesi harpy fruit bat
*C. **levicula***	*Harpyionycteris celebensis*
Black-footed shrew	Pallas' tube-nosed bat
*C. **nigripes***	*Nyctimene cephalotes*
Pale-footed shrew	Lesser tube-nosed bat
*C. **rhoditis***	*N. minutus*
Unnamed species A	Common dawn bat
C. **sp.** A	*Eonycteris spelaea*
Unnamed species B	Common long-tongued nectar bat
C. **sp.** B	*Macroglossus minimus*
Unnamed species C	**TOMB BATS Emballonuridae**
C. **sp.** C (p)	Philippine sheath-tailed bat
House shrew	*Emballonura alecto*
Suncus murinus (i)	Malaysian sheath-tailed bat
FRUIT BATS Pteropidae	*E. monticola*
Common rousette	Lesser sheath-tailed bat
Rousettus amplexicaudatus	*E. nigrescens*
Sulawesi rousette	Black-bearded tomb bat
*R. **celebensis***	*Taphozous melanopogon*
Central flying fox	Bare-rumped tomb bat
Pteropus alecto	*T. saccolaimus*
Ashy-headed flying fox	**FALSE VAMPIRES Megadermatidae**
P. caniceps	Lesser false vampire
Grey flying fox	*Megaderma spasma*
P. griseus	**HORSESHOE BATS Rhinolophidae**
Masked flying fox	Arcuate horseshoe bat
P. personatus	*Rhinolophus arcuatus* (n)
Island flying fox	Sulawesi horseshoe bat
P. hypomelanus	*R. **celebensis***
Jentink's flying fox	Broad-eared horseshoe bat
Boneia bidens	*R. euryotis*
Sulawesi flying fox	Philippine horseshoe bat
*Acerodon **celebensis***	*R. philippinensis*
Small-toothed fruit bat	**LEAF-NOSED BATS Hipposideridae**
Neopteryx frosti	Dusky leaf-nosed bat
Stripe-faced fruit bat	*Hipposideros ater*
Styloctenium wallacei	Gould's leaf-nosed bat

Species and genus names in bold indicate Sulawesi endemics, (i) = introduced, (n) = first found by EoS teams, (p) = found on Peleng Island, (s) = found on Salayar Island.

Table 1.9. (Continued.)

H. cervinus	*Cheiromeles* **parvidens**
Diadem leaf-nosed bat	Sulawesi mastiff bat
H. diadema	*Tadarida* **sarasinorum**
Fierce leaf-nosed bat	**RATS Muridae**
H. dinops	Andrew's shrew-rat
Crested leaf-nosed bat	*Bunomys* **andrewsi**
H. inexpectatus	Golden shrew-rat
Large-eared leaf-nosed bat	*B.* **chrysosomus**
H. macrobullatus	Shrew-rat
EVENING BATS Vespertilionidae	*B.* **coelestis**
Grey large-footed bat	Lesser shrew-rat
Myotis adversus	*B.* **fratorum**
Black myotis	Heinrich's shrew-rat
M. ater	*B.* **heinrichi**
Hodgson's bat	Summit shrew-rat
M. formosus	*B.* **penitus**
Horsfield's myotis	Shrew-rat
M. horsfieldii	*B.* **prolatus**
Whiskered myotis	Compact mountain rat
M. muricola	*Crunomys* **celebensis**
Sulawesi false serotine	White-tailed shrew-rat
Hesperoptenus gaskelli	***Echiothrix leucura***
Brown pipistrelle	Large grey shrew-rat
Pipistrellus imbricatus	***Eropeplus canus***
Javan pipistrelle	Sulawesi pygmy tree-mouse
P. javanicus	*Haeromys* **minahassae**
Minahasa pipistrelle	Sulawesi giant-rat
P. **minahassae**	***Lenomys meyeri***
Peters' pipistrelle	Masked long-tailed tree-rat
P. petersi	***Margaretamys beccarii***
Sulawesi yellow bat	Large-long-tailed tree-rat
Scotophilus **celebensis**	*M.* **elegans**
Asiatic lesser yellow house bat	Small long-tailed tree-rat
S. kuhlii	*M.* **parvus**
Little long-fingered bat	Dollmann's spiny-rat
Miniopterus australis	*Maxomys* **dollmanni**
Small long-fingered bat	Hellwald's spiny-rat
M. pusillus	*M.* **hellwaldii**
Common long-fingered bat	Musschenbroek's spiny-rat
M. schreibersii	*M.* **musschenbroekii**
Greater long-fingered bat	Watt's spiny-rat
M. tristis	*M.* **wattsi**
Greater club-footed bat	Short-tailed shrew-rat
Tylonycteris robustula	***Melasmothrix naso***
Eastern tube-nosed bat	Dominator/Common forest rat
Murina florium (p)	***Paruromys dominator***
Hardwicke's trumpet-eared bat	Bear-like rat
Kerivoula hardwickei	*P.* **ursinus**
Papillose bat	Rice-field rat
K. papilosa	*Rattus argentiventer* (i)
Peters' trumpet-eared bat	Lompobatang rat
Phoniscus jagorii	*R.* **bontanus**
FREE-TAILED BATS Molossidae	Beautiful-tailed rat
Sulawesi hairless bat	*Rattus* **callitrichus**

Table 1.9. (Continued.)

Polynesian rat	Sulawesi long-nosed squirrel
R. exulans (i)	***Hyosciurus heinrichi***
Mekongga rat	Mekongga dwarf squirrel
R. foramineus	***Prosciurillus abtrusus***
Hoffman's rat	Pale dwarf squirrel
R. hoffmanni	*P. leucomus*
Koopman's rat	Northern dwarf squirrel
R. koopman (p)	*P. murinus*
Monkey-tailed rat	**PORCUPINES Hystricidae**
R. marmosurus	Javan porcupine
Rat	*Hystrix javanica* (i)
R. mollicomulus	**MONKEYS Cercopithecidae**
Himalayan rat	Moor macaque
R. nitidus (i)	*Macaca maura*
Black rat	Black-crested macaque
R. norvegicus (i)	*M. nigra*
Rat	Booted macaque
R. pelurus (p)	*M. ochreata*
Roof rat	Tonkean macaque
R. rattus (i)	*M. tonkeana*
Salacco rat	**TARSIERS Tarsiidae**
R. salacco	Diana tarsier
Yellow-tailed rat	*Tarsius dianae*
R. xanthurus	Moss tarsier
Mekongga rat	*T. pumilus*
Taeromys arcuatus	Sulawesi tarsier
Gray's rat	*T. spectrum*
T. celebensis	**CIVETS Viverridae**
Sulawesi large-tailed rat	Sulawesi civet
T. macrocercus	*Macrogalidia musschenbroeckii*
Hooked rat	Common palm civet
T. hamatus	*Paradoxurus hermaphroditus* (i)
Pinadapa rat	Malay civet
T. punicans	*Viverra tangalunga* (i)
Minahasa rat	**BUFFALO Bovidae**
T. taerae	Lowland anoa
Narrow-footed rat	*Bubalus depressicornis*
Tateomys rhinogradoides	Mountain anoa
House mouse	*B. quarlesi*
Mus musculus (i)	**DEER Cervidae**
SQUIRRELS Sciuridae	Rusa
Prevost's squirrel	*Cervus timorensis* (i)
Callosciurus prevostii (i)	**PIGS Suidae**
Plantain squirrel	Babirusa
C. notatus (i,s)	*Babyrousa babyrussa*
Red-belled squirrel	Sulawesi pig
Rubrisciurus rubriventer	*Sus celebensis*

After Groves 1976; Musser 1977, 1981 in press, pers. comm.; Jenkins and Hill 1981; Hill 1983 pers. comm.: Bergmans and Rozendaal in press;

Figure 1.23. Four of Sulawesi's endemic birds. **a** - Sulawesi crowned myna *Basilornis celebensis*, **b** - White-necked myna *Streptocitta albicollis*, **c** - Sulawesi dwarf hornbill *Penelopides exarhatus*, **d** - Piping crow *Corvus typicus*.
After Meyer and Wigglesworth 1898; Goodwin 1976

1980; Watling 1983). Among the resident birds, 17 genera are endemic to Sulawesi and its surrounding islands including a large number of spectacular endemic birds such as the dark-green bee-eater *Meropogon forsteni*, the large brightly-coloured hornbill *Rhyticeros cassidix*, the crowned myna *Basilornis celebensis*, and the finch-billed starling *Scissirostrum dubium* which nests in huge numbers in holes bored out of tall dead trees (fig. 1.23). Sulawesi's best-known bird is the maleo *Macrocephalon maleo* which incubates its eggs in pits that the adult birds dig (p. 155).

Whereas 100 species of amphibians have been collected on Borneo, only 29 are so far known from Sulawesi.[17] Of this total, four frequent habitats associated with man and have probably been accidentally introduced. Of the 25 indigenous species, 19 (76%) are endemic (table 1.11). The Talaud

Table 1.10. Distribution of Sulawesi endemic birds. Endemic genera shown in bold italic type. Note that for the purposes of this table the Sula islands are regarded as part of Sulawesi (p. 5). The endemic mountain drongo *Dicrurus montanus* now appears to be regarded as the same species as the widespread spangled drongo *D. hottentotus*.

		Mainland	Sangihe	Talaud	Banggai	Muna/Butung	Togian	Sula
EAGLES and HAWKS Accipitridae								
Sulawesi serpent-eagle	*Spilornis rufipectus*	•	-	-	•	•	-	•
Sulawesi goshawk	*Accipiter griseiceps*	•	-	-	-	•	•	-
Spot-tailed goshawk	*A. trinotatus*	•	-	-	•	-	-	-
Small sparrowhawk	*A. nanus*	•	-	-	-	-	-	-
Vinous-breasted sparrowhawk	*A. rhodogaster*	•	-	-	•	•	-	•
Sulawesi hawk-eagle	*Spizaetus lanceolatus*	•	-	-	?	•	-	•
MOUND BUILDERS Megapodiidae								
Sula scrub fowl	*Megapodius bernsteini*	-	-	-	•	-	-	•
Maleo	***Macrocephalon** maleo*	•	-	-	-	-	-	-
RAILS Rallidae								
Sulawesi rail	***Aramidopsis** plateni*	•	-	-	-	-	-	-
Blue-faced rail	*Gymnocrex rosenbergi*	•	-	-	•	-	-	-
Talaud rail	*G. talaudensis*	-	-	•	-	-	-	-
Isabelline bush-hen	*Amaurornis isabellina*	•	-	-	-	-	-	-
Talaud bush-hen	*A. magnirostris*	-	-	•	-	-	-	-
SNIPE Scolopacidae								
Sulawesi Woodcock	*Scolopax celebensis*	•	-	-	-	-	-	-
PIGEONS and DOVES Columbidae								
White-faced cuckoo-dove	*Turacoena manadensis*	•	-	-	•	•	•	-
Sulawesi ground-dove	*Gallicolumba tristigmata*	•	-	-	-	-	-	-
Red-eared fruit-dove	*Ptilinopus fischeri*	•	-	-	-	-	-	-
Buff-bellied fruit-dove	*P. subgularis*	•	-	-	•	-	-	•
White-bellied imperial pigeon	*Ducula forsteni*	•	-	-	-	-	-	-
Grey-headed imperial pigeon	*D. radiata*	•	-	-	-	-	-	-
Silvery imperial pigeon	*D. luctuosa*	•	-	-	•	•	-	•
Sombre pigeon	*Cryptophaps poecilorrhoa* •	-	-	-	-	-	-
PARROTS Psittacidae								
Blue-and-red lory	*Eos histrio*	-	•	•	-	-	-	-
Ornate lorikeet	*Trichoglossus ornatus*	•	-	•	•	•	•	-
Yellow-and-green lorikeet	*T. flavoviridis*	•	-	-	-	-	-	•
Sulawesi hanging-parrot	*Loriculus stigmatus*	•	-	-	•	•	-	-
Sangihe hanging-parrot	*L. catamene*	-	•	-	-	-	-	-
Little hanging-parrot	*L. exilis*	•	-	-	-	-	-	-
Red-spot racquet-tail	*Prioniturus flavicans*	•	-	-	?	-	•	-
Golden-mantled racquet-tail	*P. platurus*	•	-	•	•	•	•	•
Blue-backed parrot	*Tanygnathus sumatranus*	•	•	•	•	•	•	•
CUCKOOS Cuculidae								
Sulawesi hawk-cuckoo	*Cuculus crassirostris*	•	-	-	-	-	-	-
Black-billed koel	*Endyramys melanorhyncha*	•	-	-	•	•	•	-
Yellow-billed malkoha	*Phaenicophaeus calorhynchus*	•	-	-	•	•	•	-
Bay coucal	*Centropus celebensis*	•	-	-	-	•	•	-

Table 1.10. (Continued.)

		Mainland	Sangihe	Talaud	Banggai	Muna/Butung	Togian	Sula
MASKED OWLS Tytonidae								
Rosenberg's owl	*Tyto rosenbergii*	•	•	-	•	-	-	-
Sulawesi owl	*T. inexpectata*	•	-	-	-	-	-	-
TRUE OWLS Strigidae								
Speckled boobook owl	*Ninox punctulata*	•	-	-	-	-	-	-
Ochre-bellied boobook owl	*N. ochracea*	•	-	-	•	-	-	-
Sangihe Scops owl	*Otus collari*	-	•	-	-	-	-	-
NIGHTJARS Caprimulgidae								
Diabolical nightjar	*Eurostopodus diabolicus*	•	-	-	-	-	-	-
KINGFISHERS Alcedinidae								
Sulawesi dwarf kingfisher	*Ceyx fallax*	•	•	-	-	-	-	-
Black-billed kingfisher	*Pelagopsis melanorhyscha*	•	-	-	•	-	-	•
Lilac-breasted kingfisher	**Cittura** cyanotis	•	•	-	-	-	-	-
Talaud kingfisher	*Halcyon enigma*	-	-	•	-	-	-	-
Large wood kingfisher	**Actenoides** monacha	•	-	-	-	-	-	-
Scaly-breasted kingfisher	**A.** princeps	•	-	-	-	-	-	-
BEE-EATERS Meropidae								
Purple-bearded bee-eater	**Meropogon** forsteni	•	-	-	-	-	-	-
ROLLERS Coraciidae								
Sulawesi roller	*Coracias temminckii*	•	-	-	-	-	-	-
HORNBILLS Bucerotidae								
Red-knobbed hornbill	*Rhyticeros cassidix*	•	-	-	-	•	-	-
Sulawesi dwarf hornbill	**Penelopides** exarhatus	•	-	-	-	•	-	-
WOODPECKERS Picidae								
Ashy woodpecker	**Mulleripicus** fulvus	•	-	-	-	•	-	-
Sulawesi woodpecker	**Picoides** temminckii	•	-	-	-	-	•	-
CUCKOO-SHRIKES Campephagidae								
Sula Cuckoo-shrike	*Coracina schistacea*	-	-	-	•	-	-	•
Sulawesi cuckoo-shrike	*C. temminckii*	•	-	-	-	-	-	-
Bicoloured cuckoo-shrike	*C. bicolor*	•	-	-	•	•	-	-
White-rumped cuckoo-shrike	*C. leucopyia*	•	-	-	•	-	-	-
Mountain cuckoo-shrike	*C. abbotti*	•	-	-	-	-	-	-
BABBLERS Timaliidae								
Sulawesi babbler	*Trichastoma celebensis*	•	-	-	-	•	-	-
Malia	**Malia** grata	•	-	-	-	-	-	-
THRUSHES Turdidae								
Great shortwing	**Heinrichia** calligyna	•	-	-	-	-	-	-
Sulawesi mountain-thrush	**Cataponera** turdoides	•	-	-	-	-	-	-
Rufous-backed thrush	*Zoothera erythronata*	•	-	-	•	-	-	-
Geomalia	**Geomalia** heinrichi	•	-	-	-	-	-	-
WARBLERS Sylviidae								
Sulawesi leaf-warbler	*Phylloscopus sarasinorum*	•	-	-	-	-	-	-
Sulawesi tailorbird	*Orthotomus* sp.	•	-	-	-	-	-	-
FLYCATCHERS Muscicapidae								
Rufous-throated flycatcher	*Ficedula rufigula*	•	-	-	-	-	-	-
Lompobatang flycatcher	*F. bonthaina*	•	-	-	-	-	-	-

Table 1.10. (Continued.)

		Mainland	Sangihe	Talaud	Banggai	Muna/Butung	Togian	Sula
FLYCATCHERS Muscicapidae (Continued)								
Blue-fronted flycatcher	Cyornis hoevelli	•	-	-	-	-	-	-
Matinan flycatcher	C. sanfordi	•	-	-	-	-	-	-
Fiery-backed fantail	Rhipidura teijsmanni	•	-	-	-	-	-	•
Caerulean paradise-flycatcher	**Eutrichomyias** rowleyi	-	•	-	-	-	-	-
FLOWERPECKERS Dicaeidae								
Golden-flanked flowerpecker	Dicaeum aurealimbatum	•	-	-	-	•	-	-
Crimson-crowned flowerpecker	D. nehrkorni	•	-	-	-	-	-	-
Grey-sided flowerpecker	D. celebicum	•	-	-	-	•	-	-
SUNBIRDS Nectariniidae								
Elegant sunbird	Aethopyga duyvenbodei	-	•	-	-	-	-	-
HONEYEATERS Meliphagidae								
Sulawesi honeyeater	**Myza** celebensis	•	-	-	-	-	-	-
Red-faced honeyeater	**M.** sarasinorum	•	-	-	-	-	-	-
WHITE-EYES Zosteropidae								
Pale-bellied white-eye	Zosterops consobrinorum	•	-	-	-	-	-	-
Sulawesi white-eye	Z. anomala	-	-	-	-	-	-	-
Streak-headed white-eye	Lophozosterops squamiceps	-	-	-	-	-	-	-
MYNAS and STARLINGS Sturnidae								
Sulawesi crowned myna	Basilornis celebensis	•	-	-	-	-	-	-
Helmeted myna	B. galeatus	-	-	-	•	-	-	-
White-necked myna	**Streptocitta** albicollis	•	-	-	•	-	-	-
Flame-browed myna	**Enodes** erythrophris	•	-	-	-	-	-	-
Finch-billed starling	**Scissirostrum** dubium	•	-	-	•	•	•	-
WOODSWALLOWS Artamidae								
Ivory-backed woodswallow	Artamus monachus	•	-	-	•	-	-	•
CROWS Corvidae								
Piping crow	Corvus typicus	•	-	-	-	•	-	-
WHISTLERS Pachycephalidae								
Chestnut-backed whistler	**Coracornis** raveni	•	-	-	-	-	-	-
Olive-flanked whistler	**Hylocitrea** borensis	•	-	-	-	-	-	-
Sulphur-bellied whistler	Pachycephala sulfuriventer	•	-	-	-	-	-	-
Sangihe Shrike-Thrush	Colluricincla sanghirensis	-	•	-	-	-	-	-

After Holmes and Wood 1979; White and Bruce 1986; K.D. Bishop pers. comm., N. Collar pers. comm.

Islands have a tree frog *Litoria infrafrenata* not found on the mainland but which is common to the east as far as Queensland (van Kampen 1923). The apparently poor amphibian fauna may be an artefact of undercollection since most of the species were found as a result of collections made in the 1870s-1890s and only three small collections have been made this century, each of which has included new species. There may, therefore, be as many species to be discovered as are already known (J. Dring pers. comm.).

There are 40 species of lizards known from the Sulawesi mainland, 13 of which are endemic, but the group is poorly known and there are certainly new species awaiting discovery. One of the most distinctive of the reptiles, is the large sailfin lizard *Hydrosaurus amboinensis*, which is usually found near water (p. 301). Better known among the reptiles are the snakes and 64 have been collected from the mainland and its coastal waters (table 1.12) compared with 136 in Peninsular Malaysia, 150 in Sumatra, 110 in Java and 166 in Borneo (Medway 1981). There are 15 endemic species and one endemic monotypic genus *Rabdion* (den Bosch 1985). Strangely, the small island of Tanahjampea south of Salayar Island has only two species of snakes and both these are endemic: the Jampea ilyssid *Cylindrophis isolepis* and the Jampea pit viper *Trimeresurus fasciatus* (de Rooij 1917). Both species were collected by a team from the Bogor Museum that visited Jampea in 1984 (Boeadi pers. comm.). Sulawesi has the distinction of being the locality of the world's longest recorded snake, a reticulated python *Python reticulatus* that measured 9.97 m in length (McWhirter 1985). These snakes are the only Sulawesi land animals that present any real threat to man.

All of the fish indigenous to Sulawesi are brackish-water species tolerant of freshwater. Some of these appear to be restricted to lakes, while the eels migrate between the lakes and the sea. Many species have been introduced deliberately or accidentally (such as the air-breathing snakehead *Channa*[18] *striata* and climbing perch *Anabas testudineus*) and it is these fish that dominate the Sulawesi freshwater fisheries (p. 330).

Until recently the invertebrates of Sulawesi were very poorly known but three expeditions have greatly increased the knowledge of what is present:

- Project Wallace (1985) organized by the Indonesian Institute of Sciences and the Royal Entomological Society of London, based in the Bogani Nani Wartabone National Park in Bolaang Mongondow, North Sulawesi;
- Operation Drake (1980) organised by the London-based Scientific Exploration Society, based in Morowali National Park; and
- a series of medical expedition teams (1970s) organized by the National Institute of Medical Research and the United States Naval Medical Research Unit, based near Lake Lindu.

A flood of papers describing new Sulawesi invertebrates appeared after Project Wallace (e.g., Hoogstraal and Wassef 1977; Bedford-Russel 1981, 1984; Hadi and Tenorio 1982; Hayes 1983; Goff et al. 1986). As an example

Table 1.11. Amphibians of Sulawesi.

TOADS Bufonidae	TRUE FROGS Ranidae
Bufo celebensis	Rana arathooni
B. biporcatus [1]	R. cancrirora [3]
NARROW-MOUTHED TOADS Microhylidae	R. celebensis
Oreophryne variabilis	R. chalconota [1]
O. sp.	R. erythraea [3]
O. celebensis	R. grunniens [2]
O. zimmeri	R. heinrichi
Kaloula baleata [1]	R. microtympanum
K. pulchra [3]	R. modesta [2]
TREE FROGS Rhacophoridae	R. macrops
Polypedetes leucomystax [3]	R. papua [2]
Rhacophorus georgei	Occidozyga semipalmata
R. edentulus	O. celebensis
R. monticola	O. laevis

[1] Species of western Indonesian forests and clearings.
[2] Species found in the Moluccas or New Guinea.
[3] Species of cultivation, secondary growth and degraded habitats, common in western Indonesia and probably introduced by man.
Other species are endemic.

From J. Dring pers. comm.

of the previous lack of information, only one species (endemic) of spring-tail or Collembola was known from Sulawesi before 1985. In a few weeks of collecting in and around Bogani Nani Wartabone National Park no less than 120 species from about 70 genera had been added to the list. Specimens were even collected of two genera that were previously known from single species in North America (P. Greenslade pers. comm.). These small animals are very important in decomposition and mineral cycling (p. 365).

There are 38 species of the large and usually striking swallowtail butterflies on Sulawesi and 11 (29%) of these are endemic. One is *Atrophaneura palu*, a large black and white swallowtail (forewing 70 mm long) known from only a few specimens collected from close to what is now Lore Lindu National Park (Haugum et al. 1980; Collins and Morris 1985). Another seemingly rare, recently-described butterfly is the 'paper handkerchief' or wood nymph butterfly *Idea tambusisiana* from Mt. Tambusisi in Morowali National Park (Bedford-Russell 1981). Further specimens have since been found by Japanese scientists in North Sulawesi (R. Vane-Wright pers. comm.).

Table 1.12. The land snakes of Sulawesi.

BLIND SNAKES Typhlopidae	*C. curta*
Rhamphotyphlops braminus	*C. muelleri*
Typhlops ater	*C. nuchalis*
T. conradi	*C. virgulata*
CYLINDER SNAKES Aniliidae	*Calamorhabdium acuticeps*
Cylindrophis melanotus	*Pseudorabdion longiceps*
C. rufus	*P. sarasinorum*
PYTHONS Boidae	*Rabdion forsteni*
Candoia carinata	*Amphiesma celebica*
Python molurus	*A. sarasinora*
P. reticulatus	*Rhabdophis chrysarga*
SUNBEAM SNAKE Xenopeltidae	*R. chrysargoides*
Xenopeltis unicolor	*Sinonatrix trianguligera*
WART SNAKES Acrochordidae	*Xenochrophis piscator*
Acrochordus granulatus	*Cerberus rynchops* *#
COLUBRID SNAKES Colubridae	*Enhydris enhydris* *
Psammodynastes pulverulentus	*E. matannensis* *
Lycodon aulicus	*E. plumbea* *
L. stormi	*Homalopsis buccata*
Oligodon octolineatus	**COBRAS, CORAL SNAKES**
O. waandersi	**and SEA SNAKES Elapidae**
Elaphe erythrura	*Bungarus candidus*
E. flavolineata	*Maticora intestinalis*
E. janseni	*Ophiophagus hannah*
Gonyosoma oxycephalum	*Naja naja*
Ptyas dipsas	*Laticaudata colubrina* #
Chrysopelea paradise	*L. laticaudata* #
Dendrelaphis caudolineatus	*Aipyurus fuscus* #
D. pictus	*Hydrophis fasciatus* #
Ahaetulla prasina	*H. melanosoma* #
Boiga dendrophila	*H. spiralis* #
B. irregularis	*Lapemis hardwickii* *
B. multimaculata	*Pelamis platurus* *
Calamaria acutirostris	**VIPERS Viperidae**
C. apraeocularis	*Trimeresurus wagleri*
C. boesemani	*T. albobaris*
C. brongersmai	

Endemic genus and species in bold italic.
* = freshwater habitats,
# = estuarine and marine habitats.

After Regenass and Kramer 1981; in den Bosch 1985; C. McCarthy pers. comm.

Occasionally insects can be seen in such unusually great numbers that the observation is worthy of record. One example is the stream of pierid white butterflies seen crossing Kalaotoa Island from dawn until dusk one day in 1936 (Doctors van Leeuwen 1937). Such migrations of pierids are well documented in Europe and it has been found that these butterflies travel in more or less straight lines, searching for suitable habitats. By so doing, they avoid returning to a place that they have just left. They feed and breed in different habitats and so, for a chort of butterflies that emerged at more or less the same time, the best place to be is somewhere else. These butterflies sometimes seem to follow rivers but it is possible that they are simply more visible there because experiments have shown that they have a sense of direction rather than a sense of location (Baker 1982).

Endangered Species

The names and status of the world's rare and endangered animals are compiled and monitored respectively by the Conservation Monitoring Centre of the International Union for the Conservation of Nature (IUCN) and Natural Resources based in Cambridge, England. 'Red Data Books' are produced detailing what is known of the ecology and threats facing each of these species, and what conservation measures, if any, are currently in force to protect the animals. Examination of the Red Data Books reveals that 16 species are considered to be at risk of extinction[19] on Sulawesi (table 1.13).

During the course of writing this book it has become clear that the above animals are by no means the animals most at risk and indeed some animals may have become extinct unnoticed.

The attractive Caerulean paradise-flycatcher *Eutrichomyias rowleyi* was discovered in 1873 by a hunter working for the German ornithologist A.B. Meyer (Meyer 1878). The label on the first specimen states that it was male but there was no verification of this (Meyer 1878), and in fact the colour is more typical of a female flycatcher than of a male (S.V. Nash pers. comm.).

That first specimen is in fact the only specimen ever collected of this distinctive bird. One may have been sighted in 1981 (White and Bruce 1986), but intensive ornithological surveys of Sangihe Island by experienced ornithologists in 1985 and 1986 failed to find any. Virtually all of Sangihe has been converted to coconut and nutmeg plantations or else is covered in patches of secondary forest from abandoned gardens. A small patch of montane forest exists on the top of Mt. Sahendaruman in the south of the island and it was felt that this was the only possible habitat for the species (Whitten et al. 1986). Surveys by Action Sampiri in 1998-99 revealed that this bird was clinging to survival in those forests but the numbers are extremely low.

The status of other animals endemic to Sangihe and the Talaud Islands, should now be an immediate cause for concern (see the Introduction). The blue-and-red lory *Eos histrio* (Sangihe and Talaud[20]), Sangihe hanging-parrot *Loriculus catamene* and Elegant sunbird *Aethopyga duyvenbodei* (Sangihe), and Talaud kingfisher *Halcyon enigma* have all been seen recently although none of them is at all common. It is possible that the blue-and-red lory is already extinct on Sangihe but it still appears to be common on Karakelang, the main Talaud Island (Whitten et al. in press). The black birdwing butterfly *Troides dohertyi* (forewing length 73 mm male, 82 mm female) is known only from Sangihe and the Talaud Islands where it lives in lowland forests in which the caterpillars probably feed on the leaves of *Aristolochia tagala*, a climbing shrub that grows in forests and

Table 1.13. Animal species considered to be at risk according to IUCN Red Data Books.

	Species	Status
Sulawesi tarsier	*Tarsius spectrum*	I
Sulawesi civet	*Macrogalidia musschenbroeckii*	R
Dugong	*Dugong dugon*	V
Babirusa	*Babyrousa babirussa*	V
Lowland anoa	*Bubalus depressicornis*	E
Mountain anoa	*B. quarlesi*	E
Chinese egret	*Egretta eulophotes*	V
Milky stork	*Ibis cinereus*	V
Maleo	*Macrocephalon maleo*	V
Estuarine crocodile	*Crocodylus porosus*	E
Leatherback turtle	*Dermochelys coriacca*	E
Hawksbill turtle	*Eretochelys imbricata*	E
Forsten's tortoise	*Indotestudo forsteni*	R
Talaud black birdwing butterfly	*Troides dohertyi*	V
Palu swallowtail	*Atrophaneura palu*	I
Tambusisi wood nymph	*Idea tambusisiana*	I

E - endangered, V - vulnerable, R - rare, I - insufficiently known.

After Miller 1977; Thornback 1978; King 1979; Groombridge 1982; Wells et al. 1983; Collins and Morris 1985

thickets up to 800 m above sea-level (Ding Hou 1984; Collins and Morris 1985). This butterfly must be considered as being under considerable threat given the greatly reduced area of lowland forest on the islands, for it is not known whether it can adapt to secondary vegetation (Collins and Morris 1985). It is probably already extinct on Sangihe but there is a hunting reserve on Karakelang, which includes hill and mountain forests and efforts should be made to determine whether the birdwing is present. The boundaries of this reserve are under pressure from illegal farmers.

The endemic fish of lakes Poso and Lindu, and of lakes Towuti, Matano, Wawantoa and Mahalona (table 4.10) are threatened by the introduction of fish to increase fisheries production. Among the endemic species are four duck-billed fish–*Adrianichthys kruyti*, about 11 cm long from Lake Poso (fig. 1.24), *Xenopoecilus poptae* and *X. oophorus*, about 10–20 cm long, also from Lake Poso, and *X. sarasinorum*, about 7 cm long, from Lake Lindu. These species used to be thought to comprise the entire family Adrianichthyidae, but recent taxonomic analyses have now placed several more species in this family, some of which are also endemic to Sulawesi, and some of which are found from India to Japan (Rosen and Parenti 1981). There are early reports that *X. poptae* does not seem to lay eggs as most fishes do, but rather voids eggs which hatch on contact with water. The young fry then swim along with their mother. The broken egg membranes, known locally as *momosonya*[21], rise to the lake surface and used to cover considerable areas of the lake (Weber and de Beaufort 1922). This has not been confirmed and the 'momoso' may be pollen or seeds. It is known that the newly discovered *X. oophorus* carries its eggs below its body until they hatch (Kottelate 1990) (see Introduction). Similar behaviour in *X. poptae* may have given rise to the other story.

Teams that visited Lake Poso in 1976 and 1983 both found the two endemic fish, but a survey of fishermen and an examination of fish catches in 1986 could not confirm the continued existence of either species. Some fishermen claimed that some fish had disappeared when Colo volcano erupted in 1983 (p. 7). This seems untenable as a reason since satellite photographs of the ash plume show how this was blown to the west rather than the south (Katili and Sudrajat 1984). At the start of the century it is said that great shoals of *X. poptae* formed at 12 -15 m between November and January, and were caught by fishermen using hooks (Weber and de Beaufort 1922).

A likely cause of the reduction in the populations of these endemic fish, and also the possible extinction of some of the snails and mussels endemic to Lakes Poso and Lindu (p. 297) (Carney et al. 1980) is the unthinking introduction of exotic fish species, particularly of the tilapia, carp and catfish, to increase fisheries production (Whitten et al. 1986). The extinction of fish species as a result of the uninformed introduction of commercial species has been reported from elsewhere, particularly

Figure 1.24. *Adrianichthys kruyti,* one of the three endemic duck-billed fish of Lake Poso.

the enormous lakes of the East African rift valley to which many species of cichlid fish are (were) endemic. Indeed, 60%-80% of the world's freshwater fisheries are based on introduced species. Not all fish introductions damage the indigenous fauna, however, and success stories have been reported from other African lakes. Major agencies now do not, as policy, advocate the introduction of new species of fish into lakes except with the most extreme caution. To avoid any possible extinction of endemic fish on Sulawesi, fisheries staff must first be aware that any introduction may result in the loss of endemic species, and second be able to justify their aims to all parties.

BIOGEOGRAPHY

Background to Biogeography

Every organism has a spatial distribution related to its ecology, behaviour, physiology, ability to travel long distances, the other organisms living in the same area, and to the geological history and climate of the area in question. A further and important factor in determining distributions is chance. Conditions suitable for an organism in terms of light, humidity, food, etc. do not necessarily occur evenly over an area and this is reflected in that species' distribution. The study of such patterns and the factors causing or limiting them is known as biogeography.

The distribution of an organism is generally bounded by unsuitable habitats, unsuitable climates or the occurrence of a species against which it cannot compete successfully. Alternatively, a species may be actively dispersing and the edge of its range may simply be the furthest point it has reached at that time. Clearly a barrier for one species is not necessarily a barrier for another, and a barrier for one life-stage of an organism is not necessarily a barrier for a different life-stage of the same species. For example, the larvae of many invertebrates and the seeds of plants are much more mobile than the adults.

Distributions can be described on various levels: thus a bird might live in forest; but perhaps only in a few types of forest; at certain altitudes; avoiding forest near rivers; and only using the tops of the taller trees, etc. Such precise micro-habitats are only available in certain locations, and so a species' distribution usually comprises a number of areas inhabited by discrete populations. If these populations are isolated from other populations over a long period, changes may occur which render the two populations reproductively incompatible and new species or sub-species[22] may arise.

Organisms disperse from one suitable area to another along one of three routes:

- corridor—a route comprising the same habitats as are present at either end of the route, thereby giving all species a chance to move through it;
- filter—a route comprising only some of the habitats available at either end of the route, thereby preventing certain species that depend on the absent habitats from moving;
- sweepstakes—a route comprising habitats which are absent from the source of species and may represent, in analogy or actuality, a sea separating two areas of land. The chances of a species crossing this 'sea' are very small but, under certain conditions of wind or the stranding of individuals on a raft of floating vegetation drifting

down a river and out to sea, a terrestrial species might successfully cross over to another area of suitable habitat. For example, after heavy rains around Manado in 1882, large forest trees were seen floating out to sea, each sufficiently large to provide a temporary home for a few small mammals such as squirrels and invertebrates. Lava can also push trees into the sea. In these two cases one might expect agile animals to jump off the trees, but a few days after the Krakatoa eruption in 1883, a monkey was found, scorched and tired, clinging to a partly-burned tree in the Sunda Straits. The monkey was kept alive for at least a year (Hickson 1889). Even if an animal is washed alive on to a new coast, if there is no member of the opposite sex available then the sweepstakes will have been lost.

So there are three ways in which an organism can disperse:
* by 'jumping' quickly over relatively large expanses of unsuitable habitat;
* by dispersing slowly across habitat which is more or less suitable; and
* by dispersing very slowly and making adaptations on the way allowing the colonizing of areas the environments of which would have been unsuitable for the original stock.

It is well understood by laymen and scientists alike that small islands support fewer species than large islands. After a certain length of time the total number of species on an island will remain more or less constant, and this represents an equilibrium between the colonizing of the island by immigrant species and the extinction of existing species. The rate of colonization is clearly higher when an island is near the mainland because more species are likely to cross the relatively narrow sea gap. Also, the rate of extinction is clearly greater when an island is smaller because the population of any species will be smaller and the chance will be greater of disease and other detrimental events reducing the population to zero or an unviable number. These relationships can be drawn graphically and represent the foundation of the Theory of Island Biogeography (fig. 1.25).

The relationship between island size and number of species is relatively constant for a given group of animals or plants, and in general reducing island area by a factor of ten, halves the number of species (figs. 1.26 and 1.27). Where an island supports fewer species than expected and so falls below the line, the reason may be that:
* the group is not sufficiently well known;
* equilibrium in species number has not yet been reached (where a volcanic island such as Una-una has been destroyed and is being recolonized);
* the island comprises a relatively restricted number of habitats, or habitats which do not support large numbers of species;
* the island is extremely remote and difficult to colonize.

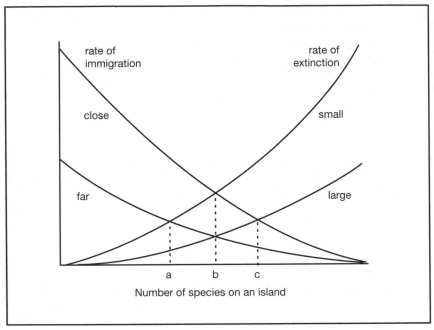

Figure 1.25. The relative number of species on (a) small, distant islands (b) large, distant or small, close islands, and (c) large, close islands.

Where an island supports more species than expected and so falls above the line, the reason may be that:

- more than the equilibrium number are present and some species will in due course be lost;
- the island is peculiarly rich in habitat types; or
- the island is a centre of species radiation in a certain group.

For total species, Sulawesi is below the line of best fit for both plants and birds. For plants this is probably because there has been insufficient collecting (p. 29), but this is unlikely to be the reason for the birds. Sulawesi falls above the line for mammals due, perhaps, to the extraordinary radiation of rat species, and is more or less on the line for snakes. Sulawesi is consistently above the line for the number of endemic species and this reflects its geological history.

These principles are useful in deciding whether an area is well-known biologically. For example, plotting known number of species of milkweed

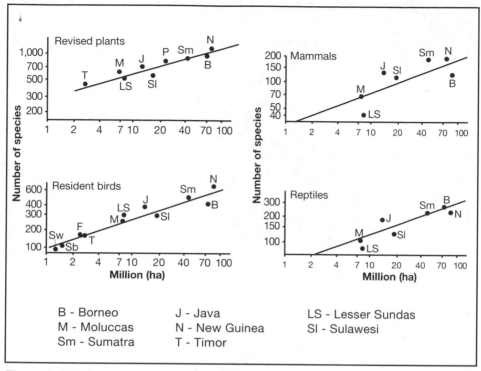

Figure 1.26. Relationship between number of species and island size for revised plants (those with a recent taxonomic review) and three groups of animals.
From Anon. 1982b

butterflies (Danaidae)[23] against island size (by rank) reveals that islands around Southeast Sulawesi and the Togian Islands have probably not been surveyed sufficiently for this group, whereas for the Banggai Islands and islands in South Sulawesi, nearly all the species of this family are probably known (fig. 1.28).

When lists of species found on large and small islands are compared, it is generally found that the species absent from the smaller islands are larger than average since these animals generally have large range requirements and low densities. A few species are more abundant and fill a wider niche[24] on islands than they do on the neighbouring mainland or larger islands where they have more competitors although this relationship does not always hold (MacArthur et al. 1972).

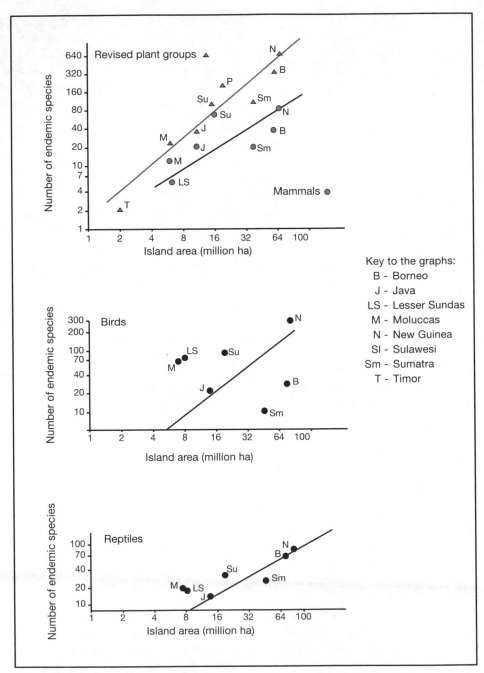

Figure 1.27. Relationship between number of endemic species and island area for revised plants and mammals, birds and reptiles.

From Anon. 1982b

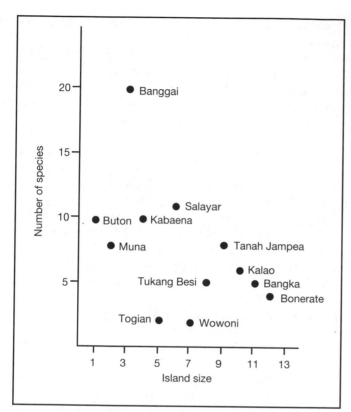

Figure 1.28. Number of species of milkweed butterflies found on Sulawesi islands.

After R. Vane-Wright (pers. comm.)

Wallace's Line

It is an enigma that educated people in Indonesia know about Charles Darwin, even adopting his surname, and yet know nothing about Alfred Wallace, his contemporary and and fellow Englishman. Darwin is usually given credit for formulating the theory of evolution by natural selection, but the primary reason he was encouraged to write down his thoughts was because Wallace had written to him from Indonesia enclosing a manuscript showing that he had reached more or less the same conclusions as Darwin. Both men had been stimulated by what they had seen while travelling, and the interpretation of their observations did not match with the contemporary wisdom concerning the creation of the world. Wallace had been particularly fascinated by a visit to Sulawesi and the first manuscript he sent to Darwin was written in Ternate after leaving Manado. The thoughts of these two men on the distribution and evolution of species turned modern thinking upside-down and Wallace's contribution should not be underestimated.

Whereas Darwin was well-educated and came from a rich family, Wallace left school at 14, in 1837, and eventually earned a living by collecting animals in remote areas of the world to sell to museums. He spent eight years in Sarawak and Indonesia (from Sumatra to Irian Jaya) and the account of his travels totalling about 22,000 km makes splendid reading (Wallace 1869).

In a letter written in 1858, Wallace expressed his view that the Indonesian Archipelago was inhabited by two distinct faunas, one found in the east, one in the west. The following year he defined these two regions, based on the distribution of birds, by placing the boundary between Lombok and Bali and between Borneo and Sulawesi. He was struck that Borneo and Sulawesi should have such different birds and yet be separated by no major physical or climatic barrier. He believed that Borneo, along with Java and Sumatra, had once been part of Asia, and that Timor, the Moluccas, New Guinea and perhaps Sulawesi had once been part of a Pacific-Australian continent. The fauna of Sulawesi seemed so peculiar that he suspected it might have been connected with both the Asian and the Pacific-Australian continents (Wallace 1859). He insisted that an explanation of the origin of the fauna of Sulawesi would have to accept that there had been vast changes in the surface of the earth, a concept which challenged the established view but which we now know to be true (p. 2). The line that Wallace drew east of the Philippines, through the Makassar Straits and between Bali and Lombok (Wallace 1863) came to be known as Wallace's Line. In 1910, three years before he died, Wallace decided that the predominance of Asian forms on Sulawesi should be reflected in the Line being moved east of Sulawesi (Wallace 1910). Many other analyses have been performed on the distribution of animal species resulting in several different Lines (fig. 1.29) (Simpson 1977). Weber's Line attempts to delimit the boundary of faunal balance, that is, where the ratio between Asian and Australian animals is 50:50 (Weber 1904). Weber used molluscs and mammals in his analysis but the exact position of the Line differs from one group of animals to another. For example, Asian reptiles and butterflies penetrate further east then do its birds and snails. Lydekker's Line delimits the western boundary of the strictly Australian fauna in much the same way as Wallace's Line delimits the eastern boundary of the Asian fauna; both these Lines effectively trace the 180-200 m depth contours around the Sahul and Sunda continental shelves respectively. The area between these two Lines has been nominated as a separate region, subregion or transition area called Wallacea (Dickerson 1928). This concept was first suggested by Wallace in 1863, but has been strongly criticized as the area does not comprise a homogenous fauna, and there is no gradual change in species composition across it; instead there are large number of

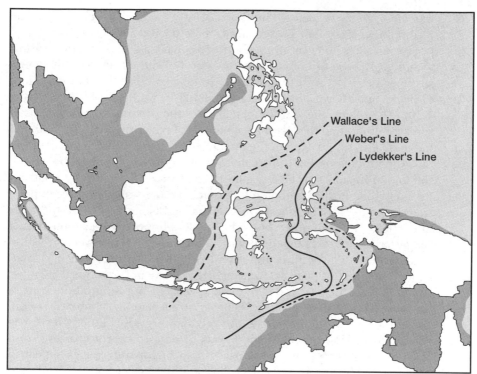

Figure 1.29. Biogeographical Lines through insular Southeast Asia.
From Simpson 1977

endemic species (Stresemann 1939; Simpson 1977). The name Wallacea should be retained, but to describe the area between the Oriental and Australian regions rather than as the name for a strict biogeographical entity.

The concept of Wallace's Line has fascinated biogeographers and it has been found that its validity differs between groups of animals and plants. For example, an early analysis of Sulawesi's flora showed similarities with Borneo, Sumatra and Java, rather than with the Moluccas and New Guinea (Lam 1945). This analysis was based, however, on a limited number (about 700) of species. An analysis at the generic level of the whole Malesian flora[25] demonstrated the existence of three provinces in Malesia, of which East Malesia comprises New Guinea, the Moluccas, and Sulawesi (van Steenis 1950). A recent analysis of 4,222 species in 540 genera that have been subject to recent taxonomic revisions revealed that the Sulawesi flora was most closely related to the floras of other relatively dry areas in the Philippines, Moluccas, Lesser Sunda Islands and Java (van Balgooy in press). There is no clear affinity between Sulawesi and the islands east or west of it, but the flora of the lowlands and of the ultrabasic soils show a stronger similarity with that of New Guinea, whereas the montane flora (1,000 m and above) shows a stronger similarity with that of Borneo[26] (fig. 1.30). The higher the altitude, the greater the distance between areas of similar altitude, and the less chance there is of receiving plants from similar habitats. This would explain the greater proportion of Bornean plants on Sulawesi mountains, but the greater affinity with New Guinea flora among the lowland plants may be a result of New Guinea having more relatively dry areas than Borneo and therefore being a more suitable source of plant species. Some of these may have been brought to Sulawesi via the 'Sula Spur' (p. 6), whereas others could have island-hopped. An examination of the percentage of taxa that do not cross imaginary lines between or within landmasses in a given direction revealed that the strongest such 'demarcation Line' was for plants of western origin between Borneo and Sulawesi. About 50% of the non-endemic plant species of Borneo do not occur in Sulawesi. This suggests that the Makassar Straits have been a barrier to dispersal for a very long time. Interestingly, however, this Line is very weak when considering non-endemic plants of eastern origin crossing from Sulawesi to Borneo. The easiest routes by which plant species appear to have entered Sulawesi are those via Java and the Lesser Sunda Islands and via the Philippines and Sangihe. The former route argues for the existence of an island chain between Java, Lesser Sundas and Sulawesi in the not too distant past.

When specific groups of animals or plants are examined, many interesting distribution patterns are found. The percentage of a total species list shared between neighbouring islands (fig. 1.31) indicates a generally closer affinity between Sulawesi and islands to the east but this is at least in part an artefact of the relatively impoverished floras and faunas to the east.

The mountain flora of Sulawesi is derived from two sources: those which originated locally (autochthonous) and those for which the centre of origin is outside the area concerned (allochthonous) (van Steenis 1972). The allochthonous flora, although a minority of the total mountain flora, allow hypotheses to be made regarding its origin. This part of the flora belongs to genera whose species are found only in cold climates (i.e., microtherm species), and in the tropics they are generally found only in the subalpine forests on mountains 2,000-2,500 m high. These genera, such as *Rhododendron* (Eric.) and *Gentiana* (Gent.), are found in many tropical and subtropical countries yet none can tolerate a hot climate. Soils seem to have little or no influence on the distributions since a single species will be found on soils originating from igneous, sedimentary or recent volcanic parent material. The age of the rocks does seem to be rel-

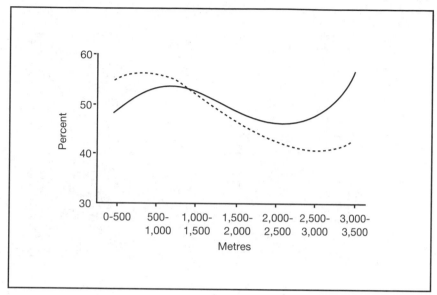

Figure 1.30. Floristic affinity of Sulawesi to Borneo (solid line) and New Guinea (dashed line) for each 500 m increase in altitude.

From data in van Balgooy in press

evant, however, with some species being absent from recent volcanic soils (van Steenis 1972).

From an analysis of the distribution of about 900 of these cold-adapted mountain species, it has been concluded that there are three tracks by which plants arrived in Sulawesi during some period or periods in the geological past (fig. 1.32) (van Steenis 1972). Continuous ranges of high mountains do not, of course, exist along the entire lengths of these tracks. During the coldest times of the Pleistocene the mean temperature dropped only about 2°C (p. 18) which, with rates of temperature change being about 0.6°C/100 m (not necessarily applicable at that period) is equivalent to a drop in the levels of the forest zones of 350-400 m. The number of suitable mountain tops would obviously have been greater during the cooler periods thereby forming more 'stepping stones' across which plants could disperse.

Palms are a useful group for biogeographical study because their genera, at least, are well known and they represent an ancient group of plants in which several genera had evolved by the Oligocene (30 Ma ago)

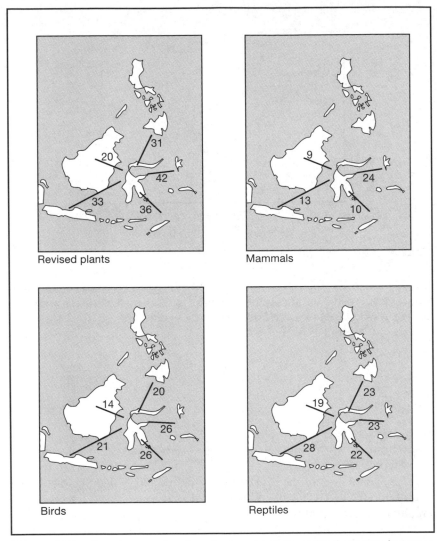

Figure 1.31. Species overlap for plants and three animal groups between Sulawesi and neighbouring islands. Figures are the percentages of total number of species recorded from a pair of areas that are shared between them.

From Anon. 1982b

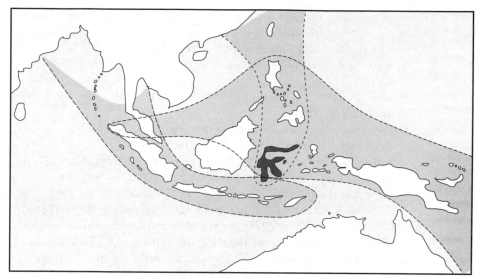

Figure 1.32. The three tracks by which mountain plants arrived in Sulawesi.
After van Steenis 1936, 1972

(Dransfield in press). Only two genera, *Gronophyllum* and the monotypic *Pigafetta*, are found in Sulawesi but nowhere further west. This is surprising in the case of *Pigafetta* because it is a tall fecund palm of secondary growth with small seeds (p. 396). There are, however, 13 genera of palms that are found no further east than Borneo illustrating again that Wallace's Line is most noticeable travelling west to east. Two genera, the spiny *Oncosperma* and the fan palm *Pholidocarpus,* cross Wallace's Line from Borneo to Sulawesi but are found no further east, and another sixteen are found in Sulawesi and to the east and west. Of these, the rattans belonging to *Calamus,* the undergrowth palm *Licuala,* and the tree palms *Cyrtostachys, Areca,* and *Livistona* exhibit a peculiar distribution having species in Sunda-land and New Guinea, but few or none in Sulawesi (Dransfield 1981 in press). Since it is known that Sulawesi was perhaps drier in the Pleistocene than were the continental landmasses (p. 18), it may be that some species became extinct, or that conditions for the evolution of species there were less suitable.

A similar distribution is found among the spiny eels (Mastacembel-idae) in which four species are found in Borneo, three in Halmahera, but none in Sulawesi. Fish are in fact a useful group to examine with respect to different opinions concerning the narrowing or closing of the Makassar Straits during the Pliocene. For this it must be understood that freshwater fish can be divided into three ecological groups:

- those confined to freshwater with no tolerance to brackish or salt water;
- those generally encountered in freshwater but showing some tolerance to salt; and
- those with considerable salt tolerance which either migrate between freshwater and marine habitats or are of marine origin and have colonized freshwater habitats (Darlington 1957). This grouping is particularly useful since there are strong correlations between the ecological and systematic groups (Myers 1949; Cranbrook 1981).

The fish fauna of Sundaland has many species from the first two groups whereas Sulawesi has no strictly freshwater fish, only a few with a little tolerance to salt, and the majority are from the third group. The Makassar Straits or Wallace's Line therefore seems to separate two distinct fish faunas and it is therefore unlikely that Sulawesi and Borneo were ever a single landmass.

Freshwater fish may not have been able to reach Sulawesi but all major terrestrial animal groups have. Amphibians, for example, have no indigenous species in common with Borneo but of the endemic species four are related to Sundaland species and two have Papuasian species (p. 47). Two of the three frogs that are found east of Sulawesi may in fact have evolved on Sulawesi and subsequently dispersed (Cranbrook 1981). Twenty-three lizard species found on Sulawesi are also found west of Wallace's Line (C. McCarthy pers. comm.). Similarly, of Sulawesi's 63 snake species, 38 are found both sides of Wallace's Line but only two are found in New Guinea and not in Sundaland (den Bosch 1985).

The tortoise *Indotestudo forsteni* was thought to be of particular interest because it was the only land tortoise known from both sides of Wallace's Line having been first found in Minahasa and Halmahera. It now appears that it is the same species as the later-described *I. travancoria* of eastern India, Thailand and Burma, and was brought from there by traders as a food animal (Hoogmoed and Crumly 1984).

The birds of Sulawesi are predominantly western with 67% of the species having affinities with the Sunda region (Mayr 1944). One, the Sulawesi Roller *Coracias temmincki*, in fact has no relatives in Sundaland and all the other members of its genus are found in Europe, mainland Asia and Africa (Klapste 1982).

Only two of the non-flying mammals on the mainland, the cuscuses, are clearly of Australian/New Guinea affinity;[27] the remainder, including the endemics, have their origin in Asia. Wallace's Line delimits the eastern boundary of the distinctive Sundaland fauna comprising moles, flying lemur, tree shrews, lorises, gibbons, pangolins, porcupines, dogs, bears, otters, weasels, cats, elephants, tapir, rhinoceroses and mouse-deer. Thus the fauna of Sulawesi has closer affinities with the Sunda than with the Philippines, Lesser Sundas or Sahul region, but cannot really be said to be part of it.

In general, Wallace's Line is not as clear a demarcation line for invertebrates as it is for vertebrates but there are many fewer genera in Sulawesi than in islands to the west (Gressitt 1961). For example, there are about 1,200 species of butterflies in the Malay Peninsula, 850 in Borneo and New Guinea, but only 450 in Sulawesi (R. Vane-Wright pers. comm.). As with the palms discussed above, the relatively low number of species may be the result of a previously unfavourable climate. Cicadas are poor dispersers and so their distributions show more detectable patterns than do many other groups. The cicadas of Sulawesi are largely of western origin but there are relatively large numbers of endemic species and some endemic genera (Duffels 1983). The same general pattern (major affinity with Borneo, relatively depauperate, but high level of species endemism) can be observed in other groups such as ground beetles, tiger beetles (N. Stork pers. comm.), and pond skaters/water striders (D. Polhemus pers. comm.), but the affinities of dung beetles Scarabaeidae, for example, are by no means clear (J. Krikken and H. Huijbregts pers. comm.). There are now about 200 species of ground beetle known from Bogani Nani Wartabone National Park and although less than half the total known from Borneo, this is still a massive total for such a relatively small area (N. Stork pers. comm.). This needs to be investigated further.

A detailed analysis of butterfly and moth relationships has shown that Sulawesi species are most strongly associated with the fauna, not in Borneo, but in the Philippines, reflecting the ancient connection through the Sangihe Islands (Holloway 1987). As shown above, the plants of Sulawesi also show a greater affinity with the north, east and south than with Borneo, and the difference in affinity of butterflies when compared with other invertebrates, may be a result of the caterpillars being dependent on specific living plants rather than on detritus. The moth families examined in most detail showed virtually no elements from the direction of Australia possibly because this moth fauna had become established by the time the connection through the Sula Islands occurred (Holloway in press). All these analyses are hampered to some extent, however, by a generally poor knowledge of the Moluccan fauna.

One recent addition to the butterfly fauna of Sulawesi was not through human agency but through natural dispersal. The range of the Monarch butterfly *Danaus plexippus* used to be restricted to eastern and western North America where it performs remarkable annual migrations between the north and south of its range covering up to 3,000 km in a year travelling up to 125 km per day. This remarkable flying ability and the relatively long life of the adult has meant that when blown off course by strong winds it has reached and successfully colonized new areas. It first reached

Hawaii in 1845, Australia in 1870, and Manado in 1873. It has not spread widely within Sulawesi, however, and was not collected by entomologists in or around Bogani Nani Wartabone National Park only 130 km from Manado (R. Vane-Wright pers. comm.).

As might be expected there is no obvious difference in the composition of coastal marine faunas either side of Wallace's Line. Corals, for example, show a great similarity between the species inhabiting reefs throughout the Indo-Pacific region and a general absence of endemic species even in remote areas. Most genera have, in fact, been present for 20-40 million years and some species are probably this old too. This is probably a result of the high-frequency fluctuations in sea-level in the Quaternary (p. 16) which alternately exposed and covered coastal regions. There was probably simply not enough time for populations of long-lived corals to complete many generations before their descendants colonized new habitats, and this may have maximised variation within a species rather than resulted in the appearance of new species. This contrasts with the marked Quaternary speciation observed for more mobile organisms such as fishes and crustaceans (Potts 1983, 1984).

Biogeographical Differences within Sulawesi

The fauna and flora of Sulawesi are far from being homogeneous and evenly distributed, but although the geological history might be expected to have influenced the distribution of species, clear patterns cannot now be seen. The flora of the east peninsula might be expected to have close affinities with the Moluccas, but it is in fact closest to the central and west regions of Sulawesi and to Borneo than to New Guinea. Indeed, all areas of Sulawesi except one are closer to each other in floral composition than to regions outside Sulawesi. The exception is the southwest peninsula whose lowland (<500 m) flora is most closely related to the Lesser Sunda Islands (van Steenis 1972; van Balgooy in press). This could be a result of dispersal from Flores via Kalaotoa, Tanahjampea and Salayar or from Sumbawa via the Sabalana and Postilyon Islands. The tip of the southwest peninsula has a dry climate similar to the Lesser Sunda Islands, bounded to the north by wetter, less seasonal climates (p. 24).

More recent geological history is reflected in the distribution of species within a particular genus. The most striking example of this allopatry (non-overlapping distribution of related species or subspecies) is perhaps the distribution of macaques. The number of species living on Sulawesi is the subject of some debate but four species and seven subspecies seems to be agreed by many (Groves 1980b). These macaques may have evolved from an ancestor of the pig-tailed macaque *Macaca nemestrina* (found now in Sumatra and Borneo), which crossed to Sulawesi, by rafting or by island hopping from Borneo or Java, possibly in the Middle Pleistocene (Fooden

1969; Takenaka 1982), but their origins and relationships have yet to be determined with certainty (Takenaka 1982; Takenaka and Brotoisworo 1982; Takenaka et al. 1985). It is interesting that *M. maura* in the southwest has the most primitive characteristics, and *M. nigra* living furthest away in Minahasa, is the most specialized (Albrecht 1977; Groves 1980b). The original condition was probably a single species but with 'clines' or gradients of character variations existing throughout the range. Disruptions to this continuous distribution would have caused populations to evolve in isolation such that when the disruption was removed the populations were reproductively incompatible (Fooden 1969).

During periods when the sea-level increased by only 4 m (p. 18) the mountainous region south of Lake Tempe would have been cut off by a narrow straits running northwest-southeast, and a narrow isthmus would have been formed between Gorontalo/Limboto and Kwandang. These inundations would help to explain the separation of *M. maura* and *M. tonkeana*, and of *M. tonkeana* and *M. nigra*. The junction of the two subspecies of *M. tonkeana* may have arisen from the formation of a narrow isthmus between Tamba and Labuhanbajo (near the prominent Tanjung Manimbaya), and the two subspecies of *M. ochreata* would have arisen because of the narrow straits between northern Buton and the southeast peninsula. It is not clear, however, how subspeciation arose between the two *M. nigra* subspecies, or speciation between *M. tonkeana* and *M. ochreata*, although unproductive forests on the ultrabasic soils (p. 457) may have formed a biological barrier in the latter case.

Other examples of allopatry are the distribution of subspecies of the large carpenter bee *Xylocopa nobilis* and of species of the large pond skater/water strider *Ptilonera* (fig. 1.33) although not enough specimens are known to be able to determine species boundaries.

Another example of closely-related species replacing each other geographically on Sulawesi is sometimes quoted, that concerning the white-eyes of lowland forests: *Zosterops anomala* in the southwest peninsula, *Z. atrifrons* in the north peninsula, the central block and Banggai Islands and *Z. consobrinorum* in the southeast peninsula (Stresemann 1939-41; Lack 1971). This distribution has now been shown to be rather less well-defined than was previously thought (Holmes and Holmes 1985).

Within a well-known group such as the birds, a number of interesting distribution patterns can be seen within those species endemic to the main island. For example, some species are known from only a single peninsula (table 1.14). Half of the 88 endemic birds are found in all regions of Sulawesi and half have partially-restricted distributions. Thus five species are known from only the north peninsula, central area and southeast peninsula, two from only the central area and southwest and southeast peninsulas, two from only the central area and southwest peninsula, etc. As a result the number of endemic species is different between the main areas (table 1.15). Some of these totals may change as more field data become available, particularly from the east peninsula, but they serve to illustrate the apparent paucity of the birds in the southwestern peninsula.

The Sula Islands, to the east of the Banggai Islands, are administratively part of the Moluccas but biogeographically they are part of Sulawesi. Of the birds, there are twice as many species with Sulawesi affinities as with Moluccan affinities (Wallace 1862) and the birds of the Banggai Islands seem to be derived almost equally from Sula and the mainland (Eck 1976). The reptile fauna of the Sula Islands is essentially a poor Sulawesi fauna (Kopstein 1927), and the flying lizards *Draco* of the Banggai Islands are more similar to those of the Sula Islands than to the mainland (Musters 1983).

Table 1.14. Endemic birds known from only one peninsula of Sulawesi.

Area		Species
North peninsula	Diabolical nightjar	*Eurostopodus diabolicus*
Southeast peninsula	Sulawesi white-eye	*Zosterops consobrinorum*
Southwest peninsula	Lompobatang flycatcher	*Ficedula bonthaina*

From Holmes and Wood 1979; K.D. Bishop pers. comm.

Table 1.15. Numbers of endemic bird species in the five main regions of Sulawesi.

Area	Number of endemic species
Southeast peninsula	65
North peninsula	60
Central area	58
Southwest peninsula	50
East peninsula	47

From Holmes and Wood 1979

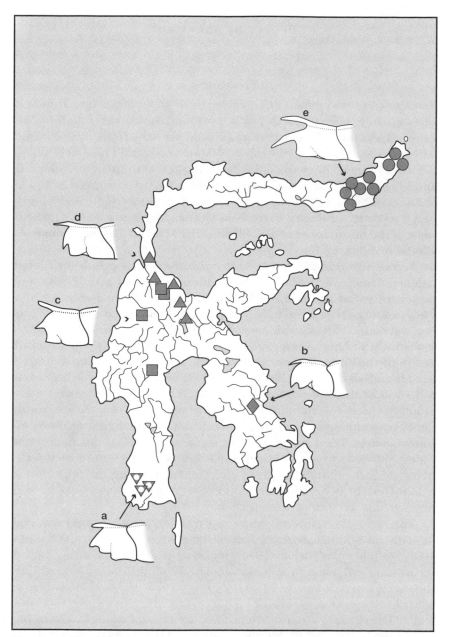

Figure 1.33. Distributions of the five known species of large pond skater/water strider *Ptilonera* on Sulawesi. Side views of terminal segments of females' abdomen illustrate the considerable morphological variation. Note the single zone of overlap, in Lore Lindu National Park. a - *Ptilonera laelaps*, b - *P. sumizome*, c - *P. oribasus*, d - *P. pamphagus*, e - *P. dorceus*.

After Polhemus and Polhemus 1986

PEOPLE OF SULAWESI

Prehistory

Remains of proto humans *Homo erectus* from about 500,000 to 1 million years ago have been found in Java but nothing similar has been found in eastern Indonesia. The first traces of modern man *Homo sapiens* in the latter area have been dated to about 30,000 years B.P.[28] These were populations of hunter-gatherer people who were present throughout Indonesia and who successfully crossed the sea between New Guinea and Australia before 35,000 years B.P. They were the direct ancestors of the Australoids, found today in Australia and the New Guinea highlands, and relatives of some of the inland forest tribes of Peninsular Malaysia and the Philippines (Bellwood 1980a).

Between 4000 and 2000 B.C., Austronesian-speaking people on Taiwan and mainland eastern Asia, with an economy based on plant cultivation, began to spread through the Philippines into eastern Indonesia and western Melanesia (the islands of the Bismark, Solomon and New Caledonia groups). This expansion was perhaps initiated by agricultural developments in southern China and Taiwan, and had, by 1500 A.D., reached over half the circumference of the globe from Madagascar (colonized from Indonesia) to Easter Island (Bellwood 1980a). The people had taken with them domestic pigs *Sus scrofa* and dogs, pottery, bows and arrows, a tradition of thatched community houses, fishing and canoe transport with sails. They cultivated taro, bananas, breadfruit, sugarcane,[29] sago and possibly coconuts. These people did not replace indigenous inhabitants but rather blended with them (Jacob 1967; Bellwood 1985), and their technological and cultural novelties were adopted. Tracing the migration is extremely difficult because racial history is very complex and some human physical characteristics are easily changed.[30]

The early setters of South Sulawesi are known from remains excavated in many caves in the southern half of the province (Sarasin and Sarasin 1905; Mulvaney and Soejono 1970; Heekeren 1972) and the most detailed records, indeed among the most detailed in Southeast Asia, come from three caves near Maros (fig. 1.34) (Glover 1976, 1977, 1978, 1979a, b, 1981; Burleigh 1981; Frank 1981; Glover, E. 1981; Mook 1981; Vita-Finzi 1981). The caves, in order of artefact age are Leang Burung 2, Ulu Leang and Leang Burung 1 and represent the period between about 30,000 and 8000 years B.P. although there is a gap of 10,000 years between Leang Burung 2 and Ulu Leang and overlap between Ulu Leang and Leang Burung 1 (Glover 1977). The culture whose remains are found is sometimes termed 'Toalean' after a tribe of hunter-gatherer and occasionally cave-dwelling people encountered by the Sarasins in the hills half-way

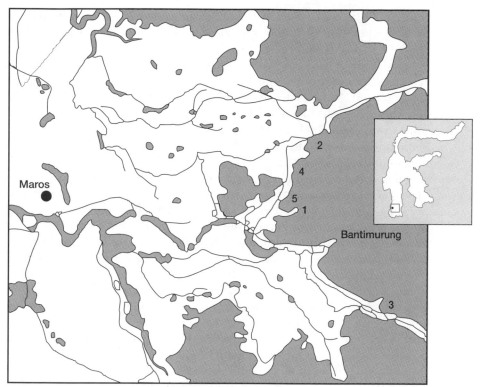

Figure 1.34. Caves with known prehistoric remains around Maros. **1**- Leang Burung 2; **2** - Leang Pattae; **3** - Leang Pettae Kere; **4** - Ulu Leang; **5** - Leang Burung 1. (Dark grey shade indicates hills.)
From Glover 1977

between Maros and Watampone. Photos of these people were printed in their book (Sarasin and Sarasin 1905). To-ala means 'forest people' and is a derogatory expression, and the To-ala may simply have been runaway Bugis debt slaves and landless villagers making a poor living in the forest. There is certainly no evidence to link them with the original cave inhabitants and in any case there is a gap of 2,000 years between the most recent, well-dated Toalean finds and the finding of the To-ala people (Mattulada 1979; I. Glover pers. comm.).

Southeast Asia during the mid-Holocene was home to two types of stone tool industry. The Hoabinhian industry was found in northern Sumatra and the Asian mainland, and is characterized by large choppers made from large split river stones. The other industry had its geographical centre in Sulawesi, typified by the 'Toalean' remains and is characterized by the production of numerous relatively small, fine flakes made from chert[31] for use as knives, scrapers, etc. These small flakes, known as microliths, are of two main types: backed flakes which appeared about 6000 years B.P. and the specific 'Maros points' which appeared about 4600 years B.P. (fig. 1.35). Similar backed flakes and blades have been found in India and western Asia from about 10,000 years B.P., and in various parts of Australia from about 5000 years B.P. (Glover and Presland 1985) and before. Some of the flakes found in the Maros caves have a gloss resulting from being polished against the materials they were used to cut or scrape. The minute scratches on the glossy surfaces have been analysed and it has been concluded that between 31,000 and 19,000 years B.P. the inhabitants of Leang Burung 2 worked wood and other plant materials with a cutting motion and that the maximum diameter that could have been cut through was only 3 cm or 4 cm. Thus these rude tools could have been used for cutting strips of stems and leaves to make string, mats, baskets and simple weapons such as spears, but would not have been used for whittling or slicing to make more refined objects such as complex wooden points, spear throwers, harpoons, etc. However the tools found in Ulu Leang (dating from 9000 to 3000 years B.P.) could have been used for this finer work. Conclusions from such analyses need to be guarded because it is not possible to be sure that the remains found in the caves represent the whole arsenal or kit of tools used by these prehistoric people (Sinha and Glover 1984). There are groups of forest people today who have taboos about bringing certain tools or weapons into the living area, and this might also have been the case thousands of years ago. Alternatively, it may be just that certain tools were stored close to the areas (outside caves) where they were used.

Animal remains from Ulu Leang indicate that the most important prey were pigs and babirusa, followed by anoa, macaque monkeys, and small animals such as snakes, bats, rodents, cuscus, lizards, tortoises and squirrels. Bird and fish remains are surprisingly rare. Another very abundant food animal which was gathered around the caves was the snail *Brotia perfecta*. Over 90% of the shells have their tips broken off whereby the small amount of flesh could be sucked out (van Heekeren 1972). These are still found in rivers around the caves which suggests that the environment has not changed dramatically during the period in question.

Plants remains found in Ulu Leang deposits are seeds of sedges, wild grasses *Panicum* (Gram.), figs *Ficus* (Mora.), *Canarium* (Burs.) and *Bidens* (Comp.), a weedy herb used by modern villagers to relieve coughs, toothaches and sore eyes and as a vegetable (Burkill 1966). In addition to the above species, remains of rice husks were found in deposits in a hearth, believed to be 1,500 years old but it is not known for certain whether cultivated forms were important in southern Sulawesi at this time (Glover

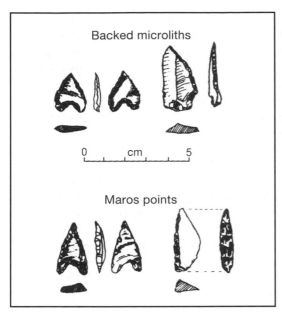

Backed microliths

0   cm   5

Maros points

Figure 1.35. Backed microliths and Maros points.

After Glover and Presland 1985

1979b, 1985). It is also very likely that the people gathered the large (up to 35 kg) surface tubers of a yam *Dioscorea hispida* (Dios.) which can still be found in the area and represented an important food for villagers during the Japanese occupation (Burkill 1966; S. C. Chin pers. comm.). The tuber and the foliage contain the poisonous alkaloid dioscorine which must be removed by rasping, pounding, grating and soaking, preferably in salty water. A piece of tuber the size of an apple is enough to kill a man if eaten raw (Burkill 1966).

Two major prehistoric sites have been excavated in North Sulawesi: a coastal cave called Leang Tuwo Mane'e in the north of Talaud's main island Karakelang, and a shell midden (rubbish dump)[32] at Paso on the southwest shore of Lake Tondano near some natural hot springs which would obviously have been attractive to early people. The cave deposits contain chert flakes from 4,500 to 6,000 years ago but made in a different style from the chert flakes found in South Sulawesi. In the upper layers, starting about 4,500 years ago, a thin and burnished type of pottery is found; about the same time that it appeared in Ulu Leang. The midden deposits date from about 7,500 years ago and contain scrapers made from obsidian[33] flakes (Bellwood 1978). The animal remains found in the midden deposits are similar to these from Ulu Leang but the proportions and quantity of the different animals are markedly different. For example, there are four times as many pigs as anoa in Paso deposits, but fifteen times as many pigs in Ulu Leang as at Paso. There were many rodents and no tortoises at Paso but few rodents and many tortoises at Ulu Leang. At Paso the long bones of animals were broken into recognizable pieces but at Ulu Leang they were chopped up and unrecognizable (Clason 1979). Interestingly, when the Paso midden was being used, the surface of Lake Tondano was higher than it is now (Bellwood 1978), indicating that either some change has since occurred to the riverbed where the water flows out of the lake, or that the rainfall and hence average height of the lake surface used to be higher.

It can be concluded, then, that eastern Indonesia had a variety of stone tool industries worked by isolated communities of people who occupied a wide range of habitats: low swampy areas, steep rocky coasts, high mountains and inland lakes (Glover 1981). The people may have cremated their dead (Boedhisampurno 1982) and buried them in a flexed position. They painted pigs, not babirusa as stated elsewhere (van Heekeren 1972), geometric designs and made stencils of their hands on the walls and ceilings of their caves. Some of these can still be seen in Leang Pettae and Leang Petta Kere caves in the Prehistoric Park near Maros (Soejono 1978; Anggawati 1985), but they are more or less impossible to date. Red ochre, of the sort used for the paintings, was found in deposits dating from 20-30,000 years ago but this might have been only for decorating the cave-dwellers' bodies. Paintings depicting hunting scenes, boats, and warriors on horseback (fig. 1.36) have been found recently in several rock shelters west of Raha on Muna Island (Kosasih 1983, 1984). These are probably relatively recent but the depiction of horses does not necessarily mark them as particularly recent. For example, horse sacrifices are mentioned by King Mulawarman of East Kalimantan in the 4th century in Indonesia's oldest inscription. Horses and goats would have been brought from western India as items of trade and diplomacy (J. Miksic pers. comm.). Horses are known to have been present in southern Sulawesi in the 16th century (Pelras 1981).

Remains from a Neolithic settlement have been found at Kamasi Hill near Kalumpang, a village 93 km upstream from the mouth of the Karama River in West Toraja (van Stein Callenfels 1951). Plain and decorated pieces of pottery were the most abundant artefact and are similar to remains found in China, Luzon, Vietnam and East Java. Human remains were found, as well as bones of anoa, wild and domestic pigs, and fish. In addition, the excavations revealed highly polished rectangular adzes, small chisels, ground oval axes (similar to those from China and Taiwan), spearheads (similar ones also found in Hong Kong) and stone rings. The minimum estimate for the age of the Kalumpang artefacts is between about 3000 B.C. and 5000 B.C. (Bellwood 1980b, 1985). A nearby site known as Minanga Sipakko is probably older but is less well known (van Heekeren 1972).

Figure 1.36. Hunters or warriors on horseback and man with spear as depicted in paintings on a wall in a rock shelter on Muna Island.
After Kosasih 1983

Shards of perhaps similar smooth, red, unglazed earthenware pots of about 40 cm diameter were found by members of Operation Drake in a cave near Kolonodale but the cave has not yet been properly excavated (Rees n.d.).

The cultural stage after the Neolithic was a form of 'Bronze Age' and Central Sulawesi is rich in artefacts of this period, notably huge stones or megaliths. These megaliths are large worked stones in the form of huge cylindrical vats, large statues, urns and mortars and are found primarily in Central Sulawesi (fig. 1.37) (Sukendar 1976; 1980a, b) but the meaning and creators of these stones are unknown. The decorations on the vats, which were probably multiple burial chambers, are of faces, figures, monkeys and lizards and most are near the village of Besoa, and are very similar to designs found in Laos.[34] The statues are mostly larger than life, usually male (only one-quarter are female), legless and armless, and set upright in the ground (Kaudern 1938).

A megalithic culture is still alive (just) in Sa'dang Toraja where the people celebrate by erecting large stones in rows or circle (Asmar 1978; Kadir 1980). Megalithic remains also abound in the hills north of Wattang

Figure 1.37. Locations where megaliths are found in Central Sulawesi.

After Kaudern 1938

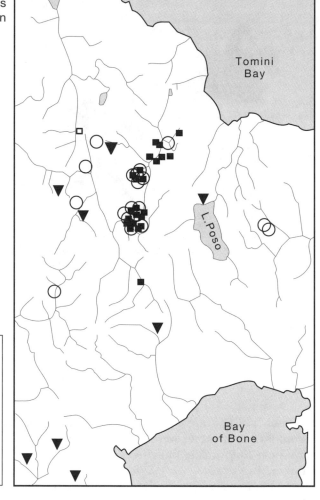

Key

▼ Menhirs

□ Large stone blocks with small grinding holes

■ Statues

○ Vats

Tomini Bay

L. Poso

Bay of Bone

Sopeng as far as Sengkang, and the traditions are still practiced by the Amparita people of Sidenreng (van Heekeren 1958b; MacKnight and Bulbeck 1985). Bronze used to be a highly valued metal and a magnificent, large (115 cm diameter) bronze kettledrum decorated with stylized frogs has been found in Salayar Island.[35] It is typical of the Dongson culture, the centre of which was in Vietnam, and may have been made about 2,000 years ago. The presence of this drum may be evidence of a significant settlement in Selayar during the first millenium A.D., possibly inland from the relatively poor soils of the west coast where most of the population lives today. The houses of the Toraja are almost identical to house motifs used to decorate some Dongson kettledrums. Bronze was regarded as a very special metal and was greatly prized. A description of Minahasan women in 1679 states that the women were seen wearing up to 10 kg of bronze. Bronze was also felt to have magical, particularly protective, powers, and many bronze axes were taken by the Dutch after they had defeated the Gowa army on Buton Island in 1667. In the Luwu and Wotu areas at the top of the Bay of Bone, bronze axes, forged generations earlier, were believed to be teeth of a spirit (van Heekeren 1958b).

Near ports or other centres, rapid cultural changes occurred perhaps 1000 years before they occurred in remote tribal areas. If prehistory is taken as referring to that period before events and thoughts are written down, then there are still groups of people in remote areas of Sulawesi who live in prehistory. The official number of these tribal people in South Sulawesi is 60,000 (e.g., the Sareung, Bentong and Towala), in Central Sulawesi 50,500 (e.g., the Tolare, Towana and Sea-sea), in North Sulawesi 10,000 (e.g., the Gorontalo), and in Southeast Sulawesi 5,800 (e.g., the Tolaki, Tooere and Koro), but only a very small proportion of these are beyond the influence of government institutions. The major trends in dealing with these people are to teach and to develop, and little or no effort is made to learn from them. They hold within their cultures more information about their various environments and about forest products (drugs, rotan, semi-domestic crops, etc.) than could be gleaned in a decade of research, yet this knowledge is being lost to Indonesia and the rest of the world. The only in-depth study of a tribal group in Sulawesi seems to have been on the Towana of the eastern peninsula (Atkinson 1979, 1985), but this concerned their sociology rather than their ecology.

Impacts of Prehistoric Man

The activities of primitive man probably affected populations of animals and plants in four ways (Rambo 1979). First, he exerted direct selection on the species of prey he hunted. In many areas of the world early man has been implicated in the extinction of the giant members of the Pleistocene fauna (these were either absolutely large or large by comparison with living relatives). In Sulawesi, this 'mega-fauna' was represented by at least Heekeren's giant pig, the giant tortoise and the stegodont (p. 34). Both giant tortoises and elephant-like animals have suffered greatly from man and these used to be found in North America, Europe, Africa and Asia. The cause of their extinction is not known but if man were present when these animals roamed Sulawesi's forests (a situation which has yet to be confirmed) no evidence has been found in cave or other deposits. There is no evidence that prehistoric man had any greater effect than causing local extinctions, probably more as a result of forest clearance than overhunting. The only animals known from cave deposits in South Sulawesi which have not been recorded in historic times are two endemic rats, one of which is known from just two specimens collected in Central Sulawesi in the early 1900s. The absence of records from South Sulawesi is probably partly a result of the very small area of lowland forest remaining (p. 98) and partly because there have been no serious studies of small mammals in the province (Musser 1984).

The relative numbers of the different animal species found in the cave deposits cannot be used to judge whether the Toalian hunters were selective or whether they simply caught and ate anything they could catch. This is largely because the excavations made before the war were not done with the rigorous techniques used in recent years and it is likely that only a proportion of the large bones were collected and many of the smaller fragments were overlooked. Many deposits have also since been mined for phosphate fertilizer (p. 553).

The earliest Australoid people would have used spears and cord snares or set traps of camouflaged pits to catch their prey, and the large ground-living species such as pigs, anoa and babirusa were, not surprisingly, the most common prey. Even with the skill and knowledge of a forest-bred hunter it would be scarcely possible to cause anything but local reduction in numbers. They would, however, have produced fear and avoidance responses in their prey. Fear and avoidance of man is a learned trait: the Tasaday, a 28-member tribe in the forests of Mindanao who had had no known contact with other humans before 1962, were able to approach deer, which they did not hunt, and to stroke them (Nance 1975). In the absence of feline predators (such as tigers, clouded leopards and jungle cats), or large eagles, man as a predator would have been a novel component of the ecosystem. The monkeys and pigs may have responded by forming large social groupings because the more eyes and ears a group has the more likely it is that a predator will be detected (van Shaik 1983). Squirrels and birds may have developed more cryptic behaviour.

Second, he dispersed trees by picking fruits in one place and discarding or voiding them in another. An early form of agriculture would have been the accidental or deliberate sowing of tree seeds in the same area, thereby reducing the distance between fruit trees. This phenomenon can frequently be seen at traditional resting places such as ridge tops or along forest paths, where rambutan, durian and other fruit trees can often be found.

Third, when he was able to fell trees, he modified habitats. The long sequences of remains from Talaud and Maros are exceptional, however, in the almost total absence of ground stone axes and adzes. They do appear in the Kalumpang deposits but these date from rather later. So, not only were these tools possibly not part of the tool-kits of the Australoid people but the immigrant Austronesians (or the culture they introduced) apparently did not use them either.[36] It is possible to conclude then that the activities of the early horticulturalists in Sulawesi did not involve large-scale forest clearance and cultivation probably took place in small fixed plots next to their dwellings (Bellwood 1980a, 1985). It was probably iron, introduced about 1,500 years ago, rather than stone axes that gave man an efficient means of felling trees and it was probably at about that time that the process of forest clearance truly began. Some of the indigenous wildlife, such as the endemic pig, may have benefitted from the greater areas of relatively succulent secondary growth in the 'edge habitats', but this would also have exposed them to greater hunting pressure. There is evidence of forest clearance from about 9,000 years ago in New Guinea, and about 4,000 to 7,500 years ago in Sumatra.

Fourth, he carried crops, pets/live foods and pests with him on his travels. Among his crops would have been plants whose edible parts were also the regenerative parts, such as yams[37] and tubers. Prehistoric man almost undoubtedly introduced the deer *Cervus timorensis*, common palm civet *Paradoxurus hermaphroditus*, Malay civet *Viverra tangalunga*, Prevost's squirrel *Callosciurus prevostii*, some of the rats, and jungle fowl or wild chicken *Gallus gallus*, as well as the later domesticated animals such as buffaloes, dogs, horses, ducks, etc.

Prehistoric man on Sulawesi had the distinction of having brought into domestication the endemic pig *Sus celebensis* some time in the early Holocene. This pig accompanied early adventurers or was traded, and morphological analyses of skulls show that Sulawesi pigs are still found in a domestic or feral state in Maluku, Flores, Timor and, amazingly, Simeulue, an island off the west coast of Aceh in northern Sumatra. Unlikely though this may seem, it is of great interest that the Simeulue language is said to be closest to a Bugis-type language! The only other pig brought into domestication was the common wild pig *Sus scrofa* which was apparently superior to *Sus celebensis* which it replaced, ousted, or with which it hybridized (Groves 1981).

History

It was in the 13th to 15th centuries that the famous Bugis kingdoms of Bone, Wajo and Soppeng, and the Makassarese kingdom of Gowa arose. During this period the power of the largely Buddhist empire of Srivijaya in Sumatra was waning, and the power of the Hindu Majapahit kingdom of East Java was in ascendency. Around this time the southwest peninsula supported a relatively small population scattered in small settlements centred on resource-rich geographical features such as lakes, rivers and estuaries to supplement the produce of swidden agriculture. Most contacts were with traders from northern Sulawesi, Maluku and Sumbawa rather than from Java (MacKnight 1983). Great quantities of ceramics were imported from China and elsewhere through these trading links, and were used in ritual, two-stage burials (still practiced by some Toraja people today) up to at least the middle of the 17th century (Hadimuljono and Muttalib 1979). Makassar, Luwu, Bantaeng and Salayar were listed as dependencies of Majapahit and their ships are described in a 14th century Javanese poem as harassing trade ships from Malacca. Bugis states were trading with Arab ships during this period. Of these states, Luwu was the first to establish a kingship, and dynasties established later elsewhere were believed to be quasi-divine though still subject to the same forces which influenced their subjects. The 'high-culture' centres of this period in Indonesia tended to be coastal and concerned largely with trade but because of its narrowness, the majority of the southwest arm of Sulawesi was affected. The history of this period is composed mainly of genealogies and was written on leaves of lontar palms *Borassus flabellifer* (Mattulada 1985) as was the custom in the drier and more seasonal parts of Asia until the Portuguese introduced the techniques of paper manufacture (Whitmore 1977).

At this time there were about 50 kingdoms in the southwest peninsula (although Bone, Wajo and Luwu were clearly dominant), and the kings and their people worshiped images–presumably a mix of the traditional animistic beliefs and traces of Hinduism. These kingdoms, most of them little more than a small town with a feudal chief ruling over a subservient hinterland, had their main settlements near river mouths. These settlements probably comprised 100-200 households and were built backing on to the river (Mattulada 1978).

The name Celebes was first used by the Portuguese historian Tome Pires, whose discerning and comprehensive accounts of eastern Asia in his *Suma Oriental* were written in India and Malacca between 1512-15. He mentions that the Portuguese sailed to the Spice Islands or Moluccas via Singapore, Tanjung (?)Puting (Central Kalimantan), Buton and sometimes Makassar. Celebes, Banggai and Siau were said to produce foodstuffs and gold for the Moluccas. 'Celebes' applied initially only to the northern point of the north arm (Punta de Celebres) and was first marked on a map in 1524. Only later did it come to apply to the whole mainland (Pires 1944).

There are four hypotheses concerning the derivation of the word Celebes. First, the Bugis word 'selihe' (the 'h' is sometimes pronounced as an 'r') means sea current, so Punta de Celebres would mean Point of Currents (Pires 1944). Second, it derives from 'sula' (island) 'besi' (iron), for the area around Lake Matano in the centre of Sulawesi is one of the richest deposits of iron ore in Southeast Asia. Third, it derives from 'Si-lebih' or the one with more islands (Crawfurd 1856); and fourth, it derives from a corruption of Klabat, the name of the impressive volcano north-east of Manado which dominates the Minahasa landscape (Sarasin and Sarasin 1905; de Leeuw 1931). The modern name Sulawesi clearly derives from Sula-besi ('b' and 'w' are frequently transposed).

The first writer to mention the northern islands was Pigafetta who wrote the journal of the first round-the-world voyage led initially by the Spanish captain Magellan. Magellan himself died in the Philippines in early 1521, but later that year his ships sought a passage from there to the Moluccas. Having left southern Mindanao, they encountered the small Kawio Islands and then sailed to Sangihe Island which was said to be "very beautiful to look at". The island was divided between four kings. The boats then passed the islands of Kalama, Kanakitang, Para, Sanggeluhang and Siau, the last of which was ruled by a king called Raja Ponto. After reaching Ruang, they turned south-east (Pigafetta 1906), never setting eyes on the Sulawesi mainland only 80 km distant.

Gowa was the first dominating power on Sulawesi, due in part to the alliance made with the Bajau or sea nomads (see below) who furnished the kingdom with a wide range of marine produce. Gowa dominated the whole Makassar-speaking region and its capital, Makassar, was a cosmopolitan trading centre. Malay traders began living there in the middle of the 16th century, followed by the Portuguese. The Dutch opened a trading post and factory in 1607, the English in 1613, and the Danes in 1618. Agents from France, the Philippines, India, Arab countries, Aceh and China were also to be found (Reid 1983).

The Dutch displaced the Portuguese from the Moluccas and tried to exercise a monopoly on the spice trade. The importance of Makassar was, however, that people were able to buy spices there without having to tangle with the Dutch. Malay and Gowa ships sailed to the Moluccas with food and foreign goods and traded them for spices which were then sold in Makassar. Makassar became even more popular when it was made a free port.

Catholicism preached by Portuguese missionaries from the 1540s did not make a great impact, and later kings sent envoys to the Malay Peninsula to find Islamic teachers. Islam was accepted as the state religion by the twin kingdom of Goa-Tallo in 1605 and this was the last major Indonesian state

to do so.[38] Gowa and the states of Bone, Wajo and Sopeng that formed a
Bugis alliance were often warring within and between themselves and
these last three later accepted Islam at different times after the powerful
Gowa. Acceptance of Islam at that time was tantamount to acceptance
that they had to be allies (or vassals) of Gowa. Gowa influence spread
slowly to south-east Borneo, Buton, Palu, Toli-Toli, Lombok, Flores, Sumbawa and Timor (Hadimuljono and Muttalib 1979).

Inland, the Toraja people of the central mountains traded with the
Bugis but their supposed headhunting traditions tended to minimize the
degree of contact. The present-day Aluk-Todolo religion of western Toraja
probably resembles the early beliefs of all the inland tribes of that region.

The Minahasans, who are physically most closely related to the Fil-
ipinos,[39] have never been subject to dynastic rule although it is believed that
there was pressure to institute kings in the early 7th century. In response to
this, a large open meeting is said to have been held in about 670 A.D.
around the stone known today as Watu Pinabetengan,[40] and the government of the independent states was discussed (Taulu 1981). In contrast,
Bolaang Mongondow and all the islands north of Manado had kings and
slaves, and were subject to the rule of the Portuguese-influenced Sultan of
Ternate (an important 'spice island' on the west coast of Halmahera,
300 km to the east of Minahasa), as were the Banggai Islands and the
mainland area of Luwuk. The first Europeans to live in Minahasa were
Spaniards from Magellan's boats escaping from the Portuguese in Ternate
in about 1524, although it seems that Portuguese ships had occasionally
called before this to buy rice (Jones 1977). Christianity was brought successfully to Minahasa and Sangihe-Talaud in the 1560s by Portuguese
missionaries, and the king and people of Siau accepted Christianity in
1568, the same year that Indonesia's third oldest church, the Evangelical
Church of Minahasa, was founded. At about the same time most people in
Bolaang Mongondow and Gorontalo were accepting Islam, although some
had been converted earlier by Bugis traders (Jones 1977). In 1574,
Spaniards from Manila sent envoys to Minahasa and in 1619 a pastor was
sent out. Eleven years earlier, the Dutch East-India Company built a
wooden fort that doubled as a trading post near Manado. In 1643 the
Spaniards attempted to impose a half-Spanish king on the Minahasans
who were adamant that they wanted to retain their independent status.
This spurred the Minahasans to call in the Dutch and ask for a defence and
trading alliance (although this was not formally concluded until 1679)
(Taulu 1981). The history of this period was written onto a rock face at
Watu Pinantik by an ousted Spanish priest, and these writings can still be
discerned (Taulu 1981). The trading agreement later turned sour and
many states actively opposed the Dutch. This came to a head when Dutch
arms and ammunition were seized and taken to Moraya Fort on the shores
of Lake Tondano. The Resident began a siege and the will of the Mina-

hasans broke only after he ordered a flock of pigeons with burning palm-fibre tied to their legs to be released; these landed on the thatched roof of the fort and razed it to the ground (Taulu 1981).

The royal court of Makassar was cultured, tolerant, and secure in its success as one of the great entrepôts of Southeast Asia (Reid 1983). It was also a thorn in the flesh of the Dutch and as a result its zenith lasted barely 50 years. In 1660 the Dutch destroyed six Portuguese ships in Makassar's deep-water harbour, captured the fort and made an alliance with the Sultan of Gowa. Later they schemed with the Bugis state of Bone, in particular with Arung (Prince) Palakka, against Sultan Hasanuddin of Gowa. They began fighting in 1666 and the Bungaya Treaty was signed a year later forbidding all foreign traders to live in Makassar, and transferring jurisdiction over Bulukumbu-Bira, Maros, Bantaeng and the Makassar fort (Hadimuljono and Muttalib 1979; Andaya 1981). There were several uprisings in Makassar during the following century and the town was totally destroyed at least twice. At this time the Dutch had little or no interest in the land they now ruled, because their main aim had been the suppression of a competitor. The main export during the 18th century had been slaves. After the subjugation of Makassar many Bugis, particularly from Wajo, emigrated and founded royal dynasties in Kutai (East Kalimantan), Johore and Selangor (Peninsular Malaysia). In 1737 the Wajo king of Kutai liberated the Wajo state from Bone, which had become increasingly more powerful, and established the most successful Indonesian maritime commercial operation of the 18th and 19th centuries.

Since most of the written early history of Sulawesi and neighbouring islands was concerned with maritime trade, the life of the traditional nomadic people of East Indonesian seas, the Bajau, is quite well known. The Bajau used to be a group of maritime hunter-gatherers and although now largely settled, they can still be met with around Sulawesi's eastern islands. In contrast to the Bugis who, although seafaring, were based on land or in beach settlements, the Bajau spent more or less their whole lives on or around boats. Another major contrast was the Bugis habit of leaving their family behind when travelling whereas the Bajau always travelled as a family.

The Bajau economy was based primarily on collecting, preparing and selling 'tripang', a collective name for a few types of edible sea cucumbers, a group of echinoderm animals (which also include starfish, brittlestars, and sea urchins) (p. 227). Sea cucumbers are bottom-dwelling (benthic) creatures and are either collected at low tides or dived for in shallow water. The most sought-after tripang were apparently painstakingly prepared by a Bajau group known as the Turijene who were based in the Spermonde Archipelago, particularly around the island of Kuring Aring, 17 km west of Ujung Pandang. All the tripang, together with tortoiseshell, pearls, birds' nests and giant clams, were sold to Bugis seafarers for the China trade in food, tonic and medicine, and this formed the basis of the interdependence of these two groups. Tripang was not eaten by any Malay people even when other food was scarce and, strangely, the Bajau rarely if ever used nets or traps for catching fish, preferring instead to use harpoons and spears.

The narrow entrance to Kendari Bay was discovered in 1839 at which time the east and southeast of Sulawesi were very sparsely populated (fig. 1.38). Inland people in this area have moved south only in the last century or so. A few Bugis and Gowan settlements were established and were dominated alternately by maritime powers such as Luwu and Ternate. Kendari Bay was the only major trading focus in the southeast but this was occasionally abandoned because of the tribal disputes and raids on traders by the inland people. The Bajau did not have a particularly cordial relationship with the Bugis and Gowan traders, but generally had peaceful relations with the inland Tolaki people. This was not always the case, however, and it seems that the Bajau were probably in frequent conflict with, for example, the Toloinang of south Tomori Bay (Sopher 1978).

The Banggai Islands, or more specifically the small islands between Peleng and Banggai, were another traditional centre of Bajau activity; in the 1840s, 100 to 150 boats gathered there at certain times of the year. The Banggai people, of the Towana group, accepted the presence of the Bajau and traded produce from their dry-field agriculture with them. Other centres of Bajau activity were Tomori Bay (west Tolo Bay), Dondo Bay (off Toli-Toli), the Sabalana Islands, and Wowoni Island (Sopher 1978).

One of the earliest dates in the history of Central Sulawesi was 1555 when the Portuguese built a fort at Parigi. In 1602 two Muslim Minangkabau traders from West Sumatra settled in Palu and Parigi to develop commercial interests and to propagate their faith. In 1680 Palu and Toli-Toli were under the influence of the Sultan of Ternate but the Dutch East India Company had growing interests and established an outpost in Parigi in 1730 to trade in gold (Davis 1976). Meanwhile the people in the hills lived in fortified stockades, and the earthworks that once surrounded these stockades are still visible in villages such as Besoa, Padang Lolo, and Bala (Anon. 1981b). Given the generally low fertility of the hill soils, it is interesting that many old villages and megalithic remains are on or around the beds of long- 'dead' lakes (Davis 1976). The small clans of people were frequently warring and when a peace was settled it was customary to smash plates. This is symbolic and represents the feeling that 'as the shards lay divided and yet together in one place, so shall we all return to our hearths knowing we belong together'. One such place is Poka Pinjang ('broken dish') on the path to the peak of Mt. Rantemario, Enrekang (p. 511), where shards of Chinese porcelain have been found (van Steenis 1937). The shy inland Towana people of the eastern peninsula also paid tribute to the Sultan of Ternate, but later came under the influence of the kings of Bungku to the southwest, and Tojo to the northwest. These rulers did not exert direct rule over this group of Towana, but indirectly through Towana

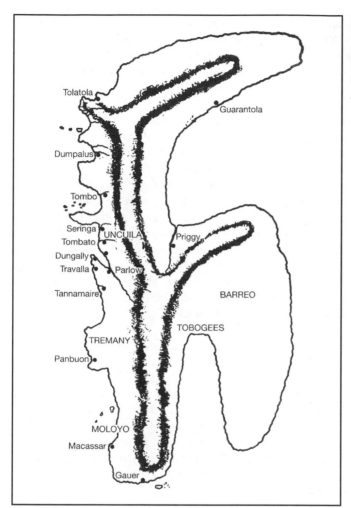

Figure 1.38. A sketch map of Sulawesi drawn in 1795 to illustrate just how poorly known the eastern coasts were.

After Woodard 1805

leaders and elders. Tributes of beeswax and small bamboo tubes of uncooked rice were given to the rulers (Atkinson 1979). A succinct description of the culture of the people around Poso and the effects on them of early western influence is available (Kruyt 1929).

The British took over the Dutch colonial possessions between 1811 and 1816 and it was not until the early years of this century that the Dutch began to exert their rule over all of Sulawesi. In fact, by the late 19th century Minahasa had already received considerable assistance from the Dutch and from Christian mission societies. The degree of support is illustrated by the fact that by 1900 there were 1,200 inhabitants for every school in Minahasa, but 50,000 inhabitants for every school in Java. The Dutch rule of Sulawesi began with much blood being shed, particularly in battles with people in Bone. Until this period remote areas such as Toraja land had withstood or had simply not experienced incursions and assaults into their territory, but now they were brought under an island-wide administration. There then followed 35 years of peace–probably the longest period in Sulawesi's history–during which roads, irrigation schemes and administrative structures and institutions were established.

The Japanese occupied Sulawesi in 1942 but the move towards independence was not entertained until the final five months of the war. Then, Dr. Sam Ratulangi, a Manadonese, and Andi Pangiran, son of the Sultan of Bone, were sent to Jakarta to assist in the preparations for the birth of the infant republic. Ratulangi became the first governor of Sulawesi but was imprisoned after the Australian allied forces handed over the administration to the Dutch in January 1946. By the end of that year a guerrilla operation against the Dutch was well under way and the Dutch responded with their bloodiest campaign anywhere in Indonesia. On 17th August 1950, the Dutch accepted the independence of Indonesia. At this time Sulawesi was a single province. In July 1950, an intermittent rebellion with Islamic and regional goals erupted in and around Luwu and its forces became allied with the Darul Islam movement of West Java. By 1958, when this movement was at its peak, only the cities in South Sulawesi were under government control, and towns such as Soroako, now the centre of an international tin-mining operation, suffered severe privations (Robinson 1983; Kirk 1986). Order was restored in the early 1960s. In the regional rebellions of 1958-61, Minahasa (together with West and North Sumatra) declared their independence from the rest of the Republic, but few lives were lost before national unity was restored.

Present-day People

Despite the relatively small size and population of Sulawesi, the number and make-up of the ethnic groups is extremely complex.[41] Early linguists used to ascribe a large number of language groups to Sulawesi. For example, one analysis recognized just one language group in each of Taiwan, the Philippines, Sumatra and Borneo, whereas for Sulawesi nine were recognized. More recent work has lessened the distinctiveness of Sulawesi languages but, even so, two of the nine language families found in Western Malayo-Polynesia are confined to Sulawesi. Linguists classify languages in a stricter fashion than do lay people, and the fifty ways of saying 'no' in Central Sulawesi alone do not necessarily represent different lan-

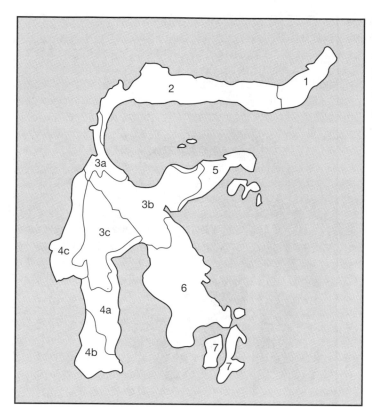

Figure 1.39. The distribution of the major ethnic groups. Numbers refer to table 1.16.

From Davis 1976

guages (Davis 1976). Rather this has arisen through small kingdoms or states being relatively isolated for long periods, and the development of dialects. The Sulawesi people themselves use region, religion and style of farming as the major criteria for determining ethnic groups. For example, someone from Central Sulawesi may refer to Christian Mamasa speakers of South Sulawesi as 'Toraja', but call Islamic Mamasa speakers 'Bugis' or 'Mandar' (Davis 1976). Similarly, the term 'Bugis' can mean seafaring Makassarese and Mandarese as well as the coastal Bugis (fig. 1.39; table 1.16). About 80% of the population of Sulawesi is Islamic and 20% Christian but there is considerable variations between regions (fig. 1.40).

The population is distributed unevenly across Sulawesi, with the area around Ujung Pandang having more than 300 people/km^2, much of the rest of the southwest peninsula, Minahasa and Sangihe-Talaud having 100-299 people/km^2, and Toli-Toli, Mamuju, the eastern arm and the east of Southeast Sulawesi having less than 30 people/km^2 (fig. 1.41).

Figure 1.40. Percentage of population following Islam by county.

Based on Anon. 1981a

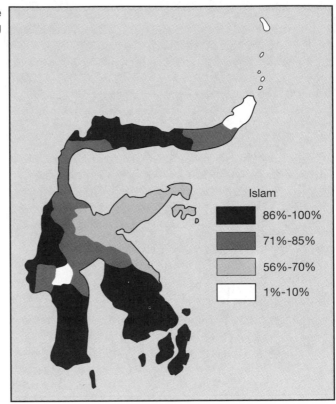

Table 1.16. The major ethnic groups.

	Dominant religion	Traditional economy
1. Minahasa	Christianity	Plantation and mixed agriculture
2. Gorontalo-Tomini	Islam	Mixed agriculture (slash and burn)
3. Toraja		
a. Kaili	Islam	Mixed agriculture
b. Upland Kaili and Pamona[1]	Christianity	Upland slash and burn
c. Sa'adang[2]	Christianity	Mixed agriculture
4. Bugis-Makassar		
a. Bugis	Islam	Lowland agriculture, commerce, seafaring
b. Makassarese	Islam	Lowland agriculture, seafaring
c. Mandarese	Islam	Lowland agriculture, seafaring
5. Luwuk-Banggai	Mixed	Slash and burn, fishing
6. Bunku-Mori	Mixed	Slash and burn, fishing
7. Muna-Buton	Islam	Slash and burn, fishing

[1] - Formerly called 'Bare'e'
[2] - The group now commonly called 'Toraja'

From Davis 1976

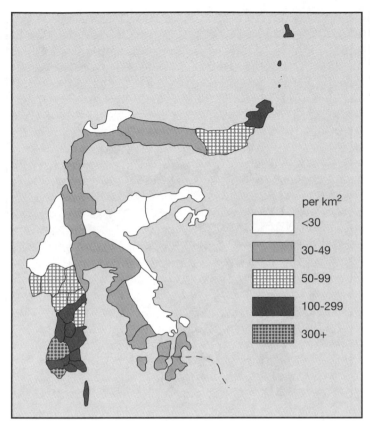

Figure 1.41.
Population
density of
Sulawesi by
county.
After Anon. 1981a

per km²

	<30
	30-49
	50-99
	100-299
	300+

PRESENT STATE OF NATURAL ECOSYSTEMS

Man began converting natural forest to other forms of vegetation many hundreds of years ago but this process has accelerated greatly since the early 1970s when commercial logging, transmigration and estate crop projects began to receive enormous government support. The Sulawesi mainland now comprises an irregular patchwork of natural forest within and between areas of cultivation (fig. 1.42). The forest cover per inhabitant in Sulawesi is more than in Sumatra, Java and Bali, or the Lesser Sunda Islands (table 1.17), but this is at least partly due to the high proportion of land in Sulawesi on slopes which are unsuitable for agricultural development projects (table 1.18). For example, 83% of Sulawesi comprises slopes of 15% or above, compared with only about half of Sumatra, Kalimantan or Irian

Figure 1.42. Extent of cultivated land (black) in 1982.

Adapted from Whitmore 1984b

Table 1.17. Forest area per person in 1982.

	ha forest/person
Sumatra	0.7
Java and Bali	0.01
Kalimantan	4.8
Sulawesi	1.0
Lesser Sundas	0.3
Moluccas	5.1
Irian Jaya	35.1

From Anon. 1982b

Jaya. Small islands have lost most of their natural vegetation: for example, Sangihe and other small northern islands were largely deforested by 1920 although Karakelang still has some forest cover (Heringa 1921). Likewise the forests of the small southern islands had all been converted to agricultural uses by 1915 with the exception of small areas of Tanahjampea and Kalaotoa (van Schouwenburg 1915, 1916a, b, c).

Calculating land areas under different forms of land use is best conducted by the interpretation of satellite images. This was done for Sulawesi (Hadisumarno 1978), but the images used were made in 1972 and the usefulness of the information is now somewhat limited to historical comparisons. More recent analyses of land use over relatively small areas have been made by a number of government departments. An excellent series of maps showing land system and suitability, land use and land status was completed in 1987 for the primary use of the Transmigration Department, but these have had great value for others concerned with land use planning and resource management. Meanwhile, different methodologies, definitions and criteria used by different agencies tend to produce somewhat different figures for the land use of Sulawesi and elsewhere (table 1.19). This makes comparisons between tables confusing but does not necessarily affect the comparability of data within tables. Thus it can be seen that Sulawesi had less forest cover than either Sumatra or Kalimantan, and also relatively less land covered with estate crops (p. 487). The percentage of the land under wet rice fields is comparable to the percentage in Sumatra. See the Introduction for more recent data.

A comparison of land area, forest area, rice production and timber exports between the four Sulawesi provinces is instructive (table 1.20). In

Table 1.18. Areas (in 1,000s ha) of different topographic classes in the main islands of Indonesia. Percentages in parentheses.

	Total swampy area	Plains to level 0%-3%	Level to undulating 3%-8%	Undulating to hilly 8%-15%	Hilly to mountain 15%	Total area
		\[Upland area according to slope spanning\]				
Sumatra	13,211 (29)	8,281 (17)	4,102 (9)	1,844 (4)	19,712 (42)	47,160
Kalimantan	12,764 (24)	2,693 (5)	4,779 (9)	3,308 (6)	29,402 (56)	52,946
Sulawesi	469 (3)	955 (5)	806 (4)	927 (5)	15,747 (83)	18,904
Irian Jaya	12,981 (30)	3,606 (9)	1,288 (3)	844 (2)	23,477 (56)	42,195

From Suwardjo et al. 1985

1985 North and South Sulawesi were similar, as were Central and Southeast Sulawesi, in their ratio of land to population (about 1 ha/person and 4 ha/person respectively), but the difference in land capability and intensity of use is reflected in the greater rice production in South and North Sulawesi compared with the other two provinces. Southeast Sulawesi generally comprises poor agricultural soils and consequently rice yields are low. Timber ceased being exported out of the natural forests of Southeast Sulawesi in 1979 and out of North Sulawesi in 1981, and it is clear that Central Sulawesi was by far the most important timber producing province. Timber revenues in fact supplied more than 95% of the provincial income in Central Sulawesi. Logging continues in all provinces, of course, to supply local needs.

Obtaining definitive data concerning the area of land covered by different categories of forest is as difficult as obtaining accurate data on land use and leads to discrepancies between the figures produced by different departments (tables 1.21 and 1.22). It is well known that some Forestry land no longer has trees growing on it by virtue of inadequate protection, and this is one of the major causes for the discrepancies; there is clearly a difference between Forestry land and forested land. In any case, the percentage of land under some form of legal protection in Sulawesi is less than the percentage on hilly and mountainous slopes which for soil conservation reasons alone should be protected. Estate crops are not allowed on suitable slopes over 25% but some of the steep land has been deforested by second-stage shifting agriculturalists who have, for various reasons, abandoned the traditional or first-stage swidden agriculture practices (p. 570). This land clearly requires legal and enforced protection.

Table 1.19. Major land uses (percent) by main island in the early 1980s. Total area in 1,000s ha. Upper row of each pair from Directorate-General of Agraria 1980, lower row of each pair from Anon. (1984b).

	Total area	Villages & towns	Wet rice fields	Dry fields & gardens	Estate crops	Forest plantation	Other forest	Grass-lands	Critical land
Sumatra	47,611	2	4	3	10	7	46	4	<1
	44,817	4	4	6	8	11	36	2	-
Java	12,609	13	30	22	8	10	6	1	2
	12,174	13	29	24	5	2	17	<1	-
Kalimantan	54,261	<1	1	2	2	3	70	4	<1
	46,765	1	2	2	3	2	63	<1	-
Sulawesi	18,840	2	4	9	4	4	30	7	1
	18,724	3	3	7	4	9	44	4	-

After Suwardjo et al. 1985

Areas with high agricultural potential have clearly been utilized more than areas with low agricultural potential. Thus, nearly all of wet lowland forest on volcanic soils has been felled compared with only 10% of similar forest on ultrabasic soils (table 1.23). Unfortunately none of the remaining wet lowland forest is within either existing or approved nature reserves. This habitat together with wet lowland forest on limestone, dry lowland forest on limestone, freshwater swamp and peatswamp forest are the habitats with the highest priority for conservation on Sulawesi.

The provinces differ strikingly in their geology (p. 5) and so the distribution of habitats between them, and the representation of those habitats within reserves, differ accordingly (table 1.24).

Sulawesi has five National Parks: Bogani Nani Wartabone National Park (formerly Dumoga-Bone) (300,000 ha), Lore Lindu (231,000 ha), Bunaken-Manado Tua Marine National Park, Taka Bone Rate National Park and Rawa Aopa-Watumohae National Park. Lore Lindu has also been chosen as a Biosphere Reserve by UNESCO in recognition of its biological, physical and cultural interest. Biosphere Reserves are areas with a protected core surrounded by utilized buffer zones, managed by a body having institutionalized relationships with the surrounding land and people, centres for management-related research, education and training, and having links with national and international monitoring schemes. Other conservation areas and the areas in Sulawesi under the control of the Directorate-General of Forest Protection and Nature Conservation include Nature Reserves (totalling 322,731 ha on Sulawesi), Wildlife Refuges

Table 1.20. Land, people, rice and forest on Sulawesi.

	Area* (A)	Population (P) (1986) (millions)	A/P (ha)	Area of forested land (F) (million ha)	Percent forest	F/P (ha)	Rice produced (t/ha) (1984)	Timber exports (1,000s tons) (1982)
North	2.5	2.4	1.0	1.3	52	0.6	3.78	0
Central	6.3	1.6	3.9	4.0	63	3.1	2.45	62.16
South	6.4	6.7	0.9	2.6	41	0.4	3.87	2.12
Southeast	3.6	1.1	3.3	2.4	67	2.7	2.24	0
Total	18.9	11.8	1.6	10.3	54	0.9	-	64.28

* - areas in million ha; provincial areas given here differ significantly from official government statistics which are quite incompatible with up-to-date provincial maps

From Anon. 1982a, 1984b, 1985a

(144,788 ha), Tourist Parks (97,000 ha), Hunting Parks (22,000 ha), and Marine Parks (none yet declared). Protection Forests (3,867,000 ha) are under the control of the Forestry Department but are also, in theory, a type of conservation forest. The regulations in force or proposed for the different categories of protected areas are shown in table 1.25.

The conservation areas in the four provinces in 1982 are shown in figures 1.43-46 and are listed with their index of conservation value in table 1.26. Descriptions of these areas, and others which have less or no importance due to exploitation and degradation are described in the Sulawesi volume of the National Conservation Plan (Anon. 1982a).

Table 1.21. Forest areas in 1,000s ha in 1982. From Forestry Department statistics.

| | Permanent forest land | | | | | | Forest land as |
	Protection/ conservation	Production forest	Total	Conversion forest	Total forest	Total area	percentage of total land area
Sumatra	10,777	14,399	25,176	5,032	30,208	46,949	54
Kalimantan	11,024	25,650	36,674	8,293	44,967	54,825	67
Sulawesi	5,273	6,018	11,291	1,587	12,878	19,661	57
Moluccas	1,991	3,106	5,097	436	5,533	8,573	59

After Suwardjo et al. 1985

Table 1.22. Forest areas (1,000s ha) in the four Sulawesi provinces, and percentages of provincial land areas. An additional category of 'other utilization' is also used.

	Protection forest	Conser- vation forest	Limited production forest	Definitive production forest	Conversion forest	Total forest	Total area
North	285	327	741	231	699	2,283	2,751
	(10)	(12)	(27)	(8)	(25)	(83)	
Central	1,157	617	1,364	1,028	335	4,501	6,803
	(17)	(9)	(20)	(15)	(5)	(66)	
Southeast	421	273	827	669	699	2,889	3,814
	(11)	(7)	(22)	(17)	(18)	(76)	
South	2,004	190	993	165	259	3,611	6,292
	(32)	(3)	(16)	(3)	(4)	(57)	
Total	3,867	1,407	3,925	2,093	1,992	13,284	19,660
	(20)	(7)	(20)	(11)	(10)	(66)	

After Anon. 1985c

Table 1.23. Original, remaining (1982) and reserved areas of, and habitat product of the different habitats in Sulawesi (in 1,000s ha). Habitat product is an index of the rarity of a habitat, its rate of loss, and its priority for conservation. 'Wet', 'Moist' and 'Dry' refer respectively to the A and B, C and D, and E agroclimates (see the Introduction).

	Original area	Remaining area (and % original area)		Area included in existing reserves (and % of original and remaining areas)		Habitat product
Upper montane forest	63	63	(100)	11	(17/17)	0.58
Lower montane forest on:						
ultrabasic soils	207	171	(83)	23	(11/13)	0.59
volcanic soils	180	144	(80)	69	(38/48)	0.35
other rocks/soils	3,446	2,284	(66)	233	(68/10)	0.49
Lowland forest						
wet areas on:						
alluvium	401	213	(53)	16	(4/8)	0.82
limestone	33	3	(4)	1	(3/33)	2.45
ultrabasic soils	636	570	(90)	29	(5/5)	0.62
volcanic soils	272	17	(6)	0	(0/0)	2.73
other rocks	1,902	1,265	(67)	78	(4/6)	0.60
moist areas on:						
alluvium	1,015	259	(26)	32	(3/12)	0.94
limestone	507	165	(33)	38	(2/3)	0.78
ultrabasic soils	520	433	(83)	12	(2/3)	0.76
volcanic soils	328	33	(10)	19	(6/58)	1.22
on other rocks/soils	3,704	2,239	(60)	130	(4/6)	0.60
dry areas on:						
alluvium	674	164	(24)	6	(1/4)	1.49
limestone	277	20	(7)	5	(2/25)	1.79
ultrabasic soils	204	99	(49)	4	(2/4)	1.10
volcanic soils	2,077	60	(3)	50	(2/83)	2.09
other rocks/soils	3,733	2,014	(54)	192	(5/10)	0.57
Other habitats						
Volcanic scrub	8	6	(75)	3	(37/50)	0.91
Freshwater swamp	282	66	(23)	25	(9/38)	2.56
Peat swamp	44	34	(77)	?	(?/?)	2.25
Mangrove forest	118	90	(76)	6	(5/7)	0.84
Beach vegetation	55	25	(45)	4	(7/16)	1.21
Freshwater lakes	96	96	(100)	3	(3/3)	0.88
Coral reefs	970	955	(98)	1	(<1/<1)	0.76

Based on Anon. 1982a, b

Table 1.24. Original and remaining (in 1982) areas (1) and percent of remaining areas in reserves (2) (1,000s ha) of different habitats in the four provinces.

	South 1	South 2	Central 1	Central 2	Southeast 1	Southeast 2	North 1	North 2
Upper montane forest	31/30	0	25/25	10	3/3	0	4/4	1
Lower montane								
forest on:								
ultrabasic soils	41/38	20	163/127	3	3/3	-	-	-
volcanic soils	61/25	64	-	-	-	-	119/120	69
other rocks/soils	1,328/825	-	1,842/1,180	161	117/103	-	159/160	8
Lowland forest								
wet areas on:								
alluvium	301/168	16	87/42	-	-	-	13/2.5	-
limestone	3/1	1	-	-	-	-	302	-
ultrabasic soils	306/280	6	270/230	20	60/60	-	-	-
volcanic soils	46/-	-	-	-	-	-	227/23	12
other rocks/soils	909/531	26	450/300	40	370/344	-	173/90	12
moist areas on:								
alluvium	643/115	-	59/31	-	181/100	65	132/12.5	5
limestone	286/27	-	-	-	221/138	39	-	-
ultrabasic soils	-	-	156/115	-	364/316	12	-	-
volcanic soils	235/13	-	-	-	-	-	93/20	19
other rocks/soils	1,191/409	-	764/475	25	1,312/1,025	65	437/330	40
dry areas on:								
alluvium	254/5	-	205/95	-	109/59	1	106/5	4
limestone	47/-	-	-	-	220/14	-	10/5.5	5
ultrabasic soils	-	-	174/87	4	30/12	1	-	-
volcanic soils	88/55	-	-	-	-	-	119/60	50
other rocks/soils	269/13	6	2,100/1,295	37	540/186	1	820/520	140
Other habitats								
Volcanic scrub	-	-	-	-	-	-	8/6	3
Freshwater swamp	225/45	1	5/3	-	19/11	-	33/5	1
Peatswamp	-	-	-	-	44/34	-	-	-
Mangrove forest	61/55	-	-	-	40/25	4	17/10	2
Beach vegetation	3/2	12	25/10	1	4/3	1	35/10	1
Freshwater lakes	82/80	69	5/5	4	-	-	9/7	-
Coral gardens	160/160	-	275/275	-	450/450	-	85/70	1

After Anon. 1982

Table 1.25. Activities permitted (•) and prohibited (-) in different categories of protected area.

	National Parks	Nature Reserves	Wildlife Refuges	Tourist Parks	Hunting Parks	Protection Forests
Growing food crops	-	-	-	-	-	-
Growing tree crops	-	-	-	•	•	•
Human settlement	-	-	-	-	-	-
Commercial logging	-	-	-	-	-	-
Collecting herbs and firewood	-	-	•	-	-	•
Hunting	-	-	-	-	•	•
Fishing	•	-	-	-	•	•
Camping	•	-	•	•	•	•
Scientific collecting with permit	•	-	•	•	•	•
Active habitat management	•	-	•	•	•	•
Non-exotic introduction	•	-	•	•	•	•
Collecting rattan with permit	-	-	-	-	•	•
Mineral exploration	•	-	•	-	•	•
Wildlife control	•	-	•	•	•	•
Visitor use	•	-	•	•	•	•
Exotic introductions	-	-	-	•	-	•

After Anon. 1982b

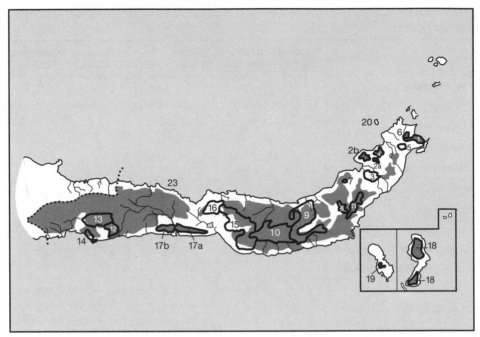

Figure 1.43. Remaining forest, conservation areas and proposed conservation areas in North Sulawesi. **1** - Tangkoko-Dua Saudara, **2** - Manembo-nembo, **3** - Mt. Soputan, **4** - Mt. Lokon, **5** - Mt. Klabat, **6** - Wiau, **7** - Tamposo-Sinansajang, **8** - Mt. Ambang, **9, 10, 15** - Bogani Nani Wartabone (formerly Dumoga-Bone), **11** - Mt. Simbolang; **12, 13, 14** - Marisa Complex, **16** - Mt. Damar, **17** - Labutodoa and Paguyaman Barat, **18** - Karakelang, **19** - Mt. Sahendaruman, **20** - Bunaken-Manado Tua.

From Anon. 1982a

Table 1.26. Major terrestrial conservation areas of Sulawesi and their conservation value. The index of conservation value is a score calculated from six factors: species richness, habitat area, rarity, rate of loss, degree of protection and degree of distinctiveness such that the higher the score the greater the conservation value.

SOUTH			CENTRAL		
Bantimurung and Karaenta	NR	600	Lore Lindu	NP	1,378
Peruhumpenai Mts.	NR	751	Lombuyan	WR	96
Lakes Matano-Mahalano	TP/NR(p)	489	Tanjung Api	NR	297
Lake Towuti	TP	367	Morowali	NR	1,553
Lariang	NR(p)	597	Mt. Dako	NR(p)	216
Mamuju	WR(p)	574	Mt. Sojol	NR(p)	632
Masapu	WR(p)	188	P. Palung	MP(p)	148
Mt. Mambuliling	WR(p)	146	Toli-Toli Mts.	WR(p)	1,198
Rongkong	PF(p)	198	Palu Mts.	WR(p)	1,540
Rompi	WR(p)	315	Morowali/Balantak Mts.	WR	1,259
Lamikomiko	NR(p)	287	Togian Islands	MP	217
Samalona Islands	WR(p)	175			
Lake Tempe	WR(p)	451	**NORTH**		
Bulu Saraung	NR	115	Tangkoko-Batuangus	NR	586
Latimojong Mts.	NR(p)	245	Manembo-nembo	WR	344
Mt. Lompobatang	NR(p)	173	Mt. Soputan	PF	1,059
			Mt. Lokon	PF	55
SOUTHEAST			Mt. Klabat	WR(p)	1,022
Gunung Watumoha	WR(p)	1,384	Wiau	PF	581
Butung Utara	WR	1,607	Tamposo-Sinansajang	PF	217
Kayu Kuku	NR(p)	745	Mt. Ambang	NR	536
Rawa Opa	WR(p)	3,255	Bogani Nani Wartabone	NP	1,954
Polewai	WR(p)	587	Mt. Simbolang	PF(p)	82
Tanjung Batikolo	WR	269	Marisa Complex	NR/WR(p)	1,224
Tanjung Peropa	WR	609	Mt. Damar	PF	402
Kakinawe	WR	343	Labutodaa/Paguyaman Barat	PF	900
Lambusango	NR(p)	464	Karakelang	HR	229
Lasolo-Sampara	WR(p)	600	Mt. Sahendaruman	WR(p)	511

NP - National Park, NR - Nature Reserve, WR - Wildlife Reserve, HR - Hunting Reserve, TP - Tourist Park, PF - Protection Forest, MP - Marine Park, (p) - proposed

After Anon. 1982a

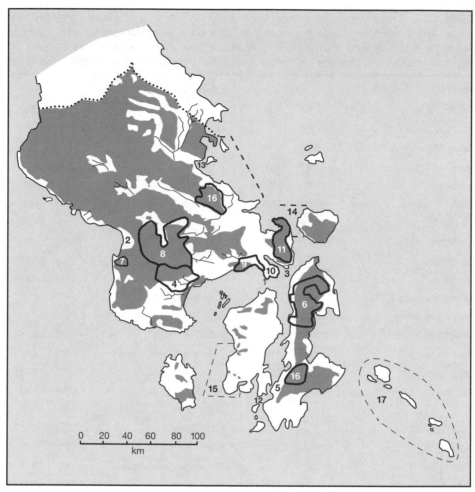

Figure 1.44.   Remaining forest, conservation areas, and proposed conservation areas in Southeast Sulawesi. **1** - Napabalano, **2** - Lamedae, **3** - Tanjung Amolenggo, **4** - Mt. Watumohae, **5** - Tirta Rimba, **6** - Buton Utara, **7** - Kayu Kuku, **8** - Rawa Opa, **9** - Polewai, **10** - Tanjung Bati Kolo, **11** - Tanjung Peropa, **12** - Wakouti, **13** - Lasolo Bay, **14** - Wowoni Straits, **15** - Muna Straits, **16** - Lasolo-Sampara, **17** - Tukang Besi Islands.

After Anon. 1982a

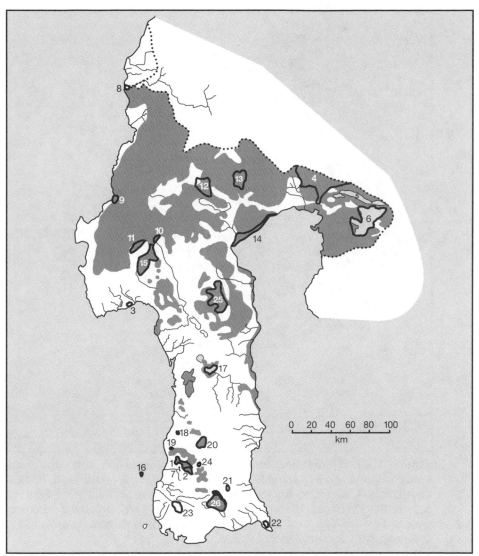

Figure 1.45. Remaining forest, conservation areas and proposed conservation areas in South Sulawesi. **1, 2, 7** - Bantimurung and Karanta, **3** - Lampuko-Mampio, **4** - Peruhumpenai Mts., **5** - Lakes Matano/Mahalano, **6** - Lake Towuti, **8** - Lariang, **9** - Mamuju, **10** - Masapu, **11** - Mambuliling, **12** - Rangkong, **13** - Rompi, **14** - Lamikomiko, **15** - Sumarorang, **16** - Samalona Islands, **17** - Lake Tempe, **18** - Torokkapai, **19** - Matanga, **20** - Bulu Saraung, **21** - Palangka, **22** - Bontobahari, **23** - Komara, **24** - Camba River, **25** - Latimojong Mts., **26** - Mt. Lompobatang.

After Anon. 1982a

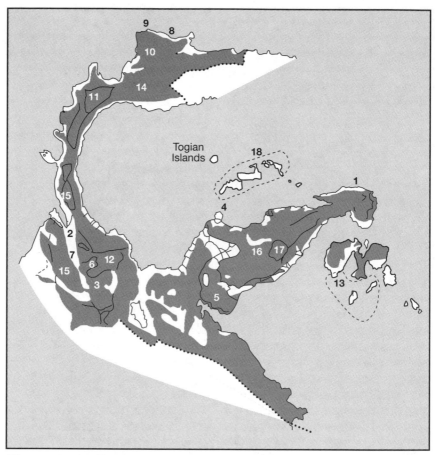

Figure 1.46. Remaining forest, conservation areas and proposed conservation areas in Central Sulawesi. **1** - Pati-Pati, **2** - Paboya, **3, 6, 12** - Lore Lindu, **4** - Tanjung Api, **5** - Morowali, **7** - Wera waterfall, **8** - Tanjung Matop, **9** - Dolongan Island, **10** - Mt. Dako, **11** - Mt. Sojol, **13** - Peleng waters, **14** - Toli-Toli Mts., **15** - Palu Mts., **16** - Morowali/Balantak Mts., **17** - Bakiriang, **18** - Togan Islands.

After Anon. 1982a

Chapter Two

Seashores

INTRODUCTION

This chapter is concerned with those habitats that lie between the reach of the lowest low and highest high tides.[1] Thus all these areas are inundated by the sea at some time. Habitats such as seagrass meadows and coral reefs that are found below the level of the lowest low tide, are discussed in chapter 3.

Sulawesi has proportionately more coastline relative to its land area than any other Indonesian island, because of its long, narrow peninsulas (table 2.1). No point on the mainland is more than 90 km from the sea and most locations are within 50 km. In addition there are over 110 offshore islands each with an area in excess of 1.5 km² within the administrative areas of the four provinces. Coastal ecosystems are of great economic and ecological importance for fisheries and other commercial activities, and those concerned with development on Sulawesi should, therefore, have a grounding in understanding the components, interactions, and mechanisms of coastal ecosystems to ensure that these resources are managed for sustainable production and benefit.

PHYSICAL CONDITIONS

Tides

The process that largely determines the characteristic features of the seashore is the ebb and flow of the tides. The tides around most of Sulawesi are termed mixed prevailing semi-diurnal. This means that each day two high and two low tides occur and that the successive tides are different in height and duration. Around the southwest peninsula the tides are termed mixed prevailing diurnal. This means that each day only one high and low tide occur and that successive tides are different in height and duration (fig. 2.1). In narrow straits and bays, such as the Gulf of Bone (Anon.

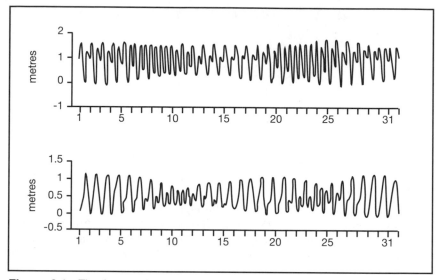

Figure 2.1. The forms of a typical mixed, prevailing semidiurnal tide (above), and a typical mixed, prevailing diurnal tide (below) over the period of a month.
After Pethick 1984

1980a), however, the patterns may become rather more complicated.

Tides are enormous waves with wavelengths of half the circumference of the earth. These 'waves' are primarily the result of the gravitational pull of the moon which acts not only on the water closest to it, but also on the mass of the orbiting earth itself, thereby pulling the earth away from the water on the opposite side.[2] This is similar to pulling someone towards you by one arm, and seeing his other arm move away from his body.

The sun also has an effect on tides but although it has 27 million times the mass of the moon, it has less than half of the moon's gravitational pull because it is 389 times as distant. Tides are greatest when the sun, moon

Table 2.1. Area, coastline length and complexity of four islands. Complexity is calculated by dividing the actual coastline length by the circumference of a hypothetical circle with the same area as the island, and is therefore a measure of the indentation of the shoreline.

	Area (km²)	Coastline (km²)	Complexity
Java	126,000	2,250	1.79
Sumatra	473,600	5,010	2.05
Halmahera	17,690	1,550	3.29
Sulawesi	189,200	6,100	3.95

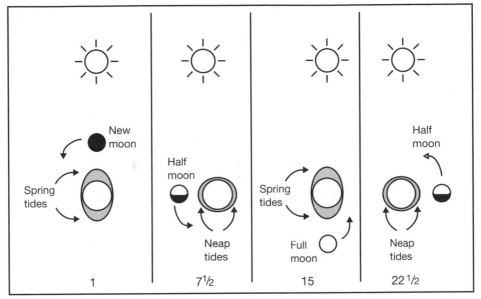

Figure 2.2. The tidal cycle.

After Pethick 1984

and earth are in a straight line (i.e., on days of the full and new moon). These large tides are called 'spring tides'. At the half-moons, when the sun, earth and moon form a right-angled triangle the combined forces are least. These small tides are called 'neap tides' (fig. 2.2). The moon rises about 50 minutes later every day so that successive high tides are 25 minutes later each day.

The arc traced by the sun changes throughout the year, being directly above the equator and in a straight line with the moon on the equinoxes: March 21 and September 21. This produces the strongest tide-raising forces. When the sun is directly above the Tropic of Capricorn or Tropic of Cancer (23.5° S and N respectively) on June 21 and December 21—the solstices—the tides are a minimum. Towards the end of the year the sun is closest to the earth and so high tides in this period, particularly around the September equinox, are among the highest of the year. For example, the extreme tides during early October in South Sulawesi each year cause the death of coral exposed to the air, and flooding in the coastal regions.

There are further complications: the moon swings 28° north and south of the equator every month, the distance from the sun to the earth varies through the year, and the gravitational pull of the other planets also has an effect. As a result there are longer cycles of tidal behaviour including the

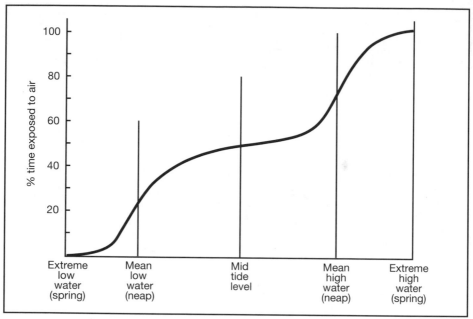

Figure 2.3. General pattern of tidal exposure up a beach.
After Brehaut 1982

18-year cycle and the 1,800-year cycle. For example, tides were particularly high in the 1500s and will reach a minimum around the year 2400.

The pattern of periodic inundation of the shore leads to a gradient of exposure (fig. 2.3). Thus the beach at the mid-tide level is covered for 50% of the time and exposed for 50%. At the mean highwater of neap tides, the beach is exposed for about 70% of the time. This exposure gradient determines to a large extent the occurrence of different species of animals and plants up a shore.

Surface Currents

Surface currents are relevant to the coastal regions because they carry detritus, animal larvae, etc., from or between coastal areas. During the north-westerly monsoon (approximately November to April) the currents run approximately anti-clockwise around Sulawesi. From May to November no such simple pattern can be discerned. The currents on the Sulawesi side of the Makassar Straits run southwards throughout the year, and there is also a year-long eastward current along the northern coast of North Sulawesi (fig. 2.4).

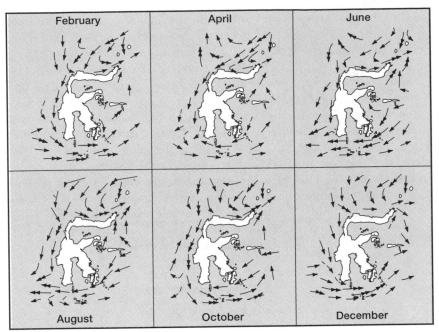

Figure 2.4. Surface currents.

After Wyrtki 1961

Salinity

As rocks weather chemically and physically, so the salts that are dissolved from them in the rain are carried by rivers, sub-surface and groundwater flows to the sea. The seas have therefore been getting saltier over time, but the rate of increase is extremely slow. The most common salt is sodium chloride.

The average level of salinity in the world's oceans is about 33.5 ppt (parts per thousand) but in coastal regions just after the onset of the wet season, the concentration falls, and the degree of variation differs between areas, being most marked off the shores of seasonal areas. Pools of seawater left on a muddy shore as the tide falls can sometimes increase their salinity to about 50 ppt as a result of evaporation, but this decreases again to 15 ppt after rain. The challenges of living in such an environment are discussed next.

For trees growing in the intertidal zone, the salinity of the tidal water is less important than the salinity of the water within the sediment where the plant roots are found. The salinity in this sediment is often less than sea water because of dilution by freshwater flowing through the soil from the land to the sea. This is an important factor in the management of mangrove forest (p. 192) and in understanding the effects of agricultural and industrial pollutants that may enter into the ground water. Some mangrove trees and other organisms are resistant to certain types of pollution, but the demise of sensitive species may upset the equilibrium of the whole system (Bunt 1980; Saenger et al. 1981).

While high concentrations of sodium and chloride ions are toxic to plants, the osmotic potential of the water is also most important as it influences the ability of a plant's roots to take up the water on which its growth depends. The osmotic pressure depends on the sediment type, being greater in fine than in coarse-grained sediments. Fine-grained sediments are capable of withholding more pure water against gravity due to the small size of the pore spaces, and therefore their osmotic pressure is higher than in coarse sediments. If it is not practicable to measure osmotic potential, then salinity and conductivity are a good second best.

Temperature

Tropical coastal waters usually have a temperature of between 27° and 29°C but can be much warmer in shallow areas. The temperature on the surface of mudflats or rocks can be so high that it is uncomfortable to walk on them in bare feet. Inside a shady mangrove forest, however, the air and soil surface temperature is much more equitable (table 2.2).

Dissolved Oxygen and Nutrients

In general, concentrations of neither dissolved oxygen nor nutrients impose any limits on productivity in coastal environments although the concentrations vary between locations. The greatest concentration of dissolved oxygen in the coastal environment is at the water's edge where wave action constantly agitates the water. The abundance of life in most coastal environments and the general abundance of nutrients in coastal environments (except sandy shores), results in a very high biological

Table 2.2. Average measurements of water quality along a transect from the sea edge into mangrove forest near Lainea, southeast Kendari.

	Metres from the sea edge							
	0	10	20	40	60	80	120	160
Temperature (°C)	25	25	25.5	24	24.5	24	23	22.5
Salinity (µmHos)	115	130	115	120	135	130	110	105
Dissolved oxygen (ppm)	9.1	9.1	8.0	8.9	9.0	8.1	8.8	8.4

Data from an EoS team

oxygen demand and this tends to lower the concentrations of available oxygen. Thus there is a gradient of increasing nutrient concentrations and decreasing oxygen concentrations moving from the water's edge through a mangrove forest. This is a result of dilution in the greater volume of water at sea, and the greater incorporation of nutrients into the sediments in the upper tidal areas where the litter is retained for longer (p. 131) (Davie 1984).

Sediment

It is a matter of debate whether new sediments should be termed soils but they can nevertheless be defined by standard soil classifications. Thus, in the Malangke mangrove forest area, the sediment is primarily a grey hydromorphic alluvium but towards the terrestrial margin merges into a gley humus reflecting alternating periods of aeration and flooding. These sediments have very low fertility but a high organic content. This was highest (4%-5.8% carbon) in the drier parts of the forest where the vegetation was older and the trees faller, in the foreshore under *Sonneratia alba* there was much less organic matter (0.5%-2.7%). The sediments are generally acidic and this increases with depth although different regimes occur under different species. Conductivity (which is directly proportional to salinity) at Malangke was highest (5.9-6.4 mhos/cm) under *Rhizophore* forest somewhat inland; a situation probably related to the fact that the percentage of sand is higher near the foreshore (due to greater wave action) and this does not bind the salt (Anon. 1981a, b).

The grain size of sediments is measured by passing the substrate through a series of sieves and calculating the percentage of the total retained by each sieve. If the grains were identical and perfect spheres then 26% of the volume of the sediment would be pore spaces (i.e., a porosity of 26%), regardless of whether the grains were large or small. In nature, of course, grains are neither spherical nor packed as closely as possible, and in many cases small grains fill the spaces between large grains.

The water content of a beach sediment depends on its grain size and porosity, but not all pore spaces will always be filled with water. When the tide is out, the water table falls faster in coarse-grained sediments than in fine-grained because the average pore size is greater. These coarse sediments therefore have a higher permeability (i.e., are better drained than fine sands or muds [fig. 2.5]). In fine sand or mud beaches the water table may stay at or near the surface, even when the tide is out, due to the massive surface tension resulting from the very large surface area in a fine-grained sediment, and to the low permeability.

When the pore spaces remain filled with water the sediment may become thixotropic, or liquid when agitated or subjected to pressure. As the water drains away, external pressures are met by increased resistance

Figure 2.5. Time taken (minutes) to drain a 50 cm column of water through 10 cm of different sand mixtures. The sand mixture is assessed by the percentage of each sample passing through a 0.28 mm sieve.

After Brafield 1972

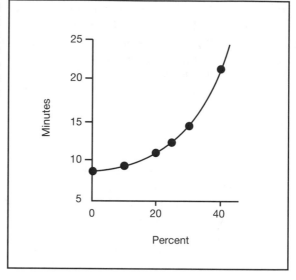

and the sediment may become dilatant or solid when agitated or subjected to pressure. This is why sand saturated with water feels soft and sloppy, whereas drier sand, whitens and becomes firm when walked upon. These properties are important to burrowing animals since thixotropic sediments are easily entered but burrows are hard to maintain whereas dilatant sediments are hard to burrow into but the burrows are easily maintained. Certain shorebirds follow the waters edge up and down the beach in order to remain in an optimal feeding zone (p. 149). In general, heavy wave action is associated with steeply sloping beaches and coarse-grained sediments. Whereas shores subjected to little wave action slope gently and are comprised of fine grains.

Oxygen within the Sediment

Oxygen concentrations are affected by three major factors. First, concentrations drop rapidly with depth. In a fine-grained sediment, oxygen concentration at 2 cm may be only 15% of the saturation concentration and virtually zero at 5 cm, and supersaturated in small surface puddles as a result of photosynthesis by diatom phytoplankton. Second, oxygen concentrations fall with an increase in temperature; for example, an increase in temperature from 25°-30°C reduces saturated dissolved oxygen concentrations by nearly 10%. Third, in coarse-grained sediments, which are relatively well oxygenated when the tide leaves, the concentration can fall rapidly in the first few hours after exposure due to the respiration of the animals within it. Should an oil slick drift onto the shore and settle on the sediment at low tide, oxygen cannot diffuse into the sediment pores with obvious extremely adverse effects on the animals below (Ganning et al. 1984).

Bacteria

As mentioned above, fine-grained sediments have a much larger surface area per unit volume of particles than coarse-grained ones thereby providing more attachment sites for micro-organisms such as diatoms and bacteria. Fine-grained sediments also tend to contain more organic debris and so it is not surprising that chemical processes involving bacteria are more complex and proceed faster in these than in coarse-grained sediments (Brafield 1972).

Only near the surface can the organic matter be broken down by oxidation processes. Below this zone, in an anaerobic environment at a level called the redox (reduction-oxidation) potential discontinuity layer, anaerobic bacteria break down organic matter by fermentation or reduction processes producing alcohols and fatty acids. Sulphate ions are reduced to hydrogen sulphide and much of this is fixed as iron (ferrous) sulphides which gives the sediment below this layer a black or dark grey colour (fig. 2.6).

The boundary of the black layer is found closer to the surface where the sediment is fine-grained and where organic matter content is high (i.e., mudflats in front of mangrove forest). Differences between the features of steep and shallow beaches are shown in figure 2.7.

Adaptations of the Fauna

Organisms in the intertidal area experience cycles of wetting and drying quite unlike those in any other ecosystem. Most animals of marine origin are unable to live in such extreme conditions because they quickly dry out, cannot breathe gaseous oxygen, can feed only on water-borne food, and are bound to the sea for reproduction. Two groups, however, the crabs and gastropod snails, have members which have met the challenges by having exoskeletons of impervious shell to restrict water loss, they are able to breathe gaseous oxygen, feeding on damp organic material or micro-organisms, and climb into trees to find food. In addition, by fertilizing eggs

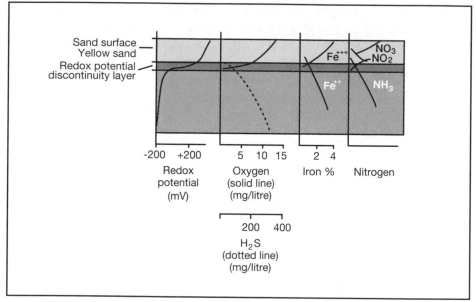

Figure 2.6. Physical changes across the black layer.

After Brafield 1978

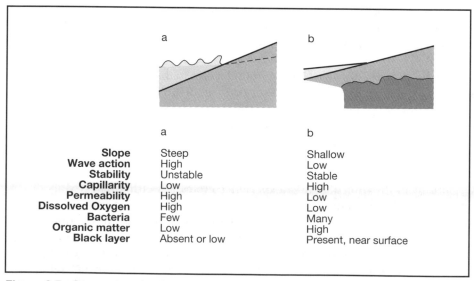

Figure 2.7. Comparison between steep, coarse-grained beaches (a) and shallow, fine-grained beaches (b).

After Brafield 1978

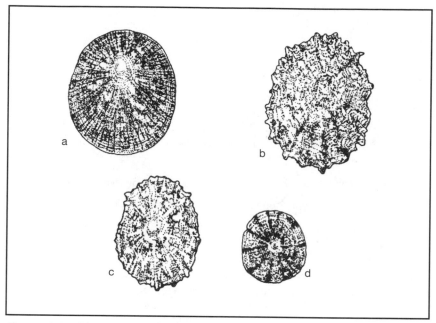

Figure 2.8. Typical limpets. a - *Cellana testudinaria,* b - *Patella exusta,* c - *Patella pica,* d - *Cellana radiata.*

internally, they can care for the young in brood pouches, or capsules, rather than having to let them take their chances among the plankton.

The primitive gastropods known as top shells or limpets (fig. 2.8) cope with exposure in at least two ways. Some species return to a 'scar' in a rock when the tide falls and the shells grow to match this scar exactly. Others are not restricted to occupying a single home site and instead secrete a mucus sheet between the shell margin and the rock surface to reduce water loss. The effectiveness of their adhesion is soon realized when attempts are made to pry them off rocks. It is water loss, rather than temperature which is the main danger to limpets even though they are more tolerant of desiccation than most animals: they can lose about 80% of their water and still recover when water becomes available. These animals, as well as mussels, and *Littoraria*[3] snails, can also lower their metabolic rate thereby allowing them to survive periods of exposure when the only oxygen available is in the water held within their shells (Brehaut 1982).

Animals and plants can survive short periods at high temperatures which would be lethal over a longer period. This may have a significant effect on the degree of exposure, or distance up a rocky shore, that an

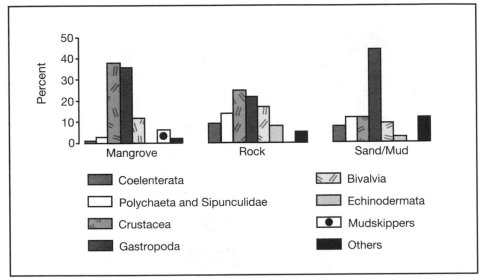

Figure 2.9. Percentages of total aquatic animal taxa recorded at three types of shore (omitting microscopic forms). Note that crustaceans and gastropod molluscs account for 74% of the total on the mangrove shore. Note that these data are not based on complete lists or on equally exhaustive surveys but they serve to illustrate the major differences.

Data from Berry 1972

animal can endure (Brehaut 1982). In addition, most marine organisms are only able to tolerate very minor variations in salinity because they do not have mechanisms of regulating the salt and water balance of their body fluids except within narrow limits. The crabs and snails cope with this problem in different ways: crabs can regulate the concentration of salt in their body tissue, whereas snails are remarkably tolerant of a wide variation in the concentration of salts in their body fluids.

The success of crabs and snails in the intertidal zone is illustrated by their predominance in mangrove, rocky and sand/mud environments (fig. 2.9).

MANGROVE FOREST VEGETATION

Mangrove forests would once have fringed much of the coast of Sulawesi (fig. 1.20) but major expanses of mangroves are now found in relatively few locations (fig. 2.10), the remainder having been largely felled and used for timber or fibre or converted into brackish fish and prawn ponds (p. 187).

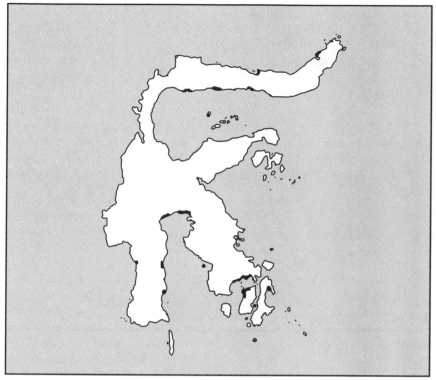

Figure 2.10. Present distribution of mangrove forests around Sulawesi (indicated in black).

After Salm and Halim 1984

South Sulawesi has more mangrove forest than the other three provinces combined (Darsidi 1982) and small patches can be found along most shallow beaches and river mouths away from centres of human habitation.

Composition

Mangrove forests are characteristic of tropical coastlines and have very similar compositions irrespective of climate. Only 19 tree species are commonly encountered in Sulawesi mangrove forest, although there are about 16 species of tree that may be found only occasionally or in the forest closest to dry land (table 2.3). In addition there may be 20 species of orchids and other epiphytes but these are generally rare. Some plants have been reported from only small areas, such as *Camptostemon philip-*

pinense (Bomb.) (fig. 2.11) from Kwandang Bay in Bolaang Mongondow (Steup 1939), but this must in part be due to inadequate collecting. A detailed list of plants found in Philippine mangrove forests (Arroyo 1979) is useful to those working in North Sulawesi. A key to the trees most likely to be encountered in mangrove forest and other coastal vegetation is given in Appendix C.

In addition to higher plants, various algae and bryophytes (mosses and liverworts) are also found. Some of the algae appear to have adaptations for living in brackish conditions and these species can be quite abundant. The algae are greenish, brownish or reddish (Johnson 1979), but are unfortunately rather difficult to identify (Teo and Wee 1983). Bryophytes

Table 2.3. Tree species recorded in Sulawesi mangrove forest. Principal species shown in bold. Old names which are sometimes still used are shown in brackets.

Avicenniaceae	Bignoniaceae
Avicennia alba	*Dolichandrona spattacea*
A. marina *[intermedia]*	Bombacaceae
A. officinalis	*Camptostemon [Cumingia] philippinense*
Combretaceae	Leguminosae
Lumnitzera littorea	*Intsia bijuga*
L. racemosa	*Pongamia* sp.
Euphorviaceae	*Serianthes* sp.
Exoecaria agallocha	Moraceae
Meliaceae	*Ficus* spp.
Xylocarpus *[Carapa]* **granatum** *[obovata]*	Myrsinaceae
X. moluccensis	*Aegiceras corniculatum*
Rhizophoraceae	Myrtaceae
Rhizophora apiculata *[conjugata]*	*Osbornea octodonta*
R. mucronata	*Eugenia* spp.
R. stylosa	Palmaceae
Bruguiera cylindrica *[caryophylloides]*	*Nypa fruticans*
B. gymnorrhiza	Pandanaceae
B. parviflora	*Pandanus* spp.
B. sexangula *[eriopetala]*	Rubiaceae
Ceriops tagal *[candolleana]*	*Morinda citrifolia*
Sonneratiaceae	*Scyphiphora hydrophyllacea*
Sonneratia alba	Sterculiaccae
S. caseolaris	*Heritiera littoralis*
S. ovata	Ulmaceae
Apocynaceae	*Gironiera* sp.
Cerbera odollam	

After Hickson 1889; Heringa 1920; Steup 1933, 1939; Anon. 1980a; Darneedi and Budiman 1984

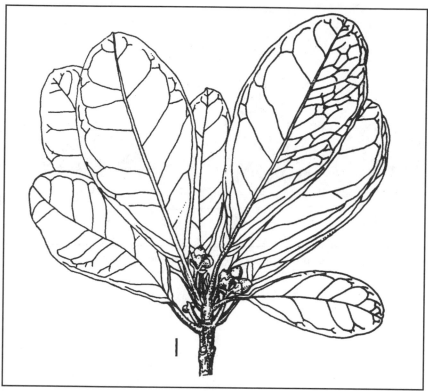

Figure 2.11. *Camptostemon philippinense* an uncommon mangrove tree known from North Sulawesi. Scale bar indicates 1 cm.

After Anon. 1968

were found on most of the major species of mangrove trees in Thailand (all of which occur in Sulawesi), but not on all trees present. They comprised of five species of moss and 21 species of leafy liverwort. *Rhizophora apiculata* bore the most species (23) but *R. mucronata* only four (Thaithong 1984). This may have been due as much to microclimate differences as to differences between the substrates provided.

Many different plant communities of mangrove trees have been identified in Southeast Asia (Chapman 1977b), many dominated by a single species, but these are not discrete. It may be that given the wide range of micro-environmental conditions occurring, there may be a virtually infinite variety of mangrove forest types. Thus any effort to classify an area of mangrove forest that is being studied as a certain 'type' is probably misguided and ultimately not particularly useful.

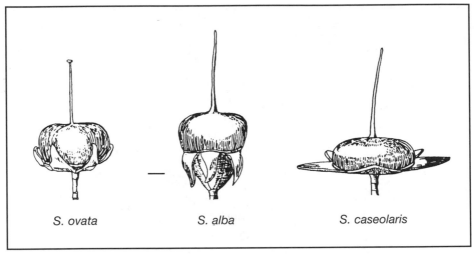

S. ovata   S. alba   S. caseolaris

Figure 2.12. Fruits of the three species of *Sonneratia* found in mangrove forests in Sulawesi. Scale bar indicates 1 cm.
After Backer and van Steenis 1951

Mangrove trees are tolerant of saline soils, that is they are halophytes (Walsh 1974), but they are facultative rather than obligate in that they can also grow successfully in freshwater. This is demonstrated by the growth, fruiting and germination of *Bruguiera sexangula, B. gymnorrhiza,* and *Sonneratia caseolaris* (fig. 2.12) in the Botanic Gardens in Bogor (Ding Hou 1958). This is typical of many plants that might be regarded as restricted to growing in certain soil conditions. In fact, there are many organisms which exist not where they fare best, their 'fundamental niche', but where they grow most successfully in competition with other species, their 'realized niche'. Mangrove trees can sometimes be seen growing at the sides of rivers in seemingly freshwater conditions but in these cases there is generally a wedge of (heavier) salt water permanently or periodically near the bed of the river which maintains saline conditions for the tree roots. The occurrence of these trees in such locations may also be due to the flooding regime of the river (J. Davie pers. comm.).

Mangrove forests, particularly those that are frequently flooded, differ markedly from dryland forests and from most inland swamp forest in the virtual absence of climbing and understorey plants (Ding Hou 1958). In essence, the only plants that grow in mature mangrove forests are trees whose crowns reach the canopy. The reason for this has yet to be confirmed, but it would seem that regular tidal (rather than seasonal) flooding is the most important factor, rather than a difference in salt tolerance between trees and smaller plants (Janzen 1985; Corlett 1986). Exactly how this flooding acts against the establishment of herbs has yet to be determined. Living in saline conditions is clearly costly in terms of energy, and this is supported by the observation that mangroves were killed after only a single spraying with defoliants in Vietnam, whereas nearby terrestrial trees had to be sprayed several times to achieve the same results (Janzen 1985).

Most trees of the mangrove forest have developed peculiar root systems to allow for gaseous exchange above a water-logged and anoxic soil (Mann 1982) (fig. 2.13). Such 'breathing roots' are known as 'pneumatophores'. The stilt roots of *Rhizophora* may also be effective in preventing the growth of seedlings too close to a growing tree. These stilt roots are generally unbranched but secondary or tertiary branching can occur due to damage of the primary root tip by scolytid beetles (Docters van Leeuwen 1911) or by boring isopod crustaceans (Ribi 1981, 1982; Whitten et al. 1984) (fig. 2.14).

The roots of *Sonneratia* and *Avicennia* are similar in gross structure and consist of a horizontal cable root held in place by anchor roots growing vertically downwards. The pneumatophores grow upwards from the cable roots and, small nutritive roots grow horizontal from these. As mud is deposited on the forest floor, so new nutritive roots are produced higher up the pneumatophores. In *Bruguiera* the cable root loops in and out of the soil and the exposed 'knee-roots' act as pneumatophores. *Ceriops* does not have special root adaptations but its bark has many large openings, or lenticels, to assist in gas exchange. If oil drifts into a mangrove forest the lenticels on the exposed parts of the pneumatophores become clogged and this is the primary reason that trees so afflicted will die. The dark oil also causes the water temperature to rise and the concentrations of dissolved oxygen to fall (Mathias 1977; Lugo et al. 1978; Getter et al. 1984). Mangrove trees that survive exhibit signs of chronic stress such as reduced productivity and gradual leaf loss (Lugo et al. 1978; Saenger et al. 1981). These effects can be quite local, however, as is evidence by the relatively small patches of dead mangrove trees around the natural oil seeps on the shore near Kabali, southwest of Luwuk.

The feathery flowers of *Sonneratia* are superficially similar to those of the Myrtaceae (such as rose apples *Eugenia* spp., and eucalypts *Eucalyptus*) and in common with many of those, are pollinated by bats which may fly up to 40 km from their inland roost when *Sonneratia* is in flower (Start and Marshall 1975). The flowers of Rhizophoraceae have a range of mechanisms by which they effect pollination. For example, the anthers of *Bruguiera* open explosively when the flowers are visited by sunbirds *Nectarinia* (in the large flowered such as *B. gymnorrhiza*), or butterflies and other insects (in the

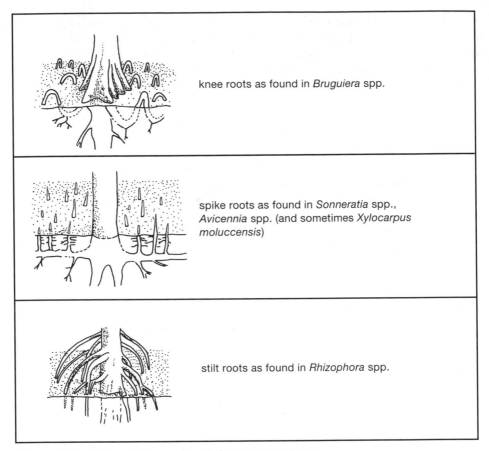

knee roots as found in *Bruguiera* spp.

spike roots as found in *Sonneratia* spp., *Avicennia* spp. (and sometimes *Xylocarpus moluccensis*)

stilt roots as found in *Rhizophora* spp.

Figure 2.13. Different types of roots in mangrove trees.

smaller-flowered species such as *B. parviflora*) in search of nectar. *Ceriops tagal* also has explosive anthers, triggered largely by moths. *Rhizophora* flowers are largely wind-pollinated but bees may also be involved (Ding Hou 1958; Tomlinson et al. 1979). Sunbirds can sometimes be seen visiting *Rhizophora* trees but this is largely to lick the sweet, sticky exudate from leaf buds or young flowers which have been slightly damaged by insects. Since the sunbirds also eat insects the birds help to reduce the damage to *Rhizophora* by scale insect Coccidae (Christensen and Wium-Anderson 1977; Primack and Tomlinson 1978; Wium-Anderson and Christensen 1978; Wium-Anderson 1981).

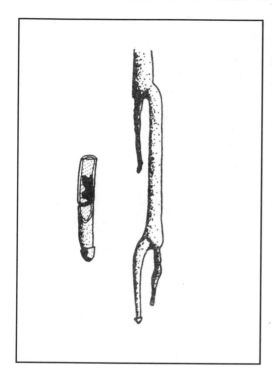

Figure 2.14. Branching of *Rhizophora mucronata* roots caused by small scolytid beetles burrowing into the pith of the root tip.

After Doctors van Leeuwen 1911

Observations of fruiting and flowering of mangrove trees in Australia and Thailand showed that there was activity in every month but that most species flowered during the dry season, and dropped ripe fruits during periods of peak rainfall. This pattern is very similar to that often found in dry lowland forests (p. 366) and is probably related to insect abundance. The production of new leaves was depressed when fruit and flower production were maximal. The time it took for a flower to form a ripe fruit varied both between and within species but was about 1-2 months for *Nypa*, 2-6 months for *Avicennia* and about 15 months for *Ceriops* (Christensen and Wium-Anderson, 1977; Wium-Anderson and Christensen 1978; Wium-Anderson 1981; Duke et al. 1984).

A few species of mangrove trees have evolved an unusual, though not unique, form of reproduction. Generally speaking, fruit develops on a plant and, when it is ripe or fully developed, the fruit or the seed inside it is then dispersed; the seed germinates when, or if, it comes to rest in suitable conditions. In most of the Rhizophoraceae such as *Rhizophora* and *Bruguiera*, however, the fruits ripen and then, before leaving the parent tree, the seeds germinate inside the fruit, possibly absorbing food from the tree. The hypocotyl (embryonic root) of the seedling pierces the wall of the fruit and then grows downwards. The cotyledons (first leaves) remain

Figure 2.15. The propagule of *Rhizophora mucronata* showing the root (often mistaken for part of the fruit) and the top of the seedling that detaches itself from the parent plant.

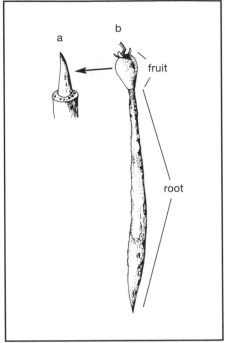

inside the fruit. Eventually, in *Rhizophora mucronata* for example, the root may reach a length of 45 cm. The seedling then drops off by separating it self from the cotyledon tube, the scar of which forms a ring around the top of the fallen seedling, and the small leaf-bud can be seen above this scar (fig. 2.15). *Bruguiera* behaves similarly, but the break occurs at the stalk of the fruit. This form of behaviour presumably allows the rapid establishment of the young plant. These types of fruit are described in more detail elsewhere (MacNae 1968).

Zonation

Mangrove tree species tend to grow in zones or belts (figs. 2.16, 2.17 and 2.18). On gently sloping accreting shores (where sediment is being actively deposited), the forest nearest the sea is dominated by *Avicennia* and *Sonneratia*, the latter usually growing on deep mud rich in organic matter (Troll and Dragendorf 1931). On firm clay sediment *A. marina* is more common whereas on softer muds *A. alba* predominates (Ding Hou 1958). Behind these zones *Bruguiera cylindrica* can form almost pure stands on firm clays which are only rarely inundated by the tide. Further inland *B. cylin-*

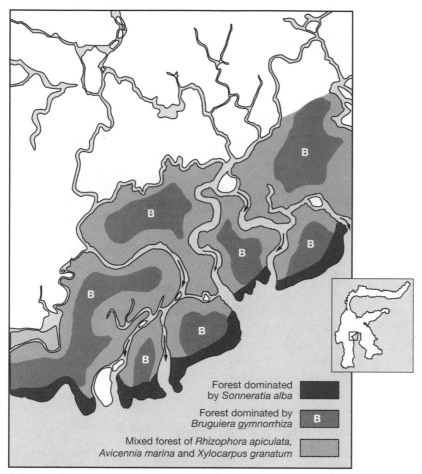

Figure 2.16. Zones of mangrove forest observed in part of Malangke.

After Anon. 1980a

drica becomes mixed with *Rhizophora apiculata, R. mucronata, B. parviflora* and *Xylocarpus granatum* (the canopy of which can reach 35-40 m). The mangrove forest furthest from the sea is often a pure stand of *B. gymnorrhiza*. Seedlings and saplings of this species are tolerant of shade but only under larger trees of other species; they are unable to grow under the canopy of their parents. This is presumably due to some chemical interaction. The boundary zone between mangrove forest and inland forest is marked by the occurrence of *Lumnitzera racemosa,* *Xylocarpus moluccensis, Intsia bijuga* (fig. 2.19), *Ficus retusa,* rattans, pandans, the

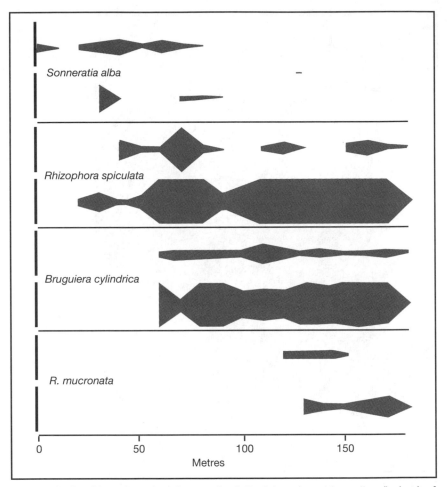

Figure 2.17. Changing abundance of adults (above) and juveniles (below) of four tree species along a transect inland from the seaward edge of mangrove forest at Lainea, Kendari. Vertical bars represent 10 trees or seedlings/10 m².
After EoS team

stemless palm *Nypa fruticans,* and the tall spiny-trunked palm *Oncosperma tigillaria.* Where the mangrove forest has been opened, the most common undergrowth plant is the fern *Acrostichum aureum.* Relatively steep-sided creeks, bays and lagoons are generally fringed with *Rhizophora* trees.

Recognizable zones may arise for two different reasons where neighbouring vegetation associations have little or no floristic affinity despite growing in the same environmental conditions (continuous variation),

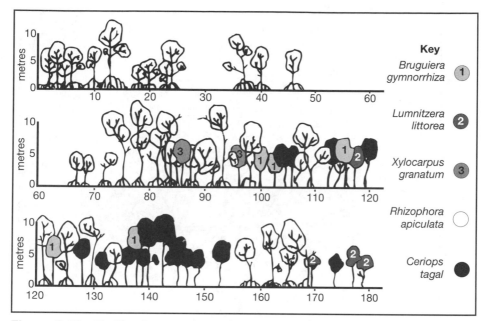

Figure 2.18. Profile diagram through somewhat disturbed mangrove forest at Lapangga, Morowali National Park. Note: the general predominance of *R. apiculata* and discrete occurrence of the other species.
After Darnaedi and Budiman 1984

and where neighbouring environmental conditions are different enough to result in sudden changes between vegetation associations (discontinuous variation) (Bunt and Williams 1981). Thus vegetational changes can be continuous, discontinuous, or a combination of both. This is why it is crucial to consider scale before attempting an analysis of mangrove forest and why, in the absence of such consideration, data from isolated transects, even from the same area, are so hard to interpret. Comparisons of transect data from different sites are useful in compiling inventories and in noting similarities, but are not a basis for a discussion of zonation which should be based instead on air photos of, and ground surveys over, parts of a number of forests.

The tendency of mangrove forests to occur in distinct zones has been interpreted variously by different authors as a consequence of plant succession, geomorphology, physiological ecology, differential dispersal of propagules and seed predation. Each of these is considered next.

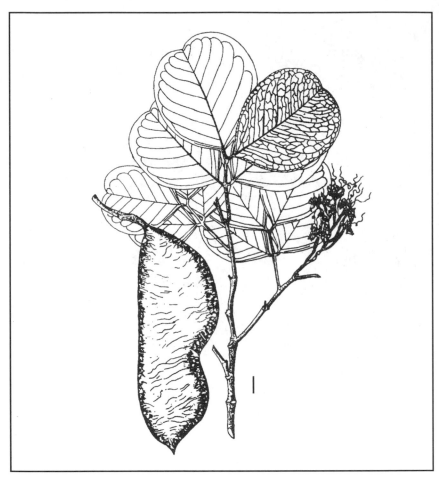

Figure 2.19. *Intsia bijuga.* Scale bar indicates 1 cm.
After Soewanda (n.d.)

Plant Succession. Plant succession is a classic ecological concept and is defined as being the progressive replacement of one plant community with another of more complex structure (p. 366). Much of the early work on mangrove forests focused on its supposed land-building role and it seemed clear from this that one species colonized an exposed bank of mud and, as conditions changed (such as an increase in the organic debris of the mud), so other species took over. For example the colonization of a new shallow or exposed substrate by *Avicennia* or *Sonneratia* trees, such as on the banks of the Rongkong River delta at the north of Bone Bay, produces a network of erect pneumatophores which have three indirect functions:

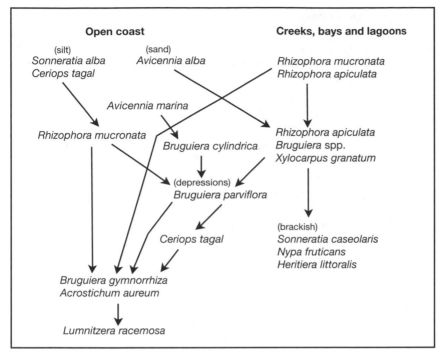

Figure 2.20. Succession in mangrove forests.

After Chapman 1970 in Walsh 1974

• they protect the young trees and germinating seeds from wave damage;

• they entangle floating vegetation which decays and becomes incorporated in the soil; and

• they provide a habitat for burrowing crabs which help to aerate the soil (Chambers 1980).

These changes, and subsequent additional sedimentation, lead to a succession of species or communities of species over a period of time (fig. 2.20). It is clear, however, that the stages of succession are not always consistent and different local environmental conditions and man's impact on those have an influence (Steup 1941; Anon 1980a).

A total of 20 characteristics have been noted for secondary succession in tropical forests (Boudowski 1963). If the characteristics are examined in relation to the succession of mangrove forests, only seven of the characters apply, nine do not, and four are inconclusive. This suggests that attributing the apparent zonation to succession is not the whole story (Snedaker 1982b).

Geomorphological Change. As stated above, early workers on mangrove felt that it was mangroves that 'built land'. However, it is clear from observations in the huge deltas of the Ganges, Indus and Irrawaddy on the coast of the Indian sub-continent, that it is the process of sediment deposition that builds land. It is now generally agreed that mangroves do not have any influence on the initial development of the land forms. Mangroves may accelerate land extension but they do not cause it (Ding Hou 1951).

From a geomorphological perspective, it is the shape, topography and history of the coastal zone that determine the types and distributions of mangrove trees in the resulting habitats. The position of species relative to tidal levels (and thus soil type) is obviously important and the pattern of tidal inundation and drainage has been considered to be the major factor in mangrove zonation (Watson 1928). This idea has been developed to include variation in the salinity of the tidal water and the direction of its flow into and out of the forest (Lugo and Snedaker 1974). Although good correlations exist between salinity and zonation, they are not proofs of direct cause and effect.

Physiological Response to Soil-Water Salinity. If salinity is an actual cause of zonation in mangrove forests it needs to be shown that the plants actually respond through their physiology to salinity gradients and not to a factor such as oxygen levels which, under certain conditions, will fall with increasing salinity. It has already been mentioned that mangrove trees are able to live in freshwater (i.e., they are facultative, not obligate, halophytes), but each species probably has a definable optimum range of salinity for its growth. Indeed, it has been found that within each zone the characteristic species had apparently maximized its physiological efficiency and therefore had a higher metabolic rate than any invading species (Lugo et al. 1975). Thus invading species would be at a competitive disadvantage due to their lower metabolic efficiency in that habitat. Similarly, it has been found that each species of the mangrove forest grows best under slightly different conditions such as the amount of water in the mud, the salinity, and the ability of the plant to tolerate shade. This means that the various species are not mingled together in a haphazard way but occur in fairly distinct zones.

Salinity can vary considerably between high and low tides and between seasons (p. 109), and thereby presents a confusing picture to a scientist conducting a short-term study. Thus, to identify the salinity levels to which the different species are optimally adapted requires long-term and detailed measurements to determine long-term averages and ranges of salinity. It is often the case that a species' ecological limits are defined by relatively rare events such as occasional extremely dry years (p. 22) or, in the case of mangroves, high salinity levels in the dry season when there is little freshwater input to the system.

Differential Dispersal of Propagules. The suggestion that differential dispersal of propagules (fruit, etc.) influences zonation of mangroves rests on the idea that the principal propagule characteristics (e.g., size, weight, shape, buoyancy, viability, numbers, means of release and dispersal, and location of source areas) result in differential tidal sorting and therefore deposition. There are as yet few data to support this hypothesis (Rabinowitz 1978; Snedaker 1982b).

Seed Predation. Experiments conducted in Australia have shown that about 75% of the propagules of five species of mangrove trees were consumed by predators, primarily grapsid crabs, and there were significant differences among species and between forest types. *Rhizophora stylosa* was preyed upon the least and *Avicennia marina* the most. As might have been expected, predation was generally highest where a particular tree did not have neighbouring trees of the same species. Predation rates seem to be associated with chemical composition because *A. marine*, selected by predators over all the other species, had the highest concentrations of protein and sugars and the lowest concentrations of fibre and tannins. The propagules with the highest tannin content appeared to be preyed upon only by a single, specialist crabs (Smith 1986). These differences will influence the pattern of seedling establishment in a mangrove forest, such that establishment is most likely to occur in areas where seed predators are least likely to occur. In the case of *A. marina* this may mean establishment in the areas most frequently inundated.

Geomorphology, physiology and seed predation thus appear to be the most relevant forest. The impact of human activities, so ubiquitous in coastal regions, plays a major role in modifying species composition and physical conditions, however, and so should be considered first in any study of zonation.

Biomass and Productivity

'Biomass' is a term for the weight of living material usually expressed as dry weight, in all or part of an organism, population or community. It is commonly expressed as the 'biomass density' or 'biomass per unit area'. Plant biomass[5] is the total dry weight of all living plant parts and for convenience is sometimes divided into above-ground plant biomass (leaves, branches, boughs, trunk) and below-ground plant biomass (roots). It appears that no study of mangrove biomass has yet been conducted in Sulawesi but several studies have been conducted in Peninsular Malaysia (Ong et al., 1980a, b; 1985). In one undisturbed forest the biomass was found to be between 122 t/ha and 245 t/ha, but in another, which has been exploited and managed for timber on a sustained basis for 80 years, the biomass of trees was 300 t/ha (Ong et al. 1980b). As is explained below, the higher biomass in managed forest is not unexpected. Above-ground biomass in Australian mangrove forests has been found to correlate with parameters of soil quality such as extractable phosphorus, redox potential, and salinity (Boto et al. 1984).

Biomass is a useful and a relatively easy-to-obtain measure but it gives no indication of the dynamics of an ecosystem. Ecologists are interested in productivity because, if the dry weight of a community can be determined at a moment, and the rate of change in dry weight measured, then the rate of energy flow through an ecosystem can be calculated. Using this information different ecosystems can be compared, and their relative efficiencies of converting solar radiation into organic matter can be calculated.

Plant biomass increases because plants secure carbon dioxide from the atmosphere and convert this into organic matter through the process of photosynthesis. Thus, unlike animals, plants make their own food. The rate at which a plant assimilates organic matter is called the 'gross primary productivity'. This depends on the leaf area exposed, amount of solar radiation, temperature and upon the characteristics of individual plant species (Whitmore 1984). Plants, like all other living organisms, respire and use up a proportion of the organic matter produced through photosynthesis. What is left after respiratory loss is called 'net primary productivity' and the accumulation over a period of time is termed 'net primary production'. Net primary productivity is obviously greatest in a young forest which is growing and it should be remembered that a dense, tall forest with a high biomass does not necessarily have a high net primary productivity. Large trees may have virtually stopped growing. Indeed in an old 'overmature' forest, the death of parts of the trees and attacks by animals and fungi may even reduce the total plant biomass while net primary productivity remains more or less constant. The major aim of silvicultural management in forests of timber plantations is to maximize productivity and so the trees are usually harvested while they are still growing fast and before the net primary productivity begins to decrease too much (fig. 2.21).

One means of assessing net primary production is to measure the rate at which litter is produced. The production of litter appears to be very similar between the sites examined in the Indo-Australian region, being about 7-8 t/ha/yr for leaf litter, and 1-1.2 t/ha/yr for all small litter (mainly leaves, but with twigs, flowers and fruit) (Ong et al. 1985; Woodroffe 1985). The total litter production of mangrove forests in Peninsular Malaysia and Papua New Guinea has been found to be about 14 t/ha/yr (Sasekumar and Loi 1983; Leach and Burgin 1985), which is similar to results obtained in Queensland (Duke et al. 1981), and therefore probably similar to that found in Sulawesi. Interestingly, these figures are similar to or higher than those obtained in lowland forest (p. 365) and support the contention that mangroves grow, reproduce, and die fast (Jimenez et al. 1985), similar to dry lowland forest on young terraces (p. 361).

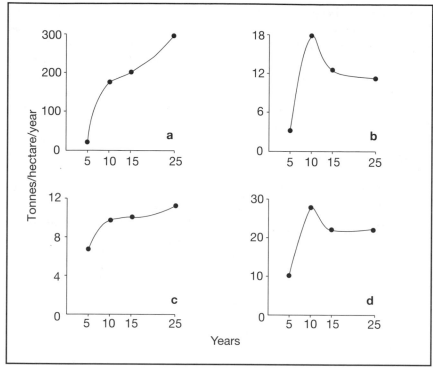

Figure 2.21. Changes in: a - biomass, b - mean annual increment, c - litter, and d - net productivity in four even-aged stands of *Rhizophora apiculata* in Peninsular Malaysia. In the forest from which these data were collected, the trees are harvested on a 30-year rotation.

After Ong et al. 1985

Variation in net primary production in mangrove forests in northeast Australia and Papua New Guinea has been ascribed to availability of phosphorus (Boto et al. 1984), which is consistent with the view that nitrogen and phosphorus are limiting in coastal marine environments (Rhyther and Dunstan 1971). The approximate annual accumulation of litter on the mangrove forest floor at Merbok was calculated to be 0.33 t/ha for leaf litter and 1.13 t/ha for total litter. An experiment at the same site revealed that 40%-90% of fallen leaves were lost after 20 days on the forest floor. The major agents in the disappearance were probably crabs which either bury them or eat them, later to be excreted as detritus. The importance of the crabs is seen when they are prevented from reaching the leaves, in which case the time needed for total decomposition was 4-6 months (Ong et al. 1980a).

The detritus becomes rich in nitrogen and phosphorous because of the fungi, bacteria and algae growing on and within it and is therefore an important food source for many 'detritivore' animals such as zooplankton, other small invertebrates, prawns, crabs, and fish. These detritivores are eaten in turn by carnivores which are dependent to varying degrees on these organisms. It is probable that most of the micro- and macro-fauna in the mangroves and surrounding coastal areas are dependent on the productivity of litter from mangrove forests (Ong et al. 1980a, b). A major initiative to study the important issue of transport of material in mangrove estuaries in the Indo-Australian region is currently underway (Ong et al. 1985).

Mangrove forests are highly productive ecosystems but only about 7% of their living leaves are eaten by herbivores (Johnstone 1981), and most of the mangrove forest production enters the energy system as detritus or dead organic matter (fig. 2.22). This detritus plays an extremely important role in the productivity of the mangrove ecosystems as a whole and of other coastal ecosystems (Lugo and Snedaker 1974: Ong et al. 1980a, b; Saenger et al. 1981; Mann 1982). Its importance to offshore ecosystems is not clear (Nixon et al. 1980).

The high productivity of mangroves and the physical structure and shading they provide forms a valuable habitat for many organisms, some of which are of commercial importance. At present the most valuable mangrove-related species in Indonesia are the penaeid prawns. The juvenile stages of several of these prawn species live in mangrove and adjacent vegetation, while the adults offshore (Soegiarto and Polunin 1980).

The influence of mangroves extends far beyond the prawn fisheries (p. 187). For example, carbon from mangrove trees has been found in the tissues of commercially important bivalves such as the cockle *Anadara granosa,* oyster *Crassostrea,* shrimps such as *Acetes,* used in the making of belacan paste, crabs such as *Scylla serrata,* and many fish (Rodelli et al. 1984) such as mullet *Mugil,* milk fish *Chanos* and barramundi or giant perch *Lates* (MacNae 1968; Moore 1982; Polunin 1983).

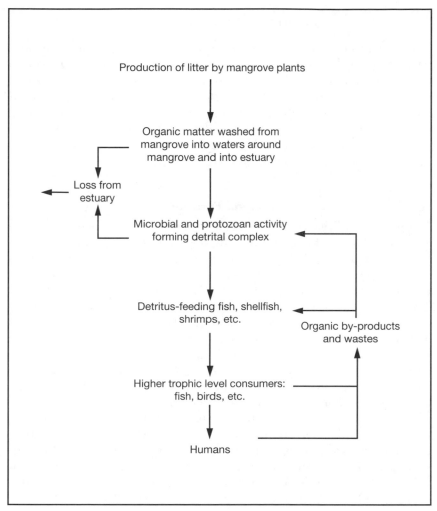

Figure 2.22. Major pathways of energy flow in a mangrove-fringed estuary.
After Saenger et al. 1981

OTHER COASTAL VEGETATION

There are three main types of beach vegetation: the *pes-caprae* formation, the *Barringtonia* formation and the vegetation of rocky shores. The removal through development of the vegetation on a sandy beach may not be regarded as a particularly serious loss in itself, but its ability to hold together a loose sandy substrate means that in its absence more or less continuous coastal erosion occurs. This, and its resultant impact on human settlements is most damaging during severe storms, because the power of the wind and waves is no longer countered by deep-rooted vegetation.

Pes-caprae Formation

The *pes-caprae* formation is found along sandy beaches which are actively accreting; that is, where sand is being deposed, or on already-developed beaches that are now being eroded. Its name is derived from the conspicuous, purple-flowered creeper with two-lobed leaves, *Ipomoea pes-caprae* (Conv.).[6] The other plants found in this formation also tend to be low, sand-binding herbs, grasses and sedges whose long, deep-rooting stems or stolons spread across or just under the surface of the sand.[7] The herbs include the small legume *Canavalia*, the herb *Euphorbia atoto* (Euph.), the sedge *Cyperus pendunculatum* (Cype.) and various grasses. Full lists are given elsewhere (van Steenis 1957; Wong 1978; Soegiarto and Polunin 1980; Whitmore 1984). Most of the species are confined to this habitat type and many are pantropical in their distribution.

The actual composition of the vegetation depends to some extent on the type of sand. There are two main forms on Sulawesi beaches: black, andesitic (volcanic) particles found in the north, and white, calcareous sand from the erosion of coral reefs as found in the southern part of Sulawesi and most of the offshore islands. The black beaches of Minahasa have a poor beach flora probably because the black surfaces absorb more heat and become extremely hot.

Plants are dependent on non-saline soil water but are tolerant of the periodic droughts, salt spray, almost constant winds, low levels of soil nutrients, and high temperatures found in the habitat. The plants also typically have small seeds which are dispersed by water, some even having air sacks around the seeds to assist floating.

The green mat formed by the *pes-caprae* formation traps leaves and other organic material blown by the wind or tossed up by waves at high tides. Small animals can also take refuge there. As a result, soil conditions improve, nutrients increase and plant succession proceeds (table 2.4).

One of the first large plants to be seen at the landward edge of the *pes-caprae* formation is the she-oak *Casuarina equisitifolia* (Casu.), which frequently forms pure stands at the top of the beach. She-oak seedlings are intolerant of shade, but even in open conditions, if there is a carpet of she-oak twigs and litter, the seedlings will not grow. This may indicate the presence of some chemical or allelopathic prevention of regrowth, but this has yet to be proven. Thus, unless the shoreline advances, the belt of she-oak will be replaced by other species.

Barringtonia Formation

The *Barringtonia* formation is found behind the *pes-caprae* formation on sandy soils. It is also found behind on abrading coasts, where sand is either being removed by unhindered ocean waves or where sand has at least ceased to accumulate; in such areas a beach wall about 0.5-1 m tall can be found and the formation is found inland of this. The plants are generally tolerant of salt spray, nutrient-deficient soil and seasonal drought and grow in a belt along the coast, usually between 25 m and 50 m wide, where the lie of the land allows it. The belt will be much narrower where the coast is steep and rocky. Large trees sometimes sprawl across the upper parts of the beach, and as the beach wall is eroded away these eventually fall over, die and become shelters for many small seashore animals.

The larger trees of the *Barringtonia* association are of three species: *Barringtonia asiatica* (Lecy.) which has huge 15 cm wide feathery flowers and unusual-shaped fruit (fig. 2.23), *Calophyllum inophyllum* (Gutt.) which has transparent yellow sap and round fruit of 3 cm diameter, and *Terminalia catappa* (Comb.) whose large leaves turn red before falling and whose boughs stand out at right-angles to the trunk in a manner similar to kapok trees *Ceiba pentandra* (Bomb.). *Barringtonia* itself is not invariably present in the formation which bears its name (van Steenis 1957) and it is sometimes found on sandy ground away from the coast. As with the *pes-caprae* formation, the plants found in this type of beach vegetation are found in similar locations throughout the Indo-Pacific region and some are typical of sandy shores throughout the tropics. Many of the species are

Table 2.4. Soil properties of sand from 2-5 cm depth every two metres inland from the seaward edge of an area of *Ipomoea pes-caprae* in Tanjung Peropa Reserve, Kendari. Base saturation was 100% in each case.

Metre depth	pH	%C	%N	C/N	P_2O_5 (ppm)
0	8.0	0.2	0.03	7	3.3
2	8.4	0.37	0.05	7	7.9
4	8.7	0.57	0.06	10	6.3
6	8.1	0.64	0.09	7	9.3
8	7.9	1.14	0.13	9	18.6

After L. Clayton pers. comm.

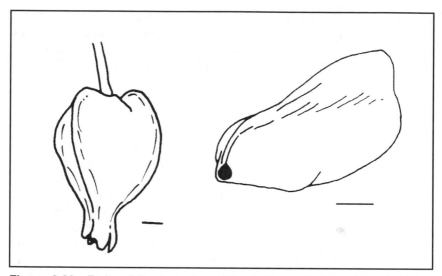

Figure 2.23. Fruits of *Barringtonia asiatica* (left) and *Heritiera littoralis* (right). Scale bars indicate 1 cm.

not found outside these formations. In addition to the trees mentioned above, other typical species include the coconut palm *Cocos nucifera*, the large bush *Ardisia elliptica* (Myrs.) with its pink young twigs and leaves, *Heritiera littoralis* with its peculiar boat-shaped floating fruit (fig. 2.23), and other trees such as *Excoecaria agallocha* (Euph.) with sticky, white sap which may cause temporary blindness (Burkill 1966), pandans *Pandanus*, the white-flowered and large-leafed *Scaevola taccada* (Good.) the fruits of which are dispersed by birds (Leenhouts 1957), and two types of hibiscus *Hibiscus tiliaceus* (Malv.) and *Thespesia populnea* (Malv.) (van Steenis 1957). Both hibiscus have large, yellow flowers with purple bases, but *H. tiliaceus* has slightly hairy lower-leaf surfaces, heart-shaped leaves which are as long as they are broad, black-coloured longitudinal glands on the leaf undersurface near the base, flowers which fall off as soon as they have dried, and smaller fruit. *T. populnea* has smooth leaves, longer than they are broad with a sharper tip, no black glands on the base of the leaf undersurface, flowers which remain on the plant for some days after they have died and larger fruit (fig. 2.24). *Hibiscus tiliaceus* is commonly planted in towns and villages.

The *Calophyllum* trees near the mouth of the Lariang River in northern South Sulawesi bear the epiphyte[8] *Myrmecodia* (Witkamp 1940). *Myrmecodia* and certain other epiphytes are able to grow where there is insufficient

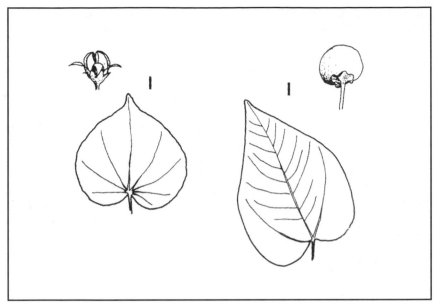

Figure 2.24. Leaves and fruit of *Hibiscus tiliaceus* (left) and *Thespesia populnea* (right). Scale bars indicate 1 cm.

organic debris for most epiphytes. This is because *Myrmecodia* shelters ants within chambers in its swollen stem, and these ants deposit organic matter in the chambers (p. 465). The cycad *Cycas rumphii* is also sometimes found in the *Barringtonia* formation. Despite their appearance, cycads are not palms, neither are they ferns, but they are related to the now-extinct seed-ferns that flourished between 280 and 180 million years ago. In addition to the species above, certain species from the *pes-caprae* association can also be found, particularly near the beach wall.

The vegetation on small islands used by seabirds as nesting sites has a peculiar composition, although such islands as Sangisangian (p. 160) have not been investigated in this (or any other) regard. Here the normally basic reaction of the calcareous soil is changed because of the large quantity of uric acid and high phosphate levels in the birds' faeces although the soil pH is about 6.5-8.5 depending on the organic matter content. One tree *Pisonia grandis*[9] with opposite leaves and reddish veins, is confined to such islands and can dominate the vegetation. Islands which lose their populations of seabirds eventually also lose this tree, whose fruit is dispersed by birds, and the more usual *Barringtonia* formation species take over (Stem-merik 1964). The status of bird islands is discussed below (p. 159).

Rocky Shores

Rocky shores occur where hard, resistant rock faces the sea in such a manner that the products of rock weathering by the waves are swept out to sea rather than deposited to form a beach. Such shores are usually steep with the rocky face often continuing down below the sea surface. There is, however, occasionally a narrow coarse sand or shingle beach. Such steep coasts and cliffs are usually formed of old limestone (e.g., Kaloatoa Island) or volcanic rock (e.g., Lembeh Island).

The vegetation clinging to the upper rock face, above the level of extreme high tides but still affected by sea spray, is similar to that found in the *Barringtonia* formation.

FAUNA OF SEDIMENT BEACHES

Open Area Communities

Most animals of sediment beaches rarely emerge on the surface and these are known collectively as the 'infauna'. Those that spend some time on the surface such as crabs and snails[10] are known as the 'epifauna'. Most of the epifauna are large (macrofauna) but the infauna can be grouped into the microfauna or protozoans, the meiofauna (defined as animals able to pass through a 0.6 mm mesh sieve but retained by a 0.05 mm mesh sieve[11]), and the large, conspicuous macrofauna such as bivalve molluscs or large worms. These are all resident animals although their larvae may have originated elsewhere. Beaches also receive visitors such as shorebirds (p. 144) and turtles (p. 151).

The meiofauna, being too small to move the grains, generally comprises elongate creatures able to wriggle between the grains. Among the most common animals are nematode worms of which hundreds of thousands may occupy a single square metre of beach. These worms are an important food source for larger animals. A new species of *Collembola* (p. 48) was recently found in the sandy beach near Tangkoko-Batuangus Reserve, Minahasa (Greenslade and Deharveng 1986). The worms and most other members of the meiofauna are most common in fine sediments since coarse sands dry out too quickly and very fine sands are too easily deoxygenated. The meiofaunal infauna is less rich in sediments comprising a mixture of grain sizes, probably because the weak waves which result in

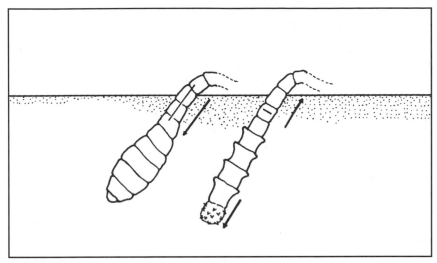

Figure 2.25. Polychaete worm burrowing into sediment. The anterior segments act alternately as a terminal anchor while the longitudinal muscles contract and the middle segments are pulled in (left), and as a penetration anchor as the circular muscles are contracted and the segments of the head are pushed further in (right).

After Trueman 1975

such sediments are incompatible with the animals in some way.

The ability to burrow is crucial to beach animals in order to avoid excessive wave action, surface predators, high temperatures and desiccation. Soft-bodied animals such as bivalve molluscs and polychaete worms pump their way through sediments (fig. 2.25).

The beach ecosystems are unusual in that the common plant-herbivore-predator structure of food chains is absent. The only 'plants' available are diatoms and bacteria, and predation on or in the sediment is difficult. Predators are found primarily among the epifauna-birds, certain mud-skippers, polychaete worms, and snails. Thus the majority of animals either filter plankton from the seawater (suspension feeders) or suck organic deposits and micro-organisms off the sediment surface or sort out edible particles after ingesting sediment (deposit feeders), although the distinction between these is not always clear. The amount of organic material on or in the sediment is generally greater in the finer sediments for these contain higher concentrations of organic carbon and nitrogen (protein in bacteria) (p. 113), and so deposit feeders flourish here. Suspension feeders are more common lower down the beach because they are covered by water for longer and so able to feed for longer.

All these animals have an affect on the environment within which they live. Burrows increase the depth to which oxygen can penetrate and the digging of them brings lower sediments to the surface; suspension feeders deposit faecal pellets on the sediment surface which become a food source for deposit feeders; the action of deposit feeders can resuspend deposits in the sea water. These suspended particles can, however, clog up the filters of suspension feeders so deposit feeders often predominate in very fine sediments while suspension feeders predominate in coarser sediments (Brafield 1978).

Animals living in finer sediments may, when the tide is out, experience oxygen deficiency. Those with burrows opening into surface pools will have few problems, and nor will those that can utilize atmospheric oxygen. Most, however, have to rely on oxygen in the pore spaces which is at low concentrations at the best of times (p. 113). Some animals reduce their metabolic rate below normal levels at low tide, some create currents of water past their gills or body by waving cilia or fine hairs, whereas others move towards the surface as soon as oxygen levels fall below a threshold. Others are able to withstand low oxygen concentration because of the structure of their respiratory pigments, such as haemoglobin, in their body fluids which have exceptional affinity for oxygen. Some, such as fiddler crabs *Uca*, have no problems at low tide and are able to respire anaerobically for short periods when the sea is covering the sediment and the crabs are in their burrows.

The sediment on a shallow sloping beach can be extremely soft, comprising about 75% very fine sand with the remainder being even finer particles. The area below the mean low water level of neap tides is covered by every tide of the year and never left exposed for many hours. The fauna here is marine with certain crab species, bivalve and gastropod (snail) molluscs[12] such as *Telescopium telescopium* (fig. 2.26), and two or three species of polychaete worms predominating. A variety of mudskippers[13]— an unusual group of fish capable of living out of water for a short periods—occur commonly along the water's edge and in burrows in the mud (MacNae 1968).

Underwater, mudskippers breathe just like other fishes, but in air they obtain oxygen by holding water and air in their gill chambers. This water to be renewed about every 5-6 minutes (Burhanuddin 1980). The oxygen they obtain in this way is supplemented by gas exchange through their skin and fins (Stebbins and Kalk 1961). On land, mudskippers move in a variety of ways: they themselves forward on their pectoral fins which move in synchrony with each other (i.e., not true walking) leaving characteristic tracks

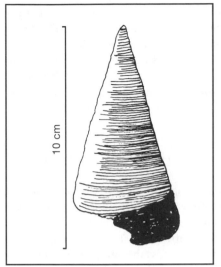

Figure 2.26. *Telescopium telescopium.*

10 cm

in the mud, they skip over the mud by flicking their tails, and they climb on vegetation using their pectoral and pelvic fins.

Although they look very similar, the different species of mudskipper have very different diets; some such as *Boleophthalmus boddarti* take mud into their mouths, retain algal material, and blow out the rest; some are omnivorous, eating small crustaceans as well as some plant material; and others such as *Periophthalmus koelreuteri* are voracious carnivores feeding on crabs, insects, snails and even other mudskippers (Burhanuddin and Martosewojo 1978; McIntosh 1979; Martosewojo et al. 1982).

Most mudskippers occupy deep burrows and some species have burrows with turretted tops (fig. 2.47). All their activities centre on these burrows which are entered briefly throughout the day to moisten the skin and to replace water in the gill pouches. They are usually evenly spaced on the mud thereby reducing the likelihood of conflicts (Stebbins and Kalk 1961), but fights within and between and species are avoided by complex behavioural mechanisms (MacNae 1968). Males dig the burrows and make courtship displays to nearby females by jumping, and by erecting their dorsal fin in the hope of being chosen as a mate. After pairing, eggs are laid on the sides of the burrow which is then defended by the male. Territorial defence is rare outside the breeding season (Nursall 1981; Hutomo and Naamin 1982).

The other main representatives of the epifauna are fiddler crabs *Uca* which can be observed by waiting patiently on the beach. Population densities of the often colourful crabs can reach 50 per m². Male fiddler crabs have one huge claw which is useless for feeding because it cannot reach the mouth, and they therefore have to pick organic deposits off the sediment surface at half the rate of the females which have two normal-sized claws. Presumably the males have to eat for twice as long. The large claw is used for warning away males and for attracting females.

Males compete with other males for females rather than for breeding sites because food resources are so rich and potential sites so abundant. Mating occurs near the burrow defended by the female, but the male-female association is very brief. Preliminary observations suggest that large and small, resident and non-resident males all have the same likelihood of breeding, but smaller males produce fewer viable larvae. Since food resources are not limiting and do not have to be defended, promiscuity is possible (Christy and Salmon 1984).

On coarser-grained beaches near the high water level, burrows about 8 cm across can be found with small piles of sand around them. Their occupants are adult, beige-coloured sand crabs *Ocypode* (fig. 2.27) which are rarely seen during the day. It is extremely difficult to dig these crabs out of their burrows or to catch them as they run across the sand. Young *Ocypode* are very numerous, and can be seen on the sand surface both by day and night. *Ocypode* feed mainly on organic material in the sand, but are sometimes predatory on small crustaceans.

On wider beaches the small ghost crabs *Dotilla* may occur in thousands with densities of over $100/m^2$ (McIntyre 1968; Hails and Yaziz 1982). Although some of the larger individuals are coloured light blue with pinkish legs, the majority are sand-coloured. As the tide rises and covers the beach, each ghost crab builds a shelter of wet sand pellets over its back. Air becomes trapped beneath the crab, and as it burrows down so the air pocket is carried down with it. The crabs emerge as the tide falls, and they are followed by a stream of small bubbles.

Isopod crustaceans can be found by careful examination of the sand and the organic material washed ashore onto the upper shore, and wading birds can sometimes be seen feeding on these and other small animals.

Lower down the beach a variety of molluscs occur but are rarely seen because they burrow beneath the surface. Examples of the bivalves are the white and pinkish *Tellina,* the large *Pinna,* and the economically important edible cockle *Anadara granosa.* This bivalve mollusc spawns seasonally, and breeding seems to be triggered by a drop in water salinity at the start of the wet season. Since plankton and algae on mud always seem to be available, the reason for synchronous breeding is probably to ensure maximum fertilization of the eggs, and to reduce the chance of any one larva being eaten by swamping the potential predators (Broom 1982).

Shorebirds

In addition to four species of resident shorebirds, at least 34 migratory species visit Sulawesi's coasts twice each year. They can be seen between

Figure 2.27. An *Ocypode* crab, a common member of the beach epifauna.

February and April and between September and November, on their way to and from their breeding grounds in northeastern and eastern Asia and their wintering grounds, possibly in northwestern Australia (White 1975). One species, the Australian courser *Stiltia isabella*, migrates from the south between February and April, and returns between September and November (table 2.5; fig. 2.28). These birds would most often be encountered on muddy rather than sandy shores (fig. 2.29).

Very little is known about the movements of these birds within Indonesia and the basic questions posed thirty years ago have barely begun to be answered. That is: What are the normal migration routes? How many birds are there (Coomans de Ruiter 1954)? EoS teams had the opportunity to work with an ornithologist from Interwader, an international shorebird study programme, during the first part of 1986, and two areas of mudflat were visited: the north of Bone Bay, and the coast north and south of Watampone. The northern site had extensive mangroves but the mud was rather sandy and, therefore, not especially suitable for waders. One exception was the muddy estuary of the Balease River where at least 18 species were seen, four of which constituted about half of the total number of birds seen. The coasts around Watampone were found to have less sand than in the north, and the shorebirds were consequently more common though of fewer species (Uttley 1986).

Table 2.5. Waders found around the coasts of Sulawesi.

PLOVERS: Charadriidae

Lesser golden plover	*Pluvialis dominica*
Grey plover	*P. squatorola*
Little ringed plover*	*Charadrius dubius*
Malaysian plover*	*C. peronii*
Mongolian plover	*C. mongolus*
Greater sand plover	*C. leschenaultii*

CURLEWS, GODWITS, SANDPIPERS, SNIPE: Scolopacidae

Whimbrel	*Numenius phaeopus*
Eastern curlew	*N. madagascariensis*
Black-tailed godwit	*Limosa limosa*
Bar-tailed godwit	*L. lapponica*
Grey-tailed tattler	*Tringa brevipes*
Common redshank	*T. totanus*
Marsh sandpiper	*T. stagnatilis*
Common greenshank	*T. nebularia*
Wood sandpiper	*T. glareola*
Terek sandpiper	*T. cinereus*
Common sandpiper	*Actitis hypoleucos*
Ruddy turnstone	*Arenaria interpres*
Swinhoe's snipe	*Gallinago megala*
Red knot	*Calidris canutus*
Great knot	*C. tenurostris*
Rufous-necked stint	*C. ruficollis*
Long-toed stint	*C. subminuta*
Sharp-tailed sandpiper	*C. acuminata*
Curlew sandpiper	*C. ferruginea*
Sanderling	*C. alba*
Broad-billed sandpiper	*Limicola falcinellus*
Ruff	*Philomachus pugnax*

STILTS, AVOCETS: Recurvirostridae

Black-winged stilt*	*Himantopus himantopus*

PHALAROPES: Phalaropodidae

Red-necked phalarope	*Phalaropus lobatus*

STONE CURLEWS: Burhinidae

Beach stone curlew*	*Esacus magnirostris*

COURSERS: Glareolidae

Australian courser	*Stiltia isabella*

* Indicates resident species

White and Bruce 1986

Figure 2.28. Some waders that visit Sulawesi shores. a - broad-billed sandpiper *Limicola falcinellus*; b - grey-tailed tattler *Tringa breviceps*; c - rufous-necked stint *Calidris ruficollis*.

After Beadle 1985

Figure 2.29. Areas of mudflats, the habitat most often visited by shorebirds (indicated in black).

After Salm and Halim 1984

From the above it is clear that the physical composition of the sediment influences the numbers of shorebirds feeding upon it, but within a suitable area of mud it is still not known precisely what attracts birds to one part of a beach and not to another. There are clues, however. Small *Ocypode* crabs, prawns, fish larvae, polychaete worms and small bivalves are among the most important foods for shorebirds and the distribution of these foods between beaches is very uneven. Differences in the fauna in mudflats can really only be determined by direct investigation (Swennen and Marteijn 1985). Where suitable prey is present, density is the most important factor, followed by prey size, prey depth and the penetrability of the substrate (Myers et al. 1980).

Tidal state, wind and disturbance all affect the density and availability of prey, and this is why certain beaches are only used by the waders at certain times (Evans 1976; Grant 1984). Casts of mud thrown up by suspension feeders and swimming movements of small crustaceans are visual clues for the birds, showing them where to feed (Pienkowski 1983), but some birds use tactile rather visual clues and have sensitive beak tips which can sense prey underground. Sandpipers, one group of partially tactile feeders, may avoid sandy mud because the sand grains are very similar in size to the poly-chaete and oligochaete worms upon which they feed (0.5-1 mm) (Quammen 1982).

The penetrability of a beach sediment depends on its water content (p. 111). This may be the reason that some shorebirds can be seen running along the water's edge on the ebbing tide pushing their bills into the thixotropic (fluid) sand. A careful examination of bill marks made in tidally formed sand ripples by dowitchers, a wading bird similar to godwits, showed that more marks were found on the crests than in the water-logged troughs. Neither the distribution of prey nor sediment grain size showed any difference between crests and troughs, but penetrating the crests required only 50%-70% of the force required to penetrate the troughs. Thus, concentrating effort on the crests reduced energy expenditure. Ripple crests are sites of active sediment transport and the arrangement of the grains is relatively unstable. This larger volume of pore space allows a higher water content and offers less resistance to penetration. Although the differences in water content between crest and trough are small, minor dif-ferences in pore volume can produce major changes in the reaction of sand grains to a shearing force (Grant 1984).

Wading birds are often seen in mixed-species flocks which might be thought to be disadvantageous to the individuals by virtue of increased competition. In fact, more often than not, the birds are taking different foods and being together has the advantage that the more birds present the more likely it is that a predator, such as a bird of prey, will be seen. One particular species is usually first to settle on a certain stretch of beach having used visual clues to make its choice. Other species follow when it is clear that food is being found. A few species act as pirates taking food away from the other species. This is disadvantageous in that the birds which lose food have to spend more time feeding to compensate for the loss, but there are advantages in that feeding birds have their heads down searching for food whereas the pirate generally keeps its head up and serves an early warning of the approach of predators (Barnard and Thompson 1985).

In addition to the waders, other common large birds of the coast include the white-bellied sea eagle *Haliaeetus leucogaster*, the osprey *Pandion halietus* and Brahminy kite *Haliastur indus* all of which fish in the shallow waters and scavenge food along undisturbed beaches. There also various storks, herons, egrets and ducks seen around the shore and roosting and

Figure 2.30. Large birds seen feeding in or near mangrove forests. a - woolly-necked stork *Ciconia episcopus*; b - little egret *Egretta garzetta*; c - Chinese egret *E. eulophotes;* d - plumed egret *E. intermedia;* e - great egret *E. alba*; f - milky stork *Ibis cinereus*; g - grey heron *Ardea sumatrana*.

After King et al. 1975

nesting in mangrove forest (fig. 2.30). Palopo Bay and the delta of the Cenrana River are good sites for seeing these birds (Uttley 1986). The milky stork *Ibis cinereus* is of particular interest because until a few years ago it was thought to be quite rare. Large numbers have now been found in Sumatra (Silvius et al. 1985) and they have also been observed, some in breeding plumage, in the Tiworo Straits between Muna Island and the mainland of Southeast Sulawesi (L. Clayton pers. comm.), near Ujung Pandang and in the Cenrana River delta (Uttley 1986).

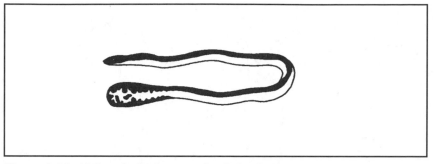

Figure 2.31. Yellow-bellied sea snake *Pelamis platurus*.
After Tweedie 1983

Such fish-eating birds might occasionally encounter venomous sea snakes (p. 232) in the shallow waters of mudflats. One species, the yellow-bellied sea snake *Pelamis platurus* (fig. 2.31), is the most widely distributed species of snake, being found from south Siberia to Tasmania, and from the west coast of America to the Indian Ocean. It is about 1 m long and is often found near the water surface and eats mainly rabbitfish and mullet-like fish (Voris and Voris 1983). Young, hand-reared egrets and herons were presented with live and dead, poisonous and non-poisonous snakes, with and without their tails. The tails of sea snakes are very distinctive. The birds were most frightened by the yellow-bellied sea snake, even if its tail had been removed. This indicates a genetically-based response; they could not have learned that the snake was dangerous from experience (Caldwell and Rubinoff 1983).

Turtles

The sandy beaches of Sulawesi are used as nesting sites by four species of sea turtle which differ in size and in the shape of the carapace (fig. 2.32):
- green turtle *Chelonia mydas* which has an olive-brown carapace about 1 m long, can weigh 100 kg and feeds mainly on seagrass in shallow seas (p. 201);
- hawksbill turtle *Eretochelys imbricata* which has a dark-brown carapace about 90 cm long, weight up to 80 kg and feeds on invertebrates on coral reefs;
- loggerhead turtle *Caretta caretta* which has a brown carapace about 1 m long, a weight of about 100 kg and feeds on crustaceans and molluscs; and

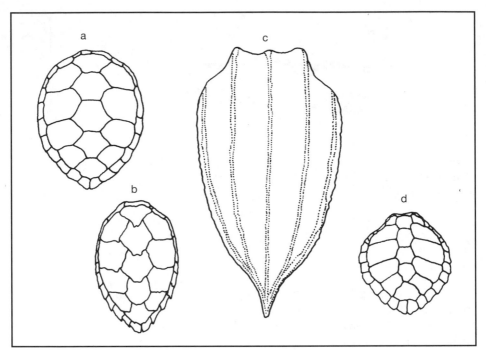

Figure 2.32. The four species of turtle found around the coasts of Sulawesi, a - green turtle *Chelonia mydas*; b - hawksbill turtle *Eretochelys imbricata*; c - leatherback turtle *Dermochelys coriacea*; and d - loggerhead turtle *Caretta caretta*.
From Anon. 1979

- the enormous leatherback turtle *Dermochelys coriacea* which has a dark brown, ridged carapace up to 2.5 m long and can weigh up to a ton; perhaps surprisingly, this species feeds solely on jellyfish. It is a strong swimmer and can maintain a body temperature 18°C above the sea water temperature. Individuals migrate over large distances, in excess of 3,000 km, and although generally found in the tropics, they are have been found feeding inside the Arctic Circle.

The green turtle is more or less absent as a nesting species in South Sulawesi except on the remote Taka Bone Rate Atoll and the Sembilan Islands near Watampone (fig. 2.33; table 2.6). This may be the result of overhunting or overexploitation of eggs but the fact that the hawksbill turtle is present on the small islands in this province and the green turtle common in the other provinces, hints at an ecological factor being respon-

Figure 2.33. Turtle nesting beaches. For key to numbers see table 2.6.

From Salm and Halim 1984

sible. It has been suggested that hawksbill turtles favour small remote islands, even those with no vegetation, with beaches rather steeper than those favoured by green turtles (Nuitja and Uchida 1983). It is important to determine what factors make a beach acceptable so that human activities can be managed or controlled. A study in Australia found that nesting beaches tended to be protected from prevailing winds and to have sand

Table 2.6 Recorded turtle nesting and hunting locations in Sulawesi. Numbers refer to locations in figure 2.33.

	Chelonia mydas		Eretochelys imbricata		Dermochelys coriacea		Caretta caretta	
	nest	hunt	nest	hunt	nest	hunt	nest	hunt
NORTH								
1. Sangihe Island	-	•	-	-	-	-	-	-
2. Tanjung Flesko	•	-	-	-	-	-	-	-
3. Tangkoko Batuangus	•	•	?	-	•	•	-	-
4. Karkaralang Islands	•	-	-	-	-	-	-	-
5. Nenusa Islands	•	-	-	-	-	-	-	-
6. Bunaken Island	-	•	•	•	-	•	-	-
7. Popaya and Mas Islands	•	-	-	-	-	-	-	-
CENTRAL								
8. Tanjung Arus/Dako	•	-	-	-	•	-	-	-
9. Simatang Island	•	-	-	-	-	-	-	-
10. Siraru Island	•	-	-	-	-	-	-	-
11. Pasoso Island	•	-	-	-	-	-	-	-
12. Togian Islands	•	-	•	-	-	-	-	-
13. Banggai Islands	•	•	-	-	-	-	?	•
SOUTH								
14. Spermonde Islands	•	•	•	•	-	-	-	-
15. Masalima Islands	-	-	•	-	-	-	-	-
16. Kalukalukuang Islands	-	-	•	-	-	-	-	-
17. Dewakang Island	-	-	•	-	-	-	-	-
18. Tengah Islands	-	•	•	-	-	-	-	-
19. Sabalana Islands	-	•	•	-	-	-	-	-
20. Tanjung Apatama	-	-	-	-	•	-	-	-
21. Kayuadi Islands	-	-	•	-	-	-	-	-
22. Sembilan Islands	•	•	•	•	-	-	-	-
23. Taka Bone Rate Atoll	•	•	•	•	-	-	-	•
SOUTHEAST								
24. Kabaena & Telaga Islands	-	-	•	-	-	-	-	-
25. Padamarang	-	-	•	-	-	-	-	-
26. Tanjung Kassolamatumbi	•	-	-	-	-	-	-	-
27. Tanjung Tamponokora	•	•	-	-	-	-	-	-
28. Menui Island	•	-	•	-	-	-	-	-
29. Wowoni Island	-	-	•	-	-	-	-	-
30. Saponda Island	-	-	•	-	-	-	-	-
31. Lintea Tiwolu Island	•	•	-	-	-	-	-	-
32. Binongko Island	•	-	-	-	-	-	-	-
33. Seira Island	-	-	•	-	-	-	-	-

? = requires confirmation

After Salm and Halim 1984

moisture with a lower salinity than beaches not used for nesting (Johannes and Rimmer 1984). How the turtles detect these differences is not known.

Many turtle research and management programs have involved the taking of newly-laid eggs and hatching them with some form of protection from predators. These eggs have frequently been taken at the start of the laying season when the intensity of predation is greatest. It has been known for some years that the sex of hatchling turtles (and other reptiles, p. 303) is determined by temperature during incubation (Vogt and Bull 1982). The weather can very over the laying season causing the sex ratio of hatchlings to also vary. In the sub-tropics the percentage of females can change from 0% in the cooler, early parts of the season to 80% in the warmest month (Mrosovsky et al. 1984a). Even close to the equator effect is still noticeable and differences can be expected in sex ratio of clutches laid in different environments—in beach vegetation, in open beach with or without inundation by cool seawater during the critical middle third of incubation when sex is determined (Mrosovsky et al. 1984b). Where eggs are hatched under conservation management programs, great care must be taken not to unwittingly distort the sex ratio of the hatchlings.

The overexploitation of turtle eggs and adults for food makes the problem of survival for the turtle population progressively harder because the mass nesting behaviour of turtles is ecologically akin to the gregarious fruiting behaviour of dipterocarps (p. 368). Both act to satiate the appetites of their predators with the result that at least some of their eggs or seeds will get the chance to develop. The smaller the population of turtles, however, the large the proportion of eggs destroyed.

Maleo Birds

The maleo *Macrocephalon maleo* is a member of a small family of mound builders, or incubator birds (Megapodiidae), which with one exception is confined to eastern Indonesia, New Guinea, Australia, and Polynesia.[14] The maleo itself, however, is found only in North, Central and Southeast Sulawesi. It is about the size of a domestic hen, weighing around 1.6 kg (Guillemard 1889), with striking black and rose-white plumage, an erect tail, and a head with a bare, helmeted cranium which may serve to keep the brain cool when it is on hot beaches (Watling 1983). The bill is pale green and red at the base. Maleo are primarily inhabitants of forest, but only lay eggs where the ground is sufficiently hot for incubation—that is, near hot springs (Wiriosoepartho 1979), near volcanic vents, or on sandy beaches. The megapodes and the Egyptian plover *Pluvianus aegyptius* are the only living birds which do not use the heat of their own bodies for incubation.

Pairs arrive at a nesting area the night before eggs are laid. The following morning, amid much duck-like quacking and turkey-like gobbling, the birds examine holes and make trial digs. When a suitable spot is found, both male and female start digging, throwing earth or sand behind them using their strong legs and claws. The toes are slightly webbed at the base which must help when scratching away loose sand (Wallace 1869). As the hole becomes deeper, so the birds take it in turns to dig and drive away other maleos that venture too close.

This digging can take over three hours, particularly where the sand is loose, after which the female lays her enormous egg, 11 cm long and 240-270 g in weight[15] (Guillemard 1889), in the bottom of the pit. Subsequent eggs are laid at approximately 10-day intervals. The refilling takes nearly as long as the digging and is lengthened by the digging of false pits near the real one to divert predators such as monitor lizards and pigs. Against humans who value maleo eggs as a delicacy, however, these precautions are of little use. During the nesting period the maleos seek food such as figs, and fruit of *Macaranga* (Euph.) and *Dracontomelum* (Anac.) in the beach forest and roost primarily in *Casuarina* (Casu.) trees (Wiriosoepartho 1980).

Maleos are communal nesters and on the largest known site at Bakiriang, on the south coast of the north-east peninsula, more than 600 birds nest early in the year with the holes only two or three metres apart. Two hundred of the birds nest on just 1 ha of sand (Watling 1983).

The surface of a sandy beach can become extremely hot, over 50°C and 80°C on white and black sand respectively (MacKinnon 1978), yet just a few centimetres below the surface, the temperature is relatively stable at about 36°C. It seems as if most eggs, on beaches or elsewhere, are laid in positions where the temperature is between 32°C and 38°C (MacKinnon 1978; Wiriosoepartho 1980). The depth of the hole might be thought to be critical, and it has been suggested that the bare head of the maleo is efficient at sensing temperature but, in reality, the exact depth and temperature (within certain limits) are not so critical. Instead it seems that the eggs are laid as deep as possible for protection against predators.

Hatching takes about three months and if the chicks survive the one-or two-day scramble to the surface, for ants are a major predator of chicks in the ground (R. Dekker pers. comm.), they are able to fly away immediately, already having adult plumage. The manner in which they 'explode' from the sand and rush away is probably an adaptation to avoid the attention of predators (Watling 1983). The great size of the egg is related to the need to produce a chick strong enough to struggle up to the surface (Guillemard 1889).

The size of the egg makes it an attractive source of food for humans and maleo nesting beaches have probably been exploited since man first arrived on Sulawesi. Unfortunately, however, over-exploitation has been a common phenomenon: for example, the beach at the Batuputih[16] just north of the present Tangkoko-Batuangus Reserve, where Alfred Wallace (p. 59) watched maleo nesting in 1859, was at one time visited by egg collectors in

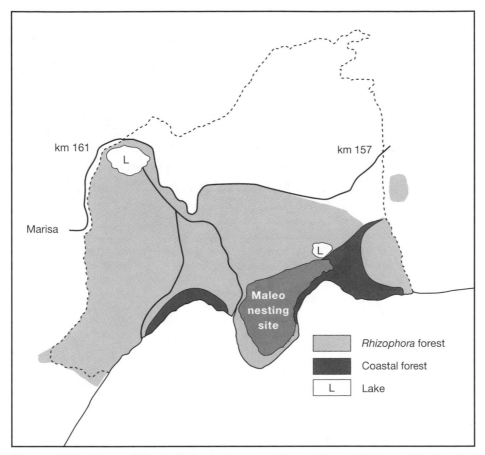

Figure 2.34. Panua Nature Reserve between Gorontalo and Marisa, showing the maleo nesting site, the large expanse of *Rhizophora* forest, coastal forest and the two small lakes.

After Wiriosoepartho 1980

an apparently more or less sustainable manner, but within six years of a settlement being established at Batuputih in 1913, maleos no longer visited the beach (MacKinnon 1978). In 1947 about 10,000 eggs were laid in 2 ha of the Panua Reserve on the coast near Marisa, Gorontalo (fig. 2.34) (Uno 1949), but the present total is less than 10% of this (Anon. 1982a). The total number of breeding hens is between 25% and 67% of the total 40 years ago (Wiriosoepartho 1980).

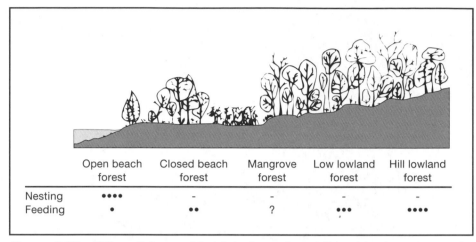

	Open beach forest	Closed beach forest	Mangrove forest	Low lowland forest	Hill lowland forest
Nesting	••••	-	-	-	-
Feeding	•	••	?	•••	••••

Figure 2.35. Differential use of habitats by maleo at Panua to demonstrate the importance of lowland forest to beach-nesting maleo.
Based on Wiriosoepartho 1980

The largest site, at Bakiriang, is only a few kilometres away from a transmigration site. The lowland forest the birds depend on behind the beach is being felled and unless this is protected the demise of this population seems almost certain (fig. 2.35). The Bakiriang site is so special that, until 50 years ago, the raja of Banggai, on Peleng Island 100 km away, determined who should collect the eggs and he received a revenue from the eggs collected. The first 100 eggs were sent to the raja and only after he had approved these could they be consumed by local people. Although the Banggai rajas were notorious pirates and unacceptable in many ways, they were among Indonesia's first resource managers. Now, however, eggs are taken despite legal prohibition and they can be found, wrapped in individual palm-leaf baskets, in the markets of Ujung Pandang and even Jakarta.

Experiments by the head of the Gorontalo Forest Service in the mid-70s showed that maleo eggs could be collected and reburied in a cage so that predation was avoided, and then hatched with significant rates of success (MacKinnon 1978). This was tried again in the Tangkoko-Batuangus Reserve and a hatching rate of 78% was achieved. This technique, together with the control of pig and lizard predators and the clearing of undergrowth to increase the area with a suitably high soil temperature, could make a significant contribution to increasing maleo populations affected by overexploitation where forest areas are sufficient (MacKinnon 1981).

Work is currently being conducted in Bogani Nani Wartabone National Park on the management of an inland population of maleo birds and results are awaited with interest.

Seabirds

The coasts and islands of Sulawesi are visited by at least a dozen species of seabirds which, unlike waders, spend months or even years at sea without returning to land. They tend to nest in large colonies, often on small islands, which are extremely sensitive to disturbance (fig. 2.36; table 2.7). Some islands, like the precipitous Batu Kapal off the northeast coast of Lembeh Island, appear to be roosting rather than nesting sites (Hickson 1889). It is likely that seabirds once nested on or near beaches on the mainland and that human disturbance is the cause of the nesting pattern seen today. Indeed, the seabird populations of Indonesia are experiencing a serious decline in numbers (de Korte 1984).

The habit of nesting in large colonies is disadvantageous because disturbance can have so serious an effect, but it has evolved for at least three important reasons: for ease of pair formation (most seabirds are solitary or live in small groups and range over vast distances when they are not breeding), for defence against predators (there is less risk to one individual of becoming the prey), and for the information shared concerning locations of the abundant food necessary for feeding young birds (Nelson 1980). Fishermen well know the value of these birds since they help to locate schools of tuna and other fish. Indeed, along the north coast of North Sulawesi frigatebirds enjoy a traditional protection because of the service they give (Polunin 1983). Since seabirds are top predators,[17] they tend to concentrate some pollutants, the effects of which only become obvious when the levels exceed a certain limit and reproduction is disrupted. Monitoring programs of pollution levels in seabird tissues can thus be extremely valuable in assessing levels of marine pollution.

Seabirds that visit Sulawesi are boobies (Sulidae), frigatebirds (Fregatidae), and terns and noddies (Laridae) (fig. 2.37). Terns can take off easily and consequently can nest directly on the ground. The red-footed booby and frigatebirds, on the other hand, are masters of soaring flight on their long wings and so they need to nest a few metres above the ground in order to take off successfully.

The nutrients which seabirds contribute to the islands on which they nest can be quite considerable. For example, colonies of the white-capped noddy *Anous minutus*, known in Sulawesi only from the small Sangisangian Island north of Kalaotoa Island in the Flores Sea, deposit in their faeces an average of about 2g dry matter/m²/day. This is equivalent to 1,030, 220,

Figure 2.36. Nesting
and roosting sites of
seabirds around
Sulawesi.

After Salm and Halim 1984

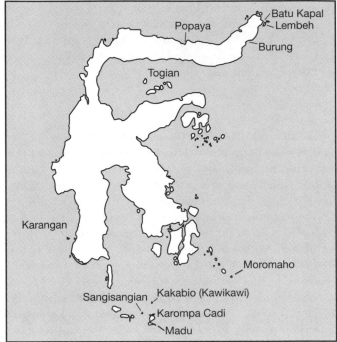

Table 2.7. Seabirds nesting or roosting on offshore islands.

	Ss	Sl	Fa	Fm	Fsp	Am	As	Sta	Stal	Stb	Sts	Stsp
Batu Kapal	-	-	-	-	(•)	-	-	-	-	-	-	•
Lembeh Island	-	-	-	-	-	-	-	-	-	•	•	-
Burung Island	-	-	-	-	•	-	-	-	-	-	-	-
Popaya Island	-	-	•	-	-	-	-	-	-	-	-	-
Togian Islands	-	-	-	-	-	-	-	-	-	-	•	-
Karangan Island	-	-	•	-	-	-	-	-	-	-	-	-
Islet off Bonerate	-	-	-	-	-	-	-	-	-	•	-	-
Karompa Cadi Island	•	-	-	•	-	-	-	-	-	-	-	-
Sangisangian Island	•	-	-	-	-	•	•	•	-	•	-	-
Madu Island	-	-	-	-	-	-	-	-	-	•	•	-
Kakabio Island	•	•	-	-	•	-	-	-	-	-	-	-
Moromaho Island	•	-	(•)	-	-	-	-	-	-	-	-	-

Ss - red-footed booby *Sula sula*
Sl - brown booby *S. leucogaste*
Fa - lesser frigatebird *Fregata ariel*
Fm - great frigatebird *F. minor*
Fsp - unidentified frigatebird
Am - white-capped noddy *Anous minutus*
As - brown noddy *A. stolidus*

Sta - little tern *Sterna albifrons*
Stal - bridled tern *S. alaetheta*
Stb - lesser crested tern *S. bengalensis*
Sts - black-naped tern *S. Sumatrana*
Stsp - unidentified tern
(•) - needs confirmation

After de Korte 1984; Salm and Halim 1984

Figure 2.37. The most common seabirds found around the coasts of Sulawesi. a - great frigatebird *Fregata minor* (adult male); b - *F. minor* (adult female); c - lesser frigatebird *F. ariel* (adult male); d - *F. ariel* (adult female); e - brown noddy *Anous stolidus;* f - bridled tern *Sterna alaetheta*; g - little tern *S. albifrons*; h - lesser-crested tern *S. bengalensis*; i - black-naped tern *S. sumatrana*; j - brown booby *Sula leucogaster*. The red-footed booby *S. sula* can be white or brown but always has bright red feet.

After King et al. 1975

140 and 50 kg/ha/yr of nitrogen, phosphorus, potassium and magnesium respectively (Allaway and Ashford 1984). It is not surprising, then, that the guano of seabirds is much sought and if the collection is conducted at times of year when the birds are not breeding (the peak breeding season is probably between February and April) and with minimum disturbance, this activity can be sustainable. The rocky rather than sandy substrate of some birds islands, however, together with the high acidity of the guano result in an unfavourable habitat for plants. On Batu Kapal, for example, the only plant found was one young fig tree *Ficus nitida* (Mora.) (Hickson 1889).

Invertebrates of Mangrove Forest

Just as the composition to the mangrove vegetation varies with distance from the sea, so also does the fauna (MacNae 1968; Berry 1972; Sugondo 1978; Anon. 1980a; Budiman 1985) (table 2.8). A number of somewhat overlapping zones can be recognised and these are described below.

Table 2.8. Changes in the composition of the mollusc fauna of a mangrove forest inland from the forest edge at Batu Gong, Kendari. All are gastropod snails except *Geloina* which is a bivalve.

	Metres inland from boundary of sea and mangrove forest										
	0-20	20-40	40-60	60-80	80-100	100-120	120-140	140-160	160-180	180-200	200-220
Strombus luhuanus	•	-	-	-	-	-	-	-	-	-	-
Gafrarium gibbia	•	•	•	•	-	-	-	-	-	-	-
Clava alugo	-	•	-	-	-	-	-	-	-	-	-
Cyclotellina sp.	-	•	-	-	-	-	-	-	-	-	-
Barbatia decussata	-	•	•	-	•	-	-	-	-	-	-
Nerita reticulata	-	•	-	-	-	-	-	-	-	-	-
Littoraria scabra	-	-	•	-	•	•	•	•	-	-	-
Monodonta labio	-	-	•	-	-	-	-	-	-	-	-
Cerithium patulum	-	-	•	•	-	-	-	-	-	-	-
Erronea caurica	-	-	•	-	-	-	-	-	-	-	-
Cyclotellina sp.	-	-	•	-	•	-	-	-	-	-	-
Cardium flavum	-	-	•	-	-	-	-	-	-	-	-
Nerita planospira	-	-	-	•	•	•	-	•	-	-	-
Terebralia sulcata	-	-	-	•	•	•	•	-	-	-	-
Natica mamilla	-	-	-	•	-	-	-	-	-	-	-
Ellobium auris-judae	-	-	-	-	-	-	•	•	•	•	-
Turbo ticaonicus	-	-	-	-	-	-	-	•	-	-	-
Semicassis vibex	-	-	-	-	-	-	-	•	-	-	-
Cassidula sulculosa	-	-	-	-	-	-	-	-	-	-	•
Geloina ceylonica	-	-	-	-	-	-	-	-	-	•	•

Data collected by an EoS team with identifications by M. Djajasasmita and A. Budiman

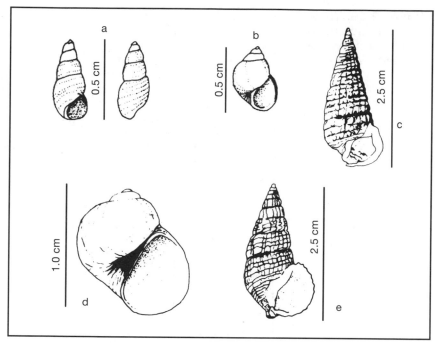

Figure 2.38. Snails of the pioneer zone. a - *Fairbankia* sp.; b - *Syncera brevicula*; c - *Cerithidea cingulata*; d - *Salinator burmana*; e - *Terebralia sulcata*.

Pioneer Zone. The pioneer zone is that area where seedlings of *Avicennia* and *Sonneratia* grow and which is covered by almost all high tides. The sediment is essentially the same as in the open areas and the fauna is also similar, although fewer strictly marine animals are found. Snails such as *Syncera, Salinator* and *Fairbankia* may occur on the wet surface of the sediment (fig. 2.38).

Mudskippers are common and large- and medium-sized fiddler crabs *Uca* can be very abundant. Various species of polychaete worms, a few bivalve molluscs and the peculiar peanut worm (fig. 2.39) which can store oxygen in its coelomic fluid when the tide is out (Brafield 1978) live permanently in the soil of this zone.

The abundance and type of fauna living on the mangrove trees depends largely on the age of the tree—older ones have denser populations consisting of more species. *Littoraria* snails (fig. 2.40) occur on almost

Figure 2.39. Peanut
worm, *Phascolosoma*
from the pioneer zone of
the mangrove forest.

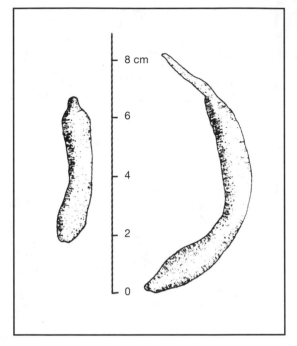

all the vegetation, sometimes up to 2 m above the soil surface. Some species are generally found on leaves while others are found mainly on bark (Reid 1986). Various authors have also noticed that individuals of a species can be pale coloured in trees and dark on the ground (Sugondo 1978; Anon. 1980a; Cook 1983) presumably as a means of camouflage. Populations of sedentary animals encrust the lower stems of trees as they grow. These animals typically include barnacles with the larger *Balanus amphitrite* below and the smaller *Chthamalus withersii* extending higher; oysters, commonly *Crassostrea cucullata*; and the small black mussels *Brachyodontes* sp. which are attached to the tree by 'byssus' threads. A total of 15,401 animals, 60% of which were mussels of the genus *Brachyodontes*, have been found on a single *Avicennia* tree (Tee 1982). These attached fauna may be eaten off the lower stems by the carnivorous snails *Thais* and *Murex* (fig. 2.41) which may move in groups decimating their prey (Broom 1982). Barnacles and mussels sometimes suffer 50% mortality in the first 10 m above the soil surface but the number of dead animals decreases upward, indicating that predation higher up the tree is less (Tee 1982).

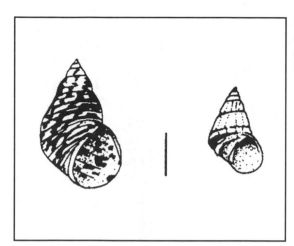

Figure 2.40. *Littorina scabra* (left) and *L. carinifera* (right). Scale bar indicates 1 cm.

Around the roots of mangrove trees can sometimes be found a bright green sea anemone which, when disturbed, buries itself very quickly, so that collecting a specimen is a very frustrating task (Hickson 1889).

Two or three species of hermit crab are found in this zone. These crabs have lost the hard protective carapace over the rear part of their bodies and depend for protection on finding empty shells. The combination of security and mobility of the adopted shells is clearly advantageous and hundreds of hermit crabs can sometimes be seen on a beach. Suitable shells are, unfortunately for the crabs, a limited resource and this affects growth and reproduction. It has been found that crabs with roomy shells put their energy into growth and do not reproduce, whereas crabs with tight shells, stop growing and put their energy into reproduction. The scarcity of shells leads to active competition between species of hermit crabs (Bertness 1981a, b).

When water inundates the pioneer zone it is sometimes possible to find the pond skater/water strider *Halobates,* one of the very few insects that has adapted to the marine environment (fig. 2.42) (Anderson and Polhemus 1976; Cheng 1976).

Eroded Banks. The edge of mangrove forest along a winding estuary is often marked by a nearly vertical bank 1-1.5 m high instead of a sloping pioneer zone. This is caused by currents sweeping away the consolidated sediment and is most obvious on the outer bend of rivers, where the current flows faster than on the inner bend. The bank may be broken in places and mangrove trees at its edge often fall into the sea as the sediment

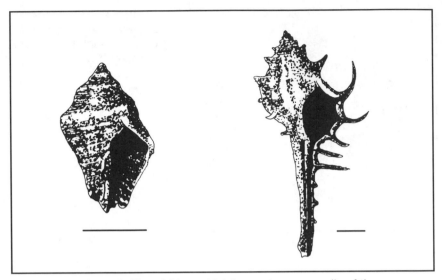

Figure 2.41. *Thais* (left) and *Murex* (right), carnivorous snails of the mangrove forest. Scale bars indicate 1 cm.

is eroded slowly away. The top of this bank is usually at, or slightly above, the mean high water of neap tides, and may sometimes be left 9-10 days without tidal cover. The sediment of an eroded bank resembles that of the mangrove forest floor behind it, with less fine sand (commonly about 65%) than in the pioneer zone. The bank is burrowed into by various crab species (Berry 1963).

True Mangrove Forest. The fauna of this zone differs in several respects from that in the preceding zones. Most of the ground is very flat and the sediment surface is exposed to the air for an average of 27 days per month. Since the trees provide heavy shade, however, the humidity is very high, and this together with the poor drainage means that the soil rarely dries out. There is also abundant leaf litter and other organic matter so that detritus feeders abound. The soil contains less sand and more of the finer clay and silt particles than in sediment near the water's edge. There is also more organic matter. A total of 15 species of mollusc and 15 species of crustaceans were found in one study around the estuary of the Malili River in the Gulf of Bone, but the study was not exhaustive and it is difficult to com-

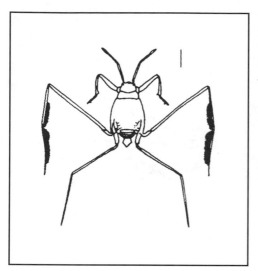

Figure 2.42. *Halobates,* one of the few insects adapted to the marine environment. Scale bar indicates 1 cm.

After Anderson and Polhemus 1976

pare these results with those of the other sites investigated. For example, two species of ellobid snails were found at Malili, compared with 12 at Way Sekampung in Lampung. The comparative study of these two sites together with one more in Lampung and one at Cilacap did show, however, that generic similarities existed between the snails found at each of the sites, but that in many cases the species were different. Snail species common to all the areas were *Littoraria scabra, Cerithidea quadrata, Telescopium telescopium* and *Neritina violacea* (Sabar et al. 1979).

About 75% of the fauna of this zone is not found in the other zones (Frith et al. 1976). The fauna is best divided into three groups: tree, ground surface, and burrowing.

Perhaps the most striking change observed in going towards the dry lands is the rapid decrease in encrusting animals on the lower stems and trunks of the trees (fig. 2.43). The remaining tree fauna is mobile, for these animals can to some extent determine their immersion in water. It is largely composed of snails such as species of *Littoraria* and *Nerita* (fig. 2.44) which are also found in the pioneer zone. Further back from the sea *Cerithidea obtusa* and *Cassidula* are found, and even further inland are species of air-breathing pulmonate snails (fig. 2.45) (Budiman and Darnaedi 1982). Most of these feed on algae growing on the sediment and move up trees when tides wet the ground, but *Littoraria* very rarely leaves

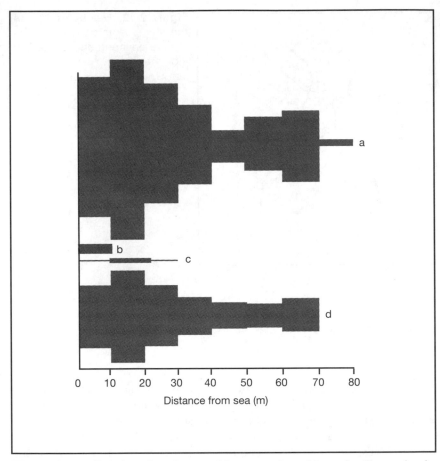

Figure 2.43. Relative abundance of four species of tree-dwelling animals along a transect through mangrove forest near Roraya, Tinanggea, on the south coast of Southeast Sulawesi. a - barnacle *Balanus* sp.; b - mussel *Brachyodontes* sp.; c - snail *Littoraria 'scabra'*; d - barnacle *Chthamalus* sp.

After L. Clayton pers. comm.

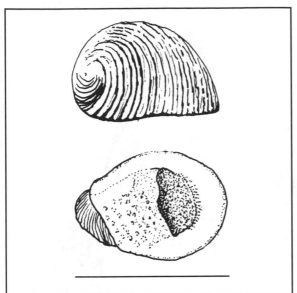

Figure 2.44. *Nerita birmanica.*
Scale bar indicates 1 cm.

the tree trunks. All these snails are able to breathe efficiently in air and the pulmonate snails, such as *Ellobium,* have lungs.

The encrusting animals on a tree also show a vertical zonation depending on their tolerance to desiccation. In a case examined in Southeast Sulawesi, the small *Chthamalus* barnacle appeared to have greater tolerance than the larger *Balanus* barnacle (fig. 2.46).

The animals seen on the sediment surface comprise mostly crabs that have emerged from their burrows, and snails, although the medium-sized mudskipper *Periophthalmus vulgaris* can be common. Among the snails, actual distance from the sea seems to matter less than details of ground conditions. In wetter areas such as where drains into small gullies, *Syncera brevicula* can be more common than anywhere else in the mangrove forest.

Molluscs at the seaward edge of the mangrove forest comprise a mixed sample of gastropods and bivalves. Further back in the mangrove forest, however, carnivorous and filter-feeding molluscs disappear. The vast majority of the molluscs in the true mangrove forest feed by grazing on algae or micro-organisms on the soil surface. Little is known about mollusc reproductive behaviour in mangrove forest but most have internal fertil-

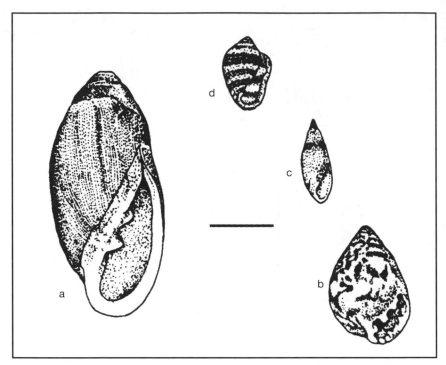

Figure 2.45. Air-breathing snails of the inland parts of mangrove forest. a - *Ellobium auris-judae*; b - *Pythia pantherina;* c - *Auriculastra subula*; d - *Cassidula sulculosa.* Scale bar indicates 1 cm.

ization of eggs and, unlike many aquatic snails, have eggs which develop directly into tiny snails rather than water-borne larvae.

Nearly all the mangrove crustaceans and worms make burrows which reach down to the water table. Many different types of tunnels are constructed (fig. 2.47) and the elliptical tunnels of the edible crab *Scylla serrata* may slope down from the bank of a river for as far as 5 m. In general, these burrows serve as: a refuge from predators at the surface, a reservoir of water, a source of organic food, a home for pairing and mating which is defended, and a place for brooding eggs and young, although no single species uses its burrow for all these purposes.

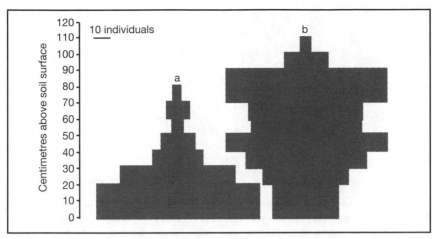

Figure 2.46. Vertical distribution of barnacles on a *Rhizophora* tree 10 m from the seaward margin of mangrove forest near Roraya, Tinanggea, on the south coast of Southeast Sulawesi. a - *Balanus* sp.; b - *Chthamalus* sp.

After L. Clayton pers. comm.

In the landward areas of true mangrove forest, the first signs are seen of an animal which itself is rarely seen, the mud lobster *Thalassina anomala* (fig. 2.48). This animal builds volcano-like mounds of mud which can reach over a metre high (fig. 2.47), and feeds on mud, digesting the algae, protozoa and other organic particles within it. The burrow below the mound is up to three metres long, extending down below the water table. The entrance leading to the main burrow is generally plugged with layers of earth. The habit of burrowing deeply in generally plugged with layers of earth. The habit of burrowing deeply in anoxic mud, closing itself off in poorly oxygenated air and water suggests that it may have evolved means of anaerobic respiration (Malley 1977).

The bivalve mollusc *Geloina* (fig. 2.49) lives buried in mud and can occasionally be found in this zone, but is more common in mangrove forests on islands in or near river deltas. *Geloina* is remarkable in its ability to feed, respire and breed so far from open water and at levels where it is sometimes not covered by tidal seawater for several weeks at a time.

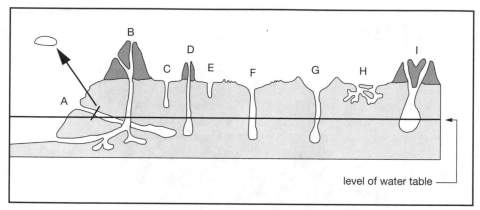

Figure 2.47. Different types of animal burrows in mangrove forest areas. The horizontal line indicates the water table. A - crab *Scylla serrata* (with transverse section of burrow); B - mud lobster *Thalassina anomala;* C - crab *Uca* spp. (burrows may reach water table nearer the low-tide level); D - crab *Sesarma* spp.; E - peanut worm *Phascolosoma;* F - large mudskipper *Periophthalmodon schlosseri;* G - smaller mudskipper *Boleophthalmus bodarti;* H - pistol prawns *Alphaeus* spp.; smaller mudskipper *Periophthalmus vulgaris.*
From Berry 1972

Rivers, Streams and Gullies. The banks and beds of water courses in the mangrove forest have a fauna which is generally distinct from that on the forest floor. For example, many of the forest floor species of polychaete worms and crabs are missing, whereas juvenile fiddler crabs and some snail species are more common. The edible crab *Scylla serrata*[18] occurs in firmer sediment near the larger streams and rivers in the mangrove forest where it is caught in traps (Anon. 1980a).

Terrestrial Margin. Far back from the sea, the soil of the mangrove forest is covered by fewer and fewer tides and suffers longer and longer intervals of exposure. Unlike other types of shore, there is virtually no wave action in mangrove forest to carry seawater higher than the true tidal level, because the wave energy is absorbed by the abundant tree trunks and roots. Thus, animals in the terrestrial margin live on a salt-impregnated soil but are covered by seawater only at irregular and infrequent intervals. Insects, snakes, lizards and other typically terrestrial animals are much more common here than further seaward. *Sesarma* and some large crabs occur here and in the *Nypa* palm swamps behind, where a small bivalve mollusc *Enigmonia aenigmatica* is found in association with the palms (Kar-

Figure 2.48. The nocturnal mud lobster *Thalassina anomala*.

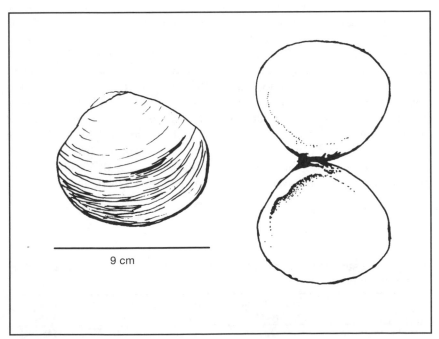

9 cm

Figure 2.49. The bivalve mollusc *Geloina ceylonica*.
From Berry 1972

tiwinata et al. 1979). The crabs' basic requirement is that their burrows must reach down to the water table. Where there is moving water some snails, such as *Fairbankia* and *Syncera,* and the large mudskipper *Periophthalmodon schlosseri,* may be found.

Only one estimate of biomass of aquatic mangrove fauna seems to have been made in the Indo-Malayan Region and that was for tree-dwelling aquatic fauna in four types of mangrove forest (table 2.9). The results show a general reduction of biomass with increasing distance from the sea. It is supposed that the turnover rate (time for one generation to replace another) must be quite rapid because most of the tree-dwelling aquatic fauna are in the lower trophic group (i.e., suspension feeders), none of which have long life spans (Tee 1982).

The biomass of molluscs in an Australian mangrove area was greater on the mudflats and pioneer zone where they tended to be filter-feeders, than in the forest where they were largely deposit-feeders. The crabs were very abundant in the open areas but they had a relatively low biomass. The biomass, density and diversity was highest in *Avicennia* forest but lower in *Rhizophora* forest (Wells 1984).

Terrestrial Fauna of Mangrove Forest

Insects, birds and mammals live chiefly in the canopy of mangrove forest. Ground-living animals such as rats, pigs and lizards only venture into the landward edge of mangrove forest for brief forays. Macaques will eat *Sonneratia* and other fruit and descend to the ground to search for crabs, peanut worms and other suitable food. Flying foxes *Pteropus* commonly roost in the mangrove canopy and others with feathery-tipped tongues such as the cave fruit bat *Eonycteris* and long-tongued fruit bat *Macroglossus* are important for the pollination of *Sonneratia*. In addition, species of insectivorous bats roost in the forest and different species feed in different microhabitats: over the canopy, just above the canopy, in the open beside the trees, and inside the forest. The composition of the guilds of bats feeding in each microhabitat can to some extent be predicted from their

Table 2.9. Estimated biomass of tree-dwelling aquatic fauna in four types of mangrove forest.

	Rhizophora forest	*Avicennia* forest	*Sonneratia* forest	*Bruguiera* forest
Distance from sea (m)	10	235	310	445
Estimated biomass (kg)	1,014	272	30	38

From Tee 1982

wing morphology and details of this from a study in northwest Australia can be found elsewhere (McKinzie and Rolfe 1986). Prediction of ecology and behaviour from gross and dental morphology can be useful tool for little known species of animal (Clutton-Brock and Harvey 1977; Kay and Hylander 1978).

The frog *Rana cancrivora* is common inland around lakes and other aquatic habitats, but is exceptional among amphibians in being able to live and breed in weakly saline water. The tadpoles are more resistant to salt than the adults and metamorphosis into adults will only occur after considerable dilution of the salty water (MacNae 1968).

Mangroves are inhabited by a variety of reptiles such as the monitor lizard *Varanus salvator*, the common skink *Mabuya multifasciata*, and the venomous yellow-ringed catsnake *Boiga dendrophila* with 40 to 50 narrow yellow rings around its body (fig. 2.50). Most snakes seen in or near mangroves are not in fact sea snakes (Hydrophidae) which live primarily in open water (p. 232). The prey of mangrove snakes comprises primarily small fish (Supriatna 1982). Potentially the largest animal of the mangrove swamps is the estuarine crocodile *Crocodylus porosus*, but persecution for centuries has reduced its number to a very low level and large specimens (they can exceed 9 m) are extremely rare around Sulawesi (p. 303).

The most conspicuous of the insects are mosquitoes, and the larvae of some species can live in water with a salinity of 13 ppt (MacNae 1977). Species of *Aedes* mosquitoes have been seen feeding on mudskippers but they are also attracted to human skin (p. 617). Mosquito collections made in an area of coconuts near the mangrove-fringed north coast of Bolaang Mongondow included large numbers of the potential vector of malaria *Anopheles subpicta* (Hii et al. 1985). Another potential insect hazard is the leaf-weaving ant *Oecophylla smaragdina* (fig. 2.51) which makes nests by glueing together adjacent leaves while they are still attached to the tree. The making of the nest is extraordinary because the adult ants have no means of producing a sticky secretion to join the leaves of the nest together. The larvae do have appropriate glands, however, and when a leaf is to be added to the nest or a tear repaired, some of the worker ants seize the leaf edges to be joined and hold them in the required position. Other workers enter the nest and collect larvae. These are held near the head in the workers' jaws and moved back and forth between the leaf edges, as they secrete the glue (Sarasin and Sarasin 1905).

Fauna of Beach Forests, Particularly Coconut Crabs

Virtually nothing seems to have been written about the fauna of beach forests, although travellers sometimes report seeing macaques in the trees.

An interesting snail *Cochleostyla leucophtalma* has been observed in bushes near the beaches of Sangihe Island. This animal, about 2 cm across

Figure 2.50. Yellow-ringed catsnake *Boiga dendrophila*.
After Tweedie 1983

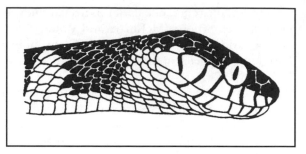

Figure 2.51. *Oecophylla smaragdina* which sews its leaf nest with larval secretions.

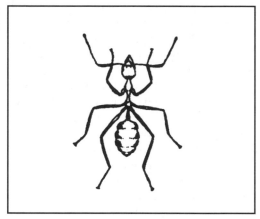

with a brown and white shell and an orangey-red body, lies across a leaf and bends its body over so the edges meet. It then starts to glue the edges together with mucus to form a bag. Before finishing, it lays large eggs inside the bag, completes the glueing and then eats a hole in the leaf blade which it covers with a very thin film of mucous. This is supposed to ensure that some air enters the leaf bag (Sarasin and Sarasin 1905).

The beach forest animal about which most is know is the coconut crab *Birgus latro* (fig. 2.52). The most recent major studies of this interesting animal, the world's largest terrestrial arthropod, were conducted in the Marshall Island and the Mariana Island in the western Pacific (Helfman 1977; Amesbury 1980, 1982; Reese 1981), and it is on the results of those that much of the information below is based.

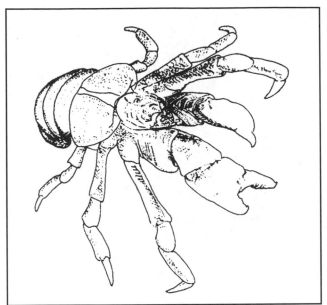

Figure 2.52. Coconut crab *Birgus latro.*

The coconut crab is a member of the land hermit crab family Coenobit-idae[19] but, unlike the other members, it does not occupy a snail shell when adult. It was once widely distributed throughout the western Pacific and eastern Indian Oceans but now is restricted to small islands, particularly those uninhabited by man. In Sulawesi it now known only from Sangihe and the Kawio, Talaud, Togian and Banggai Islands, the Togian Islands being the most westerly part of its range in Indonesia[20] (Reyne 1938; Anon. 1982b; Salm and Halim 1984) (fig. 2.53). Man has found the crab to be a desirable food item and it is easily caught even though it may be the largest animal on some of the islands it inhabits. In the Togian Islands, for example, crabs are occasionally collected by residents for their own consumption or for sale to visitors. Much more damaging though are the parties of non-residents from Gorontalo, Ampana and Poso who remove whole boatloads of crabs for sale in their respective towns as food or tourist curios. If this exploitation were regulated under an ecological management plan it could be sustained, but this is not the case (Anon. 1982b).

Most crabs are obliged to seek water for mating, but coconut crabs mate on dry land. The fertilized eggs, totalling tens of thousands, are carried under the crab's abdomen. Females release their eggs into sea water when the tides are highest; that is, around the full moon, and the larvae

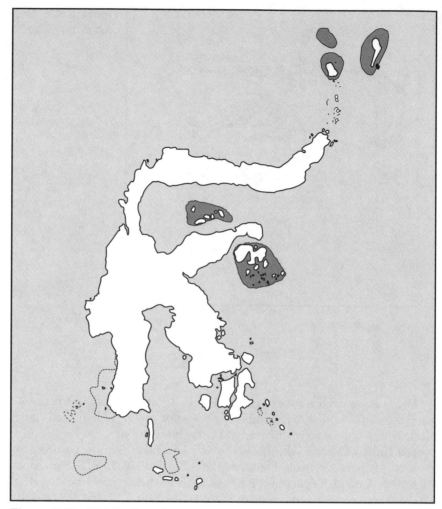

Figure 2.53. Distribution of coconut crab *Birgus latro* (dark shading). Dotted lines represent coral shoals.

After Salm and Halim 1984

hatch in response. At this time eggs are most likely to be washed out to the open sea where predation is less and availability of plankton greater than over the reef flat. The larvae spend three to eight weeks as part of the oceanic plankton before the glaucothoe or transitional larva finds a small shell to shelter inside, like a hermit crab, and migrates to land. At this stage it may face severe competition with smaller species of hermit crabs for both shells and food. This larva moults several times and then experiences a metamorphosis to become a miniature adult. When its carapace is only about 2.5 cm across, at about two years of age, it gives up living in shells, a move which brings new opportunities and problems. It is sexually mature when the carapace is only 5 cm across but the largest crabs can weigh 5 kg with a thoracic length[21] of 70 mm. Their large-clawed legs can span 90 mm. Females are smaller than males and have a maximum thoracic length of about 50 mm. Estimates of the ages of such large individuals have not been made. This large size would clearly be impossible if they were dependent on living in borrowed shells. They do need to avoid direct sun, however, so they are active primarily at night, although not all those who have studied them agree that they are strictly nocturnal (Harries 1983).

Coconut crabs generally shelter in burrows in the old coral other rocks around the coast or under coconut leaves and pandan roots. These burrows are analogous to the shells they have abandoned. They bring large pieces of food into their shelters and remain there for a few days eating. It is possible that the crabs moult in these burrows to provide protection from anything that might endanger them during this critical time. A newly-moulted crab is largely defenceless and its body is soft and easily damaged.

If a food item is too large to move, a coconut crab will defend it from all other competitors such as other crabs and rats. Apart from coconut and pandan fruits, two of their most common foods, coconut crabs will eat virtually anything organic from moulted crab exoskeletons to other crabs, and decaying wood to unwary birds. In captivity they will readily take lettuce, cabbage and live giant African snails *Achatina fulica* as well as a wide range of other food (Reyne 1939).

It was the crab's (supposed) ability to climb to the top of coconut trees, snip off coconuts, return to the ground and remove the husk, take the nut back to the tree top and drop it to the ground to crack it open, or hold it in the claw and smash it open against a rock that caught the imagination of early voyagers and biologists (Helfman 1981; Harries 1983). The original accounts were all second or third hand, and no one has yet recorded a crab opening a coconut; it may be that the crabs are only capable of opening nuts which have cracked on hitting the ground. In any case, coconuts are not essential to the survival of coconut crabs.

The relationship between coconuts and these crabs is nevertheless thought to be close. On the beaches of small coral islands the coconut palm may become the dominant natural plant form and the coconut crab the dominant animal. The coconut is so widely dispersed because the thick pericarp, or husk, allows the fruit to float in the sea without being damaged for over six months. It has recently been suggested that the glaucothoe chooses to live in cracks in the floating coconut husk in the same manner that young hermit crabs live for a period in holes in sponges and corals. The eye-end of the floating coconut would be relatively soft, moist and aerated and may attract other animals on which the young coconut crabs could feed. Other floating vegetation is also occupied, but inputs of these into the sea would occur relatively rarely, after storms for example, compared with the regular monthly flowering of coconuts and subsequent dropping of fruits through the year. This vegetable home both temporarily replaces the need for a snail shell for the crab to shelter within, and also becomes a vehicle for dispersion. Thus, before primitive man took a culinary interest in the crab and the coconut the distribution of the two were probably very closely related but that pattern has now been blurred with the one being overexploited and the other cultivated (Harries 1983).

Adult coconut crabs have no serious predators except man although the young crabs that have begun a terrestrial life suffer from predation by monitor lizards, the coastal frog *Rana cancrivora*, rats, pigs and perhaps birds. Dominance between crabs is determined almost wholly by size, and a wave of a claw can send a smaller individual scurrying for cover (Helfman 1977).

With no predation pressure or competition the larger crabs are no longer subject to the following two mechanisms by which populations are often regulated. Migration to other islands is not feasible because adult crabs have adapted to terrestrial living to such as extent that they cannot survive in seawater for more than a few hours. Another common mechanism is a negative feedback from increasing populations: the higher the population density, the less food becomes available, resources such as sites for burrows become scarce, and so reproduction is disturbed and less young are produced. Even this does not operate, however, because the new generation of coconut crabs arrive on the shores possibly from parents living on a distant island. There appears to be no normal means of regulating the numbers of coconut crabs, so they have evolved a drastic means—cannibalism. Coconut crabs have been observed to despatch ailing individuals, or smaller individuals which have approached too close. Thus the reproductive potential of the larger individuals is maximized by reducing the pressure on food and other resources, and by utilizing an important food source on a potentially food-poor island (Helfman 1977).

FAUNA OF ROCKY SHORES

Introduction

The fauna of rocky shores is not particularly rich but it has representatives from a very wide range of phyla. Many cliffs such as those on the north of Talisse Island (Hickson 1889) or the islands north of Flores (Docters van Leeuwen 1937) have clefts or small caves formed by wave action and these are often inhabited by colonies of swiftlets (p. 553), which build their nests from a mixture of hardened saliva and plant fragments (Medway 1968).

On the wet basalt rocks of Tangkoko-Batuangus Reserve, Minahasa, many thousands of rock-hopping blenny fish *Andamia* and *Alticus* can be seen (Anon. 1980b). They spend most of their time out of water but, unlike mudskippers (p. 142) they are splashed by water as each wave breaks over the rocks. They hold on by using their ventral fins like a sucker, and feed by scraping the thin film of alga off the rocks with their teeth which are set in their lips rather than their jaws (Carcasson 1977). Grapsid crabs jump between rocks like blennies (Hickson 1889).

The aquatic plant life of Sulawesi rocky shores, like that elsewhere in the tropics, is relatively poor, largely as a result of the hot and dry conditions experienced at low tide. One of the more characteristic algae is the sea lettuce *Ulva*, with thin, broad, bright-green, irregularly-shaped 'leaves', which is found between the extents of low and high neap tides. Only below the level of mean low tide is plant (and other) life particularly rich. The spray zone, the area never covered by tides but frequently wetted by spray from braking waves, is colonized by encrusting lichens and blue-green algae which, when damp, are grazed by *Littoraria* snails. At night, hermit crabs scavenge here and certain grapsid crabs seek prey. Suitable parts of the intertidal zone are covered with a thin growth of blue-green and filamentous green algae which, although not appearing particularly lush, supports an impressive range of grazing animals: limpets, chitons (fig. 2.54), winkle snails, isopod and amphipod crustaceans, and herbivorous grapsid crabs.

An examination of molluscs on a rocky shore of Tanjung Peropa Reserve, Kendari, revealed that most of the animals had discrete preferences for position although three species were found over a wide area (table 2.10; fig. 2.55). *Aemaea, Nerita* and *Drupa* are obviously quite tolerant of drought and high temperatures but appear to seek cracks in the rock where moisture may stay longer than on flat rock. *Turbo* snails generally indicate the low water mark of neap tides and this essentially is the boundary between the aquatic and terrestrial faunas (Purchon and Enoch 1954). The lowest members of the intertidal mollusc community are characterized by *Trochus* and *Pedalion*. These preferences are primarily determined by physiological and behavioural limitations of different

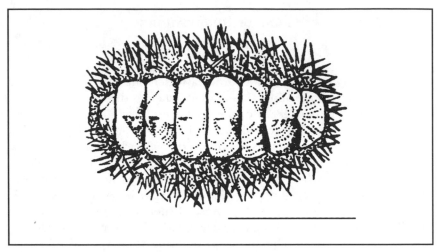

Figure 2.54. Chiton, *Acanthopleura,* a primitive mollusc found on rocky shores. Scale bar indicates 1 cm.

Table 2.10. Changes in species of molluscs present along a transect from the water's edge up a gently-sloping rocky shore at Tanjung Peropa, Kendari.

		|				Metres from the sea					
		0	3	6	9	12	15	18	21	24	27
Pedalion sp.	B	•	-	-	-	-	-	-	-	-	-
Trochus maculatus	G	•	•	-	-	-	-	-	-	-	-
Eucheus atratus	G	-	•	-	-	-	-	-	-	-	-
Cantharus sp.	G	-	-	•	-	-	-	-	-	-	-
Turbo tica-oricus	G	-	-	-	•	-	-	-	-	-	-
Nerita chamaeleon	G	-	-	-	•	-	•	-	-	-	-
Turbo setosus	G	-	-	-	•	•	•	•	-	-	-
Anadara inata	G	-	-	-	-	-	•	-	-	-	-
Gafrarium sp.	B	-	-	-	-	-	•	•	•	-	-
Clypeonous sp.	G	-	-	-	-	-	-	•	•	-	-
Nerita polita	G	-	-	-	-	-	-	-	-	•	-
N. plicata	G	-	-	-	-	-	-	-	-	-	•
Crassostrea cucullata	B	•	•	•	•	•	•	-	-	-	-
Aemaea sp.	G	•	•	•	-	•	•	•	•	•	•
Drupa margeriticola	G	-	-	•	-	-	-	•	-	•	-

B - Bivalve, G - Gastropod.

After L. Clayton pers. comm.

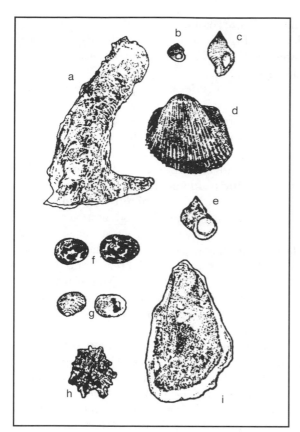

Figure 2.55. Molluscs of rocky shores.

Lower shore:
a - *Pedalion* sp.
b - *Eucheus atratus*
c - *Cantharus* sp.
Middle shore:
d - *Anadara influta*
e - *Turbo setosus*
Upper shore:
f - *Nerita polita*
g - *N. plicata*; wide-ranging
h - *Aemaea* sp.
i - *Crassostrea cucullata*

organisms, and the ability of those organisms to compete for limited resources with other species with similar optimum requirements. This competitiveness is determined by growth rate, life form, reproductive rate, aggressiveness, feeding efficiency, etc. A species is generally successful in one of these factors but this success is often achieved at the expense of success in some other factor. Different species thus perform best at different locations (fundamental niche), but in the face of competition from another species may have to shift from this optimum position to a peripheral position (realized niche). Should the successful competitor be removed through predation or disease, then the first species will be able to move into the optimum position again. Thus, zoning is the necessary outcome of the partitioning of resources between potentially competing species. The factors determining the upper zonal limits tend to be physiological, but the lower zonal limits tend to be ecological and behavioural (Barnes and Hughes 1982).

The width of the zones varies considerably depending on the tidal regime, wave force, slope and aspect. Thus, the greater the spring high tides, the gentler the slope, the greater the degree of shading and the greater the wave force, than the wider the zones. The tidal range around Sulawesi is not particularly great (p. 105) and large waves only occur off the northern part of North Sulawesi during October and November when cyclones are generated in the Pacific (Oldemann and Darmiyati 1977).

The organisms of rocky shores can be conveniently divided into two classes: sedentary and mobile. Sedentary organisms, both animal and plant, are limited by space which, on a rock, may be the rock surface itself (primary space), or the surface of another organism (secondary space). Primary space is a fixed resource but secondary space increases and decreases depending on the state of the organisms present. The space available on a rocky shore is always in state of flux—predation and storm damage can remove or dislodge organisms allowing new individuals to colonize the 'recycled' primary space.

As available space decreases so members of a population are forced to compete with their own and other species for space. The constraints this closeness imposes may be deformed growth, insufficient space for secure attachment, smothering or crushing by neighbours, reduced light or suspended food availability—each of these leading to slower growth and reduced reproductive potential. Neighbours do not have to be touching each other to become a potent factor. If sufficient food is not available then the situation of crushing and jostling may not be reached. In the same way that trees grow slower when planted closer together, so sedentary organisms growing close together will grow relatively slowly and mature at a smaller size than those growing widely spaced. Shape may also change: barnacles are tall and narrow when crowded but squat when not. Alternatively, smaller, weaker individuals in a population may die while the others continue growing at a normal rate (Barnes and Hughes 1982).

It should be stressed that the above description is meant to illustrate the existence of zonation on a rocky shore, and where more, fewer or different species are present, so the composition of the zones will change. The distribution of a species in one location is not necessarily a reflection of the sole area within which it can live. It may be, for example, that another species is better able to exploit a particular resource and can oust the first species from using such areas. In addition, distributions may depend on location—thus zones on a steep northern shore may be different from the zones on a more gently sloping southern shore.

Resources that are potentially limiting for mobile animals include food and shelter such as crevices and holes. Experiments investigating the colonization of fibreglass tubes set in rocky shores showed that the depth of the hole was important since it controlled important factors such as desiccation, grazing and competition (Menge et al. 1983). When snails are kept in cages attached to rocks, and their density increased experimentally, the growth rate of juveniles decreases although their rate of survival remains constant. At high densities the adults' body weight declines as does their survivorship. Thus, under high-density conditions, the juveniles survive at the expense of the adults because their smaller individual biomass can be maintained on less food. As in the case of coconut crabs (p. 177), young snails live as part of the plankton and there is no way the adults can regulate the number of young arriving on a rock surface.

FUNCTIONS AND PRODUCTS OF MANGROVE FOREST

If development of the coastal zone is to be both productive and sustainable, then the direct and indirect benefits provided by natural ecosystems need to be recognized. By far the most useful and yet the most misunderstood coastal ecosystem is the mangrove forest.

The transport of leaves and other debris out of mangrove forests provides an important source of nutrients for the inshore and neigbouring estuarine ecosystems (p. 197). This is of far more than just academic interest because this flow of nutrients is the cornerstone of a wide range of 'goods and services' (to use the language of economists) provided by the forests for both coastal inhabitants and commercial operations (table 2.11). In addition to the benefits derived from controlled exploitation, mangroves also efficiently control erosion of the coastline (Sato 1984), and act as nursery grounds for many commercially important species of fish, shrimp and other organisms.

Palms are among the most versatile of tropical plants and the nipa palm *Nypa fruticans* found at the inland margin of mangrove forest is no exception. The leaves are used for making roof shingles, umbrellas, hats, baskets, mats and bags. The most likely commercial prospect, however, is in its sugar production. This is tapped from the cut stem of the inflorescence in the same way as for the sugar palm *Arenga pinnata* and there are already factories producing alcohol and vinegar from the sugar in Sarawak and Papua New Guinea. Alcohol can be mixed with petrol in a 1:4 ratio and used directly in standard petrol engines. The nipa sap has a sugar content of 14%-17% which compares well with sugar cane (18%-20%). However, it is far superior in alcohol production per hectare than almost any other plant. Nipa estates in Malaysia have produced 15,000 litres of 95% alcohol per hectare per year, compared with about 7,000 litres for sugar cane, 8,000 for cassava, and 5,000 for coconut. It also has other advantages over sugar cane; it needs no crushing mills, has no crushing waste, does not compete for prime agricultural land, and requires no replanting or seasonal labour inputs (Mercer and Hamilton 1984).

Table 2.11. Uses and products of mangrove trees (above) and animals (below).

Uses and Products of Mangrove Trees	
Fuel	Firewood for cooking, heating; firewood for smoking fish; firewood for smoking sheet rubber; firewood for baking bricks; charcoal; alcohol.
Construction	Timber for scaffolds; timber for heavy construction (e.g., bridges); railway ties; mining pit props; deck pilings; beams and poles for buildings; flooring, panelling; glues; boat-building materials; fences; water pipes; chipboard.
Agriculture	Fodder; green manure.
Paper production	Paper of various kinds.
Foods, drugs, and beverages	Sugar, alcohol; cooking oil; vinegar; tea substitutes; fermented drinks; dessert topping; condiments from bark; sweetmeats from propagules; vegetables from propagules, fruits, and leaves; cigarette wrappers; medicines from bark, leaves, and fruits.
Household items	Furniture; glue; hairdressing oil; tool handles; rice mortar; toys; matchsticks; incense.
Textile and leather production	Synthetic fibres; dye for cloth; tannins for leather preservation.
Other	Packing boxes.

Uses and Products of Animals Associated with Mangrove Forests	
Source	**Product**
Finfish	Food; fertilizer.
Crustaceans (prawns, shrimps, crabs)	Food.
Molluscs (oysters, mussels, cockles)	Food.
Bees	Honey; wax.
Birds	Food; feathers; recreation (watching), hunting.
Mammals	Food; fur; recreation.
Reptiles	Skins; food; recreation.

From Mercer and Hamilton 1984

Tambak Fishponds

Given the concentration of fertile soils in coastal areas and the inherent high productivity of estuaries and inshore waters, the coasts clearly offer great promises of fulfilling development needs if, appropriate management approaches and techniques are applied. An excellent example is the development of tambak or brackish-water fishponds. Tambaks are constructed in low-lying areas on saline soils which are unsuitable for other forms of agricultural development. The culture of coastal animals probably arose in ponds that were modified salt pans and was probably operating along the north Java coast as early as the 14th century (Schuster 1950). Tambaks may have been introduced to Sulawesi one or two hundred years ago but it was only at the start of this century that they were developed there in earnest, particularly in South Sulawesi. This province now has the most extensive tambak areas in Indonesia outside Java and new areas are continually being opened.

Traditionally the major animal harvested from tambaks has been the milkfish *Chanos chanos*. This fish, which can reach 3 kg, is reared in the ponds from fry which are caught in coastal waters primarily along the west coast of the southwest peninsula (fig. 2.56). Adults are rarely caught by commercial fishermen in Indonesia except in the area around the Spermonde Archipelago and this is therefore assumed to be a major spawning area. The fry are widespread, however, tolerant of freshwater and are even found in Lake Tempe 60 km inland. About 2,500 fry/ha are required to harvest 500-800 fish which, after 8-10 months, are of marketable size (300-800 g). Under good management, the ponds nearest the sea can produce two fish harvests each year.

Milkfish are herbivorous, feeding on diatoms, green algae, blue-green algae and vegetable detritus, and production of fish is closely linked to the productivity of algae in the surface mud and water of the pond. Productivity of the blue-green algae can be stimulated by the draining and drying out of a pond although on certain soils the opposite occurs and productivity falls. Tambaks in Suppa, Pinrang, were found to support 34 species of plankton: 20 Bacillariophyceae, 4 blue-green algae, 1 green alga, 1 Dinophyceae, 6 crustaceans and 2 monogonont rotifers (Omar 1985).

Tambaks are generally constructed from behind a narrow fringe of mangrove to the limit of tidal influence. The further inland the tambaks are built the more excavation in needed, for the base of the pond should ideally be about 0.4 m below the mean level of the spring high tides. Bunds, or dykes, are built up between the ponds and a small island with a few mangrove trees is often left in the middle. Mangrove trees are sometimes planted along the bunds and these provide shade, some protection from erosion, introduce organic matter into the pond in the form of leaves, twigs, etc., and their wood can be used as fuel (Schuster 1950).

The control of water into and out of tambaks is the key to successful

Figure 2.56. Areas used for catching milkfish *Chanos chanos* fry (dashed lines).

After Salm and Halim 1984

production. The water needs to be changed regularly so that wastes can removed and new sources of food swept in. The water is controlled by sluice gates in the wall of each pond. The quantity, quality and timing of water entering is largely determined by the tides although the gates are used to allow in river water to maintain a depth of 0.3-0.6 m. The base of the pond needs to be just above the mean low tide level so that the pond can be completely drained for repair and rehabilitation. The flushes of freshwater, seawater and the effects of evaporation, cause a wide variation of salinity in the ponds, from 0 to 250 ppt. Ponds with different average salinities develop different flora of fungi, algae and diatoms. River sediments and rain introduce significant quantities of nutrients.

When the sluice gates are opened at high tide, a host of coastal organisms are swept into the ponds. Some of these are young fry or larvae of species that can later be harvested. Among these are mullet *Mugil,* snapper *Lutjanus,* ponyfish *Leiognathus,* eels *Monopterus albus,* mangrove crabs *Scylla serrata* and prawns *Penaeus.* Indeed, up to forty species of fish are found in tambaks, although some are typically freshwater species and are found primarily in the inshore ponds. The major secondary fish species which is introduced to tambak is the tilapia[22] *Oreochromis mossambicus,* although it is considerably less profitable and harder to manage than milkfish. In addition to these animals are many economically unimportant crabs, lizards, birds and insects which also exploit the tambak habitat. Among the most conspicuous of these are long-legged shorebirds many of which are migratory (p. 144), but some, such as the black-winged stilt *Himantopus himantopus* (fig. 2.57) are resident. The large birds may be viewed as pests, preying on young fish and prawns, but fishermen are generally very tolerant of their presence. An Interwader-EoS team surveyed a fishpond area north of Ujung Pandang at Lantebeong that lay behind a very narrow (maximum 25 m) belt of mangrove in March/April 1986 (table 2.12). At that time suitable coastal bird habitat (muddy fields) were provided by drained fishponds, a marsh used by grazing buffalo, a mudflat at the mouth of a small river and some marshy pools. The numbers recorded serve to demonstrate the great variation between days in the presence or abundance of the birds. This needs to be considered by those intending to make bird surveys in coastal regions.

Under traditional methods of aquaculture the milkfish is the major harvest from the ponds. In Pangkep County, for example, milkfish production averaged 800 kg/ha/yr ranging from 240-1,600 kg/ha/yr, and prawn production was only 96 kg/ha yr on average rising up to 360 kg/ha/yr (Nessa 1985). The milkfish are a largely domestic product but the prawns can join a lucrative export market. Much of the export is of prawns caught in the coastal regions. Such fishing grounds in Sulawesi are found from Mamuju around the southern coasts to the east coast of Southeast Sulawesi. None of the fishing in this area is believed to be overintensive (Salm and Halim 1984). For this reason, increased prawn production is a major thrust in the Fourth Five-Year Plan such that there are plans to open 100,000 ha of new tambak. The major prawn in Sulawesi tambaks is the tiger prawn *Penaeus monodon.* This species needs only about 3-5 weeks after being introduced into the tambak to grown from the post-larval stage, and under intensive management a yield of 2,000 kg/ha/yr is possible (Villaluz et al. 1977). It used to be believed that prawns would only spawn in the deep sea (Schuster 1950), but the technology now exists for prawn hatcheries associated with tambaks to be developed.

Figure 2.57. Black-winged stilt *Himantopus himantopus,* a resident wader seen in tambak areas.

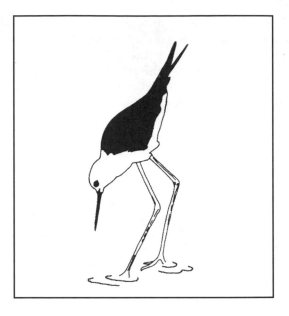

A detailed study of tambak was conducted in Pangkep County to determine the factors influencing tambak production, and to look at land use patterns in the inshore ponds in relation to the quality of irrigation water, and the impact of the ponds on the estuarine waters. The water in these ponds was changed about three times each month with half the volume being changed each time. Prawn production could be increased by acclimatization of the larvae before their introduction into the ponds, increased frequency of water change and increased pond depth, whereas the use of pesticides and increased pond area resulted in reduced yields. Water flowing out of rice fields behind an area of tambak is suitable for directing into the ponds to further support high fish production (Nessa 1985).

There are many incentives for converting mangrove to tambak but new projects are frequently proposed without regard to soil suitability or the ecological consequences of the loss of mangroves. Mangroves are the habitat of young milkfish and post-larval prawns and the loss of that ecosystem necessitates ever increasing intensity of management to maintain productivity. While responding to one part of the Fourth Five-Year Plan, most new projects fly in the face of the objectives of the same plan, namely that utilization and management of fisheries resources will be better controlled. With few exceptions, the impression is given that mangroves are an expendable resource. This is not the path to sustainable development.

Table 2.12. Numbers of water birds seen in and around tambaks near Lantebeong over three successive visits in 1986.

	7-10 March	17 March	4-5 April
EGRETS, BITTERNS and HERONS Ardeidae			
Little egret* *Egretta garzetta*	75	25	43
Great egret* *E. alba*	30	0	8
Intermediate egret* *E. intermedia*	0	0	2
Egret *Egretta* sp.	10	0	0
Purple heron* *Ardae purpurea*	8	4	7
Cinnamon bittern* *Ixobrychus cinnamomeus*	2+	0	0
Yellow bittern *I. sinensis*	0	0	2+
Javan pond heron* *Ardeola speciosa*	20	*	*
STORKS Ciconiidae			
Milky stork* *Ibis cinereus*	26	0	9
DUCKS Anatidae			
Garganey *Anas querquedula*	0	130	486
Grey teal* *A. gibberifrons*	50	*	*
STILTS Recurvirostridae			
Black-winged stilt* *Himantopus himantopus*	8	9	0
PLOVERS Charadriidae			
Little ringed plover *Charadrius dubius*	2	2	3
Mongolian plover *C. mongolus*	40	15	5
Lesser golden plover *Pluvialis dominica*	50+	25+	43
Grey plover *P. squatorola*	25	25	20
STINTS, SANDPIPERS, GODWITS, etc. Scolopacidae			
Red-necked stint *Calidris ruficollis*	270	40+	175
Long-toed stint *C. subminuta*	20+	50+	25+
Curlew sandpiper *C. ferruginea*	80	30	3
Broad-billed sandpiper *Limicola falcinellus*	260	0(?)	73
Snipe *Gallinago* sp.	50+	50+	50+
Black-tailed godwit *Limosa limosa*	750	0	0
Bar-tailed godwit *L. lapponica*	0	68	0
Whimbrel *Numenius phaeopus*	54	7+	9
Eastern curlew *N. madagascariensis*	30	32	33
Redshank *Tringa totanus*	120	100	145
Marsh sandpiper *T. stagnatilis*	30+	34	18
Greenshank *T. nebularia*	60+	20+	24+
Wood sandpiper *T. glareola*	100	40+	24+
Terek sandpiper *Xenus cinereus*	3	15	1
Common sandpiper *Actitis hypoleucos*	-	30+	34
Grey-tailed tattler *Heteroscelus brevipes*	0	0	11
Ruddy turnstone *Arenaria interpres*	0	10	5
TERNS Laridae			
Little tern *Sterna albifrons*	0	62	12
Gull-billed tern *Gelochelidon nilotica*	0	3	3
White-winged black tern *Chlidonias leucopterus*	*	*	35
Whiskered tern* *C. hybrida*	0	0	11

* = resident birds, the others are migrants from the north.

After Uttley 1986

One element of the intensification to maintain productivity is the spate of proposals to develop small-scale prawn hatcheries in South Sulawesi which are likely to have very significant impacts in the next few years. These hatcheries provide a reliable stock of prawn larvae and the tambak owners stand to make a considerable financial gain.

The hatcheries are likely to encourage the further conversion of mangrove to tambak which will lead to the loss of the beneficial roles of mangrove (p. 187), not least the provision of nursery or juvenile grounds for the very shrimps that grow up to be 'berried' females (those carrying eggs) which are caught and maintained in the hatcheries until the eggs hatch and larvae are produced. There are official policies against the uncontrolled extension of fishpond areas but this appears to be largely disregarded at the local level. The construction of the hatcheries would cause no concern if there were recognized, unambiguous policies for the sustainable management of mangrove areas and if protection regulations were fully enforced. Until these two conditions are met, however, it would probably be better if such development were postponed. This will be hard for departments to accept and so a concerted effort is required to establish the rationale for mangrove management to protect their many uses for sustainable development. This is long overdue, but no interdisciplinary study to establish such a rationale has yet been funded (Burbridge and Maragos 1985).

Mangrove Forest Management

The problems of rational management of mangrove forest are present wherever there are mangroves (van Beers 1962; Hanson and Koesoebiono 1977; Soegiarto and Polunin 1980; Koesoebiono et al. 1982; Soegiarto 1985; Darsidi and Liang 1986), and in Sulawesi they are probably most severe at Malangke. The Malangke mangrove forests are at about 12,000 ha most extensive in Sulawesi, and they are under great pressures from loggers (legal and illegal), firewood collectors, tambak developers, and establishment of settlements for local and transmigrant fishermen. As is so often the case, the forests are being removed before any detailed studies have been conducted of what is there, or of just how much disturbance a mangrove forest can sustain before its characteristic processes cease to function (Mustafa et al. 1981).

The management of mangrove forest for sustained yield forestry is possible and viable and a 30-year rotation for *Rhizophora* and replacement of *Avicennia* with *Rhizophora* for subsequent harvests have been shown to be profitable in Malaysia (Chan et al. 1982; Ong et al. 1985; Soegiarto 1985). Unless great care is taken, however, the result of large-scale disturbance is not the regeneration of productive forest but rather the appearance of a degraded type of vegetation. This is typically dominated by nipa palms, *Avicennia,* or tall mangrove ferns *Acrostichum aureum* and sea holly *Acanthus ilicifolius.*

Mangrove forest are logged primarily to supply export market for logs sent to Java and Taiwan, where they are chipped and used in the manufacture of rayon fibre. In Sulawesi itself, many mangrove trees are pulped to supply the Gowa paper mill where paper is made from 80% bamboo, 20% mangrove and some *Eucalyptus* fibre (Soegiarto 1985).

An inter-departmental debate has been proceeding for a decade concerning mangrove 'green belts' along the seaward shore but it has still not really been resolved despite a joint ministerial decree in 1984 that instituted the requirement to retain a corridor, or belt, of mangrove 50-200 m wide parallel to the shore. These figures were decided upon because they were judged to be reasonable, but they were not based on scientific criteria (Achmad 1986; Anwar et al. 1986; Darsidi and Liang 1986). They did not include, for example, the complication that forests in estuaries probably play a more important role in producing detritus than coastal forest (Woodroffe 1985). The Department of Forestry seeks to retain a narrow 50 m strip of protection forest behind which controlled felling would occur, but the Directorate-General of Fisheries, recognizing the irreplaceable role of mangroves, is seeking a much wider protected zone of 500 m.[23] It is logical to favour the Fisheries view because the value of mangrove timber is far lower than of mangrove-dependent fisheries.

Green belt arguments from just a forestry or fisheries perspective inevitably neglect the multiple uses of mangrove. Green belts do not necessarily guarantee anything since all parts of the hydrological system of a mangrove forest are connected. Changes in water flow and vegetation behind the green belt will surely dramatically affect the green belt itself. The conclusion, therefore is that mangrove areas must be managed in their entirety, not just as pieces (Woodroffe 1985).

Large-scale and uncontrolled disturbance of mangroves can lead to coastal erosion because the shoreline is no longer protected by trees. The shore may be reduced to a narrow sandy beach or to inhospitable salt pans. The coastal population centres are then more susceptible to flooding. The degradation of mangroves associated with agricultural and fisheries development can to some extent be controlled by the building of earth bunds between a belt of mangrove and the more intensely utilized land. In many areas the mangrove forest needs to be rehabilitated and selective replanting can be conducted. Young seedlings can be protected from wave action by breakwaters made of tyres. Under this system it is essential that some freshwater be allowed to flow through the mangrove from the fields or ponds so that the soil water does not become too saline (Chan 1985). Large-scale mortality of mangrove trees can almost always be traced to some human intervention such as inadvertent diversion of freshwater, blockage or reduction of tidal circulation, or increased sediment load of rivers caused by inappropriate land management (Zucca 1982; Burbridge and Maragos 1985; Jimenez et al. 1985).

1a.   Plume of ash from Mt Lokon during a new period of activity, November 1986.

1b.   Dormant Mt Klabat with the twin peaks of Mt Dua Saudara beyond.

2a.   Endemic toad *Bufo celebensis* in defensive posture.

2b.   Colony of endemic finch-billed starlings nesting in a dead tree.

3.   Endemic stripe-faced or Wallace's fruit bat.

4. Caerulean paradise-flycatcher, endemic to Sangihe Island, but now probably extinct.

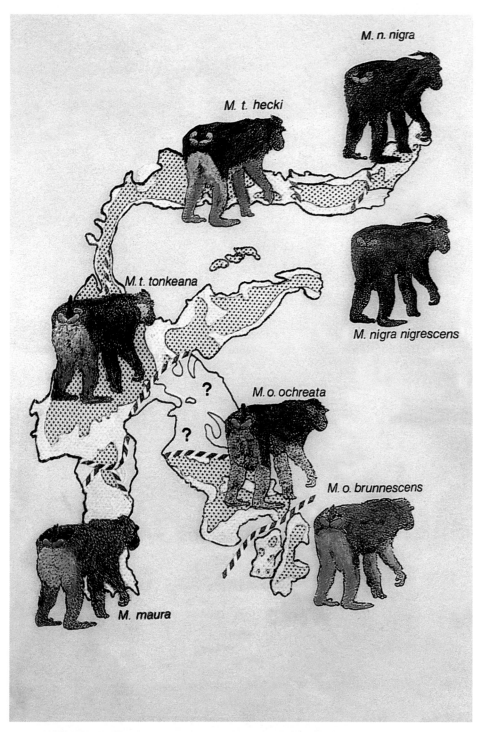

5. Distribution of the four species and seven species of macaques and approximate
    areas of remaining forest. Based on Fooden (1969).

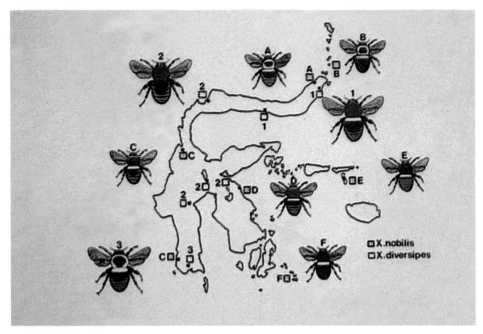

6a. Distribution of two species and nine subspecies of carpenter bees. Based on van der Vecht (1953).

6b. Traditonal Toraja houses and megaliths.

7. Megalith shaped like a human, Tamadue, near Lore Lindu National Park.

8.   Prehistoric hand stencils, Leang Petta Kere, Maros.

9a. Mangrove forest dominated by *Rhizophora mucronata*.

9b. Crabs such as this grapsid *Neosarmatrium meinerti* are important in the cycling of nutrients in mangrove forests.

10a. Pair of maleo digging a nest hole. Tangkoko-Batuangus Reserve.

10b. Maleo chick emerging from the ground.

11. Animals and plants on a rocky shore are distributed in distinct zones.

12a. Mangrove trees planted along tambak bunds provide a source of nutrients for the pond, and firewood or timber for the pond owner.

12b. Mugil fish from tambaks on sale in Rantepao market.

13a. *Enhalus acoroides* seagrass.

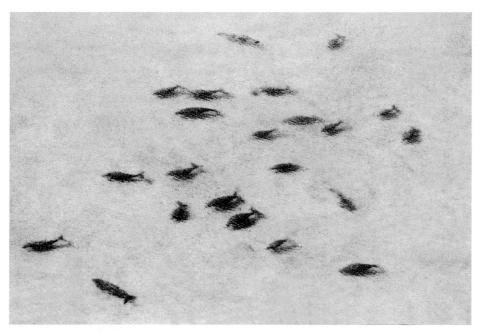

13b. Herds of dugong are now a rare sight around Sulawesi.

14a. Shrimpfish or razorfish *Aeoliscus* sp. and black coral off Bunaken Island.

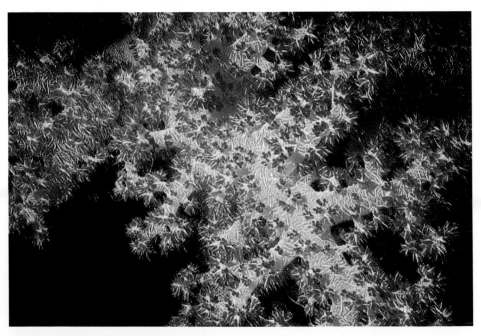

14b. Soft coral at night with its polyps extended.

15. Coral community with green and white sea squirts (above), pink coralline algae, a red sponge and green algae.

16. Bottlebrush worms *Spirobranchus giganteus*; each worm has two sets of tiered tentacles.

17a. Clear waters of the forest-fringed Tumpah River, Dumoga-Bone National Park.

17b. Flash torrent in a river arising in the deforested Palu hills.

18a. Lake Moat.

18b. Shallow water vegetation in Aopa Swamp, dominated by *Nymphoides indica* (Gent.).

19a. Large crocodiles *Crocodylus porosus* are now rare around Sulawesi.

19b. Wandering tree ducks *Dendrocygna arcuata* taking off from a pond near Dumoga-Bone National Park.

20. The decline in the fisheries production of Laka Limbota may be associated with the silting up of the lake.

21a. Fyke nets used for catching eels at the entrance to the River Poso, Tentena.

21b. EoS team collecting limnological data near Malino for the purposes of this book.

22a. Community of mosses and liverworts attached to a buttress of a forest tree.

22b. The primitive, wingless grasshopper *Karnydia gracilipes* feeds on fern leaves.

23. Leaf of the ubiquitous fan palm *Livistona rotundifolia*.

24. *Pigafetta filaris* is a fast-growing palm which produces numerous small fruit, and is associated with disturbed habitats.

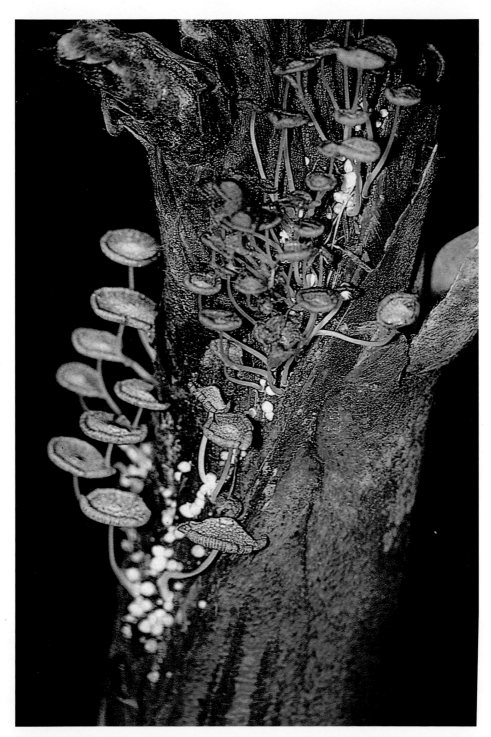

25. Fungi are major agents of decomposition in lowland forests.

26. Lowland forest with a strangling fig (background), *Licuala celebensis* palm (left foreground), young rattan palm (right foreground), and a fan palm *Livistona rotundifolia* (left background).

27. When a tree falls, light can penetrate to the forest floor, the floor growth cycle commences, and numerous temporary new habitats are formed.

28.   Morning in Lore Lindu National Park.

29a. Sweat bee taking pollen from the flower spike of the introduced and rapidly-spreading small tree *Piper aduncum*.

29b. Sweat bee returning to its tree-hole nest carrying pollen on its hind legs.

30a. Endemic rat *Maxomys hellwandii* photographed using an infrared beam.

30b. The enigmatic babirusa.

31. The pittas of Sulawesi, secretive members of the forest floor community. Above: Blue-winged pitta *Pitta moluccensis*; middle: Red-breasted pitta *P. ervthrogaster*; below: Hooded pitta *P. sordida*.

32a. Diurnal, leaf-eating bear cuscus.

32b. Nocturnal, fruit-eating dwarf cuscus.

33. Rhacophorid tree frog; there are probably as many species of frogs awaiting discovery in Sulawesi as are already known.

34a. Mountain anoa or dwarf buffalo are temperamental beasts, and stories of them charging humans in the forests are not uncommon.

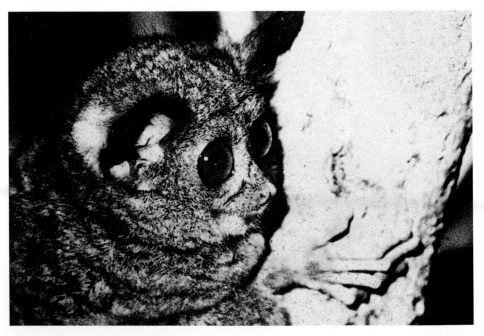

34b. Tarsiers live in small family groups and feed at night on insects, and even frogs and lizards.

35a. Crested macaques are popular village pets.

35b. Pair of endemic red-knobbed hornbills.

36. **Lowland forest and montane forest guilds of pigeons from mainland Sulawesi. Montane: a** – grey-headed Imperial pigeon *Ducula radiata*; **b** – red-eared fruit-dove *Ptilonopus fischeri*; **c** – superb fruit-dove *P. superbus*; **d** – black-naped fruit-dove *P. melanospila*. **Lowland: e** – grey-faced green pigeon *Treron griseicauda*; **f** – pink-necked green pigeon *T. vernans*; **g** – buff-bellied fruit-dove *P. subgularis*; **h** – green imperial pigeon *D. aenea*; **i** – white-bellied Imperial pigeon *D. forsteni*; **j** – black-naped fruit-dove *P. melanospila*.

37. **Open area and lowland forest parrots of mainland Sulawesi. Open area: a –** great-billed parrot *Tanygnathus megalorhynchus*; **b** – blue-backed parrot *T.sumatranus*; **c** – red-spot racquet-tail *Prioniturus flavicans*; **d** – golden-manted racquet-tail *P. platurus* (male shown, female is largely green); **e** – yellow-and-green lorikeet *Trichoglossus flavovividis*. **Lowland forest: f** – ornate lorikeet *Tr. ornatus*; **g** – little hanging-parrot *Loriculus exilis*, **h** – Sulawesi hanging-parrot *L. stigmatus*; **i** – lesser sulphur-crested cockatoo *Cacatua sulphurea*.

38. Fire kills most young trees and encourages the growth of alang-alang grass, as here south of Lambuya, Kendari.

39a. View over the northern part of Aopa Swamp.

39b. **Kingfishers that might be expected to be seen in riverine forest: a** – Black-filled kingfisher *Pelargopsis melanorhyncha*; **b** – Blue-eared kingfisher *Alcedo meninting*; **c** – Ruddy kingfisher *Halcyon coromanda*; **d** – Sulawesi forest kingfisher *Ceyx fallax*; **e** – Common kingfisher *Alcedo atthis*.

40. Colourful, peeling trunk of tree *Eucalyptus deglupta* in Lore Lindu National Park.

41. **Rails and crakes that might be encountered in wet, swampy habitats, both natural and disturbed: a** – White-breasted waterhen *Amauronis phoenicurus*; **b** – Platen's rail *Aramidopsis plateni*; **c** – Slaty-legged crake *Rallina eurizonoides*; **d** – Ruddy-breasted crake *Porzana fusca*; **e** – Slaty-breasted rail *Rallina striatus*; **f** – Rosenberg's rail *Glymnocrex rosenbergii*; **g** – Baillon's crake *Porzana pusilla*; **h** – White-browed crake *Porzana cinerea*; **i** – Band-bellied crake *Porzana paykulii*; **j** – Watercock *Gallicrex cinerea*; **k** – Purple swamphen *Porphyrio porphyrio*.

42a. Aerial view of Tonasa cement factory at the edge of the Maros karst hills.

42b. Aerial view of dry, deforested hills of the Palu Valley.

43a. Introduced cactus plants dominate the vegetation in parts of the Palu Valley such as the new campus of Taduloko University.

43b. Silt from the Palu Valley discharging into Palu Bay is suffocating coral and reducing the production of associated fisheries.

44. One of the last sandalwood trees in Paboya Reserve. This area is subject to fires set by farmers who graze their cattle in the reserve.

45. Mossy lower montane forest at 2,300 m on Mt Nokilalaki.

46a. Upper montane forest dominated by *Vaccinium* at 3,050 m on Mt Rantemario.

46b. View from the summit (3,450 m) of Mt Rantemario; fires set by hunters encourage the growth of young grass which attracts anoa.

47.   Daisy *Keysseria* sp. at 3,425 m on Mt Rantemario.

48. *Usnea* is a ubiquitous lichen which grows in somewhat exposed positions on mountains.

49a. Hairy-leaved silverweed *Potentilla parvus* at 3,400 m on Mt Rantemario.

49b. Ferns and lichens are the only plants on the 1983 lava flow on Mt Api, Siau Island.

50a. Sulawesi civet, the only carnivorous mammal native to Sulawesi, photographed in Lore Lindu National Park using an infrared 'trip'.

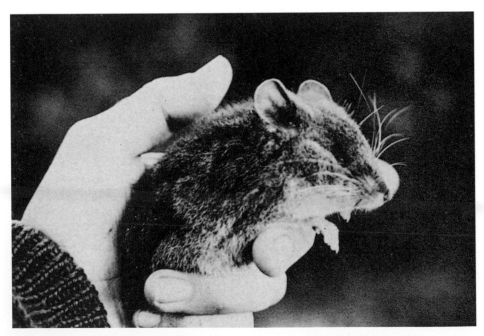

50b. Summit shrew-rat *Bunomys penitus* caught at night at 3,250 m on Mt Rantemario.

51a. Small-bodied shrew-rats from Mt Nokilalaki. Top: *Tateomys rhinogradoides*, centre: *T. macroercus*, bottom: *Melasmothrix naso*.

51b. Pygmy tree-mouse *Haeromys minahassae* eats fruit and is endemic to Sulawesi montane forests.

52a. New species of cave-dwelling atyid prawn from Salukan Kalang cave, Maros.

52b. Nursery of young long-fingered bats in Konangan cave near Kotamobagu. Note that one is being carried by its parent.

53. Greenish naked-backed bat caught in a cave mouth in Maros.

53b. Forest garden of a member of the Laoceh, a group of swidden farmers near Mt Ogoamas.

54a. Rice terrace in Tana Toraja.

54b. Cattle egrets in the Dumoga Valley are frequently seen picking up insects
disturbed by ploughs.

55a. Coconut plantation, Minahasa.

55b. Clover plantation, Toli-Toli.

56. Most coconuts are grown for their oil but failure to replace ageing trees has resulted in a fall in production.

57a. More biocides are used worldwide on cotton than on any other crops. Here, fields are being sprayed prior to germination of the seeds.

57b. Ripe cotton bolls.

58. Members of the Southeast Sulawesi Population and Environment Bureau inspecting seedlings of *Pericopsis mooniana*.

59. Spray of pigeon orchids *Dendrobium crumenatum*.

60a. The large and noisy tokay *Gecko gecko* is a secretive inhabitant of houses.

60b. Both pharoah ants *Monomorium pharaonsis* and the shy *Gehyra mutilata* have a taste for sugar.

61a. Common broad-tailed house gecko *Hemidactylus platurus*.

61b. Common round-tailed house gecko *H. frenatus*.

62.   Long-tailed garden lizard *Calotes cristatellus*.

63a. Freshwater swamp forest near Tinading, Toli-Toli, has recently been converted into rice fields by transmigrants.

63b. Self-sufficient transmigration area of Landono, Kendari, settled primarily by Balinese.

64a. Wawantobi dam near Unaaba, Kendari, being built with the aim of increasing rice production in an area of dry-land Tolaki farmers.

64b. The quality of life awaiting these children at Ameroro, Kendari, depends on commitments being made to sustainable development now.

Chapter Three

Estuaries, Seagrass Meadows and Coral Reefs

ESTUARIES

Estuaries in Sulawesi (fig. 3.1) are not as large or as many in Java, Sumatra and Irian Jaya but they are nevertheless important for fisheries and communication. They are therefore often foci of habitation and are usually the final sampling station for water quality studies of rivers. Estuaries would all, in the absence of man's activities, be fringed by mangroves.

Water Characteristics

Estuaries vary in their physical, chemical and biological properties, being affected by short- and long-term changes in river flow and tides, changing seasons, and occasional extreme weather conditions. These have considerable influence on determining salinity, temperature, nutrient levels and sediment load.

The rate of change from the freshwater riverine environment to a marine environment depends on the interaction of the volume of freshwater flowing to the sea, on friction such as wind and currents, and on tidal mixing. In a highly stratified estuary (fig. 3.2), freshwater flows over a layer of saline water with little mixing. Friction between these two layers, however, generally occurs causing varying degrees of mixing. Highly stratified estuaries are found where river flow is large compared with tidal flow. Where tidal and river flow are more equal, moderate stratification is found, and it is here that salinity increases gradually with depth. Vigorous tidal mixing can lead to a vertically homogeneous estuary in which salinity down a column of water is the same but varies with time according to the tidal state. Such stratification is found only in small estuaries. These patterns can also vary through the year depending on rainfall, and wind strength, and will also be complicated by any sand bars at the entrance to the estuary (Uktolseya 1977).

Figure 3.1. The major estuaries of Sulawesi (circled).

After Salm and Halim 1984

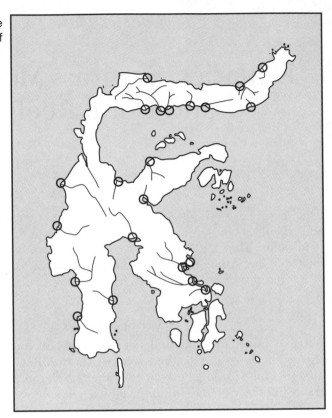

Fauna

Estuarine organisms have to contend with large fluctuations in salinity. This presents physiological challenges because adjustments must be made in the ionic composition of body fluids to ensure that the preferred ionic concentration is maintained in the body tissues. It is the responses and adaptations to changing salinity gradients of an estuary that determines the pattern or zonation of species. Since estuaries represent the transition between freshwater and marine environments, one would expect a change from freshwater to marine organisms along the length of an estuary, and this is indeed the case. Between about 5-8 ppt salinity there is a distinct dearth of species (fig. 3.3) and this may represent the threshold above which ion regulation is not necessary for marine animals but is necessary for freshwater animals (Khlebovich 1968). Not only presence of animals is affected by water salinity, but also the size to which a particular species will grow; for example, the size of certain bivalve molluscs decreases with decreasing salinity (Barnes and Hughes 1982).

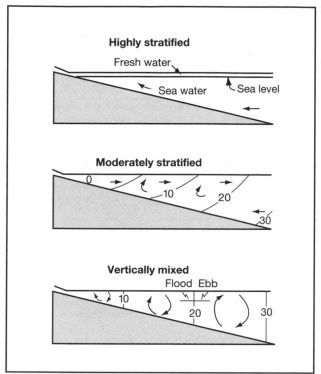

Figure 3.2. Three classes of stratification in estuaries.

After Levington 1982

The flow of freshwater out to sea might be expected to sweep organisms away from the estuary, but some members of the zooplankton are adapted to sink to the bottom of the estuary where water movement is least during the ebb tide, and to rise again when the water rises. Phytoplankton have no such adaptation and so there is no distinct estuarine phytoplankton community (Levington 1982). In the freshwater reaches of the Malangke estuaries in the north of Bone Bay, the phytoplankton is dominated by the blue-green alga *Anabaena* and the diatom *Navicula*. In the more saline locations, *Navicula* dominated the samples and *Anabaena* is more or less absent (Anon. 1980a). There appear to be no other distinct changes.

Some of the larvae of commercially-important prawns and fish, both migratory and resident, may live in estuaries for one or more years because of the high inputs of nutrients (Hadikoesworo 1977; Polunin 1983). The adults may have spawned offshore or among coastal vegetation. The most abundant fish of estuaries are those that have a broad niche, exploiting a wide range of foods and habitats. A total of 36 fish species have been found in estuaries of the Malangke region (table 3.1) some of which have

Table 3.1. Fish recorded in the Malangke estuary.

NEEDLE FISHES Belonidae *Strongylura urvilli*	Long thin fish, schooling surface dwellers. Fast swimmers, often jumping out of water. Prey on smaller fish.
HALFBEAKS, Hemirhamphidae *Zenarrchopterus buffonis*	Elongate fish with long lower jaw. Schooling often in brackish water and feeding on plankton, algae, anchovies, etc.
MULLETS Mugilidae *Liza dussumeiri* *L. vaigiensis* *Mugil cephalus*	Schooling fish of shallow coasts. Feed primarily on diatoms and other benthic algae. Commercially valuable.
WHITINGS Sillaginidae *Silago sihama*	Fish of sandy shores and estuaries, dig in sand with long snouts for worms and crustaceans.
BARRAMUNDI Latidae *Ambasis macracanthus*	Small to large carnivorous silvery fish. Some commercially valuable.
CARDINAL FISHES Apogonidae *Apogon hyalosoma*	Small nocturnal carnivores of inshore areas.
GROUPERS Serranidae *Epinephelus tauvina* *E. australis*	Large solitary carnivorous fish usually found around rocky headlands and coral reefs. Feed on fishes and benthic crustaceans.
SNAPPERS Lutjanidae *Lutjanus ehrenbergi* *L. argentimaculatus*	Medium-sized nocturnal carnivorous fish usually found near coral reefs. Feed on fish, crustaceans, molluscs, etc.
SEA PERCH Theraponidae *Eutherapon theraps* *Mesopristes orgenteus*	Small to medium carnivorous fish of shallow, brackish water.
SILVERBELLIES Gerridae *Gerres nacracanthus*	Silvery fish of shallow, sandy areas often in large schools. Feed on plankton.
EMPEROR BREAM Lethrinidae *Lethrinus ornatus*	Predators of shallow coastal areas. Commercially important.
GRUNTS Haemulidae *Podadasys maculatus*	Nocturnal predators of benthic invertebrates in shallow sandy areas.
JACKS/SCADS Carangidae *Selaroides leptolepis* *Chorinemus tala*	Solitary or school predators of fish, benthic fauna and plankton. Found in many coastal habitats.
PONYFISHES Leiognathidae *Leiognathus equula* *L. splendens* *Secutor ruconius*	Schooling carnivores found in sandy or muddy coastal waters. Feed on small benthic animals. Local commercial value.
GOBIES Gobiidae *Glossogobius celebicus* *Etenogobius suluensis* *Periopthalmus vulgaris* *Ophiocara aporos* *Eleotris macrolepis*	Small benthic carnivores or omnivores in a wide range of habitats, living in groups or as solitary animals. The mudskippers are able to walk on mud out of water.
ANEMONE FISHES Pomacentridae *Abudefduf melas*	Usually found in rocky or coral areas. Small omnivores many of which live symbiotically with sea anemones.
WRASSES Labridae *Halichoeres scapularis*	Solitary carnivorous fish, normally associated with reefs.

Table 3.1. (Continued.)

ARCHER FISH Toxotidae *Toxotes jaculator*	Group-living estuarine and mangrove fish. Able to 'shoot' insect prey off leaves with jets of water.
SCATS Scatophagidae *Scatophagus argus*	Inhabit estuaries and mangrove swamps feeding on algae or bottom detritus.
RABBIT FISHES Siganidae *Siganus vermiculatus*	Browsing herbivorous fish generally in schools in reef habitats or muddy brackish water.
PUFFER FISH Tetraodontidae *Arothron nogropunctatus* *A. aerostaticus* *A. reticularis* *Chelondon patoca*	Slow-swimming often solitary omnivorous fish of shallow inshore waters. Opportunist feeders on corals, molluscs, crustaceans, worms, sponges and seaweeds.

From Anon. 1980a. Ecological information from Carcasson 1977; Schroeder 1980

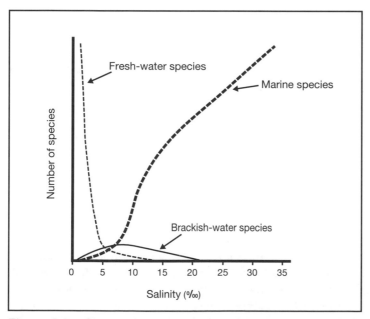

Figure 3.3. Changes in the number of freshwater, brackish water and marine species in different salinities.

After Barnes 1984a

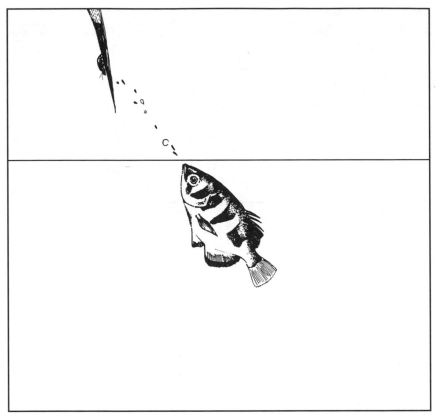

Figure 3.4. Archer fish *Toxotes jaculator* shooting a beetle off a leaf with a jet of water.

considerable economic importance (Anon. 1980a).

One of the more curious fishes found in brackish water of estuaries and mangrove creeks is the archer fish *Toxotes jaculator* (fig. 3.4). It is commonly kept as an aquarium fish because of its stunning ability to shoot jets of water from its mouth to knock insects off over-hanging vegetation. It is accurate up to about a metre and is capable of compensating for the bending or refraction of light as it passes, from air to water as well as allowing for the trajectory of the water it shoots.

Primary Productivity

Estuaries are very productive marine environments with abundant food, low predation pressure and few specifically estuarine species. The nutrients in an estuary originate from three sources: river inputs, marine inputs and bottom sediments, although rivers contain higher concentrations of dissolved and particulate nutrients than sea water. The particulate matter may be deposited on the bottom of the sanctuary becoming a food source for benthic organisms thereby entering the foodweb. A flow of deep-sea water towards the land brings dissolved nutrients into estuaries which can be utilized by phytoplankton. Plankton becomes less abundant with distance from an estuary, and primary productivity can be 20 times as great in an estuary as it is in the open sea (Doty et al. 1963). Concentrations of phosphorus, a key limiting nutrient, can be twice as great. The availability of organic matter in estuarine systems can be particularly high because it is supplemented by adjacent systems. For example, rivers bring down detached plants and freshwater phytoplankton that die on contact with sea water (Hadi et al. 1977), and colloidal matter comprising very fine sediment, aggregates or flocculates in estuaries and then sinks. Mangroves also export nutrients into estuaries (p. 135), where particulate organic material from the decomposition of mangrove forest litter is the primary source of energy for aquatic organisms.

SEAGRASS MEADOWS

Seagrasses

Seagrasses are the marine members of the aquatic plant families Hydrocharitaceae and Potamogetonaceae. In Sulawesi the former family has three genera (*Enhalus, Thalassia* and *Halophila*), and the latter four (*Halodule, Cymodocea, Syringodium* and *Thalassodendron*), and their identification is relatively easy[1] (figs. 3.5-3.8). Seagrasses bind shallow sediments and fast growing pioneer species play an important role in stabilizing land.

Seagrasses are found in shallow coastal waters or in lagoons between coral reefs and the shore. A few species extend into the intertidal or littoral zone and these can tolerate quite high temperatures as might be experienced during the day at spring low tides. In such sites, species with narrow or small leaves tend to be most common (McMillan 1984). Seagrasses generally grow gregariously and the larger ones can be said to form meadows. They often grow in associations of species, and if conditions are suitable for one species, others will probably be close by.

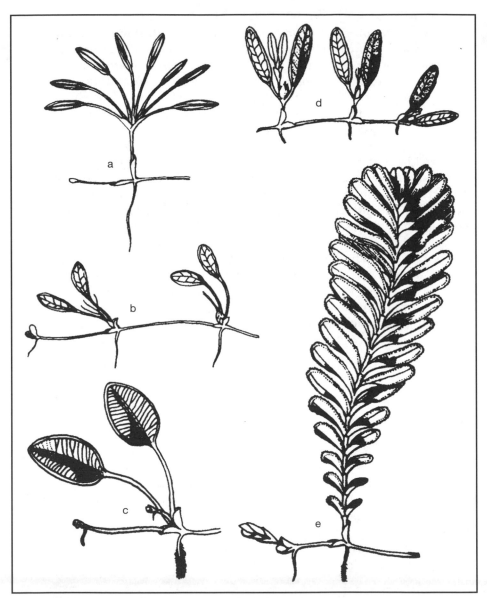

Figure 3.5. *Halophila* seagrasses. a - *H. beccarii*, b - *H. minor*, c - *H. ovalis*, d - *H. decipiens*, e - *H. spinulosa*.

After den Hartog 1957

Figure 3.6. *Thalassia* and *Enhalus* seagrasses. a - *T. hemprichii* (female), b - *T. hemprichii* (male); c - *Enhalus acoroides*. Scale bar indicates 1 cm.
After den Hartog 1957

Around northern Minahasa, seagrasses are found on all reef flats and floors of shallow bays with *Thalassia hemprichii* being the most common and conspicuous. This species frequently grows in association with *Halophila ovalis*, *Halodule wrightii*, *H. pinifolia* and *Syringodium isoetifolium*. *Enhalus acoroides* is most commonly found bordering mangroves and lagoons or in bays where the water is warm, still, slightly greenish or turbid with a relatively high organic load. In these conditions the leaf blades of this species can grow to 2 m in length. In very different conditions, on the outer reef flat close to or in the surf zone, *Thalassodendron ciliatum* is common in pure patches. Pure patches of *Halodule uninervis* also occur but on firm

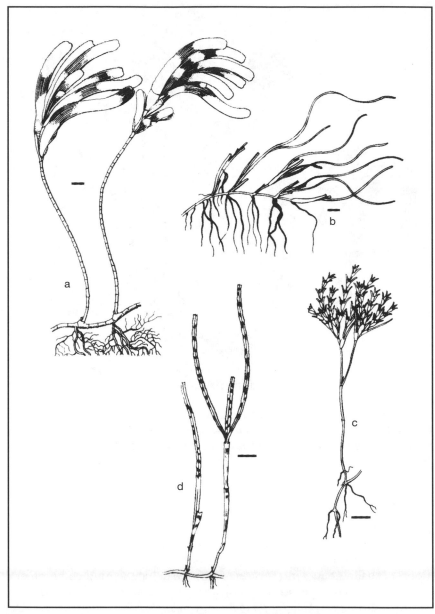

Figure 3.7. Seagrasses. a - *Thalassodendron ciliatum*, b - *Halodule pinifolia*, c - *Syringodium isoetifolium*, d - *H. uninervis*. Scale bars indicate 1 cm.

After den Hartog 1970

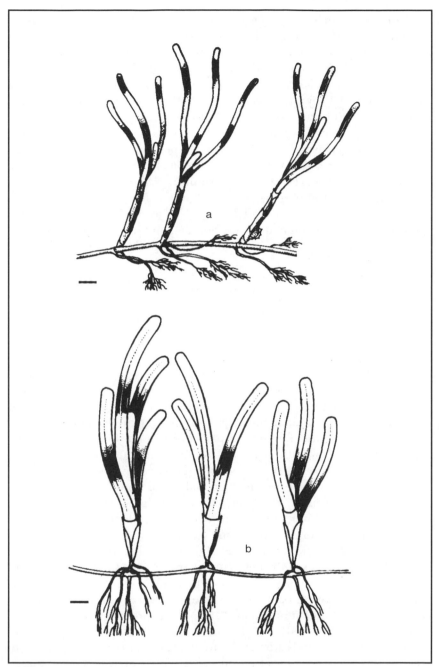

Figure 3.8. *Cymodocea* seagrasses. a - *C. rotundata*, b - *C. serrulata*. Scale
bars indicate 1 cm.

After den Hartog 1970

sand on the lower slope of beaches or between the roots of *Sonneratia* mangrove trees (p. 120) (Anon. 1981). Around the Tukang Besi Islands and the islands of Taka Bone Rate (p. 217) there are very extensive meadows of *Thalassodendron ciliatum* that stretch for several kilometres (Brouns 1985). Studies of seagrass distribution at Labuan Peropa (east of Kendari) and at Tanjung Karang (west of Donggala), concluded that the nature of the substrate was the most important factor determining the presence of a species at any particular place, although water depth and exposure clearly played a role. Transects perpendicular and parallel to the beach demonstrated that species composition changed with distance from the shore (fig. 3.9) (L. Clayton pers. comm.).

Almost as important as the seagrasses in terms of biomass and productivity are epiphytic algae found on the seagrasses. These algae vary in species composition and abundance between and within seagrass species, but there are few differences in the dominant species (Heijs 1985).

Reproduction

The reproductive biology of these higher plants is very interesting: *Halophila* and *Thalassia* are pollinated under water but *Enhalus* has flowers which reach the surface where they are pollinated by wind or insects. The flowering of *Enhalus* is gregarious and coincides, with low spring tides, a few days after the new and full moons. The male flowers are released from the parent plant, float to the surface and, due to their waxy, water-repellent coating, tend to clump together, moving around in response to the wind and currents. The simultaneous release of these small flowers is probably caused by the increase in water temperature around the plants when the tide is extremely low, although dissolved oxygen and salinity concentrations, as well as light intensity also change at such times (den Hartog 1957). The female flowers remain attached to the plant and they also have a waxy, water-repellent coating which attracts the free-floating male flowers by surface tension. Pollination usually occurs when the tide rises: the stem of the female plant reaches its greatest extension and then, when the tide gets higher still the flower can stay above the water no longer, surface tension pulls the edges of the petals together and male flowers are 'caught' between them. The fruits of seagrasses ripen only under the water and dispersal distances are probably short since the seeds are not buoyant and cannot grow in water more than 8-10 m deep presumably because the light intensity is too low. It is, however, possible that the dugong and turtles disperse the seeds after accidentally ingesting ripe fruit heads with seagrass leaves and rhizomes (Janzen 1984). The fruit, which some coastal people eat, bursts open when ripe and the developing embryo breaks out of its very thin seed coat, falls to the sand or mud, and germinates immediately (den Hartog 1957).

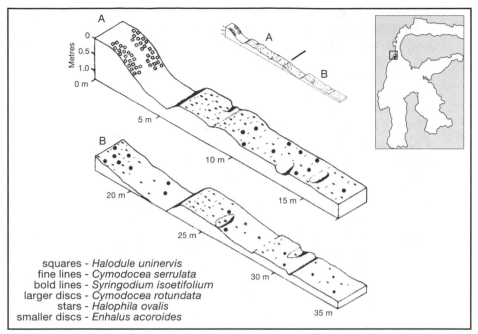

Figure 3.9. Patterns of species distribution along a 35 m transect across a seagrass meadow perpendicular to the shore at Tanjung Karang, Donggala.

After L. Clayton pers. comm.

Biomass, Productivity and Decomposition

The biomass of leaves and stems in a seagrass meadow can reach about 700 g dry matter/m^2, although most meadows studied by EoS teams had biomass densities of 80-400 g/m^2 (L. Clayton pers. comm.). About 25% of this biomass is accounted for by the epiphytic algae (Heijs 1985). The net primary productivity can be very high, and a rate equivalent to 16.4 t/ha/yr has been measured for *Thalassodendron ciliatum* meadows around one of the islands at Taka Bone Rate. This is higher than the rates for lowland forests (p. 365). There is considerable variation between sites, however, and the different methodologies used in different studies make it rather difficult to compare results. For example, the epiphytic algae can contribute at least 35% of the primary production of the system in locations where the plants are relatively widely spaced (Heijs 1985). Where the plants grow closer, less light penetrates through the seagrass stand and the algae do not thrive.

Seagrass blades contain about 30% ash or mineral matter, 10% protein, 1% fat, 60% carbohydrate and have an energy content of 3 Kcal (Dawes and Lawrence 1983),[2] compared with 8%, 8%, 2%, 82% and 18 Kcal respectively for alang-alang grass *Imperata cylindrica*, which is frequently used when young as a stock food. The main grazers on seagrasses are certain fish, turtles, dugongs and a few sea-urchins with cellulose-digesting bacteria in their guts, but only 5% of the seagrass production is consumed directly, most consumers depending more on decomposing seagrasses. This is probably because the detritus of seagrasses is rich in micro-organisms; one dry gram supporting about ten thousand million bacteria, one hundred million flagellates, and one hundred thousand ciliates with a total biomass of about 9 mg.

It is not known how much seagrass material enters off-shore food webs, but it should be borne in mind that the attempts that have been made to measure the export of material are, for practical reasons, usually conducted during relatively calm weather. A single strong storm, however, could tear up and wash away considerable quantities of plants (Barnes and Hughes 1982). In contrast to conventional wisdom concerning the role of seagrass detritus in estuarine food webs, research in the meadows in the Gulf of Mexico has demonstrated that it is the epiphytic algae, not the actual seagrasses, that are the most important component in the ecosystem. These algae are grazed intensively at night by invertebrates. In the same study carbon in the seagrasses and their epiphytes was tracked through the animal community and it was shown that animals were assimilating carbon from epiphytes rather than from seagrasses (Kitting et al. 1984). In the Seribu Islands north of Jakarta, 78 fish species have been found associated with sea-grass meadows (Hutomo and Martosewojo 1977). Molluscs such as the bivalve *Pinna*, and the gastropod snails *Lambis* and *Strombus* (fig. 3.10) sea cucumbers such as the long *Synapta* and shorter *Holothuria* (p. 228), and the brown-spotted, sand-coloured sea star *Archaster* are reported as common in seagrass meadows (Atmadja 1977).

Effects of Development

Agricultural, industrial and domestic effluents discharged into the sea, dredging, heavy boat traffic and other human activities can have detrimental effects on seagrass meadows. Where seagrasses are lost the subsequent changes that can be expected include:
- a reduction of detritus from seagrass leaves with consequent changes in coastal food webs and fish communities;
- a change in the dominant primary producers from benthic to planktonic;
- changes in beach morphology due to loss of the sand-binding properties of seagrasses;

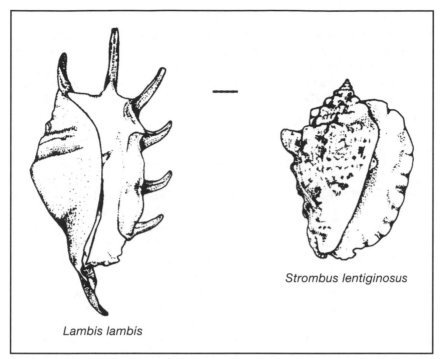

Figure 3.10. Gastropods of seagrass meadows. Scale bar indicates 1 cm.

- a loss of considerable structural and biological diversity and their replacement with bare sand (Cambridge and McComb 1984).

It may be possible to detect early stress on the meadows by monitoring changes in the communities of epiphytic algae (May et al. 1978), but any change must be judged against the quite considerable long-term changes that occur unrelated to development activities (May 1981).

In areas where seagrass meadows have been lost through mechanical disturbance, or where a once damaging effluent source has since been removed or purified, it is quite possible to restore the meadows by planting. This has been successful in a number of situations (Thorhaug 1983, 1985; Thorhaug et al. 1985). It would be reasonable to insist upon seagrass restoration if it is found that the construction of the new fishing port at Kendari, intended to become the largest in eastern Indonesia, has a negative impact on nearby meadows.

Dugongs

Dugongs, or sea cows *Dugong dugon,* are the sole species in the family Dugongidae, and are one of only four living species in the order Sirenia[3]. The dugong is the world's only vegetarian marine mammal, living in shallow coastal areas in the south-western Pacific and Indian Oceans (Nishiwaki et al. 1979). It is believed that sirenians evolved from an animal that fed in vast seagrass meadows in the west Atlantic and Caribbean during the Eocene Period (54-38 million years ago). This animal was closely related to the ungulates and was descended from an ancestor shared by elephants. Like elephants and horses, dugongs are non-ruminant herbivores[4] and consequently have a very long intestine. Bacterial activity in the hind gut assists in the digestion of the large quantities (about 10% of body weight) of the relatively high quality food they consume each day (Murray et al. 1977).

Dugongs grow up to 2.5 m in length, are mid-brown in colour, slow-moving, and have few predators except man who seeks their meat, oil and tusks. Their life span is about 50 years for females and 30 years for males, and sexual maturity is reached between about 9 and 15 years. The gestation period is about one year after which a single calf, one metre long, is born (Marsh 1981). The period between two consecutive births can be 3-7 years (Marsh 1981; Marsh et al. 1984). The calves generally swim above and slightly behind their mother. They are typical K-selected species (p. 345) and have little potential for rapid population increase if hunted too heavily. Under the most favourable conditions the maximum rate of population increase is probably about 5% per year, and so any man-induced mortality must be kept at that level or below if the population is not to decline (p. 335) (Marsh 1986). As a result of being vastly over-hunted dugongs are rare around the coasts of Sulawesi with viable populations being found only in remoter areas (fig. 3.11). Many dugongs are caught accidentally (but fortuitously for the fisherman) in gill nets set for fish, but some are deliberately hunted despite their protected status. A hunter interviewed near the Vesuvius reef southwest of the Banggai Islands reported killing one every ten days (Holdway n.d.), a rate which is surely not sustainable. Dugongs tend to aggregate in herds of about ten animals[5] and to have preferred feeding areas (Anderson 1981). The dugong needs and deserves much better protection than it receives at present and it is probably necessary to identify the major populations and their ranges[6], to ban gill nets from key areas, and to enforce the prohibition of hunting.

Dugongs are extremely hard to study, partly because the waters above the seagrass meadows they frequent tend to be turbid[7], and partly because they generally shy, fleeing at up to 18 km per hour (Domning 1977), faster than a human can swim. Even so, they are curious when danger is not perceived, approaching and watching divers (but not necessarily behaving normally) until their curiosity is satisfied. Unlike whales, they disturb the water little when surfacing to breathe, something they do about every one

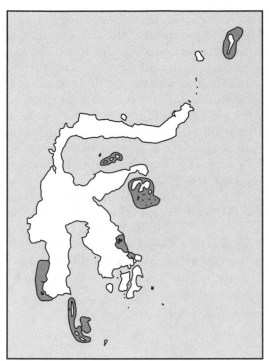

Figure 3.11. Distribution of dugongs around Sulawesi.

After Hendrokusumo et al. 1981; Salm and Halim 1984

or two minutes (Anderson and Birtles 1978; Anderson 1982).

Dugongs dig into the sediment when grazing on seagrasses because over half the biomass of the smaller seagrasses, their major food, is in the rhizomes and it is here that carbohydrates are concentrated (Johnstone and Hudson 1981; Marsh 1982). Their main potential competitor, the green turtle *Chelonia mydas* (p. 151), eats only leaf blades. Dugongs 'chew' their food between the horny pads at the front of their upper and lower jaws.

The differences in proportions of different seagrasses in the dugong diet probably reflect only differences in meadow composition rather than direct selection of certain plants by the dugongs. Seaweeds, or macroalgae, were found in half the stomachs examined in one study but only as a minor item (Marsh 1982).

Useful work can be done on dugong diets by examining either stomachs or residual samples from the mouths of dead animals. The latter technique is easier because the plant remains are relatively intact and can be identified quickly to species (Johnstone and Hudson 1981), whereas the stomach analyses require microscopic analysis of the structure of the masticated vegetable matter. The cell patterns of different seagrass genera are characteristic and so are relatively easy to identify (Channels and Morrissey 1981).

Measurements of metal concentrations in the tissues of dugong from northern Queensland provided several surprises. The livers contained exceptionally high concentrations of iron and zinc, and relatively high concentrations of copper, cadmium, cobalt and silver. Certain metals, particularly iron, increased in concentration with the age of the dugong. It seems unlikely that wastes from man's activities were having an effect on these animals from relatively remote populations and so the chemical composition of seagrasses were examined. These revealed very high concentrations of iron but low levels of copper compared with grasses from pastures. These imbalances in dietary metal intake have significant effects on dugong metabolism (Denton et al. 1980).

Heavy metal and organochlorine concentrations were examined in tissues from two female dugongs caught off Sulawesi for an exhibition in Japan,[8] and were found to be very low and undetectable respectively (Miyazaki et al. 1979). These results would have been expected since the dugong is a primary consumer (p. 559) and therefore is unlikely to accumulate pollutants. It is animals in the higher trophic levels that would be expected to accumulate pollutants from their prey.

CORAL REEFS

Importance and Species Richness

Coral reefs are of considerable importance because of their protective role and the value of the larger animals (both alive and dead) associated with the reef. Perhaps the largest coral reef fishery in all Indonesia is found in the shallow sea of the Sangkarang Archipelago within 80 km of Ujung Pandang, and thousands of people depend on this for income and protein. Coral reefs are also extremely valuable to the tourist industry, and North Sulawesi promotes diving reefs as one of its major tourist attractions. Indeed, the Nusantara Diving Centre in Manado was awarded a Kalpataru Environment Award by President Soeharto in 1985 in recognition of its work in bringing the economic role and beauty of coral reefs to the attention of the public and local authorities. It even convinced the port authorities not to allow large ships to pass the straits between Manado and Bunaken Islands where the most spectacular reefs are located. Tourists from Indonesia and many other countries now visit the reefs and find excitement, wonder and fulfillment.

Coral reefs are enormously rich in species (pp. 219 and 221) which interact with each other to form an extremely complex community. In the early days of ecology it used to be believed that complexity of a community, such as the animals and plants of a coral reef, begets stability. It was noted, for example, that crop monocultures were extremely vulnerable to invasion and destruction by pests and diseases and that forests in temperate regions were far more prone to insect outbreaks than forests in tropical regions. These and other arguments seemed reasonable but fell when it was found that natural, rather than man-made, species-poor communities were stable, and when the relationships between complexity and stability were examined mathematically. Some mathematical models indicated that stability in fact decreased with increasing complexity. Further refinements have shown that the stability of a complex community will increase if simplicity is achieved by the removal of top predators, but will decrease (as was originally proposed) when species from nearer the bottom of the food web, such as primary consumers, are removed. The stability of communities is influenced by the stability and predictability of the environment in which they live. Thus it may be that complex yet fragile communities can persist, and thereby give an impression of stability, in benign and predictable environments. The implication for management should be noted: the complexity of many communities is related to the relatively stable natural environments. When even quite minor and short-term changes are made to the environment, the members of a complex community, not used to such disturbance, may exhibit some initial resistance (an ability to avoid change), but low resilience (the speed with which the community can return to its original state) after the change has occurred (p. 569). Coral reefs, then, may be able to withstand just so much stress before the living community completely collapses.

Structure and Formation

Coral reefs form in warm water (generally above 22°C) that is relatively clear and illuminated by the sun, and has a near-to-normal seawater salinity (p. 109). Reefs are poor or absent around coasts near large river mouths because they are intolerant of lowered salinity and high sediment loads. Large rivers are rare in Sulawesi, however, and consequently much of the coasts is fringed with coral (fig. 3.12). Since coral is dependent on light, the depth to which it grows increases with distance from major landmasses where water is rather more turbid. This intolerance of turbid or brackish water results in corals growing faster the further away they are from the land. Thus reefs grow gradually away from the land, leaving shallow lagoons behind them floored with sand composed of broken, dead coral skeletons. The different structures of the steep reef edge and the gently sloping reef flat or reef platform result in changes across the reefs in physical parameters (fig. 3.13).

The coral reef structure is formed by the compacted and cemented skeletons and skeletal sediment of sedentary organisms which formerly

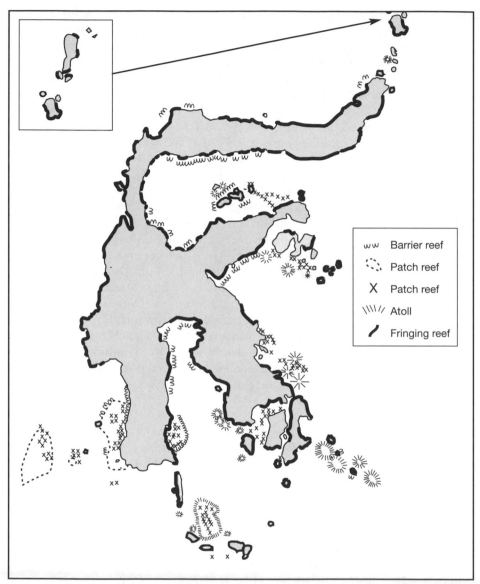

Figure 3.12. Coral reefs around Sulawesi.

After Salm and Halim 1984

lived on the reef but have since been smothered by other organisms. The outermost layer of a coral reef is living tissue comprising primarily of scleractinian (hard) corals and algae with limestone-impregnated tissues (p. 223).

It was Charles Darwin in 1842 who first distinguished the three main forms of coral reefs still recognized today—fringing reefs, barrier reefs

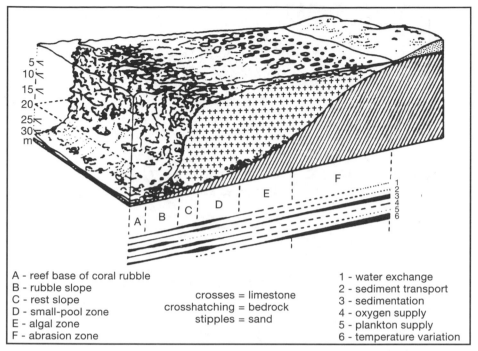

A - reef base of coral rubble
B - rubble slope
C - rest slope
D - small-pool zone
E - algal zone
F - abrasion zone

crosses = limestone
crosshatching = bedrock
stipples = sand

1 - water exchange
2 - sediment transport
3 - sedimentation
4 - oxygen supply
5 - plankton supply
6 - temperature variation

Figure 3.13. The zones of, and changes across, a typical fringing reef which grows away from the land. Zones D, E and F are together sometimes known as the reef flat, reef platform or lagoon. In this illustration, the maximum depth at which there is sufficient light is 20 m.

After Barnes and Hughes 1982

and atolls. Fringing reefs are formed close to the shore on rocky coastlines. If water depth remains the same or decreases over time growth of the reef is entirely seawards, vertical limits being set by the level of low spring tides and the depth to which light can penetrate. Between the crest of a fringing reef and the shore there is usually a shallow reef flat where coral growth is poorer because of reduced water circulation, higher sediment and slightly lower salinity due to run-off from the shore. Vertical growth can occur over geological time if the sea bed subsides or if water depth increases as a result of a rise in sea level, such as happens after an ice age (p. 16). Periods of low sea level during the ice ages would have exposed and killed most the coral reefs and wave action would have eroded their remains into platforms. At the current time, sea level is rising between 3 mm and 15 mm per year and this rate has been maintained for much of the last ten thousand years. Vertical growth of coral has been able to keep pace with this (Barnes and Hughes 1982).

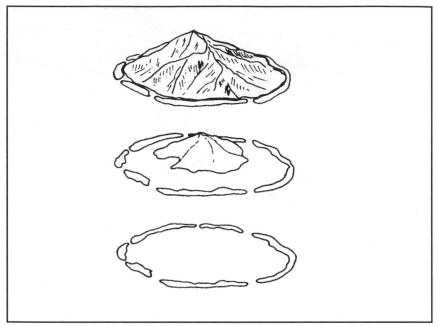

Figure 3.14. Formation of fringing, barrier and atoll reefs.
After Barnes and Hughes 1982

Barrier reefs occur where the coast has subsided and only the seaward reefs have been able to continue their growth. Barrier reefs are therefore separated from the shore by a lagoon. Atolls are formed in a similar manner where a volcanic island has subsided and disappeared (fig. 3.14). Holes drilled vertically through a South Pacific atoll passed through over 1,200 m of dead coral rock before the ancient volcanic cone was encountered (Henrey 1982). Coral reefs have been inhabited by creatures very similar to those found today for at least 200 million years and the great diversity of creatures is at least in part due to this enormous period of relative constancy although conditions in any one location have varied considerably.

These different reef forms all have the same basic biological structure, and differences between the corals found on the three types would probably be no greater than the differences encountered within any one type.

The major reef areas around Sulawesi include:
- Taka Bone Rate (formerly Tijger) Atoll which is the third largest atoll in the world, with an area of some 2,220 km? This is only 20% smaller than the largest, Kwajalein in the Marshall Islands (Anon 1982b). It is situated southeast of Salayar and north of Bonerate and comprises a complex of patch and barrier reefs with 21 sandy islands

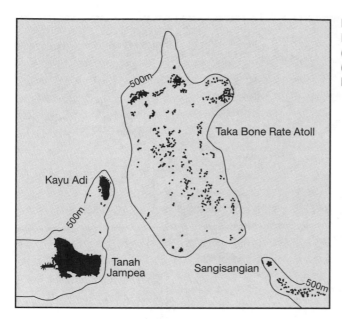

Figure 3.15. Taka Bone Rate Atoll showing land (black), coral reefs (stippled) and the 500 m bathymetric contour.

nine of which have permanent inhabitants (fig. 3.15). It is adjacent to the smaller atoll of Taka Garlarang to the northeast and both rise abruptly from 2,000 m on the side of a submerged ridge, similar to the atolls and islands of the Tukang Besi group (Umbgrove 1939, 1940). Part of the northeast corner was studied during the Snellius II Expedition (Moll 1985).

- Sangkarang (formerly Spermonde) Archipelago comprises a loose group of about 160 small islands off Ujung Pandang and covers an area of about 16,000 km² (de Nève 1982). This has been the site of the most intensive coral reef studies in Sulawesi conducted by Universitas Hasanuddin and the National Natural History Museum in Leiden, Holland.

- The Togian Islands in Tomini Bay are unique in the Indonesian archipelago in having all the major reef environments around their shores. Two atolls lie about seven kilometres off the northwest shore of Batudaka Island (p. 413) and comprise a deep (20-50 m) lagoon surrounded by shallow reef flats. The corals on the lagoon and seaward sides are dominated by different communities. The barrier reef lies close to the 200 m depth contour which in places is 15 km from the shore but elsewhere only 5 km. It reaches the surface only in certain areas and it lacks boulder or shingle ramparts and certain other features indicating that the Togian Islands enjoy calm seas in all sea-

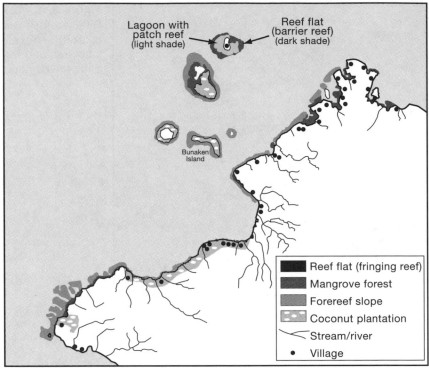

Figure 3.16. Coastal habitats near Manado.

After Anon. 1981

sons. Fringing reefs are found around virtually all the coasts and are similar throughout in their composition except in the bays where they merge on the shoreward side with seagrass meadows, which in turn are bounded by *Rhizophora* mangrove trees. Patch reefs are common within the barrier reef, and where they reach the surface are similar in composition to the fringing or barrier reefs, depending on which is closer (Anon. 1982i).

- Bunaken Island is probably one of the most-visited reefs and this and other coral areas near Manado feature in holidays promoted by international tourist agencies (fig. 3.16).

In addition, the Vesuvius reefs that lie 80 km south-southwest of the Banggai can claim some notoriety as the place where Sir Francis Drake jettisoned a canon from his ship 'The Golden Hind' during his voyage around the world in 1580. An unsuccessful search for the canon was made 400 years later as part of Operation Drake.

Zonation of corals in the Indo-Pacific is determined by exposure to winds and waves, distance from the reef edge exposed to waves, and water depth (p. 213). As with other types of ecosystems, there is no 'typical' area or 'typical' community, and the composition of communities of organisms varies from place to place as single species respond to differences in the environment. Groups of species may generally be found together but rarely in identical proportions between areas. Where sharp changes do occur in the species composition across an area, this can invariably be related to some major environmental change—such as a reef drop-off, rock outcrops, pollutants, or damage from fishing activities (Barnes and Hughes 1982).

Reef Invertebrates

There are many animals quite unknown to the average reader bearing no names that convey any concrete idea to any one who is not a specialist in the particular branch of natural history to which they belong. A long list of the Latin names of the corals of a reef, for example, conveys no impression, even to many zoologists, of the infinite variations of form, structure and colour which those corals actually present in the living state. A coral reef cannot be properly described. It must be seen to be thoroughly appreciated.—HICKSON 1889

Thus wrote one of the earliest experts on coral, the Englishman Sidney Hickson, who actually got his feet wet on a reef. In 1885 he came to live in North Sulawesi, primarily on Talisse Island off the north Minahasa coast, in order to study corals in nature, and many of his observations are still extremely useful.

There are seven major invertebrate phyla represented on a coral reef and a simple classification of these is shown in table 3.2 and short descriptions follow.

Sponges are the most primitive of multicellular animals. They are generally rather shapeless aggregations of cells whose main distinguishing feature is a maze of canals along which water and suspended food is carried by the action of flagella, or tiny hairs. A sponge the size of a tea cup can pump an astounding 5,000 litres of water per day through its canals. The holes out of which the water passes are usually visible, but the entrances are often very small. Some sponges are cup-like in shape, others are erect branches or mounds but most are encrusting forms seen as red, brown, yellow, blue or black patches on rocks. Many sponge cells are inhabited by blue-green algae, or zooxanthellae, that live symbiotically in the same way as those in corals, sea squirts and giant clams. One group of sponges, the Clionidae, are extremely important in determining the architecture of the reef because they bore into the coral rock using specialized etching cells which dissolve the limestone. Chips which fragment off are 'exhaled' in the stream of water passing through the sponge. The galleries produced in coral rock obviously weaken the structure and may result in the collapse of sections during storms (Bergquist 1978).

Table 3.2. Simplified classification of the major invertebrates occurring in the benthos of coral reefs. The phylum Cnidaria used to be known as Coelenterata. The phylum Chordata is the one to which all animals with backbones belong.

Phylum Porifera	- sponges
Phylum Cnidaria	
Class Hydrozoa	- sea firs or hydroids
Class Scyphozoa	
Order Rhizostomeae	- jellyfish
Class Cubozoa	
Order Cubomedusae	- sea wasps
Class Alcyonaria	
Order Alcyonacea	- soft corals
Order Gorgonacea	- sea whips/sea fans/sea feathers
Class Zoantharia	
Order Actinaria	- sea anemones
Order Scleractinia	- true corals
Class Ceriantipatharia	
Order Antipatharia	- black or thorn corals
Phylum Annelida	
Class Polychaeta	- marine worms
Phylum Mollusca	
Class Gastropoda	- limpets, snails, sea slugs
Class Bivalvia	- clams, cockles, mussels
Phylum Bryozoa	
Class Gymnolaemata	- marine bryozoans
Phylum Echinodermata	
Class Asteroidea	- starfishes
Class Ophiuroidea	- brittlestars and basketstars
Class Echinoidea	- sea urchins
Class Holothuroidea	- sea cucumbers
Class Crinoidea	- sea lilies and featherstars
Phylum Chordata	
Class Ascidiacea	- sea squirts or tunicates

Classification from Barnes 1984b

Cnidarians (formerly known as coelenterates) have a tube-shaped body with one opening at the upper end through which food and waste is alternately passed. The hydroids have a simple sac for a digestive chamber but this is more complex in corals and sea anemones. The two most-commonly seen hydroids are rather different. The sea nettle *Aglaophenia* grows like a pinkish-brown fern about 20-30 cm high. Fire coral *Millepora* however, looks ostensibly like an ordinary coral, but with a hand lens the characteristic pattern of five or six small holes surrounding a larger hole can be seen. The 'mouth' of all cnidarians is surrounded by tentacles armed with stinging cells or nematocysts out of which barbed harpoons are shot when touched (fig. 3.17). These can just be felt if a sea-anemone is touched but those fired from jellyfish and some hydroids can caused intense pain. Hence the common names of 'fire coral' or 'sea nettle' given to some hydroids that are capable of delivering a memorable sting to unwary divers.

An individual coral animal starts life as a small planktonic larva which comes to rest on a suitable substrate and metamorphoses into a polyp. This polyp begins dividing and forming genetically-identical polyps next to it, and this process continues until the coral dies. Within the 'skin' of the coral polyps are small yellowish-brown granules, which are in fact small plants from the phylum Dinophyta (dinoflagellates), other members of which are found in the marine phytoplankton. Only a single species, *Symbiodinium microadriaticum,* is known to live symbiotically with corals. These plants, called zooxanthellae because they live within animals, absorb waste products produced by the host polyp converting the phosphates and nitrates into protein and, with energy from the sun, the carbon dioxide into carbohydrates. Their waste product, oxygen, is used in turn by the polyp in its respiration. Coral polyps secrete their external skeleton of limestone from their bases but each polyp is connected to its neighbours by strands of tissue (fig. 3.18).

Corals mature after three to eight years and tend to breed seasonally and often simultaneously in a certain area. Until recently it was thought that most scleractinian corals were viviparous, that is brooding fertilized eggs within the polyp, often releasing larvae throughout the year. Careful observations in northeast Australia revealed, however, that most corals release gametes simultaneously at night a few days after full moons (Harrison et al. 1984).

Why should corals breed simultaneously? If breeding were not simultaneous, there would be great risks involved in spawning eggs or sperm, both of which have limited periods of viability, into the sea. The chances of egg meeting sperm if spawning were random would be very low. In addition, if only a few eggs were in the water at any given time it is likely that a relatively large proportion of the eggs would be eaten both by opportunistic or by specialist predators. Thus simultaneous spawning gives the maximum opportunity for successful fertilization, minimum chance of any

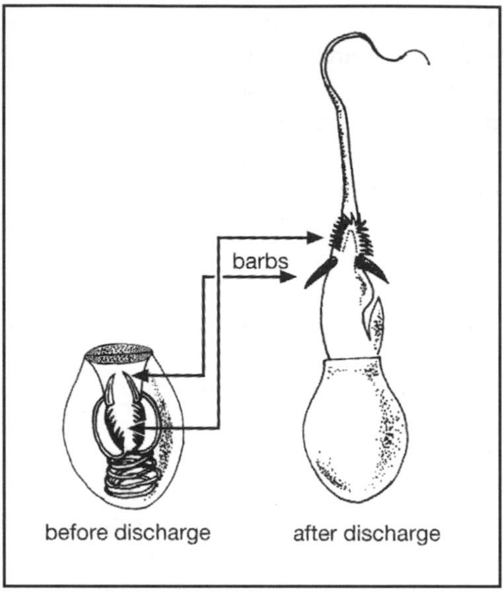

Figure 3.17. Cnidarian nematocyst before and after the firing of the 'harpoon'.

After Henrey 1982

barbs

before discharge   after discharge

particular egg, fertilized or not, of being eaten, and prevents any animal specializing on coral eggs as food.

Not all corals breed every year, though this would most often be the case for the slower-growing, longer-lived species. An individual *Porites* measuring 5.8 m in diameter was estimated to be 140 years old but was still increasing its diameter by over 4 cm per year. Monitoring of young colonies of about 1 cm diameter (about one month old) on the Great Barrier Reef revealed that five colonies per m² were recruited per year on a reef flat but that half of these died within a year (Connell 1972).

Most human visitors to a reef swim during the day and see but a pale shadow of the wonders that can be seen at night. It is then that the sharp edges of coral seen during the day are transformed into soft, fuzzy outlines as millions of colourful tentacles sweeping through the water catching minute plankton or other particles of food. Startling examples are the mushroom corals such as *Fungia* which lie loosely on the reef bed like large, upturned mushrooms.[9] During the day these are dull, lifeless and gray-brown but at night thick, bright-coloured tentacles emerge from the mushroom's gills, totally obscuring the skeleton beneath. In addition, because the wavelengths of red, green and yellow light are quickly absorbed by water, much of the deeper reef appears blueish by day, but the true colours can be seen by torchlight.

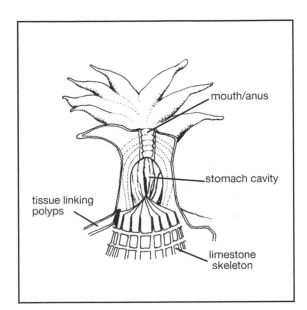

Figure 3.18. Cross-section through a coral polyp.

A total of 262 species of hard (scleractinian) corals from 78 genera and subgenera have been identified from the Sangkarang Archipelago.[10] This is more than is known anywhere else in the Indo-Pacific region. Of this total of species, six have recently been described for the first time (Moll and Borel-Best 1984).

Surveys of the Sangkarang reefs revealed that similar numbers of species were found on the reef edge, reef flat and reef slope but that only 110 species were found in all three situations. There were 24 species confined to the reef flat and 17 species to both the reef edge and reef slope (Moll 1983). However, the number and coverage of coral units or colonies is higher on the reef edge than elsewhere. The islands were divided into three strips according to their distance from the shore but little difference was found between them in terms of corals present in each strip. The northern reefs were richer, however, than those in the east or the south, and the western reefs had a greater cover than those in the east and south. Differences between the compositions that were detected, however, could be related to distance from the mainland but there was great similarity between the composition of reefs in similar situations throughout the islands. The data were obtained by repetitive transects and the method is explained more fully elsewhere (Moll 1983).

A total of 115 species from 59 genera of coral have been found around the Togian Islands. Little difference in the number of coral species was

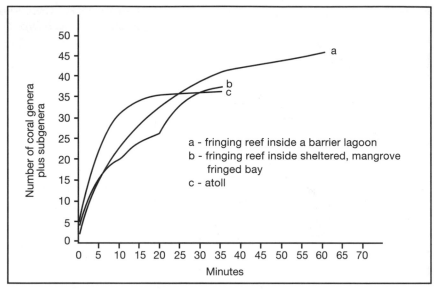

Figure 3.19. Discovery curves for coral genera and subgenera from three reefs around the Togian Islands.

After Anon. 1982a

found between the three types of reefs examined in detail but their distributions were distinct, as revealed by discovery curves[11] (fig. 3.19). Thus the initial rapid rate at which genera were encountered over the atoll reflects a well-mixed and diverse community of corals. The rate of discovering coral genera over the two fringing reefs was slower and no new genera were found after about 30 minutes. The smooth curve from the reef inside a barrier lagoon is again typical of a well-mixed community of corals but in this case the individual coral colonies are either larger or occupy larger areas. The step-like discovery curve from a sheltered fringing reef reveals that different communities of coral were distributed somewhat patchily rather than being mixed.

At a relatively undisturbed part of the Bunaken reef 58 genera of coral have been recorded, mostly in well-mixed assemblages (Anon. 1981). A total of 68 genera (158 species) have been recorded from a lagoon in Taka Bone Rate (Anon. 1982a).

The reason for the high diversity of corals in Sulawesi and certain other tropical regions is not fully understood but it is certainly related to three factors—more specialization, more time (because of lack of seasonality) and greater area. A longer time allows a greater proportion of species and/or their descendants to survive in the tropics by favouring greater specialization (Rosen 1981). The greater area is probably the most fundamental.

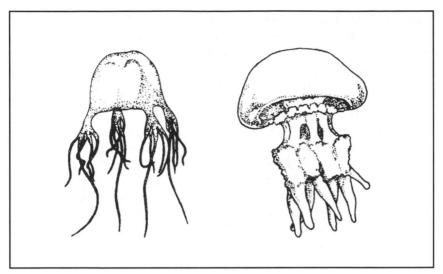

Figure 3.20. A sea wasp (left) and a typical rhizostome jellyfish (right).
After Henrey 1982

Whereas most cnidarians are colonial, some such as sea anemones and jellyfish are not. A sea anemone is a single large polyp with a thick muscular body attached to a rock. Some are over 50 cm across and the spaces between the tentacles are often occupied by clownfish. Jellyfish and hydroids pass through a life stage called a medusa in addition to the stages of the free-swimming larva and the polyp. In hydroids this is a relatively minor stage but in jellyfish the medusa is the dominant life form. Most jellyfish seen around reefs are members of the Rhizostomeae with no tentacles around the bell edge but many 'arms' around the central mouth. However, dangerous sea wasps are creatures with cuboid bells with four flattened sides and tentacles hanging from the four corners. The vicious stings from these tentacles make sea wasps among the most venomous animals in the sea (fig. 3.20; p. 232).

Soft corals do not secrete a limestone skeleton; the polyps sit instead in a gelatinous material of dull grey, brown or yellow colour. They grow into various shapes like mushrooms, fingers or just flat masses. Black corals and sea fans are similar in that they have a central axis of horny material around which the polyps live, but in black corals the axis is black and thorny and in sea fans it is brightly-coloured and branched. Black corals are generally found below 20 m, somewhat deeper than sea fans.

Bryozoans are small colonial animals looking like lacy moss attached to rocks, seaweeds or shells. Bryozoans are one of the more common organisms found growing on ship hulls leading to drag and loss of speed.

Many gastropod snails with exquisitely-patterned shells can be found around reefs, although shell hunters have drastically reduced the numbers of certain species (p. 230). For example, shells of over 170 species of snails have been found on the beaches of Tangkoko-Batuangus Reserve, Minahasa, and most of these were presumably thrown up from the fringing reef (Anon. 1980b). Other molluscs known as sea slugs (Nudibranchia) have lost their shells and evolved incredibly beautiful skin colours. The best known of the bivalve molluscs are the giant clams, most of which burrow backwards into coral rock. All that a diver sometimes sees is the intense blue, spotted flesh of the 'tips' which, like coral and sponges, contain zooxanthellae, or algae, living symbiotically within the cells. One of the species of giant clam found around Sulawesi, *Tridacna gigas*, has been found to spawn sperm during incoming tides during the full, third quarter and new phases of the moon. Plans are being aired to culture or translocate giant clams in Indonesia and it is important that the spacing and sexual behaviour of these animals be studied. In Australia it has been found that mature animals are on average 9 m apart and even where collectors have left some clams in place, this is no guarantee that the density is high enough for effective fertilization of eggs. The management of giant clams is further complicated by the ability of an adult clam to change its sex (Braley 1984).

If the word 'worms' conjures up a picture of dull, pinkish-gray creatures, then examining a coral reef will change that limited view. Here, polychaete worms live in burrows they build into the coral rock and extend their stunningly beautiful feeding structures into the water. The worms stop feeding and retract into their burrow when they detect unusual movement close by. The most common polychaetes are the intense blue, yellow, brown, orange or red bottlebrush worms *Spirobranchus* which have two sets of feeding apparatus, each of which has about five tiers of tentacles. Young bottlebrush worms in fact occupy cavities left by dead polyps and grow up as the coral grows.

The two most commonly seen echinoderms are very conspicuous in the inshore regions. They are black-spined sea urchins *Diadema*, the long spines of which are venomous and can cause considerable pain, and the blue starfish *Linckia laevigata*, which is commonly seen lying in the sun in shallow water on rocks or sand. One of the best-known starfish in coral areas is the large and very spiny crown-of-thorns starfish *Acanthaster planci*. This preys on coral by inverting its stomach over an area of coral, secreting digestive juices and sucking out the digested polyps (Chester 1969). This starfish is quite common around Bunaken Island near Manado. In the 1960s the population of this animal exploded in the Indo-West Pacific Ocean. Large areas of coral were devastated—a single adult destroys about 5 m² of coral per year—and the tourist industry at certain places along

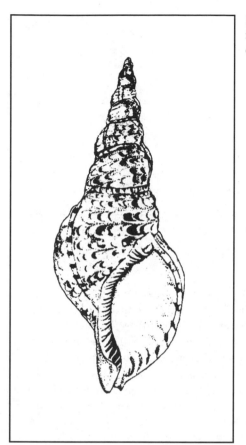

Figure 3.21. Giant triton *Charonia tritonis*, the major predator of the crown-of-thorns starfish.

Australia's Great Barrier Reef suffered considerably. Information from Indonesia is scanty. Even now, there is no satisfactory hypothesis as to why the population explosion occurred, but with an adult female producing about 20 million eggs each year the potential for growth is enormous. It is unlikely that man was directly responsible for the population explosion because the area over which it occurred was vast. It may be, however, that release from predator pressure could have caused the explosion because the only predator of any importance is the giant triton *Charonia tritonis* (fig. 3.21) which, because of its size and beauty, is much sought after by shell collectors (Anon. 1984a). Large-scale collection did not begin, however, until after about 1950 and it could be that their removal allowed numbers of *A. planci* to increase (Paine 1969).

Sea cucumbers look just like their name suggests.[12] Instead of five arms or segments they have five rows of tubefeet. When annoyed, species of

Figure 3.22. A typical sea squirt or tunicate. Arrows indicate the flow of water.

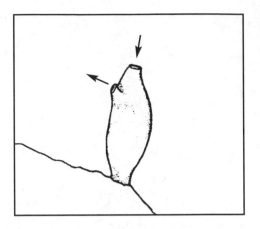

Bohadschia can exude sticky secretions to distract or entangle predators. They may even eject some of their sticky internal organs if extreme defence measures are required. The lost organs subsequently regenerate. There are probably about 30 species around the coasts of Sulawesi but the most common species are *Holothuria atra* which measures up to 60 cm in length and has sand stuck to its black body; *Stichopus chloronotus* which is smaller (about 30 cm long), dark green with orange-tipped lumps; and *Holothuria scabra* which is variable in colour but is generally yellowish cream with black and grey spots. This last species has a symbiotic relationship with a small pea crab *Pinnotheres* (similar to those found in cockle *Anadara granosa* shells and giant clams) which lives in its anus. Rather rarer are the worm-like or synaptid sea cucumbers which reach a metre in length, have a relatively thinner body and tentacles around the mouth.

Feather stars, or crinoids, have 10 or more brightly-coloured feather-like arms along which food is passed to the mouth.

Sea squirts, or tunicates, are sac-like animals with two openings—water and suspended food are drawn through the top opening and exhaled through the side (fig. 3.22). These can be found as solitary individuals or in small groups such as the green grape ascidian *Didemnum,* which is indicative of somewhat polluted water, such as near a town. Sea squirts, like sponges, corals and giant clams, have symbiotic blue-green algae growing inside their cells.

Colonial animals such as sponges, corals, bryozoans, and compound sea squirts comprise genetically identical units which together can grow to a size far greater than that possible for the individual. The modular construction permits flexibility in growth form which can adapt to the force of water currents, light intensity, silting and competitors. The colonies, or fragments of them, have considerable regenerative capabilities and large branching corals may propagate following damage caused by storms. For example, extremely large waves associated with strong winds in January 1986 bombarded the southwest coast of Sulawesi and reduced growths of *Acropora* coral to rubble even at 4 m below the surface. At one site studied intensively it was calculated that waves with velocities of 1.7 and 1.0 m/s at 0.1 m and 4 m depths respectively had caused the destruction; typical maximum wave velocities at the surface are around 0.4 m/s (P. Bloks pers. comm.). Four months later some regeneration of the *Acropora* had occurred. It would appear that *Acropora* actually benefits from this fragmentation (Bak 1981; Bothwell 1981; Highsmith 1982).

From November to April winds from the west buffet the western side of Bahuluang Island southwest of Salayar, and during the southeast monsoon it is protected from the wind's strongest effects by the much larger Salayar. The effects of this differential exposure on reef composition is quite dramatic (table 3.3). Thus, in the east (Transect 4), foliose corals contribute 40% of the cover and massive corals are very scarce. Conversely, in the west (Transects 2 and 3) there are few foliose corals but there are large numbers of massive corals. The exposed coral (Transect 1) is intermediate in its cover characteristics but it has the most species and colonies of massive corals and the least number of branching coral colonies. In each transect, however, the percentage cover of branching corals is remarkably similar. Considerable differences are found at the species level: for example, only one species of the branching *Acropora* is found on both the east and west of the island and some genera of smaller corals are found only in the east. On this eastern side, the large and fragile growth forms reflect the more sheltered conditions prevalent there (Moll 1985).

Coral transects around the small sand bar of Tinanja at the edge of the northeast part of the Taka Bone Rate Atoll also showed the effects of expo-

Table 3.3. Number of species, number of colonies, and percentage cover for four major coral growth forms at four locations off Bahuluang Island.

Transect Number	Branching			Foliose			Massive			Solitary		
	No. spp.	No. col.	% cov.	No. spp.	No. col.	% cov.	No. spp.	No. col.	% cov.	No. spp.	No. col.	% cov.
1	11	12	53	4	5	11	15	17	26	0	0	0
2	5	27	52	0	0	0	5	10	45	2	2	3
3	11	18	54	2	2	7	12	15	37	1	1	3
4	9	23	58	5	19	38	1	1	2	2	2	1

After Moll 1985

sure (Moll 1985), as did the much more detailed work in the Sangkarang Archipelago (Moll 1983). In summary, harsh exposure combined with heavy sediment transport results in low coral cover. Where water movement is moderate in a situation close to the open sea, corals flourish.

The incredible diversity of animals on a reef poses intriguing questions of how so many can co-exist, for it is generally believed that coexistence of two or more competing species can only occur if each species is exploiting a different set of limiting resources or, possibly, a similar set of limiting resources to different intensities. This is known as the 'Principle of Competitive Exclusion' or 'Gause's theorem'[13] (named after the Russian biologist who first investigated the means by which closely related species of micro-organisms coexist in the laboratory). Many closely-related species, such as those from a single genus, are often too similar in their ecology and behaviour to live in the same area because they compete for the same limiting resource and the species least able to compete is excluded—that is, the species live allopatrically (p. 69). Species that live sympatrically, that is, in the same area, exploit the available resources in different ways. An example are the cone snails *Conus* (fig. 3.23), the shells of 41 species of which have been reported from the beaches of Tangkoko-Batuangus Reserve, Minahasa (Anon. 1980b). One might reasonably ask how all these similar creatures achieve exclusive niches. That would be an interesting research topic, but some clue is given by the results of work on just eight *Conus* species conducted around Hawaii. The major food items of these predatory snails were examined and it was found that no two species had exactly overlapping diets (table 3.4). In areas where fewer sympatric *Conus* snails are found, the niche breadth or range of food in the diet would be expected to increase, and where more species coexist even more specialized diets would probably be found (Kohn 1979).

Many of the reef invertebrates, such as giant clams, are used by coastal people as food or, like sea cucumbers, have been exploited for centuries as items of international trade (p. 85). In general the inefficiency of the collecting methods have meant that the harvesting has been sustainable, and this is proven by the great age of the sea cucumber trade.

A number of invertebrates have considerable value on modern markets and the harvesting has been conducted from motorised vessels using teams of scuba divers. The most important collecting and export centre for this trade in Indonesia is Ujung Pandang which, in 1981, recorded foreign exports of 2,100 tons of reef invertebrates[14] which comprised 740 tons of pearl oysters *Pinctada*, 820 tons of mother-of-pearl shells *Trochus*, 220 tons of other shells, and 320 tons of sea cucumbers. In addition to these there were also quantities of black coral and spiny lobsters as well as catering for the growing domestic market, particularly for giant clams whose shells are now used in the manufacture of floor tiles. Some of the markets, particularly that for pearl oysters, are on the verge of collapse due to extreme over-

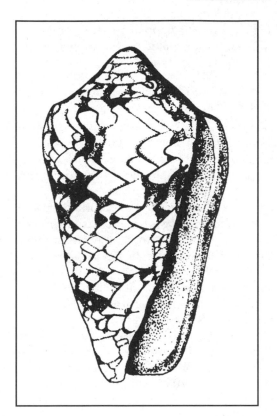

Figure 3.23. Braided cone shell *Conus textile*.

Table 3.4. Percentage of major food groups in the diet of eight species of cone shells *Conus*.

	Gastropod snails	Acorn worms	Nereid worms	Eunicid worms	Terebellid worms	Other
A	-	400	-	-	64	32
B	-	61	-	12	14	13
C	100	-	-	-		
D	-	-	-	100		
E	-	-	15	82	-	3
F	-	-	46	50	-	4
G	-	-	23	77		
H	-	-	-	27	-	73

After Kohn 1959

exploitation (Anon. 1984a). If this is not to happen to all the markets then some effective controls must be promulgated and enforced by the appropriate authorities.

Reef Fish

There is a greater density of fish species on a reef than in any other place in the sea, with 100-200 species present in a single hectare. The colours, patterns and shapes are breathtaking. The relative ease of observation and the beauty of the subjects make study of their diverse behaviour and ecology extremely rewarding.

The enormous numbers of fish species associated with coral reefs and the high diversity has led to a bewildering array of shapes, patterns and colours enabling individuals from the same species to identify one another (fig. 3.24). Two field guides are available to assist in fish identification (Carcasson 1977; Schroeder 1980).

Reef fish can be divided roughly into those found in rock pools, on the reef flat and those of the reef edge and slope, characterized by 9, 45 and 80 species of fish respectively off Salu Island, Singapore. The reef edge species tended to be larger species of the open sea belonging to the jacks or scads (Carangidae) and snappers (Lutjanidae). The more inconspicuous frog fish (Batrachoididae) and scorpion fish (Scorpaenidae) were restricted to the reef flat and rocky areas. In terms of abundance more fish were found on the reef edge and slope than on the reef flat and this may relate to the amount of available cover (Tay and Khoo 1984).

The most dangerous animals around a reef are fish. Among the fish this group would include sharks (generally small except in deeper water), stingrays (Dasyatidae), moray eels (Muraenidae), lionfish (Scorpaenidae) and stonefish (Synanceiidae). These last two could not be more different: lionfish flounce around the reef in gaudy colours, whereas the drab, grotesque stonefish lies still with its large warty head and mouth almost invisible among the corals. Spines of the dorsal fin inject a poison which can be fatal to humans (fig. 3.25). For this reason if no other, stout shoes should be worn if one is intending to walk on a reef. The other dangerous vertebrates are the sea snakes but most are rarely seen. One, however, the gray-and-black striped *Laticauda colubrina* is confined to coral reefs and the beaches where it lays its eggs (fig. 3.26). The bites of some sea snakes can be extremely dangerous but *L. colabrina* is reluctant to bite, even when handled, and has a venom of low toxicity.

Some fish, particularly those seen just off the reef edge are schooling species with large ranges. On the reef itself there is a bewildering range of social organizations. Some species feed over a 10 m wide stretch of reef reacting aggressively only to others of their species; others, such as small damselfish, occupy very small territories which they defend against every-

Figure 3.24. Reef fish of Sulawesi to show the range of shapes and patterns. **a** - Spotted seahorse *Hippocampus kuda*; **b** - Starry moray *Echidna nebulosa*; **c** - Blue-lined sea bream *Symphorichthys spilurus*; **d** - White-spot humbug *Dascyllus trimaculatus*; **e** - Pennant coralfish *Heniochus acuminatus*; **f** - Orange clownfish *Amphiprion ocellaris*; **g** - White-barred triggerfish *Rhizecanthus aculeatus*; **h** - Two-eyed coral fish *Coradion melanopus*; **i** - Toadfish *Arothron areostaticus*; **j** - Black-tailed thrush eel *Moringua bicolor*; **k** - Black-barred garfish *Hemirhamphus far*; **l** - Patterned tongue sole *Paraplagusia bilineata*; **m** - Tapefish *Anacanthus barbartus*; **n** - Boxfish *Tetrasomus gibbosus;* **o** - Red-purple parrotfish *Scarops rubroviolaceus*.
After Carcasson 1977

Figure 3.25. Two venomous fish of coral reefs: Lionfish *Pterois volitans* (right) and stonefish *Synanceichthys verrucosa* (left).
After Carcasson 1977; Schroeder 1980

thing, including divers. Some live alone, others in groups often with just one male, which results in the formation of bachelor-only groups. In some such one-male groups, when the leading male dies, the largest female in the group immediately adopts male behaviour and soon assumes the physical and physiological characters of the male.

Some peculiar niches occupied by fish are well-known but are nonetheless interesting. There are small fish called 'wrasses', some of which establish cleaning stations on the reef which are visited by larger fish to have their gills, teeth and bodies cleaned of parasites, algae and debris. Another small fish, a blenny, has very similar markings and behaviour which larger fish mistake for the cleaner to their cost because the blennies take bites of flesh instead of removing parasites (fig. 3.27).

Anemone, or clownfish (Amphiprionidae), lessen the chance of falling prey to other fish by taking refuge among the stinging arms of large sea anemones. It used to be thought that they slowly covered themselves with anemone mucus, from their tail forwards, thereby becoming indistinguishable to the anemone from its own tentacles, but it appears now that the fish itself produces its own protective mucus (Brooks and Mariscal 1984).

A study of coral reef fish off the southwest coast of Sangihe Island found seven dominant species (table 3.5). Analyses of the stomach contents showed that all except the rabbitfish ate mainly crustaceans and that the parrotfish tended to eat fewer snails than the snapper or rock cod (table 3.6). Diets also changed from month to month possibly due to changes in availability of food, but the relative contribution of the different food types remained quite similar (Tilaar 1982).

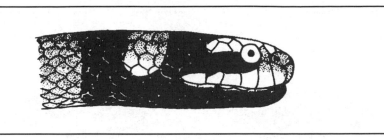

Figure 3.26. Venomous sea snake *Laticauda colubrina*. It has a grey-and-black striped body, a yellowish belly, and yellow marks above and below its eyes.
After Tweedie 1983

Parrotfish generally swim by using their pectoral fins, their tails being employed only when they have to swim faster. They are sometimes seen in small groups which can comprise of adult females and a single adult male who defends the females against other males. Older males sometimes develop a bump on their head. At night they shelter in small caves or under overhanging ledges. These are used consistently such that even if a parrotfish is taken some way from the reef it will swim directly back to its shelter, so long as the sun is visible. Parrotfish and wrasses are peculiar in that at nighttime they secrete mucus from glands in their skin which forms a shroud around them. Whether this is a protection against enemies or whether it is to prevent the gills becoming clogged with silt while the fish is asleep is not known.

Reef fish, like reef invertebrates are objects of trade (p. 230). Indeed most of the marine ornamental fish traded on the domestic and international markets are associated with reefs. The Indonesian region has more potential ornamental marine fish species, perhaps more than 250, than any other region. For comparison, the next most important areas are Sri Lanka with 165 species, the Philippines and Ethiopia with about 110, and Kenya with about 95 (Kvalvågnaes, 1980). Major collecting areas are near Ujung Pandang, Manado and Kendari, as well as around the islands of Kabaena, Salayar and Siau (fig. 3.28). Most of the fish caught are sent to Jakarta for export.

The impact on the wild populations of this collecting is hard to assess with any accuracy but it should be remembered that of every 1,000 fish caught, only about 70% survive to be sold and only half of these live more than six months (Anon. 1986). The regulation of the trade and captive breeding needs to be encouraged with economic incentives.

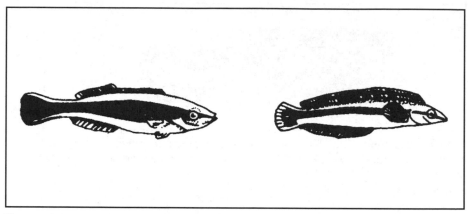

Figure 3.27. The cleaner wrasses *Labriodes dimidiatus* (left) and its mimic, the blenny *Aspidontus tractus*.

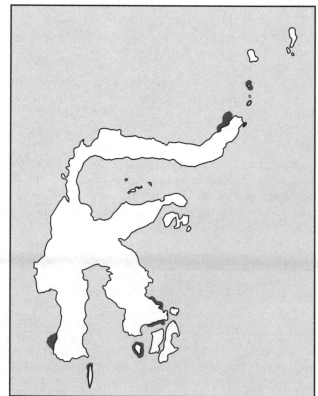

Figure 3.28. Capture sites of marine ornamental fish (indicated in black).

After Salm and Halim 1984

Table 3.5. Dominant fish over the reef edge off southeast Sangihe Island.

GROUPERS Serranidae

Estuary rock-cod, *Epinephelus tauvina* (29)

SEA BREAM Sparidae

Long-spined snapper, *Argyrops spinifer* (29)

PARROTFISH Scaridae

Purple-spot parrotfish, *Scarus dimidiatus* (27)

Green-toothed parrotfish, *S. chlorodon* (24)

Bower's parrotfish, *S. bowersi* (30)

Graceful parrotfish, *S. lepidus* (23)

RABBITFISH Siganidae

Dusky rabbitfish *Siganus furcescens* (31)

Average length (cm) given in parentheses.

Table 3.6. Percentage composition of the food of seven species of reef fish from southeast Sangihe Island.

	Et	As	Sd	Sc	Sb	Sl	Sf
Crustaceans	44	33	32	42	41	43	7
Algae	12	22	22	20	24	20	54
Gastropod snails	29	22	21	16	9	5	4
Polychaete worms	1	3	4	3	1	+	4
Protozoans	1	1	+	1	3	+	1
Arrow worms	+	+	+	+	+	+	+
Rotifers	1	+	+	+	+	+	+
Coral polyps	+	+	+	-	1	+	+
Unidentified	12	20	20	18	20	30	27

After Tilaar 1982

Epinephelus tauvina (Et)
Argyrops spinifer (As)
Scarus dimidiatus (Sd)
S. chlorodon (Sc)
S. bowersi (Sb)
S. lepidus (Sl)
Siganus furcescens (Sf)

Reef Algae and Herbivores

Although a coral reef may appear to be dominated by animals, plants can in fact constitute 75% of the biomass in an area of reef. These plants include both very fine filamentous algae growing in dense mats or turfs, larger seaweeds or macroalgae (fig. 3.29), and coralline algae. This last group are red algae that are calcified, that is they deposit limestone in their cells and are consequently extremely hard. Nearly 22% of the surface of a coral reef near Hawaii was found to be covered with coralline algae (Johansen 1981). The green algae *Halimeda* are also calcified and together with the coralline algae appear to be the major suppliers of sediment below the reef slope (C.V.G. Phipps pers. comm.). Large algae show distinct patterns in their distribution; red algae are found on inner reef flats and outer edges of coral reefs, brown algae such as *Sargassum* usually occur throughout the reef flat, and green algae tend to be found in intertidal zones. The reasons for these patterns do not seem to have been investigated.

Certain seaweeds are cultivated and harvested to produce compounds for the food and chemical industries. Red seaweeds such as *Gelidium* and *Gracilaria* produce agar used in fish and meat canning to prevent breakage, in the manufacture of ice cream, milk drinks, cakes, jams, jellies, as well as in cosmetics. Other red seaweeds such as *Eucheuma* produce gums called carrageenan which are similar to agar but are thicker and are used in the food and cosmetic industries, as well as in paint, insect-sprays and insecticides (Chapman 1970; Teo and Wee 1983). The Menui Islands in extreme southeast Central Sulawesi are a major producer of carrageenan (Hasan 1975) and areas with potential for commercial alga production are found all round the coast (fig. 3.30).

Most herbivorous mammals possess specialized, symbiotic gut organisms which assist in the digestion of plant material or secondary metabolites, producing compounds usable by the animal. Such a relationship has only recently been discovered in marine fish, specifically in the reef-dwelling surgeonfish *Acanthurus nigrofuscus*, a species studied in the Red Sea but also found around Sulawesi (Carcasson 1977). Many day-feeding fish empty their guts before resting for the night but, *A. nigrofuscus* retains a ball of undigested algae. When it begins to feed in the morning it voids this onto the reef upon which it then grazes and this may be the pathway by which it reinfects itself with the symbiotic organisms (Fishelson et al. 1985).

Sea urchins are among the most important grazing animals on the reef and exert heavy pressure on plant biomass. For example, the biomass of algae in a plot from which sea urchins were removed reached 159 g/m^2 in contrast to 12 g/m^2 in areas where sea urchins grazed, although the net productivity of ungrazed tufts may be lower than grazed turfs (N. Polunin pers. comm.).

The presence of sedentary animals seems to be inversely related to the presence of algae. It appears that filamentous algae interfere with the

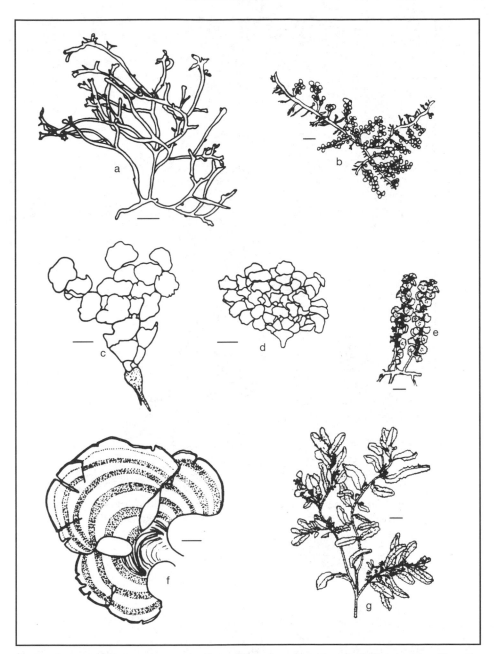

Figure 3.29. Common coral reef algae. **Red algae: a** - *Gracilaria lichenoides* (Grac.); **Green algae: b** - *Caulerpa racemosa* (Caul.), **c** - *Halimeda tuna* (Codi.), **d** - *H. opuntia*; **Brown algae: e** - *Turbinaria conoides* (Fuca.), **f** - *Padina gymnospora* (Dict.), **g** - *Sargassum polycystum* (Sarg.). Scale bars indicate 1 cm.
After Teo and Wee 1983

Figure 3.30. Areas around Sulawesi with potential for the harvesting of commercially-useful algae. Dotted line shows the Taka Bone Rate reef shoals.
After Salm and Halim 1984

feeding structures of erect bryozoans. In addition, in places exposed to grazing by parrotfish and surgeonfish, the presence of algae around sessile animals, such as sea squirts, exposes the latter to damage from those fish species that scrape algae off rocks. In this way fish 'weed out' sedentary animals from algal beds (Day 1983).

Productivity and Plankton

The gross primary productivity of coral reefs is high, about 3,000-7,000 g C/m²/yr (equivalent to 70 t/ha/yr) (pp. 207 and 311). This is balanced by very high respiration, however, so that the net primary productivity is about 300-1,000 g C/m²/yr (Mann 1982). Even so, coral reefs are about 20 times as productive as the open sea where net primary productivity is only about 20-40 g C/m²/yr. The symbiotic zooxanthellae are about as productive as the benthic algae but, because the zooxanthellae are less abundant than the algae, they probably account for less than half of the net primary productivity (Levington 1982).

Coral reefs support many fish, coral and other invertebrates, but early studies of zooplankton densities showed that the density of plankton drifting over coral reefs from the sea was too low to support the reef organisms. In fact, certain zooplankton leave the reef at night and return before dawn (Alldredge and King 1977). Another major source of zooplankton is found close to the coral. These plankton rise from their refuges at dusk and retreat at dawn and thus represent the most important food for the night-emerging corals. So, most of the zooplankton are not drifting in from the sea but are resident among the coral resulting in most of the nutrient minerals being recycled within the reef ecosystem. It is important to note that neither of the major sources of plankton would have been detected by the standard methods of plankton sampling such as net towing (Birkeland 1984).

The filamentous algae growing on the rock (dead coral) on reefs are grazed, and up to 6% of the filamentous algal biomass per day can be removed from some areas. About 50% or more of the algal production enters the grazer food chain (Hatcher 1981), primarily through parrotfish and surgeonfish. The flow of energy through a reef ecosystem is therefore primarily through grazing not through detritus feeding and this is a major factor in understanding coral reef systems (Hatcher 1983; Hatcher and Larkum 1983).

Causes of Coral Death and Reef Destruction

The major and most insidious cause of coral death is suffocation by sediment, either in suspension in the water or that has settled on the coral polyps. Both prevent light from reaching the photosynthetic zooxanthellae and the suspended matter may overburden the filter-feeding polyps with inorganic material. Thus, any plan concerning coral in coastal zone management must include considerations of the sources of that sediment: uplands subject to bad farming practice, deforestation and logging, overgrazed pastures, and public works projects that expose bare soil to the full force of the rain. To incorporate such concerns in regional planning requires dedicated and co-ordinated government action.

One of the most common causes of reef destruction is the use of explosives to catch fish and to break coral into rubble for use in construction projects. For both purposes, explosives are very efficient but it is not without cause that using explosives over reefs has been outlawed in many areas. Explosions kill fish indiscriminately without regard to food value, and more fish probably sink and decay than float to the surface for collection (Burbridge and Maragos 1985). This is because the major site of internal damage is the swim bladder, a sac of gas used to control buoyancy. This is ruptured in explosions and so many fish lose that buoyancy and sink when they die. Bottom-dwelling fish, with poorly-developed swim bladders, are less likely to be killed than other species (Wright 1982). The effects of underwater explosions on invertebrates or dugongs has not been studied, but it is known that divers can suffer perforated ear drums if diving while coral is being blasted. Despite the regulations, the use of explosives is still widespread particularly in the remote, uninhabited areas that are the most valuable for conservation purposes,

The coral itself may recover from blasting (Parrish 1980), but it would take many decades before the effects of blasting are no longer obvious to even a casual diver. The coral rubble and the algae growing on it have a relatively low productivity and the biomass of plankton over such areas is little greater than that over plain sand (Porter et al. 1977). When considering different impacts on coral it is worth remembering that most coral has a radial growth rate of between only 1 and 10 mm/yr (Soegiarto and Polunin 1980).

Although at first sight this may be difficult to understand, coral reefs play a very significant role in protecting coasts from erosion. In deep water the height of an ocean wave is one-twentieth of its length. When a wave approaches shallow water, however, the drag of the sea bed shortens the wave length and the wave is forced up into a peak. This sharp-topped wave breaks when its height is in the ratio of 3:4 with the water depth. Thus a wave 30 cm high will break in water 40 cm deep. It is for this reason that waves can be seen breaking (and have much of their energy dispelled) some distance offshore above the reef slope. In places where the reef is broken or blown apart for the collection of hard core for roads and other constructions, the concave nature of the reef lagoon is changed into a sloping shore partly because the coral 'lip' is being removed and partly because of increased sedimentation. The waves now break nearer the shore and the wave energy is now not just effective in creating water turbulence but also scours the coastline. The killing of coral reefs also greatly reduces the productivity of the coastal waters with the result that the production of reef-related fisheries is certain to fall (fig. 3.31).

It is generally true that the closer a coral reef is to a centre of human habitation, the more disturbed it will be. Thus on the eastern shore of Bunaken Island, four sites at increasing distance from the main village had

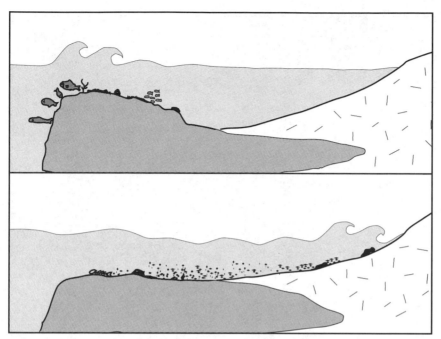

Figure 3.31. The consequences of destroying fringing coral reefs which protect the coast are that the land suffers erosion, the coral no longer supports fish and other commercial animals, the water becomes turbid, and fishing yields decline.

more or less the same number of coral species present in the study plots (about 18-21 species), but the area of dead coral decreased with distance from habitation (fig. 3.32) (Lalamentik 1985).

Coral Reef Fisheries

The largest reef fishery in Indonesia is probably found in the shallow sea of the Sangkarang Archipelago which lies within 80 km of Ujung Pandang. This fishery is based largely on the islands of the archipelago. Some of the islands of the Sangkarang Archipelago west of Ujung Pandang, and coastal areas north and south of Ujung Pandang have dense human populations many of which are dependent on reef fisheries (fig. 3.33). Fishing villages are the least developed of coastal villages and in general have the lowest household and per capita income of all types of villages. This is due to the seasonal nature of their fish harvests, the lack of relevant technology and

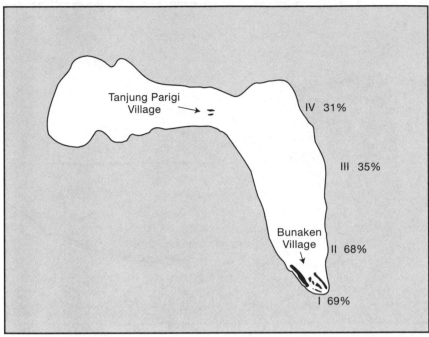

Figure 3.32. Bunaken Island, Minahasa, to show the percentage of dead coral in four study sites.
After Lalamentik 1985

skills, and the paucity of supplementary jobs available in such villages (Makaliwe et al. 1985).

This is well illustrated by a comparison of the situation on two islands 2 km apart that lie 12-15 km from Ujung Pandang. One is based on a mixed economy and the other is entirely dependent on reef-fisheries (table 3.7). Over the last two decades, the numbers of fishermen have increased and the capture per unit effort has increased for those fishermen able to exploit new technologies. Those who have been unable to exploit such changes have been locked into a situation of poverty combined with increasingly intense exploitation of coral reef fish. They can escape from this situation only if money for development is available outside the private sector for consolidating fishermen's cooperatives and possibly for the introduction of mariculture projects based on clams, pearl oysters, mussels, etc. Any such initiative will, however, have to contend with an established philosophy of resource use in a conservative and proud people (Davie and Mustafa 1984). In addition, it should not be taken for

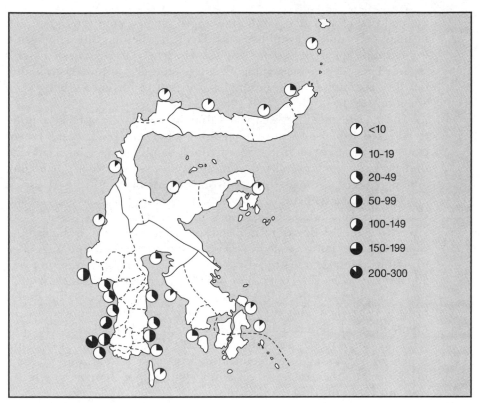

Figure 3.33. Number of fishing households per kilometer of coast by county.

After Salm and Halim 1984

Table 3.7. Comparison of two islands in the Sangkarang Archipelago.

	Barrang Lompo	**Barrang Caddi**
Area (ha)	23	10
Population	2,160	1,030
Economy	Mixed, tradition of inter-island trade, cloth, filigree silver, 10% of households engaged in fishing.	Almost entirely fishing.
Ethnic make-up	Mixed.	Makassarese.
Fishing	Higher capital investment allows pelagic fisheries with nets, relatively lucrative though fisherman income meagre.	Coral-reef based using handlines and traps, declining yields, much of income used to repay loans.
Trends	Pressure on reef fisheries increasing to avoid spending money on fuel.	Unsustainable capture methods increasingly used to compete with other fishermen.

Based on Basuni 1983; Davie and Mustafa 1984

granted that all fishing communities appreciate the connection between healthy reefs and successful fisheries. The fishermen on Bunaken Island, for example, apparently do not, and therefore need to be educated in this regard. They see tourists coming to snorkel and dive around their island, but they themselves see little financial incentive for encouraging such attention (Rondo and Sondakh 1984).

The view is sometimes expressed that, in days past, coastal communities devised systems of tenure that conserved marine resources. A recent exhaustive review of historical information relating to this subject failed to find any such systems reported from the Sulawesi (or Borneo) mainland, although there are many reports from eastern Sumatra, the Moluccas and New Guinea. The only report of marine tenure from a Sulawesi island was from Salayar where the ownership of reef areas was passed from father to son. The dearth of information from elsewhere probably indicates that other tenure systems ceased to operate before those interested in writing about such matters arrived, and that the systems were not very common anyway. The conclusion of the review was that tenure systems were instituted not through a desire to conserve resources, but rather because fishermen wanted to increase their exploitation and eventually found their neighbours doing the same thing. Traditional ownership patterns are therefore an imperfect and risky route to follow to establish responsibility among fishermen within a plan for coastal zone management, because they are based on human gain rather than on an ethic of restraint (Polunin 1984).

Coral Reef Survey Techniques

If the impacts of development activities on coral reefs are ever going to be assessed, then it is essential that work begin now on collecting data from those reefs closest to centres of development. The techniques used need to be able to provide systematic information on changes and natural variation. This repeatable collection of data on all aspects of the coral reef does not appear to have begun in Indonesia, although relevant work has been conducted in New Caledonia since 1973 (Dahl 1981b).

Sea water is an unnatural environment for man and for him to survey coral reefs requires a certain amount of skill and equipment. Firstly, those doing the survey work clearly must be able to swim. Surveys at low tide conducted by walking over the coral can cause considerable damage, puts people in danger of touching venomous fish (p. 232), and is not to be recommended. Scuba diving using tanks of pressurized air is by far the best for examining reefs but it should only ever be done in pairs and after training from qualified instructors. Snorkel and mask, particularly with belt of weights, represent a very good second best and are quite sufficient for working over the reef flat and reef edge.

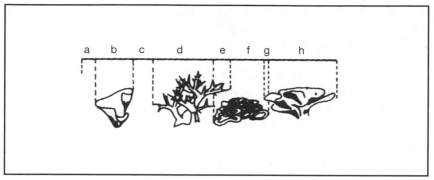

Figure 3.34. A line-transect across a patch of coral showing the measurements taken. Thus: cover by species A = b+g+h, cover by species B = d+e, cover by species C = e+f+g, overlap = e+g, total cover = b+d+e+f+g+h.

After Moll 1985

A wide variety of techniques are available for making quantitative surveys of coral but the line-transect method has been used extensively around Sulawesi and elsewhere. For example, it was used for the study of coral during the Indonesian-Dutch Snellius II Expedition in 1984-85. A series of 30-m transects are laid at set intervals from the shore and measurements are made of the coral cover, or any other parameter directly below the transect tape[15] (fig. 3.34) (Moll 1985).

Line-transects are less suitable for broader studies in a coral reef monitoring program. The methods described below are based on those recommended by the South Pacific Commission for environmental monitoring programs and are well suited to non-specialists. A single person can collect data related to all parameters after some training (Dahl 1981a). A blank data sheet is provided in Appendix N and this could be photocopied, placed back to back, and laminated in plastic. The plastic surfaces can be roughened with fine sandpaper so that they can be written upon underwater.

Circular plots of 50 m^2 can be established by fixing a permanent marker to some obvious feature and marking a circle with a radius of 4m around this marker. Plots should be established in each of the major reef habitats: inner lagoon near the shore, mid-lagoon or patch reef, outer lagoon just inside the reef, back reef where it slopes into the lagoon inner reef flat, outer reef flat, and reef slope. None of the plots need to be deeper than about four metres except on the reef slope which may have to be omitted if appropriate equipment is not available. The sites chosen as plots should appear to be typical of the area within the zone, and given this qualification, may be selected by random sampling.

The number of plots established will depend on the time and manpower available and the type of study. For a straightforward monitoring study at least two plots in each habitat should be established, but for an environmental impact analysis more plots and control plots would be necessary. Repeat surveys are necessary every year at roughly the same time but surveys every two months would provide valuable information on natural variations.

The parameters to be recorded are as follows:

Fish Counts. Since fish are easily scared, it is best to tie a floating rope to the permanent marker and to swim 100 m parallel to the reef edge counting particular fish within an imaginary 5 m transect below the swimmer. Only two groups of fish are to be counted: predators and butterfly fish. Predators to be recorded are longer than an outstretched hand[16] and belong to three families. Butterfly fish are generally smaller than an outstretched hand and are often seen in pairs around coral (fig. 3.35). They are territorial and will be seen in more or less the same location on the return swim.

Percentage Cover. Some of the 50 m² circle will not be covered with corals, etc., but by mud (grains not distinguishable), sand (grains obvious), rubble (finger-size to head-size) or blocks (larger than head-size). The percentage of the circle covered by major biological groups: live hard coral, soft coral and sponges, dead standing coral,[17] and marine plants that can be held in the hand is also recorded. Percentage cover is not easy to assess and the shapes in figure 3.36 may help.

Life Forms Present and Dominant. The major types of corals and plants present and dominant are recorded. The main groups are hard corals (fig. 3.37), soft corals and sponges (fig. 3.38), and marine plants (fig. 3.39). The dominant form (the most obvious one which covers the largest area) is noted as is its most common size: about fist-size, about the size or diameter of forearm, or equivalent to outstretched arms.

Benthic Animal Counts. Reef health can be indicated by the presence or absence of certain animals (fig. 3.40). Totals in excess of 20 need only be expressed as >20.

Visible Pollution. Objects to be recorded include cans, bottles, plastic and other synthetic materials; leaves, palm logs, and other land plant debris; and notes should be made on sediment in the water sufficient to reduce visibility, oily film on the water, or tar on rocks and sand, etc.

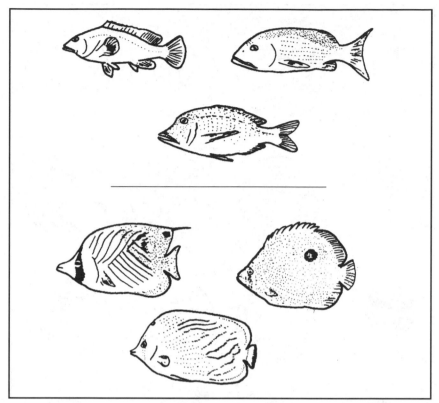

Figure 3.35. Fish to be recorded: predators (three above) and butterfly fish (three below). Not drawn to scale.

After Dahl 1981a

Other Notes. Other items worth noting are recent storms, nearby developments, proximity of human habitation, signs of fishing, etc.

In addition to the above parameters, temperature, salinity and turbidity should be measured in the different zones, preferably repeated regularly during a tidal cycle.

The data collected on their own are clearly useless without interpretation. Interpretation can either be made from a single survey, or from three or more successive surveys. Small variations between surveys are to be expected since variation is a natural condition and no observer is free from error. Important measures or changes can, however, be interpreted (table 3.8). If significant effects are noticed then survey intervals should be shortened, the results communicated to the appropriate authorities, and experts invited to make independent assessments.

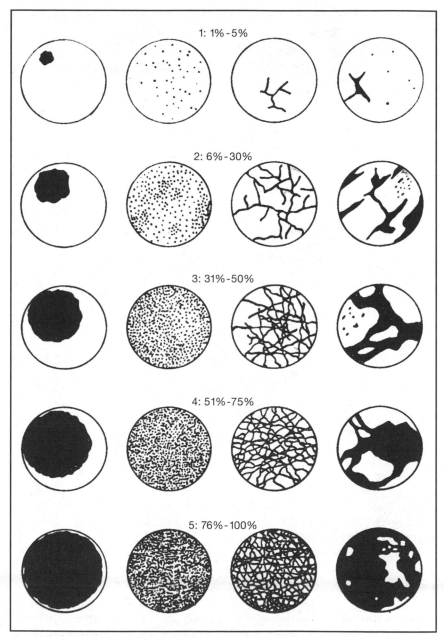

Figure 3.36. Examples of percentage cover.

After Dahl 1981a

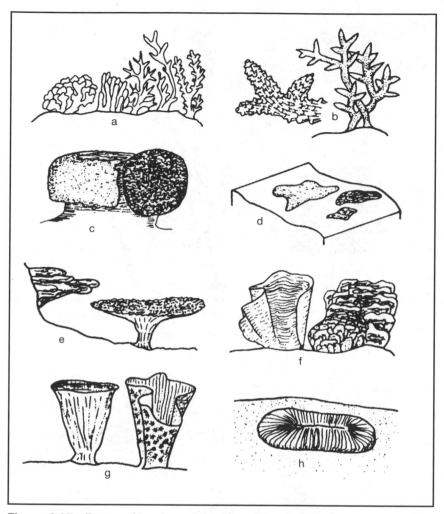

Figure 3.37. Forms of hard coral. a - branching, b - staghorn, c - massive, d - encrusting, e - tabulate/flat, f - erect foliose, g - cup-shaped, h - mushroom.

After Dahl 1981a

Figure 3.38. Forms of soft corals and sponges. Only two forms are distinguished. a - massive, b - fans and whips.

After Dahl 1981a

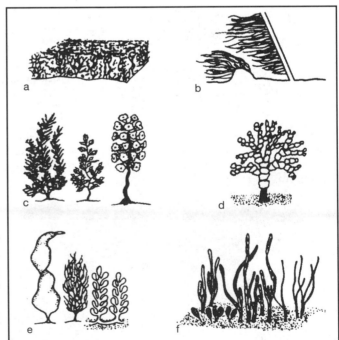

Figure 3.39. Forms of marine plants. **a** - thick turf; **b** - long filaments; **c** - large browns; **d** - *Halimeda*; **e** - other fleshy plants; **f** - seagrass.

After Dahl 1981a

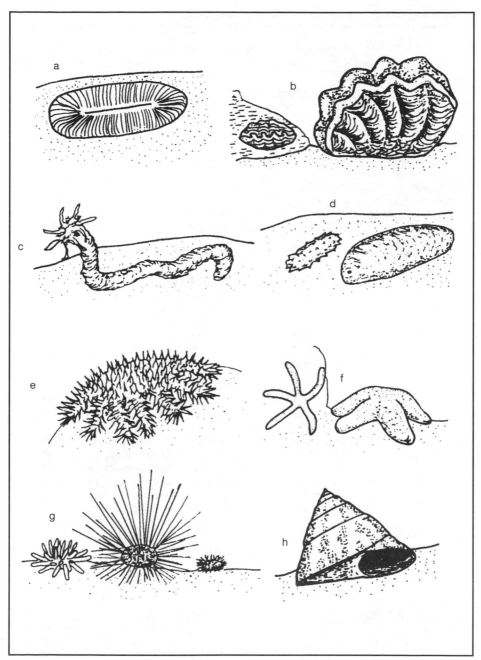

Figure 3.40. Benthic animals to be recorded. **a** - mushroom coral, **b** - giant clams, **c** - synaptid sea cucumbers, **d** - other sea cucumbers, **e** - *Acanthaster*, **f** - other starfish, **g** - urchins, **h** - *Trochus*.

After Dahl 1981a

Table 3.8. Data sheets and key for interpreting data sheets.

	Problem indicator level for any one survey	Three consecutive changes indicating problem	Interpretation
Fish Counts (100 metre line)			
predators	0 on reef front only	decrease	B
butterfly fish	10 or less, reef front or back	decrease	B
Percent Cover			
mud/sand	3 or more (except in lagoon)	increase	C
rubble blocks	4 or more (except in lagoon)	increase	C
live hard coral	0	decrease	A
soft corals/sponges	4 or more		A
dead standing coral	3 or more		A
crustose corallines	3 or more		H
marine plants	3 or more		A/D
Forms Present and Dominant			
Hard coral			
branching	disappearance or change in dominance		H
staghorn	disappearance or change in dominance		H
massive	disappearance or change in dominance		H
encrusting	disappearance or change in dominance		H
tabulate flat	disappearance or change in dominance		H
erect foliose	disappearance or change in dominance		H
cup-shaped	disappearance or change in dominance		H
mushroom	disappearance or change in dominance		E
Soft corals/sponges			
massive	disappearance		H
fans/whips	disappearance		H
Plants			
thick turf	appearance, change in dominance or disappearance		D
long filaments	appearance, change in dominance or disappearance		D
large browns	appearance, change in dominance or disappearance		D
Halimeda	appearance, change in dominance or disappearance		D
other fleshy	appearance, change in dominance or disappearance		D
seagrass	appearance, change in dominance or disappearance		D
Counts of Animals			
mushroom coral		decrease	E
giant clams		decrease	B
synaptids		increase	G
other holothurians	6 or more		H
Acanthaster		increase	F
other starfish		increase	H
urchins		change	H
Trochus		decrease	B
Visible Pollution		increase	I

Table 3.8. (Continued.)

Key for interpreting data sheets.

A - Reef is under stress or has been damaged by storms or by human activities; new developments begun during the monitoring program should be investigated; more survey circles should be established to determine extent of effect.

B - Overfishing is likely; control of fishing techniques probably necessary; fish are also adversely affected by reef damage or change in food or shelter availability; more detailed surveys required.

C - Increasing quantities of mud are caused by land-based erosion due to bad land use; rubble formed from dead coral; increasing rubble cover indicates coral death.

D - Plant growth can either increase or decrease as a result of increased nutrient load; increase in plants can also be caused by over-exploitation of herbivorous molluscs (for their shells) or other herbivores.

E - Mushroom corals are easily removed by ignorant tourists; their disappearance is a measure of uncontrolled visitor pressure.

F - A few coral-eating *Acanthaster* starfish on a reef is normal but in certain conditions they can become too abundant and seriously damage the reef; control may be necessary and the extent of reef affected should be determined.

G - Synaptid sea cucumbers increase in areas affected by urban or organic pollution; may indicate increase in pollution but they are occasionally locally common.

H - Changes in dominance is a natural phenomenon but rapid changes may be caused by man.

I - Cans, bottles, etc., are harmless ecologically (although damaging to a tourist industry), but usually indicate other forms of human activities.

After Dahl 1981a

Chapter Four

Freshwater Ecosystems

INTRODUCTION

Freshwater is an essential element of life. Its obvious value is in the daily uses of drinking and bathing, but it is also harnessed to produce electricity, used in industrial processes, diverted to irrigate crops, and utilized by animals and plants some of which have considerable commercial value. Despite the commercial importance of freshwater ecosystems to humans throughout the tropics, their nutrient cycles, carrying capacities, and ecological limits are poorly understood.

Bodies of freshwater may be flowing (lotic) as in rivers, or more or less stationary (lentic), as in lakes and swamps. The scientists who study the living organisms of these habitats are called limnologists, and their domain actually extends beyond the lakes and rivers into the drainage basins above, features of which strongly influence the chemical composition of the drainage water. For the same reason, nutrients, dust and other particulate matter from the atmosphere which reach the earth in rain water are also considered.

The term 'limnology' was coined in 1892 by a French scientist, F.A. Forel (Cole 1983), and since that time theories and research have for the most part been conducted in temperate zones. Results from tropical limnological studies have revealed some similarities to results obtained in temperate areas, but inherent differences, particularly with respect to greater solar radiation and smaller seasonal variations, suggest that categories devised to define temperate lakes may not be applicable in the tropics (C. Fernando pers. comm.). Further, higher temperatures and a continuous growing season should theoretically support a higher rate of production, but it has not yet proved possible to devise principles that would apply to all tropical situations (Anon. 1982b).

It has been suggested that the sustainable management of freshwater ecosystems requires certain steps to be taken before and after the execution of any program that might affect them (Soerjani 1985). Those steps are:

- determine the aims and the needs of any management program (the aims are generally multi-purpose);
- make an inventory of basic information related, even loosely, to the aims and needs of the management program determined in the first step;
- gather technological knowledge related to the aims of management;
- analyse the available information in the framework of an environmental impact assessment;
- determine a plan of action, accepting the need to compensate for lost options;
- execute the plan;
- evaluate the execution and impacts within a system that is sensitive and flexible enough to allow for modifications in the management goals.

It is hoped that the information provided in this chapter will allow at least some of these steps to proceed in a sounder manner than has previously been possible.

LAKES AND RIVERS

All the major lakes on Sulawesi have been surveyed within the last 10 years to determine their physical and chemical characteristics, but long-term studies to determine seasonal variations have not yet been conducted. The lakes themselves are extremely diverse, ranging from the very shallow Lake Tempe (<1m in the dry season) to the beautiful Lake Matano, at 590 m the deepest lake in Southeast Asia.[1]

In contrast to the lakes, Sulawesi's rivers are virtually unknown, and the great range of possible conditions makes generalizations difficult. The climatic regime of the tropics, particularly the high rainfall, results in rivers which differ substantially from temperate rivers in terms of temperature, substrate, chemical composition and flow regimes. Lakes may be referred to as microcosms,[2] but this term, and the principles derived from studying lakes, particularly large ones, definitely cannot be applied to rivers. Instead, rivers are systems that conduct all matter, solid and liquid, to standing water bodies. If the continuous inflow-outflow is stopped, they no longer exist. At one extreme, they may dry up; at the other, they may become swamps or lakes.

Sulawesi has 13 lakes over 5 km² in surface area and these are of widely different depths located over a range of altitude (fig. 4.1; table 4.1). Sulawesi has the second and third largest lakes in Indonesia (Towuti and Poso); the largest is Lake Toba in North Sumatra (Fernando 1984). If Aopa Swamp (p. 447) is included, the area of standing freshwater on

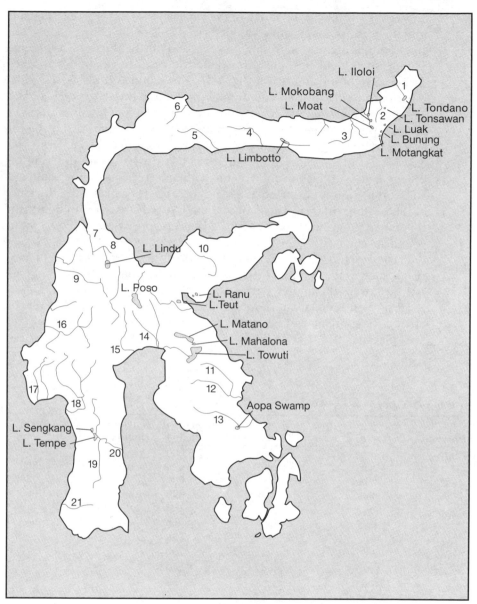

Figure 4.1. Major lakes and rivers of Sulawesi. Rivers: 1 - Tondano, 2 - Poigar, 3 - Dumoga, 4 - Paguyaman, 5 - Milango, 6 - Buol, 7 - Palu, 8 - Gumbasa, 9 - Lariang, 10 - Bangka, 11 - Lalindu, 12 - Lasolo, 13 - Konaweha, 14 - Kalaena, 15 - Kaladu, 16 - Karama, 17 - Mandar, 18 - Sa'adang, 19 - Walanae, 20 - Cenrana, 21 - Jeneponto.

Sulawesi is 191,526 ha, or 1.01% of the total land area. This area fluctuates with the seasons: for example, flooded rice fields in the rainy season create an enormous increase in freshwater area, and during the dry season, Lake Tempe shrinks to become three separate but inter-connected lakes; Tempe, Sidenreng and Buaya (fig. 4.2). The volume of freshwater is likewise not constant: for example, annual variation in the water level at Lakes Towuti and Matano is 92 cm and 56 cm, respectively. This amounts to an annual volume difference of approximately 515 million m³ for Towuti and 92 million m³ at Matano. Maps showing contours of lake depth (fig. 4.3) are useful for calculating volumes of lakes as well as for determining potential fish production and aquatic habitats. To calculate the volume of a lake, the water is envisaged as a series of layers, each layer the thickness of the contour interval. The area of these layers decreases with depth and the sum of the volumes of each of the layers is approximately equal to the volume of the lake.

The size and shape of Sulawesi preclude the development of long rivers, such as those found on Sumatra or Kalimantan, and the longest is only just over 200 km in length (table 4.2). For comparison, the Kapuas River of West Kalimantan is over 500 km in length. The area of rivers on Sulawesi is estimated to be 299,520 ha or about 1.6% of the total land area (Sarnita 1973, 1974).

Table 4.1. Lakes of Sulawesi.

Province	Lake	Type	Maximum Area (ha)	Maximum depth (m)	Elevation (m)
North	Tondano	volcanic	5,000	20	600
	Moat	volcanic	900	23	800
	Limboto	flood	5,600	2.5	25
	Dano	?	880	?	1,050
Central	Poso	tectonic	32,320	450	3
	Lindu	tectonic	3,200	100	1,000
	Teu	?	525	?	?
	Ranu*	?	750	?	c.10
South	Matano	tectonic	16,408	540	382
	Towuti	tectonic	56,108	203	293
	Wawantoa	tectonic	-	-	-
	Masapi	tectonic	-	-	-
	Mahalona	tectonic	2,440	73	?
	Tempe**	flood	35,000	9.5	5
Southeast	Aopa Swamp	swamp	38,000	6	c.100

* Ranu is in fact two adjacent lakes of 480 ha and 270 ha.
** Maximum area and depth for Tempe during the dry season are 1,000 ha and 1.0 m respectively.

After Sarasin and Sarasin 1905; Fernando 1984; Anon. 1979a, b; L. Clayton pers. comm.

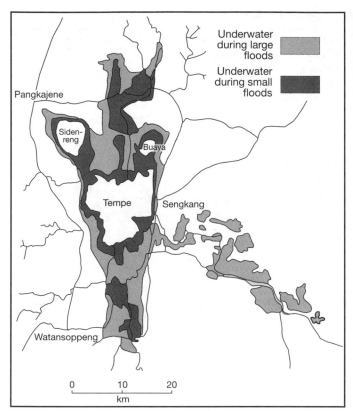

Figure 4.2. The lakes of the Tempe region with the areas under water during small and large floods.

After Anon. 1979a, 1982a

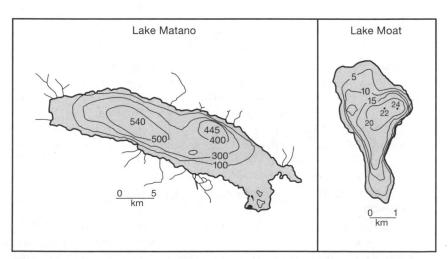

Figure 4.3. Depth contours of Lake Matano and Lake Moat of a type that can be used to calculate approximate lake volume.

After Fernando 1984; EoS team survey

Table 4.2. Major rivers of Sulawesi.

Province	River	Length (km)	Catchment area (km²)
North	Paguyaman	99	-
	Paguat	96	-
	Bone	90	-
	Dumoga	87	-
Central	La	118	2,812
	Lariang	225	7,703
	Lalenda	150	5,547
	Bangka	110	3,141
	Konaweha	190	7,438
South	Kalaena	85	1,512
	Balease	95	998
	Jeneberang	80	-
	Saddang	150	5,453
	Mambi	95	1,391
	Mandar	90	-
	Karama	150	5,574
Southeast	Lasolo	132	-
	Lalindu	145	-
	Sampara	170	-

PHYSICAL FEATURES

The formation of a lake is generally the result of geological processes. Once formed, the basin is doomed to die, following an assured sequence of stages from youth to senescence, at which time the lake has become full of sediment (Cole 1983). Examples of lakes in their final stages of 'life' are Lura (Enrekang), Mokobang (north of Mt. Ambang), Tempe and Limboto. The physical characteristics of a lake reflect the formation process as well as its stage; its chemistry is a reflection of the climate and physical condition of the surroundings. The ecology of Sulawesi's lakes are thus best understood through an initial description of the physical and chemical components.

The lakes can be divided into three major types; tectonic, volcanic and flood lakes. The majority are tectonic, that is, they were formed as a consequence of the movements of the earth's continental plates. Lakes Matano, Mahalona and Towuti are low sections of the Matano fault zone. Mahalona, Matano and Poso are flooded rift valleys (graben), while Towuti is a combination of a rift valley and more complex faulting. From deposits collected around Matano, its formation can be dated back to about 1.6 million years ago (B. Wahyu pers. comm.). Further, Miocene-age limestone on its southern boundary suggests that the area was under ocean water during the late Tertiary and early Quaternary.

From an analysis of the mollusc fauna on an island in Lake Lindu, it was concluded that the lake was formed during the Pliocene, between 5.0 and 1.6 million years ago, when part of the mountain range sank (Sarasin and Sarasin 1905; Bloembergen 1940). These dates do not seem to have had recent confirmation.

Lakes Tondano and Moat in North Sulawesi were formed as a result of volcanic activity. Moat is a high altitude depression surrounded by hills and mountains, and fumaroles near the northwest shore are evidence of continuing volcanic activity. Limboto and Tempe are flood lakes, or fairly flat depressions over which a river flowed in the past. Tempe appears to be a remnant of an ancient strait that formerly separated the southern arm of the Toraja highlands from southern Sulawesi (p. 20).

Scattered along the eastern part of the eastern peninsula of Central and Southeast Sulawesi are small lakes which can be seen clearly from the air. It is likely that these represent sinkholes, a characteristic of limestone topography (p. 470). One such lake has been described from the central peninsula in the south of Muna Island west of Lasongko Bay (Verstappen 1957) and the hills of Maros also reportedly contain such features (Baharuddin pers. comm.). Such lakes are usually shallow and circular, formed by slightly acidic rainwater dissolving the limestone, weakening and depressing the surface structure, such that the surface depression enlarges, or collapses entirely into the channel below. A lake is thus formed and may be fed by channel water or, alternatively, debris may choke off this underground supply and then surface water becomes the sole source of water.

Rivers usually arise in the headwater areas of mountains. They increase in volume and width as they flow downhill, joining other rivers to form a main river, that drains a watershed which may be thousands of hectares in area. Many temporary small rivers are created during rainstorms, particularly in headwater regions. Geology and topography determine drainage patterns and there are a number of types found on Sulawesi (fig. 4.4). Most are typical meandering channels, but the Palu River, several in Morowali National Park and the Jeneberang River, for example, are braided (Bloembergen 1940; Metzner 1981). This describes the small, shallow, interlaced rivers of lowland areas, formed after sediment is deposited during a major discharge. New channels form in the sediment when the water rises, joining, dividing, and joining again. Rivers may also originate at other water sources such as lakes (e.g., Tondano River at Lake Tondano, Guaan River at Lake Moat), limestone caves (e.g., Balangajir River, Maros) and swamps (e.g., Aopa River from Aopa Swamp).

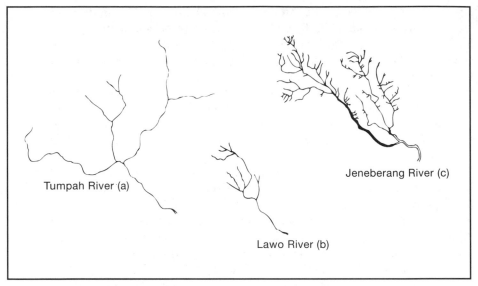

Figure 4.4. Examples of three types of drainage patterns. (a) - dendritic radial, (b) - dendritic, (c) - trellis dendritic.

A catchment is drained by a hierarchical network of channels each of which can be assigned an order depending on its position and relationship to the other channels (fig. 4.5). Thus, the Jeneberang River (fig. 4.4c) is a sixth-order river at its estuary (Pudjiharta and Mile 1981). At the headwater region the stream bifurcation ratio is 3.55; that is, there are three-and-a-half times as many streams of one order as there are of the next higher order. It must be cautioned, however, that assigning orders to rivers depends very much on the map scale used; orders can increase by up to a factor of four on a larger scale map (Dunne and Leopold 1978). Nevertheless, there are evident biotic patterns in rivers correlated with stream orders, and the drainage pattern is a signature of the topography. The trellis pattern in the lower and middle reaches of the Jeneberang River, for example, indicates topographic control effected by parallel ridges and valleys.

Water Inputs and Outputs

The ultimate source of all freshwater is rain which is just one stage in the water cycle (fig. 4.6). Water flows along three pathways: overland, under the soil surface but above the water table, or via the groundwater. Whichever path is travelled, the water is subject to a variety of physical and chemical forces, including gravity, and eventually collects in a depression, river, lake or ocean.

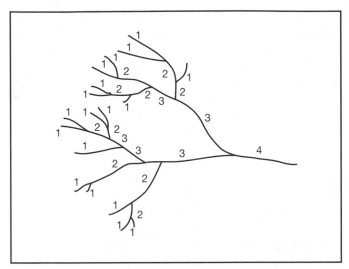

Figure 4.5. Hypothetical river basin showing the system of assigning river 'orders'. Thus: 1+1 = 2, 2+2 = 3, but 2+1 = 2, 3+2 = 3, etc.

After Strahler 1957

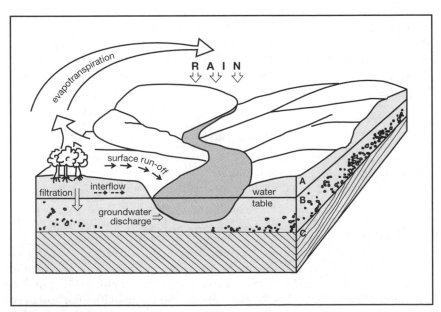

Figure 4.6. Water cycle.

After Townsend 1980

Most of Sulawesi's lakes are supplied by small inlet rivers: Towuti has 26, Tondano 25, Matano 10, Moat 7, and Limboto 6 (Achmad and Cholik 1977; Wardoyo 1978; Wardoyo and Thana 1978; Thana and Wardoyo 1980). The Lake Tempe region is fed by fairly large rivers from the south (Walanae), north (Bila) and west (Batu-Batu and others) (Anon. 1979a). At times of high river flows, water may back up in the Lawe Konaweha River which drains Aopa Swamp, causing the water level in the swamp to rise. Thus the swamp serves as a 'balancing' lake for the river system below it (Anon. 1981a).

The amount of rain reaching the ground below any vegetation cover depends on rainfall intensity and duration, and configuration and density of leaves. Some of the rain water that reaches the ground evaporates from the stomata or leaf openings of plants (transpiration), as well as from the soil, puddles, rivers and lakes (evaporation). Both transpiration and evaporation are difficult to measure but studies indicate that 21%-83% of the rainfall is lost through transpiration in a plantation forest, about 30% is lost through transpiration and evaporation in lower montane forest, and nearly 40% by both processes in the Ujung Pandang area (Wiersum 1979; Jager et al. 1984). Thus, over the period of a year only about 60%-80% of the rainfall actually enters rivers and lakes in the streamflow (Wiersum 1979; Whitmore 1984).

The response of a river to rain depends on the size, topography, geology and soil conditions of the catchment area above the river. Small rivers typically display a quick rise and fall of water levels after rain (e.g., fig. 4.35). Larger rivers are slower to respond to rainfall and the magnitude of the eventual response is less. After rain has fallen, groundwater may not appear as riverflow for hours, days, weeks or even longer. Delayed groundwater flow can therefore sustain low water flows in rivers or lakes through long, dry periods (p. 317).

Water Chemistry

The chemistry of a river or lake reflects the complex interactions of rainwater with soil, rock, plants and climate. Rain is not pure water—it contains low but measurable concentrations of dissolved gases particularly oxygen, nitrogen and carbon dioxide; positively-charged ions (cations) of hydrogen, sodium, potassium, calcium, magnesium and trace elements; and negatively-charged ions (anions), such as sulphates, nitrates, chlorides and phosphates. As the water percolates through or runs off a catchment area, it changes chemically due to the leaching of substances from soil and rocks. These nutrients infiltrate into the soil and are temporarily stored in soil water or are attached to the surface of soil colloids. This attachment or adsorption depends on the cation exchange capacity (CEC)[3] of the soil, which in turn is dependent on pH and amount of clay, humus or organic matter. The CEC may be as low as 5 meq/100 g for sandy soil and over 50 meq/100 g for a clay.[4] Water added to soil can cause some of the adsorbed cations to diffuse away from the exchange surface into solution. The ease of replacement from the complex is approximately H<Ca<Mg<K<Na; that is, sodium is more readily displaced than potassium, and hydrogen least readily of all. Leaching of cations out of the soil system is facilitated by mobile anions, particularly bicarbonate. Bicarbonate is a product of soil organism respiration and it is also a common constituent of limestone bedrock. Thus, leached soil water reaching rivers through the ground-water should contain large quantities of magnesium, calcium, bicarbonate, sulphate and have a pH of 7 or more (Freeze and Cherry 1979).

Freshwater has remarkable buffering capacity, that is, it maintains its chemical characteristics within certain close limits despite the input of liquids with a different composition such as rain-water or factory effluent. The system can, of course be overloaded. However, a study of the effects of effluent from a coconut oil factory on the Tondano River found very little to be concerned about, since the majority of organic and inorganic pollu-tants were degraded through natural processes. The only aspect that did cause concern was the high concentration of suspended matter from soil erosion (Palenewan 1984).

During prolonged rainy periods, subsurface and overland flow are active in contributing rain and soil water nutrients to rivers. Water often turns a brown colour and suspended sediment concentrations are high, perhaps over 1,000 mg/l (p. 320). At low flow periods, during the dry season, groundwater is the major contributor to river flow and the chem-ical composition of water better reflects the geology of the catchment area. Groundwater contains dissolved solids, or inorganic chemicals which eventually flow into receiving waters. These are measured by evaporating a known volume of water and weighing the dried solid residue. Freshwater has a total dissolved solid concentration of less than 1,000 mg/l,[5] brackish water between 1,000 mg/l and 10,000 mg/l, and salt water over 10,000 mg/l (Freeze and Cherry 1979).

Temperature is important in determining water quality. As temperature rises, the rate of chemical reactions increases following the laws of chemical kinetics. Oxygen solubility decreases with increasing temperature and the rate at which oxygen is consumed through chemical and biological oxi-dation of organic compounds, the chemical and biological oxygen demand (or COD and BOD respectively) rises. Thus, a river suddenly exposed to greater sunlight, due to the felling of riverbank trees for example, will experience a dramatic change in water quality associated with the increased water temperature (p. 321). In addition to a lower dissolved oxygen concentration, the warmer water is conducive to the growth of aquatic bacteria that may be pathogenic to certain fish, and will probably alter the metabolic activity of most aquatic organisms. For a given quantity of organic matter, shallow, slow-moving rivers will have lower levels of dissolved oxygen and higher BOD levels than deeper, fast-moving rivers. Rivers at high altitudes generally have higher oxygen levels than those in the lowlands (Chye and Furtado 1982), because turbulent water facilitates the mixing of layers and keeps water temperature low and constant across and through a river; oxygen may also be replenished from bubbles formed in rapids, waterfalls or waves.

PLANTS

Macrophytes

The larger aquatic plants are called macrophytes. These produce complex micro-habitats which provide shelter for animals, act as a substrate for micro-organisms and serve as a homeostatic factor in the decomposition processes of organic matter by consuming carbon dioxide and producing oxygen. Fish use the aquatic plants for shelter and food, and also graze on the small plants and animals which reside on or near the plants. Birds take refuge in emergent plants during storms, make their nests there and eat small floating plants and seeds of these and other plants. Humans use macrophytes for feed (such as *Ipomoea aquatica* [Conv.] [fig. 4.7]), traditional ceremonies, green manure, paper, cattle feed, or simply for aesthetic purposes. Aquatic plants are divisible into five groups: those that grow on wet banks and are frequently flooded, those that are rooted beneath the water but project above the water surface (emergents), those that are rooted beneath the water but whose leaves float on the water surface, those that are rooted and grow wholly beneath the water, and those that float.

The submerged and floating macrophytes of Sulawesi[6] (tables 4.3 and 4.4) include both very common species such as *Ceratophyllum demersum* (Cera.) (fig. 4.8), and very rare species such as *Aponogeton lakhonensis* (Apon.) (fig. 4.9) which is so far known from just two collections near Maros. The curly-leaved *Ottelia mesenterium* (Hydr.) (fig. 4.10) is endemic to Sulawesi and is known only from Lakes Matano and Towuti. A number of macrophytes are rare but are found over very wide areas which suggests narrow niche requirements rather than chance dispersal by aquatic birds. For example, the species of *Najas* in Sulawesi have ranges that include, North Africa, Kashmir and Australia, but in Indonesia, only one species is ever found in a single lake; they appear unable to live sympatrically. The one exception, however, is Lake Tondano which is reported to have three species (de Wilde 1962) but confirmation is needed. Freshwater at low

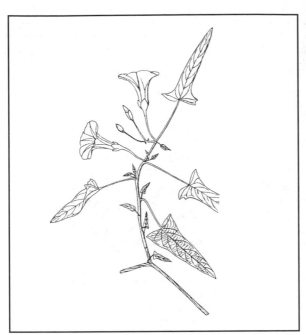

Figure 4.7. *Ipomoea aquatica.*
After van Oostroom 1953

Figure 4.8. *Ceratophyllum demersum.* Scale bar indicates 1 cm.
After van Steenis 1949

Table 4.3. Submerged and floating macrophytes of Sulawesi.

Family	Genus and species	Habitats
Alismataceae	*Sagittaria quayanensis*	Ditches, wet rice fields.
	S. sagittifolia	Swamps, wet rice fields.
Aponogetonaceae	*Aponogeton lakhonensis*	Small rivers.
Araceae	*Pistia stratiotes* (i)	Ditches, rice fields.
Azollaceae	*Azolla pinnata*	Floating on still, shallow water.
Ceratophyllaceae	*Ceratophyllum demersum*	Stagnant pools, slow rivers, lakes, ditches, shallow water.
	C. submersum	As *C. demersum*.
Convolvulaceae	*Ipomoea aquatica*	Shallow pools, ditches, rice fields.
Hydrocharitaceae	*Hydrilla verticillata*	Ditches, pools, lakes, swamps, wet rice fields, slow rivers, even tidal waters.
	Vallisneria gigantea (a)	Shallow lakes, slow rivers and rivers.
	Blyxa auberti (a)	Stagnant pools and swamps.
	B. echinosperma (a)	As *B. auberti*.
	B. japonica	As *B. auberti*.
	Hydrocharis dubia	Floating or rooted in shallow water of pools and swamps.
	Ottelia alismoides	Shallow, slow rivers, stagnant pools.
	O. mesenterium	Lakes, shallow water.
Lemnaceae	*Spirodella polyrhiza*	Stagnant ponds, ditches and rice fields.
	Lemna perpusilla (a)	Small rivers, ditches and rice fields.
	Wolffia globosa (a)	Stagnant water and ditches.
Lentibulariaceae	*Utricularia aurea*	Shallow, still water.
	U. exoleta	As *U. aurea*.
Najadaceae	*Najas indica*	Most freshwater habitats up to 5 m depth.
	N. graminea	Ditches, small rivers, rice fields.
	N. tenuifolia	Lake shores. Known in Sulawesi only from Lake Tondano.
	N. malesiana	Forested rivers and pools, occasionally rice fields.
Podostemaceae	*Cladopus nymani*	Swift, rocky, clear streams.
Pontederiaceae	*Eichhornia crassipes* (i)	Floating or rooted in all types of stagnant or slow running water.
Potamogetonaceae	*Potamogeton malaianus*	?
	P. indicus	
	P. pectinatus	
Salviniaceae	*Salvinia molesta*	Floating on still water.
	S. auriculata	As *S. molesta*.

(a) - Not yet recorded from Sulawesi but likely to be found.
(i) - Introduced.

After van Steenis 1949a, b; Backer 1951; van Ooststroom 1953; den Hartog 1957a, b; de Wilde 1962; van Bruggen 1971; van der Plas 1971; and Taylor 1977

Table 4.4. Submerged and floating aquatic macrophytes in various lakes.

Lake	Submerged	Floating
Tondano	*Ceratophyllum demersum*	*Pistia stratiotes*
	Hydrilla verticillata	*Spirodella polyrhiza*
	Najas indica	*Lemna minor*
	Potamogeton malaianus	*Eichhornia crassipes*
		Azolla pinnata
		Salvinia spp.
Matano/Mahalona	*Ceratophyllum demersum*	
	Ottelia mesenterium	
	Isoetes sp.	
	Chara sp.	
Aopa Swamp	*Ceratophyllum demersum*	
	Najas indica	
	Hydrilla verticillata	
Tempe	*Ceratophyllum demersum*	*Nymphaea* sp.
	Najas indica	*P. stratiotes*
	Hydrilla verticillata	*E. crassipes*
	O. alismoides	
		Hydrocharis dubia
		Ipomoea aquatica
		I. crassicaulis
		A. pinnata
		S. molesta
Limboto	*Hydrilla verticillata*	*Salvinia auriculata*

After Sarnita 1974; Anon. 1977a, 1979a, b, 1983a; Wardoyo and Thana 1978; Thana and Wardoyo 1980

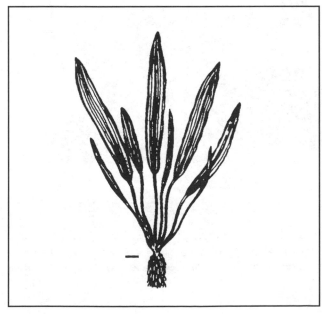

Figure 4.9. *Aponogeton lakhonensis*. Scale bar indicates 1 cm.

After van Bruggen 1971

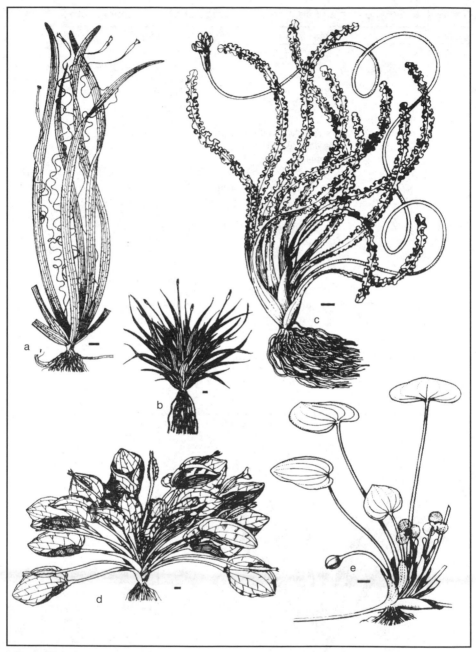

Figure 4.10. Macrophytes of the Hydrocharitaceae. a - *Vallisneria gigantea*, b - *Blyxa auberti,* c - *Ottelia mesenterium*, d - *O. alismoides*, e - *Hydrocharis dubia*. Scale bars indicate 1 cm.

After den Hartog 1957b

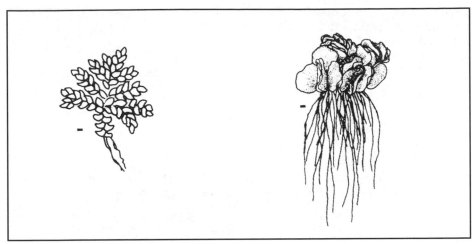

Figure 4.11. Floating ferns *Azolla pinnata* and *Salvinia molesta*. Scale bars indicate 1 mm.

altitudes generally has more species of macrophytes than freshwater at high altitudes (Bumby 1982) though this has yet to be studied in Sulawesi.

The *Isoetes* (Isoe.) reported from Lake Matano or Mahalona is the only record of this aquatic fern from Sulawesi (Sarnita 1974). The species was probably *I. philippinensis*, which has numerous grass-like leaves up to 50 cm long, 7 mm broad at the base and about 3 mm across the middle (Alston 1959). *Isoetes* is of considerable interest in plant evolution because it has the most primitive indication of seed habit in any plant (Corner 1964). Two other aquatic ferns, *Azolla pinnata* (Azol.) and *Salvinia molesta* (Salv.) are common. Both species float and are particularly common in rice fields (fig. 4.11; p. 580).

A peculiar submerged macrophyte is the tiny *Cladopus nymani* (Pods.) (fig. 4.12) which is confined to swift but relatively unshaded rocky rivers with clear water. In Sulawesi it is known only from the southwest peninsula. During the rainy season it is sterile, but when the water level falls it flowers (van Steenis 1949b). The small *Lemna* and minute *Wolffia* (Lemn.) (fig. 4.13) are very common in still waters of the tropics but the latter has not yet been reported from Sulawesi.

One of the most interesting macrophytes is *Utricularia* which has no recognizable leaves or roots, but a mass of floating rhizoids that support erect flower stalks raised above the water (fig. 4.14). Absorption of water and nutrients is conducted through the finely divided leaves. The modified

Figure 4.12. *Cladopus nymani*; single plant and a group on a stone.

After van Steenis 1949b

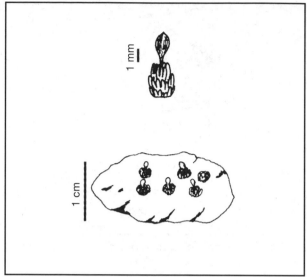

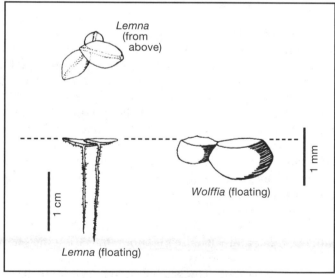

Figure 4.13. *Lemna* and *Wolffia globosa*, common, small, floating macrophytes. *Wolffia* contains the world's smallest plants; *W. globosa* measures just 0.4 mm high and 0.5 x 0.2 mm in length and width.

After Holttum 1954; van der Plas 1971

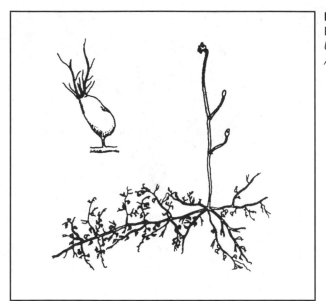

Figure 4.14.
Bladderwort
Utricularia.
After Holttum 1954

leaves bear numerous bladders, the function of which is to catch small animals and so supplement the conventional means of obtaining nutrients. Each bladder has a single opening in which sits a tightly-fitting hinged door. In their relaxed state the bladders are rounded but are usually seen with slightly concave sides because the cells of the bladder have actively passed water to the outside. If the door or the hairs around it are disturbed, it swings inwards because of the reduced pressure, sucking in with it whatever small animal caused the disturbance. The animal dies and decomposes within the now-rounded bladder and the process of passing water to the outside and setting the trap begins again.

The water hyacinth *Eichhornia crassipes* (Pont.) is a Brazilian plant that was brought to Indonesia at the end of the last century and is now considered a serious weed. It was introduced to Sulawesi relatively recently. In a review of its distribution published thirty-five years ago, Sulawesi was not mentioned (Backer 1951). Another floating, introduced macrophyte is the water cabbage *Pistia stratiotes* (Arac.) (fig. 4.15) which has caused serious problems in African lakes.

Observations of macrophyte distribution away from a shore will quickly show how different species grow in different zones. In lakes with steep sides, such as Lakes Matano (Anon. 1980), Teu and Poso (figs. 4.16 and 4.17), the zones are extremely narrow and the macrophyte community relatively small compared with the area of the lake, but in lakes with very

Figure 4.15. *Pistia stratiotes.*

After Holttum 1954

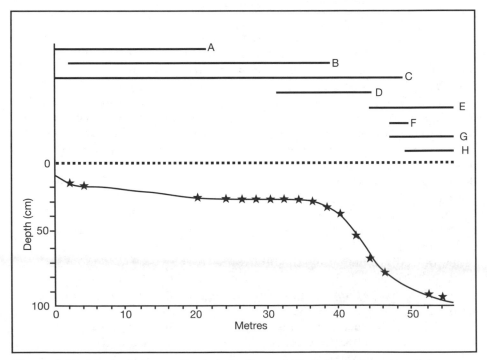

Figure 4.16. Macrophyte zonation in Lake Ranu. A, B, C, E - sedges, D, - *Polygonum*, F, G - *Chara* mosses, H - *Najas*.

After L. Clayton pers. comm.

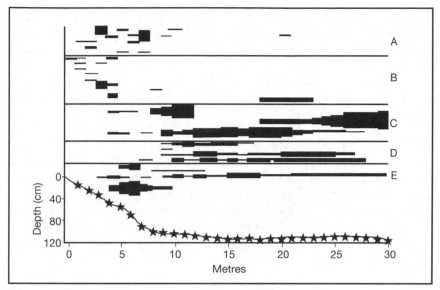

Figure 4.17. Zonation of macrophytes in Lake Poso. A - emergent dicots, B - emergent grasses, C - submerged plants, D - emergent sedges, E - mosses.

After L. Clayton pers. comm.

gently sloping sides, such as Lake Tempe, the zones can be very broad and the macrophyte community can cover the entire lake bed.

The shallow region close to shore, up to about 1 m in depth, forms the primary habitat of emergent macrophytes. These plants utilize the resources of both aquatic and terrestrial environments: there is a very rich nutrient supply in the sediments in which they are rooted, and they have an advantage over submerged plants in that they have direct access to light, oxygen and carbon dioxide. This productivity of emergent macrophytes in eutrophic lakes is consequently very high. Between 1 m and 3 m depth, plants whose rhizomes are rooted to the ground but have extended their leaves to the surface can be found: an example is the water lily *Nymphaea* (Nymp.). Deeper still, up to about 10 m depth in clear water, are found the submerged macrophytes. Floating plants can be found on water of all depths and their distribution is determined primarily by currents and wind.

At Lake Tempe, an EoS team collected and dried the above-ground parts of plants at intervals along a transect from land to about 6 m from the shore. Although the perimeter of the lake fluctuates and is disturbed by cattle and humans, it is evident that the highest biomass of plants occurs at the water's edge (table 4.5), where a mixture of terrestrial and aquatic plants grow together. Biomass decreases in both directions from the water's edge, but dense mats of the cultivated *Ipomoea aquatica* were found in other areas of the lake that had a biomass of up to 1,360 g/m², in water 0.5 m deep.

The zonation is ultimately the result of the physiological problems faced by the aquatic plants caused by the limited diffusion of carbon dioxide and oxygen, relatively low light levels, and high water pressure. Oxygen tolerance is achieved by having up to 60% of the tissue volume occupied by air spaces (Moss 1980). For example, one-third of the tissue volume of *Ceratophyllum demersum* (Cera.) consists of air spaces. Anaerobic metabolism in sediments under water produces ethanol, of which rooted macrophytes are tolerant through a compensating biochemical mechanism. Photosynthesis is also limited by shortage of carbon dioxide, particularly at high pH, but some plants, such as *C. demersum,* are reportedly able to utilize bicarbonate directly, and others are adapted to low rates of free carbon dioxide (Moss 1980).

Table 4.5. Biomass density of plants along a transect from swamp land to 6 m from the water's edge in Lake Tempe.

	Biomass (g/m²)	Plants present
12 m inland	284.6	*Echinochloa colonum* (Gram.)
		Alternanthera sessilis (Amar.)
		Eclipta alba (Comp.)
8 m inland	250.4	*Alternanthera* sp.
		A. sessilis
		Monochoria vaginalis (Pont.)
		Cyperus brevifolia (Cype.)
		Heliotropium indicum (Bora.)
5 m inland	222.1	*M. vaginalis*
		Echinochloa colonum
		A. sessilis
Water's edge	823.6	*Ipomoea aquatica* (Conv.)
		A. sessilis
		M. vaginalis
		Pistia stratiotes (Arac.)
		Unidentified submerged macrophyte
3 m from shore	265.6	*Ipomoea aquatica*
(water depth 0.15 m)		*P. stratiotes*
		Ludwigia adscendens
		Hydrilla verticillata (Hydr.)
		M. vaginalis
		Unidentified submerged macrophyte
6 m from shore	188.8	Unidentified submerged macrophyte
(water depth 0.7 m)		

Data from an EoS team

Figure 4.18. *Hydrilla verticillata.*
After den Hartog 1957b

Light is absorbed quickly by water, so submerged plants receive only a small proportion of the light reaching the water surface. Water lilies develop long petioles so that their leaves can float on the surface and therefore receive light directly. Submerged leaves assume a morphology similar to terrestrial shade plants; their leaves are thin and contain many chloroplasts thereby maximizing the use of available light energy. *Hydrilla verticillata* (Hydr.), a widespread submerged macrophyte (fig. 4.18), for example, is able to maintain a high photosynthetic rate at low light intensities (Finlayson et al. 1984), and can be found 6-7 m below the water surface (den Hartog 1957b).

Water pressure limits vascular plants to about 10 m depth[7] in lakes, but mosses and algae can live much deeper; *Chara* sp.[8] a green alga with a stout stem and whorls of slender branches which are themselves branched, has been found at 164 m in a lake in California (Moss 1980).

Aopa Swamp (p. 446) is one of the better-studied aquatic ecosystems on Sulawesi and the most recent investigation by a team from Bogor Agricultural University (IPB) divided it into four ecological zones (fig. 4.19):

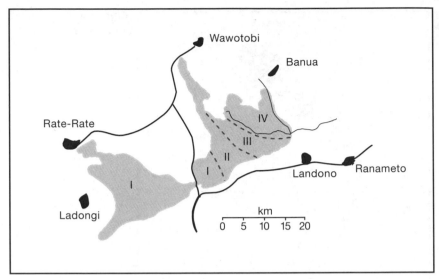

Figure 4.19. Ecological zones of Aopa Swamp.
After Anon. 1983a

Zone I or Makaleleo is the whole of the western swamp and the western portion of the eastern swamp. The vegetation cover is more or less continuous. The water is coloured as black as strong coffee and has a pH of 6.5 (Anon. 1983a) to 5.5 (Thana et al. 1981).

Zone II or Aopa extends part-way into the eastern swamp from the dike. In the rainy season some of the water comes from the Konaweha River and has the colour of strong tea and a pH of 6.6. Water plants here and in the following zones are less abundant so that open water is common.

Zone III or Muara Aopa extends further east again up to the western bank of the Konaweha River. This zone receives water from the Konaweha at all seasons. The water colour is like weak tea and slightly alkaline (pH 7.1).

Zone IV or Tangenutu covers the rest of the swamp and is influenced by water from both the Konaweha and Lahumbuti Rivers. The water has no obvious brown colour and a pH of 6.7. The last two zones are mesotrophic tending to eutrophic (Anon. 1983a).

The dominant aquatic plants differ between zones (table 4.6), but grasses and sedges accounted for 90% of those found. Among the submerged plants *Ceratophyllum* and the small *Utricularia* are the most common (Vaas 1956).

A report from 30 years ago observed that lotus plants *Nelumbium nelumbo* (Nelu.) covered 60% of the water surface (Vaas 1956). These areas may not have been enumerated by the IPB team or may have since become dominated by other species. The lotus is often confused with water lilies *Nymphaea* but the two are quite distinct (Appendix E).

The productivity of most submerged plant communities is about five times lower than that of emergent, floating or terrestrial macrophyte communities (18 t/ha/yr against 75-100 t/ha/yr) (Sutton 1985) due to the reflection of light from the water surface and suspended particles in the water, lower rates of gas transmission, epiphytic algae and protozoans that shade the leaf surfaces, and the general absence of extensive root systems. As a result, emergent rooted plants will quickly overgrow submerged ones where conditions are suitable. Productivity varies considerably depending on environmental conditions: thus in sunny pools algae has been found growing six to seven times faster than in shaded pools and this influences the density of algae-feeding fish (Power 1983).

Table 4.6. Percentage of 10 m x 10 m quadrats in Aopa Swamp covered by different plants.

	Percentage in each zone			
	I	II	III	IV
Plants of riverbanks				
Thoracostrachium pandanophyllum (Cype.)	45	30	35	5
Panicum repens (Gram.)	10	15	20	25
Leersia hexandra (Gram.)	5	10	4	20
Other	5	7	5	10
Plants rooted beneath the water but extending above it				
Nymphaea sp. (Nymp.)	10	15	5	-
*Ludwigia adscendens** (Onag.)	5	5	1	-
Other	5	5	3	-
Floating plants				
Pistia stratiotes (Arac.)	-	3	1	-
Other	-	4	4	-
Plants growing wholly beneath the water surface				
Ceratophyllum demersum (Cera.)	5	-	5	15
Chara fragilis (Char.)	3	-	10	20
Other	7	6	2	5

* This plant (shown in fig. 4.20) is sometimes referred to by its old name *Jussiaea repens*.
After Anon. 1983a

Figure 4.20. *Ludwigia adscendens,* a small emergent macrophyte found in damp locations throughout Sulawesi.
After Raven 1977

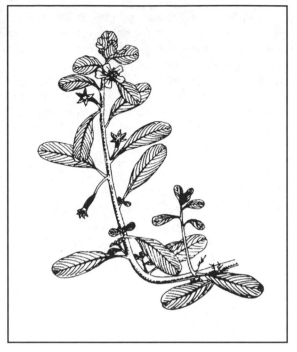

The quantity of dead aquatic plants may be so high that organic matter accumulates in the water, thus reducing water clarity and concentrations of dissolved oxygen, and increasing BOD. The decomposition half-life (the time taken for half of a leaf to disappear through decomposition) of three common Sulawesi macrophytes, *Salvinia molesta, Hydrilla verticillata* and *Eichhornia crassipes* at 26°-32°C is relatively rapid, 23.5, 20 and 11.5 days respectively, compared with leaves from terrestrial ecosystems (Sastrautomo 1985).

Aquatic macrophytes release oxygen into the water as a product of photosynthesis, and this is used for respiration by animals and decomposers on decaying or dead material. When organisms die, organic matter is released into the water, taken up by plants and animals, and thus increases the rate of nutrient cycling in the ecosystem (Sastrautomo 1985).

Phytoplankton

Green algae are the major primary producers in lakes and rivers and microscopic phytoplankton is the principal group of algae in lakes. These organisms float or drift near the water's surface, often together with microscopic animals (p. 286), and are the major photosynthetic producers. They have many unusual shapes, with spines, horns and hairs, which were originally thought to increase surface area and thus help in buoyancy, but are now believed to have evolved in response to absorption and defence against herbivory as well (Cole 1983). Apart from these different types of green algae, phytoplankton include flagellates, diatoms and blue-green algae (Cyanobacteria). Although some bacteria are photosynthetic, they are considered to be neither plant nor animal, since they combine characteristics of both.

The phytoplankton are conveniently divided into two groups on the basis of size: net plankton, which are retained by the silk of a net whose openings are normally 50 μm^9 or larger; and the nanoplankton which pass through the net as it is lowered and raised through the lake (Cole 1983). The latter group may be more numerous, up to a million in one litre of lake water, while net plankton may number thousands per litre (Moss 1980).

The phytoplankton lead a precarious existence. They are readily preyed upon by fish and zooplankton, their habitats may be destroyed by desiccation or incoming floodwaters, or they may sink to the bottom of the lake where they perish due to lack of light. These conditions have favoured the selection of rapid reproduction by simple cell division which can occur every few hours or days (Moss 1980).

Forty-eight genera of phytoplankton, in six families, have been found in Lake Tondano where there is a linear relationship between numbers of individual phytoplankton and zooplankton (Ratag 1981). At Lake Moat there were 47 genera from seven families (Buchari 1984). Although the number of genera were almost equal for the two lakes, the actual number of individuals per litre of water differed considerably: 2,413 in Tondano and only 140 in Moat. The highest count was for Aopa Swamp at over 19,000 individuals per litre (Anon. 1983a) (table 4.7). From this count, evaluations of zooplankton, algae and fish abundance, and from the distribution of aquatic plants, Moat is considered to be oligotrophic, or nutrient-poor, and Tondano and Aopa Swamp to be eutrophic, or nutrient-rich (p. 340).

In addition to the lakes of above, partial lists of phytoplankton are available for the three lakes of the Malili system (Achmad 1974), and Limboto (Wardoyo and Thana 1978).

In rivers, algae attach themselves to large rocks or boulders, and only when river flow is slow can substantial phytoplankton development occur. Factors limiting their growth and productive potential are high sediment loads, insufficient rocks, and the absence of suitable microhabitats (Chye and Furtado 1982). A survey of phytoplankton at three points along the Jereberang River by an EoS team found 16 and 14 genera of phytoplankton in the relatively slow-flowing headwaters (1,440 m altitude) and in the lower reaches (about 5 m) respectively, but only 6 genera in a swift middle reach (850 m).

Fungi, Bacteria and Blue-green Algae

The major decomposers in freshwater are bacteria and aquatic fungi. The dominant group of fungi is the Hyphomycetes which colonizes corpses, faeces, and dead plant material. Fungal spores often have four projections (fig. 4.21), which favour their attachment to solid matter, even in fairly turbulent waters (Moss 1980). Knowledge of the biology and ecology of the decomposers is confounded, however, by their minute size and the difficulty of meaningful sampling.

Bacteria are generally dominant among macrophytes where, along with algae, they live epiphytically. It has been discovered that the leaves of macrophytes such as *Najas flexilis* actually secrete organic material, including glucose, sucrose, fructose, xylose and glycine, and these are readily taken up by bacteria (Moss 1980). The strategy, if any, of this relationship remains unclear.

Table 4.7. Phytoplankton abundance and genera in major Sulawesi lakes.

Lake	Average number/litre	Dominant phytoplankton
Tondano	2,413	*Staurastrum, Cosmanium, Navicula, Synedra*
Moat	140	*Cosmanium, Staurastrum, Melosira, Echinophaurella, Nitzchia*
Aopa Swamp	19,571	*Microcystis, Oscillatoria, Chlorococcus Scnedesmus, Trachelomonas, Nitzchia Melosira, Phacus*
Lindu	-	*Microcystis, Melosira, Synedra*
Poso	-	*Staurastrum, Melosira, Botryococcus Microcystis, Navicula*
Matano	-	*Staurastrum, Cosmanium, Navicula, Synedra, Cyclotella, Diatoma, Microspora*
Tempe	-	*Rhyzoclonium, Gomphoneis*

After Sarnita 1973, 1974; Anon. 1977a, 1979b, 1980, 1983a; Pirzan and Wardoyo 1979; Ratag 1981; Umar 1984

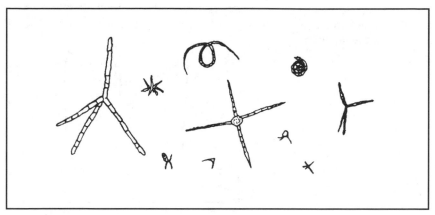

Figure 4.21. Fungal spores.

The blue-green algae, or Cyanobacteria, are present in an enormous range of habitats including hot springs hotter than 55°C (Kullberg 1982). They have evolved the use of chlorophyll 'a', the most important photosynthetic pigment in higher plants, and are thus, along with green algae, primary producers of oxygen in freshwater habitats. They are also capable of fixing nitrogen (Cole 1983). Overgrowth or blooms of blue-green algae, however, produce toxins dangerous to fish and other animals.

FAUNA

The fauna of lakes and rivers can be roughly categorized by the micro-habitat they occupy. Thus:

- the neuston comprises those animals able to live on the surface of the water supported by surface tension, such as pond skaters or water striders (fig. 4.22);
- the nekton comprises swimming animals such as fish;[10]
- the zooplankton comprises animals that drift in the water or swim weakly, such as small crustaceans and protozoan rotifers;
- the benthos are those animals closely associated with the river or lake bed, such as many insect larvae, molluscs, prawns and crabs.

Figure 4.22. A water strider or pond skater (Gerridae) a common member of the neuston.

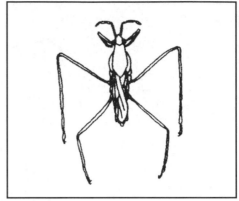

Zooplankton

The zooplankton are the most numerous animals of lakes and some rivers and are also the smallest, with sizes ranging from 0.2-5.0 mm (Moss 1980). The zooplankton of Southeast Asia has relatively few species of the single-eyed Copepods, or of the Cladocera (waterfleas) (Fernando 1980), and Sulawesi is no exception (table 4.8). It is believed that at Lake Matano the high quantity of chromium (0.04 ppm) and the presence of the other heavy metals such as cobalt and manganese (Anon. 1980) may have produced an environment unsuitable for at least Cladocera (Fernando 1984). It is important to understand that in such conditions it may not be the high concentration of any one metal but rather the combined or synergistic effects of the different metals that influence the distribution of a particular species.

The zooplankton are primarily represented by four groups; the protozoa, rotifers, waterfleas and copepod crustaceans. These animals prey on phytoplankton or smaller zooplankton in a variety of ways. Plankton such as the waterfleas grasp their prey, while copepods waft the smaller organisms towards their mouths using their thoracic limbs, rotifers filter water through their mouth, and some protozoans spread around an item of food and engulf it (Moss 1980). Certain phytoplankton can travel unharmed through the digestive tracts of zooplankton and in fact absorb nutrients during their passage (Cole 1983).

Altitudinal variation in zooplankton abundance was examined by an EoS team along the Jeneberang River system. At 1,440 m only a planktonic larva of an arguloid fish louse was found, no zooplankton at all were found in the fast-flowing middle section at 825 m (where very few phytoplankton

genera were found–p. 284), but a wide range of zooplankton were found in
the waters at a slow lowland station where populations were able to develop
and phytoplankton was relatively abundant.

Macro-invertebrates

Macro-invertebrates, invertebrates visible to the naked eye, can be categorized as:
- shredders: those feeding on large units of plant material;
- collectors: those feeding on loose organic particles either on the
  riverbed or free in the water;
- grazers: those feeding on attached algae, rotifers and bacteria; and
- carnivores: those that (usually) kill and eat other animals.

Most of the non-benthic macro-invertebrates are found associated with
macrophytes. These plants act as refuges from predators, as substrates for

Table 4.8. Dominant zooplankton of Sulawesi lakes.

Lake	Cladocera	Copepoda	Rotifera	Average number per litre
Tondano		*Keratella* *Naupilus* *Cyclops*	*Brachionus*	93
Moat		*Keratella* *Cyclops* *Naupilus*		11
Aopa Swamp	*Moina*	*Cyclops* *Naupilus*	*Brachionus* *Arcella* *Lecane* *Monostyla* *Lepadella*	295
Lindu	*Daphnia*	*Cyclops*	*Brachionus* *Polyarthra* *Monostyla* *Vorticella* *Filinia*	
Poso		*Copepoda*	*Monostyla* *Lecane* *Phormidium*	
Tempe	*Diaphanosoma*	*Macrocyclops* *Mesocyclops* *Eurycerus*	*Diaptomus* *Phyllodiaptomus*	

After Sarnita 1973, 1974; Anon. 1977a,1979b, 1983a; Pirzan and Wardoyo 1979; Ratag 1981

algae, diatoms and rotifers, but are not themselves eaten by invertebrates. At first sight this is peculiar since the leaf cuticles are thin and few possess spines or hairs that might dissuade invertebrate herbivores from eating them. It has been suggested that the plants simply are not very nutritious but in fact they contain as much protein as high-quality forage crops. In fact, the most likely reason why macrophytes are not eaten is rather because many of them appear to have significant quantities of a wide range of defensive chemicals (alkaloids) in their leaves (p. 371) (Ostrofsky and Zettler 1986). The types and diversity of macrophytes in an area play a major role in determining the abundance and diversity of macro-invertebrates present (Scheffer et al. 1984).

Freshwater molluscs are among the better-known groups of macro-invertebrates. From the distribution of species, it appears that the species on Sulawesi may be divided into two categories: those that have a few, widely distributed species, and those that have many species, most of which are endemic to the island or a single lake. A total of 45 molluscs are known from Sulawesi's lakes and of these 17 are recorded from Lake Poso (fig. 4.23; table 4.9).[11] Not all the lakes have received equal attention, however, and the total for Towuti, the largest lake, seems surprisingly small, but otherwise the ancient lakes of Poso, Matano and Towuti have the highest proportion of endemic species (67%, 76%, and 87% respectively). Lake Poso has the most distinctive mollusc fauna, including many species with primitive characters, and has two genera confined to its waters one of which, *Tylomelania*, comprises three species. As described elsewhere (pp. 52 and 297) some of these species may now be extinct.

The affinities of the different mollusc faunas are of interest. The greatest similarity (as measured by the percentage of species shared between two lakes) is, not surprisingly, between Tempe and neighbouring Sidenreng which share four (50%) of their eight species. Water from Lake Matano flows into Lake Towuti but only 10% of their total mollusc fauna appears to be shared between them. The mollusc species in these two lakes are not shared with any other lake. Conversely, Lake Poso with its distinctive fauna shares species with all the other lakes. Interestingly, the shallow and intensely-used Lake Limboto shares 10 (30%) of its mollusc species with the Tempe Lakes, although the lakes are separated by over 600 km. The similarity is probably due to the shallowness and surrounding land use common to both lakes and to the fact that many of the the snails are widespread and have probably been introduced along with fish.

Swampy lakes such as Lura, Bolano and to some extent Aopa Swamp have very simple mollusc faunas comprising relatively common species. For example, Aopa has *Ampullaria*, *Planorbis* and *Vivipara*, Lura has a single widespread snail of agricultural areas *Lymnaea auricularia*, and Bolano has two species of *Pila*, snails that have accessory breathing organs.

A bivalve mollusc of interest collected during Project Wallace in

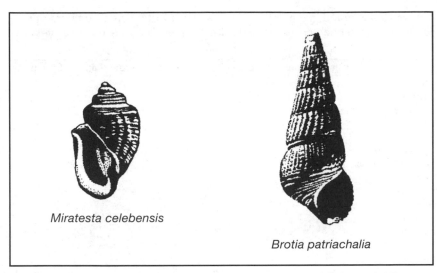

Miratesta celebensis

Brotia patriachalia

Figure 4.23. Two of the endemic freshwater molluscs of Sulawesi. *Miratesta* is an endemic genus confined to Lake Poso.

Bogani Nani Wartabone National Park was *Anodonta woodiana* found in the Toraut River. This species is a native of China and Taiwan, and was introduced to Java (Djajasasmita 1982) and has probably spread to Sulawesi along with the introduced fish tilapia *Oreochromis* on which it encysts during the glochidial larva stage (D. Dudgeon pers. comm.). *A. woodiana* is the largest freshwater mussel recorded from Indonesia: it can reach 27 cm long, 13 cm high and 61 cm wide and is of an olive to dark-green colour (Djajasasmita 1982).

Details of abundance and distribution of four snail species or species groups have been studied in Lake Tondano. Samples were taken once each month for three months and few clear patterns were found except that, in general, the snails were most common where macrophytes were most dense. In addition, some species were found more commonly over sand and others over mud (Buchari 1981). It is commonly stated that snails feed by grazing algae from stones and plants but this oversimplifies a situation in which different species prefer different foods and many ingest predominant inorganic material and detritus (Dudgeon and Yipp 1983b).

There is little information available concerning the ecology of tropical freshwater snails. The generally large, medically-important (p. 297), air-breathing pulmonate snails generally breed frequently and abundantly

Table 4.9. Molluscs found in major lakes of Sulawesi. Note: Large numbers of new, endemic species have been collected in the Malili lakes, but they have not yet been described.

	Tondano	Limboto	Lindu	Poso	Matano	Towuti	Tempe	Sidenreng
GASTROPODA								
AMPULLARIDAE								
Pila ampullacea	•	•	-	•	-	-	-	-
P. scutata	-	-	-	-	-	-	•	-
HYDROBIIDAE								
Indopyrgus **bonnei***	-	-	-	•	-	-	-	-
Oncomelania hupensis	-	-	•	-	-	-	-	-
NERITIDAE								
Neritina labiosa	•	-	-	-	-	-	-	-
THIARIDAE								
Brotia carota	-	-	-	-	-	-	-	-
B. **centaurus**	-	-	-	•	-	-	-	-
B. **germifera**	-	-	-	-	•	-	-	-
B. **insulaesacre**	-	-	-	-	-	•	-	-
B. **kuli**	-	-	-	•	-	-	-	-
B. **molesta**	-	-	-	-	•	-	-	-
B. **monacha**	-	-	-	-	•	-	-	-
B. **patriarchalia**	-	-	-	-	•	-	-	-
B. **perfecta**	-	-	-	-	-	-	•	-
B. **policolarum**	-	-	-	-	•	•	-	-
B. **scalariopsis**	-	-	•	-	-	-	-	-
B. **teradjarum**	-	-	•	-	-	-	-	-
B. **zeamais**	-	-	-	-	•	-	-	-
Melanoides amabilis	-	-	-	-	-	-	•	-
M. fontinalis	-	-	-	-	-	-	•	•
M. granifera	-	•	-	•	-	-	•	•
M. plicaria	•	-	-	-	-	-	-	-
M. tuberculata	•	-	-	•	-	-	-	•
Thiara scabra	•	•	-	•	-	-	-	•
Tylomelania carko	-	-	-	•	-	-	-	-
T. neritiformis	-	-	-	•	-	-	-	-
T. porcellanica	-	-	-	•	-	-	-	-
VIVIPARIDAE								
Angulygra costata	•	-	-	-	-	-	•	•
Bellamnya javanica	-	-	-	•	-	-	•	-
B. **lutulenta**	-	-	-	•	-	-	-	-
B. **rudipellis**	-	-	-	-	•	-	-	-
Celetaia porsculpta	-	-	-	•	-	-	-	-
Protancylus adhaerens	-	-	•	•	-	-	-	-
P. pileolus	-	-	-	•	-	-	-	-
LYMNAEIDAE								
Lymnaea auricularia	•	-	-	-	-	-	-	-
PLANORBIDAE								
Gyraulus **tondanensis**	•	-	-	-	-	-	-	-
Miratesta **celebensis**	-	-	•	-	-	-	-	-
Physastra **celebensis**	•	-	-	-	-	-	-	-
P. minahassae	•	-	-	-	-	-	-	-

Table 4.9. (Continued.)

	Tondano	Limboto	Lindu	Poso	Matano	Towuti	Tempe	Sidenreng
STENOTHYRIDAE								
Stenothyra ventricosa	•	-	-	-	-	-	-	-
BIVALVIA								
CORBICULIDAE								
Corbicula lindoensis	-	-	•	-	-	-	-	-
C. loehensis	-	-	-	-	-	•	-	-
C. matannensis	-	-	-	•	•	-	-	-
C. subplanata	-	-	•	-	-	•	-	-
C. sp.	-	-	-	-	-	-	•	-
TOTALS	11	3	6	17	8	4	9	5

*Species and genera endemic to Sulawesi are shown in bold type.
After Djajasasmita 1972, 1975; Carney et al. 1980; Buchari 1981; and EoS teams

and this appears to lead to a reduced adult life span. These snails may be regarded as r-selected (p. 345) and this probably reflects the often temporary nature of their habitats (pools, rice fields, lake fringes). The more common prosobranch snails (with no lungs) on the other hand tend to have frequent broods of relatively few young (sometimes born alive rather than hatching from a egg laid in water), and have longer adult life spans. These snails seem to be K-selected. The bearing of live young (viviparity) is clearly advantageous for a river-dwelling snail since planktonic larvae would be carried downstream and out to sea. It is not surprising, then, that the viviparous snail family Thiaridae dominates the headwaters and middle reaches of rivers in the Old World tropics (Dudgeon 1982b).

One of the most common and widespread thiarid snails is *Melanoides tuberculata* which is found in lakes, irrigation ditches and similar habitats from Africa to the Pacific Islands. Not only does this snail give birth to fully formed young, but it reproduces without the eggs being fertilized (parthenogenetically). Populations have 0%-3% males, but they do not appear to function sexually. All individuals are therefore genetically identical and this lack of evolutionary potential might be thought to doom the snail to extinction. It would seem, however, that *M. tuberculata* has all the adaptive potential required since it is an extremely effective colonizer. One particular adaptation favouring colonization is the presence in the snail's brood pouch of young of all ages: eggs, larvae and small snails, the last of which can be released throughout the year or when environmental conditions are favourable (Dudgeon 1986a).

Freshwater molluscs have been eaten by man since prehistoric times (p. 74) and are commonly eaten by villagers in certain areas even today. Only in Lake Tempe, however, does there seem to be a commercial mollusc fishery. In 1976 the lake produced 51 tons of mussels *Corbicula* and 34 tons of *Pila scutata* (Anon. 1979b).

Another group of macro-invertebrates used as food is the river prawns. The Sulawesi species are very poorly known but during Project Wallace four species of *Macrobrachium* (*latimanus, lepidactyloides, australe* and *lar*) and two of *Atyopsis* (*spinipes* and *moluccensis*) were found in the Toraut River (D. Dudgeon pers. comm.).

Among the aquatic insects the Hemiptera are represented by several families three of which are relatively often encountered: the giant water bugs (Belostomatidae) (fig. 4.24), the water boatmen or backswimmers (Notonectidae), and the pond skaters or water striders (Gerridae). The giant water bugs are attracted to lights at night and look superficially like large (about 8 cm) cockroaches. They can deliver a painful bite if handled carelessly. Their front legs are used for grabbing prey, while the middle and hind legs are used for swimming. The water boatmen swim upside down and have long hind legs which are used as oars. The bodies and legs of pond skaters are covered by a dense pile of water repellent hair and this helps them to be supported by the surface tension. They prey on other insects or catch dead ones using their front legs while the middle and hind pair of legs scull across the water. Small pools along the Tumpah River in Bogani Nani Wartabone National Park were found to have up to five genera of pond skaters, and this raises interesting questions concerning their respective feeding strategies and spacing which will hopefully be answered in due course (Calabrese 1986; D. Polhemus pers. comm.).

Ubiquitous and interesting members of the benthos in relatively undisturbed rivers are caddisfly larvae, the adults being moth-like insects, often dull-brown in colour.[12] Some groups construct a silken tube adorned with grains of sand or other material which they carry around with them. These larvae feed on plants. Another group, the Hydropsychoidea, which dominated the caddisfly fauna in the Toraut and Tumpah Rivers in Bogani Nani Wartabone National Park (A. Wells pers. comm.) spin nets of different mesh sizes which trap detritus and drifting invertebrates on which they feed. Larger mesh sizes tend to be found in the headwaters and smaller sizes in downstream region, appropriate to the differences in the sizes of suspended matter (Townsend 1980). This was studied during Project Wallace but the detailed results are not yet available (Dudgeon 1985b). Species may also be distributed differentially across a stream: the rigid nets of certain species predominate in the fast-flowing riffles, and flimsy and weaker nets of other species are more often found in the slower-flowing water of pools.

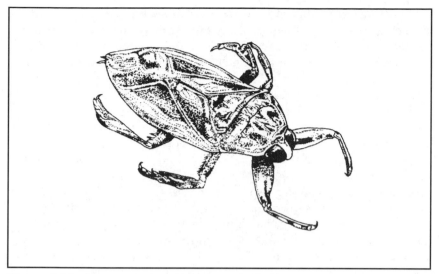

Figure 4.24. Giant water bug *Belostoma indica*.

Schistosomiasis and Echinostomiasis of the Lindu Valley

During preliminary studies of the disease echinostomiasis in Indonesia during the 1930s and 40s, a focus was found in the Lindu valley 50 km southeast of Palu. Schistosomiasis, also called 'bilharzia', is caused by a blood-fluke, or trematode flatworm *Schistosoma*, which is about 1 cm long and lives in the main veins of the middle and lower abdomen in certain mammals including humans. Early symptoms of the disease are itching of the skin caused by the entry of the aquatic larva, and skin eruption around the entry point. Four to six weeks later a high fever starts accompanied by coughing, abdominal pain and rashes. There are attacks of diarrhoea, and blood and mucus appear in the stools. As the disease progresses, the liver, heart, brain, spinal cord and pancreas can be affected and ulcers form on the skin. Victims may live for many years after contracting the disease, but become gradually weaker, and many eventually die of exhaustion or succumb to other diseases because of their weakened condition. Treatment requires expert supervision (Hadidjaja 1982).

The major fluke found in Sulawesi, *Schistosoma japonicum*, is known from elsewhere in the Oriental region and the early studies found that not just humans but also dogs and wild deer were infected. Like all other parasitic worms, the life cycle involves at least two organisms. Some trematodes such as *Alaria* require four hosts: snails, frogs, rats, and then a mammalian carnivore such as dogs or cats, and then snails again.

In the adult *S. japonicum*, the sides of the male's body fold over to form a groove in which the longer and more slender female is held. They both cling to the walls of the intestine where they suck blood using two suckers near their heads. The female lays her eggs in small blood vessels in such numbers that the vessels become congested. As a result, the eggs, which are armed with a sharp spike, rupture the walls, are discharged into the intestine and are subsequently passed out with the faeces. If these are deposited in water the eggs hatch into a ciliated or hairy larva called the 'miracidium' which is capable of swimming weakly. It can survive only about 24 hours on its own but, if it encounters a snail, it burrows into the soft body and feeds on the snail's tissues. The hairs are lost and the miracidium produces asexual buds which develop into cercaria. These are similar to the adults except that they possess a tail. The cercaria burrow out of the snail and float to the surface of the water where, if they come into contact with skin, they attach themselves by means of glands the secretions of which are also used to digest their way into a blood vessel. The larvae are carried in the bloodstream to the intestinal vessels where they feed and develop into adults (fig. 4.25).

In the 1970s there was a renewal of interest in Indonesian schistosomiasis when the intermediate host, the snail *Oncomelania hupensis* (fig. 4.26) was found in Lake Lindu (Carney et al. 1973), confirming that the organisms involved were the same as in the Philippines, Japan and China where schistosomiasis is widespread. The list of animals known to be infected by the adult worms grew to include civet cats, rats, shrews, wild pigs, water buffalo, cattle and horses, although the domestic species had relatively low rates of infection (Carney et al. 1978). Infection occurs all around the swampy edge of the lake but the distribution of the snails is focused. Both snails and flukes have also been found in the Napu valley to the east but none, at least in Sulawesi, have been found in other water courses or at elevations lower than 1,000-1,200 m (Cross et al. 1975, 1977; Carney et al. 1977a, b; Putrali et al. 1977; Stafford et al. 1980). Therefore, no other major lakes in Sulawesi should be affected. This apparent inability to disperse is fortunate since water from the Lindu River eventually flows through Palu.

The only reason schistosomiasis is not more widespread is because of the restriction of the host snail to the above areas, but the cause of this isolation is by no means clear. In the Philippines the same species of snail is found near sea level and is known from 900 m above sea level in one

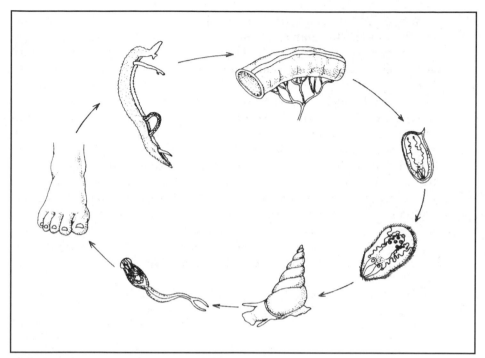

Figure 4.25. The life cycle of *Schistosoma japonicum.*
After Barnes 1968

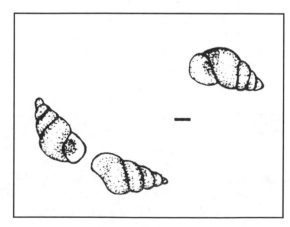

Figure 4.26. *Oncomelania hupensis* the intermediate host of the fluke *Schistosoma japonicum*. Scale bar indicates 1 mm.

province of Mindanao. Rice culture as practised in remote areas and growths of water hyacinth create habitats favourable to *O. hupensis,* but wet-rice culture was unknown in the Lindu area until the start of this century and water hyacinth was introduced into Sulawesi after the disease was confirmed (p. 275). *O. hupensis* is also found in long grass adjacent to rice fields, abandoned rice fields in swampy areas with a relatively dense cover of vegetation, and in grassy irrigation ditches. Its presence in agricultural and disturbed areas is, however, a secondary adaptation since both the fluke and snail occur in primary forests (Carney and Sudomo 1980). The natural habitats of the snail are moist areas either in the zone between forested hills and marshy lowlands, or where forest vegetation borders the lake shore. In contrast to disturbed areas, these habitats have a relatively constant and cooler temperature. It is most unlikely that the disease was introduced by humans in historical times, but then its natural dispersion from the Philippines would seem plausible only if it were known from intervening high altitude lakes such as Lake Moat and Lake Tondano, which it is not.

Despite knowledge of the disease and its ecology a transmigration site was established near Lake Lindu in a major *O. hupensis* area during the 1970s. Not surprisingly, within six months 60% of the transmigrants, about 500 individuals, were infected with the fluke (Carney and Sudomo 1980). The scale of the control problem was illustrated by counts of the snail population which found over 1,339 snails/m^2 at one abandoned rice field focus with an area of 750 m^2, giving a population of 1 million snails there alone. The land area of the Lindu valley is about 50 km^2, and if a conservative density of 500 snails/m^2 over just 10% of the area is accepted, then the number of *O. hupensis* is 2.5 billion! Of course, not all snails are infected; an infection rate of just 2.4% has been found, but that still represents 600 million snails carrying developing larvae. About 7,000 people are continuously exposed to schistosomiasis and about 2,500 of these are actually afflicted with the disease (Carney and Sudomo 1980).

A successful pilot control project was undertaken using mass treatment of human sufferers, agro-engineering, molluscicides, improved sanitation and health education (Putrali et al. 1980), but this was not maintained, and the incidence may increase. Local resettlement programs relocated people from other areas of Sulawesi in confirmed schistosomiasis areas resulting in infection. This led in turn to the departure of the migrants and, in some cases, the move to slash-and-burn agriculture in previously forested areas. Much of the credit for the relatively low incidence of schistosomiasis found around Lake Lindu today must go to Daniel Samel who works extremely hard at informing local people of the means of avoiding the disease. Recognition of his service and dedication was given in 1986 when he received a Kalpataru Environment Award from President Soeharto.

Another species of fluke *S. incognitum* is found around Lake Lindu and also further south in other river systems (Carney et al. 1977b). This is not (yet) a parasite of humans but it is common in tissues of domestic animals and commensal rodents. Its snail host *Radix auricularia* occurs in the same habitats as *O. hupensis,* and humans are frequently exposed to it. Only a small change in its genetic make-up may be necessary to allow it to exploit humans, with a consequent vast niche expansion. The fact that the two fluke species are sympatric in part of their range could also lead to hybridization; mating of the two species has been observed in rodents but the viability of the eggs was not determined (Stafford et al. 1980). If a viable hybrid were to arise and if it used the common *Radix* as its intermediate host, schistosomiasis could become a very common and widespread disease in Southeast Asia (Carney and Sudomo 1980).

Another potential danger is the accidental introduction of another schistosome snail host *Biomphalaria straminea.* This has been found in aquarium fish farms in Hong Kong, the fish from which are exported all over the world (Dudgeon and Yipp 1983a). The chances of one of these snails being infected by a dangerous schistosome are slight, but it is nonetheless important to take great care in all fish introductions, even those destined for home aquaria.

During the early surveys of schistosomiasis a related disease, echinostomiasis caused by echinostome flukes, was also found. Echinostomiasis is generally rare and of little clinical importance, but the residents of villages around Lake Lindu were exceptional in having very high rates (up to 96%) of infection. Studies indicated that various molluscs were the intermediate host of the echinostome larvae and that the bivalve mussels *Corbicula lindoensis* and *C. subplanata* were the primary source of human infection.

In the 1940s, mussel beds were common along the shore of Lake Lindu and mussels, often raw or lightly boiled, were a substantial item in the human diet. The high incidence of infection continued through to at least 1956. In the 1970s, however, no echinostome eggs were found in faeces from people living around Lake Lindu. During the intervening 15-20 years most of the mussel beds had disappeared except in one virtually inaccessible spot near the outlet of the lake in the north.

The disappearance of the mussels (and other molluscs) is almost certainly due to the introduction of fish (p. 52). The predatory snakehead *Channa striata* was present in Lake Lindu during the 1940s, but the tilapia *Oreochromis mossambicaus* was not introduced until 1951. The most likely cause of the mussel demise was the predation of young tilapia on the planktonic mussel larvae (Carney et al. 1980).

The echinostome is not extinct, however, and has been found in rats, birds and shrews, but if no new channel of infection to humans is introduced then human echinostomiasis will remain a disease that disappeared (Carney et al. 1980).

Fish

All of the indigenous fish of Sulawesi are of marine origin but are now adapted to freshwater life (p. 52). In Lake Tempe, brackish water fish such as *Mugil* sp., *Leiognathus* sp., *Terapon* sp., *Glossogobius* sp., and *Anguilla* sp. are common, even though the lake water is fresh (salinity < 3 ppt), and the connection to the sea is 70 km to the east (Suwignyo 1978). Also, eels in Lakes Poso and Moat start life in the depths of the ocean 50 km and 15 km away, respectively. The young swim upriver as 'elvers' to grow into adults in the lakes before returning to the ocean again to breed.

Details of the fish in Sulawesi rivers are hard to find but the species in the Mamasa River, southwest of Rantepao, comprised largely introduced or widespread indigenous species (Anon. 1982a). Recent collections from the rivers meandering through the forested karst areas of Maros revealed two new species of halfbeaks *Dermogenys* (Brembach 1982).

In the early 1970s it was observed that Lakes Lindu, Poso, Towuti, Matano and Mahalona were inhabited by fish that did not fully exploit the natural food available. It was therefore recommended that barbs *Puntius* spp., *Labiobarbus*[13] spp., *Osteochilus* spp., *Thynnichthys* spp., *Mystacoleucus marginatus*, Nile tilapia *Oreochromis nilotica* and *Leptobarbus hoeveni* could be introduced. At no point in the reports was thought given to possible effects on the indigenous fauna, although further studies were recommended to ensure rational fisheries management. Even today most regional fisheries staff are unaware of the existence of the indigenous fish species and no efforts are made to increase their production.

The endemic fish known from Sulawesi lakes are quite remarkable and the Malili lakes most significant 'hotspot' of freshwater biodiversity in Asia (table 4.10). Some of the endemics of Lakes Poso and Lindu may have become extinct as a result of competition and disease from introduced species (p. 52) (Whitten et al. 1987). The snakehead *Channa striata*[14] and the climbing perch *Anabas testudineus* (fig. 4.27) are often listed as being indigenous to Sulawesi but it is much more likely that they were brought by humans before scientific attention was paid to the fish fauna. Both species are airbreathers and easily transported.

Knowledge of the ecology of indigenous Sulawesi fish has increased substantially as a result of the work of Kottelat (see Introduction), and relatively little is known even of the introduced species important to fisheries. Communities of fish have been the subject of study in rivers in Sri Lanka and Panama. In the latter study, seven groups of fish based on their feeding habits were identified (algivores, piscivores, aquatic insectivores, etc.) and some fish changed their group with age. All feeding guilds except aquatic insectivores were concentrated (in terms of biomass/unit area) in the deep pools even though this was not necessarily where food was most abundant. With increasing river width the number of species increased, as did the density of algivores and other herbivores, presumably because of

Table 4.10. Endemic fish recorded from Sulawesi.

	Tempe	Sidenreng	Lindu	Poso	Matano	Mahalona	Wawontoa	Towuti	Limboto	Tondano
ATHERINIDAE										
Telmatherina celebensis	-	-	-	-	-	●	●	●	-	-
T. antoniae	-	-	-	-	●	-	-	-	-	-
T. obscura	-	-	-	-	●	-	-	-	-	-
T. opudi	-	-	-	-	●	-	-	-	-	-
T. prognatha	-	-	-	-	●	-	-	-	-	-
T. sarasinorum	-	-	-	-	●	-	-	-	-	-
T. wahyui	-	-	-	-	●	-	-	-	-	-
T. abendanoni	-	-	-	-	●	-	-	-	-	-
T. bonti	-	-	-	-	-	●	-	●	-	-
T. ladigesi *	?	?	-	-	-	-	-	-	-	-
Tominanga aurea	-	-	-	-	-	●	-	●	-	-
T. sanguicauda	-	-	-	-	-	-	-	●	-	-
Paratherina wolterecki	-	-	-	-	-	●	-	●	-	-
P. labiosa	-	-	-	-	-	-	●	-	-	-
P. striata	-	-	-	-	-	?	●	●	-	-
P. cyanea	-	-	-	-	-	●	-	●	-	-
ADRIANICHTHYIDAE										
Adrianichthys kruyti	-	-	-	●	-	-	-	-	-	-
Xenopoecilus sarasinorum	-	-	●	-	-	-	-	-	-	-
X. poptae	-	-	-	●	-	-	-	-	-	-
X. oophorus	-	-	-	●	-	-	-	-	-	-
Oryzias marmoratus	-	-	-	-	●	●	●	-	-	-
O. matanensis	-	-	-	-	●	-	-	-	-	-
O. celebensis	?	●	-	-	-	-	-	-	-	-
O. nigrimas	-	-	-	●	-	-	-	-	-	-
O. profundicola	-	-	-	-	-	-	-	●	-	-
O. orthognuthus	-	-	-	●	-	-	-	-	-	-
GOBIIDAE										
Weberogobius amadi	-	-	-	●	-	-	-	-	-	-
Glossogobius matanensis	-	-	-	-	●	-	-	●	-	-
G. flavipinnis	-	-	-	-	-	-	-	●	-	-
G. intermedius	-	-	-	-	●	-	-	●	-	-
Mugilogobius adeia	-	-	-	●	-	-	-	-	-	-
Tamanka latifrons	-	-	-	-	●	●	-	●	-	-
T. sarasinorum	-	-	-	-	●	-	-	-	-	-
HEMIRHAMPHIDAE										
Dermogenys megarrhamphus	-	-	-	-	-	-	-	●	-	-
D. weberi	-	-	-	-	●	-	-	-	-	-
D. ebrardti	-	-	-	-	-	-	-	-	-	-
D. montana	-	-	-	-	-	-	-	-	-	-
D. orientalis	-	-	-	-	-	-	-	-	-	-
D. vogti	-	-	-	-	-	-	-	-	-	-
Nomorhamphus celebensis	-	-	●	-	-	-	-	-	-	-
N. towoeti	-	-	-	-	-	-	-	●	-	-
N. hageni	-	-	-	-	-	-	-	-	-	-
N. australis	-	-	-	-	-	-	-	-	-	-
N. brembachi	-	-	-	-	-	-	-	-	-	-
N. liemi	-	-	-	-	-	-	-	-	-	-
N. ravnaki	-	-	-	-	-	-	-	-	-	-
N. sanusii	-	-	-	-	-	-	-	-	-	-
Tondanichthys kottelati	-	-	-	-	-	-	-	-	-	●
PHALLOSTETHIDAE										
Neosthetus dajaorium										
TERAPONIDAE										
Lagusia micracanthus	-	-	-	-	-	-	-	-	-	-

* The known locality of this species is "the area of Makassar." Most of the hemirhamphids, phallostethids and teraponids are not known from lakes but from rivers.

After Boulenger 1897a; Popta 1905; Weber and de Beaufort 1922; Aurich 1935a, b, 1938; Ahl 1936; Koumans 1953; Ladiges 1972; Brembach 1982; M. Kottelat pers. comm.

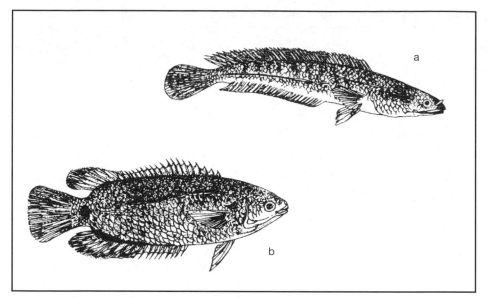

Figure 4.27. Snakehead *Channa striata* (a) and *Anabas testudineus* (b) are fish probably brought by humans to Sulawesi centuries ago.

Table 4.11. The fish found in the four zones of Aopa Swamp (p. 280). Note that only the last three species are indigenous.

		Zones			
		I	II	III	IV
Channidae	*Channa striata* *	•	•	•	–
Anabantidae	*Anabas testudineus* *	•	•	•	•
Belontiidae	*Trichogaster* spp.*	•	–	•	•
Cyprinidae	*Barbodes gonionotus*	–	•	•	–
	Cyprinus carpio	–	•	•	•
Clariidae	*Clarias batrachus* *	•	•	•	•
Aplocheilidae	*Aplocheilus panchax*	–	–	•	•
Hemirhamphidae	'*Hemiramphus*' spp.	–	–	•	•
Synbranchidae	*Monopterus albus* *	•	•	–	•
Anguillidae	*Anguilla* sp.	–	–	•	–

* Possess accessory respiratory organs: hind gut in *Monopterus*, near or around gills in the others. The ecological characteristics of most of these species are described in table 9.2.

After Anon. 1983a

the greater amounts of light striking larger rivers (Angermeier and Karr 1983).

The fish of Aopa Swamp are all widespread or introduced species (table 4.11). Those found in the blackwater swamp or Zone I have accessory breathing organs; this is not surprising considering the dissolved oxygen in its water is only 0.43 ppm compared with 2.41, 4.92 and 5.37 in Zones II, III and IV respectively. The readings were taken around midday when plants were actively photosynthesizing. Conditions at night would therefore be even less favourable. The mosquito fish *Aplocheilus panchax*[15] exploits the high oxygen concentrations in the surface exchange layer.

Aquatic Reptiles

The sailfin lizard *Hydrosaurus amboinensis* (fig. 4.28) is, when seen at all, generally resting with its feet dangling on either side of a tree branch overhanging a river. It is an impressive creature, the world's largest agamid lizard, reaching over 1 m in length, two-thirds of which is tail. Its most distinctive feature is the 12 cm tall tail crest behind the hind legs which is supported by projections from the tail vertebrae. The function of the crest is not clear, but because it is best-developed in males, it is probably connected with sparring contests for females. Sailfin lizards are always found near water and the toes have enlarged flattened scales, most obvious in juveniles, which must act like paddles. They are very able swimmers but can apparently also run across the water surface. On land they have been observed walking upright on their hind legs. Adults have no natural enemies and are not particularly favoured as food by country people, but they are wary of humans and drop off their perch into the water and swim away when approached. A particular tree limb may become a regular place for a certain animal to perch during the morning and afternoon, but around noon they seek shade in the riverside vegetation. Juvenile sailfin lizards, which measure 15 cm on hatching, bask in the sun less frequently than adults, probably because they would be easy prey for herons, eagles and snakes (Visser 1984).

Sailfin lizards are mixed feeders taking primarily vegetable matter (leaves for adults, seeds for juveniles) and some insects, and as such they are the only primarily herbivorous lizards in Indonesia. In captivity they appear to favour eating brightly-coloured fruit (Visser 1984). Lizards do not have the teeth to chew leaves and they bite them from the plant by perforating them and tearing them off. Lizards do not have a gizzard for grinding plant material but they may occasionally swallow stones to help break down food in the stomach. The plant material is probably processed slower and rather less efficiently than animal food, and so one is led to wonder why these lizards should bother to eat leaves? The answer may be in their eating habits for they do not eat continuously as do many mam-

Figure 4.28. Sailfin lizard *Hydrosaurus amboinensis*.
After de Rooij 1915

malian herbivores such as sheep. Instead, they rest motionless for long periods, basking in exposed, sunny places. Any prey they catch is the reward for waiting for suitable animals to come within reach. As a result of moving very little they have relatively small home-ranges, perhaps just a few hectares. Surprisingly, herbivorous lizards are most common where there are few leaves, such as in deserts and on mountains, and least common in tropical rain forests, but whether this is due to availability of basking sites, difficulties of resource partitioning, or some other factor, is not fully understood (Rand 1978).

Only one aquatic tortoise *Cuora amboinensis* is found on Sulawesi and nothing is known of its habits. It is likely, however, that only the young animals are wholly aquatic. It is generally brownish-black above and yellow with black spots below, although humic acids in the water can stain the yellow colour a reddish-chocolate.

Estuarine crocodiles *Crocodylus porosus* used to be very common in the lower parts of large rivers and early travellers reported seeing them in Lake Tempe (Mundy 1848), in the Maros River (Guillemard 1889), Lake

Poso, Butaioda'a River (between Buol and Marisa) and the Dumoga River (Sarasin and Sarasin 1905). Their numbers were so high at the start of this century where the Onggak and Dumoga Rivers meet that it was impossible to cross from one bank to another, and nearby villages were surrounded by bamboo stake fences to protect the inhabitants at night (Sarasin and Sarasin 1905). Elsewhere, the abundance was not necessarily as high as imagined. Last century a bounty of $2 was offered for each crocodile killed and one resident of Maros was producing numerous animals and claiming his money. His technique was eventually investigated and it was found that he had fenced off a stretch of river within which crocodiles were living and breeding happily—and providing the gentleman with a significant income (Guillemard 1889).

Crocodiles are the largest animals found on Sulawesi. A skull found 90 years ago near Maros was 73 cm long indicating a likely total length of 5.5 m (Guillemard 1889). An EoS team was shown a crocodile skull 51 cm long (probable total length about 3.5 m) at the information centre of the western section of the Bogani Nani Wartabone National Park at Lombogo (fig. 4.29). The animal had been caught by villagers in 1985 in the middle reaches of the Bone River. More recently a 6.5 m long and 80 cm broad crocodile, which was said to have had human hair in its stomach, was captured near Malili. In some areas, crocodiles were traditionally not hunted because it was believed that the soul of dead humans went to live in them. An exception was made only when someone had been attacked and killed (Sarasin and Sarasin 1905).

The estuarine crocodile is capable of entering both saline and freshwater by virtue of physiological adaptations which allow it to control the osmotic pressure of its plasma. In an estuary or in the sea, crocodiles conserve water by reabsorption in their kidneys resulting in a very concentrated urine, and by reabsorbing water from their faeces before they are voided. Most salts are excreted in the urine (Grigg 1981) but sodium chloride is also excreted through their external nasal gland and glands in the corner of the eyes. When a female crocodile is ready to lay eggs, she seeks a shady location on land where she builds up a dome-shaped nest of leaves, tall grass or peat (Greer 1971). The eggs, up to 50 or more, are laid in the middle of this nest, where they remain damp and protected from direct sunlight. The heat generated by the decomposition of the vegetable matter probably helps the incubation, but if the mother senses the eggs are becoming too hot she will spray urine over the nest to cool it down. Just before the eggs hatch, the young crocodiles make high-pitched croaks which are audible outside the nest. The mother scratches away the now-hardened surface of nest material and as the young crocodiles wrestle their way out of their shells, she (and sometimes the father too) picks them up gently in her mouth and carries them away to a secluded 'nursery' area in a swampy bank. They stay there for a month or two, guarded by

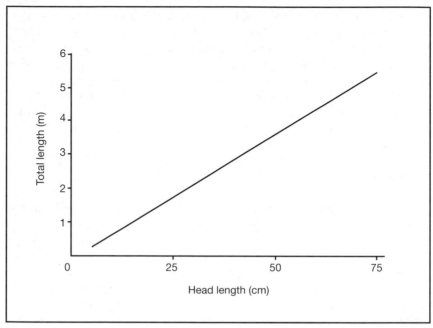

Figure 4.29. Relationship between skull and body length of estuarine crocodiles.

After Greer 1974

their parents, catching large insects and small vertebrates such as fish and frogs. As described for sea turtles (p. 155) the sex of hatching reptiles is determined by the temperature of incubation. Experiments on the Mississippi alligator, a reasonably close relative, have shown that alligator sex is fully determined and irreversible by the time of hatching. Incubation temperatures of up to and including 30°C produced all females, while 34°C and above produced all males. Temperatures in between produced clutches of mixed sexes. Since nests on exposed riverbanks or other dry areas receive more sunlight and are hotter than those nearer wet, shaded swamp, the sex ratios of hatchlings will be different between these habitats (Ferguson and Joanen 1982). It is likely that the sexes of estuarine crocodiles are also determined by incubation temperature, and this is important for management of the species. The clearing of riverine forest (p. 342) followed by regrowth of tall shrubs and grasses will reduce the amount of shading and may increase the average ambient temperature of an area. This in turn may increase the number of male crocodiles. However, the efficiency of nest temperature regulation by the female is unknown as are the social impacts of any alterations in sex-ratio.

Nest sites are exposed to the warming rays of the sun for a major part of the day but because they are generally surrounded by relatively tall vegetation they are difficult (as well as dangerous) to find and count. For these reasons, surveys of nest mounds are best conducted from the air if light aircraft or helicopters are available. Nest sites have a number of common features: access to permanent fresh or only slightly saline water, seclusion, and proximity to the type of vegetation used in the construction of nest mounds. Where plant growth is abundant, nests are built in places where the plants can resist strong winds but can be flattened by a crocodile crawling over them. Such plants break easily or can be uprooted, and have foliage from the substrate level to the tip rather than a thin stem and foliage concentrated at the top (Webb et al. 1983).

Finally, Sulawesi has four species of brownish-grey freshwater snakes: *Homalopsis buccata, Enhydris enhydris, E. plumbea,* and *E. matannensis.* The last species, which grows to about 25 cm, is endemic to Sulawesi, has an olive-brown back and a yellow-white throat[16] It is known from only two specimens: one from Lake Matano and the other from a fish pond near Raha, Muna Island (Iskandar 1979), and its ecology is unknown. If it is similar in habits to its close relatives in the Malay Peninsula, however, it feeds on fish and frogs (Tweedie 1983). Other snakes are sometimes found in rice fields but these are terrestrial species with no particular adaptation for swimming.

Water Birds

Water birds are far more common in Sulawesi than they are in western Indonesia probably because it is closer to the migration pathway of some of the species. Many of the larger birds such as the storks, egrets and herons are the same as can be seen at the coast (p. 149), but the Australian pelican *Pelecanus conspicillatus* (fig. 4.30) is found more usually near lakes. About 30 species of birds are associated with aquatic habitats and of these, most of the smaller ones are more or less confined to such areas (table 4.12).

In 1840 Lake Tempe was said to 'abound with aquatic birds' (Mundy 1848) and even now a large number of aquatic birds can be seen. An Interwader/EoS survey of lakes Tempe and Buaya in April 1986 found relatively few muddy habitats suitable for waders. A greater area becomes available in the dry season when local residents report an abundance of small waders. Even so, 29 species of water bird were found, the most for any area in Sulawesi so far (Uttley 1986).

The little grebe *Podiceps ruficollis,* a small, round, brown diving bird, is found both on large lakes and on smaller forest-fringed lakes such as those in Lore Lindu (even on Lake Tambing at 1,700 m a.s.l.) and Morowali

Figure 4.30. Australian pelicans *Pelecanus conspicillatus*, one of the largest species of birds to be seen on Sulawesi lakes.

Figure 4.31. Asian darter *Anhinga melanogaster*, a diving bird found on lakes.

After King et al. 1975

Table 4.12. Water birds observed in Aopa Swamp and around Lakes Lindu, Tempe, Buaya and Matano.

		Aopa	Lindu	Tempe	Buaya	Matano
Little grebe	Podiceps ruficollis	•	•	•	•	•
Australian pelican	Pelecanus conspicillatus	•	•	-	-	-
Little pied cormorant	Phalacrocorax melanoleuca	-	•	-	•	-
Cormorant	P. sp.	•	-	-	-	-
Asian darter	Anhinga melanogaster	•	•	-	•	•
Pond heron	Ardeola speciosa	•	•	•	•	•
Little egret	Egretta garzetta	•	•	•	•	-
Great egret	E. alba	•	-	•	-	-
Intermediate egret	E. intermedia	-	?	-	-	-
Purple heron	Ardea purpurea	•	•	•	•	•
Great billed heron	A. sumatrana	-	?	-	-	-
Cattle egret	Bubulcus ibis	•	•	•	•	-
Yellow bittern	Ixobrychus sinensis	-	-	•	•	-
Cinnamon bittern	I. cinnamomeus	-	•	•	-	-
Schrenk's bittern	I. eurhythmus	-	-	•	-	-
Black bittern	Dupetor flavicollis	-	•	•	•	-
B.-capped night heron	N. nycticorax	-	-	-	•	-
Spotted tree duck	Dendrocygna guttata	•	-	-	-	-
Wandering tree duck	D. arcuata	•	•	•	•	-
Black duck	Anas supercilliosa	-	•	•	•	-
Grey teal	A. gibberifrons	-	-	•	•	•
Garganey	A. querquedula	-	•	-	-	-
Tufted duck	Aythya fuligula	-	-	-	-	•
Wooly-necked stork	Ciconia episcopus	-	-	•	-	-
Glossy ibis	Plegadis falcinellus	-	-	•	•	-
White-breasted waterhen	Amaurornis phoenicurus	-	•	•	-	•
Common moorhen	Gallinula chloropus	•	•	•	•	-
Dusky moorhen	G. tenebrosa	-	•	•	-	-
Ruddy crake	Porzana fusca	-	-	-	-	•
Purple swamphen	Porphyrio porphyrio	-	•	•	•	-
Crested jacana	Irediparra gallinacea	•	•	•	-	-
Lesser golden plover	Pluvialis dominica	-	-	-	-	•
Long-toed stint	Calidris subminutus	-	-	•	-	-
Wood sandpiper	Tringa glareola	-	-	•	•	•
Common sandpiper	Actitis hypoleucos	-	-	•	•	•
Whiskered tern	Chlidonias hybrida	-	•	-	•	-
White-winged black tern	C. leucopterus	-	-	•	-	-

After Escott and Holmes 1980; Holmes and Wood 1980; Watling 1983; Uttley 1986

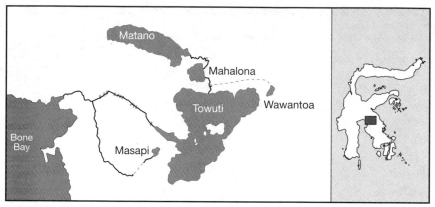

Figure 4.32. Lakes of the Malili River system.
After Brooks 1950

National Parks (Anon. 1977b, 1981b; Watling 1983). Another diving bird is the Asian darter *Anhinga melanogaster* (fig. 4.31 on p. 306). It resembles a cormorant but has a longer, more slender neck and an unhooked bill.

Malili Lakes System

Endemism in the lakes of the Malili region (fig. 4.32) has been shown to be very unusual. The entire system is connected by rivers and contains three large lakes, Matano, Mahalona and Towuti, each downriver of the other, and two much smaller lakes, Masapi and Wawantoa.

Of the 100 or so species of copepods, prawns, molluscs, and fish endemic to Sulawesi found in the system, only two (a prawn and a goby) are shared by all five lakes (fig. 4.33). This has led to speculations of species arising within the lakes and rivers due to isolation resulting from waterfalls in connecting rivers or to physiographic obstacles not evident today. It is certain that further collecting would add to the species lists of the lakes and allow better hypotheses to be made concerning the origin of the fish, and would also better establish the status of the endemic species particularly in the light of the nickel mine and other human activities. It is very important that the remarkable biology of the lakes be better known and for the lakes to be given appropriate conservation which would prevent introduction of other fish.

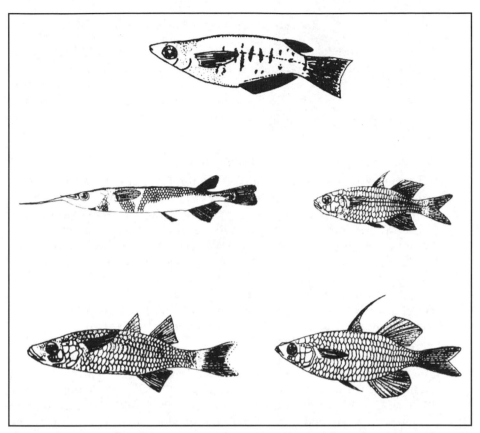

Figure 4.33. Some endemic fish of the Malili lakes. **Top row:** *Oryzias matanensis* from Lake Matano; **middle row:** *Dermogenys weberi* from Lake Matano, *Telmaterina bonti* from Lakes Mahalona, Towuti and Matano; **bottom row:** *Paratherina labiosa* from Lake Wawontoa, *Telmatherina celebensis* from Lake Mahalona. The last species is also found in Lakes Wawontoa, Matano and Towuti but these differ somewhat in shape.

After Boulenger 1897a; Aurich 1935a, b, 1938

PHYSICAL PATTERNS IN LAKES

Most lakes can be viewed as slow-moving rivers in which the riverbed has become very wide and very deep. Many of the same species of animals and plants live in both lakes and rivers, and many of the adaptations they require are also the same. So different are the physical regimes of lakes and rivers, however, that their behaviour requires separate attention.

Temperature

The sun warms the surface of a lake and this can cause density differences within the water column to produce a layering effect (fig. 4.34). The warmest and highest layer, the epilimnion, experiences diurnal fluctuations: at Lake Moat the maximum daily range of water and air is temperatures is 21.5°-23.5°C, and 16°-23°C respectively (Buchari 1984). Temperatures for the surface water of other lakes are similar but decrease with increasing altitude (table 4.13). The narrower temperature range of water than air is due to water's superior capacity to retain heat. Maximum water temperature rarely exceeds that of the air except where hot springs are present, such as at the southwest of Lake Tondano, or where water is shallow such as at Aopa Swamp and Lake Limboto.

Temperature generally decreases with depth but there are exceptions such as Lakes Moat (EoS team), Towuti and Matano (Wardoyo 1978; Anon. 1980) where temperature appears to stay more or less constant throughout the depth of the lake. In one study at Lake Towuti the temperature at 135 m depth was even found to be 1°C warmer than at the surface (Wardoyo 1978). These three lakes are in areas that are tectonically relatively active and the temperature anomalies may be due to deep hot springs (Anon. 1980). These unusual temperature profiles may cause overturns of the water (p. 315), bringing deoxygenated water to the surface, but this has not been reported. The maximum temperature of surface

Table 4.13. Temperature ranges (°C) of surface water and ambient air at several lakes arranged in descending order of altitude.

Lake	Air	Water
Moat	16-28.0	21.5-25.5
Tondano	26-33	24-28
Towuti	22-31	23-30
Aopa Swamp	26.5-29.0	27-31
Limboto	26-33	29-33
Tempe	26-33	27-31.5

After Sarnita 1974; Achmad and Cholik 1977; Suwignyo 1978; Wardoyo and Thana 1978; Anon. 1979b; Thana and Wardoyo 1980; Thana et al. 1980; Buchari 1984

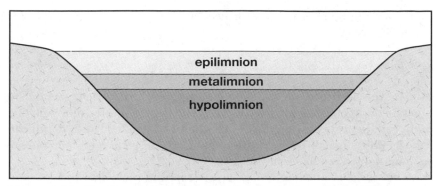

Figure 4.34. The three principal density layers of a hypothetical lake.

water on Lake Tempe was 29°C, almost 5°C warmer than that recorded at Moat, because of the higher altitude and hence higher ambient temperatures at Lake Towuti.

The metalimnion is an intermediate, relatively thin layer through which water temperature usually drops rapidly (as much as or greater than 1°C/m (Moss 1980), and this gradient is termed the thermocline.

The thermal characteristics of lakes should be examined over a longer period of time, perhaps 1-2 years, in order to assess any diurnal and seasonal variations. In deep, wind-swept lakes, a persistent stratification may never form (Moss 1980).

Oxygen

Dissolved oxygen concentrations are highest in the surface layers where photosynthetic activity of plants is greatest, and lowest at the lake bottom, where light is low and plants rare (table 4.14). An EoS team examined this in Lake Moat and found surface concentrations of about 7.9 ppm and bottom (9.6 m) concentrations of about 3.0 ppm.

Oxygen values varied with depth along all profiles, a phenomenon which could reflect micro-site variations caused by the mixing of water layers, or may possibly reflect populations of photosynthetic phytoplankton and bacteria. The deepest layer of a lake, the hypolimnion, is far removed from the two sources of oxygen—the atmosphere and green plants—and the organisms which decompose the faeces, corpses and other organic matter falling to the bottom, consume rather than produce oxygen. Oxygen concentrations in a deep lake may be reduced to zero. All lakes surveyed show concentrations of surface oxygen that were excellent for fish growth (table 4.16). The lowest concentration recorded was 1.5 ppm at the

Aopa Swamp, which is too low for fish which rely solely on their gills for obtaining oxygen.

Nutrients and Conductivity

As has been described, the hypolimnion experiences a nutrient gain due to the rain of organic matter from above, while the epilimnion and metalimnion experience a nutrient loss. A profile of conductivity with depth can reveal approximately where nutrient gain, expressed as a higher concentration of dissolved materials, starts.

The oxygen consumption (rate of respiration) in lowland lakes in Sulawesi is about 4-9 times faster than it is in temperate lakes which are 15°-20°C cooler. Carbon dioxide and other solutes are released very quickly and in the deeper lakes much of the settling organic matter would probably be reduced to mineral matter before it reached the lake bed. The high concentrations of phosphate and ammonium in the hypolimnions of some lakes have led to the suggestion that it is these deep waters, rather than surface layers, that should be used for irrigation.

Figures for the chemical composition of Sulawesi lakes indicate a large degree of variation (table 4.15) but to what extent this represents genuine differences or simply reflects different methods of analysis is not known. Almost all measurements are from only the surface layers (probably 1 m or less) and results can also be compared with water quality criteria for fish (table 4.16).[17] Lake Limboto has the lowest and highest values of carbon dioxide and pH respectively. Carbon dioxide concentrations were higher on the lake bed than in the surface layer, and in Aopa Swamp, water

Table 4.14. Dissolved oxygen concentrations (mg/l) at different depths in major lakes of Sulawesi.

Lake	Surface	Bottom	Other depths
Tondano	6.7-8.0	4.0-7.1	-
Moat	5.0-8.4	4.4-7.2	10 m: 4.6-7.1
	6.8-7.4	3.2	-
Limboto	7.0-8.2	-	-
Matano	5.5-7.9	-	-
Aopa Swamp	3.9-6.2	1.5-2.5	-
Tempe	4.1-8.4		
	4.5-4.9	-	-
Poso	5.6-8.3	-	-

After Achmad and Cholik 1977; Wardoyo 1978; Wardoyo and Thana 1978; Pirzan and Wardoyo 1979; Thana and Wardoyo 1980; Thana et al. 1980; Anon. 1981a; and EoS surveys

Table 4.15. Chemical composition of surface and bottom (b) water in major lakes of Sulawesi. All values in ppm except conductivity which is in μmho/cm and pH which is the negative logarithm of hydrogen ions.

	Tondano	Moat	Limboto	Aopa Swamp	Matano	Towuti	Tempe	Lindu	Poso
CO_2	1.6-4.2	2.5-4.1	0.0-1.9	14.1-16.9	1.4-3.2	3.0-8.2	0.0-6.0	3.5-8.8	0-2.3
(b)	2.0-4.9	1.7-5.8	-	18.4-21.9	-	-	-	-	-
BOD	0.7-1.9	2.0	-	-	0.0-2.0	0.4-0.7	-	-	-
Tot. Sus. Solids	1.45-2.05	-	-	-	-	-	-	-	-
Conductivity	205-223	-	-	201-303	-	-	-	-	-
pH	7.5-8.5	5.5-6.0	8.0-8.5	6.5-7.1	8.5	8.2-8.4	6.8-8.3	5.6-6.0	7.0-7.9
(b)	6.3-7.1	5.0-7.0	-	-	-	-	-	-	-
Alkalinity ($CaCO_3$)	77-95	-	-	-	-	-	72.0-140.0	-	-
$CaCO_3$	55-68	-	-	194-196	-	-	37.0-84.0	-	-
SO_4	0.19-0.61	-	3.3-5.8	-	1.0-1.3	-	-	-	11-82
SO_3	11.5-18.3	-	-	-	-	-	-	-	-
SiO_2	0.7-2.0	-	-	2.8-3.5	16.3-46.0	-	-	-	-
PO_4	0.04-0.10	-	-	0.08-0.09	0.1-0.5	-	-	-	-
Ortho-phosphate	0.05-0.08	-	-	0.08-0.24	-	-	-	-	-
$N-NO_3$	0.33-0.48	-	-	-	9.7-11.0	-	-	-	-
$N-NH_3$	0.16-0.17	-	-	0.4-1.8	-	-	-	-	-
NO_2	-	-	-	0.04-0.06	0.03	-	-	-	-
Cl_2	-	-	0.3-0.6	-	-	-	-	-	0.3-0.6
Si	-	-	-	2.7-9.8	-	-	-	-	-
Mg	-	-	-	-	-	-	-	0.72	0.48
Ca	32.9-37.9	-	-	81-109	-	-	-	10.4	23.2

After Sarnita 1973; Anon. 1977a, 1978, 1979a, b, 1983a; Wardoyo 1978; Wardoyo and Thana 1978; Achmad and Cholik 1979; Pirzan and Wardoyo 1979; Thana and Wardoyo 1980; Thana et al. 1980; Buchari 1981

quality was poor, according to the criteria, as might be expected.

Lakes are frequently classified on the basis of nutrient loading, phytoplankton counts, and organic productivity. In general, the lakes of Sulawesi are oligotrophic, or nutrient poor, and their water is relatively clear, with plant growth restricted to a few meters from the shore. Lakes Moat, Poso, Matano, Towuti and Mahalona are in this category. Lake Lindu has an intermediate nutrient status (Sarnita 1973) termed mesotrophic, while Lakes Tempe, Tondano, Limboto and Aopa Swamp are eutrophic, rich in nutrients, supporting an abundant fauna and flora (p. 340).

Light Penetration

The depth to which light can penetrate is called the 'euphotic zone' and below this depth primary productivity is essentially zero (Cole 1983). The zone varies daily or seasonally. A standard, simple method of estimating light penetration is with the Secchi disk. This black-and-white plate about 20 cm in diameter is lowered into the water and the depth at which the plate disappears from the viewer's sight is recorded. The disk is lowered a bit further and the depth at which it reappears is also recorded. The average of these two readings is less than the actual depth of the euphotic

Table 4.16. Water quality criteria for freshwater fish.

	Harmful	Poor	Good	Very good	Excellent
Suspended sediment (ppm)	>400	400-80	79-25	<25	<25
Conductivity (µmho/cm)	2,000-1,000	1,000-500	500-150	<500	<500
pH	>10.5+<4.0	4.0-5.5	5.5-6.5	6.5-9.0	6.8-8.5
Dissolved oxygen (mg/l)	<1.7	1.7-2.0	2.0-4.0	4.0-5.0	5.0-7.8
Carbon dioxide (ppm)	100-30	30-25	25-12	<12	<12
Alkalinity (ppm $CaCO_3$ eq) (ppm $CaCO_3$)	<10 <5	10-50 5.12	50-200 12.15	200-500 >15	200-500 >15
Calcium (ppm)	<6.25	6.25-24.9	25-62.5	>62.5	>62.5
Phosphate (ppm)	<0.02	0.021-0.05	0.051-0.1	0.10-0.20	>0.20
Ammonia (ppm)	>1.5	1.5-1.0	<1.0	<1.0	<1.0

After Alabaster and Lloyd 1980; Anon. 1983a

zone, but represents approximately 30%-80% of it (table 4.17) (Cole 1983). As would be expected, oligotrophic lakes are much clearer than eutrophic lakes in which high concentrations of phytoplankton tend to absorb the light.

Stability

Stability of tropical lakes is poorly understood, but may be very important. In general, warm waters have a greater resistance to mixing than cooler waters, and slow mixing may continue all the year round (Anon. 1982b). Overturns, in which the hypolimnion is brought to the surface, are not unknown however, and the low oxygen concentrations of water from the bottom of a lake can cause fish to perish if it comes to the surface (Green et al. 1976). Greatest stability is found in steeply walled, deep lakes with small surface areas (Anon. 1982b). Indeed, the relationship between surface area, thermocline and stability can be quantified roughly as follows (Ruttner 1931):

$$\text{Area} \quad 1 : 100 : 1,000$$
$$\text{Depth of thermocline} \quad 1 : \quad 3 : \quad 6$$
$$\text{Stability } (0\text{-}20 \text{ m}) \quad 50 : \quad 10 : \quad 1$$

Thus Lake Towuti, which has ten times the area of Lake Tondano, would have a thermocline twice as deep and require only one-tenth of the wind strength to mix the top 20 m. A strong wind-generated water current noted by an EoS team at Lake Moat could have caused sufficient mixing of layers to prevent layering. In the Philippine Lake Lanao, thermoclines

Table 4.17. Secchi disk depths in major lakes of Sulawesi.

	Depth (m)
Tondano	0.5-2.5
Moat	2.0-2.6
Limboto	0.11-0.45
Aopa Swamp	0.28-0.30
Matano	11-16
Towuti	15-22
Tempe	0.10-2.5
Poso	8-10.6
Lindu	1.7-2.3
Mahalona	11

After Sarnita 1973, 1974; Wardoyo and Thana 1978; Anon. 1979a, b; Pirzan and Wardoyo 1979; Thana and Wardoyo 1980; Thana et al. 1980; Buchari 1984; and EoS surveys

shallower than 20 m depth are easily formed but equally easily disrupted by light winds, while at greater depths secondary thermoclines are more constant, requiring squalls or storms to dissipate them (Lewis 1973).

Abiotic factors, particularly climate, are most influential in the process of change. Precipitation, for example, governs the volume and surface area of the lakes, temperature is the driving force behind chemical reactions and establishment of density layers, and wind induces layer mixing and increases surface concentrations of dissolved oxygen.

In lakes that do not mix completely, a type of chemical stratification may occur in which nutrients may be held in deep layers that resist mixing and are therefore unavailable for life in the upper lake layers. For example, in Lake Lanao, (Philippines), free nitrate levels were undetectable when the water column was stable, but seasonal circulation helped to distribute nitrate and other nutrients to the surface layers (Lewis 1973). It has been suggested that small year to year variations in the annual heat budget are the critical factors in the layer stability of lakes (Moss 1980).

BIOTIC PATTERNS IN LAKES

As would be expected, the distribution of biota in a lake is determined principally by the physical conditions, particularly the layers.

The distribution of plankton in lakes (and other bodies of water) is governed by a number of variables such as water density and viscosity, nighttime cooling, turbulence, temperature, light intensity and time of day. In addition, the form of feeding of zooplankton has an effect (Davis 1955). Differences in plankton abundance do not only occur between the epi-, meta-, and hypolimnion but considerable variation also exists within the epilimnion itself; these differences are not always easy to explain (Ruttner 1931).

Many fish obviously depend on the plankton for food and would be unable to feed on them if they were in the hypolimnion because of the low levels of oxygen found there. Benthic animals either have to be able to cope with very little oxygen (such as the red, haemoglobin-filled chironomid fly larvae) or with no oxygen, and usually no light (such as anaerobic saprophytic[18] fungi and bacteria). In the low-oxygen, dark environment these organisms have few predators.

PHYSICAL PATTERNS IN RIVERS

Variation exists across a river as well as down its length. Velocity and depth of water, and substrate composition all vary along and across a river and all influence the biota which may inhabit rivers.

Discharge

The volume of water flowing through a cross-section of river per unit of time is the discharge (Q), represented as the product of the mean velocity (V) and the cross-sectional area (A), or Q = VA. Thus, a wide, deep river with a cross-sectional area of 30 m^2 flowing slowly at 0.5 m/s has the same discharge as a narrow (A=10 m^2) river moving swiftly at 1.5 m/s (Q = 15.0 m^3/s in both cases). It may seem strange, but the average velocity of a river is lower in the steep headwater regions than in the lowlands. This was demonstrated by an EoS team that measured velocity and discharge at stations at 1,440 m, 825 m and 0 m altitude in the Jeneberang River system. The velocities at those stations were <0.005, 0.12 and 2.99 m/s and the discharges were <0.1, 0.4 and 3.3 m^3/s respectively. Discharge thus increases as the river flows downhill, though this is to be expected because of additions from tributaries and from water flowing into the river from runoff and interflow. An EoS team monitored rainfall and discharge at a small river in the headwaters of the Jeneberang River over a period of 18 consecutive days. Rain occurred in the first week only (fig. 4.35). Discharge is clearly not dependent solely on rainfall because of complicating factors such as previous rainfall, and intensity and duration of rainfall. Each watershed, and each area within a watershed, will respond differently to rainfall and so discharge is changing along the course of a river, and at any particular point (Achmad 1983).

Heavy rains can produce dramatic increases in water levels in rivers. At the Toraut River, near Dumoga, the high water mark is over 1 m higher than for low flow, representing a 300% increase in discharge. The concomitant increase in energy of the flow during a rainy season often places extra stress on benthic organisms and plants. Turbulent flow scours riverbanks and carries soil, rocks, and even whole trees far downriver. Extremely large floods may even permanently alter the river course.

Headwater streams respond quickly to short, intense rainfall, whereas the response of wide, lowland rivers are much slower. For example, the time lag between rainfall and discharge peaks at 300 m altitude on the Jeneberang River is about seven hours (Anon. 1984).

The flash flood is a dangerous event characteristic of small rivers on steep slopes or other channels which are usually relatively dry. High intensity rainfall may cause a rapid rise in river levels, which can suddenly appear as a single flood-wave travelling swiftly downriver. This happens

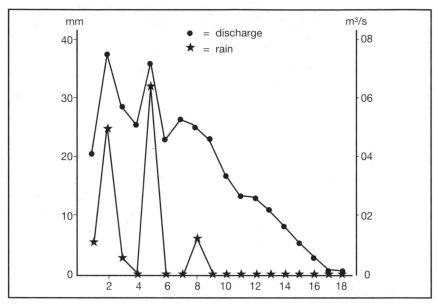

Figure 4.35. Rainfall and average discharge of a river at 1,440 m over a period of 18 consecutive days.

Data from an EoS team

because waters from small feeder channels meet and concentrate in a larger channel simultaneously, or when water pooled behind an obstacle such as a dam or a fallen tree suddenly breaks free and gushes downriver. Flash floods occasionally occur in the normally dry riverbeds leading to Palu Bay after rare heavy storms have broken over the mountains. The water surges with great force carrying enormous quantities of suspended sediment. Surprisingly large boulders are also brought down to the lower reaches of the river. The comparative rarity of these events is shown by the herbs and shrubs growing on the dry riverbeds which would get swept away in a flood. The unsuspecting traveller crossing the river can easily be swept away or hit by transported debris.

The 'lahar' is an Indonesian term that refers to mudflows due to volcanic activity. Lava, soil and water mix together to flow downslope in much the same process as the flash flood previously described. Sufficient water must be available to move the mass, either from rain, a lake or river. Lahar paths are evident on the flanks of Mt. Karangetang on Siau Island and Mt. Lokon, near Manado. On Mt. Karangetang, it appears that lahars join the main river which eventually discharges into the sea.

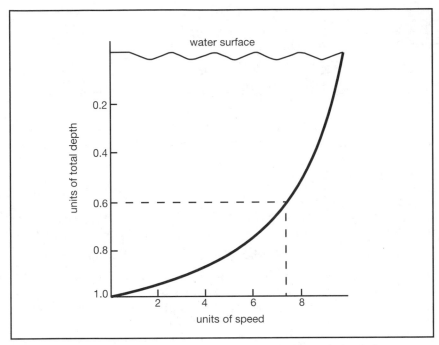

Figure 4.36. Relationship between current velocity and depth in an open channel. Average velocity is measured at 60% of the total depth.
After Townsend 1980

Shear Stress

Shear stress is the result of faster water flowing past slower water. This is described by the velocity gradient (fig. 4.36) which has a logarithmic distribution. Velocity is greater in the main body of flow just under the surface than at the riverbed or the shore boundaries. Turbulent flow near the water surface may create eddies the resultant energy of which moves particles on or near the riverbed. Pebbles or rocks are lifted up, pushed or bounced along, bounced into other rocks or thrown into faster water layers, only to settle down once again on the riverbed (fig. 4.37). The greater the shear stress on the riverbed the greater the chance that a benthic organism will be dislodged and washed downstream. Shear stress is proportional to water depth and to slope, and much greater shear stress will be experienced by benthic organisms in the headwaters, than by similar organisms in the lower reaches.

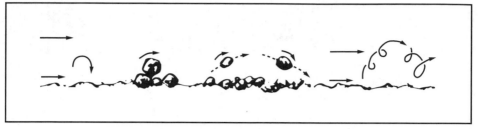

Figure 4.37. Particles on a riverbed can be pushed along, rolled, bounced or carried along.

Riverbed Particle Size

Rivers are capable of carrying sediment of all sizes ranging from the smallest clay fractions to large boulders. The particles that travel along the riverbed will be those that are larger than the flowing water is able to carry away. The smallest particles are the suspended load while the larger particles rolled along the riverbed are the bed load. As shear stress decreases downriver, so the average particle size of the suspended load decreases, larger particles having fallen to the riverbed. Thus the particle size on the riverbed also decreases downriver. Particle sizes of suspended load typically range from clay (<0.004 mm) to sand (<2 mm), although much larger rocks and pebbles may also be suspended for shorter periods of time. The diameter of a particle in suspension is, however, generally less than 0.5 mm (Dunne and Leopold 1983). The amount and type of sediment suspended in a river is related to the type and exposure of sediment sources, such as slopes, riverbanks and roads, as well as to discharge.

Discharge rates are often used to estimate suspended sediment concentrations, but caution must be exercised. This relationship is unique to each river under a certain set of conditions and cannot be transferred to other rivers or even used for the same river, if large areas in the catchment area are to be developed or disturbed. There are complications too, in that for a given discharge there may be more, or less, suspended sediment when the river level is rising than when the river level is falling. This will depend on erosion sources in the river basin and the predominant flow processes.

Bed load consists of larger particles that are rolled or pushed along the riverbed. These can become suspended briefly but the weight of the particle soon pulls it back down. Bed load is difficult to measure and no data are available for Sulawesi. Depending on rainfall patterns and watershed characteristics, bed load may account for as much as 50% of the total load (Dunne and Leopold 1983), although it is more typically suspected as being much lower. The bed load component is commonly omitted and this can result in a serious underestimation of the total sediment load.

Temperature

Altitude, rain, exposure, water source and velocity, and ambient temperatures are the major factors which influence river water temperatures. For example, the surface temperature in an exposed section of the Bantimurung River measured by an EoS team, was 27°C, whereas under shade just 5 m upriver, the surface temperature was 26°C, and in a nearby swiftly-moving tributary, water discharged from a cave was even cooler, at 25.5°C. The temperature of five different lowland rivers in the Southeast Sulawesi ranged from 26°-29°C. This is rather less than the temperature range of 25°-32°C found in Malaysian lowland rivers (Chye and Furtado 1982) but more samples would doubtless increase the range.

Temperature layering, such as found in lakes, does not occur to any extent in rivers because the water is in continuous motion and depths seldom exceed 2 m. Water temperatures can respond quickly to ambient air conditions however, particularly if the river is shallow and slow moving, a situation typical of lowland rivers. A river less than 2 m deep in Sri Lanka displayed a diurnal variation of 5°C, ranging between 25°-30°C (Benzie 1984). Night-time temperatures of water remain warmer than air temperatures due to the warming effects of surrounding earth and ground water seepage, and to the greater thermal capacity of water (fig. 4.38). Rivers at higher altitudes are cooler due to the lower mean daily air and soil temperatures (p. 489).

Dissolved Oxygen and Mineral Nutrients

Oxygen concentrations of lowland rivers are generally about 6.5 ppm to 7.5 ppm although sluggish rivers may have a concentration of only about 4 ppm. Concentrations in mountain headwaters tend to be greater because the water is more turbulent and because cooler water can hold more oxygen (Chye and Furtado 1982).

Solubility of gases, including oxygen, decreases with increasing temperature. At the same time, the rate at which oxygen is consumed through oxidation of organic compounds (BOD), rises. Thus, a river suddenly exposed to greater sunlight and hence higher temperatures due, for example, to riverbank clearing, will experience a dramatic chemical change, possibly even resulting in the death of fish (Johnson 1961; Anon. 1982a). In addition, reduced light penetration in rivers caused by high concentrations of suspended sediment levels would restrict photosynthesis of algae and other plants and would thus result in lower oxygen concentrations.

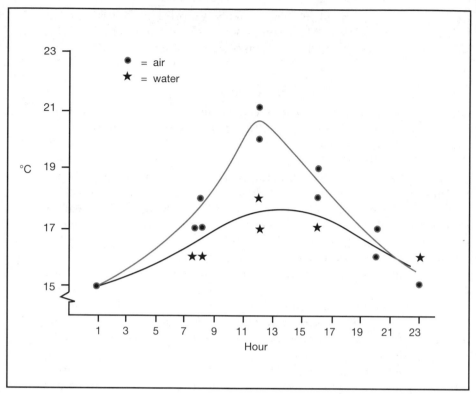

Figure 4.38. Changes in air and water temperatures in the headwaters of the Jeneberang River at 1,440 m altitude.

Data from an EoS team

In general, lowland rivers have low concentrations of chemical ions, weakly acid pH, low alkalinity (and therefore low buffering capacity), low BOD, and low mineral nutrients, especially phosphate and nitrate (Chye and Furtado 1982). Exceptions occur where human activities influence water quality (Palenewan 1984). Concentrations will vary with river volume and dilution is not always the major factor. An examination of water quality in the Jeneberang River revealed that concentrations of magnesium and potassium were much higher in the rainy period than the dry period, due to release of these cations from soil particles washed into suspension.

BIOTIC PATTERNS IN RIVERS

The abundance of many organisms changes along the length of a river, and some organisms are found only in mountain headwaters whereas others are found only in the estuary. These different distributions are largely determined by the physical factors described above. A study of fish diversity along a Malaysian river found an increase with river 'order' (Bishop 1973), but a similar study in the United States found that maximum richness was in the middle orders (Minshall et al. 1985).

Current

Shear stress is greatest in turbulent headwaters and organisms found in these waters have attachment adaptations to prevent them from being swept downriver. For example, plants of the headwaters tend to have low resistance to water flow, good anchoring ability, and high resistance to abrasion. They include encrusting algae, mosses and filamentous algae and occasionally a group of higher plants which are taxonomically diverse but show similar structural features. These plants are known as 'rheophytes' and are typically shrubs with narrow leaves and brightly-coloured fruit dispersed by water or fish (Whitmore 1984). They are not, however, particularly well represented on Sulawesi (van Steenis 1981).

As the discharge, and hence the water velocity and shear stress, varies with time and at any given point, plants are adapted for maximum flows (Townsend 1980). Plants growing near the edge of a river are subject to less shear stress and so the density of plants and composition of a plant community will vary across the river. Even so, floods also sweep away large amounts of algae attached to stones although they also bring down detritus from further upriver. It is therefore more advantageous for an animal to have a dependence on detritus than on algae growing on stones (Dudgeon 1982a). Bed load movements can drastically alter rock or riverbed environments of benthic dwellers, while suspended loads influence light penetration and photosynthetic opportunities of phytoplankton and bacteria.

Many invertebrates and fish of the headwaters have extremely flat bodies allowing them to move about easily in the almost motionless layer of water just above the riverbed, or to live under stones. Some have hooks, others suckers, and yet others have hydrodynamically streamlined shapes. Some of the fish living on the riverbed are able to maintain their position by having a smaller swim bladder (an air-filled sack in their bodies used to control buoyancy) than that found in fish that are adapted for swimming against the current.

The distribution and abundance of animals found on the riverbed across a meander, will reflect differences between the inner (slow-moving) and outer (fast-moving) bends. For example, species of net-spinning caddisflies

(p. 292) that make strong coarse-mesh nets to catch passing food are found in the fast-flowing, shallow parts of a river, whereas the species with finer and more delicate nets are found in the slower, deeper water (Dudgeon 1986b).

Substratum

The trend of decreasing particle size on the riverbed with increasing distance from the headwaters obviously influences the distribution of animals. For example, those adapted to living under stones will rarely be found on a muddy substrate, and those which burrow into mud will not be found among stones. It is interesting, therefore, to consider the effects on animal distribution of the common practice of removing small boulders from riverbeds for use as hard core in making roads and other constructions.

Rooted macrophytes may be found in parts of a river where conditions are suitable (relatively slow-moving with sufficient silt and organic material), and these in turn influence the distribution of invertebrates. Most invertebrates found on macrophytes do not feed on the plants themselves but graze on epiphytic algae growing on the leaves (p. 288) a situation which has parallels with seagrass meadows (p. 208). Other invertebrates use the leaves as an anchorage point from which to filter water for particles of suspended organic matter, and yet others are predators.

Temperature and Dissolved Oxygen

It was shown above (p. 311) that temperature and concentration of dissolved oxygen are closely related. Temperature increases with distance from the headwaters, but it is somewhat difficult to distinguish any effect this may have from the linked effect of decreasing dissolved oxygen. The fall in dissolved oxygen concentrations downstream is exaggerated because the oxygen removed from the water by organisms living in the calmer, lower reaches is less easily replaced than in the turbulent headwaters.

Studies of environmental changes and their effects on aquatic organisms have tended to concentrate on pollutants and other chemicals, whilst the importance of temperature is often overlooked. The metabolic rate and hence demand for oxygen of most animals increases with temperature, but at higher temperatures haemoglobin has a lower affinity for oxygen, and dissolved oxygen concentrations decrease.

An increase in temperature from 25°-30°C resulting from the clearing of riverine forest, for example, would cause a 9.5% reduction in dissolved oxygen at saturation and the saturation percentage will probably also fall (Crowther 1982). This will have marked impacts, particularly on animal communities, although tolerance of oxygen depletion varies between species. In general, however, indigenous species are less able to cope with changes than introduced species.

Mineral Nutrients

Most aquatic molluscs are limited in the freshwaters they can inhabit because they need calcium concentrations in the water of at least 20 mg/l for the secretion of their shells. As a result, rivers running off relatively recent volcanic debris, for example, are likely to support few, if any, molluscs although this effect needs to be quantified. Some molluscs, however, appear to get all the calcium they need from their food and can therefore live in water with much lower calcium concentrations.

Pulmonate snails, those with accessory breathing organs, tend to have greater tolerance of turbidity, low dissolved oxygen, water hardness, pH, ammonia, nitrates and phosphates than many of the prosobranch snails which are more particular in their choice of habitat (Palmieri et al. 1980). Exceptions do exist and a good example is *Melanoides tuberculata* which is tolerant of a wide range in water quality (Dudgeon 1986a).

Biotic Factors

Competition between species and predation may be the main factors constraining a particular species to its realized niche rather than its preferred or fundamental niche. Where this shift is caused by direct aggression, interference competition is said to have occurred. A second form, exploitation competition, is less easy to identify but occurs where indirect competition occurs for the same resource. Consumption of that resource by one species will reduce the amount remaining to be consumed by the other species. The species less able to convert the resource into reproductive output will either perish, move elsewhere, or specialize in other resources. Thus the diet, habitat preference and habits of a community of 20 freshwater fish species in Sri Lankan forest rivers was examined and it was found that, with a few exceptions, the niches occupied overlapped little. Where species appeared to live in similar microhabitats, it was found that their diets differed (Moyle and Senanayake 1984).

Energy Flow

At the level of communities or whole ecosystems, the study of ecology can be broadened to include the flow of energy through the ecosystem, in which organisms are regarded as transformers of energy. The energy base in most ecosystems is provided by plants converting solar radiation through photosynthesis into high-energy organic molecules. Exceptions are caves (p. 535) where there is not enough light for green plants, and rivers in which a substantial proportion of the energy base is represented by decaying organic matter. This organic matter can be divided into two components: 'allochthonous' (originating outside the system) and 'autochthonous' (originating within the system). The latter is a relatively

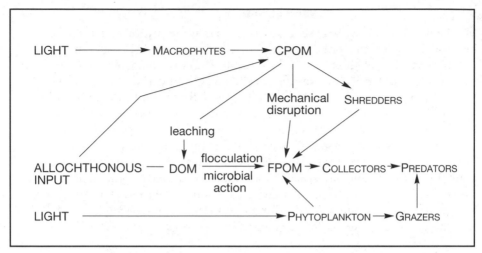

Figure 4.39. A simplified model of energy flow in a river ecosystem. To preserve clarity some arrows have been omitted. For example, all the animals contribute to FPOM in the form of faeces, dead bodies, etc.; some of the allochthonous input contributes directly to FPOM; the principal food of many fish in rivers consists of invertebrates so the 'predator' category includes fish. However, some fish feed on macrophytes and detritus.

After Townsend 1980

minor component. The available organic matter, living and dead, is processed by a wide range of organisms, which include bacteria, fungi, invertebrates and fish, all interacting in a highly complex manner (fig. 4.39) with different pathways depending on the size of the particles. The organic particles can be divided as follows:

- dissolved organic matter (DOM) arbitrarily defined as smaller than 0.00045 mm diameter,
- fine particulate organic matter (FPOM), less than 1 mm diameter, and
- coarse particulate organic matter (CPOM), more than 1 mm diameter and including whole leaves, twigs, etc. The FPOM and CPOM components also include the micro-organisms associated with them (Townsend 1980).

The path of the energy flow depends to a large extent on the nature of the energy base. The majority of river headwaters, particularly if undisturbed by people, flow through forested catchment areas and receive a substantial allochthonous input from material that simply falls into the water from the forest canopy (fig. 4.40). This decomposes extremely rapidly in

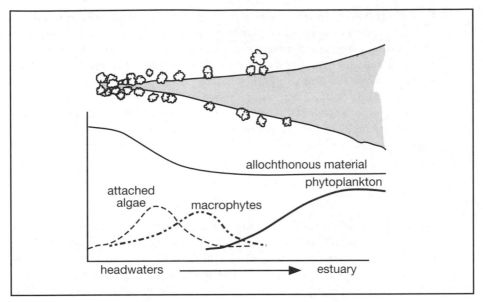

Figure 4.40. Hypothetical representation of the relative contributions of potential energy inputs to a river.
After Townsend 1980

tropical rivers, and a proportion of the nutrients released return to the terrestrial vegetation through the roots that trail in the water (Dudgeon 1982a, 1983c). Consistent with the knowledge that the fauna of shallow, stony and fast-flowing 'riffles' is different from that of deep and slow-flowing pools (p. 323), so rates of decomposition also appear to be higher in the riffles (Dudgeon 1986b). The shade provided by overhanging trees prevents, or at least hinders, the growth of both attached algae and macrophytes, but lower down the river only the riverbanks are shaded and the autotrophic component increases. The attached algae component, which is generally better adapted to extreme flows and less dependent on a substrate of sediment, would be expected to fill its maximum role nearer the headwaters than the macrophytes. Both macrophytes and attached algae should continue to make significant contributions to the river energy budget, until the depth and turbidity are such that light can no longer reach the riverbed, and then these plants will be restricted to the margins (Townsend 1980). Phytoplankton will usually make a significant contribution only where the river is long enough for this component to build up. The generation time for phytoplankton is one or two days (at least three or four days and frequently more for zooplankton), and since rivers generally flow between 20-60 km per day, it is clear that few rivers will have well-developed populations of phytoplankton (table 4.2).

Such longitudinal patterns have not been studied in Southeast Asia. The scheme illustrated is hypothetical but it seems to fit the known facts from other regions. If it applies to Sulawesi it can be seen that forest clearance in headwater areas can seriously disturb a river's energy input and therefore the life that depends on it (p. 342).

Benthos Dynamics

When a heavy fall of rain causes increased flow, the shear stress exerted on a particular area of riverbed intensifies and the substrate is scoured, often with the loss of organisms associated with it. As the flow subsides, so organisms from upriver will be deposited in their place. Even when the river flow is normal, however, benthic organisms temporarily join the plankton and move downstream. This is easily demonstrated if a net is placed in a river and held above the riverbed for a period. This phenomenon is called 'invertebrate drift'.

In just 24 hours in a major headwater river in Peninsular Malaysia, an average of 222,800 individual invertebrates would drift past a transect (Bishop 1973; Townsend 1980). This is equivalent to about 160 individuals per 100 m³ of discharge. Drift varies not just with river flow but also through the day. Studies from various parts of the world have shown that drift is highest at night, particularly just after sunset (Bishop 1973). This appears to be related to light levels rather than to chemical changes. Many invertebrates spend much of the day hiding under stones and only forage when darkness falls. It is logical that when they start to move they are more susceptible to being swept away. It has been calculated that 2.6% of benthic invertebrates shifted their position each day by drifting (Townsend and Hildrew 1976) but another study showed that 60% of drifting invertebrates travel for less than 10 m before regaining a foothold (McLay 1970).[19]

Whether or not losing contact with the riverbed and drifting downriver is accidental, there may be adaptive significance in doing so. A riverbed, as with most habitats, is composed of 'patches', some favourable for a particular organism and some unfavourable. The patch may be a food resource, a form of favoured substrate, an area experiencing a certain set of biotic and/or abiotic conditions, etc. In some cases a patch may change its suitability, for example, when a food resource is depleted or when a flood occurs (Bishop 1973; Townsend 1980). Drifting, although it has certain risks, is an energy-efficient way of moving from an unfavourable to a possibly favourable patch, for a journey of 10 m along a riverbed is not inconsiderable for many river invertebrates. For an insect larva 1 cm long it would be equivalent to 1,000 body lengths which for a man would be equivalent to about 1.7 km. If the drifting invertebrate lands on an unsuitable substrate there is a high probability of the animal re-entering the drift within 5-30 minutes (Walton 1978), suggesting that invertebrate drift is not entirely passive.

If such large numbers of normally quite sedentary animals are moving downstream, it would be reasonable to ask how the upriver regions remain populated. Do upriver movements by some organisms compensate for the downriver losses?

The displacement of organisms downstream does not necessarily lead to the extinction of those species in upriver stretches. One way to view drift is as a dispersal mechanism for removing animals (possibly as eggs or as larvae) which, had they stayed in the headwaters, would have exceeded the habitat's carrying capacity. It is obvious that not all young invertebrates could remain in the area where their eggs were laid because they would soon exhaust the initial food resource (Peckarsky 1979).

This is not the whole story, however, because an organism drifting downriver is likely to leave its zone of most suitable environmental conditions. It would therefore be reasonable to suggest that adult invertebrates that managed somehow to reach regions upriver of their optimum habitat to breed, would have an evolutionary advantage because their young would have a greater chance of developing in that optimum habitat (Townsend 1980). A 'colonization cycle' is thus envisaged with eggs being laid in the headwaters, dispersal of larvae occurring downriver and an upriver flight or other movement of adults to the headwaters to complete the cycle. This is commonly known as 'Mueller's hypothesis'.

The first two stages of this hypothesis are irrefutable but evidence for the upriver movement of adults is less convincing. Twin traps set to catch insects flying upriver and those flying downriver along a headstream river in Peninsular Malaysia revealed that the predominant direction of flight was in fact *down* river (Bishop 1973). As a rule, winged adults of invertebrate species with aquatic larvae are not strong fliers and their flight direction might simply reflect the prevailing wind direction. Strong winds occur most frequently in rainy seasons and these are the periods when insect dispersal is most common (Fernando 1963). Adults of invertebrate species which spend their entire life cycle in freshwater are not usually strong swimmers or walkers but they may travel near the river edge where shear stress is least, so that the upriver journey requires the least possible energy. Most studies have found that upriver movements represent only about 7%-10% of the individuals that move downriver (Moss 1980; Williams 1981). It must be remembered, however, that if only a single female reaches the upriver regions, she may lay hundreds or even thousands of eggs.

The study of invertebrate drift deserves more attention. If an industrial development is to be sited in the middle stretches of a river, and an environmental impact statement has been requested before it is built, one of the many problems that should be considered is the impact on the recolonization of upper stretches of the river by invertebrates (or, for that matter, fish). Is the effluent going to be poisonous or debilitating to adults moving upriver? Which months are the most critical? Since invertebrates are common food for fish, and fish are common food for humans, the problem of the mechanism of invertebrate drift is wider than that of esoteric biology.

FISHERIES

As indicated earlier (p. 298) the fisheries of Sulawesi's lakes are based not on indigenous species but on those ten introduced species that were thought to be most likely to succeed in the prevailing conditions (table 4.18; fig. 4.41).

A reflection of the different properties the of lakes is shown in the relative success of different introduced fish species. *Oreochromis mossambicus* has been extremely successful in Lake Lindu and Poso but only in Lake Poso has carp, an extremely difficult fish to introduce successfully, become the major harvested fish (Simanjuntak 1981). In oligotrophic lakes such as Matano, the carp *Cyprinus carpio* has not thrived and so there are plans to introduce culture nets moored off-shore in which the fish will be fed by hand. This method has been used with success in the Jatiluhur reservoir in West Java (Anon. 1980) and a similar method is used on Lake Tempe with another species, but the fish are released at the start of the rainy season.

The catfish *Clarias batrachus* is often regarded as undesirable in some areas because it preys on small fishes and is not favoured as a food fish. There are areas of Indonesia, such as Central Java, however, where this fish is in high demand (A. Hardjamulia pers. comm.). Catfish studied in Lake Tondano were found to feed primarily on detritus with molluscs and insects eaten occasionally. Introduced possibly only ten years ago, it has reproduced prodigiously probably because of an open niche, a wide range of suitable foods, high reproductive potential and little desire on the part of the local people to eat the fish. It has been suggested that ways to put pressure of the catfish populations should be found (Lumingas 1983).

The air-breathing *Helostoma temminckii* was so successful in Lake Tempe that it was being exported just twelve years after its introduction. But even more dramatic was the replacement of *H. temminckii* by the barb *Barbodes gonionotus*[20] and the small gurami *Trichogaster pectoralis* which, two years after their introduction in 1937 were accounting for 20% and 70% of the lake's harvest respectively (Hickling 1957). As is often the case with lakes into which exotic fish have been introduced, the production in the early years was enormous, up to 25,000 t/yr. It fell to about 5,000 t/yr but is now about 12,000 t/yr (equivalent to about 600 kg/ha/yr). This is still a large annual production; for comparison, the production of fish in Aopa Swamp is only about 3 kg/ha/yr which is very low for eutrophic/mesotrophic waters, because even the deep oligotrophic waters of Lakes Matano and Towuti have a fish production of 6.5 and 4.5 kg/ha/yr respectively. The low productivity of Aopa Swamp is at least in part because the people around the Aopa Swamp do not eat the abundant eels *Anguilla bicolor*[21] and *Monopterus alba* (fig. 4.42) or the very common large *Vivipara* snails, despite these being common foods elsewhere in Indonesia. The high productivity of Lake Tempe is primarily a result of the gently sloping banks and the shallowness of the lake, as well as the balance between herbivorous and carnivorous fishes (Anon. 1977a). Carnivorous animals are far more wasteful of primary production than herbivorous ones.

The fall in fish production of Lake Tempe stimulated considerable research into the fisheries and other aspects of this lake which has been called the 'Fishbowl of Indonesia' (Suwignyo 1979). The first study of the problem had concluded that the dramatic decrease was due to the gradual

Table 4.18. Introduced fish in Sulawesi lakes with dates of first introduction where known. The fish of Aopa Swamp are listed in table 4.11.

	Tempe	Lindu	Poso	Matano	Towuti	Moat	Limbotto	Tondano
Cyprinidae								
Barbodes gonionotus	1937	1955	+	-	-	+	+	1954
Osteochilus hasselti	1938	-	-	-	-	+	+	1941
Cyprinus carpio	1948	1955	+	+	+	+	(+)	1895
Clariidae								
*Clarias batrachus**	1944	-	-	-	-	+	-	1975
Belontiidae								
Trichogaster trichopterus	-	1950s	-	+	+	-	-	1938
T. pectoralis	1937	-	-	-	-	-	1947	1938
Osphronemus goramy	-	1955	-	-	-	+	-	1914
Helostoma temminckii	1925	-	-	-	-	-	-	-
Cichlidae								
Oreochromis mossambicus	-	1951	+	-	-	+	1942	1951
O. nilotica	-	-	-	-	-	-	+	1871
Eleotridae								
Ophiocara porocephala	-	-	-	-	-	-	+	1902

* = accidental introduction   + = date of introduction unknown
(+) = known to have been introduced but no longer present

After Schuster 1950; Sarnita 1973, 1974; Suwignyo 1978; Carney et al. 1980; Hadiwijaya 1981; Manggabarani 1981; Simanjuntak 1981; Lumingas 1983; A. Hardjamulia pers. comm.

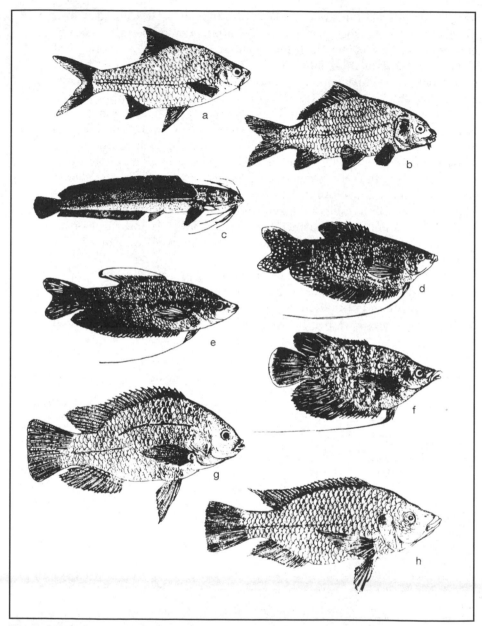

Figure 4.41. Major introduced commercial fish of Sulawesi. **a** - Java barb *Barbodes gonionotus,* **b** - common carp *Cyprinus carpio*, **c** - rice-field catfish *Clarias batrachus*, **d** - two-spot gourami *Trichogaster tricopterus*, **e** - no-spot gourami *T. pectoralis*, **f** - gourami *Osphronemus goramy*, **g** - common tilapia *Oreochromis mossambicus*, **h** - Nile tilapia *Oreochromis nilotica.*

After Mohsin and Ambak 1983

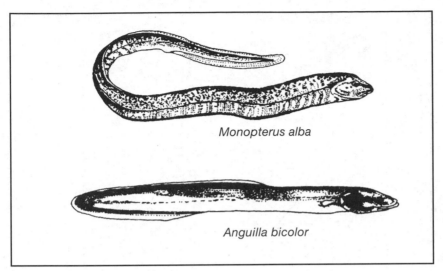

Monopterus alba

Anguilla bicolor

Figure 4.42. Common eels.
After Mohsin and Ambak 1983

shallowing of the lake (p. 258), but teams from BIOTROP in Bogor concluded that overfishing, together with some effect from shallowing, was the major cause (Anon. 1978). The local government responded to the conclusion by instituting Friday as a closed fishing day, and by proclaiming 230 ha in the centre of the lake as a fish reserve. Lack of effective enforcement meant that both these rules were flouted. More effective than such regulations would probably be the land-based enforcement of outlawing small (< 2 cm) mesh nets. Reduction of soil erosion from the water catchment area would also be beneficial. A further component in fisheries management would be to place an obstacle between the lake and the sea preventing milkfish *Chanos chanos* and other brackish water species from entering the lake. All these species (except *Mugil* sp.) are at least partly carnivorous and have a low combined production (Anon. 1978). While this course of action certainly seems reasonable from the point of view of inland fisheries, it might, when considered with the loss of mangroves, have serious effects on inshore and tambak fisheries (p. 187).

Lake Limboto has also experienced a five-fold reduction in fish harvest, at the same time as the lake has become shallower (fig. 4.43), but not enough data are available to be able to state with any certainty whether this is due to the shallowing process or to overfishing.

Overfishing can be said to occur when the rate of harvesting exceeds the rate of recruitment or production. The recruitment rate varies with the

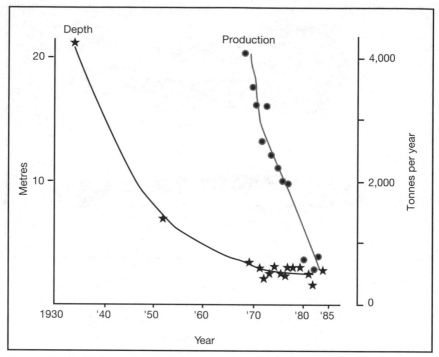

Figure 4.43. Fish production, lake depth and area of Lake Limboto.
Data from U. Alitu pers. comm.; A.H.J. Purukan pers. comm.

size of a population, being low both when there are few individuals, and also when the population is crowded and competition between individuals is intense. When the carrying capacity of a habitat is reached, recruitment is zero (fig. 4.44).

Harvesting of a population can be conducted at three general rates which for the purposes of this discussion are assumed to be constant. When the harvesting rate is high and exceeds recruitment, then the population is driven to extinction. When the harvesting rate is low, then there are two points where the number of individuals lost to the population equals those recruited to the population. At this low rate, there is an equilibrium at a relatively high population density, and another at a relatively low population density. The latter is dangerous from a management perspective because if the population gets smaller due to some factor other than harvesting, yet the harvesting rate is maintained, then the population will be driven to extinction. The intermediate case is that where the harvesting rate exactly matches the maximum population recruitment. This is

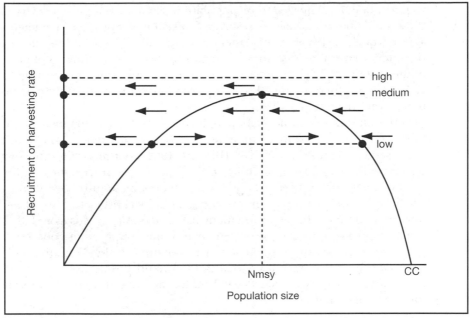

Figure 4.44. The effects of constant harvesting rates on the rate of recruitment to a population (curved line). There are three harvesting rates, high, medium and low. Arrows indicate the expected changes in population size under the respective harvesting rates. Dots indicate points of equilibria. The three rates have one, two and three points of equilibria respectively including the equilibrium of extinction. 'CC' indicates the maximum population allowed within the carrying capacity of the habitat. 'Nmsy' indicates the population size that can be harvested for a maximum yield on a sustainable basis.

known as the 'maximum sustainable yield', which is balanced between the conditions of under- and overexploitation. As with the low harvesting rate, however, whereas higher populations are pulled back to the equilibrium point, lower populations are driven to extinction if the harvesting is not adjusted accordingly.

The concept of maximum sustainable yield is central to the science of wildlife management, but it has many shortcomings. First, it does not take account of population structure such as the relative abundance of different age-classes; second, the environment varies and unpredictable events can cause reductions in populations unrelated to harvesting pressure; and third, the maximum sustainable yield is extremely hard to estimate since it requires detailed knowledge of population size and recruitment. Given these limitations it is usually necessary to make the best-possible judgement. This is true for fisheries management both in Lake Tempe and in some of the intensively-studied fisheries in North America. The usefulness of maximum sustainable yield is probably more important as a concept urging caution, and encouraging flexibility over harvesting rates that can be benign one year and catastrophic the next.

Even without the influence of people, however, fish populations do not necessarily remain static. For example, fish yields from Lake Moat increased between 1977 and 1981 (Buchari 1984), but reports given to an EoS team by local fishermen indicate that yields have since dropped, for example, eel catches have dropped from 50 per day to just 1 per month. The cause of this has not been investigated but local fishermen are of the opinion that there are simply too many fishermen. Also, yields from Lakes Lindu and Poso appear to be extremely unstable, but even in good years the yield is below that estimated as the potential yield. Migratory eels *Anguilla* spp. are indigenous to Lakes Lindu and Poso, but whereas eel fisheries are still active at Lake Poso (Taufik et al. 1980), they are now rarely found in Lake Lindu.

MANAGEMENT OF MACROPHYTES

Macrophytes can be extremely useful plants. They often provide important refuges for young fish, and they act as substrates for invertebrates and algae that are eaten by fish. Their presence in a body of freshwater is clearly important to a fishery. Some macrophytes, notably kangkung *Ipomoea aquatica* (Conv.), are eaten by people and fisherman around Lake Tempe deliberately plant this plant both for food and for the protection its hanging roots provide for young fish (Anon. 1977a). Some species, such as *Hydrilla verticillata* and water hyacinth *Eichhornia crassipes* are harvested for feeding to cattle, pigs, ducks and chickens (Swarbrick et al. 1982).

E. crassipes, *Salvinia*, *Lemna* and *H. verticillata* are also known to be efficient accumulators of nutrients and have been used in water treatment plants to lower nutrient concentrations in nutrient-rich water (Wolverton 1985). *H. verticillata* is useful as a 'sink' for nitrogen, phosphorus and zinc, whereas *Salvinia molesta* is particularly efficient at accumulating copper and lead (Finlayson et al. 1984).

Emergent and floating macrophytes can also be viewed as extremely disadvantageous. Water evaporating and transpiring from the leaves of emergent macrophytes (evapo-transpiration) may exceed the rate of evaporation from open water of the same area. This is an important consideration particularly in shallow lakes such as Tempe, Limboto and Tondano. This water loss may be expressed as evapotranspiration (Et)/evaporation (E). Et/E values for *E. crassipes* and *S. molesta* are 2.5-7.8 and 2.0, respectively (Sastroutomo 1985). That is, the active process of transpiration results in at least twice the water loss as evaporation alone.

From the viewpoint of fisheries, aquatic plants are disadvantageous in that they introduce considerable quantities of organic matter into the water which can lead to reduced oxygen concentrations. They speed the process of sedimentation by retarding water flow, thus enhancing deposition of suspended particles, and they interfere with fishing activities by snagging on fishing lines or getting caught in nets (Anon. 1983a). Since they reproduce vegetatively very quickly and are difficult to control, they are frequently referred to as 'weeds'. For example, the area covered by the floating *Salvinia molesta* can double in extent every 2.2 days under optimal growth conditions (Mitchell 1985).

The major Sulawesi lakes each have between one and four species from the list of the world's ten worst aquatic weeds (table 4.19). It must be remembered that 'weediness' is simply an attribute given to a plant by man; a weed is a plant growing where it is not wanted and the nomination of a plant as a weed is a consequence of man being thwarted in his ever-increasing attempts to manage his environment.

At Lake Tempe, *Hydrilla verticillata*, *Ceratophyllum demersum*, *Najas indica* and *Potamogeton malaianus* (Pota.), *Polygonum* spp. (Poly.) and *Ipomoea aquatica* (Conv.) are listed as potential sources of interference in the fishing industry there (Anon. 1978). Most of the above species, along with 26 emergents including *Monochoria vaginalis* (fig. 4.45) at the sides of Lake Tondano serve to slow the water current and increase sedimentation such that the lake is rapidly becoming shallower (Anon. 1979b).

In 1976 *Eichhornia crassipes* and *Salvinia molesta* were not found on Lake Tempe although both are now present. *Pistia stratiotes* was present but was not regarded as a threat because it is not adapted to surviving periods of being washed up on the shore or stranded on the mud during the dry periods. *Hydrilla verticillata* is the major threat but even this appears to be restricted to the more sheltered parts of the lake, where the sediment is not

Table 4.19. The world's ten most important aquatic weeds.

Eichhornia crassipes (Pont.)	*Panicum repens* (Gram.)
Salvinia molesta (Salv.)	*Monochoria vaginalis* (Pont.)
Pistia stratiotes (Arac.)	*Cynodon dactylon* (Gram.)
Phragmites australis (Gram.)	*Echinochloa crussgalli* (Gram.)
Ceratophyllum demersum (Cera.)	*E. colonum* (Gram.)

After Soerjani 1978, 1985

Figure 4.45. *Monochoria vaginalis.*
After Backer 1951

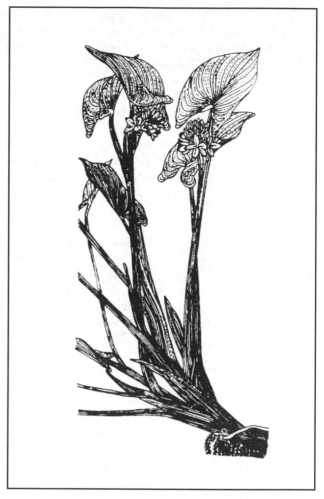

so disturbed by water movement. This is probably also the reason why *Najas indica* and *Ceratophyllum demersum* are not more common. Changes proposed by water engineers to control the water entering and leaving the lake would diminish the sediment load and slow down the process of shallowing, but this would inevitably increase the abundance and distribution of most of the potential aquatic weeds since the water level fluctuations do serve to control these plants (Suwignyo 1978).

Macrophytes clearly have potential advantages and disadvantages with respect to human uses of freshwater and to balance these informed management is necessary.

This management can be divided into three components: preventative measures, control measures and utilization (Soerjani 1985). The first of these requires monitoring areas of freshwater for new, undesirable species of macrophytes as should have been done for *Eichhornia crassipes* and *Salvinia molesta*. Control measures designed to eradicate or partially reduce the population of a particular macrophyte species, can be subdivided into mechanical, biological and chemical methods. All these can have serious side-effects associated with them. For example, mechanical control can disperse small bits of stem that are swept away and then grow in new areas; the American weevil *Neochetina eichhorniae* can control water hyacinth (Napompeth 1985) but it also feeds on arrow root *Canna edulis*[22] (Cann.) and gingers *Zingiber* spp. (Soerjani 1985). Chemical methods, the use of herbicides, can have the advantage that the dead macrophytes enter the food web again as detritus. This presents problems, however, if not all the detritus can be utilized by the fish, snails or other detritivore fauna. There are also the well-known dangers such as unpredicted toxicological effects on non-target species (Soerjani 1985) and it must be remembered that behaviour of herbicides in water is quite unlike that of herbicides in the air and on land (Robson 1985).

Considerable work has been conducted in Indonesia on the utilization of macrophytes. Much of this has concentrated on *E. crassipes* but the results can be applied with caution to other species. Macrophytes can be used as the bedding material for mushroom culture and yield up to 250 g of mushrooms per kilogram of sterilized dried plant. Decomposition of the plants to make biogas was thought to have a bright future but it does not seem to be economically and technically-feasible at the village level. Water hyacinth has some potential as the raw material for paper pulp but, again, it would only be economically feasible if competition from other forms of paper were controlled in some way. *Salvinia molesta* and water hyacinth have uses as organic additives or as mulch. Various carp feed on a wide range of aquatic plants (Sutton 1985), and adequate control can apparently be achieved with grass carp *Ctenopharyngodon idella* using a stocking rate of 79 fish/ha. The plants will also be eaten by cattle, pigs, chickens, and ducks (Soerjani 1985).

The best control is integrated management. This is not necessarily the use of all three methods (Soerjani 1985), but rather the clear understanding of the perceived problems, a clear definition of the desired goals (not likely to be complete eradication), the study of the response of the key species to possible changes, consideration of the economic and ecological costs and benefits, and attention to local socio-economics, particularly potentials for locally-organized resource management.

IMPACTS OF DEVELOPMENT

Development can affect freshwater ecosystems in a variety of ways. Industrial and domestic pollution reduce the water quality with consequent effects on the whole biota; fish poisons and bombs cause indiscriminate fish death; over-exploitation of fish stocks to supply the aquarium trade can have devastating effects; introduced fish can cause the demise of indigenous species; and forest clearance causes changes in temperature, turbidity, stream flow, water input, etc., thereby altering the habitat of many species.

Industrial, Domestic and Agricultural Pollution

Pollutants of water have been divided into four categories: pathogens, toxins, deoxygenators and nutrient enrichers (Prowse 1968). Pathogens include a wide range of bacteria, protozoa and parasitic worms harmful to man and other organisms, and most of these pathogens are associated with untreated sewage. Toxins are derived from industrial waste and agricultural chemicals. Their effects can be both dramatic and cumulative not only in aquatic animals but also plants and, of course, man. Deoxygenation is caused by bacterial and fungal decay of organic matter and by animal and plant respiration, and can also be related to certain weather conditions (Johnson 1961). Where large quantities of organic wastes are disposed of, deoxygenation and the subsequent death of animals by suffocation can be expected.

Nutrient enrichment is known as eutrophication. Eutrophic habitats are excellent for fisheries because of their high productivity. Indeed, lakes and ponds are often artificially fertilized to increase fish production. In the Philippines, for example, yields of over 1,000 kg/ha/yr have been achieved from such habitats, whereas a normal, medium-sized oligotrophic lake might produce only one-hundredth of this yield. Primary production of the phytoplankton is similarly affected.

Eutrophication is not, of itself, a danger to freshwater ecosystems. Naturally eutrophic systems are usually well-balanced but the addition of artificial nutrients can upset this balance and cause devastating results. Algae 'blooms', the most spectacular of these effects, are a result of high nutrient levels, and favourable temperature and light conditions which stimulate rapid algal growth. As the water becomes enriched so the dominant species of algae change. The blooms are a natural response to environmental change, but when the water can no longer support high algae populations, the algae that accumulated during the bloom die and decay. The ensuing rapid decomposition of organic debris by bacteria robs the water of its oxygen, sometimes to the extent that fish and other aquatic organisms suffocate.

Various studies have shown that phosphorus, rather than nitrogen or potassium, is usually the limiting factor in eutrophication (rather than nitrogen or potassium which might have been suspected) and 0.08-0.10 mg/l of phosphate has been suggested as the figure above which algal blooms are likely to occur (Dunne and Leopold 1983). Research is needed to determine the appropriate level for Sulawesi lakes, so that it can be used for the establishment of rational effluent control standards. Algal blooms may have already occurred on Sulawesi: it has been reported that Lake Limboto has appeared 'a peculiar pinkish colour' (Guillemard 1889), and from the air, parts of Lake Tempe sometimes appear brownish-red. Water samples were not taken during these events. The course of a serious algal bloom has been reported from a Malaysian reservoir (Arumugan and Furtado 1981). Reference to table 4.15 shows that Lake Matano has high phosphate levels and this is surprising considering its position and surrounding land use. Lakes Tempe and Limboto are perhaps the most likely places to experience serious eutrophication given their shallowness and the relatively dense human population in their catchment areas.

Multidisciplinary research on chemical, physical and microbiological parameters in rivers running through towns in Sulawesi have not been conducted, but one such study was made in Malaysia and its results should give stimulus and lead to future research in Sulawesi. In that study a wide range of parameters was measured at a number of sample points along the Kelang River (which passes through Kuala Lumpur) and the results were examined against fish catches at the same locations (Law and Mohsin 1980; Mohsin and Law 1980). The species diversity, richness and the number of fish per m^2 were much higher in sample points above Kuala Lumpur than in sample plots within the city itself or downstream. In the worst-polluted places only the guppy *Poecilia reticulata* was found. Downstream of Kuala Lumpur the water was unsuitable for most freshwater fishes but a few airbreathing species were found. The primary causes of the change in fish abundance and diversity were heavy siltation, low pH, and high biological oxygen demand. Heavy silt load suffocates certain fish by clogging their gills.

Poisons, Bombs and Electric Shocks

The most common traditional fish poison is 'tuba', made from roots of various species of the climber *Derris* in which the active chemical is rotenone. Other poisons, typically commercial insecticides, are also commonly used. Poisons, bombs and electric shocks are very effective and kill or debilitate large numbers of fish without regard to size or species. As a result, many urban and rural rivers have virtually no fish left in them. The effects of explosives in rivers are similar to those described for coral reefs (p. 241). Apart from any other arguments, these forms of fishing are selfish, needlessly destructive and, in the case of poisonous insecticides, can cause sickness in humans.

Forest Clearance

Forest clearance is by far the most serious threat to natural river and lake ecosystems. Many aquatic animals depend on allochthonous material (material falling into the river–p. 325) for their existence, for they feed directly or indirectly on dead leaves and other vegetable matter. If forest around the river is cleared, plant matter fails to accumulate regularly in sufficient quantities and many organisms such as fish, prawns, crabs, dragonfly nymphs, pond skaters and snails will not survive even if other conditions are still suitable (Dudgeon 1983c; Calabrese 1986).

No systematic investigation of the effect of forest clearance on the aquatic ecosystems of Sulawesi appears to have been published but it is certain that the experience in Singapore (Alfred 1966) and Peninsular Malaysia (Johnson, Soong and Wee 1969; Johnson 1973) is extremely relevant. Rivers without a riparian fringe may indeed have considerable numbers of fish but few species are represented and those present tend to be introduced or widely-distributed species. Species indigenous to Sulawesi will generally be lost. For example, the introduced gourami *Trichogaster trichopterus* and tilapia *Oreochromis* sp. were the only fish caught in the Toraut River on the edge of Bogani Nani Wartabone National Park during Project Wallace, even though the riverbanks were opened only a few years ago (D. Dudgeon pers. comm.).

Chapter Five

Lowland Forests

DIVERSITY

Most tropical forests are very rich in species and a single hectare of lowland forest may contain 100-200 species of trees.[1] The total number of plant species, including smaller trees, shrubs, herbs, climbers and epiphytes is obviously much higher, and a recent study found 233 species in a plot of just 10 m x 10 m in Costa Rica. This is equivalent to one-sixth of the British flora, for example, on an area the size of half a singles tennis court. The total of plant species in a Sulawesi forest is likely to be lower but certainly of the same order. Caution is required before enumerations of total species are embarked upon, however, because it has been calculated that to study a single hectare would take one person nearly a decade (Whitmore et al. 1986).

The only detailed species count conducted in Sulawesi was in the somewhat seasonal, and therefore rather species-poor, forest at Toraut, Bogani Nani Wartabone National Park[2] (table 5.1). The total of tree species in a hectare of forest in the ever-wet centre of the island would probably be higher. It has to be remembered that the figures are derived from relatively small, and not necessarily representative areas, and that the pattern of growth phases within the area enumerated will affect the totals.

The problem of explaining the very high diversity of plants in tropical forests, particularly lowland forests, has taxed many minds. Before detailed investigations of fossil pollen began about 15 years ago it was thought that tropical forests had climatic stability for millions of years and that this stability had allowed time for evolution of so many species. It is now realised that tropical vegetation has experienced considerable changes (p. 29), but those areas that retained a wet, non-seasonal climate have the richest floras and faunas. Some people have suggested that the formation of isolated forest blocks, or refuges, during the peaks of glacial activity, when climates were cooler and drier, would have resulted in the species of each block following a course of evolution slightly different from that followed by species in the other blocks when the climate ameliorated and the forest blocks were reunited, some of the closely-related species would have been incompatible for reproduction (i.e., would have become different species–p. 30), and thus a greater degree of diversity would exist than before the forest blocks were separated. This is generally known as the 'Refuge theory' (Haffer 1982) and has been useful in explaining the high diversity in certain organisms in Amazonia and Africa. It is less convincing as a theory in Southeast Asia, however, because the whole of Peninsular Malaysia, Sumatra, Borneo and West Java would have formed one 'refuge' and the majority of New Guinea the other (Meijer 1982).

In addition to historical factors, species richness is also influenced by the wide range of potential niches available in a forest, variations in site conditions, the changes in physical conditions through the forest growth cycle (p. 359), and the vast array of potential interactions with animals in processes such as pollination and dispersal (p. 379) (Whitmore 1984).

Table 5.1. Number of trees (roman type) and tree species (bold type) in one-hectare plots.

	Diameter at breast height (cm)				
	≥10	≥15	≥20	≥35	≥40
SULAWESI					
Toraut a[1]	408	328	237	75	44
	109	**78**	**72**	**43**	**26**
Toraut b[5]	-	270	170	82	66
Lombogo[5]	-	238	174	88	70
Dumogo (Hog's Back)[5]	-	350	-	-	-
Dumogo (Edward's Camp)[5]	-	440	-	-	-
Tanoma[6]	-	496	-	-	-
SUMATRA					
Siberut Island[2]	-	180-320	-	-	-
Ketambe[3]	287*	-	-	-	-
	148*	-	-	-	-
MALAYSIA					
Kuala Lompat[4]	-	194-307	-	-	-
	-	**110**	-	-	-
KALIMANTAN					
Wanariset[3]	338	-	-	-	-
	173	-	-	-	-
Lempeke[3]	278*	-	-	-	-
	155*	-	-	-	-
Jaro[3]	249	-	-	-	-
	129	-	-	-	-

* Figure calculated from numbers in 1.6 ha; tree numbers show a linear relationship with area, but tree species do not.

1 - Whitmore and Sidiyasa 1986; 2 - Whitten 1980; 3 - Kartawinata et al. 1981; 4 - Raemaekers et al. 1980; 5 - Bismark 1982b

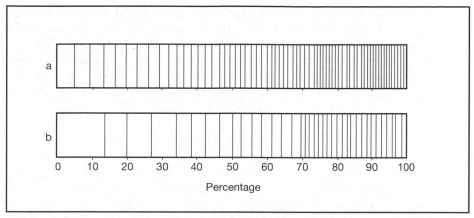

Figure 5.1. Percentage occurrence of different tree species from plots in (a) Karaenta Reserve, Maros, and (b) Mopuya, Bolaang Mongondow, to illustrate the large number of rare species. Intervals represent 10% of the trees present. In both cases half of the trees are represented by three-quarters of the species.

From data in Doda 1980; Harun and Tantra 1983

An ecosystem with high diversity has, by corollary, a large number of rare species (fig. 5.1) so, instead of asking "why is lowland forest so diverse?", one could equally well ask "why are there so many rare species?" (Flenley 1980). To answer this one has to consider the different reproductive strategies adopted by living organisms. These can be described as falling within a spectrum within two extremes: r-selected strategies in which as many offspring as possible are produced with minimal parental investment in each offspring, and K-selected strategies in which few offspring are produced but each receives a great deal of care, attention and material resources in order to ensure its survival and success. Thus, elephants, cows, and bats are K-selected, and many (but not all) rats, pigs, fish and insects are r-selected. If an organism adopts the K-strategy in a lowland forest it will almost certainly be relatively rare. It is then necessary to ask "under what circumstances can rarity be an advantage?". It is generally true that the herbivores found on a plant species in lowland forest are restricted to a few species which have evolved as the plant species evolved and have found ways to circumvent whatever form of physical or chemical defence the plants had adopted (p. 370). The chances of a herbivore finding an individual of a particular tree species are greater if that tree is common, and so the pressure exerted by herbivores on an abundant, or otherwise dominant, tree species puts it at a disadvantage and permits less common species to coexist (Janzen 1970).

There have been many attempts to devise and index that would adequately describe in a single figure both the number of species in an area of vegetation (its richness) and the way individuals are distributed between the species (the evenness). A recent evaluation of the available indices found that richness was best represented simply by the number of species, and the evenness was represented with similar degrees of appropriateness by most of the techniques available. The Shannon Wiener function or

$$H = \sum_{i=1}^{s} (p_i) (\log_2 p_i)$$

where 's' = number of species and 'p_i'= the proportion of the ith species, was the most appropriate measure of diversity for tropical rain forest[3] A measure of evenness (E) could be derived from this by

$$E = H \div H_{max}$$

where $H_{max} = \log_2 S$ where 'S' is the number of species, but the usefulness of calculating diversity has been questioned of late because its biological meaning is unclear (Stocker et al. 1985).

The high diversity of plant and other life forms in the tropics reflects the great structural diversity and an enormous range of potential niches. In an attempt to answer the question of how many species there are in an area of tropical rain forest, chemical fogs which cause insects to drop out of the canopy have been blown through a number of forests in America and Southeast Asia. One of the first attempts was in Panama and gave an estimate of 41,000 species of arthropods in one hectare of unspectacular seasonal forest (Erwin 1982), with more than 945 species of beetles recorded from just one species of tree (Erwin and Scott 1980). Animal diversity on Sulawesi, like diversity of plants, is high by the standard of temperate regions, but is relatively low compared with neighbouring Borneo or New Guinea, the largest island to the east (p. 39). Similar fogging techniques were used in the Toraut forest but the results are not yet available (N. Stork pers. comm.).

STRUCTURE AND COMPONENTS

Characteristics

Lowland forests are characterized by the huge amount of plant material they contain, and this can be measured in terms of the amount of carbon present. The tropical forests of the world (i.e., not just lowland forests but also montane, swamp and dry forests) cover 1,838 million hectares or

11.5% of the earth's land surface, yet they contain 46% of the living terrestrial carbon (Brown and Lugo 1982). Lowland forests in Sulawesi would certainly contain an even higher percentage of the carbon relative to their area.

It has often been stated that nearly all the inorganic nutrients in a tropical forest are held in plants and very little remains in the soil, a situation markedly different from forests in temperate regions. Detailed studies of various forests throughout the tropics have now shown that there is great variation between forests in the percentage of nutrients held in different parts of the forest ecosystem, and also that within a forest the distribution of different inorganic nutrients varies. In general, however the biomass of tropical forests does not hold a disproportionately large percentage of the nutrients (Whitmore 1984).

Lowland forests, particularly in the wetter regions, are characterized by the conspicuous presence of thick climbers, large buttressed trees, and the prevalence of trees with tall, smooth-barked trunks. The vast majority of trees have simple mesophyll-sized leaves (p. 498) between 8 cm and 24 cm long. The trees of the lower canopy, including the immature individuals of large tree species, often have large leaves (Parkhurst and Loucks 1972). Lower canopy saplings commonly have pronounced drip-tips (fig. 5.2). Since one theory of the functions of drip-tips is that they allow water to run easily off the leaf surface and thus prevent or hinder the growth of epiphylls (p. 358), it is reasonable that these should be most common nearer the forest floor where the relative humidity is highest (i.e., rate of evaporation lowest).

There is considerable similarity between plant families in the shape of their leaves and this, together with the high diversity of species, and the poor knowledge of the Sulawesi flora, frequently make identification, even to the genus level, somewhat difficult. Characters such as sap, bark type, buttress size, leaf-vein arrangement and arrangement of leaves on twigs are those of most use in the identification of the majority of the specimens but since plant taxonomy is based on the structure and arrangement of flower parts, final identification relies on the examination of flowers (p. 385).

Layering

The crowns of forest trees are usually said to form three 'layers'. While this is a convenient theoretical concept it has limited application in the field. An emergent layer is often quite clear but below that the supposed layers are often difficult to distinguish. Some trees may be referable to a particular layer but many more are generally not. Part of the difficulty in applying the concept of layering lies in the existence of the mosaic of pioneer-, building- and mature-phases in the forest (p. 359), each of

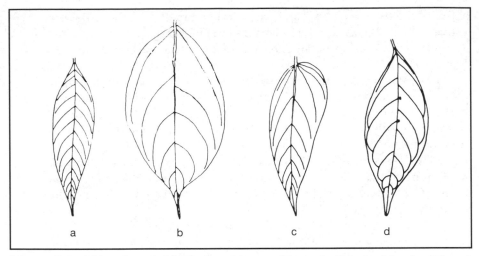

Figure 5.2. Examples of drip-tips on leaves of trees in Sulawesi lowland forest. a - *Cratoxylon celebicum* (Hype.); b - *Ficus variegata* (Mora.); c - *Pterospermum celebicum* (Ster.); d - *Vitex quinata* (Verb).

which is in a dynamic flux. The strata are clearest in forests where there is a single dominant species (Richards 1983).

Borrowing terms from limnology, two clear zones have been described. The canopy, comprising the crowns of the emergent trees together with the crowns of the main tall trees, has been called the euphotic zone (p. 314), for it is in this zone that most of the photosynthesis and production of the forest occurs. The lower level of this zone undulates between the lowest boughs of the larger trees (fig. 5.3). As in a lake, animals are most abundant in this zone which is unfortunately the furthest from a ground-bound observer (p. 416). Below the euphotic zone, in the oligophotic zone, most of the light is reflected or transmitted from above, although flecks of sunlight are found. As a result, productivity is much lower and decomposition processes predominate (Richards 1983). For the purposes of this book, the lower canopy is taken to mean the crowns of trees in this oligophotic zone.

Physical factors other than light vary through the canopy, providing a complex range of microclimates (table 5.2).

The crowns of neighbouring trees of similar heights are rarely in contact and are separated by spaces called 'crown shyness gaps'. These appear to by caused by abrasion of buds and shoots as the trees sway in the wind. These spaces are relevant because they affect movements of arboreal animals and the ability of climbers to reach a tree crown from a neighbouring tree (Putz and Parker 1984).

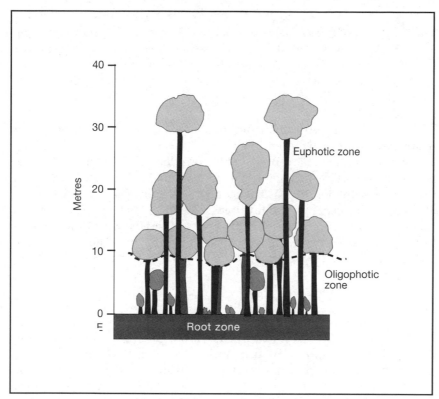

Figure 5.3. Layers and zones through a lowland forest.
After Moore 1986

Table 5.2. Differences through the forest in factors relating to microclimate.

	Humidity	Rainfall	Maximum temperature	Minimum temperature	Temperature variation	Windflow
Emergent	•	••••	••••	•••	••••	•••••
Main canopy	••	••••	•••	•••	•••	•••
Lower canopy	•••	•••	••	••	••	••
Shrubs/saplings	•••	•••	••	••	••	•••
Herbs	••••	•	•	••	•	•
Floor	••••	•	•	••	•	•

• = lowest   ••••• = highest

After Lock 1980

Basal Area and Biomass

The basal area of trees in a given area is a measure of the 'amount of tree' present. The cross-sectional area (πr^2) is calculated for each tree and the total basal area per hectare and the average basal area per tree per hectare are frequently calculated. In most lowland forests the total basal area of trees with diameters of 15 cm or more at breast height is between 25-50 m²/ha (table 5.3).

Biomass can be estimated from the volume of the trees (estimated as height x basal area x 0.5) multiplied by the specific gravity which is generally about 0.6 t/m³. There is probably considerable error for individual trees but this becomes negligible if a large number of trees are included in the sample (Whitmore 1984). The biomass of lowland forest wood averages about 380 t/ha, generally ranging between 200 and 500 t/ha (Brown and Lugo 1984). Direct measurement of above-ground biomass in two plots of lowland forest in Peninsular Malaysia resulted in totals of 475 and 664 t/ha and an estimated average of 500-550 t/ha (Kato et al. 1978).

Roots

Roots have to compete with soil organisms such as fungi, bacteria, and numerous invertebrates for the organic carbon and the other nutrients in the forest soil. It appears that almost all trees in lowland forests and almost all non-grain crops can develop a mutualistic symbiotic relationship between their roots and fungi (Janos 1980). This association is called 'mycorrhiza'. The most common form found in the tropics has the hyphae (root strands) of the fungus penetrating the plant's roots where they take up carbohydrates from the plant. For its part the fungus channels minerals into the roots (Janos 1980, 1983). Many plant species with mycorrhizae

Table 5.3. Basal areas of forests in four regions of lowland forest in Bogani Nani Wartabone National Park for trees of 15 cm diameter at breast height or more.

	No. trees and palms /ha	Basal area (m²/ha)	Basal area of trees (m²/ha)	No. palms /ha	Basal area of palms (m²/ha)
Morowali *	412	46.0	0.11	–	–
Toraut	330	29.5	0.11	106	5.2
Lombogo	248	27.2	0.12	10	1.0
Hog's Back	420	49	0.12	75	7.3
Edward's Camp	450	48	0.11	16	0.6

*After Clayton in prep.; other data from EoS teams

appear not to possess root hairs through which water and minerals generally pass, because these structures are unnecessary given the efficiency of the mycorrhizae. Mycorrhizae are probably poorly developed among plants of early successional stages which normally grow on soil enriched by fire ash, deposits left by flood water, or the decomposition products of fallen trees (Janzen 1975). In such situations a plant would be at a disadvantage if it had to wait for its roots to be infected with fungus before it could start to grow (Janos 1980, 1983). Conversely, some species cannot grow without mycorrhizae even at the seedling stage, and yet others grow better without mycorrhizae where mineral nutrients are plentiful, but better with mycorrhizae in poor soils (Janos 1980, 1983). For trees that do depend on mycorrhizae it is advantageous to have large seeds, because they provide a reserve of food upon which the seedling can draw before infection with the fungus occurs. It is not surprising then, that seedlings of these species kept experimentally without mycorrhizae, stop or slow their growth after attaining a size correlated with the average dry weight of a seed for that species (Janos 1980).

Buttresses and Trunks

Buttresses are lateral extensions to the lower part of a tree trunk and are both common and varied on tall tropical trees. Different species are relatively constant in the presence, shape and surface characteristics of buttresses and these characters can be helpful in identification of trees.

Since the decision of whether or not a tree is considered to be buttressed depends on the definition of a buttress, it is difficult to compare sets of data collected by different workers on the percentage of buttressed trees in different lowland forests. Some data were collected, however, by EoS teams in different locations in North and Central Sulawesi (table 5.4)

Table 5.4. Frequency of buttresses in three areas of lowland forest. Altitudes are the highest point of each transect.

Site	Altitude (m)	Bedrock	Trees /ha	Buttressed trees (%)	Maximum height (m)
Hog's Back	492	volcanic	350	73	2.8
Edward's Camp	664	volcanic	440	46	2.0
Binotik	35	limestone	350	20	0.9

and up to three-quarters of the trees of 15 cm diameter at breast height were found to be buttressed.

Buttresses are sometimes said to act as structural supports for trees whose roots are relatively shallow and where the substrate affords little anchorage (Smith 1972; Henwood 1973). It has been suggested that buttressing increases a tree's resistance to mechanical stress, and reduces the tensile forces on the roots. There should, therefore, be a smaller probability of the trunk of a buttressed tree snapping close to the ground. This hypothesis has not been found to be valid in a Panamanian forest, where there was no statistical difference between buttressed and non-buttressed trees in terms of the number of snapped or uprooted individuals. Further, mortality rates were no different between buttressed and non-buttressed trees (Putz et al. 1983).

Buttresses are also said to be composed of 'tension wood' to reduce the pulling strain on the roots, and so buttresses would be expected to form on the side of the trunk which experiences the greatest tension. Observations by EoS teams along a number of transects indicated that many trees do not, in fact, form buttresses at tension sites such as on the uphill side of a trunk, on the side of a trunk opposite a major congregation of heavy climbers, or on the side from which the wind blows.

Recent hypotheses on the reasons for buttress formation have been unrelated to structural problems. It has been suggested, for example, that since buttresses increase the space of forest floor occupied by an individual tree, they could be viewed as a competitive mechanism (Black and Harper 1979). According to this hypothesis the physical presence of a buttress prevents or hinders the establishment of neighbouring (i.e., competing) trees close by. If this hypothesis is correct then it should be possible to observe that:

- the distances to the nearest neighbours of buttressed trees are greater on average that from a non-buttressed tree to its nearest neighbours;
- the density of trees near a buttressed tree is less than near a tree of similar bole size, canopy form and age without buttresses;
- the density of trees around different individuals of a buttress-forming species bears a relationship to the age of the individual trees. Older buttressed trees should have both more distant and fewer neighbours in the smaller size classes;
- species diversity of trees around a buttressed tree is lower than around a non-buttressed tree because some species are likely to be less competitive against buttresses than others (Black and Harper 1979).

Data were collected in the Toraut forest from 17 mature buttressed trees and six similar trees with no buttresses. It was found that the density of trees around buttressed trees was not significantly different from the density around non-buttressed trees, but the difference in mean distance of trees from buttressed trees was significantly greater than from non-buttressed trees (table 5.5). These results suggest that of the first two predictions, the first is valid but the second is not.

Small trees sometimes establish themselves on the outer fringe of the buttress arms, thus supporting the notion that the presence of a buttress is a physical deterrent to the establishment of other trees too close to the trunk. Often, however, small trees can be found in the area between the buttress arms. It appears that these sites may indeed favour the establishment of seedlings of shade-tolerant trees. It may be that the large buttress arms serve to intercept wind-dispersed seeds, or that rats and squirrels bring seeds there to store or, in the case of rats, to take to their burrows which are often located near the base of trees. For whatever reason, the between-arms areas of buttresses are favourable micro-habitats for shade-tolerant tree species, due to low wind velocities, and increased water and nutrients from stemflow and litter fall. The virtual absence of a large tree growing between the buttresses of an older tree suggests that any saplings growing in such places eventually die, presumably as a result of competition for nutrients and water.

Living trees with rotten, empty trunks are surprisingly common in lowland forest. This may appear to be unfortunate for the tree but it has been suggested that it is in fact a strategy bringing distinct advantages. Empty cores of large trees are inhabited by bats and rats in addition to a host of invertebrates. Faeces and other products from these animals decompose and thus provide an exclusive supply of nutrients for the tree (Fisher 1976; Janzen 1976, 1981).

Climbing and Creeping Plants

Climbing plants abound in many types of lowland forest. In lowland forest at Toraut and at Lombogo,[4] 44% and 37% respectively of the trees with trunk diameters of 15 cm or more at breast height, carried climbers. Of these trees about 10% had heavy vine loads. These figures are similar to those obtained from an area of lowland forest in Panama (Putz 1984b).

Table 5.5. Mean number and distance of nearest neighbours to 17 buttressed and six non-buttressed trees at Toraut.

	Buttressed	Non-buttressed	Significance
Within 5 m	6.38	4.5	Not significant (t=1.46)
Within 10 m	10.85	9.8	Not significant (t=0.42)
Mean distance (m)	4.71	3.18	p<00.2 (t=2.29)

Data from EoS team; G. O'Donovan pers. comm.

The crown of a large climber may be as large as that of a tree, but because they do not possess any commercial value (with the exception of rattan palms), they have been little studied. Despite the value of rattans, however, the Sulawesi species are scarcely known. About 50 different kinds and qualities have been recognized by traders (van der Koppel 1928), but it is not known to what extent these refer to actual species. Given the uncontrolled and intense collection of rattans, even in reserves and national parks, and the increase in demand, a good case can be made for a period of intensive collecting and research into sustainable harvesting and domestication.

Identifying climbers is difficult but the coiled or convoluted ones with somewhat flattened stems are generally leguminous species, and the ones with regular hoops around the stem are one of the three species of the pink-fruited *Gnetum*[5] (Gnet.), best known to botanists for possessing some of the characteristics of flowering plants and some of the characteristics of conifers. *Gnetum* has been described as either an ancient type of plant or a freak of nature (Corner 1952).

When young, many species of free-hanging climbers can look very similar to young trees as they grow slowly through the undergrowth and lower canopy, waiting for an opportunity to rise to the canopy. In the Panama forest 22% of the upright plants less than 2 m high were in fact young climbers (Putz 1984b). Opportunities to rise to the upper canopy are afforded when a gap is formed and it is then that a young climber can attach itself to a rapidly growing tree and be carried upwards. Other species, during a similar period of apparent dormancy, are growing a large tuber below ground. In response to a certain cue, the energy available in the tuber is suddenly used up in a rapid burst of upward growth, whether or not a gap has formed (Janzen 1975).

A considerable proportion of the apparent 'seedlings' of climbers (and a certain proportion of some forest trees) are in fact no more than shoots growing up from a horizontal root of an established plant (Janzen 1975). In the Panama forest mentioned above, 15%-90% of the young climbers less than 2 m tall were offshoots from an underground stem rather than seedlings (Putz 1984b). This is a means by which a plant can increase the effective size of its crown, and since these shoots are genetically identical to their 'parent', the actual number of genetically distinct individuals in a given area may be extremely low.

Climbers compete with trees for light, nutrients and water and can cause mechanical damage to trees. In addition, bole diameter is often greater than average for trees with heavy climbers in their crown, indicating that trees have to divert resources away from vertical growth and reproduction in order to increase the strength of the supporting structure. Trees carrying climbers probably suffer higher mortality and cause more neighbouring trees to fall than trees that are free of climbers. The falls do not usually kill the climbers, however, for they quickly rise to the upper canopy again using the abundant supports available at a treefall site (Putz 1984a). Thus the stems of climbers can sometimes be seen to loop down towards the forest floor and back again. Even without having their growth disturbed by treefall, mature climbers are rarely supported by a single tree but rather grow horizontally through the canopy, sewing together the tree crowns. In the Panama forest, climbers connected an average of 1.6 trees (Putz 1984b) and this tying together of crowns may help to prevent trees from falling over in strong winds. If a tree is felled on purpose, however, the climbers in its crown often pull over other trees as well. For many reasons it would clearly be an advantage to a tree to shed or to avoid climbers.

In an attempt to elucidate whether some tree species do actively shed climbers, 20 trees each of 24 species in Panama were examined and their climber load was recorded. Growth rate of the trees and height of crown appeared to have little to do with load of climbers, but there were significant positive correlations between the length of tree leaves and flexibility of tree stems with absence of climbers. Flexible trees with large leaves are abundant in disturbed, well-lit habitats and these characteristics may help them to remain climber-free by shedding the climbers along with the leaves and by breaking the brittle young growing points of climbers while swaying in the wind (Putz 1984a). In addition, some trees exhibiting a symbiotic relationship with ants (such as *Macaranga*) may benefit from the ants removing vines from their branches as they have been shown to do in Central America (Janzen 1973).

Creeping plants are those whose roots are attached firmly to tree trunks as well as in the ground. They are generally shade-tolerant, and flower and fruit below the upper canopy. Between 55%-65% of trees, at least 15 cm in diameter at breast height at Toraut and Lombogo were found to have creepers and 13%-16% of these trees had considerable numbers. In Sulawesi two of the most common creepers are *Freycinetia* (Pand.), a small pandan with rough leaf edges and many-seeded, club-like fruits held within brightly-coloured bracts, and the superficially similar *Pothos* (Arac.) which has smooth-edged, two-lobed leaves and bright red fruit. Trees can have one, both or neither of these plants[6]

A sample of 234 trees with trunk diameters of at least 15 cm at breast height in Bogani Nani Wartabone National Park were scored for whether they had no, one, few or many climbers and creepers. Of these, 33% bore neither, 57% bore creepers, 38% bore climbers, and 27% bore both. It would be expected that creepers could be more common since their progress is not hampered by any of the trees' possible defences against climbers. Trees without creepers are frequently those with flaking bark such as *Eucalyptus deglupta* (Myrt.), *Tristania* (Myrt.), *Agathis dammara* (Arau.), and *Dillenia celebica* (Dill.). The trees sampled had an average diameter of 30 cm but those with abundant climbers and those with abun-

dant creepers were, not surprisingly, large and very similar in average
diameter (40-42 cm). This is presumably an effect of age and, as stated
above, the diversion of resources from vertical growth and reproduction to
increase the strength of the tree's supporting structure.

Epiphytes and Epiphylls

An epiphyte is a perennial plant rooted upon, not in, a large host, and
which does not have to produce or maintain massive woody stems and
branches to live above the forest floor. Epiphytes are common in lowland
forest, but even more common in certain montane forests (p. 507), and
they are thought to make a significant contribution to the total biomass
and species-richness in a forest (Benzing 1983; Whitmore et al. 1986).
The major higher plant families of epiphytes include Gesneriaceae,
Melastomaceae, Rubiaceae, Asclepiadaceae and Orcidaceae. In addition to
these there are numerous smaller epiphytic ferns, lichens and bryophytes
(including mosses and liverworts). The small size of bryophytes makes
them less restricted than other plants in the ranges of microhabitats they
can occupy (fig. 5.4). The most luxuriant bryophyte communities in wet
lowland forest are generally found on the bases of large trees (Pócs 1982).

Epiphytes might seem to have an easy life but they live on a substrate
which is extremely poor in nutrients. Most of them rely upon nutrients dissolved in rain, in litter fall, and in occasional mineral inputs from animals. Epiphytes also have to contend with a very erratic water supply—
being drenched when it rains, and parched when the sun is shining and
the wind is blowing.

Epiphytes cope with growing in these nutrient-poor conditions in various ways (although few species will exhibit all these features):

- their own tissues may have very low nutrient concentrations;
- juvenile stages are prolonged even though little growth is required to
  become mature;
- their vegetative parts are reduced in size;
- they may obtain nutrients from unconventional sources such as ants
  (p. 463);
- their leaves tend to be long-lasting (due perhaps to some form of
  defence against herbivores) so that the replacement of nutrient-expen-
  sive leaves is required only infrequently (Janzen 1975) (p. 370); and
- their flowers are pollinated by animals because wind dispersal of the
  pollen would result in 'expensive' losses (Benzing 1983). In addition,
  it has been observed that some epiphytes concentrate their roots on
  the underside of boughs where water and nutrients would remain
  longest (Whitten 1981).

Some epiphytes, notably the superficially-similar *Myrmecodia* (Rubi.)
and *Hydnophytum* (Rubi.) (p. 463), harbour ants in their stems and these

Figure 5.4. Bryophyte microhabitats in a lowland forest. 1 - bases of large trees, 2 - upper parts of trunks, 3 - macro-epiphyte 'nests', 4 - bark of main branches, 5 - terminal twigs and leaves, 6 - bark of climbers, shrub branches and thin trunks, 7 - pandan stems, 8 - tree fern stems, 9 - palm trunks, 10 - rotting trunks, 11 - soil surface and termite mounds, 12 - roadside cuttings, 13 - rocks and stones, 14 - submerged or emergent rocks in streams.
After Pócs 1982

may well protect their leaves from caterpillars and other arthropod herbi-vores. To test this hypothesis would not be difficult if ants on one set of plants were killed with insecticide and herbivore damage then monitored on that set and on a control set of plants.

Although epiphytes (by definition) do not take nutrients directly from the host plant, they have nonetheless been called 'nutritional pirates' (Benzing 1981, 1983). This is because they can tie up nutrients in their own biomass from the dust, leaf leachates and rain which would otherwise have fallen through to the soil to be utilized by the trees that support them. While this is true, host trees can turn the tables at least partially back in their favour by producing small roots beneath the mats of dead and living epiphyte tissue (Nadkarni 1981). These roots thus give the tree access to sources of nutrients in the canopy. The main boughs of some tree species are almost invariably covered with epiphytes but it would appear that phys-ical or chemical features of the bark of other species actively deter epiphyte growth, presumably because the relationship is disadvantageous.

Special forms of epiphytes are strangling figs. These look like large trees with latticed trunks when mature, but they started life high in the forest canopy as a seed deposited in the faeces of a bird, bat or civet. The seed germinates and grows like an epiphyte, but roots grow down to the ground. These roots and the leafy part of the plant grow in size and an increasing number of roots grow down and around the trunk. With time these roots fuse with each other, completely enclosing the tree trunk. The host tree eventually dies, probably as much by competition for light in the canopy as by strangulation.

Epiphylls are those mosses, liverworts, algae and lichens that grow on the living surface of leaves in shady situations where the air is almost con-tinually saturated, conditions also found in moss forest (p. 500). Thus they are generally restricted to the leaves of young trees, shrubs and long-lived herbs of the lowest layer of forest up to about 2-3 m above the ground, as well as near streams. Over ten species of epiphylls can occur on a single leaf and some of these are obligate; that is, they are found only on leaves (Pócs 1982). Epiphylls are restricted to certain types of leaves and are pos-sibly more common on leaves lacking drip-tips (p. 500).

DYNAMICS

Succession and the Growth Cycle

Different types of tropical rain forests, and indeed forests throughout the world, have many fundamental similarities because the processes of forest succession (the changes over time in the occurrence of species in a given area), and the range of ecological strategies of tree species are more or less the same (Whitmore 1984). These strategies form what is known as the forest growth cycle: the events of a large mature tree falling over in a closed forest thereby forming a gap[7] the gap filling with a succession of plant species until the final, large, mature tree falls over again. Many species require gaps for their growth and others are stimulated by them. The size of a gap depends on its cause; a single dead tree falling over forms a smaller gap than a fire, landslip or drought, and the biological effects of a gap are related to its structure and size. Many of the important life history features of a particular plant species are related to the strategies it employs to exploit different types of gaps (Pickett 1983). For example, vigorous, light-tolerant (shade-intolerant) species grow up in the gap and these create favourable conditions for seeds of shade-tolerant (but not necessarily light-intolerant) tree species to germinate, for the seedlings to grow and for these to eventually supersede the species which initially filled the gap. The formation of mature forest from a gap, is known as 'secondary succession'.

Three basic ecological principles can be formulated from the study of both primary[8] and secondary succession:
- succession proceeds in only one direction, with fast-growing, tolerant colonizers replaced by slower-growing species with more specific requirements, and great competitive ability where those requirements are met;
- as new species colonize an area, they will inevitably alter the environment by their presence;
- the new conditions are generally less suitable for their own seedlings but more suitable for those of other species, hence the succession continues.

The forest growth cycle can be divided into three phases—the gap-, building- and mature-phases—which together form a mosaic throughout the forest that is continually changing in state and shape (Whitmore 1984). For example, a patch of forest in the building-phase may return to the pioneer phase should a tree in a neighbouring mature patch fall across it. The floristic composition of a forest will depend on the size of former gaps because a large gap will be filled initially with tree species requiring light whereas small gaps will be filled by shade-tolerant tree species. Many of the latter may have grown slowly for many years to become part of the lower canopy but have been unable to form part of the main canopy until a small gap was formed. It should be noted that since a single species of tree can contribute to pioneer-, building-, and mature-phases, it is incorrect to refer to 'pioneer-phase' or any other '-phase' species (Whitmore 1982). Terms describing the characteristics of a species, such as pioneer, shade-tolerant or light-intolerant, are more useful. In addition, because a forest comprises a mosaic of gap, building and mature areas, it is a matter of debate whether the term 'climax' vegetation has any meaning in the field.

The regeneration of trees in a one-hectare botanical plot at Toraut seems to be in gaps caused by single tree falls. The peculiarly small crowns, absence of a middle layer and an apparent low density of climbers sewing together adjacent crowns in that forest, results in very neat tree falls causing little damage to surrounding trees. The pattern of the different phases is thus very fine-textured and in some places trees had fallen without really forming a gap at all (fig. 5.5) (Whitmore and Sidiyasa 1986).

Small gaps created by dead standing trees initiate short cycles without a gap phase. That is, immature trees beneath the dead tree are able to grow up as soon as the light intensity increases. Large gaps created by the fall of several adjacent large trees allow the full forest growth cycle to occur. The size of gaps clearly increases with the maturity of the forest (larger trees, more crowns interconnected by climbers), thus there are *more* pioneers in an area of generally mature forest than in the same area of forest in the building-phase because of the larger gaps (Brokaw 1982b).

A sample of 310 fallen trees was examined in Panama and it was found that 70% had snapped at some height above the ground, 25% had been uprooted, and 5% had broken off at ground level. Presence or absence of buttresses appeared to have no connection with the manner in which the trees fell (p. 351). Compared with trees that had snapped, uprooted trees were generally larger and shorter for a given bole diameter, and had denser, stiffer and stronger wood. The trees with less dense wood grew faster and compensated for their tendency to snap by frequently sprouting from their base, particularly those trees in the smaller size classes (Putz et al. 1983). Records of treefalls in that forest showed that most occurred in the middle of the wet season, possibly due to stronger winds or increased weight due to water in the bark and on the leaves, and this may affect germination and/or dispersal of plants that need light gaps in which to grow (Brokaw 1982b).

Determining the duration of the growth cycle in temperate regions is greatly assisted by being able to age the trees simply by counting the rings in the wood that form because of differential growth in different seasons. In the tropics, where the rate of wood growth is more or less constant through the year, other more roundabout methods have to be used to examine the dynamics of forests through time. One such study was con-ducted in East Kalimantan and it was found that it took 60-70 years after the

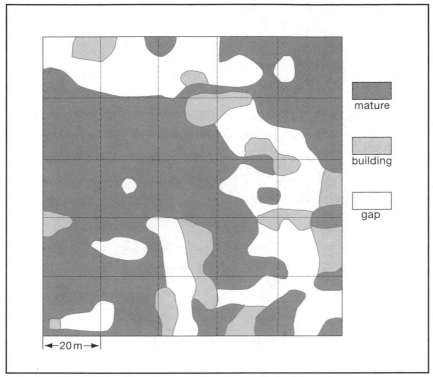

Figure 5.5. Mosaic of pioneer-, building- and mature-phase forests in 1 ha of forest at Toraut, Bogani Nani Wartabone National Park.
After Whitmore and Sidiyasa 1986

formation of a large gap for the number of growth-phase species to reach a maximum, and as long again for mature-phase species to dominate and for gap formation to begin again. Biomass continues to increase for 220-250 years after disturbance, some trees live for over 500 years, and it may take this long for a stable and dynamic system to form (fig. 5.6) (Riswan et al. 1985). Similar longevities are known from elsewhere in the tropics (Huc and Rosalina 1981; Brokaw 1982b; Lieberman et al. 1985).

Work on Sumatra has helped to elucidate variations in the length of the forest growth cycle between sites. It was found that with increasing age of river terraces (from 40-10,000 years), soil pH and fertility decrease. With decreasing soil fertility, production of leaf litter (p. 363) and fruit decreases although the forests on the poorer soils were mature phase, and the trees were longer-lived. It would appear that differences in soil fertility have caused trees to adapt different forms of nutrient economy and hence

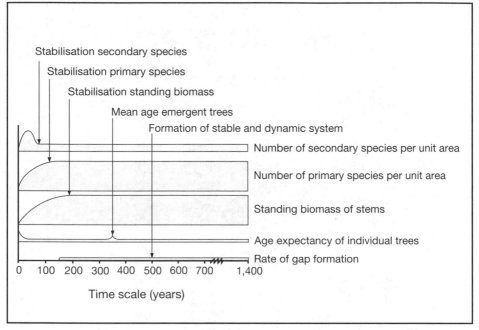

Figure 5.6. Estimates of composition, biomass, and age expectancy of trees after the creation of a large gap in lowland forest.

After Riswan et al. 1985

patterns of life history, and this is reflected in forest structure. The trees growing on poor soils clearly have an advantage if they retain scarce nutrients, and for those on richer soils there are fewer benefits from hoarding nutrients, although the cost of doing so remains similar. As trees hoard nutrients, so the soil quality will continue to decline and this in turn favours trees that hoard still further. This trend will favour later reproductions and greater longevity since it will benefit the tree to consolidate its position and retain its competitive advantage before diverting resources to the production of flowers and fruit. This helps to explain why large trees are so often found on ridges even though the soils there are generally poorer than on the slopes. Also, forests that contain tree species typical of pioneer- or building-phases, such as those on recent river terraces, are in fact merely forests that happen, by virtue of their richer soil, to have species possessing shorter cycles of regeneration than trees growing on poorer soils (van Schaik and Mirmanto 1985).

Within each of the different phases of the forest growth cycle, the trees have a number of ecological characteristics in common. For example, the pioneer species that grow up in a gap generally produce large numbers of small seeds frequently if not continuously. These seeds tend to be easily dispersed (often by wind, squirrels or birds) and, probably because of this, these species tend to have wide geographical distributions (Huc 1981). They have high viability even in full light, and the seedlings grow rapidly.

The building-phase is more or less built by the animals, such as pigeons, monkeys, bats and hornbills, that benefit most from it. This is not pure chance because the trees of this phase bear fruit which tend to be co-evolved with a particular disperser either in size, nutrient content, or phenology. For example, different individuals of *Dracontomelum dao* (fig. 5.7) fruit at different times such that the monkeys which disperse them have fruit available to them all year round. Both tree and frugivore thereby benefit by this arrangement (Anon. 1980). Many species of trees common in the building-phase are unable either to colonize open sites or to regenerate beneath their living parents (Whitmore 1982).

The forest on the young volcanic soils of Tangkoko-Batuangus Reserve, Minahasa, is largely in the building-phase of development and this, combined with the relatively rich soils and high productivity, is reflected in the composition of the fauna which comprises large numbers of relatively few species and high populations of frugivores. For example, the density of crested macaques *Macaca nigra* is about $300/km^2$ and of red-knobbed hornbills *Rhyticeros cassidix* about $60/km^2$. This is the highest-known density of macaques in any Indonesia study area (Anon. 1980).

Trees found in mature-phase forest are shade-bearers (that is, they tolerate growing in shade, but if more light is available they will grow faster), and thus can regenerate beneath a closed canopy. They tend to be found in patches but throughout an area (Lack and Kevan 1984). These species often have large seeds with rather specialized means of dispersal and a relatively restricted geographical distribution.

Changes in species composition through the growth cycle may be due to availability of nutrients, availability of seeds, chemical suppression of some species by the same or other species (allelopathy), crown or root competition, and chance, but there is still very little information available on how regeneration actually occurs. There are three ways in which trees can grow: from shoots growing out of roots, stumps or fallen trunks; from established seedlings; or from seeds. Beneath a mature forest there are clearly no seedlings of light-demanding species and, conversely, in gaps, light-intolerant species will die. Seeds can be brought into an area in the 'seed rain' (seeds dispersed from their parents) or may lie dormant in or on the soil surface. These dormant seeds are known as 'seed banks' and their numbers are quite surprising. Topsoil from four forest types in northern Queensland harboured 588-1,068 viable seeds/m^2 which is similar to other tropical lowland sites examined. The most seeds were found in coarse-grained soils, which could more easily retain a wide range of seed sizes. Very few trees characteristic of the mature phase had viable seed in

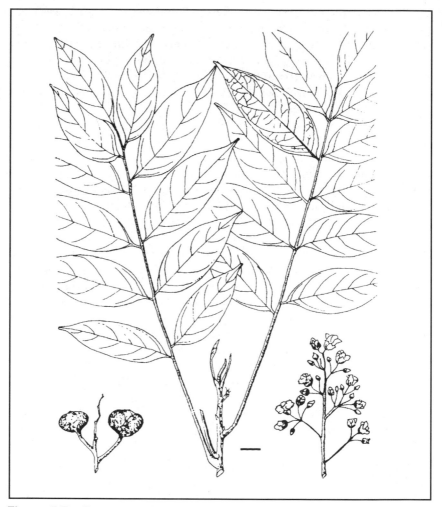

Figure 5.7. *Dracontomelum dao* whose fruit is sought by macaques and frugivorous birds. Scale bar indicates 1 cm.
After Soewanda n.d.b

the soil except those that were actually fruiting at the time of sampling (Hopkins and Graham 1983). These trees tend to fruit infrequently, have poor powers of dispersal and clearly little or no potential for extended dormancy. Their continued existence in an area therefore depends on fruit from mature trees falling into a spot with the correct conditions for immediate germination. Some seeds of pioneer species found 20 cm below the soil surface were still viable, although they must have arrived many years before.

Litterfall, Nutrient Cycling and Productivity

The first stage of nutrient cycling is the fall of leaves and other plant material from the trees to the forest floor. From various studies it would seem that the rate at which fine litter[9] falls is quite similar between tropical areas at about 7-14 t/ha/yr, of which leaves constitute about 60%-80% (Lim 1978; Ogawa 1978; Kartawinata et al. 1981; Spain 1984; van Schaik and Mirmato 1985). Five litterfall traps in the Toraut forest were monitored by an EoS team for varying periods (subject to disturbance by pigs) and extrapolation of the data indicated an annual litterfall of 12.7 t/ha/yr, about 80% of which was leaves. The methods used in different studies vary somewhat, however, and detailed comparisons are difficult. Standardized methods are proposed elsewhere (Proctor 1983).

Calculating the rate at which the coarse litter[10] falls is much harder since it occurs irregularly and unevenly. Very few studies have attempted to determine the quantity of coarse litter falling but estimates are of the order of 9-15 t/ha/yr (Yoneda et al. 1978; Brasel and Sinclair 1983).

Most decomposition in lowland forest usually occurs in the soil or on the soil surface, although a relatively large percentage of trees and branches start decaying while still standing or attached to the main trunk respectively (Edwards 1977; Kira 1978). The litter reaching the forest floor has been assumed by many to decompose much faster in the tropics than in temperate regions, although some field data suggest otherwise. The rate of decomposition is determined, at least in part, by the quality of the leaves themselves and the composition of the litter-feeding macro-fauna on and in the forest soil (Anderson and Swift 1983). Decomposition rates of tree leaves appear to depend on leaf thickness (and consequently toughness), and weight loss over six months can vary from 20% to 60% (Takeda et al. 1984). Some leaves lose only 72% of their initial weight after 14 months (Gong and Ong 1983).

The time required for leaves to decay, and thus for the minerals to be released to enter the cycling pathways, varies between species depending on the nature of the leaves. A recent study in Puerto Rico on decomposition rates of six species (three genera of which are indigenous to Sulawesi), showed that the leaves of the three species of mature-phase forest retained only 20%-50% of their original organic matter after 16 weeks and 10%-20% after 32 weeks. The three pioneer species decayed more slowly. The difference was related most closely to structural characteristics such as percentage lignin and percentage of fibre. Given that the more open habitats where pioneer species grow have potentially greater rates of nutrient loss because of the greater quantity of rain falling harder on generally more compacted soil, leaf structures which bind elements for longer periods would represent a valuable means by which trees could ensure a more sustained release of nutrients (La Caro and Rudd 1985).

Some of the nutrient cycling occurs rapidly, such as when nutrients in leaves are absorbed into a tree which grows new leaves every year, before the leaves are shed. Nutrients in leaf litter and rain may be quickly taken up by roots or they may be adsorbed into the soil particles and released slowly. Nutrients in the trunk of a tree, however, may be held for hundreds of years and even when the tree dies, a decade may pass before all the nutrients are released.

The major external input of inorganic nutrients into a forest above the flood level is from rain. These nutrients have probably been released into the atmosphere from fires. Up to 30% of the rain is intercepted by leaves and evaporates or soaks into the bark. A small amount (<1%) flows down tree trunks and the rest reaches the forest floor (Arsyad 1985). The percentage reaching the forest floor also depends on rainfall intensity, being greater when rainfall is heavier (Edwards 1982), and contains, not just nutrients from fires, but also from bark, canopy liter and leaves with which it has been in contact. As a result, the rain reaching the forest floor is significantly richer in nutrients than that reaching the soil in an open field nearby. The forest probably does not have the ideal closed or leakproof cycling of nutrients attributed to it because some nutrients are lost in the flow of water to the rivers. Potassium does seem, however, to be cycled tightly within the forest.

Very few studies of total forest primary production (p. 132) have been conducted anywhere in Southeast Asia, largely because of the many methodological difficulties involved (Whitmore 1984). At a site in Peninsular Malaysia three methods were used (Kato et al. 1978; Kira 1978; Koyama 1978; Yoda 1978a, b). The results suggested a net primary production of 30 t/ha/yr and a gross primary production[11] of 80 t/ha/yr (Whitmore 1984). The net primary productivity is lower than for rubber and oil palm plantations (Kato et al. 1978), but a mature forest would not be expected to have as high a productivity as a managed, forest plantation which is harvested before the rate of production increase begins to fall.

Flower, Fruit and Leaf Production

No studies of the cycles of flower, fruit and leaf production in Sulawesi forests have been published although the fruiting of trees was monitored in Tangkoko-Batuangus Reserve for nearly two years (Anon. 1980). Studies of canopy trees elsewhere in the tropics generally find wide variation between, and even within, species in the patterns of flower, fruit and leaf production. Some species flower at consistent but non-annual intervals, some flower more or less annually, whilst others flower at irregular intervals in excess of one year and irrespective of weather. Weather can, however, influence whether the flowers will form fruit, such that a tree may flower and yet bear no ripe fruit. Abortion of flowers and young fruit appears to be due to a lack of necessary resources such as inorganic nutrients, water and products

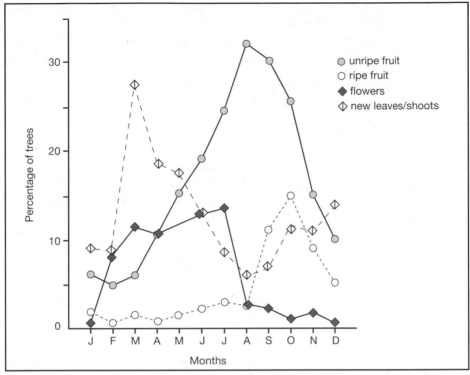

Figure 5.8. Monthly percentages of trees with unripe fruit, ripe fruit, with flowers, and with new leaves or shoots in an area of lowland forest in Peninsular Malaysia.
After Medway 1972

of photosynthesis (Stephenson 1981). Patterns of leaf production are similarly variable.

Despite these variations, the tree community as a whole does generally show relatively clear patterns of flower, fruit and leaf production (fig. 5.8). In areas with two wet and two dry seasons each year, two flushes of new leaves seem to be produced with a major peak just after the driest time of year and a minor peak just before, and extending into, the wettest time of year (Medway 1972). Flowering generally peaks after the driest time of year and fruiting, not surprisingly, peaks after that, just before the wettest time of year. The flowering of many lowland forest trees seems to be initiated by waterstress (Whitmore 1984). It should be stressed however that this is the trend in a forest and individuals may differ significantly from the above pattern.

Monitoring of cycles of flowering, fruiting and leaf production in lower canopy shrubs and small trees (<10 m tall) in Peninsular Malaysia and Costa Rica did not reveal such clear patterns as found in upper canopy species (Opler et al. 1980; Wong 1983), possibly because the microclimate near the forest floor is less seasonal that in the canopy.

The main distinguishing feature of forests in western Indonesia is the predominance of trees of the Dipterocarpaceae. Although these are more or less absent in Sulawesi (p. 386), their effects on forest ecosystems are relevant. A remarkable characteristic of the dipterocarps is that they all tend to fruit together at long but irregular intervals of five to seven years, at which times there is often exceptional flowering and fruiting activity in other tree families too (Appanah 1985). The primary advantage of this is probably similar to that of new leaf flushes in canopy trees, simultaneous spawning of coral (p. 221), or gregarious nesting habits of turtles (p. 155); that is, to satiate the appetites of predators so that at least some leaves, larvae or baby turtles escape early death. It avoids specialization of predators on those organisms and keeps the population of such predators down to relatively low levels.

It has been suggested that the large numbers of dipterocarps and the infrequency with which they produce fruit may be a factor in the relatively low biomass and diversity of animals in the forests of western Indonesia and Malaysia (Janzen 1974). In Sulawesi, the place of dipterocarps is taken by trees that do not appear to fruit gregariously but the biomass and diversity of animals is not noticeably greater than to the west, with local exceptions (p. 363), which suggests that the generally low production of fruit (when compared to the America tropics), possibly as a result of poor soils, may be the major factor in this discussion. Long-term studies (at least two years) of flower, fruit and leaf production in everwet areas of Sulawesi would help to reveal any differences with forests in Peninsular Malaysia, Kalimantan and Sumatra and would assist in understanding the role of dipterocarp trees in those other forests.

Many of the trees in lowland forest are pollinated by insects, and it is not surprising that peaks in abundance of certain insect species coincide with peaks of flower production. These peaks occur just after the drier times of the year and usually coincide with, or come slightly after, peaks of leaf production which are exploited by butterfly and moth caterpillars (Coomans de Ruiter 1951). Other insect species, particularly those associated with rotting wood, or those dependent on pools of water for breeding, become most abundant during the wetter months (McClure 1966; Hails 1982). Dramatic increases in insect numbers are occasionally observed, perhaps in response to climate and availability of food. For example, in June 1982, an expedition driving up the logging road on Mt. Roroka Timbu near Lore Lindu National Park encountered an enormous cloud of butterflies between 400 m and 1,000 m altitude. The cloud was so dense that it was necessary to use windscreen wipers in order to see the road (K. D. Bishop pers. comm.).

These variations in insect abundance are probably reflected in the behaviour of insectivorous or partially-insectivorous birds[12] which, in Sarawak and Peninsular Malaysia at least, breed and moult all year round but experience major peaks in these activities, coincident with peaks of food availability (Fogden 1972; Wells 1974). Similar patterns have been found for insectivorous bats (Gould 1978).

In East Kalimantan, birds such as large hornbills and pigeons which eat large-seeded fruits with flesh rich in lipids,[13] and seed-predating parrots, left the study area when their foods became scarce. Frugivores with fixed home ranges, such as macaques, respond to low availability of fruit by decreasing the distance they travel daily, increasing the proportion of non-fruit items such as insects, spiders, and small vertebrates such as frogs and geckos in their diet, and also by exploiting the few tree species that fruit outside the period during which most trees are fruiting. These trees, now known as 'pivotal species', and those that stagger their fruiting over a period, are extremely important in maintaining sedentary populations of arboreal frugivores (Leighton and Leighton 1983; Howe 1984).

Studies elsewhere in the tropics have shown that the peak of pregnancy in forest rats coincides with the seasonal peak in fruiting (Harrison 1955; Dieterlen 1982) (fig. 5.9) and there is no reason to believe that this would not also apply to Sulawesi rats that include fruit as a major item in their diet.

Herbivory

About 50% of a leaf consists of cellulose, a complex carbohydrate molecule which makes up the outer cell walls. Most animals lack the necessary enzymes to break down the cellulose into easily digestible molecules, but some use certain bacteria, protozoa or fungi (p. 401) to conduct this first stage of digestion on their behalf. The possession of these micro-organisms gives the animal opportunities to exploit an abundant food source. In mammals, the bacteria themselves form an important source of protein for the host and can also synthesize most vitamins except A and B.

There are two major forms of bacteria-assisted digestion: foregut and hindgut fermentation in vertebrates. Foregut fermentation is found in, for example, anoa, deer, and leaf-monkeys, and hindgut fermentation is found in, for example, rabbits, horses, and rodents (Bauchop 1978; Muul and Lim 1978) and probably cuscus. Leaves are eaten by a wide range of mammals and larval and adult insects, but not by amphibians and by very few reptiles (p. 301) or birds (Morton 1978; Rand 1978). It is estimated that 7%-12.5% of leaf production in forests is eaten by insects but only 2.4% by vertebrates (Leigh 1975; Wint 1983). These figures are usually calculated,

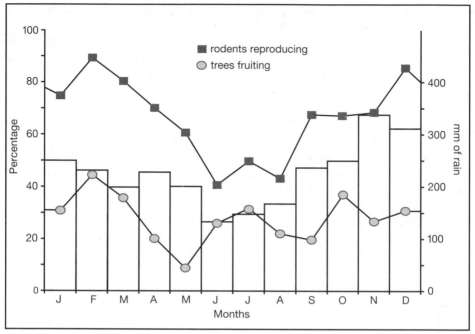

Figure 5.9. Relationship between rainfall (columns) and reproductive activity as a percentage of 13 rodent species (upper line) and percentage of trees in fruit (lower line) in a lowland forest in Zaire.

After Dieterlen 1982

however, from the remains of fallen leaves and so do not necessarily include leaves or shoots in their entirety (Janzen 1978a).

It is generally considered that ferns suffer little from grazing by insects but this view is not supported by firm data. It appears, in fact, that they suffer as much damage as flowering plants (Hendrix and Marquis 1983), and one insect specializing on eating certain Sulawesi ferns is the primitive wingless grasshopper *Karnydia* (R. Butlin pers. comm.).

In ecological studies of herbivory, the actual amount of leaf material eaten is not the most important measurement because each part of a plant is different in its value to the plant, 'cost' for replacement, food value and concentration of defence compounds. For example, the loss of a small shoot from a seedling is considerably more costly to the individual plant than the loss of a great many leaves on a mature plant. Leaves pay back their cost of production by photosynthesizing and contributing the products of photosynthesis to the plant. So, once a leaf has paid back its 'debt', its value to the plant decreases with time. Thus, it is clear that for advanced analysis of an ecosystem, a list of plant species eaten by a particular animal species is not nearly so useful as knowledge of what parts of each species are eaten and the likely cost to the individual plants of the material lost.

There is probably no plant whose defences preclude attack by all herbivores, and there is certainly no animal that can eat all types of leaves. A herbivore gains less usable food per mouthful when eating a leaf rich in defence compounds than when eating a leaf low in defence compounds. An animal which eats defended leaves probably has more plant species to choose from and experiences relatively less competition from herbivores unable to cope with the chemicals.

To the eyes of a human observer the leaves of a lowland forest, or indeed any other tropical vegetation type, vary in shape and shade of green. To an animal dependent on leaves as a food source, however, they are 'coloured' "nicotine, tannin, lectin, strychnine, cannabinol, sterculic acid, cannavanine, lignin, etc., and every bite contains a horrible mix of these" (Janzen 1981). These phenolic and alkaloid compounds (Walker 1975) are just some of the toxic and digestion-inhibiting chemicals used by plants to defend their leaves against herbivores (Edwards and Wratten 1980). To balance this, however, it must be stressed that not all food plants are chosen by herbivores on the basis of their chemistry alone; other important factors include the effects of carnivores and microclimates (Janzen 1985b).

Leaves of forest trees in Costa Rica were analysed for phenolics, fibre and alkaloids at three stages of growth: when the leaves were young but fully expanded; middle-aged (two months later); and mature (six months later). Concentrations of phenolics and fibre per gram of dry leaf remained more or less constant with growth, but water content decreased considerably. Alkaloids were much more common in leaves of deciduous species and trees appeared to defend their leaves with either alkaloids or tannins rather than both, probably because of some interaction which reduces their respective effectiveness. The distribution of two major groups of leaf-eating moth caterpillars on different trees was most striking and reflected some of these differences; large emperor moths (Saturniidae) preferred leaves with phenolics and no alkaloids in the upper canopy, and hawk moths (Sphingidae) preferred leaves with low levels of phenolics which are likely to contain alkaloids in a variety of habitats (fig. 5.10) (Janzen and Waterman 1984; Janzen 1985b).

The concentration of defence compounds may also vary between different parts of a single tree. Evidence is beginning to be gathered which shows that a tree is not really a single individual but rather something akin to a colonial organism. The tip of each growing shoot contains within it a group of undifferentiated dividing cells, called the meristem, which is capable of forming both germ cells for sexual reproduction and normal

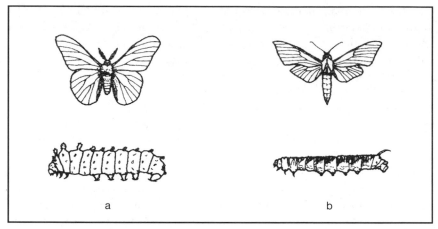

a   b

Figure 5.10. Typical adults and larvae of emperor (saturniid) moths (a) and hawk (sphingid) moths (b), major consumers of leaves in lowland forest, each of which has a different feeding strategy.

tissue cells. A large branching tree might have 10,000 to 100,000 meristems and even with conservative mutation rates at least some of the branches will be genetically distinct from the rest. Thus, if one branch contains a novel and effective cocktail of defence compounds its leaves may last longer than its neighbours' and its flowers may have a better chance of setting fruit with the chance that the genes for that defence will be passed to the next generation (Cherfas 1985).

The bear cuscus *Ailurops ursinus* (p. 429) is probably very selective in the plants it eats, choosing particular species, ages of leaves, parts (stem, tip or entire blade), and quantities. Studies of food selection by leaf-monkeys have shown that leaves selected as food are consistently lower in lignin (fibre), and usually higher in protein, than leaves that are not selected. Tannin levels do not show consistent significant correlations with food selection (Oates et al. 1977, 1980; Bennett 1984), and it is now generally felt that the role of tannins in food choice has been overemphasized (Waterman 1983). They do, however, probably remain a potent force in ecosystems where poor soil conditions produce some form of stress in the trees growing on them, such as on ultrabasic soils (p. 466).

Tannins and lignins are classed as digestion-inhibiting phenolic compounds (although lignin is 10-20 times less effective than tannins), whereas others, such as the glucosides of *Antiaris toxicaria*, are classed as toxins. The adjective 'toxic' is really a measure of the energy expended by an animal in utilizing a food relative to the energy value of usable materials gained from the digestion process. Toxicity is therefore an outcome rather than an inherent property of a chemical, although some chemicals are more likely to be toxic than others. Whether a certain percentage composition of a potentially dangerous chemical is toxic or not will depend on the food value of the material eaten and on the efficiency of the animal's detoxification system. Thus, a leaf with a certain weight and concentration of defence compounds may be more toxic than a seed of the same weight and the same concentration of defence compounds but with higher nutritive value (Freeland and Janzen 1974; Janzen 1978a; McKey 1978). Defence compounds should not be thought of as purely disadvantageous from the herbivore's point of view because they can have decidedly beneficial effects. It appears that certain large animals, at least, seek leaves or bark for the stimulatory or anti-diarrhoea effects they possess (Janzen 1978a).

Pioneer plants (pp. 359-360) are generally thought to be low in defence compounds; their growth strategy is rather to grow and reproduce quickly because, in the nature of succession, they are limited in both space and time. As a result, resources are not diverted into possibly unnecessary defence.

A study at Toraut looked at plant characteristics and leaf damage by insects in three stages of succession: a field of herbs cleared of forest three years previously with few if any pioneer trees, a forest edge with pioneer trees, and a shaded lower canopy of mature forest (table 5.6).

Preliminary results show, as have these from Costa Rica (Coley 1983a, b), that both nitrogen (in protein) and water content of mature leaves decrease through the succession sequence, while leaf toughness (determined by the concentration of lignin) increases (fig. 5.11).

Various authors have suggested that the amount a plant 'invests' in leaf

Table 5.6. Plant from three habitats used in an analysis of leaf characteristics at Toraut.

Early herbs in a field	Pioneer trees of the forest edge	Shade-tolerant plants of the lower canopy
Passiflora foetida (Pass.)	Piper aduncum (Pipe.)	Bauhinia sp. (Legu.)
Ipomoea sp. (Conv.)	Macaranga tanarius (Euph.)	Scindapsus sp. (Arac.)
Cassia sp. (Legu.)	Trema orientalis (Urti.)	Garcinia sp. (Gutt.)
Stachytarpheta jamaicensis (Verb.)	Pipturus sp. (Urti.)	Aglaonema sp. (Arac.)
Sida sp. (Malv.)	Mallotus sp. A (Euph.)	Pterospermum sp. (Ster.)
Sonchus sp. (Comp.)	M. sp. B (Euph.)	Pometia pinnata (Sapi.)
Blumea sp. (Comp.)	Unidentified A	Unidentified C
Ageratum conyzoides (Comp.)	Unidentified B	Unidentified D
Hyptis sp. (Labi.)		

After Greenwood in prep.

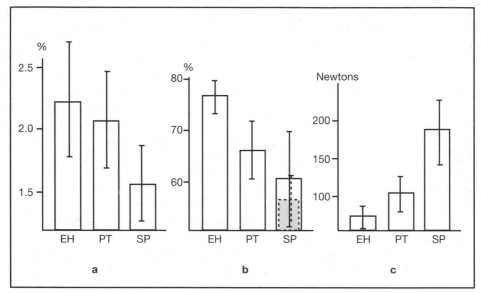

Figure 5.11. Changes in average (and range) of (a) nitrogen content, (b) water content and (c) toughness of mature leaves among early herbs (EH), pioneer trees (PT), and shade plants (SP), at Toraut. Toughness is measured as the force in newtons required to punch a rod 6 mm in diameter through a leaf. The shaded area in the water content histogram demonstrates the effect of removal of the fleshy-leaved aroid plants from the sample.

After S. Greenwood pers. comm.

defence is a function of its ability to produce new leaves. Thus a plant growing in low light or on poor soils is less able to produce new leaves and will therefore invest more heavily in defences for its leaves that a plant growing on richer soils (Janzen 1974; Coley 1983a, b). This ability, measured at Toraut as the rate of leaf production over the six-week study period, clearly decreases through the successional stages and reflects the 'quality' of the habitats. The likelihood of these leaves being discovered by a herbivore depends in part on the life span of the leaves, with long-lived leaves risking discovery more than short-lived ones. To examine this, 10 leaves of each of the 26 species studied at Toraut were tagged and after monitoring over four months it was found that life spans were as follows:

Early herbs: 43-116 days

Pioneer trees: 64-169 days

Shade plants: all leaves of all plants retained.

The shorter-lived leaves of the herbs had fewer of their leaves discovered (even partially eaten) than those in the later stages of the succession and the

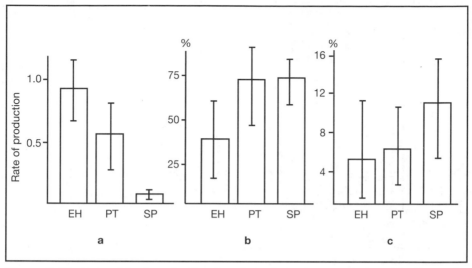

Figure 5.12. Rate of leaf production (number of new leaves produced in six weeks divided by original number of leaves) (a), percentage of leaves discovered by herbivorous insects (b), and percentage of leaf area removed (c) among early herbs (EH), pioneer trees (PT) and shade plants (SP) at Toraut.

After S. Greenwood pers. comm.

area of leaf removed was also less (fig. 5.12) (S. Greenwood pers. comm.).

The differences in percentages of leaves discovered and leaf area removed are just what one would expect given that the leaves of shade plants have a much longer life, and it was felt important to measure the actual rate of damage rather than just its occurrence, to determine whether this decreased through the succession stages. There were great differences in the rate of damage both between and within species, but it did indeed decrease for the mature leaves, although the rate remained at about 4% of leaf area for the young leaves over the six-week period. The difference between the rates for mature and young leaves is significant, however, only for the shade plants. Similarly, the rate of leaf discovery by herbivores is higher for mature leaves of the field herbs than of shade plants, whereas the opposite is true for young leaves (fig. 5.13). In a study of leaf damage in pioneer and shade plants in Costa Rica, young leaves of both types were found to suffer more damage than the mature leaves, but the rates were ten times higher for the shade plants than for the pioneer species (Coley 1982).

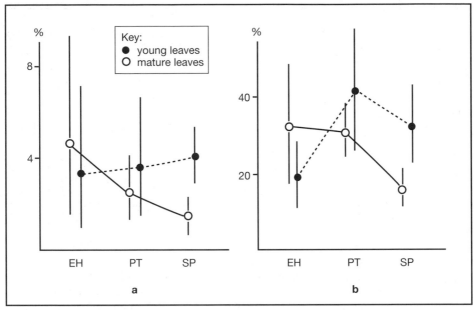

Figure 5.13. Percentage of leaf area removed (a) and leaves discovered over a six-week period (b) among early herbs (EH), pioneer trees (PT) and shade plants (SP) at Toraut.

After S. Greenwood pers. comm.

Finally the various measures of leaf damage used at Toraut were tested against both the nitrogen and water contents and the leaf toughness, but no significant relationships were found. In the Costa Rican study, however, leaf toughness, fibre content (lignin) and nitrogen content correlated significantly with damage although phenol-to-protein ratios did not (Coley 1983a, b). Clearly more work is required to understand the processes at work but it would be expected that the effects of straightforward physical and chemical factors will be complicated by, among other factors, the availability of other foods, unmeasured chemicals and microclimate. For example, some species growing in gaps have been observed to suffer less herbivore damage than the same species in shaded habitats (Maiorana 1981). Strangely, when leaves from both habitats were presented to snails[14] under laboratory conditions, it was found that the plants from gaps were preferred. It has consequently been suggested that the 'defence' used by pioneer plants might in fact be the hot, unshaded, low humidity of the gap environment itself, which is less favoured by some herbivores.

Interestingly, all the pioneer trees examined at Toraut bore abundant

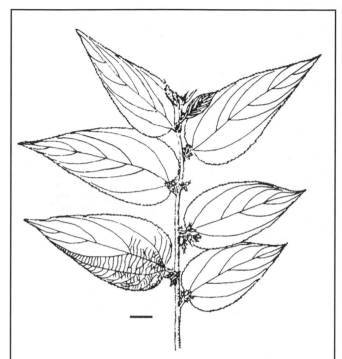

Figure 5.14. *Trema orientalis.* Scale bar indicates 1 cm.
After Meijer 1974

ants. Three of the trees had extra-floral nectaries[15] but on *Trema orientalis* (Urti.) (fig. 5.14) the ants farmed large colonies of aphids from which they collect the sticky, sweet secretion known as honeydew. Indeed, the ants do not seem to have any influence on the activities of leaf-eating insects. The insects most attracted to the extra-floral nectaries of one of the pioneer trees, *Mallotus* sp. (Euph.) appeared to be small jumping flea beetles (Haticinae), but these were also major consumers of the leaves, thereby making any adaptive argument somewhat redundant. Further research would probably prove to be rewarding.

In the Tangkoko-Batuangus Reserve at the northern tip of Minahasa, the squirrels *Prosciurillus murinus* and *P. leucomus* were common and fed entirely on bark and insects. EoS teams at Toraut also observed unidentified *Prosciurillus* eating bark from the red-coloured *Pometia pinnata* (Sapi.) (fig. 5.15), and the darker bark of *Dracontomelum dao* (Anac.) was scored by squirrel tooth marks which could easily be seen. Bark is not a rich food source but it has been recorded as a component in the diet of orangutans, deer, gibbons,

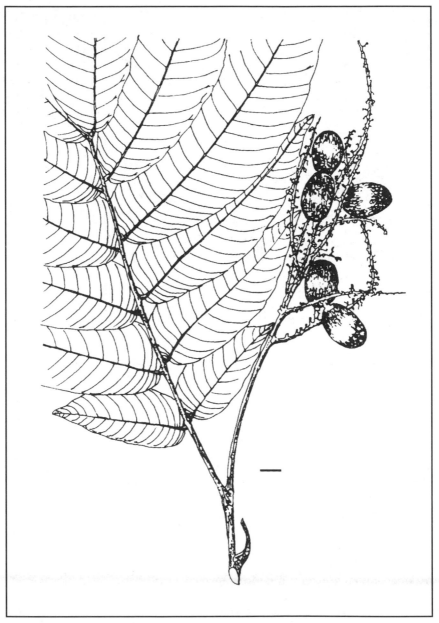

Figure 5.15. *Pometia pinnata*, the reddish bark of which is eaten by small squirrels. Scale bar indicates 1 cm.

After Soewanda n.d.b

elephants, rhinoceros, as well as squirrels. The physical and chemical features of bark eaten by small squirrels in Sumatra have been analysed and it was found that feeding tended to occur on large, smooth-barked trees which were relatively free of climbers, and in whose bark certain tannins were absent. The food value of barks eaten were, however, no different from those not eaten (Whitten and Whitten in press).

Seed Dispersal and Predation

The dispersal of a seed is not simply the event of a mature fruit being taken by an animal or being released by its parent.[16] Dispersal implies the carrying of a seed by some agent to a place where the seed will, perhaps, germinate, grow and eventually reproduce. In many cases, seeds die because the places they are dropped are unsuitable for growth; that is, they are not successfully dispersed. Seeds found directly below the parent tree have probably simply dropped from the branches. Most seeds falling beneath their parent are able to germinate and do so, but few if any have much chance of survival beyond the germination stage, because they experience severe competition from others of their species, because they represent a clumped food resource for seed predators (Becker and Wong 1985), or because environmental conditions beneath the parent simply are not suitable for the developing plant. Dispersal of seeds can be envisaged, therefore, as a means of escape from seed predators (Howe and Smallwood 1982) as well as a method of colonizing widely-separated available sites. In the species-rich lowland forest where most species are rare (p. 345), a seed dispersed away from its parent will stand a relatively low chance of being adjacent to a tree of the same species. A further advantage of dispersal away from the parent is that the biological and physical environment of the location where the parent itself germinated has, in almost all cases, changed by the time that tree has matured (Janzen 1970). This is the same as saying that the environment below the parent tree is now unsuitable for its own seeds. Pioneer, building and mature forest areas (p 359) occur in more or less predictable proportions and patterns in a forest and the means of seed dispersal will aim to maximize the colonization of suitable areas with seeds.

One particular tree, *Syzygium lineatum* (Myrt.), was found to be common in the lowland forest of Morowali National Park. All the individuals over 1 m in height were measured and mapped in 1 ha of alluvial forest and a total of 527 were found, only 10 of which were large, mature, flowering trees. The distribution of the trees was clumped but very few seedlings were found within 5 m of the parent trees despite the fact that most of the fruit fell in that area; indeed, most immature trees were at least 10 m from their parents (fig. 5.16) (Lack 1984; Lack and Kevan 1984).

The area over which seeds come to rest on the forest floor is known as

Figure 5.16. Numbers of seedlings of *Syzygium lineatum* at different distances from the parent tree.

After Lack and Kevan 1984

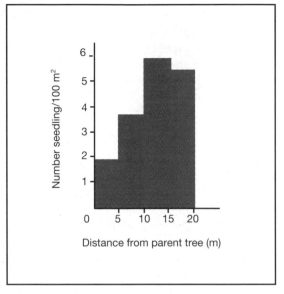

the 'seed shadow'. The seed shadow of a tree is generally most dense close to the parent tree and much less dense overall when dispersed by an animal than when dispersed by wind. Whereas a wind-generated seed shadow is quite homogeneous, a seed shadow generated by animals will usually be heterogeneous, being concentrated in dung piles, along animal paths, and in certain vegetation types. An exception may be fig trees whose fruit is fed upon by so many species of animals, and whose modes of living are so varied that a relatively homogeneous seed shadow can result. There is also likely to be variation in the seed shadows of neighbouring individuals of the same species because access for the disperser into one tree may be easier than into another, and because other attractive food sources may be close to one, but not to the other (Janzen 1970; Janzen et al. 1976). These seed shadows are not simply of academic interest but are of major importance in determining future forest structure and composition (Fleming and Heithaus 1981).

Wind-dispersed fruits are generally borne by tall trees for these are the only forest trees to be affected by wind to any extent. Wind-dispersed fruits are either very light or have wings, and in both cases their descent to the ground is delayed. Although wind does not disperse tree seeds particularly far from the parent, wind can at least be relied upon, whereas animals cannot. The fate of seeds dispersed by wind from one tree of each of nine species was followed for one year in Panama and it was clear that dispersal away from the parent tree was advantageous to all species. For eight of the nine trees, dispersal also resulted in lower mortality of seedlings, although for a variety of reasons. Pathogens or micro-organisms, such as fungi and bacteria, were the major cause of seedling mortality. Some seedlings suffered from pathogens irrespective of distance from the parent or density of surrounding seedlings, but in all cases the mortality caused by pathogens was less in light gaps largely because fungal pathogens are favoured by high humidity and low temperatures (Augspurger 1984).

The majority of forest tree seeds are dispersed by animals that eat the fruit and carry away the still-living seeds in their guts to be deposited some distance away when the animal defaecates. Birds such as pigeons, many of which are obligate fruit eaters, in New Guinea at least, appear to spend more time in a fruit tree than birds that include fruit as only part of their diet. These longer visits by birds remaining in a fruit tree between feeding bouts, probably result in many seeds being dropped beneath the parent tree since rates of gut passage are about 30-45 minutes. This behaviour might make them open to predation, but the birds tend to be quite large and cryptic in behaviour or colouration (Pratt and Stiles 1983). It is noteworthy that the backs of most pigeons in Sulawesi are green, and this is the part of the birds seen by the eagles that might prey on them.

Fruits can usually be identified as being adapted for dispersal by a particular group of animals (Payne 1980). The dispersal agent for a particular fruit depends on the nutritional requirements of the animals, the accessibility of the fruit, its shape and its size. The fruits produced by a tree vary in size and, for some species, in the number of seeds they contain. Thus different portions of the fruit crop may be dispersed by different animals over different distances (Howe and van de Kerckhove 1980; Howe and Smallwood 1982; Janzen 1982).

It is obviously to a plant's advantage to ensure that fruit are dispersed to the right type of location and numerous means are used to selectively advertise their presence and to encourage certain dispersers whilst discouraging others. These characteristics have been shaped over evolutionary time by reciprocal interactions between animals and plants but, unlike the process of pollination, few species-specific relationships exist (Wheelwright and Orians 1982; Janzen 1983b). Edible fruits are basically only seeds covered with a bait of food and this bait can vary greatly in nutritional value. Plants tend to adopt one of two strategies:

- they can produce large numbers of 'cheap' (small, sugary) fruit, most of which may be wasted or killed by the many species of obligate and facultative frugivores attracted to them, or
- they can produce smaller numbers of 'expensive' (large, lipid-rich or oily) fruit that have to be searched for by those few specialized species of frugivores which gain a balanced diet almost entirely from fruit (Howe 1980, 1981; Howe and Smallwood 1982; Howe and van de Kerckhove 1980).

An example of 'cheap' fruits is figs which grow on plants of numerous life forms, occurring as free-standing trees, epiphytes, shrubs and stranglers of other trees (p. 358). Examples of 'expensive' fruit include nutmegs (Myristicaceae) and laurels (Lauraceae).

The pollination of figs is of special interest because it directly affects seed predation. The flowers themselves are within a narrow-mouthed cup formed by the developing flower structure arching over and around itself. The hole at the mouth of a fig is partially closed by scales or modified bracts. Fig flowers are pollinated exclusively by tiny fig wasps (Agaonidae), and both figs and wasps are entirely dependent on each other for their survival since one species of fig plant will generally have just one species of pollinator wasp. Female wasps fly to the figs in which the female flowers are ready for pollinating and one or more of these wasps climb inside the fig by squeezing their way through the scales and lose their wings and antennae by so doing. Once inside, a female pushes her ovipositor down into the ovary of one of the female flowers and lays an egg (fig. 5.17). As she moves around inside the hollow fig looking for a suitable ovary, the female deposits pollen from the fig in which she hatched onto flowers in the fig into which she has crawled, and so effects pollination. Although the wasps pollinate the flowers, they are clearly also seed 'predators' and in a sample of 160 figs from four species in Costa Rica, 98% of the figs had more than 30% of their potential seeds killed by pollinating wasps (Janzen 1979).

The larvae develop and pupate and the wingless males emerge first. They search around the fig and mate with young females before the latter emerge. The males' role is not yet completed, since they then tunnel through the wall of the fig to the outside. The carbon dioxide level inside the fig is initially about 10% but when the hole is made the concentration drops to atmospheric levels (i.e., 0.03%). This change appears to stimulate the development of the male flowers, the emergence of the females, and the process of ripening. The cycle continues with the females flying off carrying pollen to a developing fig and the males dying within the fig.

A fig tree studied in Peninsular Malaysia had a crop of about 41,000 figs, some of which were ripe over a period of eight days. The consumption by bats and other nocturnal mammals was not estimated but among the diurnal animals, macaques took the most, birds took half as many, gibbons half as many as the birds, and squirrels just a quarter of this (MacKinnon and MacKinnon 1984). A more detailed study in Costa Rica monitored the fate of over 100,000 figs over five days. Birds took about 65% of the figs consumed each day (the rest were taken by bats and other mammals) and most of these were eaten by parrots which are seed predators[7] Only about 4,600 figs were eaten by seed dispersers. Taking into account the seed predation by both parrots and the pollinator wasps, there was clearly an enormous wastage and only about 6% of the enormous crop was dispersed unharmed by birds (Jordano 1983).

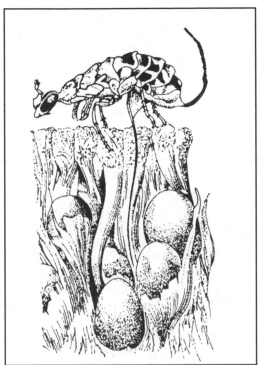

Figure 5.17. Female fig wasp inserting an egg through her long ovipositor into an ovary of a fig flower. Her wings and antennae were lost as she struggled into the fig.

Fruits advertise their ripeness to dispersers using colour, texture, taste, conspicuous shapes, and odour. Between ripening and being taken by a disperser, however, the fruit is also exposed to seed predator animals and micro-organisms. Seed predators include both animals which consistently destroy seeds of that type, and disperser animals which eat the fruit before the seeds are mature. Thus a fruit needs to be as unattractive as possible before the seed is viable and this is commonly achieved by the presence of defence compounds whose concentrations reduce as ripening progresses (p. 532).

Some fruit do not lose all their defence compounds when they are ripe, however, and this is thought to be a means of protecting fruit from being eaten or destroyed by non-disperser animals or micro-organisms (Herrera 1982; Janzen 1983). Defence compounds can be divided into two groups: toxins which can debilitate or kill, and quantitative defences or digestion inhibitors such as resins, gums, volatile oils and phenols which deter potential seed predators or cause some mildly adverse physiological effect (such as nausea) (Maiorana 1979; Waterman and Choo 1981; Waterman 1983). A review of defended ripe fruits concluded that the species producing them were in fact at a competitive disadvantage for dispersal relative to species without defended ripe fruit because of the resources which were diverted to make the defences. If the defences are viewed as a means by which undesirable seed predators can be avoided, then extreme toxicity may be a 'last resort' (Herrera 1982). One extremely toxic seed is that of the tree *Antiaris toxicaria* (Urti.) which contains poisonous glucosides. The latex from the tree contains the same chemicals and is used by many peoples in the preparation of poisoned arrows and darts, but it is actually not as deadly as popularly believed (Burkill 1966). Macaques almost certainly eat the *A. toxicaria* fruit together with the seeds which pass through unharmed, thus demonstrating that the tree probably uses its toxins to deter the 'wrong' frugivores.

A sample of 33 seed pods of *Pericopsis mooniana* (Legu.)[18] were examined in Lamedae Reserve south of Kolaka in Southeast Sulawesi. Overexploitation of the species has resulted in this being one of the few remaining *Pericopsis* stands in Sulawesi. The pods each contained 2-3 seeds and 69 seeds were examined in the sample. Of these, however, 11% had been destroyed by small beetles as evidenced by round holes in the pot shell. The average number of holes in the five pods attacked was three (L. Clayton pers. comm.). Larvae of small beetles are common predators on seeds and they can show a high level of specificity. Of over 975 species of shrubs and trees in an area of Costa Rica, at least 100 species regularly had beetle larvae in their mature or nearly-mature seeds. Three-quarters of the 110 beetles species found (primarily from the family Bruchidae) were confined to a single species of plant; if a beetle species was found on more that one plant species, the plants were found to be closely related. Of the 100 species of plants attacked, 63 were legumes and 11 were from the family Convolvulaceae (Janzen 1981). This 'preference' for legumes might be because most plant families have had strong defences against attacks from bruchid beetles whereas a few families have not. It seems more likely, however, that bruchid beetles became legume seed predators many millions of year ago and were able to counter whatever forms of defence the plant used. As the legumes evolved and diversified so the beetles evolved to meet the changes—a case of coevolution. Since plant defences appear to be so complex, the specializations required to predate on a particular species or genus are such that a beetle species would be unlikely to be able also to become a predator on seeds in another plant family.

Destructive micro-organisms (fungi and bacteria) represent a form of competition for frugivores (Janzen 1977, 1979). To a microbe, a ripe fruit is a considerable food resource that can be digested and converted into more microbes. Having started it is obviously preferable to exploit the whole of the fruit, but to do so its competitors have to be deterred. The microbe produces antibiotic substances to prevent other micro-organisms from colonizing the fruit, as well as toxic and unpleasant-tasting substances to deter frugivores. Producing tasteless toxins alone would not be a successful strategy because an animal may still eat the fruit, kill the microbe and later die itself to the benefit of neither party. For microbes to succeed, the fruits they invade have to look, smell or taste different from untainted fruit. Some fruits are known to produce their own antibiotics against decomposer micro-organisms (Janzen 1975) but this field of study has received little attention.

COMPOSITION

Composition of Mature-phase Forest

No part of Indonesia yet has a comprehensive guide to or account of its forest trees, but the first stage, annotated checklists with line drawings of the major timber species,[19] have recently been prepared by the Forest Research and Development Centre, Bogor, although the Sulawesi volume (Whitmore and Tantra in press) has not yet appeared. When it does, the way will be open for the writing of keys and detailed descriptions.

Vernacular names allow differentiation of some of the trees and represent a common and useful means of conducting a general forest survey. It should be realized, however, that vernacular or commercial names generally greatly underestimate the number of species present, although some languages have more than one name for a single species (Soewanda n.d.a, b). There is a misconception, too, that any villager employed on forest surveys from a village near the forest edge will know the names of all the trees in the forests around the village.

Numerous surveys of Sulawesi lowland forests have been conducted for forestry evaluations but these are of limited use for ecological work or for general descriptions, because they enumerate only those trees that provide profitable timber, and they concentrate on areas where timber trees are likely to be most common. Consequently, there is very little ecologically useful information concerning the composition of Sulawesi's forest arising from such studies.

Variation in the floristic composition of a given phase in the growth cycle of lowland forest can be quite considerable but, because the variation is often continuous (that is, there are no sharp boundaries, and forest of one composition changing gradually into a forest with another composition), it is very hard to study. It has been proposed that floristic composition is largely determined by chance factors, particularly at the time of fruit dispersal and seedling establishment (Poore 1968). Within a given area, however, composition is related to large-scale habitat features such as soils and topography (Ashton 1964; Franken and Roos 1981; Baillie and Ashton 1983). A detailed study of species composition and various soil and other site variables in Sarawak showed that soil texture, levels of iron and aluminum oxides, and the acidity of the soil parent material were the most important factors determining species composition (Baillie and Ashton 1983). These effects are very subtle and result more in a variation in the relative abundance of species, rather than in the presence of a species. This is partly because a particular soil type will not always exclude ill-adapted species, but adults of such species will be relatively rare and have reduced vigour.

In western Indonesia the forest trees are often dominated by members of the Dipterocarpaceae, whilst in Sulawesi this family is represented by just six species (fig. 5.18; table 5.7). Other important families in western Indonesia are the Lauraceae, Euphorbiaceae, Annonaceae, Myristicaceae, Rubiaceae and Sapotaceae. With the virtual absence of dipterocarps, Sulawesi forests are not dominated by any one family and, from what information is available, there is considerable variation between sites in the major species and families found (table 5.8). For example, the most common families at Toraut were the Lauraceae, Guttiferae and Anacardiaceae (Whitmore and Sidiyasa 1986), whereas in the Lore Lindu area Sapotaceae and Burseraceae are prominent (Meijer 1984). Some of the variation can be interpreted. For example, the presence at the Toraut site, Bogani Nani Wartabone National Park, of *Garuga floribunda* (Burs.), *Tetrameles nudiflora* (Dati.) and *Kleinhovia hospita* (Ster.) is indicative of a relatively seasonal climate (Whitmore and Sidiyasa 1986). These and other common species are illustrated below (figs. 5.19-26).[20]

The most famous lowland trees in Sulawesi are the ebonies *Diospyros* (Eben.).[21] The architecture of the trees is highly distinctive. The branches are horizontal but droop at their tips, and emerge in whorls at intervals up the trunk rather like those of *Terminalia* (Comb.) and the kapok tree *Ceiba pentandra* (Bomb.). Various species are present in the forest but the blackest and hardest timber is from *D. celebica* found mainly in the central and northern regions (fig. 5.27) (Steup 1931).

Fifty years ago in the Onggak-Dumoga area of Bolaang Mongondow, there are said to have been 50 exploitable *D. celebica* trees per hectare (Verhoef 1938). At this time there were about 1,000 tons of *D. celebica* ebony exported each year from North Sulawesi, nearly all of which was sent to Japan, and even now this very fine-grained wood is much sought after. Another such ebony-dominated area is Tanoma in Tanjung Peropa Reserve where over 50 trees/ha of *D. pilosanthera* of 15 cm diameter and over have been found (Bismark 1982b). Most interesting from an ecological perspective is that ebony trees were generally found in dense clumps over several hectares in which 90% of all trees present would be ebony (Steup 1935). In one area south of Poso, the only other species present was the *Livistona* palm (Steup 1931). This pattern is very similar to that of ironwood *Eusideroxylon zwageri*, of Sumatra and Kalimantan which also has very hard,

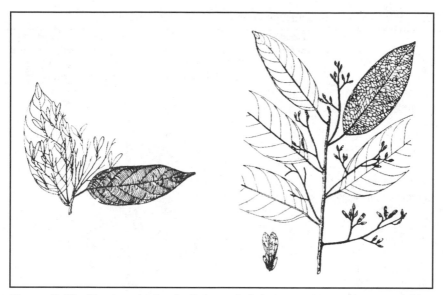

Figure 5.18. The two endemic Sulawesi dipterocarp trees: *Hopea celebica* (left) and *Vatica flavovirens* (right).

After Soewanda n.d.a

Table 5.7. The six species of dipterocarp trees known for certain from Sulawesi. Neighbouring Borneo has more than 250. *Hopea dolosa* was once recognized as a Sulawesi species but is now included in *H. celebica,* as *Vatica celebensis* is now included in *V. rassak*. Further species including new ones are almost sure to be found in due course (Jacobs 1977).

	Known distribution
Anisoptera costata	Malili
Hopea celebica	Southwest and southeast peninsula
H. gregaria	Southeast peninsula
Shorea assamica	Widespread
Vatica rassak	All regions
V. flavovirens	All regions

After Ashton 1984

Table 5.8. Species of abundant trees in lowland forests on Sulawesi.

	Bogani Nani (Toraut)	Bogani Nani	Sopu valley	Lore Lindu (a)	Lore Lindu (b)	Morowali	Bolaang Mongondow	Tangkoko Batuangus	Tj. Peropa (Tanoma)
Adina fagifolia (Rubi.)	-	-	-	-	-	-	•	-	-
Ailanthus integrifolia (Sima.)	-	-	-	•	-	-	-	-	-
Canangium odorata (Anno.)	-	-	•	-	•	-	-	•	-
Calophyllum soulattri (Gutt.)	-	-	-	•	-	-	-	-	-
Celtis sp. (Ulma.)	•	-	-	•	-	-	-	-	-
Diospyros sp. (Eben.)	-	•	-	-	-	•	-	-	•
Dracontomelum dao (Anac.)	-	•	-	-	-	-	•	•	-
Duabanga moluccana (Sonn.)	-	-	-	•	-	-	-	-	-
Dysoxylum sp. (Meli.)	-	•	-	•	•	-	-	-	-
Elmerillia ovalis (Magn.)	-	-	-	•	•	-	•	-	-
Ficus spp. (Mora.)	•	-	•	•	-	-	-	-	•
Garcinia sp. (Gutt.)	•	-	-	-	-	-	-	-	-
Garuga floribunda (Burs.)	•	-	-	-	-	-	-	-	-
Gonystylus macrophyllus (Thym.)	-	-	-	-	-	•	-	-	-
Gossampinus valetoni (Bomb.)	-	-	-	-	-	-	-	•	-
Homalium celebicum (Flac.)	-	-	-	-	-	-	-	•	-
Intsia amboinensis (Legu.)	-	-	-	-	-	-	•	-	-
Itoa stapfi (Flag.)	-	-	•	-	-	-	-	-	-
Kleinhovia hospita (Ster.)	•	-	-	-	-	-	-	-	-
Koordersiodendron pinnatum (Legu.)	-	-	•	-	•	-	-	-	-
Litsea sp. (Laur.)	•	-	-	-	-	-	-	-	•
Manilkara celebica (Sapo.)	-	-	-	-	-	•	-	-	-
Metrosideros vera (Myrt.)	-	-	-	-	-	•	-	-	-
Mussaendopsis beccariana (Rubi.)	-	-	-	•	-	-	-	-	-
Myristica sp. (Myri.)	-	-	-	•	-	-	-	-	-
Octomeles sumatrana (Dati.)	-	-	•	-	-	-	-	-	-
Palaquim obovatum (Sapo.)	-	-	•	-	-	-	-	-	-
P. obtusifolium	-	-	-	-	-	-	-	•	-
Paranephelium sp. (Sapi.)	-	-	-	-	-	-	-	-	•
Planchonella firma (Sapo.)	-	-	•	-	-	-	-	-	-
P. sp.	-	-	-	-	-	•	-	-	-
Planchonia valida (Myrt.)	-	-	-	•	-	-	-	-	-
Pterocarpus indica (Legu.)	-	-	-	-	-	-	•	-	-
P. subpetalum (Legu.)	-	-	-	•	-	-	-	-	-
Pterospermum sp. (Ster.)	-	•	-	-	-	-	-	-	-
Sandoricum sp. (Meli.)	-	•	-	-	-	-	-	-	-
Santiria sp. (Burs.)	-	-	-	-	-	•	-	-	-
Sarcotheca celebica (Oxal.)	-	-	-	-	-	•	-	-	-
Strychnos axillaris (Loga.)	-	-	-	•	-	-	-	-	-
Tetrameles nudiflora (Dati.)	•	-	-	-	-	-	-	•	-

After Steup 1933; Soeriaatmadja 1977; Anon. 1980; Bismark 1980; Wirawan 1981; van Balgooy and Tantra 1986; Whitmore and Sidiyasa 1986

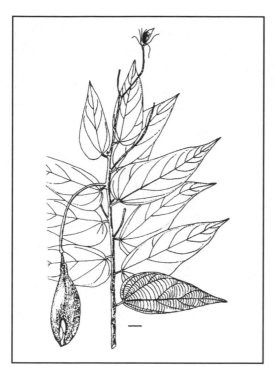

Figure 5.19. *Pterospermum celebicum.* Scale bar indicates 1 cm.

After Soewanda n.d.a

resistant, and therefore very valuable timber. It may be, too, that certain ebony species are similar to *E. zwageri* in producing relatively large, toxic seeds more or less continuously through the year (Whitten et al. 1984). The fruits of *D. celebica* measure about 4 cm x 2 cm and are amongst the largest in the genus. It would be of great interest to study fruit production and seed dispersal and predation in a population of these trees, but dense stands must now be very rare.

Certain other areas appear to be dominated by single species, although not to the same extent. In the dry hills west of Marisa, Gorontalo, for example, *Adina fagifolia* (Rubi.) is unusually common with about 12 large trees/ha (Steup 1935), and *Elmerrilia ovalis* (Magn.) appears to be extremely common in parts of Minahasa, eastern Bolaang Mongondow and the area east of Lake Poso (Steup 1931, 1932, 1933).

Palms are a common sight in lowland forest and include the black-spined *Oncosperma horridum,* the trunkless *Licuala celebensis,* the thin-trunked and spineless *Pinanga, Areca,* fishtail palms *Caryota,* and the wild sugar palms *Arenga* sp. (fig. 5.28). This sugar palm can be distinguished by

Figure 5.20.
Tetrameles nudiflora.
Scale bar indicates 1
cm.

After Soewanda n.d.b

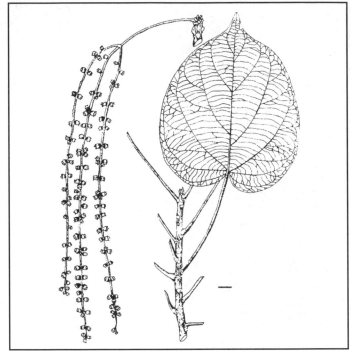

Figure 5.21. *Dracontomelum
mangiferum.* Scale bar indicates 1
cm.

After Soewanda n.d.b

Figure 5.22. *Celtis philippinensis.*
Scale bar indicates 1 cm.
After Soewanda n.d.a

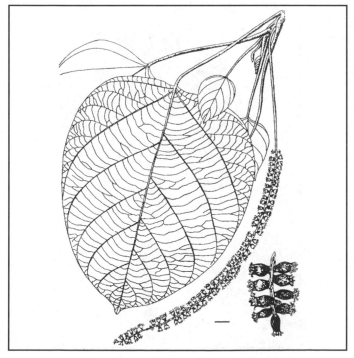

Figure 5.23.
Octomeles sumatrana.
Scale bar indicates 1
cm.
After Soewanda n.d.a

Figure 5.24.
Adina fagifolia. Scale bar indicates 1 cm.

After Soewanda n.d.a

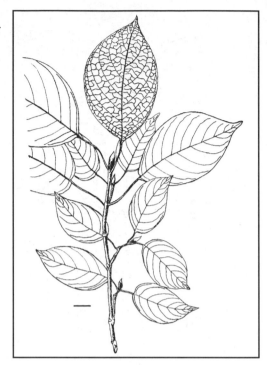

Figure 5.25.
Metrosideros vera. Scale bar indicates 1 cm.

After Soewanda n.d.b

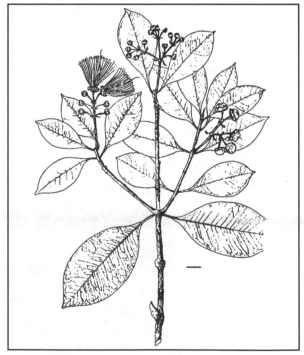

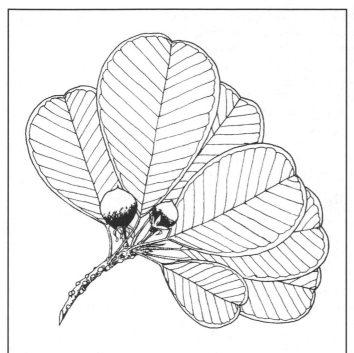

Figure 5.26.
Manilkara celebica.
After Soewanda n.d.b

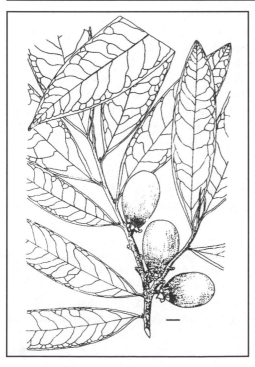

Figure 5.27.
Diospyros celebica.
Scale bar indicates
1 cm.
After Soewanda n.d.a

the pale undersurface of the leaves and the black 'hair' around the leaf bases. The distinctive *Pigafetta filaris* with its green trunk and grey rings is generally found in somewhat disturbed sites (p. 396).

One of the most common palms in Sulawesi is the tall fan palm *Livistona rotundifolia*, the leaves of which can reach 130 cm in diameter and have hard, sharp spines on the leaf stem. A total of 88 were found in one hectare at Toraut (Whitmore and Sidiyasa 1986). Observations of two seven-year-old individuals with trunks over 2 m tall in Bogor Botanic Gardens showed that new leaves were produced every 20 days (Mogea 1986). The rate of production of new leaves was in individuals with similar, shorter and no trunks monitored at Toraut, and was found to be much slower than this (EoS team). The actual rate of height increase in palms is complicated because the distance between the leaf scars (the internodes) is greater in shady conditions than in full sun, and also by the faster growth of young palms. For example, a coconut may have 6 leaf scars/m on the lowest part of its trunk but 90 scars/m at 35 m above the ground (T.A. Davis pers. comm.).

Felled palm trees such as *Livistona rotundifolia* soon develop hollows in the top of the standing stump of their trunks where the softer pith has decayed and these, like pitcher plants (p. 450), soon develop unique animal communities. These were examined at Toraut and the top carnivore was the nymph or larva of a libellulid dragonfly *Lyriothemis cleis* which fed on the larvae of beetles, mosquitoes and crane flies (Tipulidae) which also lived in the water-filled hole (Kitching 1986).

Details of the plants found in the lower layers of lowland forest in Sulawesi can be found elsewhere (Wirawan 1981; van Balgooy and Tantra 1986), but one small tree worthy of mention is *Dendrocnide* (Urti.) the leaves and stems of which have stinging hairs. The cells of the hairs contain quantities of silica which make them hard but brittle. When a hair enters the skin and breaks, both the tip and the sap, which acts on nerve endings, remain in the skin to cause irritation. The irritation is long lasting, at least in part, because the hairs remain in the skin.

Composition of Pioneer- and Building-phase Forest

As indicated earlier (p. 359), the plant species found in the younger phases of forest growth are quite distinct from those in the mature phase. The actual composition follows a certain trend, but is virtually impossible to predict because factors such as past history and treatment of a site as well as chance tend to complicate any trends predicted from the measurement of soil conditions and microclimate. Plants in these phases appear to have been studied only once in detail in Sulawesi,[22] along the boundary of Lore Lindu National Park (Wirawan 1981; Rombe 1982) and much of the information below is taken from that work.

In areas where fires occur more or less annually, the vegetation is dom-

Figure 5.28. Silhouettes of the distal ends of some Sulawesi palm leaves. a - *Caryota mitis*, b - *Areca vestaria*, c - *Pinanga* sp., d - *Gronophyllum selebicum*, e - *Arenga* sp., f - *Licuala celebensis*. Note that to date, the tree palm *Gronophyllum* is known in Sulawesi only from Southeast Sulawesi and the area between Malili and Poso.

Dransfield 1981

inated by grasses such as *Arundinella setona,* and *Themeda triandra.* The well-known sword grass or alang-alang *Imperata cylindrica* is not as wide-spread as is often stated and it is unjustly associated with the worst of land management. It is generally found on roadsides and at sites where fires are less than annual. When *I. cylindrica* does start to grow in grassland domi-nated by other species, it can be an indication of an improvement in soil conditions. Also, because it is relatively tall, certain tree seeds have a greater chance of germinating and starting growth under *I. cylindrica* than under shorter grasses. Where fires are relatively infrequent, the grasses are invaded by the tough fern *Dicranopteris linearis* and the cosmopolitan bracken *Pteridium aquailinum.* Grasslands may contain trees which are resis-tant to fires such as *Morinda tinctoria* (Rubi.), *Lagerstroemia speciosa* (Lyth.) (a tree with masses of violet flowers, frequently planted in cities), *Fagraea fragrans* (Loga.) and *Albizia procera* (Legu.) (Steup 1931, 1939a, b; Jacobs 1977). These trees are more resistant to fire than most others but the young trees are susceptible. The height to which flames reach obviously depends on the height and nature of the grass but scorch marks have been found up to 5 m on *Anthocephalus* (Rubi.) trees near Kendari (Jacobs 1977).

Some trees, such as *Ficus miquelli* (Mora.), *Nauclea* sp. (Rubi.) and *Evodia* sp. (Ruta.), can be found immediately after the garden or ladang is abandoned until the building-phase forest is formed. Other trees, like *Trema* (Ulma.), *Orophea* (Anno.) and *Macaranga* (Euph.), tend to grow from under a low shrub layer (Wirawan 1981; Rombe 1982). It should be noted that some of the trees of the early succession stages are valued and of great use to nearby villagers. Trees most notable in this respect for use as firewood, fence poles and in house construction are *Orophea* sp. and *Nauclea* sp. (Wirawan 1981).

In ladang areas that are reoccupied every decade or so a succession of species and of dominant life forms can be observed from herbs, through shrubs to trees although there is considerable overlap. Some herbs such as *Curculigo latifolia* (Amry.) and the fern *Nephrolepis biservata* are able to maintain themselves under increasing tall vegetation but most of the herbs and grasses that invaded the ladang initially are not present even under low forest. Similarly, almost all the shrubs and small trees such as *Grewia* sp. (Tili.), *Homalanthus populneus* (Euph.), *Malastoma polyanthum* (Mela.) and *Blumea balsamifera* (Comp.) die as the vegetation grows above them. Soil fer-tility is not necessarily associated with length of the fallow period, but rather with the kinds of plants growing in the area. Thus, plant species are good indicators of soil fertility and are used as such by traditional shifting cultivators (p. 570) (Wirawan 1981; Dove 1985).

Pigafetta filaris, a majestic palm of Sulawesi and the Moluccas, is dis-tinctive in having lines of shiny golden-brown spines along the bases of its leaves, and a dark-green trunk with light grey rings where the leaves have fallen off. It is a palm that favours altitudes between 300 and 1,000 m. It appears to be unusual among the palms of Southeast Asia in being adapted to growth in secondary habitats, in that it is fast-growing, tolerant of sunlight, and produces enormous numbers of small fruit (p. 363). For example, the three specimens by the front door of the Bogor Herbarium have trunks 15 m tall but were planted as seed just twelve years ago. During this time they have fruited 15 times, and new leaves appear to be produced about every 16 days (Mogea 1986). It is rarely found in mature forest, although it can be found at the site of a treefall or along a riverbank where new sediment and light penetrating to the forest floor have resulted in suitable conditions. Seedlings can sometimes be found growing together in clumps on the forest floor, as though they have been deposited in animal faeces, but these plants rarely progress beyond the one-leaf stage; it appears that the seedlings are not tolerant of shade. In the past this palm was presumably much rarer than at present, being found primarily on landslips, volcanic screes and flows, riverbanks and very steep-sided ridgetops (Dransfield 1976). Its trunk is used as supports for the traditional houses and rice barns in Tana Toraja, and is also used in the same area to make water conduits (Sneed 1981; S.C. Chin pers. comm.).

All the gaps in the forest of Toraut (p. 360) were colonized by two species of *Macaranga* and a number of gaps were colonized by *Piper aduncum* (Pipe.) and even grass (Whitmore and Sidiyasa 1986). *P. aduncum* is a wide-ranging tree of open sites and forest edges in the New World tropics from Mexico to the West Indies and down to Argentina (W. Burger pers. comm.). It was introduced to Bogor Botanic Gardens one hundred years ago (C.G.G.J. van Steenis pers. comm.), and is now common in many parts of West Java. In North Sulawesi it seems to have found conditions eminently suitable to the extent of out-competing the relatively few indigenous species of pioneer tree. In certain areas it grows in pure stands of secondary forest adjacent to undisturbed forest. Its competitiveness is quite remarkable and it will be interesting to see whether it extends its range even further. This success is partly because *P. aduncum* grows very quickly and mature specimens flower and fruit continuously both in its native habitat and in Indonesia. The minute crowded flowers on the spike-like inflorescence appear to be pollinated by small sweat bees *Trigona* (p. 418), and the ripe infructescence is taken by small fruit bats[23] (p. 421). Each tree appears to produce at least one ripe infructescence each night. As has been found in Central America, almost all the ripe fruits are taken by bats.

Other trees that can be found in almost pure stands in secondary growth are *Nauclea* and *Anthocephalus macrophyllus* (fig. 5.29) (Steup 1939a, b). Where such trees are established, reforestation is unnecessary because the shade they cast will kill the grasses and create conditions favourable to other tree seedlings. The planting of suitable shade-tolerant seedlings may be appropriate, however.

A scheme of forest succession on Sulawesi is shown in figure 5.30.

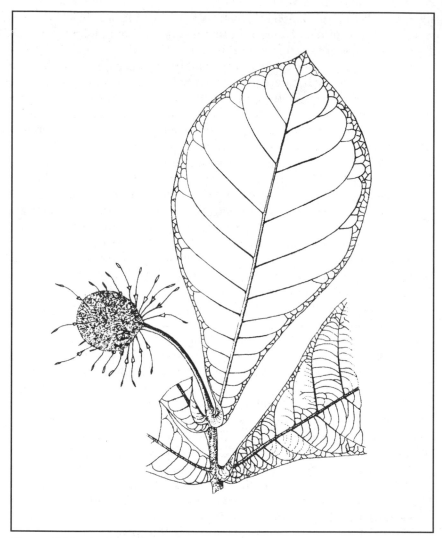

Figure 5.29. *Anthocephalus macrophyllus.*
After Soewanda n.d.

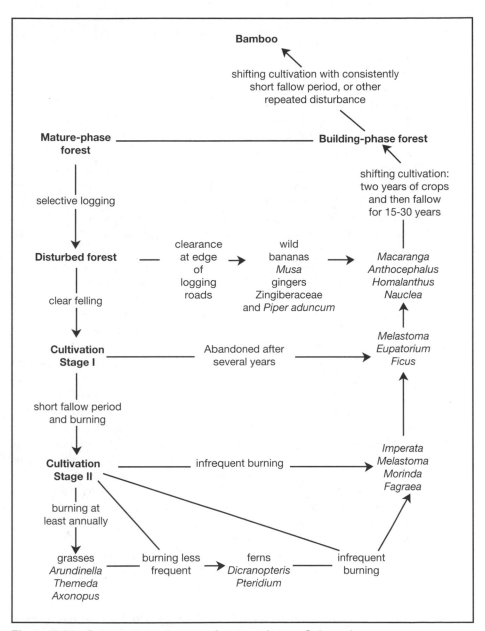

Figure 5.30. Schematic pathways of succession on Sulawesi.

After Steup 1931, 1933, 1939a, b; Bloembergen 1940; Wirawan 1981; Rerung 1983

ANIMAL COMMUNITIES

Soil and Litter Communities

The soil community lowland forest at 130 m above sea level in Sarawak comprised 2,579 visible invertebrate individuals/m² and this declined with even small increases in altitude. The dominant group of detritus feeders was the termites (62% of the total individuals) followed by earthworms (4%) and beetles (3%), and the major predators were ants (20%). The relative contribution of the groups to the biomass is somewhat different with termites accounting for 33% of the 4.1 g total, followed by earthworms (29%), ants (11%), millipedes (6%) and beetles (5%) (p. 523) (Collins 1980a). As for the animals not visible to the naked eye, 1 m² of soil in Peninsular Malaysia was found to have 10,000 mites and 56,000 nematode worms (Chiba 1978).

Termites[24] may look superficially like ants (Hymenoptera) but they are classified in a completely different order (Isoptera) which is more closely related to cockroaches (fig. 5.31). Termites are like ants, however, in that they form enormous colonies with (at least in some parts of the world) possibly a million or more members. A colony is not a simple community of distinct individuals because all except two of the colony members are siblings. The two exceptions are the 'royal pair' or parents. The parents originated in other termite nests from which they flew, along with thousands of others, to seek a mate and a new nest site. Some manage to escape the hordes of predatory ants, amphibians, reptiles, birds and mammals to which such a swarm is a food bonanza. After landing the termite wings drop off and if a male and female meet they will look for a suitable crack in the ground or a tree (depending on the species) and here they build their 'royal cell'. They copulate and the female begins laying eggs. The larvae, unlike the helpless larvae of ants, bees and wasps, are fully able to move around and undergo several moults (like cockroaches), rather than a single metamorphosis before becoming adults. These first larvae have to be fed by their parents but as soon as they are large enough to forage for food and to build walls for the nest, the royal pair devote themselves entirely to the production of eggs. The abdomen of the female, or 'queen', grows enormously and a production rate of thousands of eggs per day is common.

A termite colony can be likened to a single, although sometimes disparate, organism because none of its components is capable of independent life. The workers are blind and sterile, and the soldiers which protect the columns of foraging workers and guard entrances to the colony's nest have jaws so large that they can no longer feed themselves and have to be fed by the workers. The king and queen also have to be fed by the workers. In the same way that communication between different organs or parts of

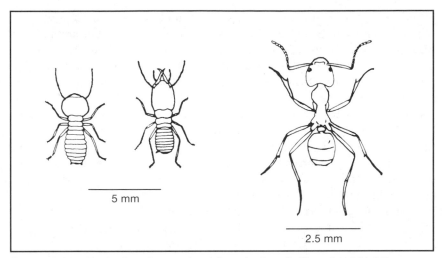

5 mm

2.5 mm

Figure 5.31. Typical worker and soldier termites (left) and ant (right).

a single body is effected by chemicals flowing through a body's tissues, so chemicals—pheromones[25]—link the different members of a termite colony into a coordinated and organized whole. All of the colony members continually exchange food and saliva which contain pheromones. Workers pass these materials from mouth to mouth and also consume each other's faeces to reprocess whatever partially-digested food remains. Workers feed the soldiers, larvae and the royal pair as well as collect the queen's faeces. There is thus a continual interchange of chemicals through the colony and this coordinates the operations of the colony. For example, all the larvae are potentially fertile termites of either sex but the food with which the workers feed them contains quantities of a pheromone from the queen which inhibits the development of larvae and produces sterile workers. The soldiers also produce a pheromone which prevents the larvae from developing into soldiers. When, for some reason, the number of soldiers falls, the level of the 'soldier repression' pheromone circulating through the colony members falls and some of the eggs develop into soldiers. On other occasions the queen will reduce her repressive pheromones and her eggs will develop into fertile winged termites which, when the air is moist, will leave the nest in a swarm and the cycle of colony formation begins again.

Most invertebrate decomposers of the lowland forest soil depend on free-living saprophytic fungi and bacteria to break down indigestible plant material into a digestible form before they can play their role. Most types of termites feed on already decomposing litter and dead wood, and their digestion is assisted by symbiotic gut protozoa or bacteria. One subfamily of higher termites, the Macrotermitinae, represented in Sulawesi by *Macrotermes* and *Odontotermes,* maintain a 'garden' of fungus in their nests which releases them from competing for partially-decomposed material outside the nest. The workers of the above genera build complex frameworks, or combs, from their own round faecal pellets. Their main wood and leaf litter, and consequently the combs, look like fragile pieces of moist, spongy rotten wood. These combs are the food of a form of fungus called *Termitomyces* which is unknown outside termite nests. If a nest is cut open, small white dots can sometimes be seen on the combs and these are clumps of asexual fungal spores (Collins 1980c).

This relationship with the fungus confers several advantages on the termites. No animal has the necessary digestive enzymes to digest lignin, a largely inert compound which is the main component of wood and thus one of the major items in a termite diet, and so lignin is excreted unchanged in the termite faeces. The fungus can, however, digest lignin. The termites are also unable to digest cellulose—the main component of plant cell walls—but the fungus produces the necessary enzymes for digesting it. When the termites eat the fungus-permeated comb, these cellulose-digesting enzymes persist in the termite gut, thereby improving the efficiency of its own digestion (Collins 1980c). The fungus releases carbon dioxide as a result of its respiration. This has the advantage to the termite that the carbon-nitrogen ratio in the combs increases roughly four-fold when processed by the fungus (Matsumoto 1976, 1978). The tremendous contribution the fungus makes to the nutrition of the termites has resulted in these termites being unable to live without *Termitomyces*—an example of obligate symbiosis. It has made them so successful that they have been reported to be responsible for the removal of 32% of leaf litter in a forest in Peninsular Malaysia (Matsumoto 1978). The role of termites in the decomposition of fallen trees was studied in the same forest and it was found that the rate of wood removal was higher for small branches (3-6 cm diameter) than for trunk wood (30-50 cm diameter) which is harder and more difficult to cut and remove. In a species of dipterocarp, 81% of the tree's dry weight of small branches was removed in the first 18 months after it fell compared with just 18% of the trunk in the same period (Abe 1978, 1979). Termites thus play a major role in the breakdown and cycling of plant material (Wood 1978).

The percentage of forest litter production consumed by termites in Southeast Asian forests appears to depend on whether the fungus-farming termites are present. For example, in an area where they are present, consumption of organic matter by termites was calculated to be 155-174 g (dry-weight)/m^2/yr, equivalent to about 15% of litter production, whereas at another site the figures calculated were only 7-36 g (dry-weight)/m^2/yr,

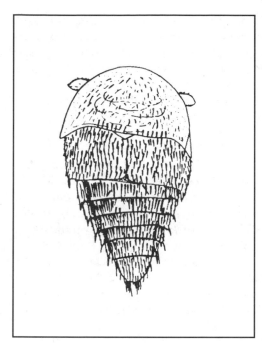

Figure 5.32. *Termitodiscus indonesiensis* beetles from a termite nest near Rantepao.
After Kistner 1984

equivalent to just 0.9%-3.4% of litter production. At the first site, where annual rainfall was 2,000 mm, the fungus-termites were responsible for the consumption of 75% of the total consumed by all the termites together, and at the second site, where about 5,000 mm of rain fell each year, the fungus termites were absent or rare. The difference may be caused by the leaf litter in the wetter area being dominated by free-living decomposer organisms which rob the fungus termites of their advantage (Collins 1983). Thus the decomposition processes should be strikingly different between the drier and wetter forests of Sulawesi.

Various animals live in termite nests, among which are *Termitodiscus* beetles (Staphylinidae) (fig. 5.32). The first species known from Sulawesi, the reddish-brown *T. indonesiensis* with a total length of about 2.5 mm, was found near Rantepao in a nest of the fungus termite *Odontotermes sundaicus*. The beetles feed on fungus in the fungus gardens and their main social adaptation is avoidance of the workers and soldiers of their termite hosts (Kistner 1984). In addition to these specialized beetles some termite nests also have phorid flies in the galleries where they breed in the fras. Ant nests also have aliens in their midst and a peculiar phorid fly was found in one such at Toraut (Disney 1985, 1986a, b, c, in prep.).

The number of termite species known from a given area of lowland forest has not yet been ascertained in Sulawesi, but there are at least 55 species of termite at Pasoh Forest Reserve in Peninsular Malaysia (Abe and

Matsumoto 1978, 1979). For population studies of these, the four most abundant species with conspicuous mound nests, part of which projected above the soil surface, were chosen (Matsumoto 1976, 1978). Approximately 100 nests were found per hectare for three of the species but there were only 15 large nests of *Macrotermes*. The population size of a *Macrotermes* nest was about 88,000, however, at least twice that of the other species.

Forest Floor Community

The forest floor community is taken to include terrestrial animals that spend a major part of their life walking on the forest floor. Among the smallest visible members are certain beetles which are exceptionally successful at utilizing dung, carcasses, rotting wood and fruit. Dung or scarab beetles (Scarabaeoidea) can comprise 50% of the arthropod biomass on the forest floor and they have an enormous diversity. The principle roles they play in the decay of organic matter are in fermentation, burial, parital assimilation of the organic matter, transportation of micro-predators and parasites, and aeration of the soil. In addition, they may be active in dispersal of seeds by rolling seeds into their balls of dung (J. Krikken pers. comm.).

The decomposer animals also include the rarely seen land crabs (p. 523). Only one species *Gecarcoidea lelandii* is known so far from the lowlands of Sulawesi (fig. 5.33) (Turkay 1974) although unidentified species have been caught in Bogani Nani Wartabone and Morowali National Parks. Four species of land crabs were caught in flood-prone alluvial forest in Sarawak where soil invertebrates in general are poorly represented, and near a river one species of crab with a carapace width of 20-35 mm reached the high density of $0.32/m^2$, suggesting it must play a major role in litter breakdown (Collins 1980b).

Beetles and other invertebrates utilizing the shrubs and herbs of the forest floor fall prey to birds and skinks. There are at least 19 species of skinks that live on the floor of lowland forests but how these divide up the available resources is not known. Perhaps the most startling species is *Emoia cyanura* with its iridescent, light-blue tail.

From the point of view of a human, two of the more unpleasant animals in the ground community of lowland forests are mites and leeches. These are locally abundant but rarely seem to occur together at the same place at the same time. Whether this is spatial or temporal (seasonal) separation is not clear. It may be that mites are relatively more active than leeches in dry conditions, or that certain habitats are generally more favourable to them.

Mites are one of the world's most ubiquitous groups of animals, being found near the North and South Poles, in deserts, hot springs, as well as in rather less extreme habitats. A handful of leaf litter from the forest floor can contain hundreds of mites from many species and many are parasitic on forest animals.

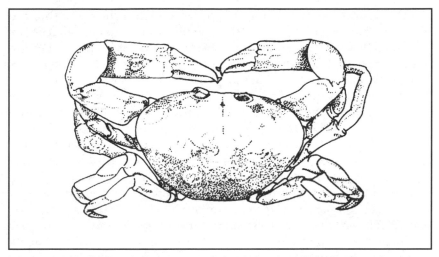

Figure 5.33. *Gecarcoidea lelandii*, the only land crab so far described from Sulawesi.

After Turkay 1974

The land leeches *Haemadipsa* are a harmless but not entirely welcome component of many lowland forests. Until recently, very little was known about their habits, mainly because little money has been expended on research concerning animals whose bite, albeit painful and liable to infection, does not transmit disease. *Haemadipsa* leeches will, if left undisturbed, feed for about 80 minutes and large individuals may feed for longer. The blood ingested exceeds the initial body weight of the leech by a factor of about 6 (although 14 times the initial body weight has been recorded), and after feeding the leeches are somewhat sluggish and only become active again after half the newly-gained weight is lost. Larger leeches tend to lose their weight more quickly but at a slower rate relative to their body weight compared with smaller leeches. A single meal may provide sufficient sustenance for at least three to eight months (Fogden and Proctor 1985). These data help to explain why leeches are so rarely found on wild animals: a one-hour feed every three months is equivalent to spending only 36 seconds feeding over 24 hours. They appear to be common on humans, but possibly only because humans tend to walk along paths. If leeches do not travel far from where they drop after a feed, and if they are repeatedly attracted to a path by warm bodies passing by, it follows that they will have a very non-random distribution.

One or two hours in a forest is long enough to hear the sound of branches, boughs or even a tree falling to the ground. Falling branches will occasionally land on a termite nest, breaking part of it open. When this happens, termite predators such as ants, devil's coachmen beetles (Staphylinidae) and skinks are rather quick to take advantage of the easy meal as are those phorid and muscicapid flies that suck up juices from termites crushed in the accident. At Toraut, females of two species of interesting scuttle flies *Diplonerva* (Phoridae) were observed at such a scene at a nasutine termite nest. The female would arrive at the damaged nest and diligently seek out worker termites which would normally not be found in exposed situations. Having found one she would prod its abdomen with her head which caused it to 'break rank' and turn toward the fly. The fly would quickly run round to the worker's head and run away closely followed by the worker. Should the worker lose interest the prod catch-me-if-you-can routine is repeated. After they have walked a safe distance from the termite nest, the worker is put into a coma by the fly which then lays her egg in the worker's abdomen. The mechanism by which she does this is not yet understood but if she killed the termite, it would putrefy and kill the larvae. The fly probably guards the comatose termite during the larva's development (Disney 1985, 1986b).

There are three main groups of birds on the floor of lowland forest—the pittas, the scrubfowl and maleo, and the ground pigeons. There are three pittas known from Sulawesi[26] and they are among its most beautiful birds. They feed by tossing leaf litter aside with their beaks and snapping up any adult or larval insects that may have been uncovered. They are rarely seen except by patient and quiet observers, but their clear 2-3 note whistles, trilled or slurred, are distinctive. Their nests, in which they lay 4-5 eggs, are constructed in tree stumps, at the base of bushes, among the roots of trees or simply on the ground within some herbaceous cover. In Papua New Guinea certain pittas decrease in abundance during the dry season (Bell 1982b) possibly because abundance of suitable food in the litter drops at that time, or because it is too hard to dig in the soil.

The habits of the maleo have been described elsewhere (p. 155) and its nest sites can be found in lowland forest. Like the nest sites around the coast, the eggs are often plundered in a quite unsustainable manner. For example, an EoS team visiting the site near Hungayono, the enclave within the western end of the Bogani Nani Wartabone National Park, found all of the 22 nest pits dug out. Its close relative, the scrubfowl *Megapodius cumingi* is a brown partridge-like bird, whose most distinctive feature is the bare red skin around the eye. They are shy and generally solitary birds but they congregate for calling displays which are followed by pairing and breeding. Their nesting habits are somewhat similar to those of maleo birds but they lay their eggs in a wider variety of sites and use not only heat from the sun and from volcanic activity, but also from the decomposition of vegetable

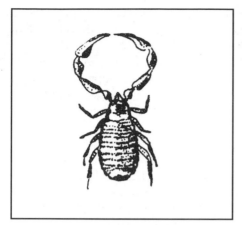

Figure 5.34. Pseudoscorpion *Megachernes grandis* from the fur of a rat.

After Durden 1986

matter which they scrape over the eggs after they have been laid at the bottom of a deep (± 1 m) hole (MacKinnon 1978). As with maleos, they disperse widely through lowland forests when not engaged in breeding activities, avoiding swamps, major disturbance and cultivated areas (Broome et al. 1984).

The ground pigeons *Gallicolumba tristigmata, Chalcophaps indica* and *C. stephani* are shy and little is known of their habits, but they appear to supplement a fruit diet with insects from the forest floor.

The other major group of animals on the forest floor are the rats such as species of *Rattus, Maxomys,* and *Paruromys.* In Tangkoko-Batuangus Reserve it has been estimated that they live at a density of about 20/ha (Anon. 1980). Studies of forest rats in Peninsular Malaysia found that the diameter of the lifetime range of individual rats was 250-500 m (equivalent to 0.05-0.2 ha) and that a 'lifetime' for a rat was an average of only 3-6 months (Harrison 1955, 1958). Some species are not confined to the forest floor but also climb.

The rats of the forest floor also support an interesting assemblage of parasites among their fur. One particular Musschenbroek's rat *Maxomys musschenbroekii* caught in the Toraut forest was infested with no less than 129 mesostigmatid mites, 7 astigmatid mites, 4 ticks, 79 sucking lice and, most interestingly, three pseudoscorpions (fig. 5.34) (Durden 1986). Pseudoscorpions usually live in or on the soil. They were first described from a living mammal when a new genus and species *Chiridiochernes platypalpus*, 3 mm long and morphologically adapted for life in fur, was found on a summit shrew rat *Bunomys penitus* on Mt. Lompobatang in 1969 (Muchmore 1972). This pseudoscorpion, and those from Bogani Nani Wartabone National Park, are incapable of feeding directly on mammals and probably prey on the smaller mites.

Different groups of rat parasites encounter their hosts in different ways. Most fleas and the larger mites are common in rat nests, but sucking lice do not form a reservoir in the nest, while ixodid ticks and chiggers search for their hosts in the forest. These differences in ecology are reflected in the rate of reinfestation for the different groups (table 5.9; fig. 5.35). The figures show a remarkable consistency in the composition of the parasite fauna, indicating that the relative numbers are dictated by conditions on the rat rather than simply by chance.

Parasites are defined as any organism that is intimately associated with, and metabolically dependent upon, a host organism for the completion of its life cycle, and which is typically more or less detrimental to the host (Lincoln et al. 1982). They are to some extent similar to herbivores for these too depend on 'host' plants for survival and cause the plants stress. Even so, there remain a number of subtle differences (table 5.10).

Frogs and toads[27] are conspicuous components of the ground community and they feed on termites, small flies and other invertebrates. When toad tadpoles hatch from pools on the forest floor they are by no means cryptic; instead, they shoal together, indicating perhaps that they are distasteful to predators. At least one of the narrow-mouthed toads *Oreophryne celebensis* is found in the lowlands. Its eggs are laid not in water but in moss on tree trunks, hollow old tree fern trunks, cavities in ant-plant stems, etc., where eggs develop directly into miniature toads without going through a tadpole stage (J. Dring pers. comm.). These and adult frogs are subject to predation by lizards, snakes and bats (p. 548).

The largest Sulawesi mammals are members of the ground community: the anoa *Bubalus depressicornis*, babirusa *Babyrousa babyrousa* and the Sulawesi civet *Macrogalidia musschenbroecki* (p. 532). The anoa is a close

Table 5.9. Numbers of different ectoparasites found on a female Musschenbroek's spiny rat *Maxomys musschenbroekii* caught over a sixteen-day period. Note the similarities in the numbers found compared with initial infestations even after gaps of just two days. This rat had no lice but other specimens of the same species did.

Day	Fleas	Ixodid ticks	Meso-stigmatid mites	Trombiculid chigger mites	Listrophorid fur mites	Lice
1	2	4	27	0	4	0
6	2	0	33	1	1	0
8	3	0	23	0	3	0
10	5	0	17	0	2	0
16	0	2	11	0	3	0

After Durden in press

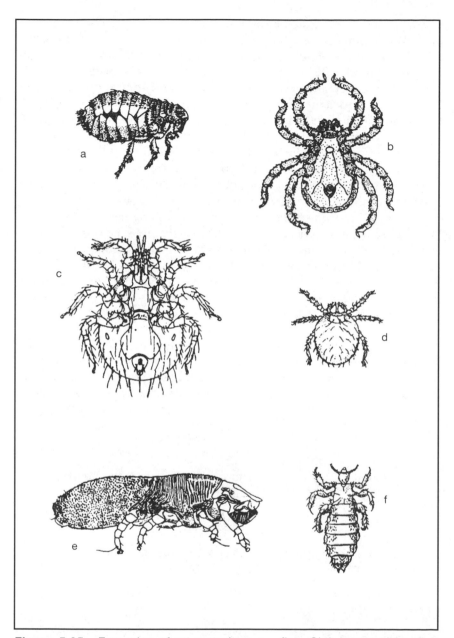

Figure 5.35. Examples of rat parasites. **a** - flea, Siphonaptera, **b** - tick *Haemaphysalis,* Ixodidae, **c** - mite, *Laelaps,* Mesostigmatidae; **d** - chigger mite, *Leptotrombidium,* Trombiculidae; **e** - mite *Listrophorus*, Listrophoridae, **f** - louse, Anoplura.

After a - Barnes 1968; b - Hoogstraal et al. 1963; c - Evans et al. 1961; e - Fain and Hyland 1974; f - Barnes 1968

relative of the domesticated water buffalo *B. bubalis* and the small buffalo of the Philippines *B. mindoroensis*. The lowland anoa is relatively large (nearly 1 m at the shoulder), has a relatively long tail, white legs and rugged horns, whereas the mountain anoa *B. quarlesi* is smaller (about 75 cm at the shoulder), has a shorter tail and smooth conical horns (Groves 1969). Their ferocity and unpredictable behaviour is attested to by villagers and scientists who have spent long periods in the forest, and some of them bear the scars made by the sharp, stout horns. The anoa used to be caught by Toraja people in attempts to breed it for its meat but its aggressiveness, even after several years in captivity, prevented it from being used directly as a domestic animal. Considering the close relationship between anoa and water buffalo, it is conceivable that they would interbreed and produce fertile offspring which might have some potential as stock animal. Such a cross has not yet been attempted. Of all the five species of wild cattle in Southeast Asia, the anoa are exceptional in being the only ones whose major habitat is undisturbed forest. Their food includes a variety of fruits, leaves of shrubs and young trees, grasses and ferns (table 5.11). Dung samples collected by an EoS team are being analysed but the results are not yet available. In and around ladangs, plants with latex, such as cassava, are said to be favoured.

Anoa appear to prefer to feed in well-drained, rugged areas where the ground is not so thick with herbaceous plants but with a variety of available food species. They generally lie down to ruminate along dry ridge tops in forest which is relatively open compared with wetter slopes (Wirawan 1981).

Table 5.10. Major differences between ectoparasites of mammals and plants.

Mammalian ectoparasites	Plant ectoparasites
Subject to little predation or parasitization while on the host.	Subject to severe predation and parasitization while on the host.
Chemical composition of blood varies little between species of mammals.	Chemical composition of tissue eaten varies widely between plant species.
Host dies if more than a small fraction of living material is removed.	Host can survive removal of relatively large fraction of living tissue.
Movement of parasites between hosts assisted by movements of host.	Host plays only a passive role in movement of parasites between hosts.
Commonly occur in large numbers on host.	Rarely occur in large numbers on host.
Spend most or all of life on host.	Spend relatively little time on host.

After Janzen 1985a

Large herbivores throughout the world seem to experience a shortage of sodium in their diets: this element is scarce in green plants (with the exception of seagrasses and other marine species) but is essential to mammals. Carnivores do not face the same problem because flesh comprises about 0.1% sodium. Many herbivores, including deer, pigs and feral water buffaloes, make up for the shortage by licking rocks or soil that are relatively high in sodium. Wildlife surveys in Central Sulawesi failed to find any such licks frequented by anoa, although anoa themselves were common enough. Licks that were found appeared to be frequented by deer and pigs rather than anoa. It seems, therefore, that anoa either did not need to make up any salt deficiency or obtained salts from other sources. For example, anoa have been seen to drink seawater. Their normal diet appears to comprise growing tips from a wide variety of plants although signs of intensive feeding in a single area were rarely seen. Anoa tracks are frequently seen, however, around springs of water (Anon. 1979; Wirawan 1981) and it might be that these supply mineral needs. Water analysis showed that although some springs had high sodium concentrations (Wirawan 1981), their mineral content in general was barely different from that of adjacent rivers which were apparently not used intensively. The frequent use of the springs is seen rather as a means by which anoa, which live mainly solitary lives at low densities, can judge the size of the

Table 5.11. Plants eaten by anoa in Lore Lindu National Park.

Species of forest gaps	Species of the forest floor
Curculigo latifolia (Amry.)	Cyrtandra leuconeura (Gesn.)
Ageratum conyzoides (Comp.)	Dysoxylum sp. (Meli.)
Axonopus compressus (Gram.)	Levieria montana (Moni.)
Setaria sp. (Gram.)	Lasianthus capitatus (Rubi.)
Melochia umbellata (Ster.)	L. stercorarius
Bochmeria virgata (Urti.)	L. spp.
Fern:	Randia pulcherima (Rubi.)
Selaginella sp.	Cypolophus ellipticus (Urti.)
	Elatostema sp. (Urti.)
	Ferns:
	Asplenium nidus
	Blechum capense
	Cyclosorus sp.
	Dennstaedtia ampla
	Diplazium sorzogonense
	Marattia sambucina
	Microlepia sp.
	Microsorium sp.

After Wirawan 1981

local population; it also helps bulls to trace oestrous females (Watling in press). Hunters take advantage of this behaviour, although anoa hunting has been prohibited since 1931.

Babirusa are primarily nocturnal although they are occasionally seen in daylight hours, too. Observations suggest that babirusa do not dig in soil to eat roots as do most pigs but rather eat fruit and break open rotten wood to obtain beetle larvae. The major fruit eaten by babirusa in the west of the Bogani Nani Wartabone National Park is the potentially poisonous pangi *Pangium edule* (Flac.) [28] (fig. 5.36). A major fruit eaten in certain coastal regions is coconut but the babirusa is not regarded as a coconut pest because it takes only sprouted coconuts or broken pieces of flesh. They can inflict great damage, however, by rooting in ladangs; it is for this reason that people set traps for them and kill them. In Islamic areas the meat is not eaten but given to dogs.

Only one study of the babirusa has been conducted and that was on Pangempan Island in the Togian Islands in 1978 and in 1979-80 (fig. 5.37) (Selmier 1983). This small island (22.5 ha) was chosen for the study after surveys were made on all the Togian Islands on which babirusa were thought to live, although it had only one small herd comprising an adult female, two adult males and two piglets. This composition was maintained over three years, suggesting that the young babirusa swam to the mainland or died. They are said to be able to swim strongly.

Babirusa have no teeth when born and young males initially develop canines in the normal way. Soon, however, the sockets of the upper pair turn upside down such that these tusks grow up through the skin (Groves 1985), and curl around towards, but not into, the skull.

The function of the curly tusks of babirusa has intrigued many people. Wallace wrote:

> It is difficult to understand what can be the use of the extraordinary horn-like tusks. Some of the old writers supposed that they served as hooks, by which the creature could rest its head on a branch. But the way in which they usually diverge just over and in front of the eye suggested the more probable idea, that they serve to guard these organs from thorns and spines while hunting fallen fruit among the tangled thickets of rattans and other spiny plants. Even this, however, is not satisfactory, for the female, who must seek her food in the same way, does not possess them. I should be inclined to believe rather that these tusks were once useful, and were then worn down as fast as they grew; but that changed conditions of life have rendered them unnecessary, and they now develop into a monstrous form.—WALLACE 1869

The question of function has recently been addressed again as a result of examining the patterns of wear on the tusks of a sample of skulls (MacKinnon 1981). On the skulls from Sulawesi, all the lower tusks had been sharpened and half of them were chipped or broken. All the upper tusks on skulls of adult animals were chipped or broken and most of these had

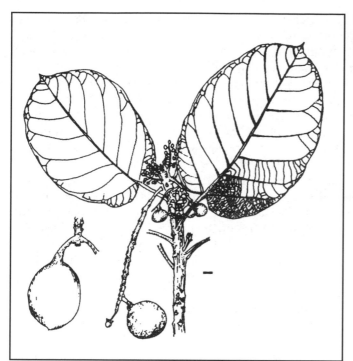

Figure 5.36. *Pangium edule.* Scale bar indicates 1 cm.
After Meijer 1974

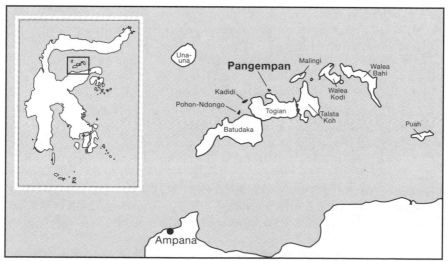

Figure 5.37. The Togian Islands to show Pangempan Island, site of the only long-term study of babirusa.

deep grooves and scratches beneath them, probably caused by collisions with hard, sharp objects. In most pigs, the sharpening of the lower tusks is achieved by the upper tusks growing across them and rubbing against them as the mandible is moved, but in babirusa the sharpening appears to be caused by rubbing against small trees, as anoa sharpen their horns and civets sharpen their claws. It must take male babirusa a considerable time to sharpen their tusks, indicating that these sharp weapons are extremely important for them. In addition, the fact that the upper tusks suffer so much damage indicates that they are highly functional, rather than simply decorative to attract mates. Since females do not have large tusks, it is reasonable to suppose that the males use them in fights between males over rival claims for females or territory. The upper tusks could not be used for butting opponents, first because the height of the lower tusks would require the head to be held very low to butt and, second, because the tusk tips would hit against the frontal bones of the skull which do not appear to be thickened as they are in the giant forest hog of Africa which does fight by butting.

When male babirusa fight they stand side by side and push each other with their shoulders. They then stand facing each other and rear up on their hind legs, jabbing their heads upwards to gore each other with the lower tusks. At this point the upper tusk can be hooked over the sharp lower tusk of the opponent, effectively disarming it, while the lower tusk of the advantaged male can be used to jab or cut the other's face, neck or throat. The eyes are to some extent protected from damage by the upper tusks which could deflect strikes by the lower tusk away from the forehead (fig. 5.38). Interestingly, since the form of the upper tusks is that of a logarithmic spiral, a form frequently found in nature, the tip of the tusk is always pointing inside the curve and so it is very difficult for a captured tusk to slip out once caught.

The only Sulawesi animals that could prey on babirusa, and then only on the young, are large pythons *Python reticulatus* and *P. molurus* and the endemic civet, but probably neither has a very potent effect on the population. This is rather different from the pressures on most pigs which have to contend with predation from wild cats, dogs and bears. Perhaps in response to this, the litter size in babirusa has no need to be large and the tusks which are used to defend the piglets in most female pigs, including the species endemic to Sulawesi, are poorly developed in female babirusa (MacKinnon 1981). The endemic pig *Sus celebensis* which arrived in Sulawesi relatively recently has a litter size of two to eight piglets with an average of five (Anon. 1983), compared with one or two slow-developing young for babirusa.

One of the reasons the babirusa is no rarer that it is, is because it is not eaten by those who adhere to Islam. The law for Moslems (and for Jews) states that animals with cloven hoofs that do not chew the cud may not be

Figure 5.38. Male babirusa locked in hypothetical combat. The upper tusk of the male babirusa on the left is used to hook over the sharp lower tusk of his opponent, thereby disarming him. The advantaged male can now jab the throat of his opponent. The long-term field study of babirusa by Lynn Clayton has never observed fighting in this way.

eaten. All other pigs have simple stomachs and do not chew the cud, but one of the many peculiarities of the babirusa is its complex stomach. Its digestive apparatus suggests that it may be able to break down cellulose with the aid of bacteria, but this has yet to be confirmed. Suggestions that babirusa actually do chew the cud are founded in wishful thinking rather that fact. It was hoped by some that babirusa would become a 'halal' or kosher pig, thereby proving its usefulness and ensuring its continued survival, but it is more likely that if religious prohibitions against eating its meat were ever lifted, then the wild populations would suffer dramatically,

Lower and Upper Canopy Communities

For the purposes of this section, the lower and upper canopy refer to the oligophotic and euphotic zones of a forest respectively (p. 348), but the lower canopy does not include the forest floor or short herbs and shrubs.

Since there are considerable differences in microclimate through the canopy
(p. 348), one would expect to find differences in the animal communities.

Traps were set up at four levels through the canopy of lowland alluvial
forest at Morowali National Park to investigate differences in insect abundance through the canopy as part of Operation Drake. Low-power ultra-
violet lamps were placed in the traps and the catches were compared. It was
clear that some of the insect groups caught were far more abundant in the
upper canopy than the lower (fig. 5.39), but the mayflies were most
common in the middle layers (Sutton 1983).

Trapping, using the same techniques and equipment, has also been
performed in Zaire, Panama, Papua New Guinea and Brunei, and the
concentration of insects in the upper canopy is more marked in Sulawesi
than in any of the other sites. The marked gradients observed at Morowali
seem to be a feature of flat sites, particularly where tree height is relatively
uniform. Insects which are dependent on leaves for some aspect of their
life would be expected to be concentrated in the tree crown and one
would also expect there to be clusters of insects focused in the different
crowns. Thus traps set to attract insects in rugged country or in a forest with
a complex canopy would not produce such a clear gradient (Sutton 1983).

In forests with marked dry seasons, arthropods in the lower canopy tend
to be most common in the wet season, but where the dry season is relatively
mild, the structure of the forest buffers the drying effects and the greater
frequency of sun flecks reaching the forest floor increases the productivity
of plants in this zone with a consequent rise in arthropod abundance.
This was the case in an everwet forest in Peninsular Malaysia, but the four-
fold variation in abundance was not correlated with the amount of rainfall
or flower, fruit or leaf production. Ants were the most abundant of these
arthropods, followed by beetles, wasps and bees. The relatively low level of
variation is probably due to the dependence of these groups on resources
other than the immediate products of plant productivity (Wong in press a).
Since foliage is available continuously in the lower canopy, however, caterpillars and their predators are not so uncommon (Wong in press b).

The relative abundance of different groups of birds also differs through
the canopy. In Queensland forest in Australia, the major factor in niche
separation of birds is differential use of the vertical strata. This is true
both within and between the different guilds.[29] Some groups of species had
similar vertical ranges and their niches were determined by differential utilization of food resources, different feeding behaviour and changes in
preferred vertical strata through the day (Frith 1984). There are also differences through the day with many species of bird feeding high in the
canopy during the early morning and moving down through the forest as
the temperature rises. In some cases, even males and females of a single
species have different preferred height strata (Bell 1983c). Among Sulawesi
birds one of the most striking divisions of canopy use is found in the

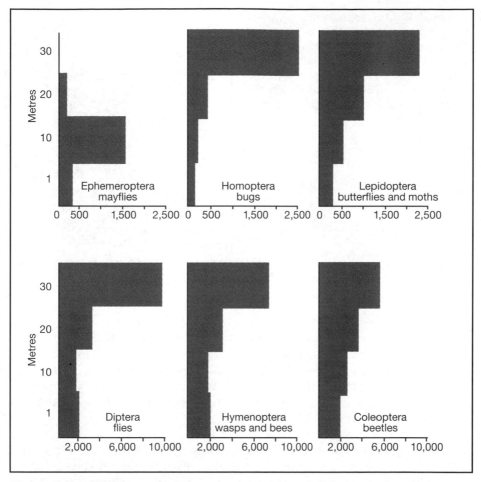

Figure 5.39. Abundance of six insect groups at four height levels through lowland forest in Morowali National Park. Note that the frequency scale on the top and bottom rows are different.

Data from Sutton 1983

hornbills. The large red-knobbed hornbill *Rhyticeros cassidix* is found in the upper canopy and the garrulous dwarf hornbill *Penelopides exarhatus* below (Watling 1983).

Seven species of hornbills were studied in an area of East Kalimantan and a wide range of social systems was found (Leighton 1982). Extrapolating from the results of that study, the large Sulawesi hornbill is probably nomadic, occasionally forming large, loose flocks of up to 50 individuals, but breeding[30] in monogamous pairs away from others of the species (fig.

Figure 5.40. Adult female hornbill in her tree hole nest, the entrance of which has been reduced to a mere slit using mud brought by the male. Until the young are able to fly, both the female and the young are entirely dependent on the male (and 'helpers' in the dwarf hornbill) for food.

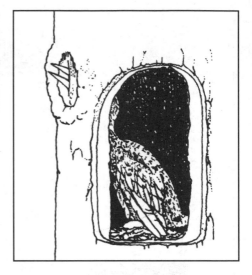

5.40). The smaller hornbill is found in small groups at all times and is probably territorial and breeds monogamously but remains in groups. In East Kalimantan such birds live at a density[31] of about 5/km? It is quite likely that young birds are fed not only by their parents, but also by other 'helper', non-breeding members of the group. Such helpers are known for many kinds of birds (Brown 1978).

Colonies of stingless sweat bees nest in cavities in moderately large trees (about 1 m diameter) with entrances at 0.5-4 m above the forest floor. In Costa Rica and Panama, the most common species have nest densities in the order of 20 nests/km² (Johnson and Hubbell 1984). Nests are sometimes built in cavities in the base of the trunk, but can also be found underground around the tree roots. The nest mouth is sometimes no more than an oval opening with black wax around its edge, but it can develop into a brittle horn perpendicular to the trunk. These bees are catholic in their choice of pollen and nectar, and on a single day twenty species of flower may be exploited. Despite this, however, a single bee tends to concentrate on a single plant species (Johnson 1983). The bees are small, less than 1 cm in length, but they can range up to 2 km from their nest and still return successfully (Roubik and Aluja 1983a, b). The workers leave the colony at dawn, and start returning in the middle of the morning. The number outside continues to decline until late afternoon when a minor burst of activity occurs. Foraging[32] has ceased by dusk.

Niche differentiation has been studied in a pair of sweat bee species which were more or less the same size, collected pollen in the same way from the same type of flowers, and lived in the same area. As explained elsewhere (p. 230) the manner in which these species are able to coexist would therefore be of considerable interest. It was found that one of the species foraged in large groups in areas with a high density of a particular flower, whereas the other foraged in smaller groups where the flower was comparatively scarce. The second species lost in aggressive encounters with the first species (Johnson and Hubbell 1975; Hubbell and Johnson 1977).

The most remarkable member of the lower canopy community is the large-eyed jumping tarsier *Tarsius spectrum*, one of the world's smallest primates, with a head-and-body length of just 10 cm, a tail 20 cm long, and a weight of 100 g. The tarsier social unit comprises an adult pair, which forms a stable, long-term monogamous relationship, together with their one or two immature offspring. They all sleep together in a tree hole within a more or less permanent territory which is defended from other tarsiers by song. Every morning, just before these nocturnal animals retire to their nest, the entire family sings a complicated territorial call. The male generally initiates the song with a regular series of squeaks, and the female joins in with a descending series of squeals which then rise in pitch to a fast climax. These calls declare to surrounding groups that the mated pair is present, is fit and claims the immediate surrounding area as its living space. When neighbouring tarsier families come close to the territory the resident group will chase the intruders away, frequently shrieking loudly as they do so.

The calls made by tarsiers are excellent clues for finding and plotting sleeping sites and so observations, at least in the early morning and evening, are relatively easy. The sleeping sites include plant thickets, tangles of vines or ferns, tree holes (with more than one exit), and the criss-cross roots of strangling figs (p. 358). Tarsiers have been found in a wide range of habitat types from urban areas, secondary growth, mangrove, lowland forest, riverine forest and montane forests, with densities ranging from 3-10/ha (MacKinnon and MacKinnon 1980b).

After leaving the nest hole, typically 20 minutes after sunset, the family spends only a few minutes interacting before the members leap off in different directions. In a single night a tarsier may leap between trees for a total of 1 km. Virtually their entire diet comprises insects; other types of animals such as crabs, shrews, small rats and frogs seen close to the tarsiers during the detailed study at Tangkoko-Batuangus, attracted little interest from the tarsiers. Most hunting occurs in the lowest 1.5 m of the forest although it also occurs on the ground and up to 9 m above the forest floor (fig. 5.41).

The home range of tarsier families at Tangkoko-Batuangus is about 1 ha, an area determined by the furthest points of ranging of the adult male who is the most active in terms of calling and marking branches with urine. Within this area, the animals regularly use the same routes. The

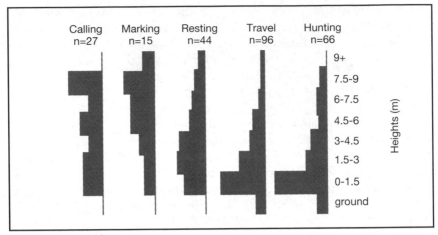

Figure 5.41. Differential use of the canopy by tarsiers for five major activities.
After MacKinnon and MacKinnon 1980b

home ranges of different families overlap to some extent but the sleeping sites are generally in areas used exclusively by the family (fig. 5.42).

The social organization of tarsiers is very similar to that of gibbons which have been studied intensively in various parts of western Indonesia. The tarsier call has considerable similarities with the structure and function of the 'great call' of gibbons, and the song structures of both vary geographically. Similarities between gibbons and tarsiers are also found in their specialized (though different) forms of locomotion, their active nature, their longevity (at least 25 and 10 years for gibbons and tarsiers respectively), their specialized diets composed of foods which are rich in nutrients and reliable, but very widely and thinly spread, and the stable nature of their environment. Their monogamous, territorial lifestyle confers advantages such as:

- minimizing reproductive wastage because only one young is born at a time and can therefore be protected and cared for to the greatest extent possible (i.e., they are K-selected [p.345]);
- minimizing time spent in conflict with neighbours because each group knows the boundaries of the area it can peacefully live within;
- minimizing time spent in searching for food and its food sources because the relatively small home range and its food sources are known well; and
- minimizing time spent competing for mates because the pair bond is more or less permanent, and is maintained outside the breeding periods by complex, coordinated duetting.

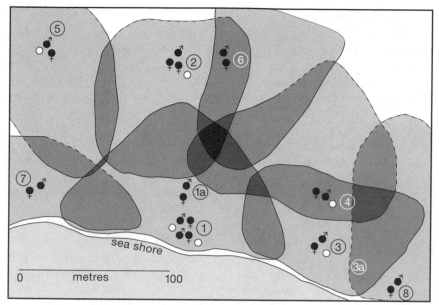

Figure 5.42. Ranges and sleeping trees of tarsier families near the shore in Tangkoko-Batuangus Reserve. White circles represent sleeping trees.
After MacKinnon and MacKinnon 1980b

There are disadvantages too, however, such as the inability to exploit rich food sources outside the home range, the small group size which cannot be exceeded without reducing the efficiency of finding food, and extreme niche specializations (MacKinnon and MacKinnon 1984).

A human observer restricted to walking on the forest floor gets a biased view of the nature of a forest. The opportunity to either climb to the upper canopy using rope slings and mountaineering ascending/-descending equipment (Perry 1978; Perry and Williams 1981; Whitacre 1981), or to walk along an aerial walkway (Mitchell 1982) brings a completely new perspective on forest life.

An aerial walkway in two spans (41 m and 70 m long) about 30 m above the forest floor was built in Morowali National Park as part of Operation Drake, and was used for a number of studies, one of which was on insect abundance (p. 425) and another on the composition and abundance of the fruit bat fauna.[33] When dusk approaches an observer standing on the forest floor frequently sees a large number of small insectivorous bats flitting erratically around the lower canopy, but rarely sees fruit bats, which fly in a much straighter, more determined fashion. Nets set on the aerial walkway and on the forest floor below revealed that fruit bats were 50 times more common above the canopy than below it, and three times more common than in any open habitat, except over water, where bats frequently come to drink water while flying along. The five most common species of fruit bats in the canopy comprised 95% of the total catch and this diversity was considerably higher than in any other habitat or even in all the other habitats of Morowali National Park combined. Netting of bats in upper montane forest showed that the three most-common species, (90% of individuals), were the same as those representing the 5% of rare species at the lowland site. Similar work was conducted in two forests in Papua New Guinea and it was found that the fruit bat fauna over the forest growing on richer soils was consistently more diverse than that over forest on poorer soils. Interestingly, the testes of adult males in one particular species were larger in those individuals caught in the open than in those caught over forest (Gaskell 1984). This difference in testis size suggests that there is some difference in behaviour between those that are reproductively active and those that are not, and that the open areas are chosen specifically, perhaps for some form of display.

Fruit bats rely more on vision than on smell when feeding, and a recent study of bat eye anatomy revealed that not only are insectivorous and fruit bats extremely different in their visual apparatus, but also that fruit bats' eyes are extremely similar to those of primates in their high-acuity binocular vision. This is a feature which, until recently, was believed to be unique to primates. This discovery encouraged further investigation and it was concluded that fruit bats are closer in many respects to primates than they are to insectivorous bats. Indeed, it has been suggested that insectivorous bats evolved from flying shrew-like insectivores in the late Cretaceous (70 Ma ago), but that later in the Tertiary (60 Ma ago) an early line of primates developed the ability to glide and then to fly. Indeed, it now seems that to exclude fruit bats from the primate order requires the redefinition of primates to include the feature that they do not fly (Pettigrew 1986a, b). It is of some interest that in the first classification of mammals by Carl von Linné (p. 669), bats were grouped together with the primates.

Flying foxes (large fruit bats) tend to roost in camps with others of their species and hundreds can be seen at a single site. This may, however, be a mere shadow of the numbers that roosted together before man began hunting. For example, a single camp extending 1 km x 7 km was reported from 50 years ago in Australia that may have contained about 30 million animals (Pierson 1984).

The majority of fruit bats feed predominantly on a range of forest and garden fruit[34] (Marshall 1985) but they are extremely selective concerning its state of ripeness. Even slightly under-ripe fruit, of a ripeness enjoyed by humans, is often rejected and for this reason the claims that fruit bats are pests of orchards are unfounded. In most cases, the fruit eaten by bats is too

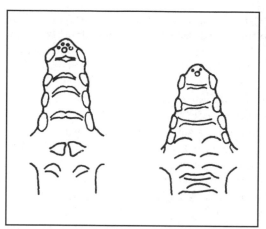

Figure 5.43. Upper palates of fruit bats showing the ridges against which fruit pulp is rubbed to extract the juice.

ripe to be picked for sale. Fruit bats will normally pick a fruit and then eat it, either while hanging in the fruiting tree, or in a regular feeding roost some distance away. These bats ingest very little other than fruit juice and pulp using their muscular tongue to rub the food across horny ridges in the roof of the mouth (fig. 5.43). If the fruit is too large to be chewed all at once, the bat will hang from one foot using the other to hold the food against its chest. Fibrous pulp and most seeds are spat out; only the smaller seeds are ingested, but these pass rapidly through the gut unharmed and are voided after only about 20 minutes at which time the bat hangs from its thumbs with its head above its feet. It is generally advantageous to the plant to have its seeds moved some distance away from the parent (p. 379), and in the case of fig seeds, germination will occur only if they have passed through an animal's gut. Fruit bats therefore serve trees not just by taking the seeds away to areas of new opportunity (or death), but also by removing pulp in which eggs or young larvae of potential seed predators may be developing before burrowing into the seed (Janzen 1982).

Fruit bats in Africa and Asia may have to ingest up to 2.5 times their body mass in fruits in a single night, although more commonly it is 1.5 times. This high intake seems to be due to their need to obtain sufficient protein because they do not include insects in their diet.[35] Most fruits contain less than 5% protein (dry weight) (table 5.12) (and are thus amongst the most protein-poor plant tissues), and because of this most frugivorous birds supplement their diets with insects, seeds or flesh. A small number of fruit have about 6-7% of their wet weight as protein and birds specializing on these do not need to seek protein from other sources (Thomas 1984).

One fruit which is eaten by bats is notably protein-rich, and is produced by the small introduced tree *Piper aduncum* (Pipe.) (p. 397) of forest edges and gaps, particularly in North Sulawesi. A similar species of *Piper* has

been found to contain sufficient nitrogen to meet not just normal maintenance metabolism, but also the needs of lactating female fruit bats. It is
not surprising, then, that fruit bats, especially lactating females, show a preference for *Piper* fruit to relieve them from the need to seek large quantities
of protein-poor fruit (Herbst 1986).

Fruits eaten by bats tend to be soft, juicy, somewhat musty or rancid and
positioned such that they can be easily taken by a bat. The units of food
taken away can be quite large but it was a surprise when a Sulawesi rousette
bat *Rousettus celebensis* was caught in a mist nest on the canopy walkway in
Morowali National Park, together with a large seed from the tree *Gonystylus
macrophyllus* (Thym.). These seeds are 'naked' in that they have almost no
aril, exocarp or appendages, they hang from the ripe fruit by 'ropes'
making their harvest easy, and are quite large (3.6 cm x 2.8 cm x 2.1 cm).
This is most unlike typical bat fruits and the possibility that it was a chance
capture was examined. A total of 297 seeds were gathered from the forest
floor and 60% of these bore marks identical in size and distance to the
marks made by the teeth of *R. celebensis*. If the bat flies to a regular feeding
roost and then drops the seed, germination may occur but very few
dropped seeds would survive the intense competition in the limited area
around the feeding roost. It may be that its awkward size and shape may
result in a high rate of fumbling by the bats and therefore a relatively
even spread of seeds around the parent plant (fig. 5.44). It seems as though
it must be the seeds which are eaten, but they taste very bitter to humans
and may contain either or both of saponins and alkaloids. Feeding trials
with captive bats and investigations of the *R. celebensis* digestive system
might provide some explanations. Interestingly, the seeds of *G. macrophyllus* from other parts of its range (most of Indonesia) are of different
shapes and sizes. This may indicate morphological adaptations in response
to different sets of dispersers (Kevan and Gaskell 1986).

Table 5.12. Percentage of dry weight of protein, fat and carbohydrate for common
garden fruits eaten by bats and for *Piper aduncum*. Ash and fibre contents would
bring row totals to 100%.

	Protein	Fat	Carbohydrate
Banana (Lady fingers)	3.8	0.5	93.8
Rambutan	4.6	0.5	92.8
Papaya	3.6	+	91.7
Mango (mangga golek)	2.8	1.1	93.8
Piper aduncum	10.0	2.2	74.4

From data supplied by the Institute for Research and Development of Agro-based Industry, Bogor, and results of
analyses conducted by that institute of samples collected by an EoS team.

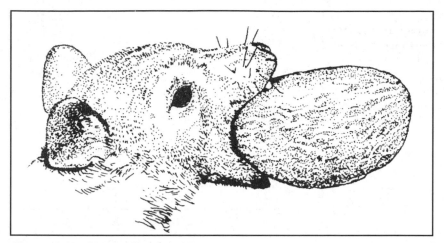

Figure 5.44. Sulawesi rousette *Rousettus celebensis* and the awkward fruit of *Gonystylus macrophyllus.*
After Kevan and Gaskell 1986

The aerial walkway at Morowali National Park also facilitated studies of the pollination of forest trees. One species, *Syzygium lineatum,* was studied in depth. The trees flowered and fruited more or less simultaneously and their small, abundant, nectar-rich white flowers predispose them to a wide range of generalist pollinators, a common feature of tropical forest trees. The numbers of flower visitors (short-tongued wasps and flies), and fruit dispersers observed, however, were few. This is not particularly surprising in a species-poor forest, but it would be interesting to observe how the tree is serviced in other parts of its range. The observations also showed a clear pattern of activity among the flower visitors with numbers increasing until about 1400 hrs., after which numbers fell. Very few visits were recorded after 1600 hrs. (fig. 5.45) (Lack and Kevan 1984).

The abundance and activity of plant-sucking insects (Hemiptera), comprising the plant bugs (Homoptera) and the leafhoppers, frog-hoppers, cicadas, aphids and scale insects (Heteroptera), in the upper canopy in the lowland forest at Morowali also varied through the canopy but this appeared to be related to differences in microclimate. The more rain that fell, the more bugs were caught and this was more noticeable within the upper canopy than below it. In addition, with increasing rain the average (modal) size of these insects increased, and the species of these insects flying on wet nights differed from that found on dry nights. Interestingly, fewer bugs were caught on moonlit nights. When the data were examined, it was clear that it was the actual occurrence of rain, not

Figure 5.45.
Abundance of insects observed feeding from marked flowers of *Syzygium lineatum* in Morowali National Park.

After Lack and Kevan 1984

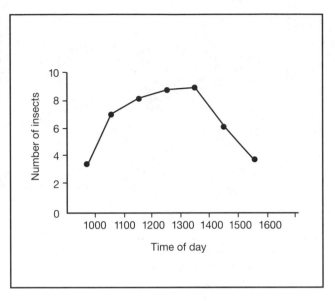

just high humidity, and of darkness that resulted in large numbers of flying hemipterans being caught. The primary predators on these insects are insectivorous bats for which raindrops would greatly reduce the efficiency of echo-location[36] because both raindrops and many hemipterans are about 3 mm in diameter and length respectively. With sound having a velocity of about 300 m/s in very humid air,[37] a sound with a 3 mm wavelength has a frequency of 100 kHz and this is in the upper range of bat calls (Fenton 1983; Hill and Smith 1984). The heavier the rain, and the denser the rain drops, make conditions for larger flying insects 'safe' from bat predation. On dry nights only the smallest (<3 mm) hemipterans fly but even these may concentrate their activity in microhabitats which bats would find hard to enter, such as among leaves and fine branches. The above explanation is elegant, but the abundance of flying hemipterans during rain might just as well be a response to low bat activity at such times because of the possibility of losing too much heat from the water covering and dropping off the wings and body (Rees 1983).

Closely-related species living in the same area frequently have different body sizes. This allows them to eat different types of food, and to use different elements of the habitat available. Large birds can take large fruit but smaller birds have access to more types of fruit by virtue of being able to perch on the more slender branches. Some years ago it was proposed that, for birds, the weight ratio between sympatric members of the same genus should be at least 1:1.2, or at least that bill size should differ by this ratio

(Hutchinson 1959). A study of two genera of New Guinea pigeons showed that the ratio of weights between one bird and the next smallest closest relative was on average 1.90; it was never less than 1.33 and never more than 2.73. Relatives with a weight ratio of less than 1.33 were too similar to live in the same area and no species pair with a weight ratio in excess of 2.73 was found, presumably because another species of intermediate size could coexist with both the larger and smaller species (Diamond 1973).

An attempt was made to determine whether the same ratios were valid for two groups of Sulawesi upper canopy birds: the parrots and fruit-eating pigeons. A simple listing of the pigeons (table 5.13) shows that most of the weight[38] ratios are in fact below the 1.33 found to be the minimum in New Guinea. Closer examination of the likelihood of overlapping niches shows that not all species should be compared in this analysis. For

Table 5.13. Approximate wing length and weight for fruit-eating pigeons of mainland Sulawesi with weight ratios for successive species.

	Wing length (mm)	Weight (g)	Weight ratios		
			All species	Lowland species	Montane species
White-bellied imperial pigeon					
Ducula forsteni (L)	260	510			
Silvery imperial pigeon			1.24		
D. luctuosa	240	410			
Pink-headed imperial pigeon			1.04	1.40	
D. rosacea	237	395			
Green imperial pigeon			1.08		
D. aenea (L)	230	365			
Pied imperial pigeon			1.24		
D. bicolor	230	365			
Grey-headed imperial pigeon			1.67		
D. radiata (M)	215	295		2.57	
Red-eared fruit-dove			1.25		1.67
Ptilinopus fischeri (M)	180	177			
Buff-bellied fruit-dove			1.35		
P. subgularis (L)	165	142			
Pink-necked green pigeon			1.78	1.35	
Treron vernans (L)	147	105			2.21
Grey-faced green pigeon			1.00	1.00	
T. griseicauda (L)	?147	?105			
Superb fruit-dove			1.31		
P. superbus (M)	135	80		2.33	
Black-naped fruit-dove			1.78		1.78
P. melanospila (L,M)	116	45			

L = Lowland species   M = Montane species
Data from Meyer and Wigglesworth 1898; Goodwin 1970; Watling 1983

example, the pink-headed imperial pigeon *Ducula rosacea* differs little in size from the species immediately larger and smaller than itself, but it is found only on the Tukang Besi Islands and the islands south of South Sulawesi and cannot really be considered a mainland species. Also, amongst the remaining species, some are predominantly montane while others inhabit lowland forests. The pied imperial pigeon *Ducula bicolor* is found primarily on small islands (and occasionally in other coastal habitats), whilst the rather similar silvery imperial pigeon *D. luctuosa* is found primarily in open woodland, cultivated areas and forest edge habitats. Thus, if the weight ratios of the four species from lowland forests, and the four from montane forests are calculated, it is found that they fall within the figures mentioned above with the exception of the two green pigeons, the grey-faced *Treron griseicauda* and the pink-necked *T. vernans*. These species are commonly seen in and around Lore Lindu National Park, and are sometimes seen feeding together in the same tree, and the ecological separation of the two is by no means clear (Watling 1983). It may be that the heavier bill of *T. griseicauda* is an adaptation to certain fruit that cannot be eaten by *T. vernans*.

A similar pattern to that found among the pigeons is found among the parrots (table 5.14). Only two species, the golden-mantled *Prioniturus platurus* and the yellow-and-green lorikeet *Trichoglossus flavoviridis*, are found in montane forest, and the two major habitats utilized are lowland forest and open areas such as secondary growth and forest edges. If these are taken into account then, once again, the weight ratios are within the range expected. As with the pigeons, one pair remains an enigma, however, and that is the two endemic racquet-tails. The red-spot racquet-tail *P. flavicans* is found only in the northern peninsula and is especially common in the Bogani Nani Wartabone National Park, whereas the golden-mantled *P. platurus* occurs throughout Sulawesi. There is virtually no difference in size and it may be that *P. flavicans* is able to hold its own against *P. platurus* but is less well adapted to dispersal. *P. flavicans* is more common than *P. platurus* in Bogani Nani Wartabone National Park (Rodenburg and Palete 1981), and Tangkoko-Batuangus Reserve (Anon. 1980). *P. platurus* probably arose in southern Sulawesi and extended its range northwards, and it would be interesting to study the ecological interactions of the two species.

The applicability of the size ratios does not seem to extend to all bird genera, and particularly not to those which are insectivorous (Simberloff and Boecklen 1981; Bell 1984). In contrast to the pigeons and parrots whose fruit food is concentrated in the upper canopy, the insectivorous birds have a wide range of options available for their feeding niche because their prey are available from the uppermost branches to the forest floor. Thus, many similar-sized species of the same insectivorous genus can coexist by specializing in taking the same or similar food as other species, but in places or by methods unavailable to the others (Bell 1984).

Mixed flocks of mainly insectivorous birds are occasionally seen in lowland forests. Such flocks in Lore Lindu National Park comprised core species which were almost always present, species which were frequent participants, and species which participated occasionally (table 5.15).

These mixed flocks have been interpreted as a means by which relatively specialized insect-eating birds can increase their food supply, particularly during times of food shortage, because of the insects flushed by the activity of the other species in the flock. The different responses of insects to disturbance and the different specializations of the species could ensure benefit for all the flock members (Croxall 1976). Studies of such flocks in Papua New Guinea also found that the participating species modified their normal behaviour to accommodate the activities of the other birds (Bell 1983).

The largest mammals of the upper canopy, although none of them is confined to it, are the cuscuses and the macaques. The large, dark-brown bear cuscus *Ailurops ursinus*[39] eats leaves especially of *Dracontomelum dao*

Table 5.14. Approximate wing length and weight for successive species.

	Wing length (mm)	Weight (g)	Weight ratios All species	Weight ratios Lowland species	Weight ratios Montane species
Great-billed parrot					
Tanygnathus megalorhynchus (O)	243	47			
Lesser sulphur-crested cockatoo			1.21		
Cacatua sulphurea (L)	232	39		1.42	
Blue-backed parrot			1.18		
Tanygnathus sumatranus (O)	208	33			
Red-spot racquet-tail			1.38	1.38	
Prioniturus flavicans (O)	184	24			2.05
Golden-mantled racquet-tail			1.04	1.04	
P. platurus (O)	180	23			
Ornate lorikeet			1.21		
Trichoglossus ornatus (L)	129	19		1.64	
Yellow-and-green lorikeet			1.36		
T. flavoviridis (O)	103	14			1.73
Sulawesi hanging-parrot			1.27		
Loriculus stigmatus (L)	94	11			
Little hanging-parrot			1.57	1.57	
L. exilis (L)	67	7			

Key:
O - Open areas   L - Lowland forest

Data from Meyer and Wigglesworth 1898; Forshaw and Cooper 1978; Watling 1983

(Anac.), *Ficus* spp., *Syzygium* spp. (Myrt.) and *Garuga floribunda* (Burs.) and is generally found in pairs which are active in daylight. The density of this cuscus at Tangkoko-Batuangus is estimated to be about one pair per four hectares (Anon. 1980). Four hypotheses have been proposed for the feeding behaviour of generalist herbivores such as these (Freeland and Janzen 1974), and they have been shown to hold for the related possum *Trichosurus vulpecula* of Australia (Freeland and Winter 1975). The hypotheses are:

- a generalist herbivore has to ingest several different plant foods in order to meet its energy requirements;
- large amounts of a single plant food are not eaten when a generalist herbivore first encounters it, the animal initially taking small samples in preference to eating a large meal;
- the amount of a single plant eaten can be gradually increased as the animal gains 'experience' with it; this probably is a result of the animal inducing enzymes to detoxify the food; and
- nontoxic foods are recognized quickly, larger amounts of them being eaten than can be eaten of single toxic foods. It is likely that these hypotheses will be found to hold true for *Ailurops ursinus* if an

Table 5.15. Three groups of bird species found in mixed-species flocks in Central Sulawesi. Many of the species are typical of montane forests, but are found above 500 m in lowland forests too.

Core species	Frequent species	Occasional species
Sulawesi leaf warbler *Phylloscopus sarasinorum*	Flyeater *Gerygone sulfurea*	Blue greybird *Dendrocopos temminckii*
Citrine flycatcher shrike *Culicicapa helianthea*	Blue-fronted flycatcher *Cyornis hoevelli*	Black-shouldered cuckoo-shrike *Coracina morio*
Yellow-bellied whistler *Pachycephala sulfuriventer*	Sulawesi fantail *Rhipidura teijsmanni*	Sulawesi mountain greybird *Coracina abbotti*
Mountain white-eye *Zosterops montana*		Spangled drongo *Dicrurus dicrurus*
Streak-headed false white-eye *Lophozosterops squamiceps*		Malia *Malia grata*
		Black-naped monarch *Hypothymis azurea*
		Mountain tailorbird *Orthotomus cucullatus*
		Maroon-backed whistler *Coracornis raveni*
		Fiery-browed starling *Enodes erythrophrys*
		Myza *Myza sarasinarum*

After Watling 1983

intensive study of it, a relatively easy task, is ever conducted.

The dwarf cuscus *Strigocuscus celebensis* differs from its larger relative by being nocturnal and frugivorous but is similar in that it lives in pairs and is predominantly arboreal. Both species use their prehensile tails as a fifth limb and the lower surface of the end of the tails is bare, presumably to improve the grip when hanging on to branches. Cuscuses are frequently described as being slow-moving and in this regard loosely resemble the sloths of Central and South America. They are also similar in that they both have relatively thick fur. Experiments on the spotted cuscus *Spilocuscus maculatus* (which has been introduced to Salayar Island) have shown that its metabolic rate is indeed relatively low for marsupials, as is the metabolic rate of sloths compared with other placental mammals (Dawson and Degabrielle 1973).

Habitat selection by cuscuses is not understood but it is interesting that in a study of their close relatives, the possums, in Queensland, the density of animals was directly related to the concentration of potassium in the foliage. This high-potassium foliage was most common in trees growing on nutrient-rich, basic soils (Braithwaite et al. 1984). This in itself is not exactly news-worthy but it serves to remind planners that if only nutrient-poor or otherwise agriculturally-unacceptable soils are left for conservation purposes, then densities of animals will probably be lower, and hence larger areas will be needed to conserve the animals present.

In terms of relative abundance, macaques are by far the most common large mammal in the forest (Sungkawa 1975). Only one of the four similar-looking species of macaques on Sulawesi (p. 69) *Macaca nigra* has been studied in detail, and that work conducted in Tangkoko-Batuangus Reserve from 1978 to 1981 has not yet been written up in detail. What is known from that study in Tangkoko-Batuangus and from shorter studies of that and other species indicate that all the species are very similar ecologically.

Table 5.16. Group size and density of macaques on Sulawesi and elsewhere.

	Group size	Density (no./km²)	Location
Macaca o. ochreata	18	32	Tanjung Peropa, Kendari
M. o. brunnescens	30+	30	Near Bau-Bau
M. nigra nigrescens	16	27	Bogani Nani Wartabone National Park, B. Mongondow
M. n. nigra	30	300	Tangkoko-Batuangus, Minahasa
M. maura	15-40	-	Karaenta, Maros
M. pagensis	15	4-15	Siberut Island, West Sumatra
M. nemestrina	15-18	20	Peninsular Malaysia
M. fascicularis	18-23	19-39	Peninsular Malaysia/Kalimantan

After Wilson 1978; Bismark 1979, 1982a, b; Aldrich-Blake 1980; MacKinnon and MacKinnon 1980a; Whitten and Whitten 1982; K. MacKinnon 1983; and EoS teams

The major exception is the very high density of the macaques at Tangkoko-Batuangus (table 5.16).

Groups comprise some adult males with a greater number of females. Individuals are very hard to recognize, although old animals of *M. maura* at least develop conspicuous white marks on their heads and forearms (Watanabe and Brotoisworo 1982) and grow somewhat bald. Group size can vary considerably, even within an area, but the social status, exclusiveness and constancy of these 'groups' is by no means clear. At Tanoma (Tanjung Peropa) groups seen ranged in size from 12 to 28 individuals. Various sites in that area were visited and it has been suggested that the relatively high density of macaques in one particular area is due to the high density of fig trees (Bismark 1982b). The few data available do indeed appear to show this but there also other trees favoured by macaques and the relationship is unlikely to be so simple. It is not stated, for example, whether there were any figs in the five locations where no macaques were seen. The group size of the Tangkoko-Batuangus macaques averages 30, but groups numbering 50 were also seen. These groups live in home ranges of about 15 ha in which the animals would move up to 0.5 km each day. The same species (though a different subspecies) in the central part of Bogani Nani Wartabone National Park[40] were found to travel about 850 m in a day (Bismark 1982a).

Macaque groups are almost always accompanied in the canopy by the endemic malkoha *Phaenicophaeus calorhynchus*, a long-tailed bird which flies weakly and more often hops or creeps around the canopy like a squirrel. Its reason for following the macaques around is to catch grasshoppers and perhaps other insects as they are flushed out by the activities of the monkeys. It also appears to follow the somewhat clumsy, endemic coucal *Centropus celebensis*.

Table 5.17. The remaining habitat and populations of Sulawesi macaques.

	Suitable habitat remaining (ha)	Percent protected	Density in good habitat	Average density (no./ km²)	Estimated total population	Population protected in reserves	Percent of in reserves
Macaca nigra	4,800	57	95 (20-300)	30	144,000	82,500	57
M. tonkeana	38,500	3	30 (10-50)	10	385,000	10,550	3
M. maura	2,800	18	30	20	56,000	9,900	18
M. ochreata	18,500	8	40 (20-155)	15	277,500	21,300	8

After MacKinnon 1983

The diet of macaques appears to be primarily fruit, with figs *Ficus* pre-dominating. Other fruit taken includes *Dracontomelum* spp. (Anac.), *Palaquium obtusifolium* (Sapo.), *Canangium odoratum* (Anno.), *Syzygium* spp. (Myrt.), *Spondias pinnata* (Anac.), and *Pangium edule* (Flac.) (Anon. 1980; Bismark 1982b; Watanabe and Brotoisworo 1982). Pet macaques are often seen to eat insects, so they probably do this in the wild, too. They may also open rotting tree trunks or disturb the litter to find beetle grubs, worms, etc., while on the ground.

The four macaque species differ somewhat in their conservation status, with *M. tonkeana* being both the most common species and the one with the least percentage of the population within protected areas. *M. maura* appears to be the most threatened with the smallest population, the smallest population in protected areas, and a distribution that includes some of the most densely-populated land outside Java, Madura and Bali (table 5.17).

Comparison of Mature- and Pioneer-phase Faunas

The high net productivity of the early phases in the forest growth cycle supports large animal populations, although the species are quite distinct from those in forests (Ewel 1983). A comparison was made of arthropod abundance and diversity between cultivated and secondary areas both near a transmigration site in Dologuo and in neighbouring disturbed areas. If the diversity of insects in the forest is taken to be 1, then the diversity was 0.4 at the forest edge and 0.1 in the rice fields (Doda 1980). In the course of collecting mosquitoes in and around Bogani Nani Wartabone National Park, it was found that there were about one-quarter as many catches in forest as there were in neighbouring farmland. Eight of the twelve mosquito species found in forest were also found in farmland but the abundance was far higher in farmland. Two species, *Anopheles flavirostris* and *A. subpictus,* are potential serious vectors of malaria and filariasis and were found in farmland but not in forest (Hii et al. 1985).

Many grasshoppers are found in and around forests but three species that might be considered agricultural pests penetrate no further than the forest edge (R. Butlin pers. comm.; K. Monk pers. comm.). Conversely no forest grasshoppers are considered to be potential pests. A similar drop in diversity between forest and disturbed areas has been observed among Collembola (P. Greenslade pers. comm.) and beetles (J. Krikken pers. comm.). The rats found living commensally with humans are rarely if ever found in forests and forest rats are never found beyond the forest edge. Not only that but the parasites they carry are quite different; both groups of rats carry fleas, but the forest species are only infested by fleas from the family Pygiopsyllidae, and the commensal species only by fleas from the Pulicidae (Durden 1986). These observations do not in any way support the contention, frequently heard, that forests are reservoirs of agricultural and human pests.

Studies of birds in forest gaps indicated that the number of species caught in nets in gaps was greater than those caught in surrounding forest. In addition, the assemblage of birds using the gaps was distinct from that in the forest, although there was, of course, some overlap in species (Schemske and Brokaw 1981).

The birds of Sulawesi have not been studied in this regard but it is likely that the situation is similar to that found in Papua New Guinea where about 15 species are more or less confined to secondary forests of the pioneer and early building-phases and about 50% of mature forest bird species might be expected to visit the younger forest. Only a very few species from the mature forest would visit, let alone feed in cultivated areas or in young pioneer-phase forests (Bell 1982d). A recent study in Central America found that migratory birds showed a preference for old pioneer- and building-phase forest due in part to the greater abundance of small fruits (Martin 1985).

There are considerable differences in microclimate between gaps and mature forest (p. 348) and this is reflected in the habits of some animals which are found in both habitats. For example, the orientation of webs of a particular spider was found to be north/south in closed habitats and east/west in open habitats. The longer a spider spends on the web, the more prey it captures (some prey manage to wriggle loose before being killed by the spider) and so presumably an orientation is chosen to allow maximum time on the web. Web orientation seems to be related to the body temperature of the spiders because both too little and too much heat can affect spider activity. Thus, in open habitats, the spiders' webs face east/west to reduce heat load and in closed habitats face north/south to increase body temperature (Biere and Uetz 1981).

THE EFFECTS OF OPENING FOREST

General Effects

The opening up of forest on a scale greater than occurs in natural ecosystems is accompanied by the following changes:

- the creation of open, hot, simple habitats containing relatively few, small, widespread species with broad niches and great reproductive potential (r-selected species–p. 345). These species are rarely, if ever, found in mature ecosystems;
- a huge decrease in biomass (± 30 kg dry weight/m² in lowland forests

to 0.2 kg dry weight/m² in alang-alang plains);
- the temporary or permanent impoverishment of the soil;
- the creation of increasingly small and isolated patches of natural vegetation whose animal and plant diversity also progressively declines (p. 442);
- virtually all the rainfall reaches the soil surface, and far more rapidly, than in a forest, but less enters the groundwater and more flows in the surface runoff. This can lead to a considerable loss of soil, a decrease in groundwater supplies and an increased propensity to flooding.

These ecological effects can be observed whether the disturbed area becomes wasteland or valuable agricultural land.

The major effect of less intense disturbance (where the forest retains at least some of its former structure) is a simplification of the ecosystem caused by deleterious changes to the soil, hydrology and microclimate as well as by the actual removal of plant material such as logs. As taller trees are removed, so the volume of living space available to the forest biota is considerably reduced (Ng 1983). There is also less substrate available for use as nest sites, aerial pathways or growing sites for epiphytes and climbing plants. In addition, there is obviously also the loss of other resources, particularly food.

It is often stated that after forest clearance the loss of soil fertility and the loss of nutrients in plant biomass pose a serious threat to the integrity of most forested ecosystems. While this is certainly true for montane forests and forests on ultrabasic soils, in other types of forest, disturbance appears to result in succession which is directed towards the re-establishment of mature-phase forest, where continued disturbance is absent. Unfortunately, once disturbed, most forests in Sulawesi experience a long series of destructive disturbances which never allow the succession to progress. Effects which are sometimes attributed to the loss of nutrients from soils may possibly be due to competitive interaction between plants. Studies have shown that the fertility of soil does indeed decline following forest removal, but the evidence of short-term loss in soil fertility is usually accompanied by evidence that the successional vegetation is remarkably good at regenerating soil fertility (Soerianegara 1970; van Baren 1975; Harcombe 1980; Uhl and Jordan 1984). This is not an argument against taking all possible precautions to reduce nutrient loss, but it does indicate that further studies are required to elucidate the situation.

A detailed comparison of soil chemistry, soil respiration, seed storage and plant growth rates between soils of primary forest, cut-over forest, and cut-over and burned forest both before and after rain, has produced many interesting results (Ewel et al. 1981). Amongst these it was noted that after the disturbed vegetation had been burned, 57% of the initial amount of nitrogen and 39% of the initial amount of carbon remained in the soil.

The number of viable seeds in the forest soil (the seed bank) was 8,000 seeds/m^2 (67 species) but this was reduced to 3,000 viable seeds/m^2 (37 species) after the burn. Thus vigorous and relatively diverse growth followed the burn because only a proportion of the seeds were released from the burnt material. The 5,000 seeds that did not survive the fire may have either been killed in the heat of the fire or, as might be expected for primary forest trees, be dependent on high humidity for germination (Ng 1983). If this is not available for a certain length of time, the seeds will die.

A 1 ha plot of Queensland lowland forest (which contains many of the genera found in Sulawesi lowland forest) was first felled and then burned shortly afterwards. Two years later the regeneration of the trees that had reappeared was studied (Stocker 1981). All of the 82 tree species present had regenerated in one or more ways; 74 species had formed shoots from the base of the old trunk, 10 had formed shoots from roots, and 34 had germinated from seed. This last category appeared to have the greatest growth rate, but frequency of shoots from old trunks shows that this is also an extremely important mode of regeneration. The ability of these trees to form shoots is probably limited, however, and repeated cutting or repeated burning may give smothering, light-demanding creepers the advantage and prevent regeneration from taking this form.

When forest is cut selectively for timber, recovery is relatively rapid because the seed bank contains many types of viable seeds and many tree species have the ability to sprout from stumps. When forest is cut, dried and burned and then abandoned, the succession proceeds more slowly because part of the seed bank has been destroyed and because some of the sprouting species are killed by fire (Uhl et al. 1981).

Thus when forests are cut, burned, farmed, weeded or burned and farmed again, recovery of the forest (if allowed) is extremely slow. In places where the only remaining stand of forest (source of seeds) is some kilometres away, full regeneration may take hundreds if not thousands of years. Even if a few mature trees have been left to stand, there is no guarantee that these will regenerate successfully because some species have seeds which have to pass through an animal's intestine before they will germinate (p. 381) (Ng 1983).

Seed banks are large in *Imperata* grasslands which have not been burned for five years or more, but small in similar but regularly-burned grasslands. In both cases, however, the seeds tend to be of agricultural weeds rather than of pioneer trees. Interestingly, *Imperata* seeds are not present to any great extent and this suggests that these grasslands are propagated by fresh seeds and vegetative growth from rhizomes (Hopkins and Graham 1983).

Wider Implications of Forest Conversion

The conversion of forests and changes in land use are sometimes suggested as a cause of the rise in levels of atmospheric carbon dioxide. The argument is that forests fix a great deal more carbon dioxide than grasslands or other vegetation types. Most of the Sun's energy passes through the atmosphere to warm the Earth beneath. The warm ground reflects some of this radiation and radiates some upwards at longer, invisible wavelengths in the infra-red. Carbon dioxide absorbs this radiation which is in turn radiated back at the Earth's surface. The more carbon dioxide, the more radiated energy is absorbed, and the warmer the atmosphere becomes. This is known as the 'greenhouse effect'. Levels of carbon dioxide have been increasing over the last 100 years or so and, if present trends persist, the concentration will have doubled from 0.03% in the middle of last century to 0.06% by the middle of the next century. Although still a small proportion of the atmosphere, the increase could well cause major climatic changes.

The global effects of atmospheric pollutants was brought into focus in mid-1986 by a report from the United States Department of Energy. This carried the full weight of establishment respectability behind an idea which, ten years ago, was thought to be no more than an eco-scare story. The fact now has to be faced that the world is getting warmer because of the carbon dioxide greenhouse effect. Rainfall patterns, temperature and sea-levels throughout the world will change, as they have indeed been changing for the last 100 years. Sea level is currently rising by about 0.1-0.25 cm/yr and may become even more noticeable in the next few decades.

Not all the carbon dioxide that should be in the atmosphere, according to calculations, is actually there, indicating that a large amount is being absorbed into natural reservoirs or 'sinks'. No one is certain where these are or how they work, but since plants absorb carbon dioxide from the atmosphere, it is reasonable to suggest that tropical forests may be one of the major sinks. As the area of these forests gets smaller, and the burning of fossil fuels increases, so the situation is aggravated.

Another possible deleterious effect on the atmosphere arising from forest conversion is increasing levels of methane produced as a by-product of termite metabolism. It has been estimated that termites are 8-10 times more common in land disturbed by human activities than they are in forests. Recent research has shown that termites produce about 150 million tons of methane each year, compared with a total global input of between 350 and 1,210 million tons. The level of atmospheric methane is increasing at about 2% per year, and most of the increase appears to originate from tropical areas undergoing development (Zimmerman and Greenberg 1983). The scale of the problem has been challenged (Rasmussen and Khalil 1983) and potential environmental impact of the increased level of atmospheric methane has yet to be assessed, but it is clearly important to be aware of the possibility and to encourage its study.

Effects of Selective Logging on the Forest

Selective logging is the removal from a forest of trees of designated species which exceed a designated size.

While selective logging clearly has many advantages, it also removes the best individual trees, leaving only inferior or younger ones to produce seeds for future 'crops'. Each time logging occurs, plant succession is set back a step, the forest becomes progressively poorer in desirable species and progressively richer in 'weed' species (Wyatt-Smith 1963; Whitmore 1984). For this reason, foresters examining regrowth in the Malili area, for example, suggested that timber trees be planted for the next cropping (Sutisna and Soeyatman 1984). This genetic erosion is potentially extremely serious for forestry in the future (Ashton 1980; Kartawinata 1980; Sastraprada et al. 1980; Whitmore 1984) but in many logged areas it is doubtful whether in fact a full cycle of regeneration will ever be allowed to occur.

Selective logging is, in theory, a repeatable exercise in any given area. After the largest commercial trees have been extracted the area should be left for about 70 years until the next timber crop is harvested. Logging of a forest at intervals of less than about 70 years is unsound forestry practice but as more species become economically worth exploiting and when timber prices increase, repeated logging occurs. Over fifty species of Sulawesi trees are probably exploited commercially; for example, in an area near Malili, species such as *Santiria laevigata* (Burs.), *Kalappia celebica* (Legu.), *Agathis alba* (Arau.), *Vatica flavovirens* (Dipt.), and *Calophyllum soulattri* (Gutt.) are the major trees felled (Sutisna and Soeyatman 1984), but of these only *Agathis* is of any great value. In certain areas of Sulawesi the logging is extremely selective and only the large valuable *Agathis* are removed, because it is not felt to be economically worthwhile to take poorer species. As a consequence, the 'value' of the depleted forest to a logging company is greatly reduced and the overall damage and effects of logging in those areas will be less than in areas where a wider range of species is taken.

The term 'selective logging' sounds a very mild and benign activity to many people who have not worked in or visited a logging concession, and it comes as a surprise to learn just how much damage is often caused to the forest as a whole. One estimate is that five times more timber is destroyed or badly damaged than is extracted (Burgess 1971). Another illustration is that for every large tree felled, 17 similar or smaller-sized trees are destroyed (Abdulhadi et al. 1981).

The felling of trees and extraction of logs are the primary causes of disturbance during a logging operation. Felling damages tree crowns, boles and saplings, exposes wood, making it susceptible to fungus damage, and also covers seedlings. Extraction of the logs exposes bare soil and damages large areas of the forest floor (Kartawinata 1980).

Much disturbance is also caused by the access roads which remove all cover from the soil and from channels that are further scoured by runoff during storms. It has been estimated that 20%-30% of a logged area is completely bare, being composed of roads and logyards (Meijer 1970; Kartawinata 1980). Some of the roads cut across small rivers, acting as dams, and the subsequent flooding kills most of the inundated trees and other plants (Anon. 1980; Kartawinata 1980). The lorries used to drag logs out of the forest cause compaction of the soil surface and this disturbance is traceable in some forests by the occurrence of mature pioneer trees even decades after logging.[41] By this time other pioneer trees have died or become rare as different genera take over dominance in the succession. The soil compaction caused by lorries is, however, only a fraction of that caused by the bulldozers used to make the roads. Additional disturbance is caused when trees are felled on both sides of the main logging roads to hasten the drying of roads. It is estimated that these 'daylighting strips' accounted for 8 ha/km of logging roads (Hamzah 1978).

There does not seem to have been a detailed study of the effects of logging on Sulawesi forest, but the results would probably be very similar to those found on South Pagai Island in the Mentawai Islands, West Sumatra (Alrasjid and Effendi 1979). A total of 2,416 trees (20 cm in diameter and over), half of which were commercial species, originally stood in the 15 1 ha plots that were enumerated. A total of 194 trees (13/ha) were felled and extracted. On average, about half of the remaining trees had been noticeably damaged, broken or knocked down, and the other half had escaped damage (fig. 5.46). This proportion seems to be relatively consistent between studies (Burgess 1971; Tinal and Palinewan 1978; Abdulhadi et al. 1981).

The microclimatic changes caused by logging are obvious—it is hotter, lighter and drier in logged-over forest. These changes result in the dieback of crowns, scalding of sensitive trunks and branches, water stress and even an increased likelihood of insect attack, any one of which may lead to the death of a tree (Kartawinata 1980).

There are many ways in which logging operations and their supervision could be improved, such as by yarding or dragging the logs up, rather than down, slopes in order to reduce erosion, avoiding forests adjacent to rivers, felling trees in a direction such that their extraction causes minimal damage, and by mulching areas to be abandoned with forest litter (Marn and Jonkers 1982; Hamilton and King 1983). These do not necessarily increase the overall costs to the contractor.

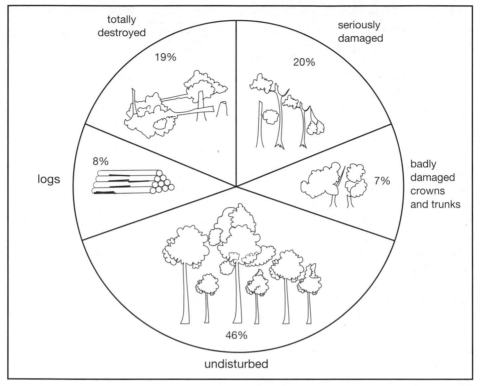

Figure 5.46. The effect of extracting 8% of the trees by selective logging on an area of forest on South Pagai Island (Mentawai). The other categories are: totally destroyed; seriously damaged; badly damaged crowns and trunks; and undisturbed.
From data in Alrasjid and Effendi 1979

Effects of Selective Logging on Soil and Hydrology

The damage that logging causes to soil has barely been studied in Indonesia but the general subject has been reviewed (Kartawinata 1980). The major problems are soil removal by bulldozers when making roads, soil compaction caused by heavy vehicles, soil loss due to rain striking the ground with its full force, and the greater quantity of rain reaching the soil. The removal or destruction of trees also reduces litter fall and hence the organic inputs to the soil, but in areas where the soil has not been too greatly damaged, soil nutrient levels and micro-organisms can recover within a few years. Since only the larger portions of the tree trunks are removed, and most of a tree's nutrients are in its leaves and roots, the loss of biomass after logging is proportionally greater than the loss of nutrients.

It is obvious from a brief visit to a logging site that a great deal of water runs off the roads, carrying soil with it. This has been quantified by the measurement of the silt content of a river near, but not directly influenced by, a particular logging area; from a river in the logging area, and from a ditch by a logging road. The silt contents were 0.01%, 0.05% and 0.1% respectively—a ten-fold increase (Hamzah 1978). The increased silting of the major rivers in Sulawesi can probably be traced to logging as well as other forms of unprotective land use. In the lower stretches of major rivers the actual volume of water is probably not greatly affected by forest clearance, and it is in the higher river regions that dramatic effects are noticed (p. 630).

Soil erosion and subsequent sedimentation can have serious and expensive impacts on irrigation schemes. For example, the concession granted in 1978 to a logging company in the catchment area of the Gumbasa River south of Palu threatened to reduce the effectiveness of the Gumbasa irrigation scheme. This scheme was begun in 1973 with the intention of irrigating 11,500 ha of rice fields in the Palu valley, in order to make that region into a net rice exporter rather than a rice importer. Thirteen years later, only 5,000 ha were being irrigated, at least in part because of the highly erodible soils in the catchment area which had been washed downstream after the forest cover had been disturbed by logging and inappropriate farming practices. The irrigation canals now have to be dredged every year when about 30,000 m^3 of soil is removed (Mattulada pers. comm.).

Many accusations are bandied about concerning whether concession loggers, illegal loggers or upland farmers are to blame for the continual degradation of forest lands (Anon. 1986a). Illegal loggers are significant not just in the opening up of the forest, but also in the loss of taxes and fees to the government. Illegal loggers are also guilty of encouraging the complicity of villagers by buying the wood they extract from the forest (Anon. 1986b). An example of the scale of this activity is given by the seizure in Palu in October 1986 of 668 lengths of ebony *Diospyros celebica* (Eben.) (p. 386) that had been taken illegally from a transmigration area on the pretext of establishing a cocoa plantation. By the time the wood was seized, it is estimated that 150 other lengths had already been taken to Poso (Anon. 1986c).

The concession holders are supposed to work within the limits imposed by the Department of Forestry but supervision of the operations is difficult and the regulations are open to abuse. Logged forest is easier to clear and burn than undisturbed forest. In this situation the farmer who takes advantage of this in production and other logged forest is most at fault, but the absence of any likelihood of being caught, let alone prosecuted, in these inevitably remote areas, means that he is likely to proceed with scarce worry or concern. The small chance of reprimand is also relevant in the clearance of protection forests. Where social pressures are such that farmers begin to endanger forests and soils, then the agriculture of the area needs close appraisal and assistance. This would not inevitably lead to the adoption of intensive rice culture (p. 572).

Effects of Selective Logging on the Fauna

The majority of mammal and bird species found in Sulawesi are dependent on mature forest. Considering the very large number of insects which are restricted to a single species of plant (p. 346), the percentage of insects dependent on undisturbed forest is probably also extremely high. There is clearly a need to know what effects different types of disturbance have on the fauna, although the trends and general patterns, at least for mammals and birds, are now quite well understood. Thus, the forest fauna is affected by disturbance in at least three ways:

- the noise and shock of disturbance may cause an immediate change in behaviour;
- the actual removal of parts of the forest canopy will alter ranging patterns and diet which may in turn affect social behaviour and population dynamics;
- the slow regeneration rate may cause permanent changes in population density (Johns 1985).

Different species show different degrees of tolerance to disturbance and at least some animals will usually be found even in the most disturbed areas. A study of terrestrial mammals in different habitats ranging from mature forest to an area of alang-alang grass in Peninsular Malaysia showed, not surprisingly, that the total number of species (rats) increased from 0%-100% (Harrison 1968) (fig. 5.47). A lowering of richness and evenness (p. 346) of small mammals species has also been reported from tropical Queensland where the effect was most marked on relatively infertile soils (Barry 1984).

Properly executed selective logging is not disastrous for much of the forest fauna although certain squirrels, bats and birds may fare badly (Marsh and Wilson 1981; Wilson and Johns 1982; Johns 1985; Johns et al. 1985). 'Properly-executed' in the ecological sense used here means an average of 8-10 and an absolute maximum of 15 trunks removed per hectare, no relogging for at least 50 and preferably 70 years, replanting with tree species found in the area, and the preservation of an adjacent area of mature forest from which fruits can be dispersed into the logged area. If such practices are adhered to, there is no reason why nature conservation and forestry should conflict. Logged forest supports a lower species diversity of animals but is able to maintain viable populations of many species.

The initial effects of selective logging on the fauna are probably the most serious and this is illustrated by a lower birth rate and a greater mortality amongst infant primates in logged areas (Johns 1985). These effects seem to be temporary, however, and the populations return to normal

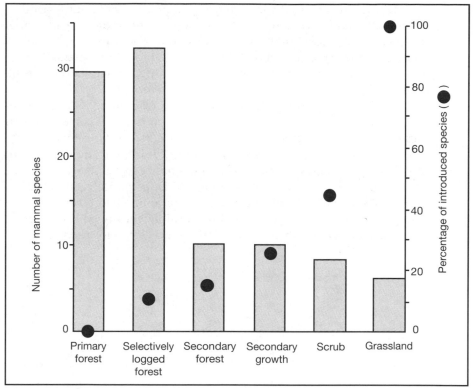

Figure 5.47. The total number of non-flying mammal species in six types of vegetation, and the percentage of introduced species.

After Harrison 1968

eventually. It is likely, however, that the effects of disturbance will last several decades because it takes that long for the lost resources such as food sources and nest sites, and for microclimatic conditions to recover (Johns 1985; Wong 1985).

The effects of disturbance depend in part on the social system and diet of the species concerned. Territorial species are worst affected because they are unable, even temporarily, to move out of the disturbed area because they will probably encounter defended territories of other groups of the same species. Non-territorial species may be able to withstand some temporary crowding before moving back into the disturbed forest. Species with relatively specialized diets would be expected to fare rather worse in disturbed forest than unspecialized species.

When forest is cleared completely for large-scale agricultural or other purposes, the animals that once lived in it will eventually die. The territorial species are very unlikely to be able to find unoccupied areas in the neighbouring undisturbed areas, and while the non-territorial species may be able to move into adjacent forests (assuming there are any), sooner or later the food resources will limit the population to more or less its original level.

Some of the displaced animals frequently enter gardens and steal food. The macaques are a major pest, particularly on crops of corn, although they will eat a wide range of other crops. It is clearly hard for them to resist food that has been planted within their range. It is these animals that are caught to become pets (Anon. 1978), and it has been estimated that there are at least 100 pet *Macaca nigra* along the Domoga valley alone (Mac-Kinnon 1983). Babirusa are one of the first animals to become locally extinct after logging or land opening, possibly because village or feral dogs kill the piglets, a form of predation to which babirusa are poorly adapted (p. 414).

Chapter Six

Specific Lowland Forest Types

INTRODUCTION

The composition and ecology of lowland forests are extremely variable due to a variety of physical factors. The more obvious forest types on distinctive soils or in dry areas are described below, together with any noteworthy aspects of their animal life.

PEATSWAMP FOREST

Formation and Location

Peat is a type of organic soil which is at least 50 cm deep and which contains more than 65% organic matter (that is, it loses more than 65% of its dry mass when burned). Peat deposits can be of two forms although only the second is found in Sulawesi:

- Ombrogenous peat: the most common type in Southeast Asia which has its surface above the surrounding land. Plants that grow on this peat utilize nutrients only from within the plants themselves, the peat or directly from the rain; there are no nutrients entering the system from the mineral soil below the peat or from rain water flowing into it. This type of peat is usually found near the coast behind mangrove forest and can be extremely deep (about 20 m). The peat and its drainage water are very acid (pH <5.0) and poor in all nutrients (oligotrophic) particularly in calcium. Outside Sulawesi, vast areas of ombrogenous peatswamps occur in South and West Kalimantan and eastern Sumatra.
- Topogenous peat: this peat occurs in depressions. Plants growing on this peat can extract their nutrients from the mineral subsoil, river water, plant remains and rain. Topogenous peat can be found behind coastal sand bars and at other sites where free drainage is hindered. The peat is usually found in a relatively thin layer (less than 4 m).

445

Peat accumulation in topogenous swamps is relatively slow and the high acidity (low pH), characteristic of ombrogenous swamps, is less pronounced. Instead, the peat and drainage water are only slightly acid (pH 5.0) and nutrients are relatively abundant (mesotrophic).

The only major area of peatswamp in Sulawesi is Aopa Swamp, 100 km west of Kendari, and it forms much of the northern portion of the Rawa Aopa-Watumohae National Park. The major attractions of this park are fishing and watching the traditional hunting of deer by horsemen using lassos[1] (Anon. 1985). Some years ago there were plans to mine the peat of Aopa Swamp and to burn it for electricity generation but companies have been far more attracted by the massive volumes of peat available in Borneo and Sumatra. Aopa Swamp is all that remains of an ancient lake which is reaching the final stages of its life as an aquatic ecosystem. In contrast to the extensive ombrogenous peat swamps which arose because of the inability of decomposer organisms to break down organic material in saline conditions, the Aopa peat was formed because of the continuous, water-logged conditions which are unfavourable, but not anathema, to decomposer organisms. A contour map shows how the swamp is almost entirely surrounded by high ground (fig. 6.1).

Aopa Swamp covers a maximum area of about 31,400 ha, and this is surrounded on all sides by grasslands, cultivation or transmigration settlements. The area of the swamp varies through the year depending on rainfall. It is at its maximum between May and September, falling to 20,000 ha between November and December, about 15,000 ha in January, rising to 28,500 ha between March and May. Its depth varies between 1 m and 2 m depending on location and time of year. The rivers draining the swamp, particularly in the western half, are dystrophic, that is, the water is shallow, nutrient-poor and dark-coloured because of suspended plant colloids. The abundant organic matter is decomposed by microorganisms, most of which require oxygen, and as a result the concentration of dissolved oxygen is low (Anon. 1983). Small areas of coastal peatswamp also exist such as in the west of Muna Island (Anon. 1982) but these have so far received no attention from ecologists.

Vegetation

The extent to which the vegetation of Aopa Swamp is known is thanks to a small team which in 1978 made preliminary botanical observations there[2]. Aopa Swamp is in fact two swamps partially separated from each other by a ridge extending southwards from Mt. Makaleleo. The swamp to the east is relatively disturbed and has only a few small trees of she-oak *Casuarina* (Casu.) scattered all over the area beneath which grow sedges, such as *Scleria* (Cype.). The next most abundant tree is a *Eugenia* (Myrt.) which has aerial roots produced from the trunk up to 2 m above the ground. Other

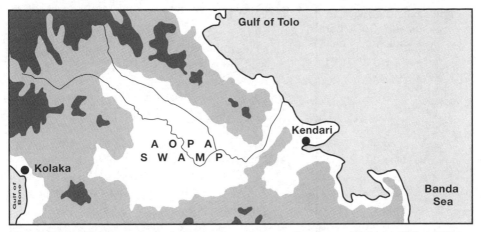

Figure 6.1. The location of the enclosed Aopa Swamp with land 400-1,000 m (medium grey), and 1,000+ (dark grey).

plants include sago palms *Metroxylon sagu*, a large tree palm *Pholidocarpus*, occasional *Corypha* tree palms (p. 481), 5 m-tall pandans, at least two species of climbing rattan, and epiphytic *Lecanopteris* ant ferns (p. 509). Epiphytes in general, however, are uncommon.

When approaching the western swamp from the road which crosses the Aopa River, one passes expanses of open water. The most spectacular plants at the edges are lotus lilies *Nelumbo nucifera* (Nelu.) (p. 281). Before reaching the forest, there is a floating mat of vegetation dominated by a tall papyrus-type sedge *Cyperus* sp., and in the drier parts the even taller grass (>2 m) *Saccharum spontaneum* (Jacobs 1979; Susanto 1984). The floating mat of vegetation is probably extending its area by growing over the water surface; one floating runner, for example, was 7.5 m long.

The forest in the western swamp is more luxuriant than in the east. The trees grow to 35 m tall, but no conifers or she-oaks were found by the team in 1978. Large buttresses are common but aerial roots are rare perhaps because the substrate is more stable than in the eastern part. Strangling fig trees are a distinctive component. There are many palms: the trunkless *Licuala*, the tree palms *Livistona*, sugar palm *Arenga*, the spiny *Oncosperma*, rattans, but no sago. Climbers are common as are creepers such as the small pandan *Freycinetia* and the superficially similar *Pothos* (Arac.). Epiphytes are comparatively rare but the herbs of the undergrowth are varied

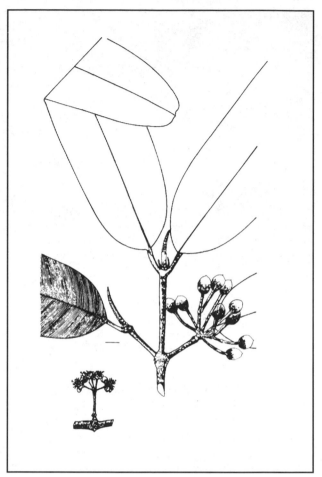

and interesting (Jacobs 1979). Other trees recorded are *Hopea gregaria* (Dipt.), *Baeckia frutescens* (Myrt.), *Saccopetalum horsfieldii* (Anno.), *Planchonia valida* (Lecy.), *Diospyros malabarica* (Eben.), *Artocarpus teysmannii* (Mora.) and *Calophyllum soulattri* (Gutt.) (fig. 6.2) (Anon. 1978; Susanto 1984). This last plant is particularly easy to identify because its leaves, in common with other members of its genus, have numerous straight veins, all running parallel to each other on both sides of the midrib.

The two most common trees in the peatswamp in west Muna are members of the teak family Verbenaceae, that is *Geunsia paloensis* and *Premna foetida* (Anon. 1982).

Fauna

Elsewhere in Southeast Asia it has been found that densities of mammals in peatswamps are generally lower than in most dryland lowland forests. Detailed studies of animals in Aopa Swamp have not been conducted but it appears that most of the larger mammals are present with the exception of the Sulawesi civet *Macrogalidia musschenbroeckii* (p. 532) which has never in fact been reported from anywhere in Southeast Sulawesi. Wallows surrounded by tracks of the lowland anoa *Bubalus depressicornis* are found in the forest although this animal is not particularly common. In addition to the indigenous fauna, wild water buffalo and deer are found, particularly in the grassy areas of the eastern swamp (Anon. 1984).

A single day's bird watching in Aopa Swamp resulted in 80 species being seen (Anon. n.d.), a relatively large total for Sulawesi, and many of these were associated with the water (p. 305).

Three large reptiles are reported from the swamp: the sailfin lizard *Hydrosaurus amboinensis* (p. 301), crocodile *Crocodylus porosus* (p. 302) and the reticulated python *Python reticulatus* which is hunted nowadays by some of the Balinese transmigrants for food.

FRESHWATER SWAMP FOREST

Physical Conditions

Freshwater swamp forest grows where there is occasional inundation of mineral-rich freshwater of pH 6 and above and where the water level fluctuates such that drying of the soil surface occurs periodically. The floods may be a result of heavy rain or of river water backing up in response to particularly high tides. The major difference from conditions of the peatswamp forest described above is that the peat layer is more or less absent although a thickness of a few centimetres may be found on the soil surface. Freshwater swamps are normally found on riverine alluvium[3] but also occur on alluvium deposited in lakes such as in the southeast of Lake Lindu, the southeast of Lake Poso (Sarasin and Sarasin 1905), and the south of the Ranu lakes (p. 259) (L. Clayton pers. comm.).

The alluvial soils are more fertile than those on adjacent slopes and have great agricultural potential when drained. The soils are very variable but are generally young and therefore do not have clearly differentiated horizons. They comprise relatively fine particles and the general colour is grey turning mottled grey or orange-grey where occasional drying occurs as a result of gleying.[4] Soil animals such as termites and earthworms, which normally take organic matter into the soil, are not tolerant of the water-logged anaerobic conditions and so there is little mixing of soils.

The agricultural potential of these swamp soils has meant that fresh-water-swamp forest have suffered greatly from human activities (p. 97).

Vegetation

The vegetation of freshwater swamps varies according to the wide variation in its soils and the proximity to free water. Few if any plants are restricted to freshwater swamps and these forests are relatively poor in species. The fringes of the swamp near open water are grassy but, where the ground is firmer, palms and pandans can be common. Behind these, a forest very similar to normal dryland lowland forest is generally found. The relative instability of the soil, by virtue of its periodic inundation with water, is probably one reason that supportive structures associated with trunks such as long, winding buttresses and stilt roots are common, and certain species have accessory breathing structures such as pneumatophores (p. 121).

Only four areas of freshwater swamp appear to have been examined in Sulawesi:

- a patch of coastal freshwater-swamp forest on the west coast of Sulawesi across the hills from Donggala and Palu was found to contain *Barringtonia racemosa* (Lecy.), the rare *Quassia indica* (Sima.) (fig. 6.3), *Terminalia copelandii* (Comb.), *Polyalthia lateriflora* (Anno.) (with pneumatophores like *Sonneratia*), *Elaeocarpus littoralis* (Elae.) and *Horsfieldia irya* (Myri.) (Meijer 1984);
- an EoS team examined a freshwater-swamp forest adjacent to the transmigration settlement of Tinading, Baulan, Toli-Toli, and at least 90% of the trees were *Terminalia copelandii;*
- the freshwater-swamp forest in the southeast of Lake Lindu comprised, in 1934, almost pure stands of *Nauclea* sp.[5] (Rubi.) trees and tall grasses (Bloembergen 1940), but it is not clear whether these still exist;
- freshwater-swamp forest near the shores of Lake Ranu in Morowali National Park contains the trees *Palaquium* (Sapo.), *Manilkara* (Sapo.), *Mimusops elengi* (Sapo.), *Calophyllum soulattri* (Gutt.), *Parinari corymbosa* (Rosa.) and *Haptolobus celebicus* (Burs.) (Anon. 1980).

Pitcher plants *Nepenthes* (Nepe.) appear to be ubiquitous in freshwater-swamp forest. They are generally climbing or scrambling plants with the climbing stem arising from a rosette of young leaves. A single plant can bear two or three distinct forms of pitchers; the pitchers on the ground generally have a rounded base, whereas those somewhat higher up are more slender and have two frilly vertical wings on their front. The uppermost pitchers are longer and more funnel-shaped and their wings are reduced to prominent ribs. The variety of forms has led to confusion in identification, and the number of species on Sulawesi is not yet known

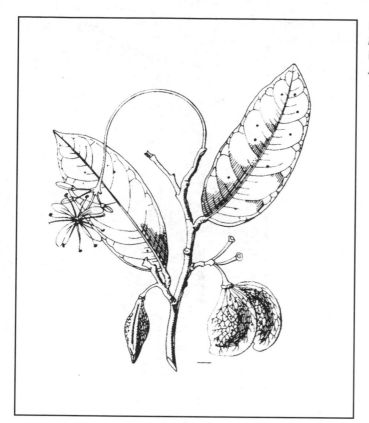

Figure 6.3. *Quassia indica*. Scale bar indicates 1 cm.
After Nooteboom 1962

since new species continue to be recognised (Kurata 1983; Turnbull and Middleton 1984). The lip of the pitcher is ribbed and often toothed around the inner edge where nectar glands are situated. The lid, which also has nectar glands, projects above the pitcher but does not prevent rainwater from entering. The nectar and perhaps the colour of the pitchers attracts insects many of which slip on the pitcher lip and fall into the liquid below.[6] The inner wall of the pitcher is divided into two zones: the upper half is waxy, very smooth, and more or less impossible for insects to climb up, and the lower zone which has numerous glands that secrete digestive enzymes. These enzymes digest unwary insects that fall into the deadly brew having been attracted by the nectar and the blood red colour. The fluid in the pitcher has a pH of 3 to 6 and this favours the action of the digestive enzymes, and the plant then absorbs the available nutrients.

It is well known that insects are lured into the pitchers, but it is less well known that the water inside a pitcher supports a community of winged-

Figure 6.4. A pitcher plant showing the animal community within it.

After Kitching and Schofield 1986

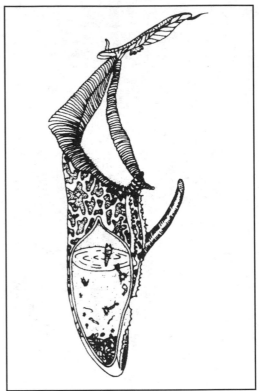

insect larvae and other organisms that are resistant to the digestive enzymes (fig. 6.4). Most of these animals feed directly on the drowned insects or feed indirectly on them by being predators on bacteria and other microorganisms that have themselves fed on the corpses. Since many of these species spend only the early part of their life cycle inside pitchers, they represent an export of nutrients which would otherwise have been used by the plant (Beaver 1979a, b), but on the other hand they enhance the plant's digestive process by breaking down the accumulating detritus of dead animals into nitrogenous compounds (Kitching and Schofield 1986).

The community of animals in a pitcher is relatively poor in species, because of the specialized nature of their habitat, and as such is amenable to study. Most of the animals exploit the detritus: larvae of biting midges and mosquitoes filter out and feed on fine particulate and suspended material, while mites feed on the very finest material and fungal strands. These animals are protected from many potential predators, but not from all. Predatory larvae of syrphid hoverflies crawl among the detritus to catch fly larvae but are generally seen at the water surface with their posterior breathing tubes poking just out of water (Kitching and Schofield 1986).

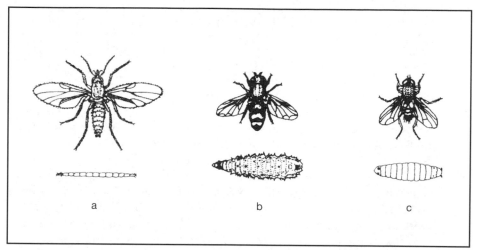

Figure 6.5. Members of the pitcher plant community. a - adult and larva ceratopogonid biting midge, b - adult and larval syrphid hoverfly, c - adult and larval anthomyiid root maggot fly.

Root maggot flies lay eggs in the unopened pitchers but by the time the larvae are ready to feed the first prey will be available (fig. 6.5) (Kevan unpubl.).

The food webs in pitcher plants in the seasonal Bogani Nani Wartabone National Park are relatively simple compared with those found in Peninsula Malaysia and in Morowali National Park (L. Clayton pers. comm.). This agrees with the hypothesis that complex food webs tend to occur only where the environment is relatively stable and predictable (Kitching and Schofield 1986; Kitching in press).

Around Lake Tempe, in those areas that are little disturbed by people, a form of swamp forest is dominated by *Gluta renghas* (Anac.) (Anon. 1977), which has a pinkish-brown, warty, puckered fruit. The seed can apparently be eaten after roasting (*Gluta* is in the same family as the cashew *Anacardium occidentale*). Otherwise its parts should be regarded as poisonous and it is unwise even to shelter under these trees during rain, since the resin washed from the leaves and branches can cause severe itching. The trees can usually be recognized by the black stains of dried sap on the bark, dippled scaly bark, and spirally-arranged leaves (Burkill 1966; Ding Hou 1978).

RIVERINE FOREST

Vegetation

Riverine forest is found along rivers, influenced by both the wet conditions and by the new sediment deposited by rivers on the inside of meanders. The only location in Sulawesi where riverine forest has been studied in any detail is in the Sopu valley northeast of Lake Lindu and Mt. Nokilalaki (van Balgooy and Tantra 1986). The forest here was dominated by the distinctive *Eucalyptus deglupta* (Myrt.) (fig. 6.6), some of which reached 60 m in height with a 1.5 m girth but in some areas (see below) it grows even taller[7]. A 20 m x 100 m transect in the Sopu valley contained 115 trees of at least 10 cm girth belonging to 47 species, 38 genera and 28 families, but very few of these are restricted to riverine forest. Some trees such as the sugar palm *Arenga pinnata* and the pandan *Pandanus sarasinorum*, are common along the main stream but not the tributaries, but others such as *Pigafetta filaris* (Palm.) and *Erythrina* (Legu.) were common along the tributaries but not the main stream. Other large trees in the Sopu riverine forest were *Duabanga moluccana* (Sonn.) (fig. 6.7), *Ficus* (Mora.) and *Elmerillia ovalis* (Magn.). One of the smaller trees *Saurauia oligolepis* (Acti.) is of special interest because its inflorescences grow from the base of the tree, but, descend into the soil and the white flowers are finally seen up to two metres away from the tree. This phenomenon of geocarpy—the pushing of fruit into the soil by the inflorescence—was observed in a number of species present in this forest (van Balgooy and Tantra 1986), and is also known in riverine species of figs (S. C. Chin pers. comm.). One possible advantage of geocarpy is that a buried fruit is less likely to be swept away by water when a river is in flood, although dispersal of fruits away from the parent plant is the more usual strategy (p. 380).

Riverine forests in Lore Lindu National Park are reported to contain *Octomeles sumatrana, Eucalyptus deglupta* and *Duabanga moluccana* (Wirawan 1981), whereas *Dracontomelum dao* (Anac.) and the red-barked *Pometia pinnata* (Sapi.) are common along both the Tumpah River near Toraut (Whitmore and Sidiyasa 1986) and the Boliohuto River in the west of Bogani Nani Wartabone National Park (EoS team).

In addition, it is reported from 50 years ago that the Dumoga valley, now the site of a major irrigation scheme (p. 629), was often under water and much of this area may once have been extensive riverine forest. The forest even then was in a disturbed or secondary state but the trees mentioned include *Octomeles sumatrana* (Dati.), *Koordersiodendron pinnatum* (Anac.), and *Eucalyptus deglupta* (Myrt.) (Steup 1933).

Eucalyptus deglupta dominates almost all riverine forests which are more or less undisturbed. It has the distinction of being the first-described

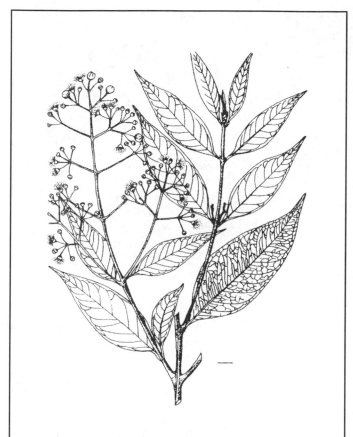

Figure 6.6.
Eucalyptus deglupta.
Scale bar indicates
1 cm.
After Soewanda n.d.b.

species of *Eucalyptus,* and it is now know from Mindanao, Sulawesi, Seram, New Guinea, and New Britain, but it is grown as a timber and reforestation tree throughout the tropics. In Sulawesi it has been collected in all provinces except Southeast Sulawesi (Davidson 1977; Cossalter 1980) but it is almost certainly found there. Efforts have been made by foresters since the 1970s to make full use of existing *E. deglupta* genetic resources in order to assist in the domestication and improvement of the species. Evaluation of early experiments show that *E. deglupta* from Sulawesi are among the best for reforestation of wet lowland sites (Davidson 1977).

Eucalyptus deglupta is the only one of its huge genus (about 500 species) that is regularly found in a rain forest. It has the thinnest bark (3 mm) of all eucalypts and is, as a consequence, very sensitive to fire which is of course extremely rare in the moist habitats it inhabits (Pryor 1976). Stands of *E. deglupta* are generally dense and comprise individuals of similar size.

Figure 6.7. *Duabanga moluccana.* Scale bar indicates 1 cm.

After Soewanda n.d.b

Mature stands at Tobou-Tobou on the Tomini River, south of Lake Poso, have 30-40 trees per hectare with the average tree reaching 65-70 m, 40-45 m of which is clear bole, and a diameter at breast height of 115 cm. The tallest trees reach 75 m to 80 m, beneath which a dense lower stratum of forest has developed. Occasionally, however, mixed-age stands are found, such as at the confluence of the Sopu and Lindu Rivers, where the river-banks are unstable, and a mosaic of different-aged stands from seedlings to giant trees can (used to be) be found on the banks of both rivers. Similar stands can also be found near Sedoa, Wuasa and Torire on the upper part of the Lariang River. It is found in small groups at the base of steep mountain slopes, such as along the Tumpah River in the area used by Project Wallace in Bogani Nani Wartabone National Park, whereas it is found only occasionally on the banks of the five broad rivers flowing into the north of the Gulf of Bone (Cossalter 1980).

The flowering of *Eucalyptus deglupta* occurs at the peak of the wet season when vegetative growth also reaches its maximum. The feathery flowers are pollinated by beetles, flies (syrphids and callophorids) and bees. Twelve weeks later the buds have developed into mature fruit or capsules but it is two or three more weeks, usually at the end of the wet season, before the fruits open and the seeds are dropped onto the newly-emerged gravel or silt along the riverbed which has been swept clear of the vegetation unable to withstand high river flows. The diminishing river flow strands seeds on these banks and the seeds are therefore given the best possible start (Cossalter 1980).

Eucalypt seeds are dispersed predominantly by wind although for *E. deglupta* dispersal by water is probably also important. In fact, the distance a eucalypt seed falls from its parent tree can be calculated from the equation $D = V_w H/V_t$ where 'D' is the distance a wind[8] of velocity 'V_w' will blow a seed falling with a terminal velocity 'V_t' from a height 'H'. Of 15 species of *Eucalyptus* used in experiments in Australia, *E. deglupta* had the lightest seeds (4,210 seeds/g), the lowest terminal velocity (2.1 m/s), and the greatest dispersal distance: (52.9 m) if released from 40 m with a wind of 10 km/h. At the other extreme, *E. globolus* had only 71 seeds/g, which travelled at a terminal velocity of 5.54 m/s and dispersed only 20 m under the same conditions (Cremer 1977).

Fauna

Animals of the riverine forest would include certain frogs[9] such as *Occidozyga* spp, near small rocky rivers, and *Rana arathooni* and *R. leytensis* on the banks of somewhat broader rivers. The last two species can include both fish and large crickets in their diets (J. Dring pers. comm.). No birds are restricted to riverine forest although a number of species of kingfishers might be seen along the riverbanks more commonly then in the forest away from the river. The shy endemic rail *Aramidopsis plateni* is found in disturbed riverine forest near small rivers (Coomans de Ruiter 1946).

FOREST ON ULTRABASIC SOILS

Soils

Ultramafic rocks are dense, igneous in origin and are composed of magnesium and ferric minerals (hence the term 'ma' 'fic'). Ultrabasic rocks are those that contain less than 45% silica. Most ultramafic rocks are ultrabasic, and both are rich in iron, magnesium calcium, aluminum and heavy metals (Parry n.d.). In the following discussion only the term ultrabasic is used.

The soils that develop on ultrabasic rocks are notoriously infertile due to combinations of the following factors: high levels of exchangeable magnesium and a skewed calcium : magnesium ratio, a deficiency of calcium, nitrogen, phosphorus, potassium, molybdenum and zinc, and toxic concentrations of heavy metals such as nickel, cobalt and chromium. The infertility of these soils is well known to indigenous people and their distribution in many areas is more or less delineated by the boundary between cultivated and uncultivated soils (Parry n.d.). Examples of this are the area south of the Loa River near Kolonodale, and parts of Morowali National Park on the other side of Tomori Bay (D. Holmes pers. comm.). It is also of interest that the five-armed island of Padamarang, off the west coast of Southeast Sulawesi, comprises ultrabasic rocks and has little water, and as a consequence of both, the island has virtually no inhabitants. The low agricultural potential of ultrabasic soils has resulted in few studies being made of the soils, a singular disadvantage when plans for transmigration and other land development are considered.

Although patches of ultrabasic rocks occur in Sumatra, Borneo, Halmahera, Timor, Sumba and Irian Jaya, the most extensive, both in Indonesia and in the world, are in the east of Central and South Sulawesi and in Southeast Sulawesi where they cover some 8,000 km^2 (fig. 6.8). These rocks are largely composed of peridotite and serpentinite minerals. The residual soils are highly weathered, well-drained, friable when moist, and of a reddish colour. The concentrations of metals are generally high, exhibiting a considerable range, but the amounts available to plants are relatively small and appear to bear no significant relation to the total content (table 6.1). Alluvial and colluvial soils[10] also have high concentrations of heavy metals and magnesium although these are lower than for residual soils (table 6.2).

Little is known about the action of chromium in the soil except that it is a non-essential element for plants and relatively inert. Some authors regard it as being severely toxic, particularly in the form of chromates, but others regard it as only moderately toxic. Concentrations of nickel in ultrabasic soils can be ten or more times greater than in other soils. As indicated above, however, the amount of nickel available to a plant will be much less, but even as little as 3 µg/g of soil can be toxic to certain plants. Other plants, however, are able to tolerate much higher concentrations and to concentrate the metal in their tissues such that 5%-25% of their mineral (ash) content is nickel (Lee et al. 1977). Samples of the shrub *Psychotria dourrei* (Rubi.) from New Caledonia have revealed the highest known concentration of nickel in plants—2.2% by weight of its dry leaves (Baker et al. 1985). The ingestion of such plants by humans can lead to heavy metal poisoning. These accumulator plants represent one end of a spectrum of responses to high metal concentrations, the other extreme being plants that actively exclude the metal. Some other plants may act as indicators; that is, there is a correlation between metal levels in the soil and concentrations in the plants. Some other plants, however, may act as an accumulator, indicator and excluder across a range of soil metal concen-

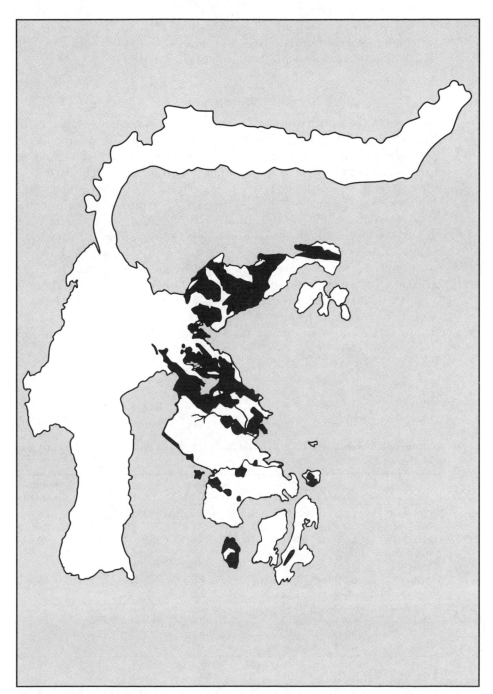

Figure 6.8. Areas of ultramafic (ultrabasic) soils (in black) on Sulawesi.

After Parry n.d.

trations. Accumulators tend to be those species that are confined to soils with high metal concentrations, and excluders tend to be those with both metal tolerant and metal-intolerant traces (Baker 1981).

Vegetation

Nowhere in the tropics with the exception of New Caledonia (Jaffre 1976) has forest on ultrabasic soils been examined in detail; a consequence, in part, of its low agricultural potential and the virtual absence of marketable timber. It would be expected that a fairly high proportion of the plants growing on these soils would be confined to them, having become adapted to the exceptional soil conditions. These would be of two types: those that have evolved on the islands of ultrabasic soil, and those that were once widely distributed but are now found only on ultrabasic soils. Some of the plants confined to ultrabasic soils may actually require high concentrations of magnesium, nickel or even chromium (Proctor and Woodell 1972). In a 500 m x 10 m transect through an ultrabasic area on the western shore of Lake Ranu in Morowali National Park the trees of 15 cm diameter or more at breast height had a basal area equivalent to 31 m^2/ha which, with 348 trees/ha, shows an average basal area of 0.09 m^2/tree which is only

Table 6.1. Summary of soil analyses from residual ultrabasic soils in Southeast Sulawesi.

	Units	Mean	Standard deviation	Range	Number of samples
Total Ni	µg/g	3,943.0	1,972.0	1,028-9,000	24
Avail. Ni	µg/g	85.1	65.3	0.4-217.5	24
Total Cr	µg/g	5,911.0	2,788.0	1,080.0-11,630	24
Avail. Cr	µg/g	1.4	1.2	0.1-4.3	24
Total Cu	µg/g	68.0	27.0	22-108	19
Avail. Cu	µg/g	1.8	1.2	0.3-4.5	19
Total Zn	µg/g	212.0	173.0	31-547	20
Avail. Zn	µg/g	1.9	1.2	0.1-5.3	20
Total Mn	µg/g	2,950.0	2,200.0	868-9,295	20
Avail. Mn	µg/g	439.0	486.0	2.4-1,200	20
Avail. P	µg/g		1.5	0.1-4.8	18
Exch. Ca	µeq/g	29	40	1-133	19
Exch. Mg	µeq/g	138	138	17-490	19
Ca : Mg	-	0.3	0.3	0.01-13.3	19
pH H$_2$O	-	5.3	0.5	4.5-6.7	20

From Parry n.d.

slightly less than trees in other lowland forests (p. 350). A histogram of tree heights in the 0.5 ha transect shows how few understorey (<10 m) trees were present and that more than half the trees were between 10 m and 20 m in height (fig. 6.9) (L. Clayton pers. comm.). A 2 m x 30 m transect in this forest was enumerated and its profile shows that the trees are relatively short and scrubby compared with trees on, for example, alluvium (fig. 6.10). Most of the species were either *Ficus* or members of the Sapotaceae. Given the high concentrations of metals and low inherent fertility, it is not surprising that the vegetation on ultrabasic soils tends to be relatively low and scrubby, and as such it is easily distinguishable on satellite images or air photographs.

The ultrabasic forest around Soroako has a great deal of local iron wood *Metrosideros* (Myrt.), some *Agathis* (Arau.), much *Calophyllum* (Gutt.), various Burseraceae and Sapotaceae and at least two dipterocarps (*Vatica* and *Hopea celebica*). This forest has a relatively regular, low canopy and the only emergents are the *Agathis* but most of these have now been felled. The dominant family reaching the upper canopy is the Myrtaceae such as *Eugenia*, *Kjellbergiodendron* and *Metrosideros*. The nutmegs (Myri.) such as *Horsfieldia*, *Gymnacranthera* and *Knema* (including the endemic *K. celebica*

Table 6.2. Summary of soil analyses from alluvial ultrabasic soils in Southeast Sulawesi.

	Units	Mean	Standard deviation	Range	Number of samples
Total Ni	µg/g	2,104	839	825-4321	24
Avail. Ni	µg/g	52.7	46.8	1.7-185	24
Total Cr	µg/g	3,068	3,537	290-11,923	24
Avail. Cr	µg/g	2.3	2.1	0.1-0.7	24
Total Cu	µg/g	54	34	23-124	24
Avail. Cu	µg/g	1.8	1.1	0.4-0.4	24
Total Zn	µg/g	101	314	29-946	24
Avail. Zn	µg/g	1.8	0.9	0.5-4.0	24
Total Mn	µg/g	1,901	1,747	440-4,783	29
Avail. Mn	µg/g	236	229	1.4-724	29
Avail. P	µg/g	3.4	2.8	0.01-12.0	24
Exch. Mg	µeq/g	107	133	4-542	20
Ca : Mg	-	0.14	0.12	0.1-0.4	20
pH H$_2$O	-	6.1	1.1	3.9-7.3	24

From Parry n.d.

Figure 6.9. Histogram of tree heights along a 10 m x 500 m transect of forest on ultrabasic soil.

After L. Clayton pers. comm.

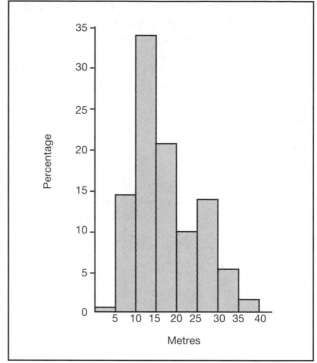

with forked [emarginate] tips to its leaves) are common below the tallest trees (de Wilde 1981; Meijer 1984; van Balgooy and Tantra 1986). A conspicuous tree is *Deplanchea bancana* (Bign.) which has dense heads of yellow flowers (which attract large numbers of sunbirds), and long, oblong pods of winged seeds (van Balgooy and Tantra 1986).

At 700 m on Mt. Konde, west of Soroako, the forest on ultrabasic soils has a regular closed canopy reaching 30-35 m above the ground. Few trees have trunks more than 60 cm in diameter and few have buttresses. The tree crowns are compact and the leaves are generally dark and leathery. No single family predominates and the most common large trees are *Eugenia* (Myrt.), *Ficus* (Mora.), *Kjellbergiodendron* (Myrt.), *Lithocarpus* (Faga.) and *Santiria* (Burs.). A 0.2 ha plot contained 234 trees of 10 cm diameter or more at breast height belonging to 36 genera and 27 families. It is not possible to calculate an index of diversity to compare with other sites because some of trees have not yet been identified to species.

Around the south shore of Lake Matano, to the east of Soroako, the ultrabasic vegetation comes right to the water's edge. It is dominated by just one family, the Myrtaceae, and most of the forest trees are less than 15 m tall. The main species are *Metrosideros petiolata* (fig. 6.11), with cream-

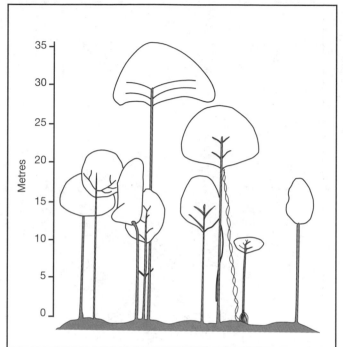

Figure 6.10.
Profile of 2 m x 30 m of forest on ultra-basic soils in Morowali National Park. Horizontal and vertical scales are the same.

After L. Clayton pers. comm.

coloured flowers and opposite leaves, and *Xanthostemon confertiflorum,* with bright red flowers and alternate leaves. Fruit measuring 8 cm x 15 cm found floating in the water belong to a lakeside *Kjellbergiodendron* with a gnarled trunk and spreading branches, which is probably not the same as *K. celebicum* found in inland forests. The spongy flesh around the single seed is apparently eaten by bats. The tallest trees in the community are the relative of the she-oak *Gymnostoma sumatrana* (Casu.), *Planchonella* (Sapo.), and the endemic *Terminalia supitiana* (Comb.). The most conspicuous among the smaller is an *Elaeocarpus* (Elae.) with yellowish-green fruit and white flowers. Epiphytic on the trees are *Drynaria* and *Lecanopteris* ferns and *Hydnophytum* (fig. 6.12), and the similar *Myrmecodia* (Rubi.), all of which are served by ants bringing organic frass into the enlarged, chambered stem (van Balgooy and Tantra 1986). Such ant-plant interactions appear to be relatively common in low-productivity vegetation. This is possibly because the organic frass dropping into or around epiphytes from the tree canopies is inadequate or of too low a quality for most epiphytes to grow. In such locations plants in which ants can deposit large quantities of insect (mainly ant) parts will be relatively successful (Janzen 1979). Although not yet investigated for Sulawesi, there is generally a preponderance of plants on

Figure 6.11.
Metrosideros petiolata.
Scale bar indicates 1 cm.
After Soewanda n.d.a

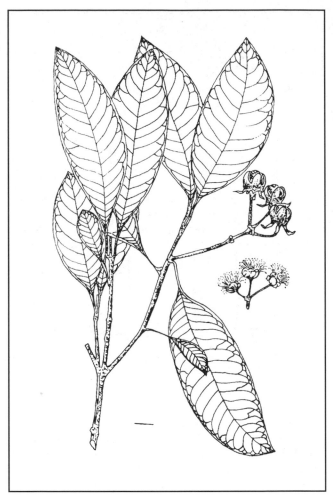

infertile soils such as ultrabasics that have seeds adapted for dispersal by ants; this may be due to ants being particularly common in such forests or to some additional factor.

Analyses of leaf protein, fibre, ash, carbohydrate and major minerals from four species of trees growing nearby, but on ultrabasic soils, were compared with results from analyses of leaves from four different tree species growing on alluvial soils in Morowali National Park. No significant differences were found, but when analyses were made comparing the composition of leaves from four species found on both ultrabasic and alluvial soils it was found that protein concentrations were consistently lower in leaves from the ultrabasic sites (medians of 6.00% and 8.25%), and fibre

Figure 6.12. *Hydnophytum* with the enlarged stem cut away to show the chambers inside.

concentrations were generally higher (medians of 23.0% and 19.7%) (L. Clayton pers. comm.).

One area in which an ecological understanding of ultrabasic vegetation and its succession processes could be of significant benefit is in the regreening of industrial sites on ultrabasic soils. The major examples of this are the P.T. INCO mining and hydropower sites near Soroako south of Lake Matano where revegetation of the bare areas caused by these operations has been generally slow. Many of the plants used in regreening programs have reflected a certain degree of tunnel vision among the proponents: plants such as *Hibiscus* and pines, neither of which is found naturally on ultrabasic soils, are unlikely to be much value.

In flat areas where the top soil is cleared the tree small *Alphitonia* (Rham.) appears to be one of the best colonizers, but on steeper areas revegetation is very slow although eventually the stiff grass *Miscanthus sinensis* (Gram.) forms a dense cover, sometimes with the yellow-flowered

Scaevola oppositifolia (Good.) (fig. 6.13). Also there are fast growing trees on exposed soils in the mining area such as *Alphitonia, Macaranga gigantea* (Euph.), *Homalanthus* (Euph.), *Callicarpa* (Verb.) and *Trema amboinensis* (Urti.), and these pioneers could protect young trees of more mature forest growing beneath them. *Baeckia frutescens* (Myrt.) with *Gleichenia* ferns together form an effective shrub cover. Many garden plants do not thrive on ultrabasic soils without expensive dressing, but suitable plants can be found in forest on ultrabasic soils that have been or are soon to be logged. One such plant is the endemic gardenia *Gardenia celebica* (Rubi.) (Meijer 1984 pers. comm.).

Fauna

There is very little known about animals on ultrabasic soils in Sulawesi or anywhere else in the world (Proctor and Woodell 1968), but the little information which does exist indicates some certain peculiarities. For example, a butterfly and its larvae are restricted to ultrabasic soils near San Francisco Bay, California. The plant used as food by the caterpillar is not restricted to such soils, however, nor has any chemical difference between the plants in the butterfly area and those outside it been detected. In the laboratory the larvae will eat plants from either soil type with no ill effects. The butterfly distribution may, therefore, be a result of microclimate differences caused by different vegetation structure (Johnson et al. 1968).

Differences in vegetation have also led to the occurrence of a large population of a gopher[11] in another area of California. Its principal food comprises the corms of a widespread plant that grows abundantly on ultrabasic soils, possibly because of reduced competition for light. The corms in question have three to four times as much magnesium as calcium in their tissue but how the animals are adapted to this is not known (Proctor and Whitten 1971).

The reports of experienced field workers who have visited ultrabasic areas in Sulawesi agree that the density of vertebrate animals is, in general, very low. Whether this is due to low productivity, high levels of defence compounds in the leaves, low concentrations of major nutrients in leaves (Jaffre 1976), or toxic concentrations of metals in the plant parts is not yet known (p. 458). An understanding of the mechanisms are important if appropriate conservation policies are to be made for Sulawesi.

A brief survey of birds in different areas around Soroako found that the least number of species was found in vegetation on ultrabasic soils although to some extent this may have been due to difficulties of observation. Five species, there of which were endemic to Sulawesi, were found only in ultrabasic areas. It also appeared that some species of flycatchers and starlings had a preference for ultrabasic areas but, again, this may be an artefact of the brevity of the survey (Holmes and Wood 1980).

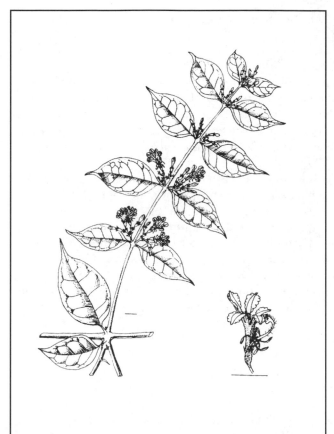

Figure 6.13. *Scaevola oppositifolia.* Scale bar indicates 1 cm.
After Leenhouts 1957

Five pitfall traps baited with shrimp paste were set 10 m apart in both ultrabasic and alluvial forests in Morowali National Park to compare the invertebrate fauna. Each trap was left for 20 hours before collecting the animals caught in the water at the bottom of the trap. One of the major differences was that many more small fruit flies were caught in the alluvial area than in the ultrabasic area, possibly due to a greater abundance of fruit in the more productive alluvial forest. Ants were visibly more conspicuous in the ultrabasic forest, reflected somewhat in the pitfall trapping, and this may be explained by the large number of plants possessing mutualistic associations with ants. The greater abundance of ants was also demonstrated by comparing the times it took ants to find prices of shrimp paste in ultrabasic and alluvial forest. In the former it took just 2.9 minutes for the paste to be found whereas in the latter forest it took over 8 minutes (L. Clayton pers. comm.).

FOREST ON LIMESTONE

Physical Conditions

The limestone areas of Sulawesi (fig. 6.14) comprise rocks of various origins. There are Quaternary-Tertiary reef limestones found over much of Banggai, Togian, Muna, Buton and the coast north of Palu, Tertiary limestones formed in Eocene to mid-Miocene times in southwest and northeast Sulawesi, and Cretaceous limestone found near Lake Matano and on Buton (Sukamto 1975).

Most but not all limestone areas are described by geomorphologists as karst landscapes that have arisen from the abnormally high solubility of the bedrock (Jennings 1971). Calcium carbonate, or calcite (the major constituent of limestone), is not particularly soluble in pure water but is more soluble in weak acid. Carbonic acid arising from the solution in rainwater of carbon dioxide from the air and, to a lesser extent, humic acids arising from the soil as a result of vegetation decay, are thus effective in dissolving the rock. The process is rather complex with reversible reactions and ionic dissociations but in its simplest form can be summarized thus:

$$CaCo_3(\text{rock}) + H_2O + CO_2(\text{dissolved}) \longleftrightarrow Ca_2^+ + 2HCO_3^-$$

$$\updownarrow$$

$$CO_2(\text{in air})$$

The karst landscapes of Sulawesi are of two major forms typical of the humid tropics: conical hill karst, such as found in the north of Bone, on Buton and Muna (fig. 6.15) and tower karst such as the famous hills around Maros and Tonasa (fig. 6.16). As in many categorizations of the natural world, intermediates are found, but in general the two forms are distinguishable.

The hills east of Maros and Tonasa (fig. 6.17) comprise about 300 km^2 of Eocene and Miocene coral limestone. These overly the limestones of former backreefs and lagoons (fig. 6.18). The limestones were uplifted and exposed briefly to erosion in the early Miocene before being covered by the products of volcanic eruptions from a later period in the Miocene. Since then the volcanic rocks have eroded away and the exposed limestones have been dissolved by intense rainfall (up to 400 mm per day) and dissected by rivers (Balazs 1973).

The Maros hills are mostly between 150 m and 300 m high but the tallest reaches 575 m. The area receives an annual rainfall of about 3,500 mm which causes a great deal of surface runoff. The morphology of the hills was probably caused by rivers which arise in the east and flow to the west through caves which the rivers deepened and widened. The roofs

Figure 6.14. Limestone areas (in black) of Sulawesi.

of the caves eventually collapsed forming steep-sided valleys. The steepness of the valleys is emphasized further by undercutting caused by the erosive action of meandering rivers, some distance beneath the surface of the plain, which tend to flow around the hill bases rather than across the alluvial plain. The undercutting forms rock shelters some 2 m to 3 m high, tens or hundreds of metres in length, and usually 1 m or 2 m deep although some passages extend far into the hills (Jennings 1976; MacDonald 1976).

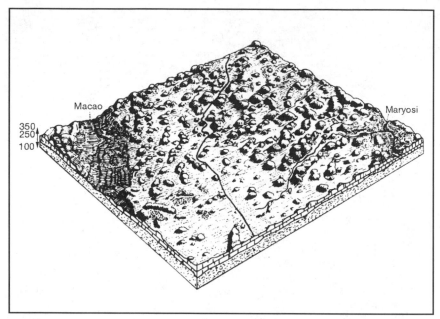

Figure 6.15. Diagrammatic representation of conical hill karst in north Bone between the Maryosi and Macao Rivers.
After Sunartadirdja and Lehmann 1960

Whereas tower karst may have arisen from conical hill karst, not all conical hills necessarily progress to towers. The processes by which conical hills are formed are less clear, but they are possibly formed by streams flowing between blocks of limestone. The process is quite rapid; even the coral reefs raised during the Quaternary to form much of Muna Island have conical hills (fig. 6.19) and various factors have combined such that these hills are better developed than those in other limestone areas. In adjacent areas of raised coral lagoons, sink holes or dolines up to 150 m across can be found. Dolines also occur in the Maros tower karst, but these are not as wide nor as frequent (Balazs 1973; Anon. 1986). These are more typical of karst in temperate regions and look like pock marks on an otherwise relatively flat surface. In some areas of the north Bone conical-hill karst, such as the Macao River, steep canyons have formed where a previous subterranean river has dissolved away the roof of its course until it collapsed (Verstappen 1957).

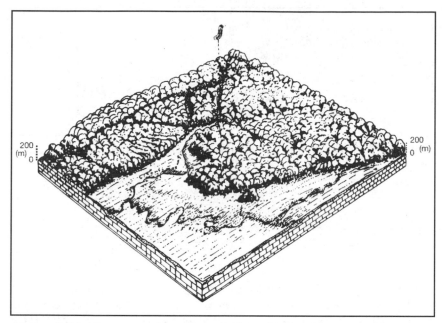

Figure 6.16. Diagrammatic representation of lower karst near Maros and Tonasa showing the position of the Bantimurung waterfall.
After Sunartadirdja and Lehmann 1960

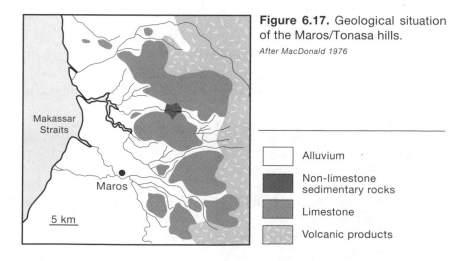

Figure 6.17. Geological situation of the Maros/Tonasa hills.
After MacDonald 1976

Makassar Straits

Maros

5 km

☐ Alluvium

■ Non-limestone sedimentary rocks

▨ Limestone

▦ Volcanic products

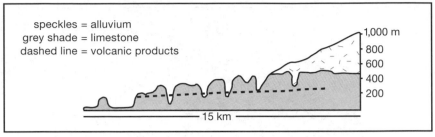

Figure 6.18. Cross-section through the hills east of Maros.
After MacDonald 1976

Soils

Soils on limestone, not surprisingly, are often richer in bases particularly calcium and magnesium, with a higher cation exchange capacity than soils in similar situations on different parent materials. On moderate slopes and hollows, clay-rich, leached brownish-red latosols are formed (Burnham 1984). On crests and shelves an acid, humus-rich, peat-like soil can sometimes develop on top of the limestone. On steep slopes and craggy hill tops the soil is very shallow. The composition and depth of soil can be variable depending on the purity of the parent material and the topography (Crowther 1982, 1984). The soils of the conical hills of Kambara on Muna Island were rendzinas,[12] 5-25 cm deep, well-drained and with a pH of 7-7.5 (Anon. 1982). Those examined by an EoS team at Hanga-Hanga waterfall, Luwuk, had high concentrations of calcium and magnesium (82 and 44 meq/100 g soil respectively).

Hydrology

Rain penetrating any soil over permeable rock passes downwards through an unsaturated (or vadose) zone, where rock pores are only temporarily filled with water, into the saturated (or phreatic) zone. The upper surface of the phreatic zone is called the watertable. The watertable is more or less parallel to the land surface and the groundwater flows according to the slope of the watertable. Springs are encountered where the watertable intersects the surface. There is an intermediate zone in which the watertable rises and falls. Thus the same opening that act as spring in the rainy season act as drainage openings in the dry season (Jennings 1971).

The actual nature of water circulation in karst areas is not fully understood and hypotheses vie for acceptance. An understanding is of great

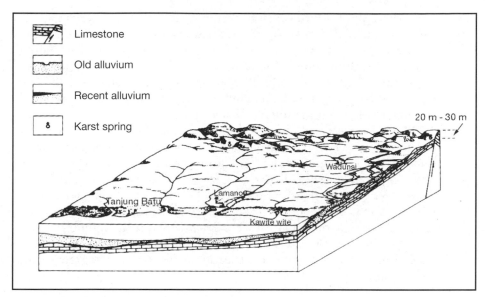

Limestone

Old alluvium

Recent alluvium

8 Karst spring

20 m - 30 m

Wadunsi

Tanjung Batu

Lamano

Kawite wite

Figure 6.19. View east over the west coast of Muna Island to show conical hill karst.

After Anon. 1982

importance to the development of karst regions because very little water is found near the surface due to the extreme porosity of the rock. For people living in karst areas the supply of water can be critical and the famines in the Mt. Kidul area of Yogyakarta in the early 1970s were a direct result of bad land management in a critical karst area. Also, water issuing from karst springs is extremely pure and less variable in flow compared with rivers outside karst areas. The phreatic zone thus represents a significant reservoir of groundwater recharge for domestic, agricultural or industrial supplies although significant groundwater recharge may only occur after heavy rainfall (Crowther 1984).

It is extremely difficult to determine the exact drainage area of an underground river because topographical and geological catchments are often not identical. Even so, the mean annual runoff from the 300 km^2 Maros karst has been estimated as 35-40 l/s/km^2, which removes in solution about 80 m^3/km^2/yr of limestone (Balazs 1973).

Despite being relatively stable in its flow, the discharge of karst springs in the Maros area varies according to seasonal patterns of rainfall. Some cave rivers often flow during the rainy season and are dry for the rest of the year, whereas others with a larger catchment are permanent. In several areas around the Maros karst, villagers have dammed cave springs or rivers to create reservoirs so that outflow can be controlled and availability increased. Such impounded water is used mainly for domestic purposes. Most of the irrigation water in the Maros region originates from the karst region and, loaded with calcium and magnesium cations it may serve the additional role of raising the pH of acid soils.

The seasonal variation of karst water chemistry around Maros is unknown but even at low flow periods there are evident, albeit slight, daily fluctuations in pH, alkalinity and total hardness in the spring water emerging from Baharuddin cave. Organic debris was found trapped 3-4 m above the surface of the river in Salukan Kalang Cave during a dry period indicating that floods sometimes occur that fill virtually the entire cross-sectional area of some active cave passages. The large volume of water during the wet season and the shorter time water would spend in the cave would presumably dilute the concentrations of bicarbonate, magnesium and sodium.

Vegetation

The appearance of the vegetation around the hills of Bantimurung was described by Wallace as follows:

> Such gorges, chasms, and precipices as here abound, I have nowhere seen in the Indonesian Archipelago. A sloping surface is scarcely anywhere to be found, huge walls and rugged masses of rock terminating all the mountains and inclosing the valleys. In many parts there are vertical or even over-hanging precipices five or six hundred feet high, yet completely clothed with a tapestry of vegetation. Ferns, Pandanaceae, shrubs, creepers, and even forest trees, are mingled in an evergreen network, through the inter-stices of which appears the white limestone rock or the dark holes and chasms with which it abounds. These precipices are enabled to sustain such an amount of vegetation by their peculiar structure. Their surfaces are very irregular, broken into holes and fissures, with ledges overhanging the mouth of gloomy caverns; but from each projecting part have descended stalactites, often forming wild gothic tracery over the caves and receding hollows, and affording an admirable support to the roots of the shrubs, trees, and creepers, which luxuriate in the warm pure atmosphere and the gentle moisture which constantly exudes from the rocks. In places where the precipice offers smooth surfaces of solid rock, it remains quite bare, or only stained with lichens and with clumps of ferns that grow on the small ledges and in the minutest crevices.

Compared with forests on deeper soils, forests on limestone generally have few trees and tree species (Crowther 1982; Proctor et al. 1983a, b), although the total number of plant species present is probably not dissim-ilar. The basal area of trees on moderate slopes (<45%) can be quite sim-ilar to other forests and only on steeper slopes and rocky hilltops do soil conditions seriously affect tree growth. The shallow limestone soils may be able to support relatively high basal areas of trees because of the relatively fertile condition of the soils. A representative patch of forest on steep limestone 30 m x 2 m was enumerated along a gorge cut by the Tomasa River near Kuku, Poso, to produce a forest profile (fig. 6.20). It can be seen that in the area sampled there were few large trees, but this may at least in part be due to the steepness of the gorge (>100% or >45°). Only five of the 13 trees were buttressed (L. Clayton pers. comm.).

The relative paucity of tree species probably arises because some trees of lowland forest cannot tolerate the high calcium levels in the soil. The different tolerance of trees to calcium and the unique physical habitat makes the composition of these forests rather different from other lowland forests, giving rise to a specific community of trees on limestone. On steep limestone cliffs with bare rock faces, clefts and shelves, a distinctive herbaceous flora occurs. Many herbaceous plants are known to be endemic to the limestone hills of Peninsular Malaysia: for example, of 1,216 plants recorded as growing on them, 254 (21%) are found only on such hills and over half of these are endemic to the Peninsula (Chin 1977, 1979, 1983a, b). A species of grass *Cymbopogon minutiflorus* appears to be endemic to limestone areas of Central Sulawesi (Dransfield 1980), and many others may be found when intensive collecting is undertaken. It has been predicted that rare plants occur on the Maros hills although these are much younger geologically than the limestone hills of Peninsular Malaysia (van Steenis 1933).

Most of the information regarding limestone flora on Sulawesi concerns parts of Maros and Rantepao/Makale (Toraja) largely due to visits paid by an EoS team led by Dr. S. C. Chin of Universiti Malaya. Brief surveys showed that virtually all of the accessible parts of the Maros and Toraja limestone hills have been exploited to some degree, but accessibility has to be measured by the standards of the ingenious villagers who manage to reach rugged outcrops and steep gullies using ropes and bamboos in their search for firewood and other forest products such as rattan and fruits of *Pangium edule* (Flac.) (p. 412) and candlenut[13] *Aleurites moluccana* (Euph.) many of which are planted (N. Wirawan pers. comm.). In addition, sap is collected from the sugar palm *Arenga*.[14] Leaves of *Garcinia* are added to the fresh sap apparently to delay fermentation. When the sap is boiled down, these leaves are removed and candlenut seeds and other hard objects are added to stop the sugary brew from boiling over (Burkill 1966; S.C. Chin pers. comm.). From the evidence available it is assumed that all the hills within several kilometres of the surrounding villages have experienced some significant degree of disturbance.

The upper parts of the limestone hills of Maros are clothed in small trees growing 7-10 m tall out of cracks and small pockets of soil. The scrub vegetation likewise depends on the availability of these cracks and soil

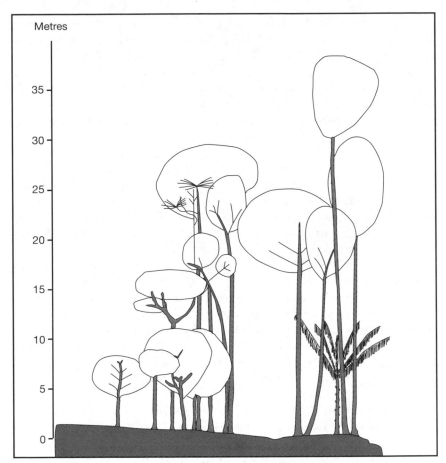

Figure 6.20. Profile diagram of 30 m x 2 m of forest on limestone from Kuku, Poso. Only trees of 15 cm diameter at breast height are shown. Horizontal and vertical scales are the same.

After L. Clayton pers. comm.

pockets. On the steepest cliffs the most common trees clinging to the rocks are figs *Ficus* (Mora.) the roots of which clamber over the rocks, in and out of crevices, for as far as 70 m. Such an extensive root network allows these plants to exploit wide areas for water and nutrients.

In undisturbed valleys, trees grow to have a diameter at breast height of up to 50 cm and common trees are *Pangium edule* (Flac.), *Artocarpus* (Mora.) and figs *Ficus*.

The composition and development of the vegetation of the Maros hills is determined not only by the concentrations of calcium, relative abundance of soil, and the peculiar biogeographical position of Sulawesi as a whole, but also the relatively long dry season (p. 22). Many of the plants in the area have clear adaptations to this climate—some are deciduous, losing their leaves in the dry season and remaining more or less dormant until the rains begin; some have fleshy stems in which they store water, some have very stiff, coriaceous leaves, and some have their leaves reduced to spines, both of which also reduce water loss. Some plants, such as the yams *Dioscorea*, have large storage organs (rhizomes, roots and corms), and some have the ability to lose most of their water (except that within their protoplasm which resists drying out), and then to revive on rewetting, a phenomenon known as 'poikilohydry' (Whitmore 1984).

When a severe dry season occurs the tolerance of these plants may be exceeded and they may be forced to survive as annual plants; that is, they must complete their life cycle during the wet season so that their seeds can germinate when the following drought ends. Among those plants that appear to be annual in at least parts of Maros are the single-leafed *Monophylla* (Gesn.) and the maidenhair fern *Adiantum*. Fleshy plants there include a cactus-like *Euphorbia* (Euph.), the red-and-green *Kalanchoe pinnata* (Cras.) and the dragon tree *Dracaena* (Lili.).[15] *Kalanchoe* is an example of a plant that avoids severe water stress during droughts by closing its stomata during the day. This prevents normal photosynthesis from occurring and instead a pathway called CAM (Crassulacean Acid Metabolism) is adopted in which important photosynthetic processes can be carried out at night. In the relatively high limestone areas of Tana Toraja, such adaptations are also seen, especially on exposed rocky hill tops. Although the rainfall here is greater and more evenly distributed, and the humidity higher due to the clouds that frequently swathe the hills, there is also a greater intensity of sunlight and exposure to wind. The presence of plants with thick fleshy leaves such as *Kalanchoe* suggests that there is periodic water stress (Chin 1986).

The forest on limestone at 1,000 m on Mt. Wawonseru, west of Soroako, has an uneven canopy reaching a maximum of 40-45 m. This only occurs where the soil is relatively deep, however, and in places where the rocks are covered with thinner soil, only shrubs are found. No single family predominates although Lauraceae and Annonaceae are common. Interestingly, neither of these families is well represented on nearby Mt. Konde which is on ultrabasic soils, and none of the trees from Mt. Konde is important on this particular limestone hill. Among the largest trees on Mt. Wawonseru are *Bischofia* (Euph.), *Eugenia* (Myrt.), *Podocarpus* (Podo.) and *Vernonia* (Comp.). In the lower canopy there are many *Polyalthia* (Anno.) and *Antidesma* (Euph.) and below this the undergrowth is rather sparse. Orchids, wild pepper *Piper* (Pipe.) and a *Rynchoglossum* (Gesn.) are able to live on the bare rock (van Balgooy and Tantra 1986).

The forest on limestone near the tip of the northeastern peninsula visited by another EoS team differed structurally from many lowland forests on unexceptional soils by having no palms in a 0.5 ha transect (there are between 80-100/ha at Toraut). The forest had a generally small stature having a basal area of only 19.8 m²/ha and a basal area per tree of 0.056 m² (both values about half those obtained at Toraut) (p. 350).

The plants growing on limestone along the shores of Lake Matano are of mixed composition and also do not show dominance by any particular family among the trees, shrubs or herbs (de Vogel 1986).

An EoS team that visited Luwuk attempted to determine any differences there might be between insect damage on plants of different succession stages and on two different soil types. Insect damage was examined in the vegetation on steep limestone soils near Luwuk and in vegetation on richer soils near the edge of Bogani Nani Wartabone National Park (p. 373). According to prevailing theory, plants will make greater investments in leaf defences where the environmental conditions are relatively harsh. The steep, thin soils of the limestone area may be said to represent a harsher environment than the flat, alluvial soils at Bogani Nani Wartabone. The percentage total amount of leaf area removed by insects did indeed appear to be lower, and leaf toughness greater, in the limestone area than in Bogani Nani Wartabone, although concentrations of nitrogen and water showed no consistent pattern (S. Greenwood pers. comm.).

Effects of Disturbance

Limestone vegetation is destroyed in the process of limestone quarrying for the Tonasa cement factories (p. 563), but even the enormous quantity of limestone quarried (the factory supplies almost all the cement needs of eastern Indonesia), cannot be considered a threat to limestone-related ecosystems in Sulawesi because the area affected is very small compared with the total area of limestone. It should be stressed, however, that new quarries and extensions to the existing ones should be sited such that the environmental impacts are minimal. The secondary vegetation in these limestone areas is dominated by the introduced *Eupatorium* (Comp.) and *Lantana* (Verb.) the dense thickets of which probably hold back the succession process by some years. Trees found in the succession include *Homalanthus* (Euph.), *Lagerstroemia* (Lyth.), *Pterospermum* (Ster.), *Kleinhovia* (Ster.) and *Villebrunea* (Urti.). Given their aggressiveness in competing with *Eupatorium* and *Lantana*, some of these trees could reasonably be used in programs for the reclamation of weed-dominated lands on limestone soils.

Fauna

No vertebrates are restricted in their distribution to limestone hills but certain species of snails, which use calcium to form their shells, have been found to be restricted to very small areas of limestone elsewhere in Southeast Asia. From 108 species found on 28 limestone hills of different degrees of isolation in Peninsular Malaysia, 70 were known from single hills, and one hill had no less than seven species known only from its slopes. At the other extreme, however, one snail species from a genus that included some very restricted species, was found on 19 of the hills (Tweedie 1961). There are clearly major differences between species in their environmental tolerances.

The snails of limestone hills on Sulawesi have not been studied. All one needs to start such a study is a bucket of water. Leaf litter from the base of a limestone hill is stirred around in the bucket and after waiting for a few minutes it will be seen that the soil and much of the vegetation will sink while small snail shells, which usually have a bubble of air trapped inside them, float on the surface where they can be collected. Identification to species is a problem better left to a specialist and representative samples of unnamed but recognizably different species can be sent to a major museum.

One of the more spectacular butterflies known largely from Sulawesi limestone, presumably because its food plant is restricted to a limestone habitat, is the large swallowtail *Graphium androcles*[16] (fig. 6.21) made famous in the following description of the butterflies of the Bantimurung waterfall by Alfred Wallace:

> As this beautiful creature flies, the long white tails flicker like streamers, and when settled on the beach it carries them raised upwards, as if to preserve them from injury. It is scarce, even here, as I did not see more than a dozen specimens in all and had to follow many of them up and down the river's bank repeatedly before I succeeded in their capture. When the sun shone hottest around noon, the moist beach of the pool below the upper fall presented a beautiful sight, being dotted with groups of gay butterflies—orange, yellow, white, blue and green—which on being disturbed rose into the air by hundreds, forming clouds of variegated colours.

Other collectors followed Wallace and 25 years later in 1882 *G. androcles* could no longer be found although thousands of other species remained (Guillemard 1889). This may have been an effect of the seasons (p. 368), however, because 45 years later the butterfly was found to be numerous again as were other swallowtails and less spectacular butterflies (Leefmans 1927), but now the butterfly is more or less extinct around the waterfall and many others have fallen prey to commercial collectors (Anon. 1985). A farsighted gentlemen in Bantimurung has managed to raise butterfly larvae to adults but his enthusiasm has waned because so many of the adults are caught (illegally) and sold near the reserve such that population levels in surrounding areas have been decreasing (Soetjipto et al. 1982).

Figure 6.21. The upper surface of a fore and hind wing of *Graphium androcles*.

After Haugum et al. 1980

Monsoon Forest

For the purposes of this book, monsoon climates are defined as those areas falling within the climatic zones 'E' to 'H' in the Schmidt and Ferguson classification (p. 22) that are characterised by a long dry season.

Vegetation

It is a matter of debate and definition whether primary lowland monsoon forest actually still exists in Sulawesi or indeed anywhere else in Indonesia (van Steenis 1957; Whitmore 1984). Its original extent has been much reduced and the primary reason for this has been fire. During the dry season, monsoon forest trees and other plants are easily burned and repeated burning eventually results in a persistent grassland vegetation. None of the grasslands or savannas of Sulawesi are natural but are instead the result of human activities. In Sulawesi, as in Java, "fire for hunting, for pleasure, for pestering neighbours or neighbouring villages, by carelessness, for clearing land, for making land passable, for converting forest into pasture land, in short for innumerable purposes, has played havoc with the monsoon forest" (van Steenis and Schippers-Lammertse 1965).

Only one area of monsoon forest appears to have been studied in any detail, and detail, and that is the much-disturbed forest of Paboya Reserve east of Palu (Sidiyasa and Tantra 1984). The Reserve was originally established to safeguard the sandalwood trees *Santalum album* (Sant.) (fig. 6.22). The oil in the heartwood of this tree is extremely valuable and wild stands of the tree have suffered considerably in dry areas of Indonesia because of uncontrolled exploitation. There are relatively few sandalwood trees remaining in Paboya, partly because of illegal felling, and partly because of fire. Fires are set by cattle farmers who wish to encourage the growth of young grass. The fires kill the sandalwood seedlings (as well as *Acacia fernesiana* trees planted in a regreening effort some years ago), although sandalwood will shoot from its base when the above-ground parts are burned. In addition to sandalwood, the trees of the Paboya forest include *Duabanga moluccana* (Sonn.), *Ficus* spp. (Mora.), *Canaga odorata* (Anno.), *Harpulia* sp. (Sapi.), *Alstonia angustifolia* (Apoc.), *Elatostachys verrucosa* (Sapi.) and *Buchanalia arborescens* (Anac.). Where the forest has been disturbed, the most common trees are *Casuarina sumatrana* (Casu.), *Pittosporum ferrugineum* (Pitt.), *Mallotus philippensis* (Euph.) and the introduced *Cassia siamea* (Legu.) (Sidiyasa and Tantra 1984).

The forest comprises a simple, lightly-closed community of trees containing a large proportion of deciduous or leaf-shedding species. It contains relatively few species found in the everwet lowland forest and one of the few emergents in certain areas is *Salmalia malabarica* (Bomb.)[17] (fig. 6.23). One tree of monsoon forest which is also found in slightly wetter areas, particularly those near the coast, is the large legume *Pericopsis mooniana* (fig. 6.24) which has a hard, dark-red timber greatly sought after for making furniture (p. 679). Its pods are flat, up to 10 cm long and 4 cm wide and contain only two to four seeds.

In some strongly seasonal areas, repeated burning has given rise to a savannah with relatively fire-resistant trees such as *Morinda tinctoria* (Rubi.) (also found in wetter areas), *Acacia tomentosa* (Legu.) *Phyllanthus emblica* (Euph.) (fig. 6.25), tamarind *Tamarindus indicus* (Legu.)[18] teak *Tectona grandis* (Verb.)[19] paperbark *Melaleuca* (Myrt.), *Timonius sericeus* (Rubi.), *Garuga floribunda* (Burs.) (fig. 6.26), and the two large fan palms *Borassus flabellifer* and *Corypha elata* (Steup 1936, 1939; Metzner 1981; van Balgooy and Tantra 1986). The two large palms are easy to confuse but *Borassus* has blue-green or greyish leaves, a smooth trunk with a diameter similar to that of a coconut palm and no large spines on the leaf stalks. *Corypha* is generally more massive, its leaves are nearly 2 m in diameter and twice as large as *Borassus*, and it has small fruits up to 5 cm in diameter compared with the massive 18 cm fruits of *Borassus*. *Corypha* can be identified from a distance by the lines formed by the persistent leaf bases spiralling up the trunk. *Corypha elata* has the second largest inflorescence of any flowering plant[20] with the flowering branches projecting nearly 4.5 m above the top of the palm and setting hundreds of thousand of fruit. After fruiting, the tree dies in the same manner as the more familiar sago palm *Metroxylon sagu*.

Figure 6.22. *Santalum album*. Scale bar indicates 1 cm.

After Soewanda n.d.b

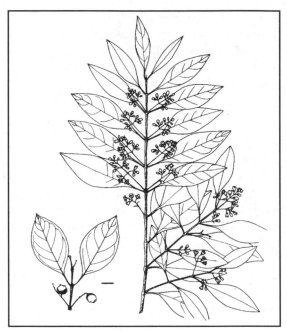

Figure 6.23. *Salmalia malabarica.* Scale bar indicates 1 cm.

After Soewanda n.d.

Figure 6. 24. *Pericopsis mooniana.* Scale bar indicates 1 cm.

After Soewanda n.d.a

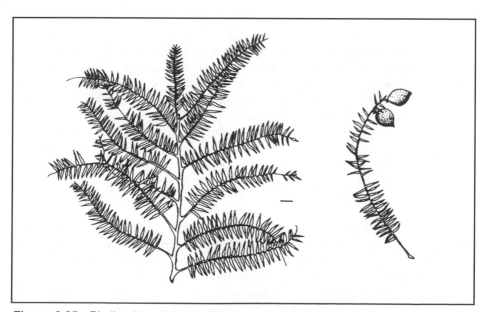

Figure 6.25. *Phyllanthus emblica.* Scale bar indicates 1 cm.

After Whitmore 1972

Figure 6.26. *Garuga floribunda.*
Scale bar indicates 1 cm.
After Leenhouts 1956

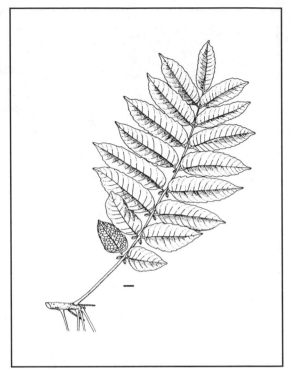

The Palu valley is the only locality in Indonesia that has the 'H' rainfall type of the Schmidt and Ferguson classification (p. 22) and it is not known exactly what its original vegetation would have been. Even in its present, greatly disturbed, condition it has many characteristic plants indicative of water stress such as *Acacia farnesiana* (Legu.), the dragon tree *Dracaena* (Lili.), *Capparis* (Capp.) and introduced plants such as the succulent-leafed, red-and-green flowered shrub *Kalanchoe pinnata* (Cras.), brought from Africa a very long time ago (Backer 1951), the tamarind *Tamarindus indica* (Legu.), and the prickly pear cactus *Opuntia nigricans* (Cact.) brought from South America. This plant is also found at the southern end of the southwest peninsula.

When the Palu district was visited in 1902, large stands of *Dracaena* were conspicuous and only a few *Opuntia nigricans* were seen (Sarasin and Sarasin 1905), but in 1911 dense stands of the cactus were abundant, particularly in abandoned rice fields (Grubauer 1923). By the late 1920s *O. nigricans* had become the dominant plant in the eastern foothills of the Palu valley and bay (Steup 1929) and in the mid-1930s the plant had taken over the entire northern part of the valley (fig. 6.27). Then, in 1934, the

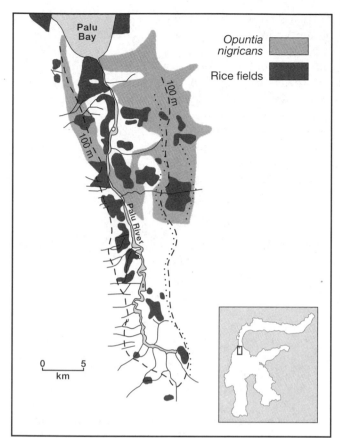

Figure 6.27. Distribution of rice fields and the cactus *Opuntia nigricans* in the Palu valley in 1935.
After Metzner 1981

cochineal scale insect or mealy bug *Dactylopius tomentosus*[21] was introduced from Australia where it had been used against *Opuntia*[22] (van der Goot 1940). By 1939 it had done its work and stands of the cactus were restricted to just a few localities and *Leucaena glauca* (Legu.) trees had taken over where the cactus had been killed (Bloembergen 1940). One would have expected the cactus and the scale insect to remain in small numbers, as is the case in Australia, but *O. nigricans* is now abundant once more, particularly around the new campus of Tadulako University, and it is possible that the scale insect has become extinct. If so, then this most unpleasant plant will increase until controlled, hopefully with a biological rather than a chemical agent. In parts of America *Opuntia* are exploited and domesticated for their fruit, and in dry periods the spines are burned off with flame sprayers which makes them acceptable as cattle food (C. E. Russell pers. comm.).

The extensive grasslands of the Palu valley comprise species such as *Cynodon dactylon* (Gram.), but the grasses are mixed with a few small legumes such as the yellow-flowered *Tephrosia*. Sword grass *Imperata cylindrica* is rarely seen except in areas which have been disturbed relatively recently. Some of the shrubs have spiny branches or thorns such as the introduced *Alternanthera* (Amar.), but their predominance may be an artefact caused by the presence of many goats and sheep which favour other elements of the vegetation (Meijer 1984). Another plant that seems not to be eaten by goats is the giant milkweed *Calotropis gigantea* (Ascl.) with distinctive large, white and wooly leaves, and this avoidance may be due to the copious white sap which has emetic properties (Burkill 1966).

Fauna

Nothing has been published specifically on the fauna of monsoon forests; it probably does not differ much from that found in other forest, except that the species would have to be able to withstand the long dry season, either by some behavioural or physiological adaptation of by moving away from the area.

One bird that favours drier areas is the Asian palm swift *Cypsiurus batasiensis,* which was first recorded from Sulawesi in 1978 (Escott and Holmes 1980). Its unusual nest is built in the centre fold of the upper surface of those leaves of large fan palms that are ageing, folded and hanging vertically. A delicate nest is built from feathers and feathery seeds (such as kapok *Ceiba pentandra* [Bomb.]) glued onto this vertical surface using saliva. The normal clutch is two eggs on which the adults sit in an almost vertical position. The diet of these birds can be investigated by examining the insect wings that are voided in the faeces dropped below the nest by the nestlings. Their food appears to comprise primarily of flying ants, termites and beetles caught by the adults about 10-15 m above the ground and which is brought to the nesting every 15-75 minutes through the day. The interval is shorter at the start than the middle of the day and this is related to the availability of suitable food (Hails and Turner 1984).

Certain species of tropical butterflies are represented by different adult forms in the wet and dry seasons, particularly in strongly seasonal areas. Common differences shown are that the wet season forms are active, more colourful and may have conspicuous 'eyespots' on their hindwings, whereas the dry season forms are very cryptic both in pattern and movement. It is assumed that these differences reflect changes in predation pressure (Brakefield and Larsen 1984; Janzen 1984). This has not yet been recorded from Sulawesi.

The Palu Valley: Past and Future

In the early 18th century the Palu valley was described as 'a blessed place' with fertile rice fields (Valentyn 1724), but two centuries later an agricultural engineer gave the opinion that "I also think that nowhere in the Archipelago has deforestation had such a fatal influence as in this place which otherwise, being richly endowed with water flowing down from the forest-covered mountains, could have been a little Egypt" (quoted Steup in 1929). By that time much of the former rice fields had been abandoned, partly because of problems of internal security and partly because the vast amounts of soil debris washed down from the deforested slopes had made the irrigation systems inoperable. Even in 1896 the south and southeastern slopes of the Palu valley were virtually devoid of trees up to 1,700 m.

Trials to determine appropriate means of reforesting the hills began over 60 years ago but the vast majority of the land to the east of Palu Bay and Palu River is still bare (Hadisumarno 1977). It is used as pasture for sheep and cattle but its carrying capacity is extremely low at about 5 hectares per head of cattle per year (Rauf 1982), at least in part because of the scarcity of small leguminous plants (Bismark et al. 1978). The general comments made about reforestation on the ultrabasic soils of Soroako apply here: the best species to choose are those that clearly excel in the natural or semi-natural setting. However, it should be noted that the area has been so degraded that it is necessary to find appropriate plants to enrich the soil before planting trees such as *Leucaena farnesiana* that have some social and agricultural uses.[23] This is particularly important because resettlement schemes in the region have been devised numerous times over the last few decades to encourage farmers in the hills to abandon their inappropriate forms of land management but almost all have met with some degree of failure (Metzner 1981).

One course of action that might be followed is to aim to recreate a monsoon forest on the Palu hills. A similar scheme, to recreate tropical dry forest, has begun in Costa Rica where a sufficient level of knowledge of this forest type, its components and their interactions has been reached such that judicious forest and animal management can be used to extend the area of this very rare forest type (Janzen 1986). All is not lost for the monsoon forest of Sulawesi—the Paboya Reserve, so degraded and small, probably contains most of the plant species that would be needed in the initial stages of forest development and disperser animals could be encouraged to return or reintroduced if necessary. The sandalwood is by far the most important monsoon forest species and provision of seedlings could be assisted by tissue culture techniques developed for this species by Dr. L. Winata of Bogor Agricultural University.

If a plan to recreate the monsoon forest could be put into effect, it would probably pay for itself in the long run given the huge amount of soil currently lost from the hills, the low productivity of the land, the potential harvesting of goods from the forest at all stages of its growth and the consequent improvement to the irrigation schemes in the lowlands. All that is required is start-up money and political will.

Chapter Seven

Mountains

INTRODUCTION

The physical and biological differences between the hot, luxuriant lowlands of Sulawesi and its cold, stressed mountain tops are very striking and parallel some of the changes found when travelling from equatorial to temperate and arctic regions. As such, mountains are stimulating subjects for study and can be instructive in understanding the physical limitations to growth and reproduction in both plants and animals.

Most of Sulawesi lies above 500 m and about 20% of the total land area is above 1,000 m. The highest blocks of land, those over 2,000 m, lie in Central and northern South Sulawesi (fig. 7.1). With lowland forest all but disappearing due to the many pressures upon it, mountains are important refuges and sources of genetic diversity. An understanding of what does or can live at higher altitudes is important to regional planning.

CLIMATE

Temperature

To spend a night on a mountain top is to learn some fundamental physical principles. Mountains do not warm up or cool down in the same way as the lowlands primarily because there is a shorter, and therefore less dense, column of air above them. Infra-red (warming) radiation from the sun excites gas molecules in the air causing them to collide with each other and thus to give off heat. Where the air is less dense, fewer collisions occur and hence less heat is generated. There is less water vapour and particulate matter in mountain air and this results in there being less reflection and absorption than at sea level even though the actual intensity of radiation is greater.

Radiation of heat from the ground occurs both by day and night, but at night there is no compensating radiation from the sun. Heat is lost quickly

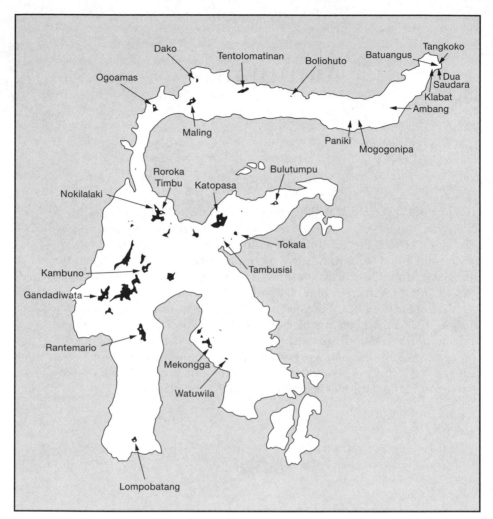

Figure 7.1. Land over 2,000 m (in black) and the major mountains.

from high altitudes at night causing the daily temperature range there to be as much as 15°C to 20°C (fig. 7.2). As the surfaces of plants, soil or rocks cool down at night so the air around them also cools. This, being heavier than the warmer air, will become progressively colder if there is no slope down which it can flow, for example in hollows and valleys. Frosts are most likely to occur in flat hollows, called frost pockets, on clear, calm nights when long-wave radiation is lost to the skies. The fastest rate of cooling occurs on non-conductive surfaces such as dead plant material

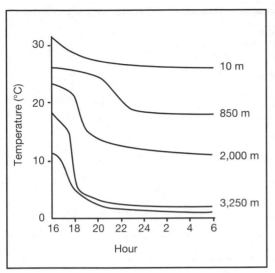

Figure 7.2. Drop in temperature from the afternoon to early morning at sea level (Ujung Pandang) and at three altitudes on Mt. Rantemario.

Data from an EoS team

and dry soil rather than conductive material such as living plants, wet soil, and water.

The rate of temperature change with altitude (the lapse rate) is generally accepted to be about 0.6 °C/100 m but this depends on factors such as cloud cover, time of day, and amount of water vapour in the air. Readings of minimum temperature from Mt. Rantemario, Sulawesi's highest mountain (3,450 m), taken during an EoS expedition demonstrate the decrease in minimum temperature with increasing altitude, and from these few readings the rate of decrease appears to be 0.7°C/100 m (fig. 7.3). It is quite likely that frost would occasionally form on the summit of Mt. Rantemario and a few of the other highest peaks, and during the cooler periods of the Quaternary permanent snow would probably have been present.

Relative Humidity

The percentage saturation of a mass of air increases as its temperature falls. The dew point (the temperature at which condensation occurs and clouds or drops of dew form) at different altitudes depends therefore on the temperature and the initial moisture content of the air. The forests of higher altitudes experience a very high relative humidity, particularly at night when the temperature falls, and the dew point is frequently passed so that water condenses on the leaves. Relative humidity has been recorded

Figure 7.3. Minimum temperature readings on successive nights during the ascent of Mt. Rantemario.

Data from an EoS team

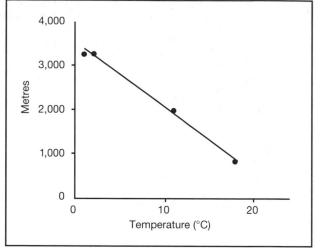

over 60 days at 1,700 m on Mt. Nokilalaki and results ranged from 86% to 100% (mean average 94%) at dawn, 68% to 98% (mean average 87%) in the early afternoon, and 86% to 100% (mean average 94%) after dark (Musser 1982). During dry spells at higher altitudes above the main cloud layer (above about 2,500 m), however, the relative humidity can reach 20% during the day, leading to wider daily extremes.

Clouds

Clouds originate where ascending air reaches its dew point and where there are the necessary dust or other particles for the water vapour to condense upon. Once water droplets have been formed they tend to act as 'seeds' and the droplets may grow as they bounce around in the clouds. During the wettest months, the slopes and peaks of high mountains can be enveloped in clouds for days on end. During the drier months, however, when the air is not saturated with water vapour, it is common for a belt of clouds to form around a mountain, often at about 2,000 m. As experienced mountain climbers know, a low, even continuous layer of cloud at 2,000 m does not always mean that clouds are present at 3,000 m or above.

Rainfall

Although rainfall is generally higher on the side of a mountain facing the prevailing wind (fig. 7.4), there appear to be no guiding principles relating

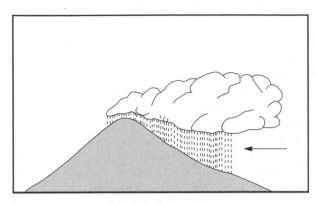

Figure 7.4. Rainfall on mountain slopes is generally greater in the lowlands and on the side facing the prevailing wind.

After van Steenis 1972

altitude to rainfall except that rainfall on mountain slopes up to 1,500 m will generally be greater than on the surrounding lowlands. At the cloud level, however, rainfall measurements are not especially useful for ecological studies because much of the water used by plants is in the form of water droplets in the clouds which adhere to the surfaces they touch. Studies on Mt. Pangrango (3,025 m) in West Java, showed that the summit received more frequent rain than the slopes but this amounted to a lower total rainfall. There is, however, great variation between years: the summit of Mt. Pangrango has in one year received only 13 mm of rain between July and September but 460 mm during the same period in another year (van Steenis 1972). When the EoS team reached the summit of Mt. Rantemario in October 1985 a note written by climbers the previous July appeared almost pristine despite being open to the elements for the intervening months.

Ultra-violet Radiation

It has been suggested that ultra-violet radiation on tropical mountains is probably more intense than on any other mountainous region on earth (Lee and Lowry 1980). This is because the amount of ozone, which absorbs ultra-violet radiation in the stratosphere, is appreciably less near the equator than elsewhere, and the atmosphere at low altitudes is more turbid and dense and thus more capable of absorbing or reflecting this radiation. Climbers must therefore protect their skins or rapidly suffer the effects of sunburn. One possible effect of high levels of ultra-violet radiation is to cause cell mutation, so it is possible that the rate of cell mutation on these mountains is particularly high, and this may be one reason why the levels of endemism are relatively high on mountains.

SOILS

The nature of mountain soils changes with increasing altitude, becoming more acid and poorer in mineral nutrients, and this is particularly the case where acid peat is present. Peat accumulates in wetter places, in the cloud belt, or upper montane zone for example, where decomposition processes are generally slower. Ridges, knolls and summits only receive water from the atmosphere and so their soils are continually leached because soil and water in the interflow (p. 265) cannot be received from above. The soil in these places is therefore drier and more nutrient-poor than soil in valleys or the lower slopes (Burnham 1984). Differences in bedrock composition and climate are the major factors influencing soil formation up a mountain, although steepness of slope and openness of the vegetation cover are also important. Low temperatures slow down the processes of soil formation for various reasons. For example, reduced evapotranspiration causes less movement of water through the soil, chemical reactions occur 2-3 times slower with every 10°C drop in temperature, and the reduced abundance of soil organisms means that biological processes affecting soil formation are also slower. With increasing height the soil is less well formed, and the roots of the plants growing in it are shallower. The differences in the soil obviously have their effects on the vegetation but whether the soil is more important than direct climatic effects and the changes in communities of disperser animals is not known.

Soils were collected on EoS expeditions and results of the analyses (table 7.1) illustrate some of the principles described above.[1] Mountain

Table 7.1. Results of soil analyses from a series of soil samples taken from 2-5 cm depth in different zones of Mt. Rantemario (3,445 m) and a single sample from Mt. Kara on Siau Island.

	Lower montane	Upper montane	Mt. Rante-mario subalpine	Summit (south)	Summit (west)	Mt. Kara 1,000 m lava flow
pH	5.4	5.0	3.8	4.1	4.1	4.3
% C	3.81	5.45	5.05	1.4	1.3	2.11
% N	0.25	0.35	0.31	0.14	0.11	0.11
C/N	15	16	16	10	12	19
K meq/100g	0.18	0.28	0.16	0.18	0.86	0.1
Na meq/100g	0.33	0.73	0.47	0.8	0.7	0.1
Ca meq/100g	7.1	4.5	2.4	3.8	1.4	0.7
Mg meq/100g	12.8	0.0	0.47	2.8	0.2	0.2
C.E.C.	16.5	16.0	16.1	9.0	16.9	4.7

From Burhanuddin 1986; and EoS teams

soils are generally deficient in calcium (<1.8 meq/100g) (Whitten et al. 1984) but this does not seem to be the case for most of the soils examined in Sulawesi. The high potassium (and relatively high cation exchange capacity) on the western (more open) slope of the summit of Mt. Rantemario may be a result of fires set by people. The soil collected from Mt. Kara was on lava that was a product of an eruption in 1983, and although the low concentrations of cations are what would be expected, the percentage of carbon seems surprisingly high.

VEGETATION

Structure

The forest on mountain slopes below 1,000 m is very similar to other low-land forests although on small mountains changes with altitude are evident (fig. 7.5). With increasing altitude, however, the trees become rather shorter and less massive, and epiphytes such as orchids become more common. This is called lower montane forest. Further up, the tree canopy becomes more uniform, the trees are even shorter, squat and gnarled, the leaves are small and relatively thick, and moss abounds. This is called upper montane forest. Beyond this, if the mountain is tall enough, is the subalpine forest of yet smaller trees with smaller leaves, whose branches bear epiphytic lichens in profusion but virtually no orchids (fig. 7.6; table 7.2). Valleys and other depressions in the subalpine zone are generally devoid of trees, and it appears that no trees have adapted to high altitude, water-logged conditions. Here and elsewhere in this zone there is, instead a covering of shrubs, colourful herbs and tough grasses. The woolly hairs of some subalpine plants have been variously attributed to the ability to protect the plants against high temperatures (Lee and Lowry 1980), intense ultraviolet radiation (Mani 1980) and frost/freezing (Smith 1970). Maybe they play a role in all three. For plants with leaf rosettes on the ground such as the silverweed *Potentilla* (Rosa.), however, the wooly hairs are concentrated on the lower surface and petioles only. This would make sense as frost protection but not against ultra-violet radiation or high temperatures. Other plants provide their growing buds and perennial parts with thermal insulation by retaining old leaves, and by having tufted branches and persistent scales. Such protection is necessary because the 'cost' to a subalpine plant of replacing a leaf is much greater than for plants growing in the more hospitable environments lower down the mountain or in the lowlands. Adaptations of leaves to intense ultraviolet radiation are thick cuticles, wax deposits, and high concentrations of red,

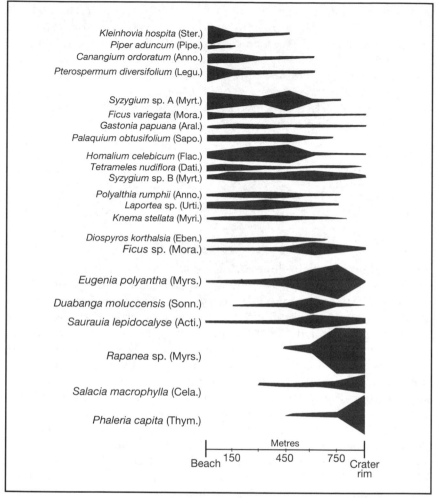

Figure 7.5. Changes in relative frequency of dominant tree species with altitude on Mt. Tangkoko (1,109 m).

After Anon. 1980

protective leaf pigments called anthocyanins in young leaves (Lee and Lowry 1980).

Where the slopes in the sub-alpine zone are treeless, the cause is probably hunters having set fires in order to make hunting anoa easier (p. 494).

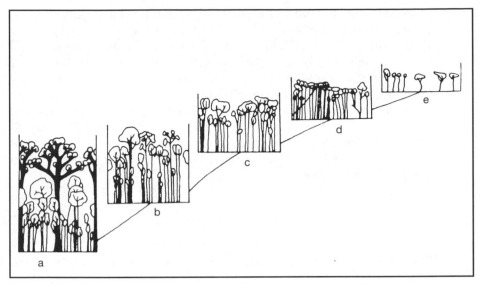

Figure 7.6. Cross-sections through five forest types to show the decrease in tree height and simplification in structure with an increase in altitude. a - lowland forest, b - lower montane forest (lower levels), c - lower montane forest (upper levels), d - upper montane forest, e - subalpine forest.

Adapted from Robbins and Wyatt-Smith 1964

Two transects (200 m x 10 m), one at 3,100 m in upper montane forest and the other at 2,100 m in lower montane forest, were enumerated on the slopes of Mt. Rantemario (table 7.3). Trees in the upper montane forest clearly have thicker trunks for their height compared with trees in the lower montane forest: average height in the former was 6.5 m whereas, for trees of the same trunk diameter range (15 cm to 20 cm), the average height in lower montane forest was 11.5 m or nearly twice as tall.

Trees in exposed, windy sites tend to have a greater wood density than species characteristic of more sheltered habitats. The denser wood would to some extent counteract the damaging effects of wind on exposed sites, but since investment in wood must be made at the expense of investment in leaves, shoots and trunk thickness, this may be a reason for the dwarf stature of upper montane forests. This shorter forest with dense wood probably grows slower than forests at lower altitudes, and together with its greater structural stability would result in the rate of gap creation being less than in the lowlands (p. 360). This must have significant effects on forest dynamics and composition (Lawton 1984). Morphological differences within a species of tree in the upper parts of lower montane forest have been shown to be related to a gradient of wind stress, which reaches a maximum on ridge crests. For individuals of a given height, trunk girth and branch thickness, increases with proximity to a ridge crest (fig. 7.8). It would seem, therefore, that gnarled trees may be a direct result of wind stress (Lawton 1982), but factors such as periodic drought and nutrient stress might be involved.

Table 7.2. Characteristics of four types of forest found on mountains. The most useful characters are shown in bold type.

	Lowland forest	Lower montane forest	Upper montane forest	Subalpine forest
Canopy height (m)	25-45 m	15-33 m	1.5-18 m	1.5-9 m
Height of emergents (m)	67 m	45 m	26 m	15 m
Leaf-size class[1]	**mesophyll**	**notophyll or mesophyll**	**microphyll**	**nanophyll**
Tree buttresses	**common and large**	**uncommon, small or both**	usually absent	absent
Trees with flowers on trunk or main branches	common	rare	absent	absent
Compound leaves	**abundant**	present	rare	absent
Leaf drip-tips	**abundant**	present or common	rare or absent	absent
Large climbers	**abundant**	usually absent	absent	absent
Creepers	usually abundant	common or abundant	very rare	absent
Epiphytes (orchids, etc.)	common	abundant	common	**very rare**
Epiphytes (moss, lichen, liverwort)	present	present or common	**usually abundant**	abundant

[1] - the leaf-size classes refer to a classification of leaves devised by Raunkier (1934) and modified by Webb (1959). The definitions are: mesophyll: 4,500-18,225 mm^2, notophyll - 2,025-4,500 mm^2, microphyll - 225-2,025 mm^2, nanophyll: less than 225 mm^2. An approximate measure of leaf area is 2/3 (width x length). See figure 7.7.

After Grubb 1977; Whitmore 1984

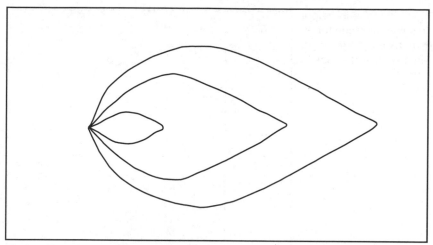

Figure 7.7. Hypothetical leaves of 4,500 mm², 2,025 mm² and 225 mm² to assist assessment of leaf-size class.

As can be seen from table 7.2, average leaf size changes with altitude, but so also does leaf thickness. The thickness of ten leaves of each of the four most common plants around the summit of Mt. Rantemario were measured and the median thickness was found to be 0.45 mm (range 0.30-0.85). The equivalent figures for a sample of leaves from lowland forest (Bogani Nani Wartabone National Park) were 0.2 mm (range 0.1-0.3). The thicker leaves, or pachyphylls, characteristic of montane habitats (Grubb 1977; Tanner and Kapos 1982), have a thick palisade layer (the

Table 7.3. Differences in forest structure at two altitudes on Mt. Rantemario.

Altitude (m)	No. trees /0.2 ha	Basal area m²/ha	Maximum diameter (cm)	Maximum height (m)
3,100	6	0.69	20	8
2,100	29	8.3	63	18

Data from an EoS team

Figure 7.8. Changes in height of trees in lower montane forest in relation to average wind velocity on slopes and exposed ridge crests.

After Lawton 1982

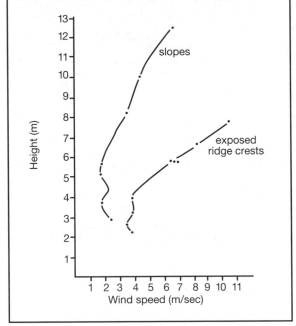

long, vertical photosynthetic cells) and frequently a strengthened, protective hypodermis (the layer below the epidermis). The thick hypodermis may be an adaptation to resist undesirable penetration by fungi, bacteria, moss and lichens, some of which grow on leaves where they are called epiphylls (p. 358). The high humidity provides ideal conditions for the growth of these plants and they probably represent a greater threat to leaf life in the frequently cloud-swathed upper montane and subalpine forests than does insect damage, the major factor in the lowlands (Grubb 1977).

Lower down the mountain, the growth of epiphylls is not so luxuriant and different adaptations seem to have evolved to cope with a constant film of water on the leaf. Under such conditions, transpiration is very slow because water cannot evaporate from the openings or stomata on the leaf (Leigh 1975). In addition water on leaves can leach nutrients from leaf tissues and can also reflect sunlight reducing the rate of photosynthesis and hence growth. It would therefore be reasonable to expect plants to have adaptations to ensure that water drips off, rather than sits on, their leaves.

One common leaf feature noticed last century is the long, narrow leaf tip which, even within a species, is longer and narrower in wetter microhabitats: leaves possessing long tips shed water faster than blunter leaves. Leaves would not need such 'drip-tips' if they could be held vertically but they obviously need to be orientated in such a way as to make the most of the little sunlight received in the understorey.[2] The optimum position for photosynthesis would be horizontal but then the water would not drain away. Assuming the majority of light comes from directly overhead (the shortest distance to the 'outside'), 100% of the light is reflected if the wet blade of the leaf is tilted down to 40° from the horizontal yet only 3% is reflected when tilted at 30° from the horizontal. Thus orientation between 30° and 40° would be optimal. If the leaf tip is narrow and downward-pointing, however, the leaf can be held more horizontal and yet still lose its water (Lightbody 1985). Leaf orientations are not rigidly fixed and in fact can vary through the day, being steeper at night when precipitation is greatest and evaporation least, and more horizontal during the day to intercept sunlight except when rain falls when the leaves tilt further downwards (Dean and Smith 1978).

The high winds, the rocky, nutrient-poor soils, and relatively low rainfall (at least seasonally) cause the plants of summits, and other exposed locations, a certain amount of stress. One means by which a plant can cope with such a poor, drying environment is by developing a relatively extensive root system to support a relatively small above-ground component. The ratio of dry weight of below-ground parts to above-ground parts (often called the root : shoot ratio) within a given species of plant or among communities of plants tends to increase with altitude. In lower montane forest in Papua New Guinea, the ratio was about 0.1 (Edwards 1982a), meaning that the trunks, branches and leaves were together ten times as heavy as the roots. In harsh habitats, such as exposed locations in the subalpine zone, the ratio is generally considerably higher. This was examined on Mt. Rantemario by the EoS for two plants: *Potentilla leuconata* (Rosa.) and *Hedyotis* sp. (Rubi.). Roots were carefully excavated from the soil and later dried with the above-ground parts. The results demonstrated that although there was considerable variation, the median ratio of the eleven plants was 0.7 (table 7.4).

Zonation

The altitude at which the various forest types occur is determined largely by temperature and cloud level. Various schemes have been devised to differentiate between the forest types[3] and a useful scheme for Sulawesi would be:

lowland and hill forest	0 - 1,500 m
lower montane forest	1,500 - 2,400 m
upper montane forest	2,400 - 3,000 m
subalpine forest	3,000 m +

Such zones are compressed on small mountains or on mountains on the periphery of large mountainous areas. This is known as the 'Massenerhebung' (literally 'mass uplift') effect (Grubb 1971), and should be borne in mind when reading about zones on mountains in, for example, Peninsular Malaysia (highest peak 2,200 m) or New Guinea (highest peak 5,031 m), where zones will be lower and narrower, and higher and broader respectively. In Sulawesi itself the effect is hard to observe because most of the vegetation on the smaller and outlying mountains has been felled. Zone compression may be thought to occur when dwarfed, moss forest is found at the top of even quite small mountains such as Mt. Tangkoko, but the apparent similarity of these forests to true upper montane forest is due to similar physical conditions; steep slopes, high winds and a relatively low cloud level (p. 492). The botanical composition of moss forest on peaks of 1,000 m or so is completely different from that of moss forest on high mountains of 3,000 m+. For example, the dominant trees in the elfin moss forest on Mt. Tangkoko are *Rapanea* sp. (Myrs.), *Phaleria capitata* (Thym.), and *Saurauia lepidocalyse* (Acti.) (Anon. 1980) normally found in lowland forest.

There are other reasons why the distribution pattern of a particular species of montane plant does not necessarily correlate with altitude. For example, the physical conditions experienced at high altitudes can occur in frost pockets (p. 489), beside mountain streams, on bare soil and near waterfalls. High-altitude plants in these relatively low locations may not flower or may have infertile flowers. To maintain a population, therefore, they are dependent on seeds or spores being dispersed from above. A similar but opposite mechanism occurs up the mountain whereby sterile or

Table 7.4. Root : shoot ratios of two subalpine plants examined by the EoS expedition to Mt. Rantemario.

	Dry weight below-ground parts (g)	Dry weight above-ground parts (g)	Root : shoot ratio
Potentilla leuconata	2.97	2.83	1.05
	2.06	3.12	0.66
	0.16	0.15	1.06
	0.83	2.41	0.34
	1.20	3.52	0.34
	0.14	0.17	0.82
	0.11	0.16	0.69
	1.20	2.49	0.48
Hedyotis sp.	6.08	8.85	0.69
	8.41	40.69	0.21
	16.72	19.87	0.84

From O. Maessen pers. comm.

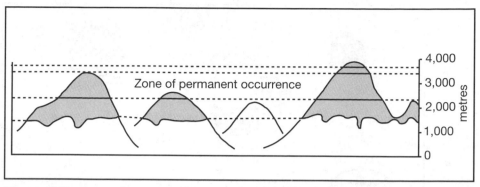

Figure 7.9. Zone of permanent occurrence with zones of temporary occurrence above and below it, showing the distribution of a hypothetical species (shaded areas). *After van Steenis 1972*

non-flowering plants are found higher above the species' normal range, for example, around volcanic fumaroles. Thus, for a given species there is a zone of permanent occurrence which supplies seeds or spores to adjacent zones of temporary occurrence. For this reason, a mountain may lack a species of plant even though at the same altitude on a higher mountain (with a zone of permanent occurrence) that plant may be present (fig. 7.9) (van Steenis 1972).

Characteristic Plants

Whereas lowland forests are not dominated by any one family of trees (p. 386), the lower montane forests are characterized by the large numbers of oaks *Lithocarpus* and chestnuts *Castanopsis* (Faga.). This has been reported, for example, from Donggala and Gorontalo between 800 m and 1,200 m where the *Castanopsis* is in constant association with *Phyllocladus, Agathis* and *Eugenia* (Steup 1931). Only four species of *Lithocarpus* and two of *Castanopsis* are known to be present in Sulawesi (compared with 50 and 21 respectively in Borneo) (Soepadmo 1972) and these can be identified quite easily from their fruits (fig. 7.10).

Oaks and chestnuts have large, relatively heavy fruit which are poorly fitted for long-distance dispersal (Soepadmo 1972) (p. 379). Pigs and anoa eat the seeds and those which pass through the gut without being chewed or otherwise damaged may remain viable. Squirrels certainly predate on the seeds but they also take fruit out of the parent crown and store and hide some of them in the ground or in other 'safe' places (Becker et

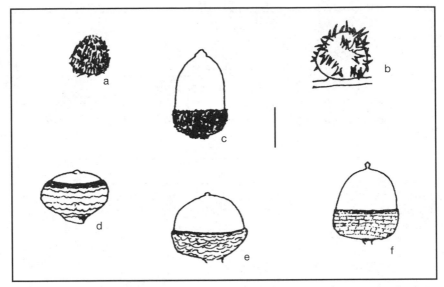

Figure 7.10. Fruits of Sulawesi oaks. a - *Castanopsis acuminatissima*, b - *C. buruana*, c - *Lithocarpus havilandii*, d - *L. glutinosus*, e - *L. elegans*, f - *L. celebicum*. Scale bar indicates 1 cm.

Drawn from specimens in the Bogor Herbarium

al. 1985). A proportion are forgotten and subsequently germinate.

Others trees common in the lower and upper montane forests are members of the Coniferae (pines and related trees) such as *Podocarpus* (Podo.), *Dacrycarpus* (Podo.), *Dacrydium* (Podo.), *Phyllocladus* (Phyl.) (fig. 7.11) and the magnificent and commercially important *Agathis* (Arau.).

Agathis is a genus of tall trees, commonly known as 'kauri' after their Maori name in New Zealand. The trees found in Sulawesi are *A. dammara* which is also known from the Philippines, Moluccas and parts of Borneo (Whitmore and Page 1980). Although found in lowland areas, they are dealt with in this chapter because the stocks of lowland specimens have been seriously depleted through general forest clearance and over-exploitation for the resin, the tapping of which has led to the death of many trees. It is the resin, known in the world of commerce as Manila copal, which has given *Agathis* much of its commercial interest because it can be tapped from the bark on a sustainable basis. The resin is used in spirit varnishes, lacquers and in making linoleum, but in villages it has been used as fuel for torches (Burkill 1966). The wood itself is finely-grained, pale and uniform and commands a price above most other hardwoods. The value of the wood has encouraged foresters to establish *Agathis* plan-

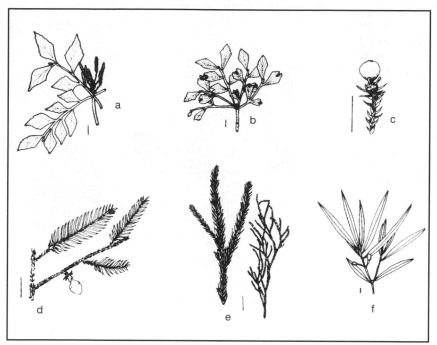

Figure 7.11. Some conifers of Sulawesi montane forests. a - *Phyllocladus hypophyllus* male, b - *P. hypophyllus* female, c - *Dacrycarpus steupi*, d - *D. imbricata*, e - *Dacrydium nidulum* juvenile leaves left, adult leaves right, f - *Podocarpus neriifolius*. Scale bars indicate 1 cm.

After Wasscher 1941; de Laubenfels 1969; Keng 1972, 1978

tations with varying degrees of success, and in Java yearly increments of 30 m³ timber/ha have been achieved. It can grow up to 45-50 m in height with a diameter of over 1.5 m. The bark is composed of round scales which flake off, and the leaves are broad, flat blades (fig. 7.12).

Agathis is recorded from all over Sulawesi except the southeast peninsula and the region between Parigi and Toli-Toli. The former absence may be a result of a climate which is too seasonal, whereas the latter may be an artefact due to inadequate collecting. It is affected by climate, such that only in the central block of Sulawesi is it found in the lowlands; on the four peninsulas it is known only from the mountains. A general survey of distribution published in 1940 indicates that it is found in discrete but extensive pockets of forest where it is usually the largest tree present, although young trees are abundant in the undergrowth and lower canopy (van der Vlies 1940). Many of these pockets are as yet too isolated for

Figure 7.12. *Agathis dammara.* Male branch with pollen cones (left), female seed cone (right). Scale bars indicate 1 cm.

From Keng 1972

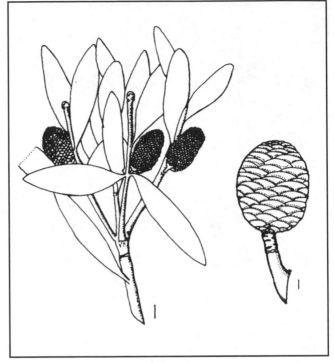

logging companies to consider felling them. It has been suggested that the occurrence of *Agathis* in the forests of Mt. Kinabalu, Sabah, is correlated with the presence of podzol soils (Askew 1964).

The price of resin was very high in the first few decades of this century, and in the rush to harvest the resin many trees were killed. Attempts made to control the harvest met with little success (van der Vlies 1940). *Agathis* is not a strong pioneer species: forest dominated by *Agathis* on the limestone Mt. Malindu between the lakes of Matano and Towuti was felled in 1969 but five years later no seedlings or saplings were to be found (Whitmore 1977).

Rattans occur in lower and upper montane forest and some species are excellent indicators of altitude. Until they are collected properly, however, the genuine relationships will remain obscure (J. Dransfield pers. comm.).

Tree ferns[4] are a common component of montane (particularly lower montane) areas in both disturbed and primary forests. Tree fern trunks can grow vertically one metre in 15 years but, like palm trees, some years elapse between the first growth of a young tree fern and the start of the ver-

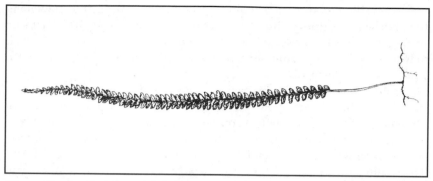

Figure 7.13. *Lindsaea pulchella* an epiphytic fern of tree ferns and trees in montane forests.
After Kramer 1971

tical growth of its trunk (Tanner 1983). The rough trunks evidently provide an unusual niche since it was noted on Mt. Roroka Timbu that a species of the fern *Lindsaea* (fig. 7.13) was found exclusively on tree ferns (van Balgooy and Tantra 1986). At present twenty species of tree ferns are known from Sulawesi, almost all of which are found on mountains (table 7.5). The species are distinguished by the form of the reproductive parts, and by the shape of the fine brown scales found covering the young, coiled leaves and the bases of older leaves.

One genus of epiphytic ferns *Lecanopteris,* found in lower montane forest, has developed a large, multi-chambered, spiky rhizome in which ants live. Sulawesi appears to be the centre of evolution for these plants with more species found here than anywhere else. An expedition to Mt. Rantemario in 1969 discovered a new species (fig. 7.14) which was inhabited by *Crematogaster* ants. These enter the chambers through holes formed when a frond dies, falls off, and leaves a raised leaf base with a soft, vulnerable centre which breaks down rapidly, providing access to the inside (Jermy and Walker 1975). The relationship between different ants and another species of *Lecanopteris* in Sarawak has been described (Janzen 1974), and the principles involved have broader application. Apart from the obvious benefit of a secure shelter, the ants probably also receive food in the form of highly nutritious spores, or young or aborted sporangia. The latter are full of globules which could be oil bodies (Holttum 1954), but this has yet to be confirmed. The plant probably benefits from organic and mineral material deposited in the chambers in the form of dead bodies, faeces and discarded food. The fern certainly grows roots into the chambers but whether these absorb anything more than water has yet to be proven. It is possible also that the ants defend the fern from animals which may seek to eat the rhizome or fronds, but this too, has yet to be proven (Jermy and Walker 1975).

Pitcher plants *Nepenthes* (Nepe.) are also found in montane habitats as they are in other nutrient-poor sites (p. 450), where their unusual means of obtaining nutrients gives them a competitive advantage. The nutrient concentrations in pitcher leaves from montane forests were compared with the concentrations in leaves of nearby plants that were not able to catch insects. The results showed that pitchers had three times as much phosphorus and twice as much potassium as the other leaves (Grubb 1974).

Characteristic of the upper montane forest are members of the Ericaceae such as the large and colourfully flowered *Rhododendron*, bilberries *Vaccinium*, and wintergreen *Gaultheria*. There are 24 species of *Rhododendron* known from Sulawesi of which 19 are endemic. The bilberries *Vaccinium* are represented by 16 species of which 13 are endemic. The

Table 7.5. Tree ferns known to occur in Sulawesi.

		Distribution	Altitude (m)
Cyathea	*?perpelvigera*	N	1,200-1,800
	pallidipaleata	SW Mt. Rantemario	3,000
	saccata	C Mt. Topapu	1,300-1,700
	inquinans	SW Mt. Lompobatang	2,000-2,800
	oosora	N	1,500-2,000
	oinops	SW	2,000-2,500
	?amboinensis	C, N	Lowland
	dimorpha	C Mt. Bohaa, SE	C - 1,500-1,700
			SE - 125-650
	strigosa	SW Mt. Lompobatang	2,800
	sangirensis	N (Sangihe)	Lowland
	elmeri	N (Sangihe + Talaud)	500-1,400
	tenggerensis	SW	1,500-2,300
	sarasinorum	C Mt. Sibaronga	Montane
	contaminans	Common	200-1,600
	?celebica	-	100-1,750
	moluccana	SW, ?C	0-900
Dicksonia	blumei	C	1,500-2,500
	?mollis	C	1,500-2,900
Cystodium	sorbifolium	N	Lowland
Culcita	villosa	N, Mt. Tampusu	1,600-2,230

Endemic species shown in bold type, ? = doubtful record needing confirmation.
After Holttum 1963

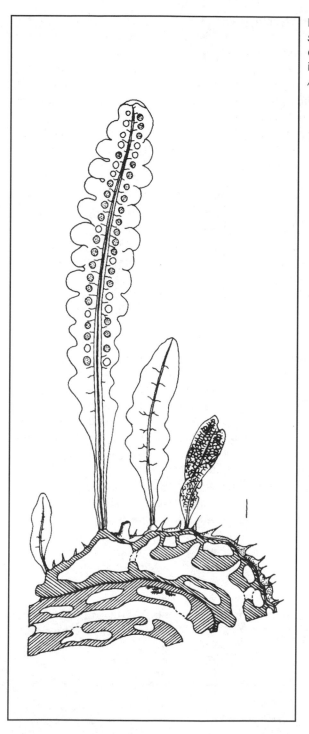

Figure 7.14. *Lecanopteris spinosa*, an ant fern discovered on Mt. Rantemario. Scale bar indicates 1 cm.

After Jermy and Walker 1975

leaves, flowers and large black berries of *Gaultheria* taste strongly of winter-green oil (methyl-salicylate) a substance which is often used to alleviate the symptoms of rheumatism. Only two species are known and both are endemic, *G. celebica* with white, pink or red flowers, and *G. viridiflora* with greenish-white, red-based flowers. A lesser-known genus *Diplycosia* with smaller berries than *Gaultheria* has 17 species all of which are endemic and some of which also smell slightly of wintergreen. *Rhododendron* has very small seeds each of which has a small tail or wing at both ends which is pre-sumably an adaptation for dispersal by wind. The fleshy fruits of the other species are eaten by birds (particularly thrushes and white-eyes), and mam-mals (Sleumer 1966-67).

One of the most obvious characteristics of the wetter parts of the upper montane forest is the enormous quantity of moss covering the ground and adorning every twig, branch and bough up to 2-3 m above the ground. In the upper parts of lower montane forest garlands of moss, typically *Aerobryum*, are frequently seen hanging from branches. Above the cloud zone the most common epiphyte is the beard 'moss' *Usnea*. This is in fact a lichen, a composite plant composed of a complex spongy framework of fungal mycelium strands within which sit algal cells. The algae photosyn-thesize and the fungus feeds on a proportion of the food produced. The algae can in fact lead a completely independent life but not so the fungus and so new generations have to re-establish the relationship. Thus *Usnea* and other lichens are given a name of their own although they are com-posed of two separate organisms. Interestingly, *Usnea* has been used as a medicine since classical Greek times. People living near mountains on Sulawesi use it for a number of purposes, but primarily exploit its astringent properties for curing intestinal troubles (Burkill 1966). It has been found also that the usnic acid found in the tissues of *Usnea* has antibiotic prop-erties against tuberculosis (Fitting et al. 1954).

In the subalpine zone the myrtles *Leptospermum* and *Decaspermum* pre-dominate, although it seems that they are never found together. In the open areas, many beautiful herbs can be found, many genera of which are typical of temperate regions, as well as grasses, sedges and rushes. Many subalpine plants form rosettes of leaves just above the ground. These rosettes are formed by the distance between successive leaves being extremely short, in order to reduce water loss and to keep the plant near the warmer ground. Only the flower stalk rises any distance above the ground, to facilitate pollination and seed dispersal. The under-surface of the leaves in some species with rosette leaves are covered with dense, silky, white hairs (p. 495).

Selected Mountains

Mt. Rantemario (3,450 m). The southwest slopes of this mountain have been converted to agricultural land up to 1,650 m where the lower montane forest begins. Elements of upper montane forest are found from about 2,150 m but oaks, the jagged-leaved *Phyllocladus hypophyllus* (Phyl.), the yew *Taxus sumatrana* (Taxa.)[5] and the conifer *Podocarpus steupi* of the lower montane forest are still dominant at that altitude. At 2,650 m the forest is clearly upper montane with *Rhododendron* bushes and some *Phyllocladus* predominating. The first subalpine herbs are found here, notably the deep-purple, closed flowers of the gentian *Gentiana laterifolia* (Gent.), and the large, bright-yellow flowers of *Hypericum leschenaultii* (Hype.) described as 'the jewel' of the mountain flora of Java (van Steenis 1972) and equally beautiful in Sulawesi. This plant is one of the rare examples of a wild plant being used directly as as a garden plant without modification through selective breeding. Also common in the undergrowth is the erect, dark-green giant moss *Dawsonia* and the small magnolia *Drimys piperata* (Magn.) with its attractive hanging white flowers (Buwalda 1949).

By 3,200 m the small trees comprise mainly *Rhododendron, Decaspermum* and *Hedyotis* sp. (Rubi.) with tightly stacked, thick leaves. Beneath these are found small shrubby *Gaultheria, Styphelia suaveolens* (Epac.) with small pink berries, and colourful flowered herbs such as the daisy *Keysseria,* a small ginger *Alpinia,* the little violet *Viola kjellbergii,* and two silver-leaved *Potentilla* (Rosa.): *P. leuconata* with large yellow flowers, and *P. parvus* with small yellow flowers. Along a little river at about 3,200 m a distinctive community of plants thrives including a ragwort *Senecio* (Comp.), a *Swertia* (Gent.), and clumps of the small spreading *Epilobium prostratum* (Onag.) with leaves less than 10 mm long but with small pale purple flowers atop a 8 cm flower stalk. Several grasses such as *Poa, Monostachys,* and *Agrostis* and sedges were also found as were the erect stems of the club moss *Lycopodium.*

In places where the subalpine forest opens into glades and where the bedrock of biotite schist is more or less visible, various species of 'cushion'-forming plants are found such as *Centrolepis philippinensis* (Cent.) (Ding Hou 1957), *Monostachys oreoboloides* (Gram.), *Eriocaulon celebicum* (Erio.), and *Oreobolus ambiguus* (Cype.). The shoots of these plants are densely packed; this is an adaptation against strong winds, keeping transpiration to a minimum and the temperature inside the cushion to a maximum. When these plants are relatively young they form single round cushions, but with age the centre (the oldest part) tends to die and a ring is formed as the plant grows outwards. Some rings found during the EoS expedition measured over one metre across although the rings were not complete. These rings are still single plants which have extended vegetatively and must be of considerable age. Most rings were, however, only 15-30 cm in diameter. Analyses of soil taken from inside these rings showed high nitrogen and phosphorus levels (0.23 and 48.6 mg/l respectively) relative to outside the ring (p. 494).

The plant collection, made by a brief botanical expedition to this mountain in 1981, demonstrated how poorly known the Sulawesi flora is. The collection comprised 57 high-altitude species, and more than 10% of these were new to science while others were previously known from just one or two specimens (Smith 1981; van Steenis and Veldkamp 1984).

Mt. Lompobatang (2,871 m). This mountain, whose name literally means 'swollen belly' (van Zijll de Jong 1934), had its first recorded ascent in 1840 when James Brooke, later Rajah Brooke of Sarawak, reached the summit (Mundy 1848). Sited as it is in a sea of densely-populated agricultural land east of Ujung Pandang, it is not surprising that the lower slopes of this mountain have been utilized for firewood and felled for farms and plantations. Reports from 1905 (Sarasin and Sarasin 1905) reported the encroachment of farmland and in 1933 plantations of exotic pines were established above 1,650 m. In the lower montane forest that remains, various species of the native conifer *Podocarpus* as well as the native maple *Acer caesium* can be found (fig. 7.15) (Meijer 1983)[6] It is one of the few trees in relatively unseasonal areas to loose its leaves regularly and when its branches are bare it generally flowers. The layer of recently fallen leaves is one clue to its presence, another is the buzzing of bees and wasps in the canopy visiting the flowers for honey (Bloembergen 1948).

The upper montane forest is dominated by twisted *Leptospermum,* but towards the first summit (Mt. Asumtatumpang or Buluasumpolong) this is found together with *Vaccinium, Styphelia* and *Rhododendron.* Above 2,500 m the only trees found are oaks and *Leptospermum.* In 1931 the tree trunks were all charred, the result of a fire 13 years earlier, and the situation now is similar to that described for Mt. Rantemario above. The *Leptospermum* seems to regenerate better than the oaks perhaps because the relatively large acorns are a favoured food of anoas and pigs, or perhaps because the *Leptospermum* is better able to shoot from its base. The summit of Mt. Lompobatang has blackberries *Rubus* (Rosa.), *Hypericum,* bilberries *Vaccinium* and many herbs and grasses.

Interesting orchids also used to be present (Bouman-Houtman 1926), but their current status is unknown.

Mt. Tambusisi (2,422 m). In the lower montane zone of this mountain in Morowali National Park, from about 1,300-1,700 m, the forest has an average height of 15 m with occasional 25 m-high *Agathis* emerging. The undergrowth has more palms than found in the lower forests. In the upper montane forest, between 1,700 m and 2,500 m, tall shrubs up to 4 m high are found, mainly *Vaccinium.* The epiphytes are numerous. Nearer the top, between the patches of bare rock, the shrubs are barely 1 m tall and were

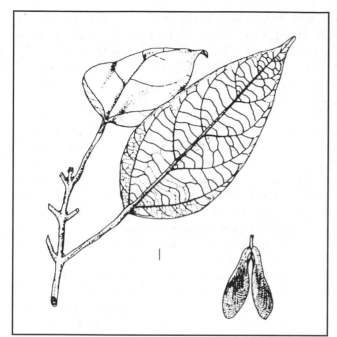

Figure 7.15. *Acer caesium* branch and fruit. Scale bar indicates 1 cm.
After Whitmore 1972

interspersed with herbs including the sticky-leaved, insect-catching *Drosera peltata* (Lack n.d.). This plant appears to have a wide altitudinal range, having been collected between 800 m and 3,225 m (van Steenis 1953).

In the valleys and in depressions in the ridge, elfin forest about 5 m tall is found. The gnarled and twisted trees have trunks of less than 6 cm in diameter, yet are surrounded by a 10-15 cm layer of brown and orange moss and liverworts. These also cover the ground to a thickness of 0.5 m or more. In this, as well as all the other montane areas, pitcher plants were common and at least eight species have been collected (A. Lack pers. comm.).

Mt. Nokilalaki (2,280 m). Mt. Nokilalaki can be seen from Palu but it is generally climbed from Lake Lindu (Bloembergen 1941). *Agathis* (Arau.) used to be a characteristic tree of the lower montane forests but many have now been felled. Formerly they were 40 m to 50 m tall with diameters in excess of 1 m. At about 1,600 m trees also include other conifers such as *Dacrydium* and *Podocarpus,* magnolias *Aromadendron, Mastixia* (Corn.), the maple *Acer caesium* (Acer.), *Elaeocarpus* (Tili.), all four species of oak *Lithocarpus* (Faga.) known so far from Sulawesi (p. 504), *Calophyllum* (Gutt.), and the endemic *Macadamia hildebrandii*[7] (Prot.) (fig. 7.16), but the most common tree is the chestnut *Castanopsis acuminatissima* (Faga.). Higher up, above 2,000 m, *Agathis* is absent and the most common trees are *Tristania* (Myrt.), *Lithocarpus* and *Castanopsis* which are progressively smaller and more covered with moss with increasing altitude. The actual summit was cleared, as was the case for almost all Indonesia's major mountains, when the Triangulation Service set up concrete pillars in the early decades of this century for surveying and mapping—the Mt. Nokilalaki pillar was set up in 1919[8]. In the 200 m below the summit the most common tree is the conifer *Dacrydium* which here grows to 3-5 m tall, with *Rhododendron, Ardisia,* and *Psychotria* beneath it. Below these are the giant mosses *Spiridens* and *Dawsonia,* peat moss *Sphagnum,* pitcher plants, scrubby *Saurauia,* and many ferns. A collection of these in 1976 yielded nine species previously unknown from Sulawesi (Meijer 1983). Among the other plants present is a climbing pandan *Freycinetia,* blackberry *Rubus* (Rosa.), the small magnolia *Drimys piperata,* and *Trachymene* (Umbe.) (Bloembergen 1941).

Mt. Roroka Timbu (2,450 m). The lower montane forest is dominated by the oaks. For example, *Castanopsis acuminatissima* is first found at 1,500 m and was extremely abundant up to about 1,700 m. In October the abundant acorns are apparently collected by local people and are roasted like peanuts. The tallest trees were *Lithocarpus glutinosus, L. elegans,* figs and an occasional *Macadamia* sp. In the understorey the dominant trees are species of *Eugenia* and a species of climbing bamboo *Racemobambos* (Dransfield 1983; van Balgooy and Tantra 1986).

Between 1,900 m and 2,000 m the vegetation changes quite abruptly with *Agathis* becoming dominant up to 2,250 m. These are the straightest and tallest trees (about 40 m) and beneath them grow other conifers such as *Dacrycarpus, Dacrydium* and *Phyllocladus.* Among the understorey trees are giant pandans reaching 25 m in height with stilt roots 10 m long. The dense crowns of the *Agathis* and *Pandanus* make the forest floor very dark, and small trees, shrubs and herbs were only found near small streams above which the canopy is not so thick. The only common climbers are pitcher plants, and a new species of small climbing pandan *Freycinetia micrura* (Stone 1983). The branches and trunks of trees are covered with abundant epiphytes.

Above 2,000 m the forest becomes lower and more crooked, the canopy is more open, and hence the undergrowth is denser. By 2,250 m *Agathis* has been left behind and the dominant trees are various conifers and *Lithocarpus havilandii.* The understorey is rich in Myrtaceae (*Leptospermum* and *Eugenia*), and *Rhododendron* and *Vaccinium* are found growing epiphytically and on the ground. At the summit the woody plants are almost all *Leptospermum, Rhododendron, Dacrycarpus, Phyllocladus* and *Podocarpus,* and all are covered with thick cushions of moss. Various herbs grow in the moss such as pitcher plants and orchids (van Balgooy and Tantra 1986).

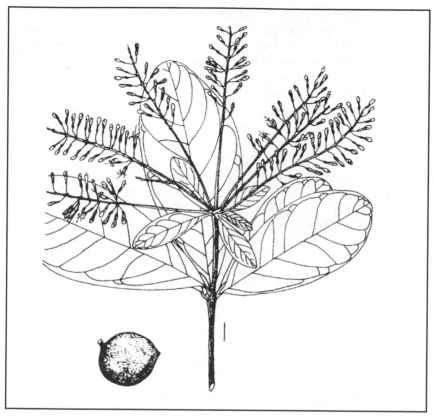

Figure 7.16. *Macadamia hildebrandii.* Scale bar indicates 1 cm.
After Soewanda n.d.

Pollination and Dispersal

Studies of flowering, fruiting and new leaf production have not been made on Sulawesi mountains but relevant information is available from Java. Most species have peaks of flowering relating to climate, but there are generally at least a small number of individuals from any one species flowering at any one time. Some plants appear to be indifferent to climatic patterns, others flower most in the dry season, while others appear to concentrate flowering in the wettest season; these differences may be due to the availability of different agents of pollination.

The pollination of montane plants has been little studied and simple observations of insects or other animals visiting flowers do not necessarily constitute observations of pollination. Also, plants which look as though they are adapted for pollination by a certain group of animals or by wind, may in fact be self-pollinated, even though they are visited by beetles, flies and other insects. On Mt. Rantemario one obvious self-pollinating flower is the purple gentian *Gentiana laterifolia* whose flower never fully opens. The plants of montane vegetation are visited by a range of animals such as:

- bats on bananas in lower montane forest;
- moths (noctuids and sphingids) on whitish or greenish long-tubed flowers with nocturnal scent;
- butterflies on colourful trumpet-shaped flowers such as the orange form of *Impatiens platypetala* (Bals.) endemic to Sulawesi;
- carpenter bees (up to about 1,400 m) and bumble bees (above this) on *Vaccinium, Gaultheria, Hypericum* and many other plants;
- flies and beetles on flowers such as certain orchids and aroids with a smell resembling decaying flesh;
- wasps on figs and, perhaps;
- birds such as flowerpeckers on long-tubed red flowers, but observations from Sulawesi mountains do not seem to have been recorded.

There are four main ways in which plants disperse their fruit and seeds (or, in the case of ferns, mosses, fungi, etc., their spores):

- in water;
- by wind;
- by terrestrial animals onto which seeds stick, or which eat seeds but do not destroy them, or which remove seeds for later ingestion but then forget them; or
- by sticking to, or being eaten and not destroyed by, flying birds or bats. An interesting example of a Sulawesi plant with a double adaptation for dispersal is the herb *Triplostegia glandulifera* (Dips.) (fig. 7.17) whose fruit has both sticky hairs and hooks (van Steenis 1951; van Steenis and Veldkamp 1984).

Dispersal by water or mammals are clearly of no significance in explaining the regional distribution of high-altitude plants (p. 62) because water only flows downstream to unfavourable habitats, the distances between high mountain tops are too great for terrestrial mammals to cross before the seed is either excreted or rubbed off, and bats are scarce at high altitudes. Wind would seem a likely candidate, therefore, for the dispersal of the minute, dustlike seeds of orchids, fern spores, and the light, parachute-like, plumed seeds of daisies, but the expected wide distributions of such species is not always found to be the case (van Steenis 1972).

Many of the shrubs and small trees in the upper montane and sub-alpine forests produce berries which are an important source of food for birds and mammals. Many of the seeds in these juicy fruits pass through bird intestines without harm. Since the time taken by a seed to pass through the intestine is little more than an hour, however, the possibility of transferring seeds to mountain tops more than about 30 km away is remote

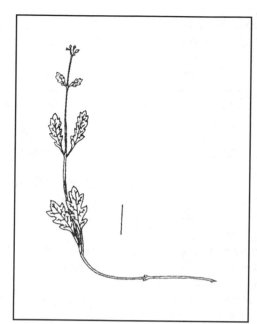

Figure 7.17. *Triplostegia glandulifera*, a plant with two means of animal dispersal. Scale bar indicates 1 cm.
After van Steenis 1951

and, in addition, birds of mountain tops do not generally fly far. Sticky or adherent seeds and fruits may attach themselves to feathers or legs but, before a long flight, birds will generally preen themselves thoroughly to ensure maximum flying efficiency, and by so doing dislodge any attached seeds. A further complication to understanding dispersal is the observation that communities of plants having a variety of dispersal mechanisms have exactly overlapping ranges on different mountains as though means of dispersal were immaterial (van Steenis 1972).

As stated previously (p. 62), the lowering of forest zones during the Pleistocene would have made chance dispersal easier if only because more 'islands' of suitable habitat would have become available. But even then, large gaps between these habitats would have existed and it is clear that the means by which plants or groups of plants effected their dispersal is not yet fully understood.

Biomass and Productivity

The trends in biomass and productivity change from lowland to upper montane forests are as follows (Grubb 1974):

- biomass decreases proportionately less than height, resulting in shorter, stockier trees in the upper montane forest;

- production of woody parts declines from about 3-6 t/ha/yr to about 1 t/ha/yr;
- production of litter, particularly leaf litter, decreases proportionately much less than biomass or production of woody parts;
- biomass of leaves decreases proportionately much less than total biomass;
- mean life span of leaves increases only slightly;
- the leaves become thicker and harder and the leaf area per gram decreases from up to 130 cm^2/gm to about 80 cm^2/gm.

Thus, with increasing altitude, total production decreases but plants invest a greater proportion of their production in making leaves although these more 'expensive' leaves may last no longer than the 'cheap' leaves of the lowlands (p. 370).

Only two studies of litter production have been conducted on Indonesian mountains and both of those were on Java. The annual rate of total litter production in lower montane forest was between 6,000 kg/ha/yr and 7,500 kg/ha/yr, about 75% of which was leaves (Yamada 1976; Bruijn-zeel 1984), which is not so different from results from lowland forest in Peninsular Malaysia (Lim 1978). Leaves fall throughout the year but the main peak is during the dry season. Branch fall is greatest during storms which occur most often during the wetter seasons.

Mineral Cycling

There are two main classes of mineral cycling in a forest. First, there is the rapid cycling of minerals in the fallen leaves and twigs, and in the rain falling through the forest canopy to the floor. Second, there is the much slower cycling of minerals held in the tissues of large woody parts of trees.

Lower quantities of minerals cycle through leaves in montane forests than in lowland forests, and fewer still through woody parts. This is a consequence of lower production of woody parts compared with leaves, and the concentration of nutrients in leaves of montane plants being about half that of their counterparts in lowland forest. The concentrations of minerals in the montane leaf litter are proportionately even less than in lowland litter because montane plants absorb into their permanent tissue about half of their leaf minerals before shedding them, whereas lowland plants absorb only about one quarter of their leaf minerals.

Assuming results from New Guinea to be broadly applicable to the forests of Sulawesi, the major external input of minerals is from the rain (Edwards 1982b) (fig. 7.18) which has been contaminated by ash from fires (Ungemach 1969) and volcanic eruptions. Not all the rain falling over a forest reaches the floor because it evaporates, is intercepted by epiphytes and other leaves, or is absorbed by bark, root mats, etc. When the rain is

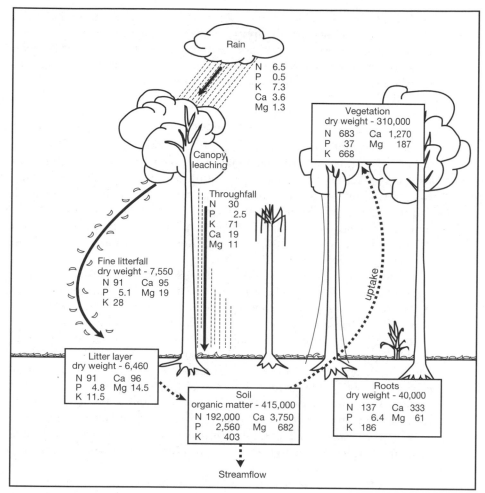

Figure 7.18. Mineral cycling in a lower montane forest. Figures in boxes are in kg/ha. Complete arrows represent major pathways of mineral transport, and figures alongside them are in kg/ha/yr. Dashed arrows are presumed pathways that have not been quantified.

From Edwards 1982b

light, little may reach the forest floor, and even in the heaviest rain only about 25% is intercepted (Edwards 1982b). The rain that does complete its journey to the ground will have also picked up minerals from animal droppings, plant exudates, and humus around epiphyte roots and will be far more mineral-rich than rain falling in a clearing.

As indicated above, leaf litter does not have the same mineral content as the living leaves because of absorption by the plant. Some minerals, such as calcium and magnesium, do not seem to be resorbed while others such as nitrogen and phosphorus are. It is possible that minerals that are absorbed may be relatively scarce and limiting in the soil.

Volcanoes

Volcanic eruptions are major geological and biological events. Not only does a great deal of subterranean material come to the earth's surface, but also many plants and animals die as a result of the intense heat of the lava, and the smothering by ash. The major volcanic product in Sulawesi is ash and this can interfere with living leaves by adhering to them and preventing photosynthesis. A layer of ash just 1 mm thick can reduce light penetration to the leaf by 90% or more. These are short term effects, however, and rain soon washes off the dust.

Around active craters or sulphur vents the soil is usually rocky, very pervious, sterile, acid, and lacking in organic matter. For example, soil collected from near a fumarole above Lake Moat, Bolaang Mongondow, had a pH of 3.6. The atmosphere is generally a choking mixture of sulphur dioxide (SO_2), and hydrogen sulphide (H_2S) mixed with smaller quantities of chlorine (Cl_2), carbon monoxide (CO), and nitrous oxide (NO). In addition, the ground is frequently heated from beneath. These are conditions which are hardly favourable to plants, but it is amazing how tolerant some can be. Those closest to the sulphur vents tend to be dwarfed, prostrate, and to grow extremely slowly. The closest plants are often *Vaccinium*, *Rhododendron* and ferns. Lichens may be found closer still and some blue-green algae are able to live in hot, muddy, sulphurous pools (Kullberg 1982).

It may take centuries before new lava flows have been sufficiently weathered to support closed forest but, if there is sufficient moisture, the first plants can be found virtually as soon as the lava cools down. For example, an EoS team examined the lava that erupted from Mt. Api on Siau Island two years after the eruption in 1983. The only plants that could be found were ferns and some patches of lichens.

The vegetation on volcanic ash develops relatively quickly. For example, the island volcano of Ruang (fig. 1.4), erupted dramatically in 1874, and 11 years after that the ash slopes were covered in sedges and leguminous herbs (Hickson 1889). The rate of colonization is obviously affected by the proximity of a source of seeds and spores and also of disperser animals (p. 55). The warm, volcanic ash of Ruang had, in 1885, already been marked by abundant footprints of maleo birds *Macrocephalon maleo* (p. 155) (Hickson 1889) which may have brought seeds on their feathers or in their guts from the neighbouring island of Tahulandang. Other colonizers

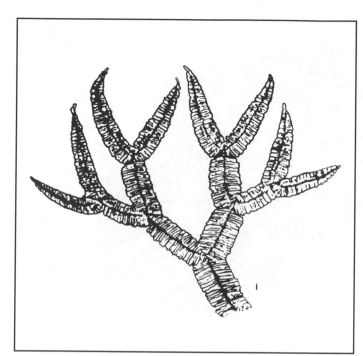

Figure 7.19.
Gleichenia truncata.
Scale bar indicates
1 cm.

After Holttum 1959

of ash slopes include grasses, *Vaccinium, Gaultheria* and *Rhododendron,* but the fern *Gleichenia* (Glei.)[9] (fig. 7.19) is common where landslips have occurred. An interesting and conspicuous plant found on Mt. Awu, on Sangihe Island and on volcanoes in Minahasa is *Gunnera macrophylla* (Halo.) which has coarse and strongly-veined roundish leaves measuring up to 50 cm in diameter. At the base of the leaf stems are three warts, immediately below which a root emerges. These warts contain colonies of the blue-green algae *Nostoc* which presumably help in nitrogen fixation. This would be distinctly advantageous in nutrient-poor, young volcanic soils. It is believed locally that the flowering of the *Gunnera* on Sangihe means that an eruption of Mt. Awu is imminent. The plant produces large amounts of yellow pollen and it may be by association that the relationship is held to exist (van Steenis 1972).

A comparison between the number of tree species and the percentage of ash in the topsoil over an area north of Mt. Batuangus shows that, as expected, fewer trees are found on soils containing high quantities of ash (fig. 7.20). Due east of the Batuangus crater the ash content is very high, but some ash blew into the neighbouring Tangkoko crater and even over the crater wall to the north of Tangkoko. The very high quantity of ash half

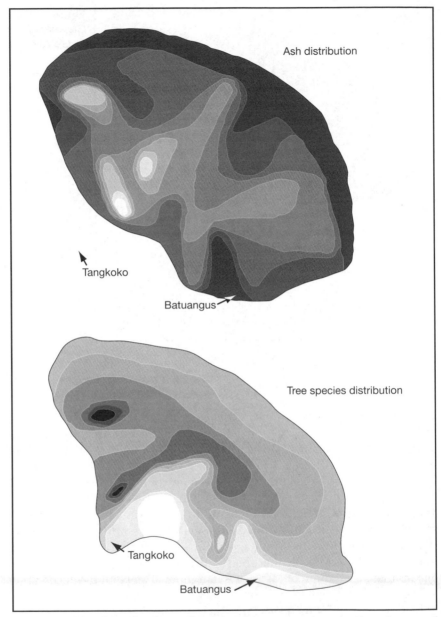

Figure 7.20. Distribution of ash and tree species on the slopes of Mt. Tangkoko (left) and Mt. Batuangus (right). For ash: darkest shading indicates at least 80% ash in topsoil, lightest shading 40%-44%, intermediate shadings in increments of 5%. For tree species: darkest shading indicates at least 45 species per 0.5 ha, lightest shading 5-9 species, intermediate shadings in increments of 5 species.

After Anon. 1980

way along the coast is due to a separate centre of volcanic activity. This correlation indicates that the last eruption in 1839 must have killed large numbers of trees where most of the ash fell and the number of species is still increasing in the impoverished areas (Anon. 1980). Parts of the cone of Batuangus are still bare but others are sparsely forested with groves of *Casuarina* (Casu.). This same tree is also present in the parts of the Tangkoko crater worst affected by the Batuangus ash (Anon. 1980).

ANIMALS AND THEIR ZONATION

Most tropical animals are unaccustomed and unadapted to cold temperatures. This is because the majority of animals, including insects, amphibians and reptiles, draw their body warmth from their surroundings. Some of these will raise their temperatures by basking in the sun and as long as nighttime temperatures do not reduce body temperatures below a critical minimum, then these animals can live successfully in relatively cold climates. No lizards or amphibians seem to have been recorded on mountain tops in Sulawesi but this may in part be because they have not been specifically sought. The cold experienced even on summits would not totally exclude the possibility of these animals living there: for example, a skink found in sub-alpine vegetation in south-east Australia had a critical minimum temperature of -1.2°C (below the freezing point of the animal's blood) and was able to move around below a cover of snow (Spellerberg 1972).

A number of animals are restricted to high mountains, but even on Mt. Tangkoko large animals of lowland forest show preferences for certain altitude ranges (fig. 7.21).

Invertebrates

The number of invertebrates declines with altitude. On Mt. Mulu, Sarawak, for example, the decrease in soil macro invertebrates was accounted for largely by the gradual reduction in the abundance of ants and termites. Biomass, however, reached a maximum in the peaty soils of the taller parts of upper montane forest where beetle larvae took over from termites as the major detrivores. At higher altitudes, centipedes and spiders took over from ants as the major predators (Collins 1980a). Land crabs and detritivore noctuid moth larvae were found on the top of '1,440', one of the peaks in Bogani Nani Wartabone National Park, indicating a distinctive detritivore community. Land crabs have also been reported from 1,350 m in the Quarles Mts., South Sulawesi (Coomans de Ruiter 1950).[10] Land crabs found in alluvial lowland forest in Sarawak were found to have veg-

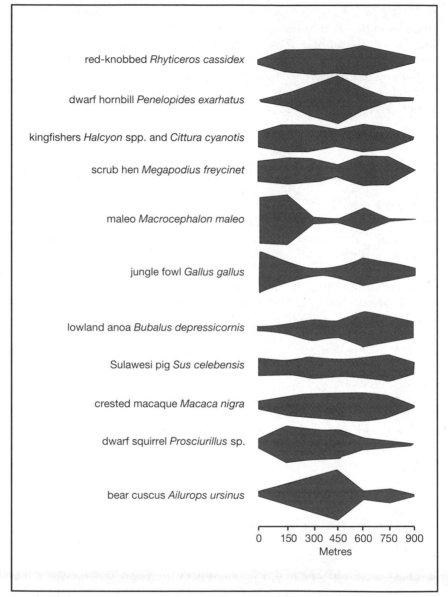

Figure 7.21. Changes in relative frequency with altitude of common birds and mammals on Mt. Tangkoko.

After Anon. 1980

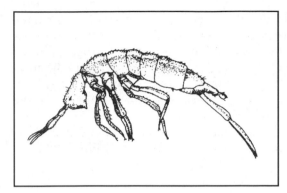

Figure 7.22. Springtail.

etable, mineral and insect remains in their guts (Collins 1980b). Ten pitfall traps[11] were set at 2,000 m and 3,250 m on Mt. Rantemario and were cleared after two days and nights. The ones at 2,000 m contained a variety of tiger beetles, ground beetles, a few spiders and springtails (fig. 7.22) whereas those at 3,200 m caught mainly large numbers of melolonthine scarab beetles, some ground beetles, and one spider. Earthworms were absent above 2,100 m on Mt. Mulu and dousing 1 m² of ground with weak formalin[12] resulted in no earthworms being found at 3,250 m on Mt. Rantemario. At 2,200 m, however, 32 of at least 10 cm length were caught with this same method and on the very summit of Mt. Rantemario, large earthworms were found under every sizeable rock and these must be significant detrivores in the subalpine zone. Huge 50 cm long earthworms *Amyntas* sp. have been reported from 1,350 m in the Quarles Mts. (Coomans de Ruiter 1950). Spoonfuls of honey were placed on the ground at 3,250 m on Mt. Rantemario and not a single creature was seen to visit them, a dramatic contrast to the rapid attention honey attracts in lowland forest.

Communities of animals change with altitude. For example, a disjunction in the distribution of beetle species was found up Mt. Mulu such that no beetle species found below 500 m (lowland forest) also occurred above 1,500 m (upper montane forest) (Hanski 1983). The highland beetles were thus a distinct guild of species and preliminary results from Lore Lindu and Bogani Nani Wartabone National Park suggests that the same general pattern is found in Sulawesi (J. Krikken pers. comm.).

Dawn-to-dusk captures of insects were conducted in Morowali National Park using very bright mercury vapour lamps set against a white sheet at sea level, 1,200 m and 2,000 m, but unfortunately results are not yet available.

The flying insect fauna on the peak of Mt. Rantemario was dominated by large sciarid/syrphid flies which were most abundant on the flowers of *Decaspermum*. No bumble bees or butterflies were seen. Insects on the relatively open peak were quite abundant, and one reason may have been the quantity of anoa dung which many insects use as a food resource for their larvae. Insects (bugs, flies and beetles) have also been reported from the near sterile summit of the active volcano Mt. Soputan, Minahasa, in December 1985 (C. Bennett pers. comm.) This phenomenon has been observed elsewhere and the term 'summit-seeking' has been given to the way they fly up mountains and sometimes congregate in large numbers. The basic factors underlying this behaviour are not known and a bewildering variety of explanations have been proposed (Mani 1980).

Birds

Although numerous species have quite a wide altitudinal range, others are found within quite narrow limits (table 7.6). Because of this, knowing one's altitude on a mountain allows one to predict which species are likely to be encountered or, conversely, hearing or seeing a number of bird species enables one to determine one's approximate altitude (Heinrich 1939).

The lowland zone, up to about 1,200 m, is characterized by large or conspicuous birds such as hornbills, malkohas *Phaenicophaeus calorhynchus* and coucals *Centropus celebensis*, maleo *Macrocephalon maleo*, bee-eaters *Merops* spp., pittas *Pitta,* and by the relative paucity of songbirds. The middle zone is characterized by various doves and pigeons, drongos *Dicrurus hottentotus*, and song birds such as the serin *Serinus estherae* (Coomans de Ruiter 1950; Schuchmann and Wolters 1982; Bishop and King 1986). The ranges of these birds extend down to the lowland zone and upwards, not to a common upper limit as in the lowland fauna, but rather extending to varying degrees into the upper zone. The doves of the middle zone are replaced at higher levels by the fruit dove *Ptilinopus fischeri* and the very rare dusky pigeon *Cryptophaps poecilorrhoa*, songbirds are far more common, and Sulawesi woodcocks *Scolopax celebensis* may be found.[13] A few birds can be found in the subalpine zone but only one species, the non-endemic Island thrush *Turdus poliocephalus* is confined to this habitat where it would eat the abundant fruit and earthworms. The others are more at home below this relatively harsh zone. Strangely, perhaps, on an island with so many endemic species, only about half of the high montane birds are endemic, and the Island thrush is widely distributed in Indonesia.

There are also a fair number of birds which can be found from, more or less, sea level to upper montane forest. Some species including the rare endemic hawk-eagle *Spizaetus lanceolatus* and the racquet-tailed parrot *Prioniturus platurus* are strong fliers, but other wide-ranging montane species such as the little fulvous woodpecker *Mulleripicus fulvus,* the pink-breasted cuckoo-dove *Macropygia amboinensis* and the ground-dove *Gallicolumba tristigmata* are not known for their flying prowess.

Table 7.6. Altitudinal distribution of selected bird species.

The horizontal bars indicate the altitudinal range (in metres) over which each species occurs, read against the scale: 0, 500, 1,000, 1,500, 2,000, 2,500, 3,000.

Species	Approximate altitudinal range (m)
Trichastoma celebensis	0 – 1,000
Pitta erythrogaster	250 – 1,000
Halcyon monachus	0 – 1,000
Cittura cyanotis	400 – 1,000
Meropogon forsteni	250 – 1,250
Cuculus crassirostris	400 – 1,000
Centropus celebensis	0 – 800
Phaenicophaeus calorhynchus	250 – 1,000
Rhyticeros cassidix	250 – 1,000
Penelopides exarhatus	0 – 1,000
Ptilinopus subgularis	250 – 600
Tyto inexpectata	400 – 1,000
Ninox ochracea	0 – 500
Gymnocrex rosenbergi	250 – 1,000
Macrocephalon maleo	0 – 1,100
Dicrurus hottentotus	400 – 1,250
Macropygia amboinensis	0 – 2,250
Ducula forsteni	600 – 1,750
D. radiata	500 – 2,000
Ptilinopus temmincki	600 – 1,400
Accipiter nanus	600 – 2,250
Muscicapa panayensis	1,000 – 2,250
Pachycephala sulphuriventer	850 – 2,200
Cryptophaps poecilorrhoa	850 – 1,850
Lophozosterops squamiceps	1,250 – 2,750
Scolopax celebensis	1,100 – 2,500
Phylloscopus sarasinorum	1,000 – 3,000
Zosterops montanus	1,500 – 3,000
Cyornis hoevelli	1,500 – 2,500
Ficedula melanoleuca	850 – 3,000
F. hyperythra	1,500 – 3,000
Heinrichia calligyna	1,500 – 3,000
Myza sarasinorum	2,000 – 2,900
Turdus poliocephalus	1,500 – 2,900
Cacomantis sepulcralis	0 – 2,750
Prioniturus platurus	0 – 2,750
Otus manadensis	0 – 2,700
Spizaetus lanceolatus	0 – 3,000

After Heinrich 1939; Marshall 1978; Klapste 1982; Watling 1983

The abundance of most species of birds varies gradually with increasing altitude but there are also pairs or trios of species which replace one another relatively abruptly with altitude. The precise altitude would vary from place to place depending on exposure, slope and rainfall. One such pair is Fischer's fruit dove *Ptilinopus fischeri* and the dark-chinned fruit dove *P. subgularis* which meet at more or less the boundary between upper and lower montane forests but not necessarily at the boundary of vegetation types. A trio of species with different altitudinal ranges is found among the white-eyes, at least near Lore Lindu National Park. Thus *Zosterops atrifrons* with a black forehead is found in lowland and hilly forest,[14] *Z. montana* in lower montane forest, and *Lophozosterops squamiceps* in upper montane and subalpine forest (K. D. Bishop pers. comm.). Such transitions may not be heeded by juvenile birds and if a bird is found at the 'wrong' altitude, it is generally found to be a juvenile or immature individual (Diamond 1973).[15]

As a result of the different altitude preferences, the proportion of birds occupying different niches changes with altitude such that:

- the proportion of tree-nesting insectivorous species increases;
- the proportion of tree-nesting frugivorous species decreases;
- the proportion of tree-nesting omnivorous species stays about the same;
- the proportion of predatory birds decreases; and
- the proportion of ground-living birds increases (Kikkawa and Williams 1971; K. D. Bishop pers. comm.).

Mammals

Birds and mammals generate their own heat and heat loss poses enormous problems for the smaller species because their surface area is very large compared with their weight, and their heat loss must be compensated for by consuming large quantities of high-energy food. It is therefore surprising to learn that the smallest Sulawesi mammals, the shrews, are found at high altitudes.

Shrews are almost entirely carnivorous and eat small insects, snails, earthworms, millipedes, centipedes and spiders; resources which are scattered, cannot be stored and whose abundance is often unpredictable. Shrews therefore have to spend a great deal of time searching for food, an activity which in itself uses up energy. It is not surprising, then, that shrews eat about their own body weight in food each day and are active almost the whole day and night although they rest briefly every few hours. If more than a few hours pass without a meal a shrew will starve to death, although larger ones such as *Crocidura elongata* (head and body 94 mm) can survive hunger better than smaller ones such as *C. lea* (head and body 60 mm). The response to food shortage also differs with size: smaller species increase and larger ones decrease their foraging activity (Hanski 1985). Shrews make underground burrows in which they build a nest such that during the cold nights the nest temperature is probably significantly higher than outside.

The little red tree mouse *Margaretamys parvus* (head and body length 10 cm) and the small rat *Haeromys* (head and body length only just over 7 cm), experience similar heating and eating problems but the diet of the former is primarily composed of insects, like the shrews, and of the latter of fruit, roots and seeds, the sources of which are often relatively large and storable.

The mountains of Sulawesi are home to a large number of rats which have been studied intensively on Mt. Nokilalaki in Central Sulawesi (Musser 1982). Among the species present were three species of shrew-rats, so called because of their relatively long snouts and superficial resemblance to shrews. Their head and body lengths are between 11 cm and 15 cm and they are known only from mountainous areas in Sulawesi and the Philippines.

The shrew-rats live in mossy upper montane forest mainly amongst the wet undergrowth of wild gingers, young rattans, ferns, and small shrubs. In clearings they live mainly among the various sedges. Runways, regular routes used by the shrew-rats, are frequently found along, or partly under, rotting trunks or boughs. These runways provide a certain amount of security from predatory owls and may be shared by other species; for example, one particular runway was used by all the shrew-rats, a ground squirrel *Hyosciurus heinrichi*, a rat *Rattus hoffmanni* and shrews. One shrew-rat *Tateomys rhinogradoides* on Mt. Nokilalaki had a very limited range. The mountain's summit is 2,280 m yet no specimens of this animal were caught below 2,210 m despite intensive trapping.

Melasmothrix naso was found to be diurnal and most were caught in the middle of the morning. This shrew-rat shares with other diurnal rats in Southeast Asia the characteristics of a chunky body, tail shorter than head and body combined, and a dark chestnut colour. The two *Tateomys* species are nocturnal and are grey or greyish-brown, a characteristic of nocturnal rats. These differences presumably relate to camouflage effectiveness.

The dietary range among the various rats in the upper montane forest was determined by food trials with captive animals and stomach analyses (table 7.7). Both *Tateomys* species seem to eat only earthworms whereas *Melasmothrix naso* eats mainly earthworms supplemented with fly larvae. Animals favoured by other rats living in the same habitat included moths and cicadas. The food resources are divided between the rats and shrew-rats by virtue of the animals' different activity periods, and by their different means of finding food: for example, *T. rhinogradoides* digs into soil, moss and rotten wood on the ground, whereas *T. macrocercus* climbs over mossy boulders and tree trunks seeking worms in the moss. None of the other rats in the moss forest eat earthworms; the squirrels eat *Lithocarpus* acorns, and the shrews eat insects. There are, however, other earthworm-eating rats at lower altitudes such as *Maxomys hellwandi, Echiothrix leucura, Bunomys andrewsi* and *B. chrysocomus*.

A number of other small mammal species were found near the summit of Mt. Nokilalaki: two long-nosed ground squirrels *Hyosciurus,* the small tree squirrel *Prosciurillus abstrusus,* and the very small squirrel *P. murinus.* The large red squirrel *Rubrisciurus rubriventer* was found up to 1,500 m but no higher. Various small insectivorous bats were seen occasionally, and the most common bat was the frugivorous tube-nosed bat *Thoopterus nigrescens.*

Some of the rats found around the summit of Mt. Nokilalaki, such as *Paruromys dominator, Rattus marmosurus, R. hoffmanni,* and *Maxomys musschenbroekii,* are also found all the way down to sea level but others such as *Bunomys chrysocomus, B. penitus,*[16] *R. arcuatus, R. callitrichus, R. hamatus, R. facetus, Margaretamys parvus,* the little *Haeromys* sp., *Eropeplus canus* and three species of shrew-rats are found only in lower or upper montane forests.

A study of rats in Morowali National Park showed that fruit-eating climbing species were more abundant at higher than lower altitudes. Also none of the species trapped in lowland forest was found in the montane forests. Only two rats were caught in the upper montane forest, a new *Maxomys* and a new *Bunomys,* but only the latter was restricted to this forest

Table 7.7. Food preferences of rats and shrew-rats between 1,950 m and 2,300 m on Mt. Nokilalaki.

	Fruit	Leaves	Insects	Snails	Fungi	Earthworms
Haeromys sp. (a)	+	-	-	-	-	-
Paruromys dominator (b)	+	-	-	-	-	-
Rattus hoffmanni	+	-	-	-	-	-
R. marmosurus (b)	+	-	-	-	-	-
Taeromys hamatus	+	-	-	-	-	-
T. callitrichus	+	+	+	-	-	-
Eropeplus canus	+	+	+	-	-	-
Margaretamys elegans (a)	-	-	+	-	-	-
M. parvus (a)	-	-	+	-	-	-
Maxomys musschenbroekii	+	-	+	+	-	-
Bunomys penitus	+	-	+	+	+	-
Melasmothrix naso	-	-	+	-	-	+
Tateomys macrocercus	-	-	-	-	-	+
T. rhinogradoides	-	-	-	-	-	+

(a) = Arboreal rats nesting in trees but occasionally caught on the ground.
(b) = Ground-nesting rats which seek food in understorey and canopy trees.
Others = mainly ground dwellers but occasionally found above ground.
After Musser 1982

type. After a period of intensive trapping it appeared, however, that just 10 m down the slope, it was replaced by its congener *B. chrysocomus*—a neat example of allopatric distribution of sister species (C. Watts pers. comm.).

In contrast to the commensal rat species of towns and fields whose high reproductive output (p. 584) and relatively low inherent survivorship mark them as r-selected species, the montane rats of the much more stable forest environment appear to be K-selected species with much smaller litters of just one or two young (C. Watts pers. comm.).

The largest mammal of mountain forests is the mountain anoa *Bubalus quarlesi*.[17] These are generally solitary but a herd of five ran past an expedition climbing Mt. Nokilalaki (Meijer 1983). Dung samples collected near the summit of Mt. Rantemario were analysed to determine the plant species.[18] The results showed that, as expected, the anoas were browsing animals for which grassy plants are relatively unimportant (table 7.8). Most interesting was the significant quantity of moss in the dung. Moss fragments are very easy to identify and thus may have been overrepresented when fragments were chosen, but there are enough present to suggest that it was certainly not ingested accidentally. Anoa may eat moss as much for its water content as for its food value. The suggestion that anoa eat moss was also made from observations of moss damage on Mt. Roroka Timbu (van Balgooy and Tantra 1986).

Anoa dung has often been reported as being common along the logging road constructed up to 2,225 m on Mt. Roroka Timbu in the northeast of Lore Lindu National Park. On a recent visit, however, very few signs of anoa were found. This might have been a seasonal phenomenon but many snares and traps set for anoas were found and these quite correctly were destroyed since the anoa is protected by law (K. D. Bishop pers. comm.). A ranger, Alex Alisi, of the Forest Protection and Conservation Service in Central Sulawesi, lost his life on that mountain and it is possible that his death was somehow connected with the poaching of anoa in and around

Table 7.8. Results of analyses of anoa dung from the summit of Mt. Rantemario.

Food item	Percentage
Woody plants with thick cell walls	38
? Ferns	9
? Broad-leaved plants	16
Grasses, sedges	17
Other monocots	5
Moss	13

From M.J.B. Green pers. comm.

Lore Lindu National Park. He was posthumously awarded one of the Kalpataru Servants of the Environment awards in 1985.

Also known from upper montane forest is the Sulawesi civet[19] *Macrogalidia musschenbroekii*, one of the world's least-known carnivores. The animal itself is rarely seen and its presence is more often indicated by tracks,[20] isolated faeces, and scratches on trees up to 2.5 m above the ground. The Sulawesi civet is a master of acrobatic climbing and like some of its relatives, its hindfeet can be rotated to allow head-first descent of vertical surfaces (Wemmer et al. 1983). In the last five years positive signs or sightings of the civet have been made on Mt. Ambang (Minahasa), Lore Lindu National Park (Wemmer and Watling 1986), and Mt. Rantemario (EoS expedition). In the more than 100 years since it was first described for science, very few have been seen by scientists or collected for museums despite intensive searches. In the past, however, the Toraja people are said to have trained these civets for hunting, although they were unreliable (Anon. 1977).

Analysis of 47 faecal samples from Sulawesi civets showed that small mammals and fruit were common components of the diet although they may also take ground-living birds. Most of the small mammals were rats but the small nocturnal cuscus *Strigocuscus celebensis* is evidently taken occasionally and feathers were found in two of the samples. Local informants report that young piglets, farmyard chickens and bananas are also eaten. From the fruit remains it was clear that Sulawesi civets ate the pulp-covered seeds of the wild sugar palm *Arenga* sp. but only when the fruit was ripe (Wemmer and Watling 1986). This is not surprising because when still unripe the pulp contains water-soluble sodium oxalate which, if it comes into contact with the lips, mouth or other mucus-covered tissue, absorbs calcium from the mucus and forms water-insoluble, needle-sharp crystals of calcium oxalate which are extremely painful. During the ripening process, however, the sodium oxalate is broken down and the ripe fruit can be eaten safely (Whitten 1980). Observations suggest that the Sulawesi civet walks around a particular area of forest, not necessarily on obvious paths, taking 5-10 days to complete the circuit of its home range measuring about 150 ha (Wemmer and Watling 1986).

Macaques have been observed up to 2,000 m; for example they have been seen on the peak of Mt. Klabat (1,998 m) (Sarasin and Sarasin 1905), but they are probably relatively uncommon above 1,500 mm.

EFFECTS OF DISTURBANCE

Most people in Sulawesi live in climates where plants grow extremely quickly. It is important, therefore, that those people who climb mountains for recreation or study should remember just how long-lasting the effects of disturbance can be in a much less productive environment.

As mentioned earlier, montane meadows are almost certainly engineered by people in attempts to attract and thence to kill anoa (p. 531). When Mt. Rantemario was climbed in 1937 it was remarked that a fire had quite recently burned the vegetation around the summit (van Steenis 1937; Steup 1939). When the EoS team climbed to the summit in 1985, most of the dominant *Decaspermum* trees were only about 2-2.5 m tall but there were isolated trees of the same species reaching 4 m. These are probably those that escaped the worst affects of the fire 50 years earlier. Lowland forests are not easily ignited and none of their trees have any fire-resistant characters. The vegetation is generally wet or moist and there is no pile of easily-dried material that could act as tinder. In upper montane and subalpine forests, however, periods of drought do occur and the oil-rich leaves of *Rhododendron, Vaccinium* and *Gaultheria,* can catch alight even in the rain. In addition, people are able, through successive attempts to burn an area, to increase the abundance of grasses and sedges, the dead leaves of which can serve as potential tinder for the next fire (van Steenis 1972). Members of the Ericaceae, such as *Vaccinium* and *Rhododendron,* may be fire resistant to some extent for in places in the Latimojong Mts. and near Rantepao, nearly pure stands of the endemic, 3-4 m tall, yellow-orange flowered *Rhododendron vanvuurenii* can be found in fire-generated grassland (Sleumer 1966-67).

The succession of plants in montane areas up to 2,000 m altitude is quite similar to that lower down. *Eupatorium inulifolium* (Comp.) and *Lantana camara* (Verb.) quickly dominate abandoned fields, form a protective cover for the soil and for young tree seedlings. Where the area is subjected to repeated burnings, herbs such as *Lobelia nicotianaefolia* (Camp.) (Moeliono and Tuyn 1960) and the white-leaved 'edelweiss' *Anaphalis longifolia* (Comp.), and bracken fern *Pteridium aquilinum* (Polp.) increase in abundance. With further burning alang-alang grass *Imperata cylindrica* would become dominant. In moist places the wild sugarcane *Saccharum spontaneum* dominates. The course of succession above 2,000 m has not yet been investigated.

The regeneration of Costa Rican subalpine vegetation following a fire[21] has been studied and the response of two shrubs, *Vaccinium* and a *Hypericum*, both of which have species on Sulawesi mountains, was monitored for three years after the fire. The fire killed the above-ground parts of the shrubs but suckers grew from their bases. The suckers grew less than an average of 50 cm in the three years. The soil surface exposed by the fire did not, as one would expect in lowland areas, become covered with fast growing pioneer species. Instead, much of the bare soil remained unvegetated. In other places liverworts and mosses started to grow. Very few seedlings of the surrounding plants were found. Three years after the fire, dead stems of plants were still standing and even those that had fallen to the ground showed no signs of decay. This extremely slow rate of decomposition is caused partly by the low temperatures but also by the absence (or extreme paucity) of most of the lowland decomposers such as ants, termites, and earthworms. This in turn may be due to the extremely wet condition of the soil, branches, and logs which never warm up to any great extent (Janzen 1973).

Regeneration of Costa Rican oak forest, which is similar to that found on Sulawesi mountains, was also found to be extremely slow. Repeated cutting would deplete the seed stocks irrevocably and it was concluded that montane forests are 'truly the tropics most fragile ecosystems' (Ewel 1980).

Chapter Eight

Caves

INTRODUCTION

Caves[1] are simple natural ecosystems of great value for understanding ecological inter-relationships as well as for their own intrinsic interest. They have advantages over many other ecosystems in terms of potential for research, both theoretical and applied, in that the boundaries are discrete, and most of the species inhabiting them can be easily studied and manipulated in the cave or laboratory.

Caves are enormously varied: they range in length from a few metres to over 100 km in a cave network in the U.S.A., and in depth from a few metres up to 1,311 m in France. The largest chamber, indeed the largest enclosed space, natural or synthetic, in the world, is the Sarawak Chamber in Mt. Mulu National Park, Sarawak. It is nearly two kilometres in circumference, and the floor area is equivalent to 17 football fields. Caves may have small entrances opening into large chambers, or massive entrances behind which the cave is only penetrable for a short distance. Some caves have rivers flowing through them, and are called active, and others are more or less dry having been formed in the past, and these are called non-active. Caves of sorts were created during the Japanese occupation when tunnels were dug into hillsides. These have many of the same features as natural caves and are not without interest. For example, the tunnels at Kawangkoan near Tomohon, Minahasa, are home to both bats and swiftlets.

All of the large areas of karst limestone in Sulawesi (fig. 6.14) may contain caves but hardly anything is known about their abundance, size or ecosystems. To stimulate interest, an attempt has been made to make an preliminary inventory (table 8.1). The longest cave is the 11 km Salukan Kalang Cave in Maros which was penetrated and investigated for the first time during the preparation of this book (Anon. 1985b). It is the second longest cave known so far in Indonesia (R.V.T. Ko pers. comm.).

One of the earliest accounts of a Sulawesi cave was by James Brooke (later Rajah Brooke of Sarawak) who visited Mampu Cave, Bone, in 1840 hoping to confirm reports of the presence of ancient statues carved by animistic peoples. The statues turned out to be nothing more than fallen

535

Table 8.1. Preliminary list of Sulawesi caves.

Province	County	Name		Map available	Type	Prehistoric site	Length (m)
South	Maros	Burung		b	N	-	210
		Bembe		b	A	-	30
		Batu Karope		-	N	•	?
		Sampeang		-	N	•	?
		Ulu		-	N	•	?
		Ulu Wae		-	N	•	?
		Pettae		-	N	•	?
		Petta Kere		-	N	•	?
		Lambatorang		-	N	•	?
		Jarie		-	N	•	?
		Saripa		-	N	•	?
		Wattonong		b	A	-	440
		Putih		e	N	-	150
		Towakkalak		e	S	-	80
		Bantimurung		e	N	-	150
		Anggawati		e	N	-	50
		Baharuddin	1	b	S	-	140
		Baharuddin	2	b	N	-	170
		Baharuddin	3	b	N	-	210
		Baharuddin	4	b	A	-	30
		Baharuddin	5	-	N	-	90
		Baharuddin	16	b	N	-	2
		Baharuddin	17	e	N	-	20
		Baharuddin	25	b	N	-	20
		Ancing		b	N	-	60
		Bengo		b	N	-	50
		Salukan Kalang		b	A	-	11,000
		Kappang	7	b	N	-	130
		Kappang	8	b	N	-	150
		Kappang	9	b	N	-	150
		Patunuang	2	b	S	-	80
		Patunuang	1	b	S	-	557
		Rumbia		b	S	-	20
		Rumbia	2	b	N	-	50
		Sambukeaja		b	S	-	200
		Semanggi		b	N	•	780
		Samanggi	1	b	N	-	610
		Samanggi	2	b	N	-	60
		Samanggi	3	b	N	-	90
		Samanggi	4	b	N	-	140
		Semanggi	5	b	N	-	100
		Semanggi	6-10	b	N	-	275
		Semanggi	11	b	N	-	80
		Semanggi	12	b	N	-	60
		Semanggi	13	b	N	-	25
		Semanggi	15	b	N	-	130
		Semanggi	16	b	N	-	55
		Semanggi	19	b	A	-	30
		Semanggi	20	b	A	-	220
		Semanggi	24	b	N	-	100
		Semanggi	25	b	N	-	25
		Kado		b	N	-	110

Table 8.1. (Continued.)

Province	County	Name	Map available	Type	Prehistoric site	Length (m)
South	Maros	Jaria	b	N	•	150
		Apah	b	N	•	450
		Sinjai Karampuang	-	?	?	?
		Bappajeng	-	?	?	?
	Barru	Kassi	-	N	•	?
		Macinna	-	N	•	?
		Sassang	-	N	•	?
		Patennung	-	N	•	?
		Sumpang Bita	-	N	•	?
		Sakapao	-	N	•	?
		Saripa	-	N	•	?
		Batang, Lamara	-	N	•	?
		Elle Masigi	-	N	•	?
		Garunrung	-	N	•	?
	Bone	Mampu	a	N	•	135
		Cakondo	-	N	•	-
		Ulu Leba	-	N	•	?
		Bola Batu	-	N	•	?
		Parisi Ta'butu	-	N	•	?
		Tomatoa Kacicang	-	N	•	?
	Soppeng	Codong	-	N	•	?
	Jeneponto	Batu Ejaya	-	N	•	?
		Panganreang	-	N	•	?
	Bulukumba	Ara	-	N	•	?
	Pangkajene	Karrasa	-	N	•	?
		M2	b	N	-	535
		M3	b	N	-	30
		N1	b	A	•	550
		N3	b	S	-	200
	T. Toraja	Lando	b	N	-	>800
		Kete	b	N	-	>100
		Loko Malilin	b	N	-	480
		Mangana	b	N	•	100
		Lapia	-	A	-	>150
		Lompo	-	N	•	120
Central	Poso	Peda	c	N	•	170
		Tadula	-	A	-	1,000+
		Permona	f	N	?	80
		Togian Is.	-	N	-	-
Southeast	Kendari	Tomba Watu	-	N	-	?
		Menui	-	N	-	?
North	Bol. Mong.	Konongan	a	S	-	150
		Solok	-	N	-	10

N - Non-active cave, **A** - Active cave, **S** - Spring at or near cave entrance.

Maps: a - this book; **b** - Anon. 1985b, 1986; **c** - Rees n.d.; **e** - Anon. 1986; **f** - L. Clayton pers. comm.

After Glover 1978; Anon. 1985; Anggawati 1986; Anon. 1986; Effendi n.d.; D. Owen pers. comm.; and EoS teams.

Figure 8.1. Habitats associated with limestone caves. A - soil and litter in dense forest; B - superficial underground compartment; C - evoluted soils; D - resurgence; E - underground stream; F - pools inside cave; G - sump; H - speleothems, clay, etc.; I - guano; J - river sink; K - river; L - soil and litter in dense forest; M - superficial underground compartment.

After Ko 1986

stalactites with various names arbitrarily bestowed on them. These 'statues' can still be seen and are said to be the members of the court of the Mampu who were turned to stone through a curse. This curse was set when the princess dropped a spool and asked her dog to pick it up because she was too lazy to do it herself. Brooke found the cave impressive enough:

> Mampu Cave is a production of nature, and the various halls and passages exhibit the multitude of beautiful forms with which nature adorns her works; pillars, and shafts, and fretwork, many of the most dazzling white, adorn the roofs or support them, and the ceaseless progress of the work is still going forward and presenting all figures in gradual formation. The top of the cave, here and there fallen in, gives gleams of the most picturesque light, whilst trees and creepers, growing from the fallen masses, shoot up to the level above, and add a charm to the scene.—MUNDY 1848

Caves have potential as tourist areas. For example, part of the Banti-murung Nature Reserve has also been designated a Tourist Park because of its beauty, its geological interest as a classic example of tower karst, its prehistoric interest and its caves. A survey by a member of the Federation of Indonesian Speleological Activities,[2] however, regarded the known caves as too small, too short and too poor in stalactites and stalagmites to hold any real touristic value (Anggawati 1986). Due at least in part to this, the Ujung Pandang Forestry Office has recommended the area south of Bantimurung-Patunuang as an extension to the Tourist Park. This has some longer caves (up to 375 m) with attractive cave formations. In the future the Bantimurung-Patunuang area may be proposed for National Park status (Anon. 1985a). The newly discovered Salukan Kalang Cave could certainly become an important object of specialist tourism.

The characteristic features of a cave are its definable limits, its enclosed nature, low light levels, and comparative stability of climatic factors such as temperature, relative humidity and air flow (Bullock 1966). Variation in these characteristics between caves creates a surprisingly wide range of habitats which determine the type and number of animals that can inhabit a cave. Habitats associated with caves include soil and litter in limestone forest, the superficial underground compartment[3] cave streams, sump zone, and cave floor habitats (fig. 8.1).

CAVE FORMATION

Cave formation is a subject full of special terminologies and opposing theo-ries (Jennings 1971) but until the geology of the various karst areas in Sulawesi is known in more detail, only the simplest of outlines is given below.

Rainwater contains carbon dioxide absorbed from the atmosphere and is therefore slightly acid (p. 468). This weak acid dissolves calcium car-bonate (the main constituent of limestone) and forms channels which, in time, achieve the dimensions of caves, often with a stream running through them.

Two of the commonest features within a cave are stalactites and stalag-mites, which together with other cave decorations are known as speleothems. They are columns of calcium carbonate containing various impurities (the cause of the wide range of pale colours found) and are formed by the repeated evaporation of water containing calcium car-bonate, leaving thin layers of mineral deposits (fig. 8.2).

Evaporation in caves is very slow because there is no solar radiation to excite the water molecules, air movement is absent or minimal, and the air is virtually saturated with water vapour. At a point 1 km into Salukan Kalang Cave, air temperature at about midday was only about 26°C, relative

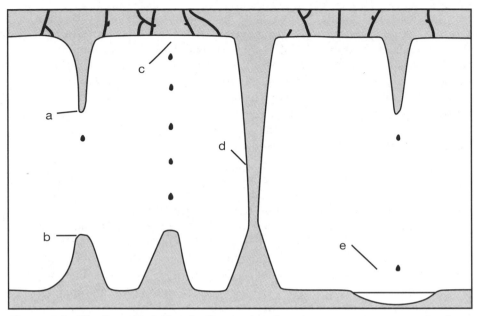

Figure 8.2. The formation of stalactites and stalagmites. **a** - water evaporates from drips precipitating calcium carbonate and impurities thereby forming a stalactite; **b** - water evaporates also precipitating calcium carbonate but forming a squatter stalagmite; **c** - where water drips quickly no stalactite is formed; **d** - stalactite and stalagmite eventually join to form a single column; **e** - where the drips from a stalactite fall (or fell) into a river, no stalagmite is formed.

humidity was 95% (when outside it was about 65%), and there was no detectable wind. In such conditions, the daily evaporation from small pans is 0.044 mm (Dunne and Leopold 1983), and so a column of water only 1 mm deep would take 23 days to evaporate. This explains why speleothems take so long to grow. Precipitation rates are greatly affected by air movement and by impurities in the limestone but the figures above serve to illustrate the slow growth. It has been suggested that small stalactites commonly grow in length by only 0.2 mm per year.

There are many types of speleothems in addition to the well-known stalactites and stalagmites. Water dripping onto the cave floor may form a splattered calcite formation, similar to the shapes formed momentarily when a raindrop falls into thick mud; concentric spheres of calcite may form under special conditions; calcite may precipitate out of water as it flows over cave walls and rocks thereby forming 'frozen waterfalls', often of beautiful colours because of the impurities. In non-active caves eerily-shaped, fragile, hair-thin speleothems called helictites may project from the walls. Lastly, on floors once covered with shallow water, slow evaporation may result in the formation of coral-like spikes, fans, and glistening crystals. Another physical feature of caves which can sometimes be found is 'moon-milk', a soft whitish mass of carbonate minerals some of which are associated with certain species of bacteria (Poulson and White 1969). All speleothems act to increase the surface area of a cave and therefore also the living area available to the cave inhabitants.

Some growths on cave walls, loosely referred to as wall-fungus, are not of mineral origin but are colonies of actinobacteria which look like white or coloured pendants of lichen or fungus. It is possible that these organisms give some caves their characteristic musty smell (Williams and Holland 1967). Growths resembling encrusting lichens found on the wall of a Sumatran cave in 1983 (Whitten et al. 1984), have been identified as an Arthrobacter (Actinobacteria) in association with a *Penicillium* fungus (H. Reid pers. comm.).

The chemical composition of water seeping into a cave depends on the capacity of the percolating water to dissolve rock and sediment, the rate at which minerals dissolve, and the rate and nature of deposition of calcium carbonate by evaporation. Water seeping into Anak Takun Cave in tower karst in Peninsular Malaysia was collected from 63 sites and the results of analyses show considerable variation. Specific conductance varied from 135 to 1,399 µmho/cm (25°C) and potassium concentration from 0.10 to 26.01 ppm. This variation was determined by whether the water had come into contact with guano (table 8.2). The higher calcium hardness[4] is probably largely due to the more acid water which passed through old guano dissolving more calcium carbonate (table 8.3) (Crowther 1981).

A unit of tower karst generally has an insufficient catchment area to support permanent streams above the water table. Thus, apart from seasonal or short-term inputs from diffuse-flow seepages, the caves within it are generally dry. As a result enlargement of existing caves may be minimal and they are instead subject to gradual infilling. It appears, however, that limestone may be dissolved below the floor deposits due to renewed aggressiveness (table 8.3) imparted to seepage water after passing through deposits of bat guano (Crowther 1981).

Table 8.2. Comparison of seepage waters on a cave floor between those which had and had not passed through guano. Samples totalled 46 and 17 respectively and all the differences were statistically significant.

	Seepage water without guano contact	Seepage water after guano contact
pH	7.85	7.69
Specific conductance (μmho/cm)	214.1	794.7
Total hardness (ppm)	111.0	351.6
Calcium hardness (Ca) (ppm)	103.4	336.5
Magnesium hardness (Mg) (ppm)	7.6	15.1
Mg:Ca	0.08	0.04
Potassium (K) (ppm)	0.32	10.68
Sodium (Na) (ppm)	0.92	2.24
K:Na	0.25	2.43

After Crowther 1981

Table 8.3. Chemical properties of saturation extracts from samples of fresh and partially decomposed guano. That is, water and guano were mixed to form a saturated solution which was then analysed.

	Fresh guano	Partially-decomposed guano
pH	7.03	5.68
Specific conductance (μmho/cm)	1906	1640
Ca (ppm)	23.6	25.3
Aggressiveness* (ppm $CaCO_3$)	12.0	156.3
Mg (ppm)	32.5	19.0
K (ppm)	2.27	1.24
Na (ppm)	286.0	194.0
K:Na	16.3	10.2

* Aggressiveness is a measure of the amount of calcium carbonate a solution will dissolve.

After Crowther 1981

TEMPERATURE, HUMIDITY AND CARBON DIOXIDE

The insulating role of the walls and roofs of caves effectively buffer the rel-atively wide daily variations in temperature and humidity of the outside world. Conditions thus remain fairly stable day to day, but there are still sea-sonal changes which can greatly alter conditions in caves. For example, during a rainy period the humidity and amount of free water within a cave tends to increase.

Air movement is also buffered by the cave walls, but still occurs as air is drawn out of the cave during the day when the air outside is warmer and lighter. This air movement follows a regular pattern, but leaves pockets of stagnant air in deep caves where spiders can weave delicate and complex webs, and preserves pockets of high humidity. Under such stable conditions the minute disturbance of air caused by the approach of predators or prey can be detected.

The constant high humidity in the deeper parts of caves appears to have led many cave invertebrates to become morphologically similar to aquatic arthropods. Many have lost cuticle pigments and wings, while some have developed a thinner cuticle, a larger and more slender body than their rel-atives outside the cave, adaptations to their feet allowing them to walk on wet surfaces and also a lower metabolic rate (Barr 1968; Howarth 1983). In these deeper areas, the concentration of carbon dioxide increases if there is no inflow of air except from the cave mouth; this was detected in Salukan Kalang Cave. It has been suggested that the lower metabolic rate of some cave invertebrates may be a physiological response to this high carbon dioxide concentration (Howarth 1983).

CHARACTERISTIC ANIMALS AND FOOD CHAINS

Cave animals can be divided into three ecological groups:
- troglobites or obligate cave species unable to survive outside the cave environment;
- troglophiles or facultative species that live and reproduce in caves but that are also found in similar dark, humid microhabitats outside the cave; and
- trogloxenes or species that regularly enter caves for refuge but nor-mally return to the outside environment to feed.

In addition, some other species wander into caves accidentally but cannot survive there (Howarth 1983). It is now realized that some small 'cave-adapted' animals also live in superficial underground compartments even in locations distant from caves.

It used to be thought that there were no troglobites anywhere in Southeast Asia (Leefmans 1930), but exploration of caves in Mt. Mulu National Park, Sarawak, have confirmed the existence of at least 27 species there (together with about 70 species of troglophile) (Holthuis 1979; Roth 1980; Peck 1981; Chapman 1984). A troglobitic crab and a prawn are known from the Mt. Sewu region of Central Java (Holthuis 1984; Ko 1986). Small, white atyid prawns with much reduced eyes found in the deep Salukan Kalang Cave by a APS[5]-EoS team have recently been examined and found to be not just a new species but also the first troglobitic atyid known from Indonesia (L. B. Holthuis pers. comm.).

All cave dwellers are dependent for food on material brought into the cave from outside. Some animals feed on plant roots attached to the cave roof, wood and other material washed in during floods (if the cave has a river running through it), or the organic matter percolating through from the surface. The major providers of food, however, are bats and swiftlets[6] that roost and breed in the cave but feed outside. Bats and swiftlets supply food in a number of ways:

- during their lives they produce faeces, collectively known as 'guano', which has nutritive value to the various animals that feed on it (coprophages), and is also a source of nourishment to fungi and bacteria;
- as live animals they are hosts to many parasites, both internal and external, and provide food for predators;
- they moult hair and feathers and shed pieces of skin;
- they produce progeny which may be susceptible to different predators and parasites;
- when the bats and swiftlets die, their bodies form a source of food for various corpse-feeding organisms (necrophages).

Almost all animals provide food for others in these five ways, but within the cave ecosystem, in the absence of green plants, these are the only major sources of food. The bats themselves roost on the roof and walls and form the primary basis of one community; their faeces and dead bodies fall to the cave floor and form the basis of another. Thus there is a distinct division of the animals into a roof community and a floor community (Bullock 1966).

EFFECTS OF DARKNESS

The cave environment can be divided into three zones according to the degree of darkness and other physical conditions:

- the twilight zone near the cave entrance where light and temperature vary, and in which a large and varied fauna can be found;
- the middle zone of complete darkness but variable temperature, in

which a number of common species live some of which make sorties
to the outside; and

- the dark zone of almost constant temperature and complete darkness[7] in which obligate, cave-adapted species are found (Poulson
  and White 1969). Since light is essential for photosynthesis, green
  plants are not found in the dark parts of caves. Plant roots can penetrate fissures leading to a cave and these can commonly be seen
  attached to or hanging from the cave roof. The most important
  effect of this virtual total exclusion of green plants is to make all cave
  dwellers dependent on material brought in from the outside and to
  exclude all animals that feed directly on the above-ground parts of
  green plants.

Most animals have a clearly-defined daily cycle of activity, being most
active at night (nocturnal), during the day (diurnal), or around dawn
and dusk (crepuscular). Such cycles are obviously associated with daylight
and darkness and thus might be thought to be absent in a cave. There are,
however, certain events which may impose a daily rhythm on cave inhabitants. The most important of these is the departure and subsequent return
of the bats and swiftlets. In their absence, food is not available for the free-
living ectoparasites in the roosts, and there is a halt to the rain of fresh
faeces from the roof (Bullock 1966). In addition, air draughts probably
have a daily rhythm caused by the heating and cooling of air outside the
cave. Therefore, although data are lacking, it is probable that a daily
rhythm does exist in a cave.

The departure of the bats is an event which may continue for two
hours after the first bat flies out of the cave. The different timing and pattern of flight activity of bat species in a single cave have been interpreted
in different ways, such as avoidance of competition for food, avoidance of
predators, and optimizing energy budgets (Fenton et al. 1977; Erkert
1982). It should be stressed, however, that bats are not out of their caves
from dusk to dawn, nor are they are necessarily roosting or asleep while in
the cave. Instead, even if not disturbed by voices or carelessly-aimed flash-
lights, individuals or groups can often be seen flying around the caves and
there appears to be a constant high-pitched chatter during the day.

In total darkness, a cave-dweller becomes reliant on senses other than
sight to detect food or enemies. This is not peculiar to cave animals, however, because many nocturnal and cryptic animals found only above
ground and animals living in soil also depend almost exclusively on
hearing, smell and touch (Bullock 1966). Some, for instance, have very
long appendages, such as the legs of scutigerid centipedes and the
antennae of cave crickets (fig. 8.3), although antennae can also function as
chemoreceptors and may be sensitive to relative humidity (Howarth 1983).
Diurnal animals can live in a cave provided their other senses are
sufficiently acute, and sight may even be useful to animals ranging into the

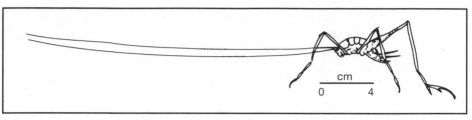

Figure 8.3. A cave cricket *Diestrammena gravelyi* showing its extremely long antennae.

twilight areas, and also to those that range outside the caves. The presence or absence of eyes in troglobites is immaterial, since they are restricted to the dark zone. In fact, the lack of evolutionary selection to maintain good sight permits deleterious variation to appear resulting in the total blindness of some species (Holthuis 1979; Roth 1980; Peck 1981). The relative scarcity of food in the deepest parts of caves has led to the few animals that live there being able to withstand long periods of starvation, to gorge when food is available and to store a large amount of fat (Howarth 1983).

Echo-location

Many bats and swiftlets (p. 553) have developed the ability to echo-locate; that is, a sound is produced and the echoes which reflect back from solid objects are interpreted to give a 'picture' of the surroundings. Echo-location in bats has evolved in at least two ways, but in all cases it is characterized by high frequency sounds (10-200 kHz) which are produced in the larynx or speech-box, mostly far above the threshold of human hearing (15-20 kHz), and by the reception of echoes in complex and often large ears. The principle can be appreciated if you stand close to and facing a wall and then speak; then compare the sound of your voice with the sound when you turn around and speak into the middle of the room. The two sounds are quite different. This ability to sense the proximity of large objects can become quite well-developed in blind people. Rain disturbs echo-locating bats because very humid air absorbs high frequency sounds, and the raindrops confuse the echos received by the bats (p. 426).

Mouse-eared bats (Vespertilionidae)[8] (fig. 8.4) use a predominantly frequency-modulating (FM) system; that is, the frequency of the sound they emit through their mouth varies and is given in very short pulses. When cruising[9] in the open a pulse is emitted and some time is spent listening for echoes. When closing in on a flying insect, however, pulses are emitted

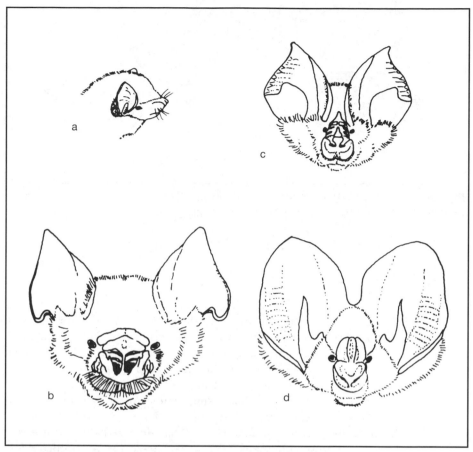

Figure 8.4. Representatives of the four most common families of insectivorous bats. a - *Miniopterus schreibersii* (Vespertilionidae), b - *Hipposideros diadema* (Hipposideridae), c - *Rhinolophus arcuatus* (Rhinolophidae), d - *Megaderma spasma* (Megadermatidae).

After Payne et al. 1985

rapidly so that the exact locations of the prey can be determined. The system used by a flying mouse-eared bat is sufficiently refined for it to detect objects less than 1 mm across.

Horseshoe and leaf-nosed bats (Rhinolophidae and Hipposideridae) use mainly a single rather than a variable frequency and each species uses a characteristic frequency. Instead of emitting the sound through their

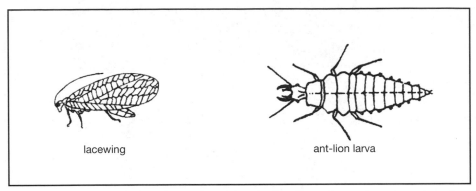

lacewing   ant-lion larva

Figure 8.5. Lacewing (Neuroptera) and its ant-lion larva. Lacewings have 'ears' on their wings that can detect the approach of a calling bat.

mouth like mouse-eared bats, these bats keep their mouth shut and emit the sounds through their nostrils which are positioned half a wavelength apart to give a stereo impression when the echos are received. The peculiar 'horseshoe' around the nostrils has the function of a megaphone, causing the sound to be emitted in a concentrated beam.

It used to be thought that bats using echo-location had no difficulty catching their insect prey, but it now appears that some moths can detect bats from 40 m away and before the bat has detected the moth, using 'ears' on their chests, abdomens or mouths. Lacewings, the adults of ant-lions (fig. 8.5), have 'ears' on their wings and those that have been artificially deafened have a 40% greater chance of being caught by a bat than those that can still 'hear' (Fenton 1983). Some moths have developed the ability to utter clicks that confuse the bat, while others have a variety of behavioural responses which make it difficult for the bats to predict their flight pattern. Some bats in their turn do not keep their echo-locating system 'switched-on' continuously so as to give as little warning as possible to the moths, while others emit frequencies above the hearing threshold of moths (Fenton and Fullard 1981; Fenton 1983).

Whereas the bats mentioned above catch flying insects, false vampires (Megadermatidae) feed by picking lizards, frogs and small rodents off the ground, insects off leaves, or fish from the surface of water. They have also been known to eat other bats (Medway 1967). To avoid swamping the echos from their prey they 'whisper' their sounds which are FM like those of mouse-eared bats. False vampires also sometimes hunt like owls and locate their prey solely by homing in on sounds made by the prey itself.

This is similar to a frog-eating bat which has been studied in Panama and which can differentiate between the calls of edible and poisonous frogs and between the calls of small frogs and frogs that are too big to capture. Its efficiency at catching frogs has probably led to adaptations in the frogs' calls so that the males still call to attract females but in such a way as to reduce their chances of being caught (Tuttle and Ryan 1981).

The only fruit bats to echo-locate, the cave-dwelling rousette bats *Rousettus,* use a low-frequency (1.5-5.5 kHz) tongue-click like swiftlets, which is audible to humans and reminiscent of a wooden rattle (Yalden and Morris 1975; Fenton and Fullard 1981). For this bat and the swiftlets, the echos enable them to detect large objects, or rock walls such that they are able to navigate, nest and breed within a totally dark cave but the system is not sufficiently accurate to enable them to catch insects at night (Medway 1969). *Rousettus celebensis,* at least, appears to learn the route in and out of its cave and does not always use its echo-locating ability. These bats were caught by EoS teams when they collided with the team members in the caves.

It might be thought that the different activity periods of bats and swiftlets would represent temporal partitioning of a common food resource. In fact, swiftlets feed mainly on small wasp-like insects (Hymenoptera) (Medway 1962; Hails and Amiruddin 1981), whereas insectivorous bats concentrate on various moths and beetles (Yalden and Morris 1975; Gaisler 1979; Fenton 1983). Bats do interact ecologically, however, with nocturnal birds (Fenton and Fleming 1976).

ROOF COMMUNITY

The roof community includes bats and swiftlets as well as all those animals that feed on or parasitise them. Over half of the insectivorous bat species and three or four of the fruit bat species probably use caves as permanent or temporary roosts. The large flying foxes *Pteropus* spp. rarely roost in caves although this is not unknown (Stager and Hall 1983).

Cave-roosting bat species differ in their preference for certain conditions. Some, such as the dawn fruit bat *Eonycteris spelaea* are found in chambers near to the cave mouth. Some have wings that allow them to maneuver in tight spaces such that they are found roosting in narrow crevices or 'chimneys' (Goodwin 1979). Others, such as the long-fingered bats *Miniopterus* tend to be found in the dark zone. It is quite common to find at least two species of *Miniopterus* living in the same cave system or even the same chamber. The species differ little in size and in total darkness they must use olfactory and vocal cues, tactile behaviour and physiological responses to distinguish their own species.

EoS teams visited various caves in the course of collecting information for this book and one, the winding, ascending cave of Konangan, north of Kotamobagu, Bolaang Mongondow, was of particular interest because a nursery of common long-fingered bats *M. schreibersii* was found. Over a thousand naked pink bats hung so close to each other on a projection from the cave roof that the rock beneath could not be seen. Nursery colonies such as this are known in relatively few species of bats and in Sulawesi they are formed only by long-fingered bats. Pregnant females congregate once each year to give birth at 'traditional' locations after a 4.5-5 month gestation period (Medway 1971), and the nurseries are not necessarily used only by the bats that normally inhabit a particular cave, but may also be used by pregnant females from other caves in the area. In India a nursery of *M. schreibersii* is known to service an area of 15,000 km^2 (Hill and Smith 1984). The dense cluster of babies huddle together and this increases the thermoregulatory potential of the group. The mothers roost away from the babies visiting them only to suckle. In *M. schreibersii* communal care has even extended to the mothers nursing the young indiscriminately. The young may start to fly only a month after birth by which time they are almost the same size as their parents (Krishna and Dominic 1983), but they may continue suckling for a couple of months after this.[10]

Little is known about bat predators, but pythons are sometimes seen in or around the cave entrance (Sarasin and Sarasin 1905). The bat hawk *Macheiramphus alcinus* was recorded for the first time on Sulawesi in 1981 near Lake Lindu; it was previously known from Africa south of the Sahara, Madagascar, the Malay Peninsula, Sumatra, Borneo and southeast New Guinea. It is rarely seen, possibly because it is only active around dawn and dusk, but is quite distinctive with the white throat contrasting with the rest of the plumage which is black, a short crest, long and pointed wings, and a square-cut tail. It is about the size of a male peregrine falcon *Falco peregrinus* (Bartels 1952; Eccles et al. 1969; Klapste 1982). Owls may occasionally take bats, and this is one possible reason why bats are not very active during the times of full moon. Birds of prey such as owls characteristically evacuate pellets of undigested food such as hair and bones through their mouth. This is often done from a regular roosting or nesting site. Pellets were collected by an EoS team from the floor of the furthest chamber in Mampu Cave and were probably produced by a barn owl *Tyto rosenbergi*. Careful dissection of the dense pellets revealed the skulls of some of the owl's prey: rice-field rats *Rattus argentiventer*, shrews *Crocidura* sp., but only one bat—the relatively large diadem leaf-nosed bat *Hipposideros diadema* (Boeadi pers. comm.). Of additional interest was part of the leg and claw of what can only have been a young owl, it is known from studies elsewhere that when food supply is limited, the stronger of two young owls in a nest may kill and eat the weaker.

Bats are hosts to parasites, some internal, and many external which bite

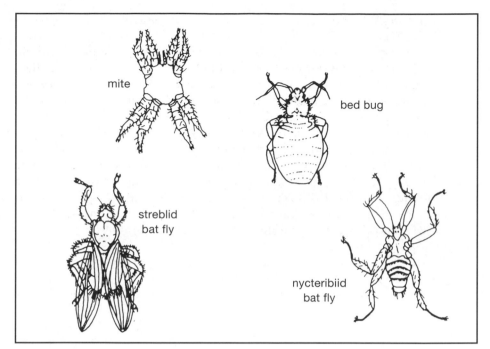

Figure 8.6. Common parasites of bats.
After Fenton 1983

their host to suck blood. Some, such as the spider-like wingless nycteribiid
bat flies[11] (Marshall 1971), live almost their entire lives on bats, while
others such as streblid bat flies, bed bugs (Cimicidae) and chigger mites
(Trombiculidae) spend only part of their life cycle on bats (fig. 8.6). It has
been suggested that the tropical bed bug *Cimex hemipterus* may have begun
its association with humans when they used caves as habitual shelters.

The parasitic insects on bats are diverse taxonomically—six families
from four orders are represented—but they show considerable convergence in many characteristics. They are generally flattened, some vertically
and some horizontally, to ease their movement between the bat hairs.
Most of the insects have tough but expandable 'skin' which allows for the
consumption of large meals of blood from the host. The skin often bears
backward-pointing spines which lessen the chance of being dislodged by a
scratching bat. They also have well-developed grasping claws. Wings have
been lost in species belonging to all but one of the insect families concerned. The loss of wings and general lack of light in caves have led, not

surprisingly, to the loss or reduction of eyes in some of the species (Marshall 1971).

Most species of parasitic insects are found on only one or two closely related host species (table 8.4). For example, each of 28 species of wingless bat flies were found on only a single bat species. Within a species of host, however, not all individuals are necessarily crawling with these parasites. A detailed study of one nycteribiid bat fly on its major host showed that male bats were more often infested than females. Of the males, most were not infested at all and of those that were, most had only one individual parasite. The reason for this low population is not fully understood but it may be because of the relatively low density of bats in the abundant roosting sites (Marshall 1971). There is some evidence to suggest that animals on low-protein diets carry larger populations of ectoparasites (Nelson 1984), and this may be due to the host being less able to scratch the ectoparasites out of its fur.

It is frequently stated that bats carry rabies. This is certainly true in Central and South America but there is only a single report of rabies in a bat from Asia—a dog-faced fruit bat *Cynopterus brachyotis* in Thailand (Hill and Smith 1984).

Table 8.4. Numbers of insect species from six families found as parasites on different totals of bat species and genera in Peninsular Malaysia. Numbers in parentheses indicate the number of species known from the bats examined.

	1 species	2 species 1 genus	3-7 species 1 genus	2 species 2 genera	4 species 3 genera	Unknown
Earwigs						
Arixeniidae (2 spp.)	2	-	-	-	-	-
Bed bugs						
Cimicidae (3 spp.)	1	-	-	1	1	-
Polyctenidae (3 spp.)	3	-	-	-	-	-
Wingless bat flies						
Nycteribiidae (44 spp.)	28	11	1	-	-	4
Winged bat flies						
Streblidae (29 spp.)	21	4	3	1	-	-
Fleas						
Ischnopsyllidae						
(4 spp.)	2	1	-	1	-	-

After Marshall 1980

Swiftlets

Three species of swiftlets are known from Sulawesi, all of which inhabit caves, though not necessarily exclusively (table 8.5). The moss-nest and white-rumped swiftlets emit rattle sounds indicating that they are able to echo-locate, but the white-bellied swiftlet does not. This last species is found, not surprisingly, relatively close to the cave mouth, and once the birds roost at nightfall they do not move again until dawn. All swiftlets use saliva produced by exceptionally large salivary glands to cement the nest material together, and to anchor the nest firmly on to vertical or over-hanging surfaces. The species of swiftlet that uses only saliva to build its nest, a delicacy sought after for Chinese cookery, is not known from Sulawesi.

Floor Community

The organic matter on the floor of most dry caves is composed largely of material formed from waste products and bodies of animals. Samples of this guano were taken from three caves, and the analyses showed that although the composition varied (table 8.6), there was generally a low level of carbon, a moderate to high concentration of nitrogen, a very low carbon:nitrogen ratio, and an extremely high level of phosphorus. Not surprisingly, local villagers are extracting the guano to sell as fertilizer.

On the cave floor the coprophages and necrophages predominate. It is often difficult to distinguish between them, because whereas a few animals are exclusively necrophagous, many of the coprophages will include dead bats or swiftlets in their diet. The majority of cave-dwelling bats are insect-eating and the faeces they produce are hard and dry and readily exploited by coprophages such as woodlice, caterpillars of *Tinea* moths which carry a cocoon around with them, flies and beetles, although the primary decomposers are bacteria. The faeces produced by the few species of fruit-eating bats that roost in caves, however, are soft and rich in carbohydrates and not generally utilized by coprophages. In this case, cockroaches ingest the faeces and the general coprophages feed in turn on the faeces of the cockroaches (Doyle 1969). Cockroach density can exceed $100/m^2$ (Ko 1986). Fungi (and some bacteria) digest these faeces and some coprophages such as crickets and small Psocoptera flies also exploit this food resource (McClure et al. 1967).

The bacteria found in caves are in no way unique, but are a selection of the species found outside the cave. It is believed that they may produce antibiotics that exclude fungi and moulds from the foods on which they live. The clayey floor deposits on which the bacteria grow may be eaten by small invertebrates which may in fact be dependent on them. This would

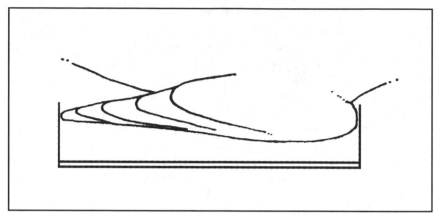

Figure 8.7. The measurement of standard wing length.
After King et al. 1975

Table 8.5. Characteristics of Sulawesi swiftlets.

	Description	Rattle call	Nest
White-rumped swiftlet *Collocalia spodiopygia*	Wing 115-120 mm,* black upper parts with slight blue/green gloss, whitish band across rump, grey under parts.	Yes	Primarily vegetable material with quite copious cement often laid down in layers alternately with vegetable matter.
White-bellied swiftlet *C. esculenta*	Wing less than 107 mm, black upper parts with strong blue/green gloss, whitish tips to feathers of greyish under parts.	No	Moss, roots, lichens, twigs, etc., bound with saliva cement.
Moss-nest swiftlet *C. vanikorensis*	Wing 118-128 mm, black upper parts with brown gloss, pale grey under parts.	Yes	Green moss only cemented with little saliva except where attached to cave wall where it is copious.

* See figure 8.7.

After Hartert 1896; Medway 1966, 1975; Wells 1975

Figure 8.8. Tineid moth, the caterpillars of which carry cocoons around with them and feed on guano.

explain why certain cave invertebrates are difficult if not impossible to rear away from cave deposits (Poulson and White 1969).

The floor community includes many predators such as long-legged scutigerid centipedes, assassin bugs *Bagauda,* and medium to large spiders which feed on the coprophages and the small *Tinea* moths. Some of these predators live on the walls and wait for wandering coprophages to come to them, or only venture to the ground when hungry. Small predators may also form part of the diet of larger visiting predators such as shrews *Crocidura.*

Tineid moths are dull-coloured with a wingspan less than 30 mm across and with furry scales on the wings (fig. 8.8). Of the many species known, most are associated with guano although only twenty are recognized as being strictly cave-dwelling.[12] The larvae, which carry cocoons around them, do not eat plants but bat and bird guano as well as dead bats, birds

Table 8.6. Results of analyses of guano from three caves.

	pH	C	N	C/N	P-av	Exchangeable bases (meq/100g)			
	H$_2$0	(%)	(%)		(ppm)	K	Na	Ca	Mg
Mampu	5.1	1.39	1.04	1.34	127	0.1	0.2	0.6	10.0
Simangi	4.4	1.33	1.55	0.86	57	0.1	0.1	0.9	8.5
Tadula	3.3	1.83	0.28	6.54	305	0.1	0.0	3.5	0.5

Data from EoS teams; L. Clayton pers. comm.

and invertebrates. Where more than one species of tineid moth is found in a cave it is likely that they exploit different types of food (Robinson 1980). For example, it is important to remember that the composition and appearance of the guano produced by the different bats and swiftlets differ markedly, and these different types are generally found in different locations according to the main roosting sites of the bats and birds. In this way, competition between species is to some extent avoided.

Tail-less whip scorpions (Amblypygi) are common in caves and look dangerous and nightmarish, but the mouthparts do not contain poison glands. The pedipalps (appendages just behind the mouth), end in a simple claw but they look formidable, with long spikes on their inner surface. They are used for seizing and holding prey in a similar fashion to the legs of a praying mantis. The front pair of legs can be mistaken for antennae (which whip scorpions, spiders and related creatures do not possess), for they are long and thin and are usually held in front of the animal. One of these whip scorpions is *Stygophrynus* (fig. 8.9) which is known from the floor and walls of Mampu Cave where they possibly hunt the large *Rhaphidophora* crickets (Leefmans 1930).

An examination of the 'stomach' of these crickets revealed pollen and other plant tissue, bits of small insects such as wasps and silverfish (Thysanoptera), and other animal tissue such as bat hair. Some of this could have been eaten if the crickets had ventured outside the cave to feed, but the contents could equally well have come from the bat dung on which the crickets are commonly seen (Leefmans 1930).

Fifty years ago invertebrates were extremely abundant in Mampu Cave. The upper layer of the guano consisted almost entirely of living insects such as click beetles (Elateridae), ground beetles (Carabidae) (fig. 8.10), scarab beetles (Scarabaeidae) (larvae of these live in the guano as do the caterpillars of tineid moths). Also present were at least three species of cockroaches as well as crickets, earwigs, centipedes, millipedes, and even some ants (Leefmans 1930). A French speleological team that visited Mampu Cave in 1986 reported an abundant invertebrate fauna, most of which are dependent on the guano produced by the thousands of bats that roost there.

Members of Operation Drake made a detailed survey of Peda Cave, near Taronggo in Morowali National Park (fig. 8.11). The most conspicuous invertebrates in the dark zone were the very large *Rhaphidophora* crickets with bodies 50 mm long and antennae 90 mm long, whip scorpions and large hairy spiders. The crickets appeared to consume virtually all the organic debris available and even ate through a wax candle placed in the cave for surveying (Rees n.d.). The cave moth *Tinea microphthalma* was found (M. Brendell pers. comm.) which was described only recently from specimens collected in the Philippines. This species is noteworthy in that it has reduced eyes, the first tineid to have lost some of its sense of sight. This

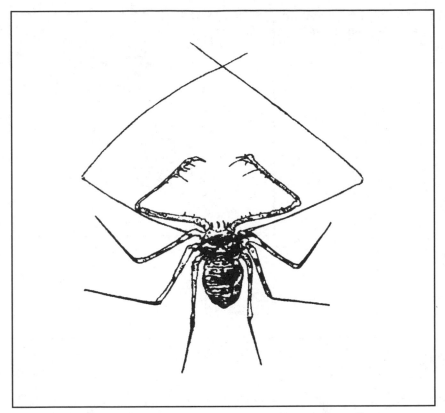

Figure 8.9. Whip scorpion *Stygophrynus* from Mampu Cave, Bone.
After Leefmans 1930

appears to go against the conventional cavers' wisdom that guano-depen-
dent invertebrates tend to retain the morphology of their terrestrial relatives, and non-guano invertebrates are those that tend to change.

The various relationships described above may be drawn as a food
web (fig. 8.12). As the understanding of a cave ecosystem grows, so its food
web becomes more and more complicated. As food webs become more
complicated (and it should be remembered that caves probably represent
Sulawesi's simplest ecosystem), there is good reason to produce generalized representations or models. The animals at the base of a food web are
relatively abundant while those at the top are relatively few in number, with
a progressive decrease between the two extremes. This 'pyramid of numbers' is found in ecosystems all over the world and provides a useful means

Figure 8.10. Carabid ground beetle, one of the animals that feed on guano.

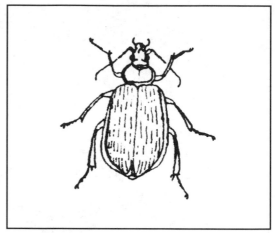

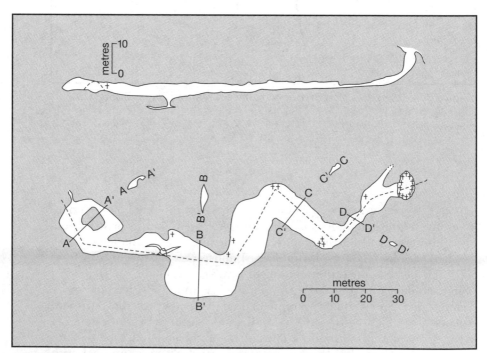

Figure 8.11. Elevation (above) and plan (below) of Peda Cave, Morowali. Crosses indicate guano deposits; the dotted line is the line surveyed for the elevation; cross sections are illustrated at A, B, C and D.

After Rees n.d.

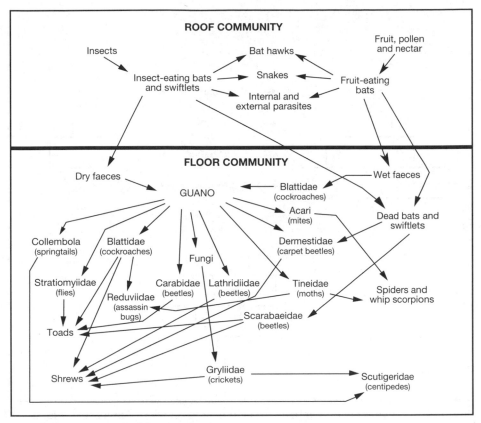

Figure 8.12. Simplified food web of a cave ecosystem.

of comparing communities. To construct the pyramid, species are grouped together according to their food habits. Thus all the plants are called primary producers, herbivores are called primary consumers, predators on herbivores are called secondary consumers, predators on secondary consumers are called tertiary consumers, etc. In general, a large number of primary producers support a smaller number of primary consumers which support an even smaller number of secondary consumers supporting one or two tertiary consumers (fig. 8.13a). Variations of the pyramid shape occur when, for instance, a single tree is considered (fig. 8.13b). Even an inverted pyramid can be formed if one considers a single animal, such as a

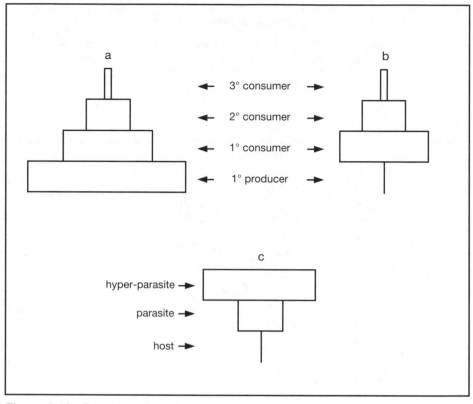

Figure 8.13. Pyramids of numbers: a - with a large number of primary producers, b - with a single primary producer, c - the case of parasites and hyperparasites.
After Phillipson 1966

bat, or a plant, which carries a large number of parasites, which are themselves parasitized by an even larger number of hyper-parasites (fig. 8.13c).

The information contained in a pyramid of numbers permits us to state the number of herbivores supported by a certain number of plants and so on. For comparisons between ecosystems, however, a better approach is to use biomass rather than numbers, and to show this as a 'pyramid of biomass' which usually has a similar shape to the pyramid of numbers (Phillipson 1966).

DIFFERENCES WITHIN AND BETWEEN CAVES

It is clear from the descriptions above that although at first sight a cave appears to be a fairly uniform habitat, this is certainly not the case. One of the chief factors causing variation in the cave habitat is the distribution of bats, the producers of guano. Cave maps (figs. 8.14-8.16) show that the occupation of a cave roof by bats is very patchy, and the distributions of guano and invertebrates on the floor reflect this. Mampu Cave comprises nine major chambers all of which, except for Chamber 9, have either swiftlets or bats roosting in them. The total bat population is in excess of 8,000 but these are divided into distinct colonies consisting of between 100 and 4,000 bats. Physical factors also vary from place to place in the cave; for example, during a rainy period, standing water will accumulate in one part of a cave but not in another, and some parts are subject to air movement, while others are not.

Conditions between caves clearly differ far more than conditions within a cave, or even within a single area of limestone, and when comparing caves of broadly similar conditions, the differences are striking. The invertebrates in various Maros caves and in Mampu Cave show considerable differences (table 8.7) as do the species of bats (table 8.8).

EFFECTS OF DISTURBANCE

Little is known about the detailed effects of disturbance on caves. At one extreme, it is clear that opening up a cave by mining the limestone and allowing in the sunlight will utterly destroy the specialized cave communities. There is no such thing as regeneration of cave life without rebuilding the cave. Instead, a succession of plants from the limestone flora (p. 474) will colonize the rocks where light newly penetrates. At the other extreme, moderate extraction of old guano need not have catastrophic effects since cave organisms utilize only relatively fresh guano as food. Even so, the resilience of the cave fauna to disturbance is unknown and this sort of exploitation would obviously be better undertaken after an ecological survey and assessment of the possible impacts of different collection techniques and sites.

Bats are sensitive to disturbance and, when caves are visited during daylight hours, strong lights and loud noises should be avoided in the darker chambers. Catching bats requires skill and practice and should not be attempted except with good reason. Scientific collecting of bats should be conducted in moderation and other studies should be planned to cause as little disturbance as possible. Some of the bats that pollinate flowers of commercial fruit trees roost in caves, and if they abandoned a

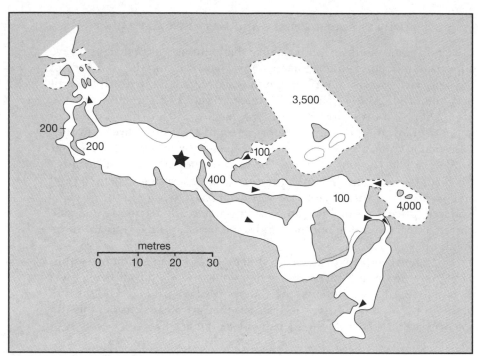

Figure 8.14. Sketch plan of Mampu Cave, Bone. **Dotted lines** - approximate dimensions, **grey lines** - roof openings to the outside, **star** - 'King of Cave' shrine, **arrows** - point up slopes, **numbers** - indicate approximate numbers of roosting bats.

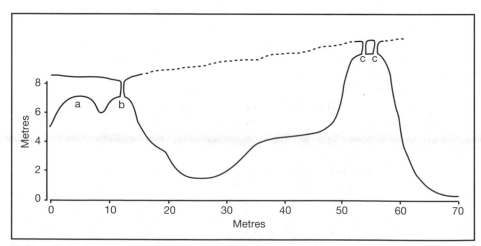

Figure 8.15. Elevation of the central part of Tadula Cave, Kuku, Poso. Note that the vertical scale is exaggerated. Bats: a - 500 *Hipposideros cervinus*, b - 30 *Rousettus amplexicaudatus*, c - 1,000 *Hipposideros diadema*.

After L. Clayton pers. comm.

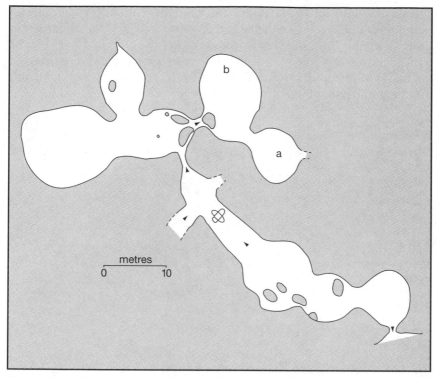

Figure 8.16. Sketch plan of Permona Cave, Tentena, Poso, a National Cultural Reserve. a - 20 *Emballonura monticola*, b - 30 *Rhinolophus arcuatus* and *E. monticola*.

After L. Clayton pers. comm.

disturbed cave and left the area this could cause considerable financial loss. Some caves are being considered as tourist sites (Anon. 1985a; Anggawati 1986) and unless this form of development is carefully controlled it could have significant negative impacts. For example, Permona Cave near Tentena, Poso, is a National Cultural Reserve, and suffers considerable disturbance from visitors and consequently not many bats are found (L. Clayton pers. comm.).

Limestone mining can cause a variety of disturbances, even if it is not immediately adjacent to a cave. For example, the vibrations caused by blasting necessary to break up the rock can cause shock waves that break stalactites and stalagmites, or cause thin cave roofs to collapse. Even so, there are still quite a number of bat species roosting in caves near the three

Tonasa limestone and cement factories[13] but it is not known what species were present before quarrying began. Limestone is vital for construction projects and for the agricultural productivity of the marginal soils of most transmigration settlements, but to avoid unnecessary negative impacts of quarrying, it is important that the blasting be conducted with care and informed understanding.

Table 8.7. List of invertebrates found in various caves in South Sulawesi. All the scientific names are of orders except Diplopoda and Gastropoda which are classes.

	Baha-ruddin	Salukan Kalang	M2	N1	Watanang	Mampu
Snails (Gastropoda)	-	-	-	-	-	•
Woodlice (Isopoda)	•	•	-	-	•	-
Shrimps (Decapoda)	-	•	-	•	-	-
Crabs (Decapoda)	•	•	-	•	-	-
Scorpions (Scorpiones)	-	-	•	-	-	-
Tartarides (Schizomida)	-	•	-	-	-	•
Whip-scorpions (Amblypygi)	-	•	-	-	-	•
Spiders (Araneae)	-	•	-	•	-	•
Millipedes (Diplopoda)	-	-	-	-	-	•
Campodeids (Diplura)	•	-	-	-	-	-
Springtails (Collembola)	-	•	-	•	-	-
Crickets (Orthoptera)	•	•	-	-	-	•
Cockroaches (Blattaria)	-	•	-	-	-	-

After L. Deharveng pers. comm.

Table 8.8. Bats collected in recent years from various Sulawesi caves. The Toraja naked-backed bat was first collected by an EoS team and appears to be a new species (W. Bergmans pers. comm.).

	Solok	Konangan	Mampu	T. Watu	Taronggo	Lando	Tadula	Permona
Sulawesi rousette *Rousettus celebensis*	•	-	-	-	•	•	-	-
Common rousette *R. amplexicandatus*	-	-	-	-	-	-	•	-
Toraja naked-backed bat *Dobsonia sp.*	-	-	-	-	-	•	-	-
Cave fruit bat *Eonycteris spelaea*	-	-	-	•	-	-	-	-
Philippine sheath-tailed bat *Emballonura alecto*	•	-	-	-	•	-	-	-
Sheath-tailed bat *E. monticola*	-	-	-	-	-	-	-	•
Black-bearded tomb bat *Taphozous melanopogon*	-	-	-	•	-	-	-	-
Arcuate horseshoe bat *Rhinolophus arcuatus*	-	-	-	-	-	-	-	•
Sulawesi horseshoe bat *R. celebensis*	-	•	-	-	-	-	-	-
Broad-eared horseshoe bat *R. euryotis*	-	-	-	•	•	-	-	-
Gould's leaf-nosed bat *Hipposideros cervinus*	•	-	-	•	•	-	-	-
Diadem leaf-nosed bat *H. diadema*	-	-	-	-	-	•	•	-
Fierce leaf-nosed bat *H. dinops*	-	-	-	•	-	-	-	-
Common long-fingered bat *Miniopterus schreibersii*	-	•	-	-	•	-	-	-
Little long-fingered bat *M. australis*	-	•	-	-	•	-	-	-
Horsfield's bat *Myotis horsfieldii*	-	-	-	-	•	-	-	-

After Hill 1983; Boeadi pers. comm.; R. Dekker pers. comm.

Chapter Nine

Agroecosystems

CHARACTERISTICS

The founding of agricultural practices on ecological principles, whether deliberate or not, is certainly not new; two thousand years ago the Romans had considered ecological factors in agricultural management, and groups of swidden farmers in remote areas today farm in a sustainable and environmentally benign manner. Agricultural development is marked by periods, however, when linkages between components of agroecosystems (agricultural ecosystems) and the fragile ecological processes have been broken. Indonesian examples of the recent past would include the use of increasing quantities of the biocide[1] DDT during the 1970s to control increasingly resistant bagworms (larvae of psychid moths) in oil palm plantations, the planning of some transmigration settlements in totally unsuitable locations, and the adoption of inappropriate farm models. Inattention to ecology has resulted in recurrent outbreaks of pests and diseases (p. 293), soil erosion, declining soil quality, and pollution. It is now realized that these problems cannot be dealt with one by one, for they are linked by ecological, social and economic processes. For this reason systematic ecological analyses of agroecosystems are a new and effective means of research and development (Conway 1985a, b).

Agroecosystems are in some sense simpler than natural ecosystems because there is limited influence from other systems, the boundaries can be defined, and the numbers of species present and physical processes are fewer, but the large number of management and control decisions (both correct and incorrect) by the farmer introduces great complexity.

Any ecosystem can be described in terms of certain measurable characteristics, and those of most concern to agroecosystems are productivity, stability and resilience (or sustainability). Productivity is the net gain of a sought item over time and, in the agricultural sense, can be measured as annual yield of a crop or net income. Stability is a measure of the variation of production caused by normal, small-scale variations in environmental or economic conditions. Resilience, or sustainability, is the ability of a system to maintain or regain its productivity after a period of stress or a sudden major environmental or economic change. These three can be assessed for

567

any ecosystem, but a fourth characteristic, equitability, is really only of relevance to agroecosystems. Equitability is the evenness with which the products, or the income derived from it, of the agroecosystem are distributed among the members of the farm, village, region or nation (fig. 9.1) (Conway 1985a, b).

Different agroecosystems can be described in terms of the above four parameters. Agricultural developments tend to be aimed at increasing productivity but, as can be seen, this is achieved at the expense of equitability and sustainability. Thus the different agroecosystems examined briefly in this chapter have different ecological properties (table 9.1).

Table 9.1. Characteristics of major farming systems on Sulawesi.

	Productivity	Stability	Sustainability	Equitability
Swidden	*	*	***	***
Shifting	**	*	***	**
Settled gardens	**	**	**	**
Rice	***	**	*	**
Coconut	**	***	*	**
Clove	***	***	*	**
Cotton	***	*	*	*

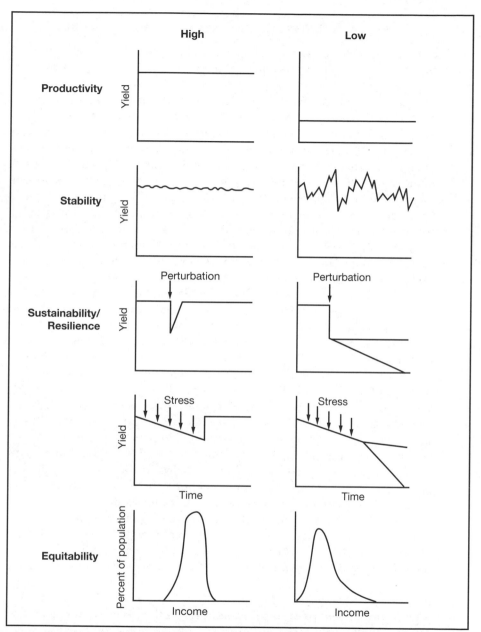

Figure 9.1. Performance of agroecosystems under conditions of high and low productivity, stability, sustainability and equitability.

After Conway 1985a, b

Swidden, Shifting and Intensive Agriculture

Swidden agriculture,[2] in its purest sense, is the repeated use of a patch of forested land for the cultivation of crops, and is characterized by long fallow periods between short periods of intensive production. One of the few groups of Sulawesi farmers which still practices traditional swidden cultivation is the Laoceh people who live on the slopes of Mt. Ogoamas, Donggala. Swidden agriculture is an extensive form of shifting agriculture and is not the same, despite what one might be led to believe, as the generally unsustainable slash and burn or intensive shifting cultivation practised in many areas (fig. 9.2). Swidden agriculture proceeds as follows: an area of forest is cleared, the remains are left to dry and these are then burned. Crops are subsequently sown in the ashes between the remaining large tree trunks. Some nutrients are lost in the smoke from the fire, pH is lowered and rain washes much of the ash into rivers before it has had a chance to become incorporated into the soil (Geertz 1963). Other nutrients are made available to plants by burning the organic matter (Andriesse and Koopmans 1984). What erosion does take place (due to the mosaic pattern of surrounding fallowed land), is usually trapped by vegetation at the field boundary (Hamilton and King 1983). After one or two crops, yields decrease, weeds become a serious problem (Chapman 1975) and the area is abandoned for 15-30 years during which time the soil recovers (Soerianegara 1970; van Baren 1975). The farmer then moves to an area of building-phase forest (p. 359). This is generally the site of a former field, because this type of forest has a smaller stature and is easier to clear than primary forest. This old field may contain mature fruit trees, such as durian, the fruit of which would have been harvested in the intervening years.

The use of cleared and burned land for swidden agriculture alters the early stages of forest succession (p. 359); weeding removes shoots growing from tree stumps and although efforts are made to remove herbaceous plants, their numbers increase. These herbs dominate the area soon after the land is abandoned, but fast-growing pioneer trees begin to smother the smaller plants after a year (Symington 1933; Uhl et al. 1982). Herbs and pioneer trees dominate these cultivated areas because they have many viable seeds in the soil and/or the seeds are easily dispersed (p. 362). It is clear that when the swidden agriculture cycle is long enough (e.g., 30 years), species of herbaceous weeds are not present in the secondary forest felled prior to cultivation. When the cycle is reduced to, say, just five years and succession has not been allowed to progress very far, domination by herbaceous weeds during even the first crop can become a serious problem.

Swidden agriculture, of the type described above, can support 30-40 individuals/km² and depends for its success on the farmers harvesting for subsistence rather than for cash crops and on cultural restraints that have

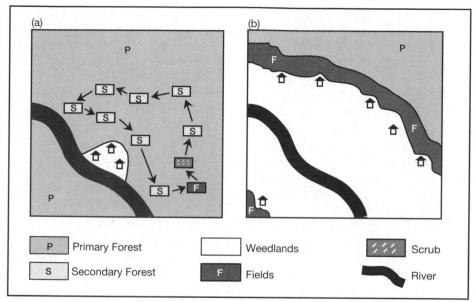

Figure 9.2. The distinction between traditional swidden agriculture (a), and the now prevalent shifting agriculture (b) which is endangering so many forested areas.
After Rijksen 1978

evolved as part of the traditions bound up in animistic religions. It is possible, however, to have a rather over-romantic idea of swidden agriculture and there is a mistaken belief that all people distant from missionary or government influence are good stewards of their environment. For example, in land occupied by the Kayu Merangka[3] group of the tribal Towana people in the remoter parts of the Morowali region of Central Sulawesi, there are vast areas of grassland caused by shifting cultivation on poor soils (Powell n.d.). Similarly, the Pipi Koro region west of the headwaters of the Palu River, also in Central Sulawesi, is desolate grassland, but in and around the nearby Lore Lindu National Park, where settled, rice-based agriculture was introduced at the start of the century, large areas of magnificent forest remain (Anon. 1981). It is not known what 'went wrong' in these areas but it may be that the farmers originally worked land in the richer lowlands but were forced by various pressures to to move to the poorer soils of the hills where their techniques were inappropriate.

With increasing populations and closer contact with the outside world, the restraints of swidden agriculturalists are eroded (Geertz 1963), and it is clear that the environmental care implicit in Christianity and Islam (Schaeffer 1970; Sutopo and Suayeb 1980) does not seem to find its way into the lives of new generations of converted shifting cultivators. Indeed, traditional, extensive swidden agriculture in its true sense has virtually disappeared from Sulawesi and a very small percentage of the population is involved. Published figures for areas under swidden or shifting agriculture do not as a rule include definitions of the terms used or differentiate between the two systems and are therefore of little use.

Shifting agriculture, as a term, thus covers a wide range of farming practices causing varying degrees of damage to the environment but, in general, it is practised with almost no cultural or social restraints and the cultivation of cash rather than subsistence crops is the major use of cleared land. Rested land is scarce and mature-phase forest is usually cleared when the former cultivated area is exhausted or when alang-alang grass can be kept at bay no longer by simple (cheap) weeding. Regeneration to mature-phase forest in such areas would take many centuries (p. 360).

Irrigated rice culture, introduced to Indonesia from India about 1,000 years ago is clearly regarded at most levels of society as the ultimate, the most rational, and most advanced form of agriculture for Indonesia. While recognizing the major role that rice plays in feeding the people of Indonesia, it is instructive, however, to stand back and to examine the applicability of rice growing for all people in all places. The agricultural system seen as the opposite end of the spectrum of modernity, is swidden agriculture, yet when this, and even some shifting agriculture, is subject to detailed analysis, they appear to be rational uses of manpower and land. For example, a typical Indonesian upland rice field may yield more rice per man-day than a typical wet rice field. Thus the rejection by such hill swidden farmers of incentives and invitations to change their agricultural system may be based on very sound economic motives, not just on backward thinking or fear of change.

Detailed study has shown how some swidden farmers are rational and sophisticated users of their natural environment (Dove 1985a). This is important to realize, since most of the blithe, blanket statements concerning swidden and shifting agriculture indicate that it is based on ignorance and inevitably results in misuse of the environment. It is this view that supports resettlement programs of remote groups of people (Satjapradja and Mas'ud 1978; Mile and Semadi 1981), but there is no compelling environmental reason to move swidden agriculturalists if they are engaged in stable, sustainable farming (Hamilton and King 1983). Among one group of Towana practising swidden/shifting agriculture between Luwuk and Ampana, there are former kampong leaders, former Muslims and Christians, and men who had been educated in Dutch schools. These people know about the methods of intensive agriculture but have decided to keep their options open and maintain an agricultural system whose benefits and advantages in their hilly environment they know (Atkinson 1979; Hidayan 1980).

The intensive nature of rice farming, promoted since the days of the early Javanese kingdoms, actually makes poor use of manpower and limits farmers' opportunities for planting cash crops. It is interesting, for example, that in the authoritative book on Indonesian rice production (Mears 1981) there are abundant data on yields per hectare, but not a single set of data on yields per man day. High yields per unit area are clearly favourable in areas of high population density, where labour is abundant and suitable land is scarce, but they are less important where those conditions do not apply. It is argued that swidden agriculture is incapable of absorbing continual increases in population density, but the production of rice from irrigated rice fields tends to decline as the population density increases (Geertz 1963). Absorption of manpower is not necessarily a superior goal or a better answer to demographic and economic problems than productivity of manpower (Dove 1985b).

So much could be done to breed improved strains of crops grown by shifting agriculturalists, to introduce new species, and to discover the best species for reforestation of fallow slopes. Agricultural intensification is neither a natural nor an inevitable development even in areas where environmental conditions make it possible. For example, alluvial soils in many Sulawesi valleys and along many coastal strips were not used for intensive agriculture until this was effectively imposed by the colonial and national governments. Conversely, agricultural intensification can and does occur as a result of government programs in areas whose physical, biological and social environments are not suitable. For example, the large Wawotobi dam being built near Una-aha, Kendari, is intended to allow the irrigation of 18,000 ha of existing and future rice fields. The Tolaki, the dominant group of people in the Kendari area, are dry-land farmers or shifting agriculturalists and have shown little inclination in the past to change to wet-land rice culture. In addition, other irrigation schemes in Kendari County have foundered because the actual water availability was not as great as was predicted. This experience is shared by certain irrigation schemes in seasonal parts of Gorontalo (Anon. 1986).

PEST ECOLOGY AND CONTROL

It is generally true that in their natural ecosystem, species that become agricultural pests are present at relatively low densities, tend to have specialized and ephemeral niches, and have populations that never reach pest proportions. With a change in land use, however, a niche that was once restricted can become available continuously over wide areas, and natural predators of the pest may not be able to hold the growing population in check if suitable habitat or other vital resources are not available. An insect predator may be dependent, for example, on certain plants on which its larvae have to feed.

The spraying of biocides onto crops is the conventional modern way of controlling pests[4]. Continued use can lead to resistance as evidenced by the hundreds of insects, mites and ticks that have developed resistance to one of more biocides. The resistance spreads through populations because the individuals that survive the effects of the biocide by behavioural, bio-chemical or physiological means, pass on their genetic adaptations to the next generation. Repeated applications of a biocide may thus lead to resistant populations. There is a greater tendency for species with a high intrinsic rate of population increase (the r-strategists–p. 345) to develop resistance, but the main determining factor is frequent application of high concentrations of biocides. Once a species has developed resistance to a biocide—and resistance has been documented for all chemical groups—the state is more or less irreversible. It is therefore quite conceivable that resistance to all known biocide groups may one day be found in populations of certain insect strains (Conway 1982). Resistance does not pose an immediate threat to world food production but it can cause extremely serious problems on a local level. In one of the latest reviews of biocide resistance, Asia is identified as the area most likely to suffer crises of pest resistance, particularly if the use of biocides grows to the high levels currently used in countries such as Japan (Conway 1982).

During the 1960s and early 1970s the potential dangers and undesirable effects of biocides were brought to the notice of the general public. Much was overstated or biased and it was impractically suggested that controlling pests by biological means was the only acceptable path to follow. Biological control is a term used for pest control by predators, parasites, pathogens, behaviour-changing hormones, sterilization of male pests, or by using resistant crop varieties. Most biological controls are highly selective and do not cause environmental damage. Other advantages are that a controlling organism can seek out its prey in a way that chemicals cannot, can increase in number and spread, and resistance to the predator by the prey is either slow or unlikely to develop. Disadvantages are that control is slow, pests are almost never wholly exterminated, patterns of control are often unpredictable, techniques are expensive to develop properly and to apply, and expert supervision is required. In addition, problems can occur when pathogens are imported along with a controlling predator or parasite or when the agent is less specific than was hoped. These disadvantages were not emphasized during the early chemical scares and society was presented with an unrealistic choice between biocides on the one side and

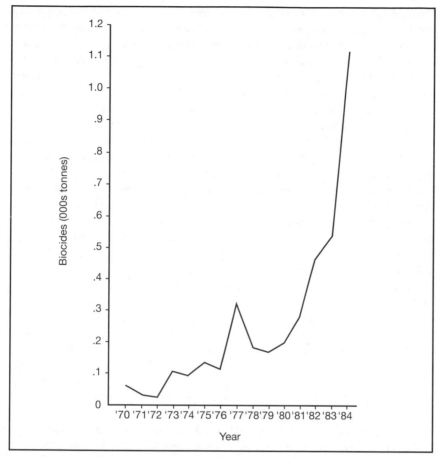

Figure 9.3. Increase in the use of biocides (000s tonnes) in South Sulawesi between 1970 and 1984.

From Anon. 1982

apparently perfect biological control on the other. Fortunately, agriculturalists did not make such simplistic choices and instead have developed the concept of integrated pest control.

Integrated pest control was defined over 25 years ago as "applied pest control which combines and integrates biological and chemical control. Chemical control is used as necessary and in a manner which is least disruptive to biological control" (Stern et al. 1959), and despite much research it does not receive the committed support it deserves (Conway 1985a). This is reflected in the huge increase in the use of biocides (fig. 9.3). One problem in pest management was that until recently only very few types of biocides were available and these were incapable of controlling the full range of insect pests. Improper use has led to an unimpressive impact and a consequent lack of enthusiasm for biocides on the part of the farmers. Certain pest problems or certain stages of those problems can only really be overcome using chemicals. Other methods can serve to keep most pest populations within limits for most of the time. These methods include making the agricultural environment as favourable as possible to predators and as unfavourable as possible to pests, and the various forms of biological control described above. When pest populations reach a certain predetermined threshold level judged capable of precipitating a serious economic loss, however, selective chemicals need to be used (p. 599).

It is notable that accounts of pests and the problems they cause often close with the remark that if the ecology of pests predators and crops (or related wild relatives) (Salick 1983) were better understood, better control measures could be developed. In theory, if the pest and its position in a food web is thoroughly understood, then accurate predictions can be made concerning the probable effects of different forms of control. The gathering of data which could genuinely achieve such ends for just a single pest species could take many years (van Emden 1974; Heckman 1979), during which time serious increases in population levels could occur. Thorough documentation and analysis of the different ecological responses of pest populations to different forms of control should lead to better selection of control programmes and the effects of such programs could be predicted with increasing accuracy (Altieri et al. 1983). In this way, ecological knowledge of the organisms concerned will lead to effective and environmentally sound pest control.

Predatory spiders have been all but forgotten in discussions on biological control, because little is known of their ecology and so they have not received the attention they deserve. Their main characteristic is that they are generalist rather than specialist predators but agricultural entomologists have rather sought *the* parasite or predator to control the pest under scrutiny. Evidence suggests that spiders can reduce pest populations significantly in rice and cotton crops, but it is the community of spiders which needs to be encouraged, not just a single species. Safeguarding the spiders can be achieved simply by paying attention to their habits; for example, spraying at mid day when spiders tend to be inactive and under some form of shelter (Riechert and Lockley 1984).

Most of this section concerns the control of insect pests, but rats and birds are ubiquitous pests which also cause considerable damage (p. 583). Rats are considered to be the most important group of pests hindering agricultural production in Southeast Asia (Soerjani 1980). Rats are known to occur in almost all agricultural crops and to damage many stored products (Anon. 1976, 1980; Myllymaki 1979; Estioko 1980; Soekarna, et al. 1980). They present the additional hazard of being a health risk to humans; the organisms causing scrub typhus, meningo-encephalitis, plague and lepto-spirosis are all known to be carried by parasites of agricultural or urban rats (Lim et al. 1980; Sustriayu 1980). For example, the mite *Leptotrombidium deliense* known as a vector of scrub typhus has been found on rats caught in Central Sulawesi and Bogani Nani Wartabone National Park (van Peenan et al. 1974; Durden 1986). Luckily, in this and many other cases, even though the disease and the organisms may be found close to where humans work or live, the disease does not necessarily break out. The environmental conditions required for the disease to be transmitted are not known, but it is clear that detailed ecological knowledge of rats should be collected now, rather than once an epidemic has begun.

The number of bird species inhabiting rice fields, plantations and other areas of permanent agriculture is quite considerable but none of them is an inhabitant of mature- or building-phase lowland forests. This serves to demonstrate how few of the forest birds can survive in disturbed areas (p. 434).

RICE FIELDS

Introduction

Much of this section is concerned with, and uses data from, South Sulawesi. This is because South Sulawesi is a major rice-growing area with over three times as much land under rice as all the other Sulawesi provinces combined. Indeed, South Sulawesi is one of only two Indonesian provinces that produces a yearly rice surplus. Just over half its total arable land (about 1,100,000 ha) is committed to rice, and the next most planted crop is maize which covers just 13% of the rice area. About 50% of the rice is rain-fed, 45% is irrigated by various means, and just 5% is unirrigated in upland areas. Of the rain-fed and irrigated rice[5] most is grown in the alluvial coastal plains which have the most suitable soils but even these may be deficient in nitrogen and sulphur (van Halteren 1979).

In some karst areas of Indonesia, such as the famous Mt. Sewu area of the Special Region of Yogyakarta, rice culture is more or less impossible because of the porous nature of the bedrock. In the Bone cockpit karst area (p. 470), the topography is superficially similar to that of Mt. Sewu but not far below the surface are impermeable rocks which can cause ponds to form on the surface. In this area very simple irrigation can produce two crops of rice per year (fig. 9.4). The hill slopes are used for tobacco and after just one year under this crop, the fields are left fallow for three to five years and so there is less soil exposure and hence less erosion. The human

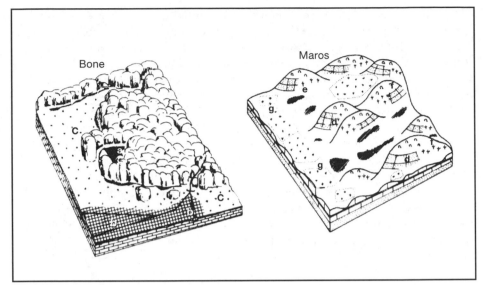

Figure 9.4. Bone and Maros karst regions showing their agricultural patterns. **a** - favourable agricultural area with karst springs (irrigation, two rice harvests, or one rice and one tobacco harvest, per year); **b** - rice fields with technical irrigation (one rice harvest per year); **c** - rice fields with impounded rainfall or simple ditch irrigation (one rice harvest, during the rainy season); **d** - terraces with permanent dry-field cultivation; **e** - shifting cultivation; **f** - pioneer-phase forest, grassland; **g** - rice fields with impounded or simple ditch irrigation (one rice harvest, during the rainy season).
After Uhlig 1980

population densities of Mt. Sewu and the Bone karst area are more or less the same (about 330 people/km²) but the environmental conditions in Bone are far more favourable thanks to the underlying rock (Uhlig 1980).

The plain around the tower karst of Maros is densely settled and at first sight the settlement and cultivation of the area is very similar to most other rice growing areas of South Sulawesi. The main difference though, is the karst springs at the base of the cliffs which provide a year-round supply of water and two crops per year can be grown using simple irrigation ditches. Water from the relatively large Bantimurung River issuing from the karst of Maros also has a reliable year-round flow and has been routed to provide technical irrigation to 70 km² of rice fields—one of the largest irrigation schemes in Sulawesi (Uhlig 1980).

Rice Fields as an Ecosystem

Rice fields are essentially modified swamps and the animals and plants they contain are typical of such ecosystems. Rice fields as an ecosystem are maintained by humans orchestrating the yearly or twice-yearly cycles of flooding, ploughing, removal of organic matter (harvesting) and enrichment with fertilizers. If people were to relax their efforts the condition of the rice field would very rapidly change and enter the forest growth cycle (p. 359). A rice field is generally by no means a monoculture. Apart from the rice crop, people also harvest kangkung *Ipomoea aquatica* (p. 269) as a vegetable, snails, prawns, crabs, fish and frogs. In certain areas, such as Tana Toraja, the centre of rice fields are deliberately deepened so that, after harvesting when the field is allowed to dry out, a pond is formed. These ponds are a refuge for aquatic organisms and a source of food for ducks.

Rice fields as an ecological system have hardly been studied at all and work has concentrated instead on the important subject of pest control and fish culture. Fish culture could most easily be enhanced, however, by increasing the productivity of the aquatic organisms on which they feed and this is only possible through an understanding of how rice fields function in an ecological sense.

Comprehensive analyses of the fauna[6] and flora have been conducted, however, in Peninsular Malaysia (Fernando 1977) and northeast Thailand (Heckman 1979). The site chosen for the detailed Thai study was near Udorn Thani at about 17°N which has a very seasonal climate with three months when virtually no rain falls and minimum temperatures in the dry months fall to 5°C. Consequently, many species common in more stable environments in Thailand and elsewhere in Southeast Asia are absent or rare in this study area, but the rice fields are by no means depauperate in species. One reason for this may be the fact that rice had probably been grown in the area for at least 5,500 years. In areas where rice has been grown for only a 100 years or so the number of animals and plants is somewhat low but, with time, adaptations by other species are likely to occur and the species total will increase. Some animals, such as certain fish, are deliberately introduced into the ecosystem and a number of these have been cultured for food for at least 2,000 years.

The study in Thailand found that a single field was inhabited by a staggering 589 species of plants and animals, 209 of which were considered rare. This species total included 38 flowering plants, 173 algae, 120 protozoans, 52 rotifers, 12 molluscs, 33 crustaceans, 21 dragonflies, 39 beetles and 18 fish. No real plankton are present, possibly because the water is simply too shallow and too full of plant material to allow such species to survive. Mosquitoes are rare in rice fields where the water is more than a few centimetres deep because predatory fish will soon consume them (Heckman 1979). The manner in which these organisms divide up the available resources is scarcely understood although it is easy enough with fishes (table 9.2).

Most of the larger aquatic animals are adapted to living in water with little or no dissolved oxygen. The fish have accessory breathing organs, the snails have lungs and gills, some snails live in close contact with the oxygen-rich surface layer, some insects and spiders carry a bubble of air with them below the surface, while others may store oxygen in their tissues.

The periodic drying of the fields means that newly-wet fields have to be recolonized or that plants or animals need resistant, dormant stages to endure the 'drought'. Some species such as snails, crabs, the swamp eel *Monopterus alba* and even an *Enhydris* snake (p. 305) bury themselves in the mud and remain inactive during the dry period, some move to permanent water bodies and many die. Those that die may leave resistant eggs, spores or seeds on the mud. Recolonisation can occur from nearby swamps, rivers or lakes, but in areas where virtually all swamps have been converted, the ricefield fauna is relatively poor (Fernando et al. 1980).

Most agronomists are charged with increasing the yield of rice per unit area. Where this involves the use of chemicals in pest control, insufficient attention is usually given to the other productive components of the rice fields. Great care must be taken in the application of chemicals so as not to distort or break the food web in a way that would reduce the capacity of a rice field to produce protein. A few organisms may even be encouraged to live alongside the rice: an example is the small fern *Azolla pinnata* (Azol.) which has clearly useful nitrogen-fixing blue-green algae *Anabaena azollae* in its leaves and is also used as animal feed. This inattention is regrettable since low species richness and high population densities of few species are often regarded as symptoms of an ecosystem under stress.

Table 9.2. Ecological characteristics of rice field fish.

	Ecological Characteristics
Trichogaster pectoralis	Vegetarian, grazes on periphyton (algae and rotifers attached to plants, stones, etc.).
T. trichopterus	Deeper water, groups of 3 or 4, searches for invertebrate prey.
Aplocheilus panchax	Feeds in large schools on microscopic particles.
Betta splendens	In small pools or shallow water, sit-and-wait predator on invertebrates.
Monopterus alba	Nocturnal, lives in burrows, eats quite large prey.
Clarias batrachus	Large predator on insects and other relatively large animals; also scavenges.
Channa striata	Large predator on fishes and frogs.

After Heckman 1979

In Peninsular Malaysia improved rice yields achieved through the use of biocides and double-cropping have been shown to be associated with reduction of the associated aquatic fauna (Yunus and Lim 1971; Fernando 1977).

This may seem defensible where there are pressures to over produce and to export but, for many farmers subsistence is the major goal. Chemicals may result in high yields, but if the farmers are unable to benefit from what can be, and perhaps should be, a multi-culture agroecosystem, effort and time must be spent producing alternatives to the varied protein and vitamin sources of a diverse and healthy rice field.

Insect Pests on Rice

All parts of a rice plant (except roots under water) at all stages of growth are attacked by insects (fig. 9.5) and the roots are attacked by nematode worms. Some, such as grasshoppers, feed on rice and many other plants while others, such as certain leafhoppers are known only from rice. The insect pests on rice in South Sulawesi have been studied in detail and the white stem borer *Tryporyza innotata* is the most important followed by the rice seedbug *Leptocoria oratoria* and the rice leaf folder *Cnaphalocrosis medinalis*. Other important insect pests include the striped stem borer *Chilo suppresalis*, the green leafhopper *Nephotettix virescens*, and the brown planthopper *Nilaparvata lugens*. It is not the actual damage done by the leafhoppers to the plants that causes concern, but rather the virus diseases they spread. The green leafhopper transmits the tungro virus which devastated large areas of rice around Palu and in South Sulawesi from 1972 to 1974. The brown planthopper transmits grassy stunt virus and had long been identified as a species with great potential for becoming a very serious pest. The brown planthopper is considered to be Indonesia's major insect pest and the greatest challenge for plant protectionists. When it was clear toward the end of 1986 that Indonesia was experiencing another set of brown leafhopper outbreaks, the government acted with great determination. First, it recognized that this pest was now resistant to certain pesticides and so the use of 57 biocide brands was outlawed and, second, the President announced that approved biocides were to be used only to control serious outbreaks, not as a preventative tool against possible pest attacks. In this way it was hoped that predator populations would be encouraged. There is, therefore, much greater government support at last for integrated ecological approaches paying attention to cultural controls (synchronized planting, etc.), natural enemies, pest surveillance, migration monitoring, different modes and patterns of insecticide use, and resistant varieties of rice. In short, the government will go to all lengths to ensure that the hard-won self-sufficiency in rice achieved in 1985 is maintained. The present situation shows just how

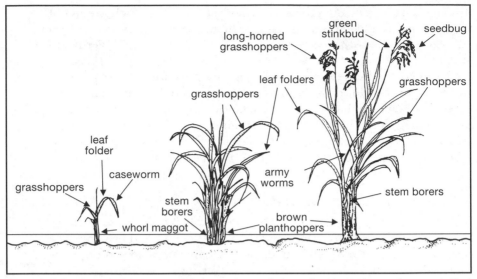

Figure 9.5. Insect pests on different parts of rice plants at the early vegetative stage, late vegetative stage and mature stage.
After van Halteren 1979

relevant ecological approaches are to the management of agroecosystems.

The stem borers are caterpillars of pyralid moths which lay eggs on the young rice leaves. On hatching the caterpillar moves into a leaf sheath from where it bores into the stem. Depending on which part of the stem is bored into, the rice will either die or recover, but in the latter case will always produce a lower yield than an undamaged plant. *Tryporyza innotata* feeds only on rice. The rice leaf folder is also a pyralid moth caterpillar but this folds one or more leaves around itself and then feeds on the inside. This results in whitish streaks on the leaves and a badly infested field can take on a whitish appearance. Both adults and larvae of the rice seed bug damage rice by puncturing the seed with sucking mouthparts. This causes both physical damage and the introduction of bacteria and viruses which lead to both a reduction in yield and a reduction in grain quality. The areas heavily infested[7] by the major pests in South Sulawesi vary between years in response to climatic conditions and effectiveness of biocides and other pest management (fig. 9.6). Some pests thrive on plants given high applications of nitrogen, others are favoured by the particular microclimate afforded by short varieties, while still others may be encouraged by a second crop of rice in a year (van Halteren 1979).

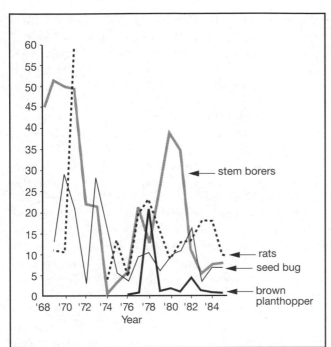

Figure 9.6. Changes in areas (000s ha) of rice fields in South Sulawesi during April-September, heavily infested with different pests.

Rats and their Control

As can be seen from figure 9.6, rats cause at least as much damage to rice crops as do the major insect pests. Rats cause the loss of at least 12 million tons of rice each year worldwide, and it is possible that rats and birds combined may have more serious impacts on rice crops than all other types of pests (Anon. 1976). Rats in Indonesia are probably responsible for a 5%-6% overall loss of the annual rice yield, but losses in individual fields can be much more than this. The rats live in tunnels they construct in the bunds between rice fields. These tunnels comprise main and secondary entrances and exits, blind tunnels for emergencies, and nest chambers.

The most serious rat pest in Sulawesi is the rice field rat *Rattus argentiventer* which damages rice crops throughout Southeast Asia, Two further species, the little rat *Rattus exulans* and the house rat *Rattus rattus* are generally found in peripheral areas but will enter rice fields to feed when other food is scarce. These species are usually only found close to buildings or farms.

Analyses of the stomach contents of rice field rats have shown that they all eat insects and snails, and that the amount of plant material they ingest depends on the type of habitat, and varies between species (Lim

1974). Rats eat rice at all stages of its development and although the theft of ripening rice grains is a direct loss from the harvest, the shredding of growing stems to eat the tender growing shoot does more absolute damage. A single, ripe seed head may represent a rat's daily food requirement, but the same rat may eat 100 growing shoots in a day before it is satisfied (Anon. 1976).

The reproduction of rats clearly follows an r-strategy (p. 345). The gestation period of the female rice field rat is 21 days, a litter generally numbers seven, and about eight litters are produced per female per year. Females could theoretically have more litters but reproduction tends to occur only when an abundant food supply is available, that is, towards the end of the rice-growing cycle. Male rats are capable of breeding at two months of age and female rats at just 1.5 months (Lam 1983). The mean life span of the rats is only about 4-7 months (Harrison 1956), but even by seven months a single female could have raised about 20 young.

The very high intrinsic rate of growth among rats is the major reason why attempts to exterminate them using traps or poisons are effective only locally or for short periods. If any rats are left, their litter size is likely to be larger than normal (up to 11) because competition for food is less and so the population will increase quickly. If, as rarely happens, all the rats in an area are killed, the unoccupied rice fields represent a wonderful opportunity for rats from other areas to colonize. If one thinks of an animal's niche as its 'profession', then when rats have been exterminated from a field, hundreds of 'jobs' become vacant for which the remaining rats are eminently qualified. It should be remembered that rats do not produce large numbers of offspring for the good of the species, or because that many are needed to fill all vacant 'jobs'. They are produced because rats are locked into their r-strategy and each female has to produce as many fast-growing, healthy young as she can, so that her hereditary line is not swamped by others. Many of the rats die young (as is shown by the life span figures above) but this is a consequence of the r-strategy. If the mother rat produced just a few offspring to whom she devoted her time for the fetching of food and for their defence, the available 'jobs' would have been filled by the larger number of other rats from other mothers.

Thus, control of rats in the long-term has to be centred on land management. Rats can live for only a few days on a diet of just rice stems (Anon. 1976) and this may be the only food at certain times, particularly if insect populations are also controlled in some way. Thus, if rice is planted and therefore harvested at the same time over large areas, and if scrub and other neighbouring habitats where rats could find alternative foods when rice grain was not available, were removed or utilized, rat populations could be kept within bounds. These are thus two of the main methods of rat control encouraged by the Department of Agriculture, the others being the digging out of rat holes and poisoning (Soekarna et al. 1980).

Attempts have been made recently at the national level and with the help of the armed forces to control rats but, despite the impressive piles of rats photographed for the newspapers, these will lead to only temporary improvements. Perhaps a more positive approach, as demonstrated in part of Central Java, is to consider rats as a resource which can be harvested on a sustainable basis and converted into animal feed.

Rice Field Birds

Among the more serious bird pests on rice are the Java sparrow *Padda oryzivora*, chestnut munia *Lonchura molucca* and the spotted munia *Lonchura punctulata*. The first of these was probably introduced into Sulawesi during the last couple of decades.

Birds generally range far wider and into more habitats than do insect pests and so their control poses considerable problems. Poisoning does not seem to be a viable control method because effective poisons are usually toxic against other animals, including man. The most common means of controlling birds is by scaring them with moving strings, cloth or scarecrows (fig. 9.7). In an account of bird pests, it is repeatedly stated that more ecological information is required for effective control programs (Adisoemarto 1980).

Rice fields are also favoured habitats of various birds whose feeding habits suggest they may assist in keeping the populations of certain insect pests under some degree of control. The birds can be divided into two ecological groups: the rails and crakes, and the herons and egrets.

Rails and crakes are very secretive birds and are hard to see except at dawn or dusk when they are more active and come to the edge of thicker vegetation. They fly weakly, with rapid beating of short rounded wings and their legs hang limply below their bodies. They usually flee from danger by running or swimming back to the cover of dense vegetation. When walking they frequently cock their tails and they typically jerk their heads while swimming. Because of their secretive and possibly nocturnal habits little is known of their ecology but their diet certainly includes small insects.

The most conspicuous large rice field birds are the herons and egrets. The three major species are:

- The black-crowned night heron *Nycticorax nycticorax* (fig. 9.8) which, because of its nocturnal habits, is more often heard[8] than seen.
- The cattle egret *Bubulcus ibis,* a white bird with yellow bill and greenish-yellow or blackish legs stands about 40 cm high. During the breeding season it develops a rufous-coloured head and neck. In Tana Toraja it is believed by many to bring bad luck if white birds are harmed and large numbers of these birds can be seen in

Figure 9.7. An ingenious bird scarer from the beginning of this century.
After Sarasin and Sarasin 1905

the rice fields.[9] There is a roost of over 300 of these birds in a bamboo grove by the road from Makale to Rantepao.

- The Javan pond heron *Ardeola speciosa*, a rather shorter and stockier brown bird with white wings. In the breeding season it develops a black back and a reddish brown breast but at other times it has brown and black speckled plumage.

The above species are rare or completely absent over many rice field areas of Sulawesi as a result of shooting, poisoning with biocides, or destruction of roosts. Both egrets and herons roost and nest in colonies and it is possible that these assemblies serve as information centres from which the members will gain knowledge of the location of the inherently patchy food sources (Fasola 1982). This is shown by the manner in which birds follow one another when flying from the roost to the feeding ground and the aggregations of birds foraging on the ground.

Although both cattle egrets and pond herons feed in rice fields, they do not compete with each other for food because they feed at different times and on different prey (table 9.3). Both birds are primarily insect eaters although the pond heron also takes frogs and fish, and the egret also eats spiders. The quantities involved are quite surprising. One Javan pond heron examined had over 500 individual prey items in its stomach, while

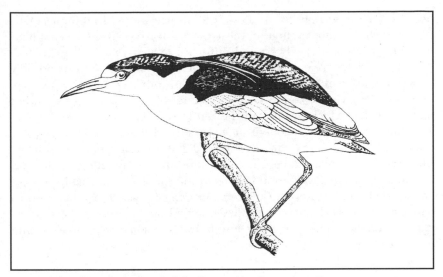

Figure 9.8. Black-crowned night heron *Nycticorax nycticorax.*

Table 9.3. Ecological differences between Javan pond herons and cattle egrets in the Dumoga valley, Bolaang Mongondow.

	Javan pond heron ***Ardeola speciosa***	**Cattle egret** ***Bubulcus ibis***
Feeding habitat	Most abundant on recently ploughed and harrowed wet fields. Also found in fields with young rice, on tree branches over lakes, and on stones by rivers.	Most abundant on recently harvested fields and fields being ploughed. Also found along roadsides, dry fields being ploughed, in rice seedling nurseries and near cattle.
Foraging strategy	Stands still and waits.	Hunts actively, eating insects disturbed by themselves, ploughs or cattle.
Diet preferences	1. Beetles 2. Dragonflies 3. Grasshoppers	1. Grasshoppers 2. Spiders 3. Beetles
Prey habitat	Most of prey live in water permanently.	Almost all prey live on dry land.

After J.W.C. Vermeulen pers. comm.

one cattle egret was observed to catch 89 insects in just 15 minutes (Vermeulen 1985). However, neither these birds nor frogs appear to eat plant-hoppers or stem borers (Berry and Bullock 1962; Berry 1965; Yap 1976).

Grasshoppers are minor pests of rice fields in South Sulawesi and they are not regarded as troublesome on rice or any other crop in the Dumoga valley (K. Monk pers. comm.). One common genus *Oxya* has expanded pads on its feet which presumably help it to run across the water surface. The 3,000 cattle egrets resident in the rice fields between the villages of Toraut and Dumoga prey on an estimated 250,000 grasshoppers each day which must be a major factor in controlling their populations. Similarly, each day the 2,500 Javan pond herons in the some area probably prey on about 50,000 mole-crickets *Gryllotalpa arachnoidea* which are believed to damage rice roots. These are relatives of grasshoppers with reduced wings and an incredibly loud song made louder by the resonance provided by the tunnel from which they call.

COCONUT AND CLOVE

Coconut and clove are the two most important industrial crops[10] in Indonesia after oil palm and rubber. Unlike the oil palm and rubber, they are both grown extensively throughout Sulawesi. North Sulawesi produces more cloves than any other province in Indonesia, and has nearly 10% of its land area under coconut. This compares with about 2% in Central and South Sulawesi, and 1% in Southeast Sulawesi (Bennett and Godoy 1986).

Dried cloves sell for about Rp 5,000-7,500/kg, and the yield from one hectare of mature clove trees can provide a smallholder farmer with Rp 5-10,000,000/yr[11] In contrast, one hectare of mature coconut palms provides a gross income of barely Rp 200,000. Mixed plantings of cloves and coconuts reduces the yields of both because of root competition and shading, but a farmer who intercrops his coconuts with cloves could increase his annual income to Rp 2,100,000 (Davis et al. 1985). During the 5 or 6 years before the clove trees begin to flower, he still has income from the coconuts which he can harvest every 3-4 months. The regular coconut harvests also provide income between clove harvests which usually occur annually, or sometimes once every two years for either physiological or climatic reasons which are not fully understood. This mixed cropping buffers the farmer against fluctuations in commodity prices and the effects of pest and disease outbreaks.

Coconut

The coconut is justifiably known around the world as the 'tree of life', 'mankind's greatest provider in the tropics', or 'one of Nature's greatest gifts to man'. The centre of origin of the coconut is probably in eastern Indonesia or Melanesia. The dispersal of coconuts to other tropical regions may have occurred naturally because the buoyant fruits can be carried great distances by ocean currents. However, the almost ubiquitous presence of coconuts on tropical sandy beaches owes a great deal to the movements of people (Purseglove 1968; Ohler 1984). Various parts of the coconut are used for food, drink, oil, medicine, fibre, timber, thatch, mats, fuel and to make utensils.

Coconuts can grow on a wide range of soils—sandy, coralline volcanic, alluvial, latosolic, lateritic and swamp soils (Ohler 1984). They require good soil aeration and drainage, and a constant supply of groundwater, which are all available in their characteristic location at the top of sandy beaches. Similar conditions are found in the volcanic soils. This is a major reason for the extensive planting of coconuts throughout Minahasa. Long dry periods adversely affect palm growth and nut production. For example, between July 1982 and April 1983 rainfall in Manado was just 369 mm compared to a previous average of 3,400 mm. Many coconuts died. It took about 2 years for nut production in the surviving palms to return to normal. This delay in recovery corresponds to the time it takes for a flower to develop and emerge from the heart of one palm and to develop into a mature bunch of nuts ready for harvest.

Coconuts were the world's major source of vegetable oil until the 1960 when soybean and palm oil became more important (Purseglove 1968). Coconut oil is principally made from dried kernel or copra which contains 65% oil. The oil was used primarily as cooking oil but the production from enormous coconut plantations in Minahasa, Toli-Toli and elsewhere provided the impetus for its use in soap and margarine manufacture in Europe during the later half of the 19th century. Secondary uses are in detergents, cosmetics, resins, wax candles, and confectionery.

Indonesia once obtained considerable export earnings from copra, reaching a maximum of US$31.4 million in 1970. Failure to replant ageing plantations established before Independence, together with an increase in the domestic demand for cooking oil since 1976, have led to a decline in the coconut export industry. About 60% of the cooking oil now uses oil from the oil palm *Elaeis guineensis*. It has been estimated at decreasing production in Central Sulawesi alone has cost Rp 6 billion over the last four years (C. Bennett pers. comm.). Since 1979, however, high-yielding hybrid coconut palms have been planted by farmers under the Smallholder Coconut Development Project. This hybrid, PB-121, is a cross between female Malaysian Yellow Dwarf and the male West African Tall.

A stand of mature coconut palms allows considerable light to reach the

Figure 9.9. Characteristic 'V'-shaped notches caused by rhinoceros beetles eating the leaflets of the spear leaf.

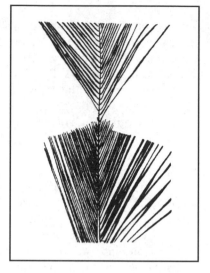

ground and multistorey intercropping can be practiced with coffee, cocoa, clove, bananas, sapota, pineapple, ginger, beans, maize or rice. Intercropping can both safeguard and consistently increase the income of coconut farmers considerably. The example of clove intercropping has been given previously (p. 588); another example is ginger which can produce 6,000 kg of rhizome/ha. At Rp 1,500/kg of rhizome, the farmer's additional gross income is Rp 9,000,000/ha. Coconut production is not reduced, but rather may subsequently increase because the palm benefits from the tilling, weeding and application of fertilizer for the ginger (Davis et al. 1985).

A review of insect pests on coconut palms forty years ago listed 751 species of which 165 (22%) were not known to attack other plants. Among the best-known insect pests are the large, handsome rhinoceros beetles. The adults burrow into the spear leaves which consist of the unopened, folded young leaves. When these leaves open they have characteristic 'V'-shaped notches where *Oryctes* has fed on them (fig. 9.9). The beetle breeds in decaying organic matter, such as the pith of old coconut stumps, dung and decomposing leaves and husks. Rhinoceros beetle is a vernacular name given to two similar dynastic beetles: the adult male of *Oryctes rhinoceros* has a single backwardpointing 'horn' on the top of its head, whereas *Xylotrupes gideon* has two, forked, forward-pointing 'horns' (fig. 9.10). The first of these species is the more serious pest although even its effects are rarely fatal to the tree unless it reaches the single growing point beneath the terminal bud.

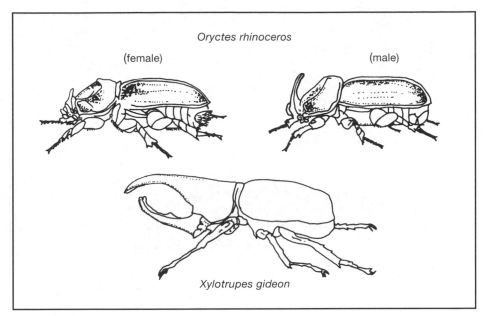

Oryctes rhinoceros

(female) (male)

Xylotrupes gideon

Figure 9.10. Rhinoceros beetles.

The palm weevil, *Rhynchophorus ferrugineus,* lays its eggs in soft or damaged parts of the palm crown. The larvae often burrow into leaf bases and the trunk through old *Oryctes* wounds. Attacks by the weevil are often fatal because their larvae usually destroy the growing point. Weevils and *Oryctes* can be controlled by the removal or burning of old trunks and their stumps. Biological control of *O. rhinoceros* using bacilovirus and *Metarrhizium anisopliae* has met with some success (Davis et al. 1985).

Coconut plantations in the Sangihe-Talaud Islands (and various locations in the Moluccas and Irian) suffer serious depredations from attacks by the long-horned (tettigonid) grasshoppers *Sexava* spp. with extremely long antennae (fig. 9.11). These 8-10 cm, green and brown insects seem to have very weak flight. This, in part, accounts for their absence on the North Sulawesi mainland. They lay most of their eggs in the soil at the base of the coconut palm. The eggs hatch in the evening and the first instar[12] larvae climb the trunk and start eating the older leaf blades. The larvae generally spend the day in the canopy and the night on the ground. Infestations are more severe on older leaves, possibly because of the toxic

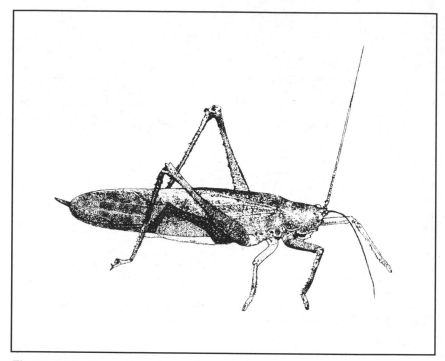

Figure 9.11. *Sexava* long-horned grasshopper.

trichomes on the very young and expanding leaves. Trichomes are minute hairy outgrowths concentrated on the lower surface and edges of the leaves. They contain chemical defences and may also discourage insect attack simply by physically obstructing their access to the leaf blade itself. Defoliations of coconuts on Sangihe-Talaud can be so severe that whole communities of smallholder farmers have had to abandon their coconut gardens and move to other areas (Kalshoven 1981; Davis et al. 1985).

Trichomes apparently also discourage leaf-mining hispid beetles *Plesispa reichi* and *Brontispa longissima* (fig. 9.12) (Davis et al. 1985), but serious outbreaks of the latter have been recorded in all provinces of Sulawesi. They attack the spear leaf particularly on young palms and are most abundant during dry periods. Plantations in South Sulawesi in the 1920s and '30s were brought under control using the eulophid wasp

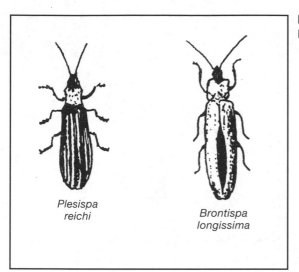

Figure 9.12.
Hispid beetles.

Plesispa reichi

Brontispa longissima

Tetrastichus brontispae as a parasite. The wasp can attack 60%-90% of the pupae and 10% of the larvae; about 20 wasps will emerge from a single beetle pupa. This parasite was also successfully released in North Sulawesi including the Sangihe Islands. *Plesispa reichi* is a blacker relative distinguished by a much darker carapace. It is generally found on seedlings and young coconut palms.

Nettle and slug caterpillars (Cochlidiidae/Limacodidae) are amongst the most serious pests, causing considerable damage to coconut palms and other crops in Southeast Asia. They are easily recognised by their gelatinous bodies and slug-like appearance. Some species have colourful, barbed spines on their back and sides. If touched, these barbs enter the skin and release a toxin which can elicit an acutely painful allergic reaction. The brilliant colouring of the nettle caterpillars, as in many insects, warns potential predators of the unpleasant consequences of an attack. In contrast, the adult moths are often small and dull coloured. In 1978 one widespread species *Setora nitens* devastated large areas of coconut plantations throughout Indonesia. Observations at the Coconut Research Institute in Manado showed that the dwarf palms suffered less than tall varieties probably because of the higher density of leaf trichomes in dwarf palms. Some farmers reacted to the outbreak by pruning the infested leaves but this led to young fruits being scorched by the sun and to flower shedding. The

Figure 9.13. *Parasa balitkae* in characteristic resting position.

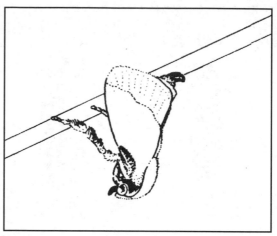

pruning led to a 76% loss in yield compared with a 14% drop for unpruned palms. In the same year there were serious outbreaks of *Chalcoscelis, Darna* and *Parasa* nettle caterpillars. Individual coconut leaves were attacked by as many as 700 caterpillars at once.

Darna catenata is a limacodid species apparently restricted to Sulawesi (including Muna and Sangihe) and Irian Jaya. It is found primarily in dry areas and is an important pest around Palu (pp. 484 and 487), where outbreaks seem to be triggered by drought. One major outbreak occurred there in 1950 when more than 100,000 coconut palms were badly damaged (Metzner 1981). In 1982 the same area was devested and 700 ha of palms appeared to have been scorched by fire. By 1984 an additional area of 9,000 ha was affected.

A new coconut pest species which is locally important, *Parasa balitkae*[13] has recently been described from North Sulawesi (fig. 9.13). This and the other nettle caterpillars are parasitized, sometimes with 100 wasps of several species and families emerging from a single caterpillar. These have considerable potential as control organisms.

One of the most destructive coconut pests is the endemic pig *Sus celebensis*, which uproots newly-planted seedlings in order to eat the tender heart or cabbage. Other mammals, including rats, may cause as much loss to the coconut grower as insect pests, but fewer resources are devoted to their control. Rats live in the crowns of coconuts and are a persistent problem because they damage young leaves and nuts of all ages. They can also consume stored copra. Rats can cause losses of 20% and occasionally as much as 50%. Rat poisons have been used with some success but rat resistance and cunning have reduced their efficacy.

It is very unlikely that rats could ever be exterminated and so a major aim should be to keep populations as low as possible using sustainable and natural means of control. One method used in some oil palm plantations is to encourage the widely-distributed barn owl *Tyto alba* to establish its territory among the palms. Analysis of pellets of bones regurgitated by barn owls showed that these birds feed almost exclusively on wood rats *Rattus tiomanicus*, and an adult pair of owls could be expected to kill 1,300 rats/yr within a territory of about 20 ha (Lenton 1983). In oil palm plantations the wood rat commonly lives at a density of 250/ha. Thus an owl territory could contain a standing crop of 5,000 rats, one quarter of which would be killed each year. It should be remembered, though, that the total number of rats living in an area during a year would be greater than the standing crop because the rats are continually reproducing and dying. A study of owl predation in England found that 20%-30% of the population of mice was taken every two months. The barn owl *T. alba* is not found in Sulawesi, but the grass owl *T. capensis* occurs in Central and Southeast Sulawesi, the endemic Minahasa barn owl lives in North Sulawesi, and the endemic Rosenberg's barn owl *T. rosenbergi* in all provinces. These owls have not been reported from coconut plantations but they could possibly be encouraged if nest boxes were erected at 6 m on electricity, telephone or other poles (fig. 9.14), because nest sites appear to be a limited resource (Lenton 1984). There is no guarantee that these owls, which typically inhabit forest rather than open areas, would adopt these nest boxes, but they are cheap to make and erect and their low cost would compare very favourably with control programmes using chemical poisons. A simple pilot study would show if this were a useful course to pursue.

The territory of barn owls in oil palm plantations appears to be a minimum of about 20 ha. Territory sizes are determined, in part, by abundance of essential resources but there are also strict upper and lower limits. The minimum size of territory is probably determined by the area required during periods of lowest prey density and the maximum by the area that can be effectively monitored and defended without diverting too much time and energy from other important activities. The population of rats in a coconut plantation is likely to vary through the year, but the resident population of owls would not continually shift territorial boundaries. Once established territories are probably quite stable, and the owls respond to fluctuations in prey density by producing more or less young. Their breeding strategy appears to be to produce large clutches and broods as frequently as possible in habitats with high potential carrying capacity (Lenton 1984).

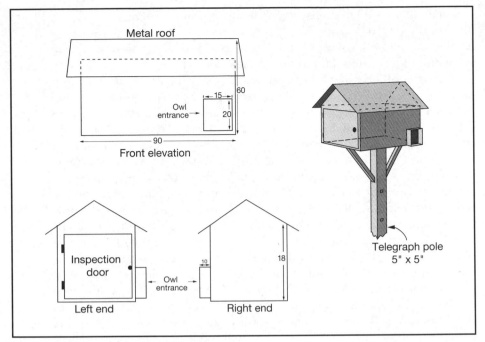

Figure 9.14. A nest box likely to attract owls to coconut plantations to control rats. The box dimensions are 0.5 m x 0.5 m x 1 m. Numbers in above illustrations indicate measurements in centimetres.

After G. Lenton pers. comm.

Diseases are not generally a problem in coconut palms, but some coconut diseases are locally important. One of these is premature nutfall disease outbreaks of which are caused by the fungus *Phytophthora palmivora* and are correlated with the peak of the rainy season in December to January (Bennett et al. 1985). Another disease affecting the leaves is often caused by another fungus *Pestalotiopsis palmorum* invading leaf wounds made by the limacodid pests.

Clove

Clove trees *Syzygium aromaticum* (Myrt.) are indigenous to the Moluccas and for more than 2,000 years the dried, unopened flowers (the cloves) have been used in east Asia as a spice, as a cure for tooth decay and as a breath sweetener. Indonesia is the world's largest producer, consumer and importer of cloves. Numerous parts of Sulawesi have caught what is often referred to as 'clove fever'. In areas such as Toli-Toli, hillside after hillside is covered with a regular forest of clove trees. Those people who own cloves receive a good or excellent income.

Indonesia consumes 25-30,000 tons of cloves each year of which 10%-30% is imported, primarily from Zanzibar. About 85% of dried cloves are shredded and added to tobacco in the ratio of about 1 : 2 to make 'kretek' cigarettes. Kreteks were already being smoked in the Moluccas at the start of the 17th century (Alauddin 1977). North Sulawesi produces 29% of the Nation's home-grown cloves, followed by Central Java (18%), Lampung (15%) and West Sumatra (13%).

The clove tree grows best at low altitudes where there is a marked dry period. The annual rainfall in its natural habitat is 2,000-2,500 mm and the agroclimatic zone is 'D' or 'E' (p. 22). Good drainage on deep acid loams seem to be important, as is shade in the early stages of growth (Purseglove 1968). Clove seeds remain viable for only a short length of time, and all these characters reflect its origins as a small (<14 m) forest tree unable to germinate and grow in full sun and unable to wait for favourable conditions (p. 359).

Pests of clove trees include a number of longhorn beetles (Cerambycidae). For example, the larva of *Nothopeus fasciatipennis* burrow into the trunk through a 3mm x 5mm hole which eventually seals itself with a white calcareous crust. The larvae tunnel both in the heartwood and just below the bark. Where these tunnels extend right around the trunk they may ring girdle the tree and kill it. A tunnel may be several metres long. The wasp-like adult (fig. 9.15) emerges higher up the tree through kidney-shaped holes.

Other groups of borers include moths of the family Metarbelidae whose larvae live in tunnels in stems, often near injuries, and feed on bark. It was estimated recently that 30,000 of the 5 million clove trees in Toli-Toli are infected with these stem borers. They pose a considerable threat to clove production in Central Sulawesi since there are only another 2 million clove trees in the province (Anon. 1986b). Scale insects (Coccidae) are quite common on cloves but they do not have serious deleterious effects. Many of these pests are controlled successfully using biocides. Mammalian pests on clove trees are restricted to squirrels which chew bark off the young twigs.

Most serious of all is a disease which was first detected among cloves in the Sonder region of Minahasa, and which is known simply as 'clove leaf drop'. Although it has been suggested that it is result of inappropriate methods, its pattern of spread is not consistent with this. The disease appears to be triggered or exacerbated by drought but the causal organism and its mode of action are still unknown. Leaf blight associated with a fungus *Phyllosticta* sp., has recently affected hundreds of hectares of cloves

Figure 9.15. *Nothopeus fasciatipennis.*

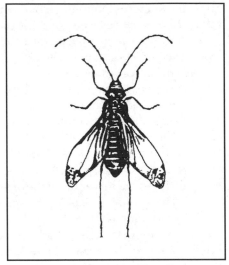

in Tana Toraja to which it was introduced from either Sumatra or Java. Still unknown is the cause of Sumatra disease which was first recognized with certainty in 1961 in West Sumatra where it may have been present since the 1920s. The disease is caused by a xylem-limited bacterium (Bennett et al. 1985). It is now found in North Sumatra, West Sumatra, Bengkulu, Lampung and West Java. Thousands of hectares of clove trees have been destroyed by the disease and its appearance in Sulawesi would spell ruin for many people.

COTTON

Cotton *Gossypium hirsutum* (Malv.) is the world's major fibre crop and covers about 30,000,000 ha or 5% of the world's cultivated land area. Conversion of land to cotton can bring considerable economic benefits[14] particularly in Indonesia where textile exports are so important, but there are a number of potentially serious negative impacts on the environment, the causes of which need to be continuously monitored.

Cotton has been grown in Indonesia for many years but interest started to increase during the Japanese Occupation at the end of World War II. In Sulawesi, cotton is grown by the State Plantation Estate Company XXIII in

South Sulawesi, and by P.T. Kapas Indah in Southeast Sulawesi. The second company runs a nucleus estate and small holder project near Tinanggea which began in 1977. It currently covers an area of about 5,200 ha, of which the mechanized nucleus covers 1,200 ha, and 4,000 ha is operated by smallholder, contract growers. The nucleus used to cover 1,700 ha but 500 ha were turned over to producing hybrid coconuts because the competition from American cotton is so great (Moeharram pers. comm.).

The first obvious danger is that, for any crop grown intensively over large areas, pests can wreak havoc and financial disaster. To counter this it is tempting to use large quantities of biocide and it is revealing that more biocides are used worldwide on cotton than on any other crop (Goodland et al. 1984). The problems encountered by pest control programs are well illustrated by the experience of P.T. Kapas Indah. It seemed unlikely that a new crop in a new area would suffer much from pests, but the entire crop of the first year was wiped out. A leafhopper *Empoasca* found the crop early on and reached such alarming densities that a major and successful spraying programme was undertaken. Unfortunately, however, it seems as though this programme exterminated insect predators and what remained of the cotton crop was destroyed by the earworm *Heliothis armigera*, which is the caterpillar of a noctuid moth which could not be controlled with the chemicals available. In the second and third years the number of biocide sprayings rose from between 3-5 to 5-7 respectively. In the fourth year, even though the crop was sprayed thirteen times, almost all the crop was lost to *Heliothis*, largely because of unexpected rains which prevented the biocide from reaching its targets. Over the following three years, the biocides used were changed and the sprayings were reduced from seven, to six to five. For the crop just harvested (1985-86), however, seven sprayings were necessary, presumably due to pest resistance, and next year the biocides will be changed again (R. Iskandar pers. comm.). Interestingly, the smallholder fields consistently suffer less insect damage than the fields of the nucleus estate and this is probably due to the position of the cotton between other crops and the dilution of the cotton in its field due to intercropping with other plants (M. Wade pers. comm.).

The control of cotton pests needs to be based on integrated control programmes that take into account the ecology of the pests and their enemies. For example, it is common practice to have four to six months (during the dry period) with no cotton crops in the fields in order to break the breeding cycles of pests. It is less common for surveys to be conducted to determine what other plants the pests may be using as alternative foods for the larvae. The growing plants need to be monitored for eggs and larvae of pests but their presence does not automatically require the initiation of a spraying programme. Many crops can tolerate heavy (<50%) loss of their leaves while young, and early spraying may retard the development of pest predator populations.

The development of resistant crop varieties would undoubtedly be the best approach (hairy varieties of cotton are, for example, resistant to *Empoasca*), but until these are developed for the pests of cotton in Sulawesi, the best approach is to maintain a range of biocides with different strengths to counter a range of pests under different conditions. The use of the biocides would be determined by regular monitoring of both pests and predator populations. For example, at P.T. Kapas Indah, the threshold of pest density is an average of 1.5 larvae per 10 cotton plants. Only when this density has been found will spraying be allowed.

Cotton has the potential to be quite lucrative for the producers. It will be viable, however, only if sound pest management can be introduced that reliably minimizes pest damage while safeguarding the profitability of the project and the quality of the environment.

Chapter Ten

Urban Ecology

INTRODUCTION

There is a school of thought that defines urban ecology as the interaction of the urban organism with its environment. The organism is the town or city itself which consumes the resources of surrounding areas, excretes its waste, and has its own circulatory and digestive system. This organism has been further described as a parasite on its host, the natural environment (White 1985). In areas where urbanization, and associated land loss, food supply, migration, waste, energy and water shortages are serious problems, this concept together with a systems analysis approach can be extremely important in the development of urban areas. Jakarta, as one of the world's largest cities, deserves this kind of attention. It may be appropriate to analyse the different forms of growth, resource use and waste[1] from urban centres in Sulawesi (Waworoentoe 1984), but this chapter treats urban ecology as the study of interactions between organisms and the urban environment.

Urban ecology is a vastly underexploited field of study in Indonesia. Budgetary and time constraints act against conducting ecological studies in remote areas but neither of these can be used as excuses against the initiation of ecological projects in any school grounds or university campus. It is not enough to learn about ecology from books: direct and personal observation of ecological interactions is essential. More attempts at teaching and learning about the ways urban organisms react to changes in their environment would lead in time to a greater familiarity with ecological concepts. Presentation of the results from studies on urban ecology to an increasingly well-informed planning sector could promote changes that would make towns and cities increasingly dynamic, attractive and hygienic and improve the well-being and mental satisfaction of the urban dwellers.

Towns comprise numerous habitats and opportunities for adaptable species:

- rubbish heaps provide nocturnal scavenging civet cats with a dependable food resource;
- these tips and houses are occupied by rats which are potential carriers of disease;

- eaves of roofs provide alternatives to tree holes for nesting birds;
- drainage water with high organic loads are exploited by toads, fish and mosquitoes, the last of which are important disease vectors;
- ponds around fountains are exploited by various small animals and plants;
- bare plots of rubble in building plots are colonized by plants with distinctive life histories;
- and house walls provide specialized niches for a few pioneer plants and animals.

Numerous ideas for imaginative research by dedicated school and university teachers can be found in urban ecology textbooks and guides (Collins 1984; Smith 1984; Hammond and King 1985) which, although written to promote urban studies in Europe, would serve as valuable guides to work that could be achieved in Sulawesi.

GARDENS AND STREETS

Vegetation

Urban trees provide aesthetic pleasure and shelter for humans, as well as a habitat for many animals such as birds, that use them as sites for nesting, feeding and roosting. Relatively few species of trees are planted and many of the species planted now are more or less the same as those planted over the last one or two hundred years. There have been some changing fashions in species, however; for example, many of the older streets in the major towns are dominated by mahoganies *Swietenia* planted about 70 years ago, but the modern fashionable tree is the quick-growing *Acacia auriculiformis*.

Few of the species planted are indigenous[2] (table 10.1) and perhaps more efforts could be made to plant more Sulawesi species. There are indeed no indigenous species that could compete with the beautiful shape of the rain tree *Samanea saman* or the stunning red flowers of the flamboyant *Delonix regia*[3] but it is certainly not true that Sulawesi has no species to offer, it is rather that landscape gardeners rarely make the effort to find appropriate trees and bring them into cultivation. Examples of under-exploited indigenous trees with great ornamental potential are *Elaeocarpus teysmannii* (Elae.), *E.* cf *macroceras*, *Deplanchea bancana* (Bign.), *Knema celebica* (Myri.), *Terminalia supitiana* (Comb.) and *Macadamia hildebrandii* (Prot.) (p. 513) (van Balgooy and Tantra 1986). It is probably correct that indigenous plants would be more attractive to birds than introduced plants, but some of the introduced species of trees have greater potential for attracting birds than some local species (table 10.1).

Legumes and *Casuarina* are obvious choices of trees because both possess the capability of fixing atmospheric nitrogen in root nodules. This means that the trees can grow on young or infertile soils and help to improve the soil for other plants. The choice of tree is determined by the aim of the planting. Parks personnel and horticulturalists are largely concerned with whether a particular species is aesthetically pleasing. Ecologists, on the other hand, are more concerned with the appropriateness of the plants in relation to other components of the environment.

Students and teachers should be encouraged to identify urban trees (Corner 1952) and other plants (van Steenis 1981) in order to become acquainted with plant characters and the classification of plants. Identifying trees in natural ecosystems will then be much easier. Most of the epiphytic ferns are also quite easily identifiable (Piggott 1979).

Table 10.1. Common urban trees of Sulawesi, their areas of origin, and their attractiveness to different groups of birds.

		Original distribution	Insecti-vorous birds	Frugi-vorous birds	Nectari-vorous birds
Anac.	*Mangifera indica*	E. and SE Asia	-	-	-
Apoc.	*Plumeria acuminata*	C. America	-	-	-
Bign.	*Spathodea campanulata*	Africa	-	-	-
Burs.	*Canarium vulgare*	Sulawesi and E. Indonesia	-	-	•
Casu.	*Casuarina equisetifolia*	SE Asia including Sulawesi	•	-	-
			-	-	-
Legu.	*Acacia auriculiformis*	Eastern Indonesia not Sulawesi	-	-	-
	A. farneciana	Tropical America	-	-	-
	Cassia siamea	S. America and Asia	-	-	-
	Delonix regia	Madagascar	-	-	•
	Erythrina variegata	Origin unknown	-	-	•
	Leucaena glauca	Tropical America	-	-	-
	Pithocellobium dulce	Tropical America	-	•	-
	Pterocarpus indica	SE Asia including Sulawesi	•	-	-
	Samanea saman	C. America	•	-	•
	Tamarindus indica	Africa and Middle East	-	•	-
Malv.	*Hibiscus tiliaceus*	SE Asia including Sulawesi	-	-	-
Meli.	*Swietenia macrophylla*	C. and S. America	-	-	-
	S. mahogani	C. and S. America	-	-	-
Mora.	*Ficus benjamina*	Indo-Australia	•	•	-

EoS teams made a survey of epiphytic ferns on roadside trees in a 1 km² area of residential Ujung Pandang, centred on the junction of Jalan Hasanuddin, Jalan Botolempangan and Jalan Arief Rate. Each tree was marked and the following were recorded: fern species present, their substrate (branch, trunk, etc.), and density. In addition, the species of each tree was noted together with its height, crown cover and diameter at breast height.

A total of 175 trees from six species were examined and four species of epiphytic fern were found (table 10.2; fig. 10.1). No more than two species were found on one tree although epiphytic pigeon orchids *Dendrobium crumenatum* were found on 6% of the trees.

The ferns were found most commonly on the trunks, and these were in fact the only sites for *Drymoglossum piloselloides*. *Drynaria sparsisora* was found on all parts of the trees. All bark-covered parts of the trees had some epiphytes (table 10.3), but the densest growth of ferns was on the trunks 4-8 m from the ground.

Exposure brings two problems: rapid drying and few sources of nutrients although exposed habitats are colonized by air-borne algae particularly the nitrogen-fixing blue-green algae which colonize bark (Wee 1982). *Drynaria sparsisora* clearly has advantages over other ferns growing in exposed places because its dried nest-leaves collect small pieces of decomposing organic material which break down releasing nutrients and retaining moisture. It is interesting that *D. piloselloides* is found beneath branches, where water and the nutrients it contains are likely to remain longest. Also, the observation that the densest growths are on the trunks 4-8 m above the ground may be because of the relatively nutrient rich water flowing down the trunk. *Davillia denticulata* is most common in moist, humus-rich microhabitats such as the axis between bough and trunk. The creeping rhizome is very thick and can store water.

Table 10.2. Epiphytes found on a sample of trees in Ujung Pandang.

	Number of trees	*Drynaria sparsisora*	*Drymoglossum piloselloides*	*Davillia denticulata*	*Phymotodes* sp.
Swietenia mahogani	131	131	9	1	0
Mangifera indica	18	18	3	0	1
Samanea saman	18	18	2	0	0
Tamarindus indicus	5	5	1	0	0
Casuarina sumatrana	3	3	0	0	0

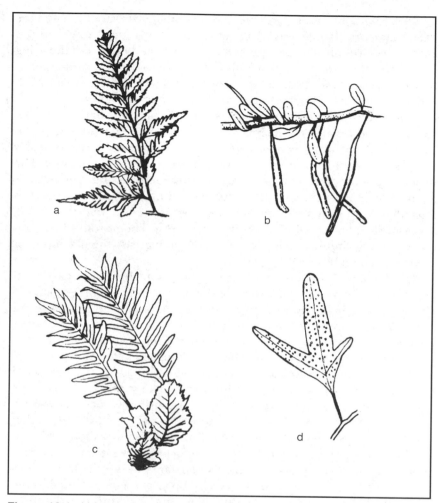

Figure 10.1. Urban epiphytic ferns. a - *Davillia denticulata,* b - *Drymoglossum piloselloides*, c - *Drynaria sparsisora*, d - *Phymotodes* sp.

Table 10.3. Percentage occurrence of epiphytic ferns on different parts of urban trees in Ujung Pandang.

	Trunk	Base of bough	Branch	Under branch	All parts	Dead branch	Dead tree	Stump
Drynaria sparsisora	54	30	6	–	6	2	0.5	4
Drymoglossum piloselloides	93	–	–	7	–	–	–	–

Of great interest is the fact that *Acacia auriculiformis* was never, and the flamboyant *Delonix regia* was hardly ever, found bearing epiphytic ferns. Half of the ferns on *D. regia* appeared to be dead, and these and the living ones were sparsely distributed on only the younger specimens. Both species have relatively smooth bark and none of the urban ferns is able to establish itself. Bark texture is unlikely to be the whole reason, however.

One of the most common epiphytic orchids on big urban trees is the pigeon orchid *Dendrobium crumenatum,* all plants of which tend to flower at the same time. Similar gregarious flowering in dipterocarp trees is probably a response to some environmental cue, possibly water stress (p. 367). The major trigger for pigeon orchids is also environmental but in this case it is a rapid fall in temperature, usually associated with a rainstorm. Nine days after that drop in temperature all the pigeon orchids will bear white, sweet-smelling flowers (Burkill 1917; Coster 1926). The reasons the pigeon orchids flower simultaneously is presumably to increase the likelihood of cross-pollination by insects.

If the crowns of urban trees, particularly fruit trees, are examined closely it may be noticed that parts are dominated by leaves other than the tree's leaves. These may be the leaves of mistletoes *Scurrula, Viscum* and *Dendrophthoe* (Lora.) (fig. 10.2), which are parasites whose roots penetrate the host bark and grow into the living tissue beneath. They are thereby able to absorb water and nutrients for their own use. Unlike certain other parasites that have no green leaves, the mistletoes do have green leaves and photosynthesize. Mistletoe flowers are relatively long and the four or five petals remain closed until touched by a visiting flowerpecker *Dicaeum celebicum.* As the bird sucks nectar from the newly-opened flower, pollen from the anthers is deposited on the base of the bill, and is then carried to another flower where it may be transferred to a receptive stigma. This same bird also eats the mistletoe fruit which comprises a single seed covered by a sticky 'glue' enclosed in a sweet pulp. The pulp is digested by the flowerpecker but the 'glue' is not, so that when the seed is voided it sticks to the branch on which the bird is sitting. It has been reported that the seeds are very large for the intestinal tract of the bird (whose major food is spiders) and the bird has to hop along a branch knocking the seeds from its anus (Dammerman 1929). It appears that germination only occurs after passage through a bird's gut.

One garden shrub indigenous to forest edges of Sulawesi is the white-flowered *Costus speciosus* (Zing.), a relative of ginger. Like ginger, its rhizome has a host of traditional uses particularly as a medicine (Burkill 1966). The lanceolate leaves are arranged spirally on a stem which itself grows in a spiral manner. The flowers open one by one from an elongate mass of red bracts, with one opening each day. In the centre of the flower the fine stamens fuse to form an upturned yellow-white tongue or lip which hides the throat of the flower. The flowers are visited by large female

Figure 10.2. Mistletoes. Left - *Dendrophthoe* growth habit, right - germination of *Macrosolen*. Scale bars indicate 1 cm.
After Holttum 1969

carpenter bees *Xylocopa* at about 0600 hrs. to 0800 hrs., from which the bees gather nectar for their nests. To obtain the nectar they have to push up the flower's tongue, and in so doing their backs are dusted with pollen, which in turn is rubbed on to the stigma. Males are rarely seen at flowers.

Carpenter bees are solitary bees and the males are said to be territorial (Frankie and Daly 1983; Louw and Nicholson 1983). Females visit the same flower locations every day (called 'trap-lining') and will generally visit different species in a regular order. Such trap-lining species tend to be large, have long life-spans, exhibit plasticity in their behavioural patterns, have powerful flight abilities, and clearly have a good knowledge of their surroundings (Janzen 1983). The nest tunnels which the females bore into timber are over 1 cm in diameter and 30 cm long and are bitten out along the grain of dead timber that is in a vertical or horizontal, rather than diagonal, position. The female deposits a ball of nectar and pollen at the end of the tunnel and then lays a large egg next to it. She then closes off this cell with a portion of chewed wood (Frankie and Daly 1983).

One of the largest garden spiders is the black and yellow *Nephila maculata,* whose huge webs of tough silk are constructed between trees, telephone wires and bushes. Most spiders eat their webs in the morning to avoid them being broken by larger animals such as birds. The web is rich in protein and represents a considerable investment on the part of the spider. *Nephila* does not eat its conspicuous yellowish web but remains instead in the web's centre, strikingly obvious to passing birds. Some smaller spiders also do not eat their webs, but these advertise them against birds by constructing zigzag or other patterns near the centre thereby making the web much more obvious (Eisner and Nowick 1983).

Birds

In general the bird fauna of towns tends to have a lower species richness and diversity than nearby forests, but the biomass and density are higher, and there are a very few dominant species. In addition, the major guild[4] shifts from the bark- and canopy-insect eaters to ground feeders (Ward 1968; Beissinger and Osborne 1982; Yorke 1984). These changes are not surprising since relatively few urban trees exceed 10 m in height, and they are often widely dispersed between large areas of grass or hedges. The total number of individual birds and bird species is low in Sulawesi's urban centres. Part of the reason is the predatory habits of young children with catapults, but ecological reasons include the lack of fruit suitable for birds and the few insects able to utilize the 'foreign' trees, leading to less food being available for insectivorous or partially-insectivorous birds (p. 369).

Lists of birds were collected for Manado and Ujung Pandang[5] forty years ago (Coomans de Ruiter and Maurenbrecher 1948) and although there have been some changes and additions as species have spread or been introduced (Escott and Holmes 1980), they provide useful baseline information. The totals counted are very similar to those reported for Kuala Lumpur and Singapore (McClure 1961; Ward 1968) (table 10.4).

An examination of the natural habitats of urban and suburban birds of Ujung Pandang (table 10.5) suggests that over half originate from coastal habitats, only about 5% from lowland forest and a similar percentage are normally cliff or cave-mouth nesters. About 25% have been introduced or are recent immigrants. The similarity between cliffs and buildings are obvious, and the swifts and swiftlets have clearly taken advantage of that. The similarity between coastal habitats and towns is not so clear, however, although a common factor is their simple plant communities containing relatively few species (Ward 1968). Because of this, generalized foragers rather than species with specialized niches are the major town invaders. In parts of Africa where savannah vegetation is common, birds in the relatively open urban environment are far more diverse. Sulawesi has no large areas of natural open country or savannah which might be expected to form a source of urban birds and so the number of urban birds originating from indigenous natural habitats is limited (p. 480).

The tree sparrow has spread widely in Indonesia. It is a native of Europe, Russia and China, and probably first arrived in Ujung Pandang aboard ships in the 16th and 17th centuries. It has now found throughout most of the southwest peninsula, to Palu, to Manado and elsewhere in Minahasa and Bolaang Mongondow, but has not so far been reported from Poso, Kolonodale or Kendari (L. Clayton pers. comm.).

The most frequent feeding preference is for insects, accounting for 24 of the 39 species above. Some of these eat insects as a major part of their diet, while others eat insects as an important component of their diets when a higher protein intake is required such as during moulting or breeding, or when they are feeding young. Insects are more abundant on trees with finely-divided leaves such as *Samanea, Casuarina, Delonix, Parkia* and *Albizia,* probably because of the greater number of potential resting places. Insect-rich microhabitats are also created by the growth of epiphytes, creepers and climbers on the trees.

The proportion of nectarivorous birds is very small, but they are attractive to urban dwellers because of their bright colours and their pleasing songs. In addition, the flowers from which they suck nectar are generally large and showy. To attract these birds (as well as butterflies) to heights at which they can be easily seen, plants such as *Hibiscus rosasinensis* (Malv.) (the wild red rather than the cultivated forms), *Ixora* (Rubi.) and *Calliandra* (Legu.) can be planted.

Birds need not only food, but also places and materials for nesting. Two of the most favoured nesting materials are the fluffy seeds from kapok *Ceiba pentandra* (Bomb.) and long grass. Certain swiftlets also use the dead leaves of *Casuarina*. Thus, these plants must be available in any area where birds are being encouraged to live. Hole-nesting birds can also be encouraged by leaving dead boughs in place, but only where human life is not endangered. In summary, the best way to attract birds to urban areas is to provide areas of heterogeneous vegetation—tall and short trees, shrubs and undergrowth, including long grass—and protection from catapults.

Birds represent excellent subjects of urban ecology studies. Observation conditions are as near ideal as one could wish for and the number of food species and competing birds species are relatively few. The study of such topics would not have island-wide environmental significance, but it would

Table 10.4. Numbers of bird species in Ujung Pandang, Manado, Kuala Lumpur and Singapore.

	Total	Resident breeders	Occasional visitors	Northern migrants	Southern migrants
Ujung Pandang	74*	22	42+	8	2
Manado	36	?	?	?	?
Kuala Lumpur	62	24	30	8	-
Singapore	53	25	20**	8	-

* This includes 13 species of raptors and 10 species of coastal or rice-field birds.

** This is low partly because relatively few birds of prey and coastal/rice-field birds were seen.

After Coomans de Ruiter and Maurenbrecher 1942; McClure 1961; Ward 1968; Escott and Holmes 1984

Table 10.5. Birds of Ujung Pandang with their natural habitats and feeding preferences.

	Introduced, recent arrival or cage bird	Man-grove, estuaries	Sandy beach vegetation	Cliffs, rocky shores	Forest edge	Low-land forest	Natural habitat not known
Blue-breasted quail (G, I)							
Coturnix chinensis	-	-	-	-	-	-	•
White-breasted waterhen (I)							
Amaurornis phoenicurua	-	-	-	-	•	-	
Pink-necked green pigeon (F)							
Treron vernans	-	•	-	-	-	-	-
Black-naped fruit dove (F)							
Ptilonopus melanospila	-	-	-	-	-	•	-
Superb fruit dove (F)							
P. superbus	-	-	-	-	-	•	-
Peaceful dove (G)							
Geopelia striata	•	-	-	-	-	-	-
Spotted dove (G)							
Streptopelia chinensis	•	-	-	-	-	-	-
Ornate lory (F)							
Trichoglossus ornatus	?	-	-	-	-	-	-
Sulphur-crested cockatoo (F)							
Cacatua sulphurea	?	-	-	-	-	-	-
Sulawesi hanging-parrot (F)							
Loriculus stigmatus	-	-	-	-	-	•	-
Plaintive cuckoo (I)							
Cacomantis merulinus	-	-	-	-	-	-	•
Common koel (I)							
Eudynamis melanorhyncha	-	-	-	-	-	-	•
Lesser coucal (I, C)							
Centropus bengalensis	-	-	•	-	-	-	
Sulawesi scops owl (I, C)							
Otus manadensis	-	-	-	-	?	-	
Sulawesi barn owl (C)							
Tyto rosenbergi	-	-	-	-	?	-	
White-bellied swiftlet (I)							
Collocalia esculenta	-	-	-	•	-	-	
House swift (I)							
Apus affinis	-	-	-	•	-	-	
Asian palm-swift (I)							
Cypsiurus batasiensis	-	-	-	-	-	•	-
Grey-rumped tree swift (I)							
Hemiprocne longipennis	-	•	-	-	-	-	-
Collared kingfisher (I, C)							
Halcyon chloris	-	•	-	-	-	-	-
Sulawesi pygmy woodpecker (I)							
Dendrocopos temminckii	-	-	-	-	•	-	-
Pacific swallow (I)							
Hirundo tahitica	-	•	-	-	-	-	-
Pied triller (I)							
Lalage nigra	-	•	-	-	-	-	-

Table 10.5. (Continued.)

	Introduced, recent arrival or cage bird	Man-grove, estuaries	Sandy beach vegeta-tion	Cliffs, rocky shores	Forest edge	Low-land forest	Natural habitat not known
Sooty-headed bulbul (I, F)							
Pycnonotus aurigaster	•	-	-	-	-	-	-
Yellow-vented bulbul (I, F)							
P. goiaveri	•	-	-	-	-	-	-
Spangled drongo (I)							
Dicrurus hottentotus	-	-	-	-	-	•	-
Black-naped oriole (I, F, C)							
Oriolus chinensis	?	-	-	-	-	-	-
Flyeater (I)							
Gerygone sulphurea	-	•	-	-	•	-	-
White-breasted woodswallow (I)							
Artamus leucorhynchus	-	-	-	-	-	-	-
White-vented myna (I, F, C)							
Acridotheres javanicus	-	-	-	-	-	-	•
Brown-throated sunbird (I, N)							
Anthreptes malacensis	-	-	•	-	-	•	-
Black sunbird (I, N)							
Nectarinia jugularis	-	-	-	-	-	-	-
Black-sided flowerpecker (I, F, N)							
Dicaeum celebicum	-	-	-	-	-	•	-
Mangrove white-eye (I)							
Zosterops chloris	-	?	-	-	-	-	-
Tree sparrow (G)							
Passer montanus	•	-	-	-	-	-	-
Java sparrow (G)							
Padda oryzivora	•	-	-	-	-	-	-
Moluccan munia (G)							
Lonchura molucca	-	-	?	-	-	-	-
Scaly-breasted munia (G)							
L. punctulata	-	-	?	-	-	-	-
Pale Sunda munia (G)							
L. pallida	-	-	?	-	-	-	-
TOTALS	6(3)	6(1)	2(3)	2	3(2)	7	4

(I) - Insectivore, (G) - Granivore, (F) - Frugivore, (C) - Carnivore, (N) - Nectarivore.

After Coomans de Ruiter and Maurenbrecher 1948; Nisbet 1968; Ward 1968; K.D. Bishop pers. comm.

furnish those involved with an invaluable awareness of ecological complexity and principles (Ward and Poh 1968; Ward 1969, 1970; Hails 1984), of great use when required to work in more complex ecosystems on an environmental impact assessment or similar study.

Bats

At first sight, bats in flight all look the same but, with a little time and patience, different groups or species can be distinguished (Gould 1978). The medium-sized, roof-dwelling, long-winged tomb bat *Taphozous longimanus* is usually the first to start flying at dusk and can be seen flying with sharply bent wings about 25 m above the streets. Before it leaves its roost it becomes quite vocal and is easily heard in the house below. It and other smaller insectivorous bats fly rather erratically as they swoop to catch insects. Frugivorous bats generally fly in a more direct manner.[6] Soppeng, in the middle of the southwest peninsula, has a famous roost of fruit bats near the Bupati's office in the centre of town. Local legend states that if the bats leave the roost the town will fall.

When watching bats forage at night it is clear that many of them fly slower than birds. The amount of lift produced by a flying animal, or indeed an aircraft, depends on the speed of the animal relative to the air (the airspeed), the wing area and the coefficient of lift, which in turn depends on the shape or efficiency of a particular wing shape. Bats and birds can change their coefficient of lift by altering the shape of their wings in the same way that the change in shape of an aircraft wing affects the airspeed. An animal that flies fast therefore needs less wing area for a given wing shape and weight than one that flies slowly and needs more lift. Fruit bats often carry quite heavy fruit from of a fruit tree to a feeding roost and consequently need a large wing area to increase lift. If one compares two urban flying animals of similar weight considerable differences can be seen (table 10.6).

Thus the sparrow carries more weight for its wing area than the fruit bat, and has a relatively long, thin wing typical of relatively fast-flying creatures. The bat therefore has some leeway to enable it to carry a 20 g guava, for example, to a feeding roost.

Bat wings are often a mosaic of scars, cuts and holes and urban bats are sometimes found with missing thumbs or feet. Exactly what causes these injuries is not known.

The usefulness of bats in producing fertilizer (p. 553), controlling fruit flies (p. 423), pollinating fruit trees (p. 561), and consuming vast numbers of insects (p. 549) has been described elsewhere and it is possible that some of these benefits, certainly the first, could be engineered by the building of bat towers or bat boxes. The first bat tower was built in 1908 in Texas but it took six years to design a tower that was truly attractive to bats. The purpose behind it was to control mosquitoes and hence improve the health of the surrounding population. Malaria did become less prevalent, and the roosts created a reasonable income from the sale of guano. Local governments and the Ministry of Health encouraged the building of towers and they were even exported to Europe but with little success. The American free-tailed bat *Tadarida brasiliensis* which occupied the towers has a relative, *T. plicata* in Sumatra, Java and Borneo. Bat towers were used with some success in Sumatra decades ago but it is not known which species colonized them (Anon. 1985). Sulawesi has a single species of *Tadarida* (p. 41) which does not seem to be particularly common, but it could probably be encouraged.

Building a tower is probably overly ambitious, and bat boxes are rather more practicable. Bat Conservation International, an interest group based at the University of Texas, U.S.A., recommends two types of boxes of slightly different sizes (fig. 10.3), that can be hung on trees 3 m or more above the ground. The different sized gaps should accommodate bat species of different sizes, but it may take several months before the boxes are colonized. Bat boxes could be erected in the grounds of schools or university departments, and the bats inhabiting them should be monitored.

Table 10.6. Weight, wing loading and aspect ratio for two flying urban animals. Wing loading is a measure of weight divided by wing area, whereas wing aspect ratio is the square of wing length divided by area.

	Male weight (g)	Wing loading	Wing aspect ratio
Dog-faced fruit bat, *Cynopterus brachyotis*	28	0.12	3.2
Sparrow, *Passer montanus*	30	0.30	6.3

After Yalden and Morris 1975; B. Gaskell pers. comm.

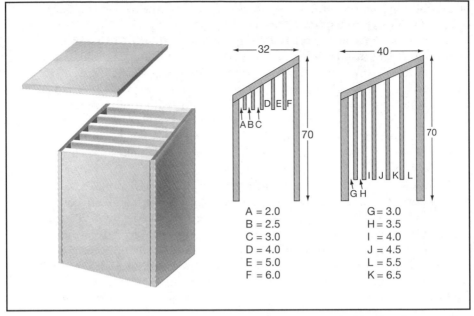

Figure 10.3. Bat boxes adapted from the design recommended by Bat Conservation International. Numbers indicate measurements in centimetres. Bats enter and leave through the bottom which is open. The wood used should be as rough as possible so that the bats can grip the surface with their feet. The top of the box can be covered with heavy plastic tacked down firmly at the edges. Boxes should be hung in relatively shady situations to prevent the bats getting too hot during the day.

WALLS

Walls are an integral part of the urban environment and are not utilized as much as they might be for the teaching of ecology. They represent a major habitat: it would be instructive to measure the area of exterior walls present in a defined area of town, remembering that many boundary walls have two external faces.

A wall has essentially four zones: the base, the lower level, the upper level and the wall top. Of these, the wall base is the wettest, most shaded and most nutrient-rich since rain and other material from the upper zones run into it. It is onto the soil adjacent to the wall base that seeds are excreted by birds or are blown by wind. Many of the seeds grow into plants that climb up walls forming a number of micro-habitats. The wall top differs from the other zones by being generally horizontal, and this has profound effects on the rate and type of colonization.

Numerous factors influence the colonization of a wall such as aspect (the direction it faces), shading, age and type of building material, and age and type of paints (if any). New walls are not immediately colonized because cement and red brick are quite strongly alkaline and unfavourable to all but the most specialized algae. Gradual decomposition, the roughening of the surface and the formation of cracks set the stage for colonization. If the process of succession is left undisturbed, algae tend to colonize first, followed by mosses, ferns and then flowering plants.

White-washed or emulsion-painted walls quickly discolour. Areas of green, black and orange appear in patches or streaks, some in shade, some in the open. The 'stains' are caused by green and blue-green algae and by diatoms. The black stains are usually blue-green algae that have caught airborne dust particles in their mucilaginous sheaths. The distribution of these microscopic plants on buildings has been investigated in Singapore (Chua et al. 1972) and an identification key is available (Lee and Wee 1982). Algal cells and filaments are dispersed in the air, and samples of Singapore air have been found to contain 21 species of algae, some of which are known to cause allergic reactions in humans (Wee 1982). Some of the algae are able to fix atmospheric nitrogen and as more dust and soil particles become attached, so a favourable habitat is formed for flowering plants to colonize. This obviously takes time, and regular maintenance of the wall prevents succession from proceeding.

Some of the oldest standing walls in Sulawesi are those of Fort Rotterdam in Ujung Pandang. The first walls were built in 1545 during the reign of the first Gowa king. The walls were rebuilt in 1634 and the Fort was surrendered to the Dutch in 1667. The walls are formidable: up to 6 m high and over 10 m thick in places. The Fort was partially restored in the 1920s and after the Japanese Occupation, then in the early 1970s the whole Fort underwent major repairs to house a provincial museum and offices of the Department of Education and Culture. There are 25 lengths of wall, seven of which have been repaired recently. The others comprise a mixture of the original blocks, coral blocks and cement.

Thirty species of moss, lichens and flowering plants were found growing on the walls during the rainy season in February 1986, and the distribution of the 23 species most commonly found was investigated (table 10.7). Eleven of these species were present on all the walls and the distribution of the other twelve did not seem to relate to aspect (the direction faced).

By far the most common plants during the wet season are an unidentified moss and the small South American herb *Pilea microphylla* (Urti.) each of which covered nearly 20% of some walls. The next most abundant plant is *Hedyotis corymbosa* (Rubi.), which was found at densities of $20/m^2$ in the sample quadrats, followed by *Peperomia pellucida* (Pipe.), *Urtica urens* (Urti.) and *Eragrostis amabilis* (Gram.) all of which averaged 2-3/m². Only one plant, the fern *Cheilanthes farinosa* (Polp.), appeared to have any clear preference for microclimate; it was found only on the lower parts of walls close to other buildings, where shade and humidity were high.

By the middle of the dry season in June, only 12 of the 30 species found in February were still present, but even these were less abundant (table 10.7). Two species, the widespread fodder plant *Alysicarpus vaginalis* (Legu.) and seedlings of the small tree *Muntingia calabura* (Elae.) whose seeds are dispersed by birds and bats, were relatively rare in the wet season but, because they persisted, were dominant in the dry season although their numbers did not increase. Some of the species of the wet season were clearly annuals, setting seed and then dying, but the EoS team watered the 'dead', brown remains of some other plants, particularly the common moss and *Pilea microphylla* and found that within five minutes the plants had become green and were becoming turgid. This ability to recover from droughts, known as poikilohydry, is also known from natural habitats such as limestone cliffs which are also subject to periodic droughts (p. 477).

Table 10.7. Flowering plants found growing on the walls of Fort Rotterdam in the wet and dry seasons. Rare plants indicated by parentheses.

		Wet	Dry			Wet	Dry
Apoc.	*Lochnera rosea*	(•)	-	Gram.	*Digitaria timorensis*	•	-
Comp.	*Eclipta alba*	(•)	-		*Eragrostis amabilis*	•	-
	Emilia sonchifolia	(•)	-	Labi.	*Leucas lavandulifolia*	•	•
	Ageratum conyzoides	•	•	Legu.	*Acacia auriculiformis*	(•)	-
	Tridax procumbens	•	•		*Desmodium triflorum*	•	-
	Vernonia cinerea	•	•		*Alysicarpus vaginalis*	(•)	•
Cype.	*Cyperus compressus*	•	-	Mora.	*Ficus benjamina*	(•)	-
Elae.	*Muntingia calabura*	(•)	•	Pipe.	*Peperomia pellucida*	•	-
Euph.	*Acalyptha indica*	•	•	Polp.	*Cheilanthes farinosa*	•	-
	Euphorbia hirta	•	•	Rubi.	*Hedyotis corymbosa*	•	-
	E. microphylla	•	-		*Spermacoce glabra*	•	•
	Phyllanthus niruri	•	•	Scro.	*Scoparia dulcis*	•	-
	P. urinaria	•	-	Urti.	*Pilea microphylla*	•	•
	P. zollingeri	•	•		*Urtica urens*	•	-
	Bridelia ovata	(•)	-				

DITCHES

Urban roadside ditches are regarded by many as simply a means of preventing floods or of removing household waste water to larger water
courses. To an ecologist, a ditch is a simple, small river, the life in which
can give indications of water quality.

Where ditches are obstructed and where water flows sluggishly, mosquitoes may breed. Their preference for still water is the major reason that
they are more common in the dry season than in the wet season. In the
latter the eggs and larvae are swept away. The frequency of mosquito-
borne diseases is usually higher in urban areas than elsewhere because of
the availability of suitable breeding habitats and biting targets.

There are about 125 species of mosquito in Sulawesi but only the
genera *Anopheles, Culex, Mansonia* and *Aedes* (represented in Sulawesi by 38,
21, 5 and 35 species respectively) have species known to spread debilitating diseases such as dengue, filariasis and malaria (fig. 10.4) (O'Connor
and Sopa 1981; Hii et al. 1985). The disease organisms are spread when
female blood-sucking mosquitoes[7] inject a small quantity of saliva into the
bloodstream of the host. If the mosquito is not yet carrying the disease
organism, she may pick it up in the blood sucked from an infected host.
The identification of those species that spread disease is far from easy, and
suggestions that posture or leg colour alone can be used to distinguish
species accurately are ill-founded. There are considerable differences
between and within genera in behaviour, and the habitat preference of
larvae and adults (table 10.8). In the towns of southern South Sulawesi, for
example, *Anopheles sundaicus, A. subpictus* and *A. barbirostris* are the most
important vectors of malaria; the peak biting period for *A. sundaicus* is
between midnight and 0100 hrs, for *A. subpictus* the biting peak is between
2200 hrs. and 2300 hrs., but *A. barbirostris* bites through the night until 0500
hrs. with no clear activity peak. *A. sundaicus* prefers human targets over
cattle and buffalo, whereas *A. subpictus* prefers to take blood from cattle and
buffalo. Both these species are tolerant of brackish-water and so are
common in locations close to coastal tambak fishponds (p. 187). The
normal method of controlling mosquitoes in Indonesia is by spraying with
a 75% solution of DDT at a rate of 2 g/m^2 (Collins et al. 1979). No one
appears to have studied the effect of the use of this infamous chemical on
other aquatic organisms or on organisms, such as humans, higher up the
food chain in Sulawesi.

Ditches are commonly inhabited by two species of small fishes, the
mosquito fish *Aplocheilus panchax* and the guppy *Poecilia reticulata* (fig.
10.5). Both species have been introduced to Sulawesi from South America
by people in attempts to control mosquito larvae. Mosquito fish are primarily eaters of small aquatic insects whereas guppies are primarily eaters
of algae, but will also prey on insects.

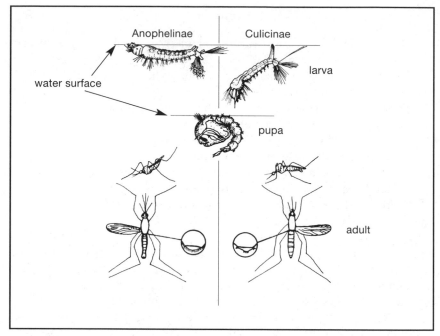

Figure 10.4. Differences in larval shape and position in water, adult posture, wing-spotting, mouth parts, and shape of scutellum between anopheline (*Anopheles*) and culicinine (*Aedes, Culex, Mansonia*) mosquitoes.
After Anon. 1967

Table 10.8. Major disease-carrying mosquitoes of Sulawesi and their major breeding habitats.

	Breeding sites	Diseases carried
Anopheles	Clean or dirty, fresh or brackish water	Malaria
Culex	Clean water	Malaria and filariasis
Aedes	Clean water	Dengue fever
Mansonia	Clean water	Filariasis

After M. Rachmat pers. comm.

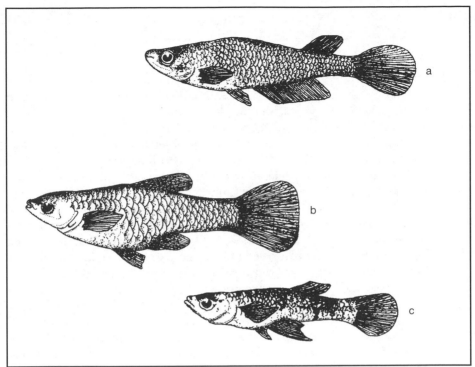

Figure 10.5. Mosquito fish *Aplocheilus panchax* (a), guppy *Poecilia reticulata* female (b), and male (c). Note the different positions of the dorsal fins, and the long anal fin of *A. panchax*. Also, *A. panchax* has a shiny silvery spot on the top of the head.

Both these fish are most obvious in urban ditches when they mouth at the surface, looking as though they are breathing air. Some fish can use atmospheric oxygen (pp. 301 and 580) but the two species are in fact taking in water from the air-water interface where the concentration of dissolved oxygen is greatest (Kramer and Mehegan 1981). This 'aquatic surface respiration' is generally only used where oxygen levels are low (such as slow-moving ditch water with a high organic content), and when the oxygen levels of water are raised the fish respire normally. Aquatic surface respiration does not allow them to stay alive for long periods in highly deoxygenated water but it does confer an advantage such that these fish can survive where other fishes cannot, and they can thus be used as indicators of water quality.

HOUSES

Spiders and Ants

Spiders are a ubiquitous group of predators found in houses. Some of the most common are the often brilliantly-coloured jumping spiders which form the world's largest spider family (Salticidae) with over 4,000 species. These bold beasts can leap up to 40 times their body length, and not surprisingly have more acute vision than any other spider. The large middle pair of eyes at the front that give these animals such a surprisingly endearing look, act like telephoto lenses with a narrow field of vision but high resolution. There are remarkably like human eyes but, since the eyes are unable to move, the retina rather than the eyeball is moved to change the field of view. The other three pairs of small eyes (two at the back and one at the front), have a much wider field of vision and the pair at the front give these spiders some degree of binocular vision, thus enabling them to judge distances when jumping and hunting. The small eyes are probably the first to pick up movements of potential prey and the main eyes are used for the fine hunting techniques. Since at least one pair of their eyes look behind them, jumping spiders are very hard to catch. They do not make webs but, like all spiders, they trail behind them a strand of 'silk' which is attached to the ground at intervals so that they are saved from falling after an unsuccessful leap or pounce on flies and other prey (Preston-Martin and Preston-Martin 1984).

Kitchens often harbour a number of ant species. The smallest is the yellowish Pharaoh's ant *Monomorium pharaonis* with a black tip to the abdomen. It has been carried unintentionally all over the world and is quick to find sugar if left in open containers. Once it has found a food source it may keep other species away with a repellent odour secreted from the poison gland.

Geckos

There are few houses or other buildings in towns without a resident community of geckos and they are unusual amongst the animals living alongside man in that they are not, for the most part, regarded as dangerous or undesirable.

Five gecko species[8] regularly inhabit houses but the two larger species, *Gekko monarchus* and *G. gecko*, are relatively rarely seen. The 'to-kay' call of the latter species is quite frequently heard, however, particularly in suburban villages where palm-leaf roofs are still used.

The other three geckos are remarkably similar in size and diet although *Gehyra mutilata* appears to shun brightly-lit locations (Church 1962). In many suburban areas *Hemidactylus frenatus* is the most common species followed by *H. platurus* and *G. mutilata* although *H. platurus* is often the most common species on well-lit walls inside houses, with *H. frenatus* most common on outer walls. The females of each species lays two hard, white eggs at a time in cracks, behind pictures, etc., and these take about three months to hatch. At least a year is required for the hatchlings to reach adult size although they become mature after the snout-to-vent length exceeds 3.5 cm (Chou 1978). Their primary food is insects but the stomach of one *H. platurus* examined in Bandung contained a *H. frenatus* tail. *H. platurus* also appear very partial to rice, honey, juices from fruit, bread crumbs, meat and fish. Most lizards develop fat bodies under their skin the size of which is determined by the abundance of food, but none of the geckos in the Bandung study were ever found to contain such fat bodies. This may be due to the year-round supply of scraps from the humans with which they share houses, and this also results in the absence of breeding seasons (Church 1962). Geckos have relatively short tails and this is typical of lizards that are habitat specialists (the ability to run upside down is taken to be a speciality), relatively cryptic, and have a sit-and-wait strategy for hunting (Vitt 1983). The bright green, diurnal, and very long-tailed lizard *Calotes cristatellus* of suburban gardens would be placed at the opposite end of the spectrum.

It is a frequent matter of debate how geckos manage to climb on vertical surfaces and run upside down on a ceiling. The feet do not have suckers but instead have small overlapping flaps of skin. These flaps are covered with minute, closely-set hairs which make contact with the slight irregularities of a surface and enable geckos to cling where other animals would fall. Geckos are also known for their ability to shed their tails when caught—a response designed either to enable them to break free if caught by the tail, or to distract the 'predator' with a wriggling tail, or perhaps both The shedding of the tail is obviously extremely important for geckos since they divert a great deal of energy into the growth of a new tail, and the majority of geckos appear to have lost their original tail. Regenerated tails are rather larger than the original but often not so long or so symmetrical (Vitt et al. 1977). Tail shedding is also known in some other lizards and in snakes, and in all cases it involves muscular contractions which causes a fracture to occur across a vertebra, rather than between two vertebrae.

Possible Urban Ecology Studies

Plots of land awaiting development, roadsides, the gardens of unoccupied houses, and neglected corners of towns represent opportunities to study the process of plant succession. Which plants colonize open ground? How does exposure affect colonization rates? Which plants are the first to arrive? How quickly does the organic content of the soil increase? Assuming the first plants are grasses and herbs, how long until the first woody plant appears? Measure samples of the above-ground biomass at different times. Is the rate of increase constant? Does it show changes related to the seasons? What animals are associated with different stages of the succession?

Investigate different urban ditches, and try to catch one of each of the fish species present. Remember that some fish are mainly nocturnal in their habits. It is more informative if the fishes caught can be given a scientific name (Schuster 1952; Alfred 1961, 1966; Mohsin and Ambak 1983), but simply the total number of different species provides useful data. Plot a graph of the number of species against the biological oxygen demand or the oxygen concentration of the water. Repeat in a different section of town and in irrigation canals. Is it possible to show a correlation between the results? What other parameters are important? Use the same procedure for other aquatic organisms.

Select a number of walls of different ages (either absolute age or age of most recent paint), and quantify the differences in algae and higher plants growing on them. Is exposure more important than aspect? Pay attention to the base of the walls. Are there plants which probably grew from seed carried by birds? Drill holes of different sizes and lengths into a wall and monitor the rate at which these are colonised by wasps and other animals (Darlington 1981).

Stand outside at dusk and watch for bats, noting the times the different species appear. Where do they seem to come from? Finding roosts is sometimes easier at dawn when bats fly around them before settling. Over a series of days, try to pin down the actual roosts. Do the bats emerge at the same time each day? What effect does rain have? Are roosts available all over the town or are they concentrated in one area? How many bats emerge from each roost? Is it the same number day after day? Take a series of 1 ha plots in a town and collect data to estimate the number of resident bats. Attempt an estimate of the quantity of fruit and insect food the bats ingest each night. If bat boxes have been made (p. 614), attempt to determine why some boxes are inhabited and others are not.

The three most common species of house gecko are more or less the same size. What forms of competition and niche separation allow them to coexist? Collect several of each species at intervals of several weeks over a period and analyze the stomach contents. Without necessarily identifying the animal food remains to genus or even family, is it possible to detect differences in composition or size of prey between species. Watch the geckos. How do their hunting strategies and prey differ? Do any of them catch food during daylight hours? How do the species space themselves around light bulbs? Are resting places (in cracks, behind pictures and mirrors, under stones, in thatch roofs, in corners of ceilings or behind cupboards and curtains) different between species?

Ants frequently take advantage of scraps of food left in kitchens. Deliberately leave small quantities of sugar, meat, oily nuts or other food and notice which ants are attracted to which food. If the foods are mixed, which ant reaches the pile first and which dominates? Place a known, weighed quantity of food in a small pile—monitor how quickly it is removed. Is there any difference between the rate at which food piles are found between day and night time? Follow a trail of ants. How far does it extend? How long does it last in any one pattern? In what ways can ants be regarded as useful?

Chapter Eleven

Resources and the Future

FORESTRY

Any country with a valuable, renewable, but exhaustible natural resource should strive to behave prudently and aim for sustainable production. Unfortunately, the prevailing economic paradigm thwarts such policies because the proximal and easily-achieved aim is to maximise the present profits that can be gained from the investments it has made. The same money invested in the forestry industry could of course be invested in some alternative industry or business with a higher rate of return on the money invested. The difference between the size of the two returns is known as the opportunity cost and represents benefits forgone. Benefits, particularly environmental benefits, may continue well into the future from an investment made today. The value of the future benefits are to be discounted by a certain percentage, currently 9% in Indonesia, which takes into account the cost of borrowing money, the likely rate of inflation, as well as attendant risks. Thus, benefits 20 years from now are worth very little under the prevailing economic theory.

Bodies concerned with forestry, such as FAO, quote figures for timber production, whereas forestry production can only really be said to exist in plantations. Figures represent instead the harvest of accumulated past production, the supply base of which is rapidly becoming smaller. Tropical timber is outside the rules of supply and demand economics since the most valuable and often traded trees take about 100 years to grow (Guppy 1983, 1984).

Any resource manager is faced with one basic decision: to exploit on a sustainable basis, or to mine the resource with no aim to renew the resource in order to take an early profit to invest in some other project or business.[1] If you conserve your stock you might expect an annual return of Rp x, but you might decide to convert the entire forest into cash totalling Rp y which you can immediately invest elsewhere with a rate of return of i% per year. The annual income from the second strategy is thus Rp yi. Your decision will depend on the relative values of x, y and i. Investment on monetary capital usually provides a return of at least 10%, but an area of forest will take *at least* 70 years to recover its full value even after careful selective logging (i.e., an average return of 1.4%/yr that can be cashed only at the end of the period), and so it is clear that care and long-term conservation are extremely hard to argue in an economic sense. Plantations produce a greater rate of return than forests, but since this is still lower than normal financial rates, there is little enthusiasm to invest in them (Anon. 1986a). This is why there are such pressures on the forests from licensed and unlicensed loggers. The rapid consolidation of forests into tangible monetary gain is quite the most rational path to follow under prevailing economic theory.

One of the main drives behind the policies controlling the forestry industry is the generation of foreign exchange, outside the oil and gas sector, in order to pay back loans given by multi- or bi-lateral agencies for development projects. It is also very important to those provinces with no oil or gas reserves; for example, over 95% of the income generated in Central Sulawesi is from forestry. The timber, in various forms, is sold relatively cheaply (certainly cheaper than its true cost which should include the replacement cost) to developed countries and so the consumers must accept some culpability in the loss of forests.

In 1984 the International Tropical Timber Agreement (ITTA) was signed in Geneva under the umbrella of UNCTAD[2] which brought together both the producers, of which Indonesia is one of the most significant, and the consumers, Japan, United States and Europe. The cornerstone of that agreement is the intention to "encourage the development of national policies aimed at sustainable use and conservation of tropical forests and their genetic resources, and maintain the ecological balance in the regions concerned". The headquarters of the International Tropical Timber Organization was hoped to have been sited in Indonesia, but Yokohama, Japan, was eventually chosen. Japan consumes over half of all the tropical timber traded on the international market.

The ITTA recognized, therefore, that a balance could be found between the economic and ecological aspects of forest development. Two likely means by which this balance will be sought are:
- the use of little-know and little used species of tree that could increase the economic returns from an area of forest that is likely to be felled in succeeding years, and
- the development of plantations (Johnson 1985).

In addition to the timber species already exploited, there are many tree species whose potential for timber, fibre or cellulose remains unexploited (Whitmore 1980). People of rural areas generally have a deeper understanding of the uses and value of forest plants, but knowledge of these can be lost within the timespan of a generation (Sastrapradja and Kartawinata 1975).

Plantations might succeed from a number of points of view, but if they comprised just a few species of trees then the problems of genetic resource depletion and wildlife requirements will not have been addressed. In the past the hardwood trees of tropical forests have been unsuitable as a raw material for paper products. New technology has changed this, however, and as literary rates increase, so more paper for books, and magazines is required. If future generations of schoolchildren and students are not to suffer because of paper shortages, it is important that extensive pulp plantations be established, preferably on land that is currently deforested and unproductive. Certain machines are able in just one minute to reduce an entire forest tree to chips for use in pulp, paper, particle or fibre board (Myers 1984a). The efficiency of these machines clearly makes their use in normal logging operations desirable, although smaller versions that could utilize the wood wasted and otherwise burned at small sawmills would be a bonus. The quantities so produced would be insufficient for export but would make a sound base for local carton manufacturing. The processing of all timber in land clearance schemes is neater and tidier, but considering that much of the nutrient capital of the forest is in the living plants, their removal could lead to long-term agricultural problems on the already marginal lands.

Sustainability is desired but this is a serious problem in the exploitation of all forest products. Sustainability is the key but to achieve it will require significant and possibly painful economic and social changes in priorities and policies. There were fears voiced over fifty years ago that rattan canes were being exploited too rapidly (van de Koppel 1928). The enormous national and international trade in rattan and rattan products has resulted in over-exploitation of rattan stocks and there are probably few areas, reserves and national parks notwithstanding, that have not been visited by rattan collectors. Sustainable collection is possible but much more needs to be known about growth rates under different light conditions, the importance of supports, the difference between those species that sprout from their base when the stem is cut and those that die. Sprouting success is also controlled to some extent by the age of the rattan and perhaps whether it is flowering or fruiting. Under present conditions certain species that do not sprout are generally cut before they fruit, may be in serious decline.

Compared with the single-operation timber operations currently sanctioned, the continuous harvesting of timber and minor products is ecologically sounder, much less destructive and of a greater advantage to local economies but, as stated above, the support of local economies is not a primary objective of forestry policy. Much greater emphasis could be placed on research into polyculture forests combining commercial hardwoods, softwoods, fruit trees, rattans and game.

WATERSHED MANAGEMENT

One thousand years ago the great Maya civilization in what is now southern Mexico, Guatemala and northern Belize collapsed after enjoying some 1,700 years of plenty. The collapse was rapid and was precipitated by the agricultural patterns imposed by the Spanish conquerors and the soil erosion resulting from this.

A high rate of soil loss is not only a long-term threat to agriculture, it also reduces water quality necessitating extra treatment if it is to be used in water supplies, reduces the diversity and richness of aquatic life (p.342), accelerates sedimentation in check dams, reservoirs and rivers, and this in turn raises the level of riverbeds which aggravates flooding, and deleteriously affects coastal fisheries, particularly those associated with coral (p. 241).

As populations increase, so more and more people are forced to move onto increasingly marginal land. Such people are generally from the poorest sections of society without the knowledge, or even the motivation, to prevent the degradation of their environment. Ultimately these people will move, either because the government offers them job opportunities in the lowlands, or because the land they farm becomes exhausted. The land they leave is unproductive, and its rehabilitation would benefit upland, lowland and coastal environments. Watershed rehabilitation is generally left behind, however, while shorter-term development priorities in the lowlands are met.

If job opportunities cannot be offered, the practices and productivity of the upland farmers have to be improved, erosion and floods must be controlled and certain areas may have to be replanted and designated as protection forests. These different initiatives can take at least a decade, however, before they show any real positive impact on the environment (Spears 1982).

In the 1950s and '60s many villages in Sulawesi were suffering from the cumulative affects of bad land management. The land was hot, bare and, due to alang-alang grass, was difficult to cultivate. In one such village, Tabbingjai, in Gowa near the southern tip of South Sulawesi, a man named Solle with no formal schooling took it upon himself to plant hillsides at 1,200 m a.s.l. with the tree *Acacia decurens*.[3] He prepared a nursery and explained the long-term benefits of what he was attempting to do to other villagers who were, at first, skeptical and mocking. He later organized the building of two 20 m x 4 m check-dams across rivers to capture sediment and retard water runoff, and planted *Eucalyptus* seedlings around the *Acacia* trees in order to supply the wood the villagers needed (Anon. 1986b). All this was achieved with no government assistance or subsidy. In all he has been responsible for the planting of over 1,000 ha of unproductive uplands, and he was presented with a Kalpataru 'Environmental

Pioneer' Award by President Soeharto on World Environment Day 1986. The previous year the same award was given to La Ode Muhammad of Wantimoro, Southeast Sulawesi who had organised villagers to replace alang-alang with coconuts and coffee.

Of considerable significance to watershed management is the role the World Bank had in encouraging the establishment of the 300,000 ha Bogani Nani Wartabone National Park from the Dumoga, Bone and Bulawa reserves to protect the watershed of the Dumoga valley. The wide, flat Dumoga valley is a former lake bed and the potential of its 13,000 ha of irrigable fertile soils could only be realized if the quality and quantity of the water supply could be maintained. Never before had the World Bank set such stringent conditions for an irrigation project or made such a sizeable loan for the development of a National Park although various parks, reserves and endangered species have benefited from World Bank projects (Goodland 1985). The protection of the forest also safeguards a large proportion of the animals and plants of North Sulawesi (Goodland et al. 1984; Wind 1984). The danger posed by illegal occupants of the Park was recognized and the local government and the Conservation Office were actively involved in their eviction in 1983, and subsequent resettlement on the south coast. Various recommendations for a system of buffer zones have been made (e.g., Anon. 1983; Wegman 1983) but no Acts, amendments or regulations have been passed by the central government that would legalize such demarcations.

The environmental conditions set by the World Bank brought together the needs of development and conservation. The World Bank loan included funds for the protection of the forests as well as for the building of a research laboratory.

The argument is often heard that the replacement of wildland forest with rubber or fruit trees preserves water and soil. This is certainly not always the case. In one study of the hydrology of a plantation area it was found that storm flows doubled, low flows decreased to one quarter and soil loss at least tripled compared with the conditions under the original forest. Average rates of soil loss on undulating land covered with forest in West Java were found to be only 0.2-10 t/ha/yr, yet the figures from under dense but uniform tree plantations were 20-160 t/ha/yr, and from pasture 200 t/ha/yr. The most dramatic impacts are experienced while the forest is being cleared and before the replacement trees have produced a continuous canopy (Wiersum 1979; Hamilton and King 1983).

Watersheds can be managed such that low and high flows are more predictable and water yield better meets human needs. Ephemeral streams in degraded areas can be converted to permanent streams through a well researched rehabilitation programme. Excess water can be reduced by planting more vegetation and improving soil infiltration. This sort of manipulation requires the very careful assessment of at least ten years of rainfall and flow data.

Suspended sediment loads from watersheds generally increase with the amount of disturbed land, or land under cultivation. For example, the Toraut watershed encloses a well-forested National Park and the maximum suspended sediment concentrations recorded is 350 mg/l with a total suspended sediment load equivalent to a soil loss of 14 t/ha/yr (Walang 1984). By contrast, the maximum suspended sediment and total suspended load for two disturbed watersheds near Manado, the Tondano and Kinilow-Malalayang, were 1,290 mg/l and 5.2 t/ha, and 1,100 mg/l and 2.8 t/ha respectively (Molenaar 1984; Mantalalu 1985). The sediment discharged from the Tondano River into Manado Bay settle over 1 km from the shore (Molenaar 1984).

Rivers flowing into Lake Tempe have long been contributing large amounts of sediment, causing the voyager James Brooke, to write in 1840 that. "it (Lake Tempe) is filling up." This is still the case today, but when sediment outflow via the Cenrana River is also considered the net sediment accumulation is only 1 cm/yr (Anon. 1979a). At this rate, it will be approximately 320 years before the lake disappears at low water season and 950 years at high season.

Lake Tondano itself is silting up at the rate of 20 cm/yr due to sediment inflow from the many inlets and erosion from surrounding, agricultural activities (Anon. 1980a).

The massive reforestation programs in many tropical countries including Indonesia were founded to a large extent on the popular belief, generated in the West, that these would inevitably benefit groundwater aquifers and create more reliable flows during the dry season. A careful recent review of watershed studies found, in fact, that there was overwhelming evidence that groundwater levels are lowered and stream yields reduced following reforestation, and both these effects are more pronounced in the dry or growing season. It is true that the severity of local flooding is reduced and delayed, but as the actual stormflow volume may not be any less, there may be no impact on downstream flooding (Hamilton and King 1983).

This is not to say that reforestation is a wasted effort. Its major contribution is in the reduction of rates of soil erosion but if attention is not paid to erosion from road building, site preparation, site maintenance, and the eventual harvesting of the wood products, then the benefits will be far less marked. Reforestation is most frequently conducted on steep land but at least as valuable is the establishment of forested strips some 25 m wide along rivers, although this can cause considerable problems of land ownership, compensation and maintenance.

IMPLICATIONS OF ISLAND BIOGEOGRAPHIC THEORY

As the disturbance of forests proceeds, so smaller and smaller pockets of relatively pristine vegetation are formed and consequently these contain smaller populations of animals and plants. The study of island biogeography (p. 52) is not yet twenty years old and academic arguments continue over what predictions the mathematical models can actually make about the extinction and colonization of species in different-sized areas of land, and therefore what shapes and forms of management are theoretically best for nature reserves (Cole 1981; McCoy 1982; Harris 1984).

As far as Sulawesi is concerned, the subject of nature reserve design is largely one for academic debate. The question of what percentage of species will become extinct in different-shaped reserves in 50, 500 or 5,000 years' time is not of great relevance in Sulawesi when it is by no means certain how much of its reserves will be intact in even 25 years' time. The shapes of protected areas do not generally conform to the neat models of the theories (Gorman 1979) because the design of reserve boundaries has been set primarily by patterns of human settlement. In Sulawesi today the major priority is simply maintaining the integrity of reserves against the pressure of legal and illegal forms of habitat disturbance. Each day, the mature-phase forests are being destroyed at a vastly faster rate than they are being added to by the process of succession from building-phase forests (p. 359). Anyone acquainted with the major Sulawesi reserves knows that not only nonreserved areas are being destroyed, but that incursions into reserved areas are frequent, in the form of logging, settlement, and other exploitation.

During the last few years, attention has begun to focus on the rather more pragmatic problem of extinction. Extinction is now no longer viewed as an event, but rather as a process that can be observed and understood. To avoid extinction, the population of a species must be large enough to maintain a sufficient proportion of its genetic variation to avoid the deleterious effects of inbreeding. Thus the concept of 'minimum viable population' has evolved, and this is of most concern for large, wide-ranging animals, and for plants that provide some important or critical resource for a significant number of other species. It is true, of course, that any population, whatever the size, is subject to the possibility of extinction through chance events such as epidemic diseases or rapid environmental deterioration, but given the insurmountable problems of quantifying the unpredictable, these factors are not generally considered.

It has been ascertained that the effective size of an absolute minimum viable population is about 50 randomly-mating adults living in a population with a one-to-one sex ratio. With an effective population of 50, however, the genetic variation remaining after 100 generations will be only 36% of the original, and the additional variance due to mutation would be negligible. It has therefore been proposed that an effective population of 500 randomly mating individuals is the minimum size of a population that would enable the species to adapt to environmental changes (Wilcox 1986).

An 'effective population' of 50 is not the same as 50 individuals nor necessarily the same as 25 breeding animals of each sex. Many animals live in social groups which are not simply pairs of adults. The effective number, N_e, in a group or population can be calculated from the actual number present as follows:

$$N_e = \frac{4\,N_m N_f}{N_m + N_f}$$

where N_m and N_f are the number of breeding males and breeding females respectively (Frankel and Soulé 1981).

To see how this theory can be used in practice, one can take the moor macaque *Macaca maura* of southern Sulawesi as an example. This is probably the Sulawesi primate most threatened by forest loss. Only about 10,000 live in about 500 km² of protected forests (p. 433) (MacKinnon 1986). A group of these macaques comprises about 16 individuals of which two would be breeding males and six would be breeding females. Using these numbers in the equation above gives an effective group size of six. If one desires to protect an effective population of 500, then a total of 83 (500/6) groups with 1,333 (83x16) individuals are required. The macaques live at a density of about 20/km² and so the area needed for 1,333 of them is 67 km² (1,333/20). This is clearly less than the area already protected but, lest complacency take root, it is important to remember that the discussion concerns populations not just total numbers. A road between two forest areas is obviously not an important barrier to a population, but an expanse of agricultural land is sufficient to separate different populations. The next step is, therefore, to determine the size of each contiguous, or essentially contiguous, forest block to check that areas of 67 km² still exist and are being adequately protected. Since this macaque is the largest of the endemic animals found in the forests of southern South Sulawesi, adequate forests for its survival should be adequate for most of the others. Some plants, particularly dioecious species with separate male and female individuals, may need still larger areas to shelter their minimum viable populations.

The approach described above provides a guide to the areas of forest required by species, but it is based largely on theory rather than the results of scientific experiments. Rather firmer support will in the future be provided by the results of the world's largest environmental experiment. An area of 600 km² in Brazil has been guaranteed protection for 30 years, and a system of different-sized patches of forest is being established. Some patches will be just 1 ha in area, others will be 10 ha, 100 ha, 1,000 ha, and the largest will be 10,000 ha. The rate, order and means of extinctions of the species within these patches are being monitored by teams of scientists as the patches are gradually isolated from neighbouring forest areas. This pioneering work will provide a firm basis from which recommendations regarding the area of forests required to safeguard the principal ecological charcteristics of an area can be made. Luckily, it is not necessary to wait 30 years for all the results. Already it is clear that the 'edge effect' or the influence of the disturbed ecosystem outside the forest penetrates very deep, about 0.5 km, inside the forest (Lewin 1984). Thus, in the case of the macaques above, it might be necessary to enlarge the areas required for the minimum viable population to allow for a 0.5 km boundary of edge-affected forest.

EXTINCTION AND CONSERVATION

The planet Earth supports about 35 million species of animals and plants (Erwin 1982; Myers 1983, 1984a) each of which has a certain range of genetic variation. It is likely that at least two-thirds of all species occur in tropical regions and most of these are confined to the rain- or moist-forests. So about 40% of all living species live in these forests which themselves account for just 7% of the earth's surface. One of the other most species-rich environments is coral reefs.

Periods of mass extinctions have dotted the earth's history and these mark boundaries in the geological time scale (p. 3). In the last 500 million years there have been six mass extinctions of which the worst occurred 240 million years ago and resulted in the loss of perhaps 95% of all marine organisms. These events were sudden when viewed against the age of the earth. For example, the loss of the dinosaurs 65 million years ago took about 2 million years. After the extinctions there was no genetic evolution of species. Instead, there was a lag of several million years before species diversity climbed up towards previous levels.

The rate of mammal species extinctions has increased from about 0.01/century during most of the Pleistocene, to about 0.08 during the late Pleistocene, to 17 from 1600-1980. The higher rates were caused by neolithic hunters or climatic change and the hunting and commerce encouraged by European expansion. The extinction rate of mammal species in the years between 1980 and 2000 could be about 100 per century and one of the predominant causes is habitat disruption (Wolf 1985).

The destruction of forests, coral reefs and other natural ecosystems by their conversion to other land uses or the over-intensive or careless exploitation of resources that they support, must be leading to a significant loss of species. Figures frequently quoted are that one species per day is being lost at present, and that an average of 100 species per day will be lost between now and the end of the century as the human population increases. If people have no option other than to follow uncontrolled and unrestrained extensive agricultural techniques, then the natural ecosystems will suffer. If the government will underwrite and support sustainable intensive agriculture, however, then the impacts on natural ecosystems will decrease. Agricultural development and conservation of species will, in this case, proceed hand in hand.

No one would dream of driving a road through the National Archives or a major museum or library, yet the loss of the majority of lowland forest, for example, and the resultant loss of species is akin to that. A major difference, however, its that the items in a Sulawesi lowland forest have not been catalogued let alone classified and studied with a view to development or their intrinsic significance.

The arguments for conservation can be grouped in two categories as follows:

Economic
- the maintenance into perpetuity of commercially valuable renewable resources such as timber and fish;
- the protection of soil and water supplies;
- the stability of climate and atmosphere;
- maintenance of plants and animals of use in the improvement of present and future food and medicinal needs;
- the needs of research and education; and
- the development of tourism and recreation.

Social and Moral
- to maintain the quality of life;
- the acceptance of a moral responsibility not to leave less to future generations than we ourselves received; and
- national pride (Anon. 1982a).

At the beginning of the conservation movement, the arguments used to seek support and money for conservation work tended to concentrate on the social and moral matters. Later, particularly in Africa, the economic benefits of tourism were stressed. Nowadays attention is drawn to the wide range of economic benefits from conservation policies. The subject was included in an influential document, *The World Conservation Strategy*, published by the IUCN[4] The Strategy has three key components:
- the maintenance of essential ecological processes which move energy, materials and nutrients through ecosystems;
- the preservation of genetic diversity; and
- the sustainable use of species and ecosystems (Anon. 1980).

Thus, species are seen as an insurance and an investment for the future; the options for which must be kept open. Where the use of a species is unknown, it is argued that a future use may be found, or that the species may be a vital component in an ecosystem.

Species can have unexpected uses. For example, the horseshoe crab, an ancient relative of spiders sometimes found around the coasts of Sulawesi, has bright blue blood which congeals rapidly when exposed to even minute quantities of bacterial endotoxins. This is used to test the purity of fluids intended for injections, and also in identifying a serious form of meningitis. The crested macaque *Macaca nigra* of North Sulawesi is the best human model known for diabetes research. The structure of the skull and vertebrae of woodpeckers have given insights to the design of crash helmets. A woodpecker generally feeds by hammering on tree trunks with its bill to find insect larvae, and the speed of the head on impact with the tree has been calculated to be about 300 km/hr; enough to kill any human. Also, the wing movements of chalcid wasps have assisted the design of improved helicopter rotors (Myers 1983). How much effort has really been made to determine the potential of other species? Is anyone bold enough to state that any species will, with all certainty, prove to be of no practical benefit to humankind?

These arguments related to economics are based on the known uses of certain wild species, but they cannot be used in regard to all species (Collar 1986). The potential (and admittedly unlikely) utility of the elegant sunbird *Aethopyga duyvenbodei* endemic to Sangihe Island (p. 51), for example, will not be accepted by those charged with increasing the domestic production of coconuts on Sangihe. This small bird may indeed have been an essential species in the functioning of the Sangihe forest—but the forest has all but been replaced by coconut groves. Economic values are all relative values, and even if we could put a value on the sunbird, its value may be less than that of coconuts. Thus, relying on economic arguments provides no security for the survival of a species. It is clear, however, that the Government of Indonesia does not only listen to economic argument. It supports the large Department of Religion because it recognizes the central importance of spiritual health.

It can be concluded from the above that the ultimate reasons for species conservation are moral and religious.[5] No one wants species to become extinct, and no religious code would sanction planetary stewardship that resulted in extinction. The only secure value for a species is an absolute value, in the same manner that the sanctity of human life is recognized as having an absolute value (Collar 1986; Ehrenfeld 1986; Naess 1986). The desire to conserve species for their own sakes is sometimes regarded as outmoded, but there is a growing reaction against the economic imperatives of some conservation arguments because, if these alone are followed, a large number of species will not be with us much longer.

When a plant becomes extinct it is likely that a number, perhaps tens, of species that depend on that plant at some stage of their life cycle will become extinct, too. As the areas of natural ecosystems are reduced so the equilibrium number of species decreases (p. 56). Some of the doomed species may not become extinct for a century or more but the damage will have been done (Myers 1985). The loss of half of an area is likely to lead to the extinction of 10% of its species.

Conflict arises, of course, when the absolute value of an animal or plant species comes up against human life. Indeed, the necessity of choosing between species—which ones will be made target species of conservation programs and which will be passed over—may be closer than we think. The term used for this choice is triage[6] (Lovejoy 1976; Myers 1984b).

This book brings into planners' hands, perhaps for the first time, the information that, for example, some of the birds endemic to the Sangihe-Talaud Islands, or fishes endemic to lakes in Central Sulawesi are on the brink of extinction. Any plan to develop the last areas of scrub on Sangihe, or a fisheries project for Lake Poso, could push some species over the brink into the oblivion of extinction. Hopefully the arguments for species' conservation will persuade the planners that the approval of plans that would relegate species to the history books would be unacceptable, with the sole exception of the planned project alleviating human suffering in a manner that could not be achieved by any alternative scheme.

The basis for conservation in Indonesia is enshrined in the 1945 Constitution of the Republic of Indonesia; specifically Article 33 in Chapter XIV which states that: "Land and water and the natural waters therein shall be controlled by the State and shall be made use of for the greatest welfare of the people". It should be noted that the term 'welfare' in the spirit of the Constitution, concerns not just economic benefits, but also spiritual health and intellectual life. The Broad Guidelines on State Policy are developed by the People's Consultative Assembly and are held to be the expression of the desires of the Indonesian people as interpreted by that Assembly. Part of the section of those guidelines concerned with long-term development states that: "Rational use of Indonesia's natural resources is necessary in the execution of development. Exploitation of these natural resources should not destroy environmental living conditions and should be executed by an overall policy that takes into account the needs of future generations". Finally, the Environmental Management Act of 1982 states that forests are the primary means of maintaining harmony between people and the environment, and in recognition of that, "Everyone is obliged to conserve the environment and prevent as well as intervene against damage and pollution". Lists of specific legislation from both the colonial and the independence eras can be found elsewhere (Anon. 1985a; Hardjasoemantri 1985). A major problem, however, is that such statements have not been tested in courts of law, and so the legal interpretation is not certain.

GENETIC RESOURCES

Destruction of tropical forests, and particularly lowland forest, represents an extremely serious global loss of plant and animal genetic resources. Like the extinction of species, the loss of genetic resources is irreversible. This is of concern because this variation is the raw material not just for evolution, but also for the improvement of domestic and cultivated animals and plants, the development of industrial and medicinal products. As such, genetic resources play an essential role in world economic productivity. If their availability and diversity are reduced or lost, the effects on humans and their growing needs will be severely felt. Anyone who has the slightest doubt that conservation of genetic resources is one of the highest environmental priorities should consult some of the many papers and books on the subject (Anon. 1980; Frankel and Soulé 1981; Ehrlich and Ehrlich 1982; Jacobs 1982; Prescott-Allen and Prescott-Allen 1982; Myers 1983, 1984a, b, 1985; Ayensu et al. 1984; Wolf 1985; Ehrenfeld 1986).

While complete disappearance is the most dramatic fate for a species, the loss of races or unique populations is also serious since the genetic pool available for further evolution is eroded. This is important for species of known or potential economic use, since genetic diversity is the property of a species upon which improvement in crops and domesticated animals is based. The preservation of genetic diversity is essential if continued improvements in nutritional quality, taste, pest resistance, etc., in domesticated crops and animals are to be sought, if new species are to be domesticated and if natural compounds useful in the drug and chemical industries are to be found. Crop breeders cannot work without stocks of genetically diverse wild species.

Natural (rather than synthetic) plant compounds are extremely important in the medicines taken by Indonesians. This does not just apply to the herbal remedies most commonly encountered in villages, but also to the drugs prescribed by doctors. About 40% of all drug prescriptions in the U.S.A. are compounds of natural plant origin and the figure for Indonesia is probably similar to that.

The role of wild plant species in the drug industry is somewhat enigmatic. The role of plant compounds in current drugs is well known, so why are the powerful drug companies not campaigning at least as loudly as conservation organizations for the wise management of watersheds to save the wealth of wild species? To answer this question for the purposes of this book, letters were sent to the heads of the ten major pharmaceutical companies operating in Indonesia requesting clarification. Only four replies were received. One company, Pfizer, was established on the success of a world-wide screening of soil samples that resulted in the discovery of the antibiotic oxytetracycline (Terramycin) (I.R. Young pers. comm.). The opinion was generally held that screening plants for potentially useful compounds is very expensive and has a very low success rate. For these reasons and because such screening is both complex and interdisciplinary, the research needs to be centralized. For most companies this means research outside Indonesia. P.T. Burroughs Wellcome Indonesia alone appears to be actively engaged in the collection, analysis and appraisal of medicinal plants from Indonesia (R.C. Young pers. comm.).

The potential value of certain genetic resources has resulted in a somewhat bitter debate within the United Nations Food and Agriculture Organization. Representatives of the lesser-developed countries, which have contributed most of the world's valuable food plants, take the view that genetic resources are a common heritage for all nations. Representatives of the more-developed countries, where many high-producing hybrids or mutations have been produced, refuse to give away what they regard as the added value for which they have been responsible. The debate continues.

FUTURE SCENARIOS

Indonesia is a resource-rich country with enormous potential for becoming a major international political force, and a centre for pragmatic research in many fields. The future is, however, by no means certain and the course taken over the next few decades will be determined to a large extent by the prevailing attitude towards natural resource conservation (table 11.1). Either of the visions of the future is possible. The pessimistic vision will arise if sustainable or eco-development is not seriously adopted as the main tenet for economic development.

The management of forest resources is crucial in this context. Compared with the single-operation timber operations currently sanctioned, the continuous harvesting of minor products is ecologically sounder, much less destructive and of far more advantage to local economies. Much greater emphasis should be placed on research into polyculture forests combining commercial hardwoods, softwoods, fruit trees, rattan and game.

DRAWING THE LINE

Land is a finite resource. It is possible of course, to open marginal land, and in the future it may be technically possible to grow crops on almost all of Sulawesi's lowland soils with increasingly expensive technological and financial support to supply the needs of the population. One day, however, it is certain that there will be no more (even marginal) land available and that other courses will then have to be followed. At that point it will not be

Table 11.1. Two visions for the future of Indonesia.

	VISION I Unbridled resource exploitation	VISION 2 Sound resource management
Population	500 million and growing.	250 million and stable.
Lowlands	All lowland forest converted to agricultural land, much of it very marginal with poor yields. Lowlands frequently damaged by floods with economic and human losses: critical lands increasing.	All quality lowlands under intensive, high-yielding agriculture based on wide variety of crops and farm models. Former critical lands being used for productive agriculture or social forestry.
Land Use	Little land under forest; reserves and protection forests exist on paper only.	30% of land under natural forest, 20% under fruit trees; boundary encroachment in reserves under control, many with utilized buffer zones.
Mountains	Montane forests being stripped for firewood with consequent serious hydrological impacts in lowlands.	Major watersheds actively protected and managed, thereby sustaining agriculture in the lowlands.
Transmigration	Programme halted because of serious regional unrest, and poverty, and because extremely marginal soils had begun to be settled.	Programme halted because aims achieved and no longer regarded as necessary.
Economy	Government unable to tax enough, borrow enough or sell enough natural capital to pay off interest on development loans much less the loans themselves or to tackle increasingly wretched social and economic problems. Increasingly dependent on 'soft' loans. Living standards eroding. Dwindling foreign currency reserves.	Government reaping benefits of sustainable development and an example to other countries. Living standards comparable to, or above, those of today. Healthy foreign currency reserves from export of manufactured goods.
Tourism	Few, low-quality, tourists.	Tourism active in many areas.
Defence	Military defences unstable and backward due to enormous budget reductions. Attempted schisms in resource-rich, outlying provinces.	Strong military defence.
Pollution	Frequent pollution of waterways as cost-cutting increases and supervision decreases.	Environmental regulations enforced and improved.
Industry	Industrial sector declining, relying increasingly on imported goods.	Manufacturing of industrial goods thriving.
Research	Moribund.	Important centre for research in tropical forestry, ecology and environmental problems.

Adapted from Anon. 1982a

possible any longer to draw the line that preserves all the options regarding natural genetic diversity and to safeguard soil and water. Unless the line is drawn deliberately prior to this point, it will inevitably be drawn for us at a time and place which we will no longer be able to choose.

The loss of habitat is closely linked to development projects but it has been shown in recent years that the narrow economic argument does not always win over conservation. For example, permission to exploit a copper deposit within the Bogani Nani Wartabone National Park has not been granted, and there appears to be a feeling in mineral companies that it is not economically worthwhile to undertake explorations in parks or reserves. It has been demonstrated to them that the line has been drawn. In 1979, the Morowali Plain in Central Sulawesi was being considered by the Department of Transmigration for the settlement of 10,000 families, by the Department of Mines and Energy for the mining of chromite, and by the Department of Forestry for the establishment of a National Park. Morowali was still largely forested, and had a complex pattern of relatively fertile alluvial soils interspersed with decidedly infertile soils. Neighbouring areas also had reasonably fertile soils, but they were much more uniform in distribution, and there was very little forest remaining. In Indonesia as a whole, alluvial forest has generally been converted to agricultural land, and eventually the argument for protecting this area was shown to be the strongest, and a National Park was established. It was shown that for the conversion of alluvial forest, the line had been drawn. So also with species. No one should be placed in a position in which a decision has to be made whether or not a species is to be allowed to live naturally into perpetuity. Breeding programs are no guarantee for any rare species and can only be justified as an adjunct to some other form of management in the natural habitat. Enough species have become extinct, or are endangered, such that the line for premeditated extinction must be considered to have been drawn. The underlying pressure on forests and other natural resources is, of course, the growth of human populations that have outgrown economic opportunities. Unless this continues to be tackled with every ounce of political and popular will, then sustainable development is just a dream.

Appendices

Appendix A - Interpretation of Soil Analysis Data.[1]

	Very low	Low	Moderate	High	Very high
C (%)	<1.00	1.00-2.00	2.01-3.00	3.01-5.00	<5.00
N (%)	<0.10	0.10-0.20	0.21-0.50	0.51-0.75	<0.75
C/N	<5	5-10	11-15	16-25	>25
P_2O_5 (mg/100 g)[2]	<15	15-20	21-40	41-60	>60
P_2O_5 (ppm P)[3]	<4	5-7	8-10	11-15	> 16
P_2O_5 (ppm P)[4]	<5	5-7	11-15	16-20	>20
K_2O (mg/100 g)[5]	<10	10-20	21-40	41-60	>60
CEC (meq/100 g)	<5	5-16	17-24	25-40	>40
Ca (meq/100 g)	<2	2-5	6-10	11-20	>20
Mg (meq/100 g)	<0.3	0.4-1.0	1.1-2.0	2.1-8.0	>8.0
K (meq/100 g)	<0.1	0.1-0.3	0.4-0.5	0.6-1.0	>1.0
Na (meq/100 g)	<0.1	0.1-0.3	0.4-0.7	0.8-1.0	>1.0
Base saturation (%)	<20	20-40	41-60	61-80	80-100
Mineral content (%)	<5	5-10	11-20	20-40	>40

	Very acid	Acid	Rather acid	Neutral	Rather alkaline	Alkaline
pH (H_2O)	<4.5	4.5-5.5	5.6-6.5	6.6-7.5	7.6-8.5	8.5

Appendix B. Plant family abbreviations used in the text.

Acer.	Aceraceae	Gnet.	Gnetaceae
Acti.	Actinidiaceae	Good.	Goodeniaceae
Amar.	Amaranthaceae	Grac.	Gracilariaceae*
Amar.	Amaryllidacaeae*	Gram.	Gramineae
Anac.	Anacardiaceae	Gutt.	Guttiferae*
Anno.	Annonaceae*	Halo.	Haloragaceae
Apon.	Aponogetonaceae	Hydr.	Hydrocharitaceae
Arac.	Araceae*	Hype.	Hypericaceae
Aral.	Araliaceae	Isoe.	Isoetaceae
Arau.	Araucariaceae*	Jugl.	Juglandaceae
Aris.	Aristolochiaceae	Junc.	Juncaceae
Ascl.	Asclepiadaceae	Labi.	Labiatae
Azol.	Azollaceae	Lecy.	Lecythidaceae*
Bign.	Bignoniaceae	Legu.	Leguminosae*
Bomb.	Bombacaceae	Lemn.	Lemnaceae
Burs.	Burseraceae	Lent.	Lentibulariaceae
Cact.	Cactaceae*	Lili.	Liliaceae
Cann.	Cannaceae*	Loga.	Loganiaceae
Capp.	Capparaceae	Lyth.	Lythraceae*
Capr.	Caprifoliaceae	Magn.	Magnoliaceae*
Casu.	Casuarinaceae*	Malv.	Malvaceae
Caul.	Caulerpaceae*	Moni.	Monimiaceae*
Cela.	Celastraceae	Mora.	Moraceae*
Cent.	Centrolepidaceae	Myri.	Myristicaceae*
Cera.	Ceratophyllaceae	Myrs.	Myrsinaceae
Char.	Characeae*	Myrt.	Myrtaceae*
Codi.	Codiaceae*	Naja.	Najadaceae
Comb.	Combretaceae	Nelu.	Nelumbonaceae*
Comp.	Compositae*	Nepe.	Nepenthaceae
Conv.	Convolvulaceae	Nyct.	Nyctaginaceae
Corn.	Cornaceae	Nymp.	Nymphaceae*
Cras.	Crassulaceae	Olac.	Olacaceae
Cyat.	Cyathaceae	Onag.	Onagraceae
Cype.	Cyperaceae	Palm.	Palmae*
Dati.	Datiscaceae	Pand.	Pandanaceae*
Dict.	Dictyotaceae*	Pass.	Passifloraceae
Dill.	Dilleniaceae	Phyl.	Phyllocladaceae*
Dios.	Dioscoreaceae	Pipe.	Piperaceae*
Dips.	Dipsacaceae	Pitt.	Pittosporaceae
Dipt.	Dipterocarpaceae	Podo.	Podocarpaceae*
Dros.	Droseraceae	Polp.	Polypodiaceae*
Eben.	Ebenaceae*	Poly.	Polygonaceae*
Elae.	Elaeocarpaceae*	Pont.	Pontederiaceae
Epac.	Epacridaceae	Prot.	Proteaceae
Eric.	Ericaceae	Rham.	Rhamnaceae
Erio.	Eriocaulaceae*	Rhiz.	Rhizophoraceae
Euph.	Euphorbiaceae*	Rosa.	Rosaceae*
Faga.	Fagaceae	Rubi.	Rubiaceae*
Flac.	Flacourtiaceae	Ruta.	Rutaceae*
Fuca.	Fucaceae*	Salv.	Salviniaceae
Gent.	Gentianaceae*	Sant.	Santalaceae*
Gesn.	Gesneriaceae*	Sapi.	Sapindaceae*
Glei	Gleicheniaceae	Sapo.	Sapotaceae*

* Families not revised in *Flora Malesiana I* up to Volume 10 part 2, or *Flora Malesiana II* up to Volume 1 part 3.

Appendix B. Plant family abbreviations used in the text.

Sarg.	Sargassaceae*
Scro.	Scrophulariaceae*
Sima.	Simaroubaceae
Sonn.	Sonneratiaceae
Ster.	Sterculiaceae*
Thym.	Thymelaeaceae
Umbe.	Umbelliferae
Urti.	Urticaceae
Verb.	Verbenaceae*
Viol.	Violaceae
Xyri.	Xyridaceae

NOTE - *The following keys are for use on Sulawesi only. If they are used elsewhere, incorrect identifications are likely to be made.*

Appendix C. Key to the trees of mangrove and estuarine areas.[1]

1. Bole with prominent stilt roots; leaf undersurface usually with black dots . .2
   Bole without stilt roots (or inconspicuous if present); leaf undersurface
   without black dots 4

2. Tree of sandy shores and coral terraces . . . .*Rhizophora stylosa*
   Tree of muddy shores (with perhaps a little sand) 3

3. Bark with deep horizontal cracks, grey to almost black; inflorescence with
   2-16 flowers arising from the leaf axils; stipules and young buds green . . .
   . . . .*R. mucronata*
   Bark with shallow vertical fissures, greyish; infloresence with two flowers
   arising from below leaves; stipules and young buds (and midrib in mature
   leaves) frequently reddish . . . .*R. apiculata*

4. Bole fissured 5
   Bole smooth, scaly or dimpled 12

5. Trees with breathing roots projecting from the ground 6
   Trees usually without breathing roots (although present in *Lumnitzera
   littorea*) 10

6. Breathing roots conical, spike-like, stem often crooked, lenticillate; flowers
   large with numerous stamens 7
   Breathing roots knee-shaped, stem not crooked, prominently lenticellate;
   flowers smaller with few stamens 9

7. Twigs hanging down, leaves almost without petioles, base of midrib red, along
   banks of tidal rivers . . . .*Sonneratia caseolaris*
   Twigs stiff, leaves with distinct petioles, base of midrib green 8

8. Leaf blade broadly ovate with rounded base, surface corrugated; uncommon
   small tree . . . .*S. ovata*
   Leaf blade obovate, base narrowed, surface not corrugated; pioneer species on
   new mud in sheltered places . . . .*S. alba*

9. Stipules with white resinous sap inside; calyx tube slightly ribbed at top; leaves
   15-22 cm x 5-7 cm; germinating radicle up to 15 cm long . . . .
   . . . .*Bruguiera gymnorhiza*
   Stipules without white resinous sap inside; calyx tube ribbed to base; leaves
   11-16 cm x 4-6 cm; germinating radicle up to 7 cm long . . . . . .*B. sexangula*

10. Undersurface of leaf copper or silver coloured . . . .*Heritiera littoralis*
    Not as above 11

11. Flowers red; large tree with breathing roots . . . .*Lumnitzera littorea*
    Flowers white; shrub or small tree without breathing roots . . . . . .*L. racemosa*

12. Trees with breathing roots 13
    Trees without breathing roots 20

13. Buttresses present, bole lenticillate, breathing roots knee-shaped 14
Buttresses absent, bole without lenticels, breathing roots spike- or pencil-like
....18

14. Bark brownish; leaves obovate 15
Bark greyish; leaves elliptic 16

15. Trees up to 9 m tall; radicle up to 30 cm long, warty throughout .*Ceriops tagal*
Shrub or small tree up to 5 m tall; radicle up to 15 cm, warty at apex only ..
....*C. decandra*

16. Flowers solitary ....*Bruguiera sexangula*
Flowers 2-5 on a stalk 17

17. Lenticels small, inconspicuous, blade yellowish-green, usually on silty mud
    B. parviflora
Lenticels large and conspicuous; buttresses inconspicuous ......*B. cylindrica*

18. Bark scaly ....*Avicennia marina*
Bark smooth to cracking 19

19. Leaves elliptic, apex pointed, white beneath; pioneer species especially within
    rivers ....*A. alba*
Leaves ovate, apex rounded, grayish-green beneath; along banks of tidal rivers
....*A. officinalis*

20. Cut bark producing white sap; leaves with slightly notched margins, yellow-
    green withering scarlet ....*Excoecaria agallocha*
Cut bark without sap, simple leaves 21

21. Leaf margin faintly toothed, finely hairy below ....*Hibiscus tiliaceus*
Leaf margin entire, glabrous below ....*Thespesia populnea*

Appendix D. Key to seagrasses.[1]

1. Leaf blade emerging from the petiole through a thin 'tongue' or leaf sheath
   2
   Leaf blade not emerging from petiole through a leaf sheath 8

2. Rhizome with single main growth point, herbaceous, with short lateral shoots.
   Leaf sheath persisting longer than blade 3
   Rhizome with more than one main growth point, with elongate and erect
   shoots. Leaf blade shed with sheath . . . . *Thalassodendron ciliatum*

3. Leaves flat 4
   Leaves not flat . . . . *Syringodium isoetifolium*

4. Veins 3 5
   Veins 7-17 7

5. Leaf tip rounded, leaves 0.5-1.2 mm wide . . . . *Halodule pinifolia*
   Leaf tip not rounded 6

6. Leaves 0.5-1 mm wide, 5-6 cm long . . . . *H. ciliata*
   Leaves 0.5-1.0 mm wide, 6-15 cm long, midrib conspicuous . . . . . *H. uninervis*

7. Scar of leaf blade continuous giving stem annular appearance, veins 9-15, leaf
   blade 2-4 mm, leaf tip more or less smooth . . . . *Cymodocea rotundata*
   Scar of leaf blade not continuous, veins 13-17, leaf blade 4-9 mm, leaf tip
   toothed (fig. A.1) . . . . *C. serrulata*

8. Plants coarse, leaves in tufts on firm rootstocks, ribbon-like, no differentiation
   between petiole and blade 9
   Plants delicate, leaves opposite in spaced pairs on thin creeping stems, ovate to
   lanceolate 10

9. Leaves 30-150 cm by 13-17 mm; rootstock covered with persistent stiff hairs .
   . . . . *Enhalus acoroides*
   Leaves 10-30 cm by 4-10 mm; rootstock without such hairs . . . .
   . . . . *Thalassia hemprichii*

10. Up to 5 pairs of petioled leaves at the apex of lateral shoot 11
    Lateral shoots with 10-20 pairs of sessile leaves . . . . *Halophila spinulosa*

11. Short lateral shoots (<1 cm) with only one pair of leaves, midrib joining intra-
    marginal vein, veins present. 12
    Longer lateral shoots (1-2 cm) with 6-10 leaf blades at top; petiole with broad
    sheath, midribs crossing intramarginal vein reaching apical margin; veins
    absent . . . . *H. beccari*

12. Lateral shoots hardly developed, glabrous; leaf margin entire 13
    Lateral shoots 0.5-10 mm, hairy; serrulate leaf margin . . . . *H. decipiens*

13. Leaf blade 10-40 mm long with 12-25 pairs of veins at angle of 45°-60°. . . . . .
    . . . . *H. ovalis*
    Leaf blade 7-14 mm long with 3-11 pairs of veins at angle of 7°-90° . . . .
    . . . . *H. ovata*

Figure A.1. Leaf tips of *Cymodocea rofundata* (left) and *C. serrulata* (right).

Appendix E. Keys to submerged and floating freshwater macrophytes.

Key to the families of submerged macrophytes including those with leaves floating flat on the water surface.[1]

1. Internodes of main stem unicellular or with unicellular medulla surrounded by sheath of cortical cells, 'fruit' with spirally-ridged wall . . . Characeae (algae)
   Internodes of main stem distinctly multicellular; fruit without spiral ridges. 2

2. No recognizable leaves, sterile specimen comprising only rhizoids and bladder-like traps . . . . Lentibulariaceae
   Not as above 3

3. Leaf tip round in cross-section, flattened above the base, leaf bases swollen, often bearing spores . . . . Isoetaceae (fern)
   Not as above 4

4. Leaves about 2 mm long, resembling a moss or liverwort, attached to rocks in shallow, fast-flowing rivers . . . . Podostemaceae (*Cladopus nymani*)
   Not as above 5

5. No obvious petiole 5
   Obvious petiole 9

6. Whorls of narrow, toothed and forked leaves 7
   Long, linear flat or wavy-edged leaves. . . . . Hydrocharitaceae (Key C)
   . . . . (*Vallisneria, Blyxa, Ottelia*)

7. Leaves with auricled sheath, 1-nerved, toothed . . . . Najadaceae[2]
   Not as above 8

8. Leaves toothed along one side, 2-4 times forked, rootless . . . .
   . . . . Ceratophyllaceae (Key D)
   Leaves relatively broad in whorls of 3-8 leaves . . . . Hydrocharitaceae (Key C)
   . . . . (*Hydrilla*)

9. Large (generally over 12 cm), round leaves either peltate (the petiole attached to the centre of the leaf like an umberella) or cordate 10
   Smaller, round cordate or lanceolate (fig. A.2) leaves 12

10. Peltate leaves, sometimes held above the water, lotus flower . . . .
    . . . . Nelumbonaceae
    . . . . (*Nelumbium nelumbo*)
    Cordate leaves, never held above the water unless water level abnormally low
    11

11. White flowers with 5 petals 1.5 cm long . . . . Gentianaceae
    . . . . (*Nymphoides indica*)
    Pink, purple or white flowers with 8-30 petals over 5 cm long . . . .
    . . . . Nymphaceae (Key E)

12. Round, oval, or arrow-shaped leaves with cordate leaf base 13
    Lanceolate leaves 14

Figure A.2. Examples of peltate, cordate and lanceolate leaves.

13. Base of petiole not bearing a sheath or stipules, leaf base deeply cordate . . .
. . . . Alismataceae (Key F)
Sheath or stipules at base of petiole, leaf base not deeply cordate . . . .
. . . . Hydrocharitaceae (Key C)

14. Variable leaf size up to 25 cm by 6 cm, petiole 7-35 cm long, main nerves 7-9
. . . . Aponogetonaceae
. . . . (*Aponogeton lakhonensis*)
Leaves less variable, about 15 by 3 cm . . . . Potamogetonaceae

Key to the families of floating macrophytes[3]

1. Plants tiny, not differentiated into a stem with leaves nor a main axis with
branches. . . . . Lemnaceae (Key G)
Not as above 2

2. No recognizable leaves, sterile specimens consisting of only rhizoids and
bladder-shaped traps . . . . Lentibulariaceae
Not as above 3

3. Leaves less than 5 cm long, or minute, bright green or reddish, pinnately
arranged along an elongate horizontal stem 4
Not as above 5

4. Leaves simple, 1-2 cm long, bright green, arranged in opposite pairs along the
stem . . . . Salviniaceae
. . . . (*Salvinia molesta*) (fern)
Leaves 2-lobed, 1-2 mm long, red-tinged, overlapping and scale-like . . . .
. . . . Azollaceae (fern)
. . . . (*Azolla pinnata*) (fern)

5. Cabbage-like, leaves 5-15 cm long, gray-green, radial on compact erect stem.
. . . . Araceae
. . . . (*Pistia stratiotes*)
Not as above 6

6. Leaf blade greater than 7 cm long and wide, petiole swollen and spongy, some-
times rooted in mud, lilac-blue flowers . . . . Pontederiaceae
. . . . (*Eichhornia crassipes*)
Leaf blade generally less than 7 cm long, stem hollow or very spongy, stem
rooting at internodes, funnel-shaped pink, pale-lilac or white flowers. . . . . .
. . . . Convolvulaceae
. . . . (*Ipomoea aquatica*)

Key to the Hydrocharitaceae known or to be expected from Sulawesi.[4]
(# = not yet recorded from Sulawesi)

1. Leaves radical or in rosettes connected by stolons 2
Leaves on a distinct stem either in whorls or spirals 7

2. Leaves linear, margins parallel, flat not crisped. 3
Leaves lanceolate, crisped if linear 6

3. Leaves 15-100 cm long, 1-2 cm wide . . . . *Vallisneria gigantea* #
Leaves narrower 4

4. Leaves 10-150 cm long, .5-1 cm wide, 5-9 nerves. . . . . *Blyxa auberti* #
Leaves 10-120 cm long, .5-1 cm wide, 5-7 nerves. . . . . *B. echinosperma* #

5. Stolons absent, spathe with 6 ribs or 2-10 longitudinal wings, leaves submerged
or partly floating 6
Stoloniferous, spathe not ribbed or winged, leaves floating. . . . .
. . . . *Hydrocharis dubia*[6]

6. Leaves broad-ovate with cordate or truncate base, petiole triangular in cross-
section. . . . . *Ottelia alismoides*
Leaves linear, strongly crisped . . . . *O. mesenterium*

7. Leaves in whorls of 3-8 . . . . *Hydrilla verticillata*
Leaves spirally arranged . . . . *Blyxa japonica*

Key to the Ceratophyllaceae known from Sulawesi.[5]

1. Leaves 1-4 cm long in whorls of 7-10, dark green, plant grows to 3 m. . . . . . .
. . . . *Ceratophyllum demersum*
Leaves 1.75-2 cm in whorls of 6-8, bright green, plant grows to 1 m . . . .
. . . . *C. submersum*

Key to the Nymphaeaceae known from Sulawesi.[6]

1. Leaves glabrous, flowers open in the day time, leaves 10-23 cm by 8-18 cm,
green or light purple beneath, flower light violet or blue, sometimes white
. . . . *Nymphaea stellata*
Leaves hairy beneath, flowers open at nighttime 2

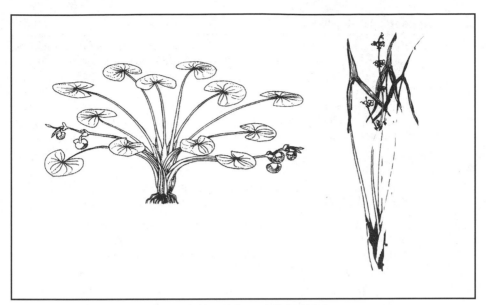

Figure A.3. *Sagittaria guayanensis* (left) and *S. sagittifoila* (right).

2. Leaf margins entire or slightly toothed, leaves 15-50 cm by 12-45 cm, dark purple beneath, flower white, sometimes pink or red, often cultivated . . . .
. . . . *N. nouchali*
Leaf margins distinctly sinuate, toothed, often cultivated . . . . *N. lotus*

Key to the species of Alismataceae known from Sulawesi.[7]

1. Leaves floating, rounded leaves with deeply cordate base, 1.5-6 cm long, 2.5-10 cm wide. Found in ditches and wet rice fields . . . . . *Sagittaria guayanensis*
Leaves erect, emergent, like narrow arrows (fig. A.3), 4-9.5 cm long (along midrib). Found in swamps and wet rice fields . . . . *S. sagittifolia*

Key to the Lemnaceae known or to be expected from Sulawesi.[8]

(# = not yet recorded from Sulawesi)

1. At least one root present 2
Roots absent, leaf less than 1 mm in any direction . . . . *Woffia globosa* #

2. One root (rarely none), fronds without ventral or dorsal scales, 1-3 often indistinct nerves. . . . . *Lemna perpusilla* #
One to many roots, fronds with ventral and dorsal scales, 3-15 nerves . . . . . .
. . . . *Spirodela polyrhiza*

*van Steenis, C.G.G.J. 1949a, b, 1981; Backer, C.A. 1951; van Ooststroom, S.J. 1953; den Hartog, C.
1957a, b; de Wilde, W.J.J.O. 1962; van Bruggen, H.W.E. 1971; van der Plas, F. 1971; Taylor, P. 1977;
Leach, G.J. and Osborne, P.L. 1985; Pancho, J.V., Tjitrosoepomo, G. and Megia, R. 1985*

Appendix F. Key to the tree ferns.[1]

1. Plant with scales around stem apex, and on the base of the petiole and on the veins and/or the leaflet midribs on the underside (abaxial surface) of the leaf; spore-producing organs (sorus) situated well away from the margin . . . . (*Cyathea*) 2

   Plant lacking scales as above but with flexible or rigid hairs at stem growing point and especially on the petiole and usually on the main vein and its branches (rachis and costae), at least beneath; sorus near the margin . . 13

2. Leaves once pinnate, pinnae (leaflets) entire or with shallow rounded teeth near apex . . . . *C. moluccana*

   Leaves more compound, pinnae (leaflets) at least pinnate again and usually deeply lobed pinnules 3

3. Undersurface of leaves covered in a web of finely fringed interlacing pale scales . . . . *C. celebica*

   Undersurface of leaves without such a covering although midribs and veins may be hairy or scaley and leaf tissue may be bluish (glaucous) 4

4. Scales at base of petiole and at trunk apex pale straw coloured 5

   Scales at base of petiole and trunk apex red-, mid- or gray-brown. 7

5. Petiole purplish with glaucous bloom and substantial spines; most petiole scales up to 2 mm wide; underside of costules (leaflet midribs) with long whitish hairs and very few bullate (balloon-like) scales . . . . . *C. contaminans*

   Petiole not as above, if spiny then scales rarely more than 1 mm wide; underside of leaflet midrib (costule) lacking long whitish hairs but with pale bullate scales 6

6. Sporangia lacking a membraneous covering (indusium) but containing branched filaments (paraphyses) amongst the sporangia; pinnules 80-120 mm long . . . . *C. sangirensis*

   Sporangia completely covered, at least when young, by a membraneous indusium which breaks at maturity and persists as a cup or flaps around the sporangia; paraphyses lacking; pinnules up to 50 mm long only . . . . *C. strigosa*

7. Sporangia completely covered (when young) by a membraneous indusium which usually breaks, the lower part persisting and conspicuous 8

   Indusium lacking even on young specimens (although in *C. amboinensis* the young sori may be covered but then the indusium is soon lost completely) 10

8. Petiole and lower part of rachis with dull gray-brown scales; sharply pointed hairs present on lower (abaxial) side of the midribs of leaf segments . . . . *C. oinops*

   Petiole scales dark or red-brown, usually shining; underside of leaves lacking hairs on the midribs, although often very scaley 9

9. Petiole densely covered with bright red-brown scales up to 2 mm wide; rachis scaley at least on the underside . . . . *C. inquinans*

Petiole covered at base only with deep brown narrower scales (up to 1 mm wide); rachis smooth and glabrous .... *C. oosora*

10. Petiole scales red-brown, stiff, less than 2 mm wide, with long setose teeth on margin; upper surface of pinnule midribs with pale stiff hairs 11
Petiole scales pale or dark mid-brown, at least some wider than 1 mm, with pale irregular margins with occasional short setose teeth; upper surface of pinnule midribs glabrous 12

11. Scales on costule dark brown, flat, with pale setae on margin; petiole and lower rachis with small conical warts .... *C. sarasinorum*
Scales on costule bullate, pale coloured, with dark setae on margin; petiole and lower rachis scabrid from persistent scale bases................ *C. elmeri*

12. Pinnule midrib densely scaley, especially in lower half; pale scales with long setose teeth seen amongst sporangia .... *C. tenggerensis*
Pinnule midrib with few isolated scales only; dark branched scales seen amongst sporangia .... *C. amboinensis*

13. Upper (adaxial) surface of petiole and rachis raised; rachis hairy, hairs of two sizes; stem a distinct erect trunk .... (*Dicksonia*) 14
Upper (adaxial) surface of petiole grooved; rachis not hairy; stem short, usually creeping 15

14. Hairs of undercoat of petiole rigid, their walls not collapsed; spores smooth
.... *Dicksonia blumei*
Hairs of undercoat of petiole flaccid, their walls collapsed when dry; spores verrucose .... *D. mollis*

15. Leaves bipinnate, lanceolate (lowest pinnae shorter than those above).....
.... *Cystodium sorbifolium*
At least the larger mature leaves quadripinnate, deltate (lowest pinnae longer than the next)...................................... *Culcita villosa*

Appendix G. Key to termities[1]

Key to families.

1. Head without fontanelle (shallow depression on the surface of the head) or frontal gland (gland which enters into the fontanelle); eyes present; mandibles with prominent marginal teeth. . . . . Kalotermitidae
   Head with fontanelle and frontal gland; eyes absent; mandibles with or without marginal teeth. 2

2. Pronotum (dorsal, shield-like section of the prothorax) flat . . . .
   . . . . Rhinotermitidae
   Pronotum shaped like a saddle . . . . Termitidae

Key to genera of Kalotermitidae.

1. Head brown to black, (plug-like), cylindrical, truncated anteriorly mandibles short and robust (fig. A.4a) . . . . *Cryptotermes*
   Head yellow to orange, not plug-like, mandibles long 2

2. Forehead steep, somewhat lobed; antennae with 11-12 articles (fig. A.4b) . .
   . . . . *Glyptotermes*
   Forehead not steep or lobed, antennae with 15-17 articles (fig. A.4c) *Neotermes*

Key to genera of Rhinotermitidae.

1. Mandibles without prominent teeth 2
   Mandibles with prominent teeth 3

2. Large opening of frontal gland on front of head with short tubular extension (from which a white sticky secretion exudes when handled) (fig. A.4d) . . . .
   . . . . *Coptotermes*
   Opening of frontal gland small, in normal, central position (fig. A.4e) . . . . .
   . . . . *Heterotermes*[2]

3. Mandibles very finely serrated or toothed at base; only one form of soldier (fig. A.4f) . . . . *Parrhinotermes*
   Mandibles without serrations or teeth at base; two forms of soldiers differing greatly in size (fig. A.4g and 4h) . . . . *Schedorhinotermes*

Key to genera of Termitidae.

1. Mandibles degenerate and non-functional; head drawn out into long 'nose' with frontal gland opening at tip . . . . (Nasutitermitinae) 2
   Mandibles well-developed and functional; head not shaped like a nose . . . 5

2. Head not constricted behind antennal sockets 3
   Head slightly to strongly constricted behind antennal sockets 4

3. Antennae with 13-14 long articles; apical projection of mandible with small tooth; head length (including 'nose') >2 mm (fig. A.4i) . . . . *Havilanditermes*
   Antennae with 12-14 short articles; apical projection of mandible without tooth; head length with 'nose' <2 mm (fig. A.4j) . . . . *Nasutitermes*

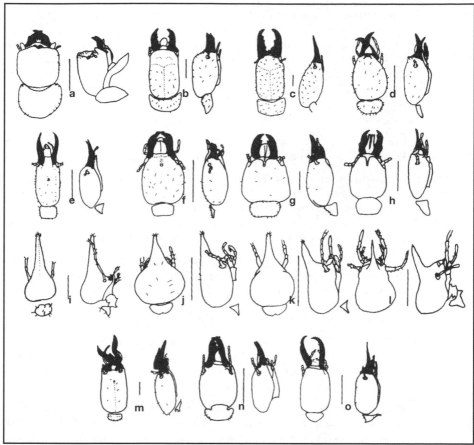

Figure A.4. Dorsal and side views of head of Sulawesi termite soldiers. a - *Cryptotermes*, b - *Glyptotermes*, c -*Neotermes*, d - *Coptotermes*, e - *Heterotermes*, f - *Parrhinotermes*, g - *Schedorhinotermes major* soldier, h - *Schedorhinotermes minor* soldier, i - *Havilanditermes*, j - *Nasutitermes*, k - *Bulbitermes*, l - *Hospitalitermes*, m - *Pericapritermes*, n - *Odontotermes*, o - *Microcerotermes*. Scale bars represent 1 mm.

4. Head not extending far beyond neck (fig. A.4k) . . . . *Bulbitermes*
   Head extending far beyond neck; head strongly constricted behind antennae;
   legs greatly elongated with hind femora as long as or longer than abdomen
   (fig. A.4l) . . . . *Hospitalitermes*

5. Mandibles asymmetrical and twisted (fig. A.4m). . . . . *Pericapritermes*
   Mandibles symmetrical, curved at tip 6

6. Postmentum broad, strongly arched centrally (fig. A.4n). . . . . *Odontotermes*
   Postmentum narrow, flat, not arched (fig. A.4o) . . . . *Microcerotermes*

Appendix H. Key to mudskippers![1]

1. Flattened snout with mouth close to ground; eye stands well above the top of the head; attachment of posterior dorsal fin to the back not much longer than attachment of anterior one; frequently seen on land; burrows into mud using mouth 2
Snout somewhat prominent; eye stands above the top of the head but not as much as above; warty skin; attachment of posterior dorsal fin to the back considerably longer than attachment of anterior one; seen commonly at water's edge; burrows into mud tail first 7

2. Length greater than 18 cm 3
Length less than 18 cm 4

3. Front margin of anterior dorsal fin conspicuously white with 3-4 spines, tail rounded; body dark brown, lighter below, blue spots on head and body, dark brown band from head to shoulder; about 27 cm (fig. A.5a) . . . .
. . . . *Periophthalmodon schlosseri*
Front margin of dorsal fin pale, 4-7 spines on anterior dorsal fin. . *P. barbarus*

4. Tail rounded, body light brown to blackish brown, pale below; dorsal fins brown at base, blackish in the middle with row of white spots and broad white margin; other fins also spotted white; burrow turreted, shorter than 12 cm (fig. A.5b) . . . . *Periophthalmus vulgaris*
Tail truncate or round; length greater than 12 cm 5

5. Tail truncate; body brownish to blackish brown, paler below, with irregular transverse bands; white margin on anterior dorsal fin narrower than on posterior dorsal fin; about 15 cm . . . . *Periophthalmus koelreuteri*
Tail round 6

6. Dorsal fins nearly continuous at the bases; anterior dorsal fin a little convex; first ray shorter than second; body bluish brown with dark blotches and shiny spots; dorsal fins yellowish orange; tail bluish brown above, yellow below; about 12 cm . . . . *Periophthalmus sobrinus*
Anterior dorsal fin relatively short and low; body greyish brown with oblique dark stripes (also on head and gill covers); about 6 cm . . . . *P. cantonensis*

7. Rounded tail, no long free spines in anterior dorsal fin, no barbels on lower lip, darkish green with 6-7 dark spots or oblong bands, head with blue or brown spots, first dorsal fin with blue spots, about 22 cm (fig. A.5c) . . . .
. . . . *Boleophthalmus boddarti*
Long, pointed, asymmetrical tail, upper half with dark bands, lower half white long, free spines in anterior dorsal fin, short barbels on lower lip, banded blue above, pale below, black spots on head, base of pectoral fin, back and dorsal fins, about 14 cm (fig. A.5d) . . . . *Scartelaos viridis*

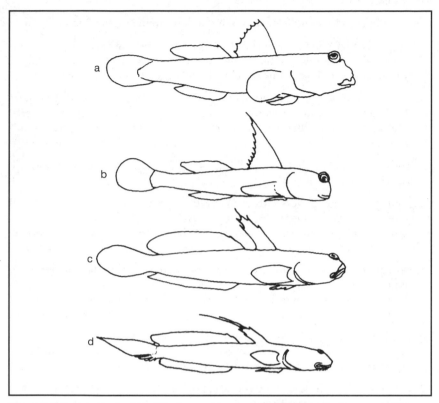

Figure A.5. a - *Periophthalmodon schlosseri*, b - *Periophthalmus vulgaris*, c - *Boleophthalmus boddarti*, d - *Scartelaos viridis*. Not to scale.

Appendix I. Key to toads and frogs.[1]

(Figures in parentheses indicate approximate snout-vent length of adults.)

1. Large parotoid gland present (fig. A.6); back rough with small swellings (tuberculate) belly coarsely granular; adults 30-110 mm (toads) 2
   No parotoid gland; back smooth to tuberculate, belly smooth or areolate . 3

2. Adults with a pair of longitudinal bony crests between eyes; a small crest or knob between eye and parotoid. . . . . *Bufo biporcatus* (80)
   Adults without crests between eyes; crest behind eye swollen and merging with parotoid . . . . *B. celebensis*[2] (110)

3. Toe webs basal or absent . . . . (microhylids) 4
   Toes half to fully webbed. 8

4. Adults 45-75 mm; fingers with truncate discs, toe tips blunt; inner metatarsal tuberculate and spade-like, outer metatarsal . . . . (*Kaloula*) 5
   Adults 15-40 mm; fingers and toes with oval discs; inner metatarsal tubercle flat, outer tubercle absent . . . . *Oreophryne*[3] 6

5. No pale dorsolateral stripes; web reaches middle subarticular tubercule on inner edge of fourth toe. . . . . *Kaloula baleata* (75)
   Broad pale dorsolateral stripes; web does not extend beyond basal tubercule of fourth toe . . . . *K. pulchra* (75)

6. Finger discs very small, disc of first finger not wider than digit . . . .
   . . . . *Oreophryne* sp. nov.[2]
   Finger discs two or three times as wide as digits 7

7. Belly dirty white or dusted with brown . . . . *O. celebensis*[2] (30)
   Belly 'uniform orange-yellow' . . . . *O. zimmeri* (?)[2]
   Belly in adult with glandular yellow patches . . . . *O. variabilis*[2] (30)

8. Tympanum hidden; lower jaw with a single median cusp; adults 20-45 mm . .
   . . . . (*Occidozyga*) 9
   Tympanum distinct; lower jaw with two or three cusps 10

9. Toes half webbed; fingers and toes with small discs . . . . . . *O. semipalmata* (?)[2]
   Toes fully webbed; fingers without discs, toes with small discs . . . .
   . . . . *O. laevis/celebensis* (?)[4]

10. Finger discs absent. 11
    Finger discs present, sometimes small 18

11. Toe tips blunt or pointed; back with irregular short ridges; adults 50-90 mm
    . . . . *Rana (Euphlyctis) cancrivora* (85)
    Toe tips with minute or small discs . . . . *R. (Limnonectes)*[5] 12

12. Blackish stream-frogs; dorsum with numerous short ridges and white warts; adults 60-120 mm . . . . *R.(L.) microtympanum* (100)[2]
    No white warts, adults smaller 13

13. Adults 40-50 mm, tibia length 0.59-0.71 of snout-vent length 14
    Adult tibia length 0.48-0.59 of snout-vent length 15

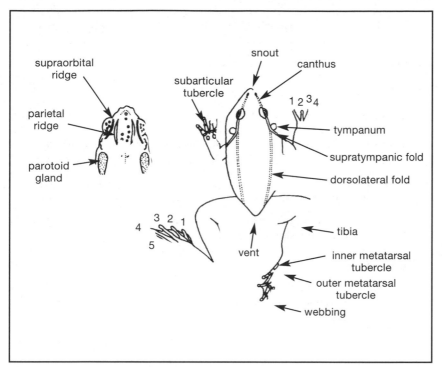

Figure A.6. Features of toads and frogs used in the key.

14. Rear thigh plain; tympanum small, would fit in space between dorsolateral and supratympanic ridges . . . . *R. (L.) arathooni* (?)[2]
Rear thigh dark mottled; tympanum larger, diameter about twice distance between dorsolateral and supratympanic ridges . . . . *R. (L.) sp. A*[2]

15. Back smooth between narrow unbroken dorsolateral ridges; rear thigh dark mottled; male with extensive dark markings on throat and chest; adults 30-40 mm . . . . *R. (L.) sp. C*[2]
Back weakly tuberculate, dorsolateral ridges broken 16

16. A sharp-edged black supratympanic marking and no canthal stripe (?); rear thigh plain; throat with few dark markings; adults about 37-40 mm . . . .
. . . . *R. (L.) sp. B*[2]
Supratympanic dark marking not as above; rear thigh dark mottled; throat strongly dark marked 17

17. Adults 24-32 mm; fourth toe broadly webbed to middle tuberculate only . . .
. . . . *R. (L.) cf. leytensis*
Adults 47-72 mm; fourth toe webbed to distal tuberculate or beyond . . . .
. . . . *R. (L.) modestal/heinrichi* (100)[2]

18. Dorsolateral glandular ridges present, sometimes indistinct; digital discs large or small .... *R. (Hylarana)* 19
    Dorsolateral ridges absent; all digits with large discs (rhacophorid tree frogs) 23

19. A broad yellowish stripe on dorsolateral ridge; discs small; adults 31-75 mm . ... *R. (H.) erythraea* (80)
    No pale dorsolateral stripes 20

20. First finger longer than second; outer finger discs less than twice finger width; adults about 40-70 mm 21
    First finger as long as second; outer finger discs two or three times width of digit 22

21. Dorsolateral ridges narrow .... *R. (H.) papua* group[6] (65)
    Dorsolateral ridges broad .... *R. (H.) celebensis*[2] (50)

22. Snout longer than upper eyelid; adult males 44-51 mm, females about 75-80 mm, male without distinct humeral gland .... *R. (H.) chalconota* (70)
    Snout shorter than upper eyelid; adult male about 30 mm long, with large gland on upper arm .... *R. (H.) macrops* (45)[2]

23. Fingers unwebbed; adults 37-75 mm .... *Polypedates leucomystax* (80)
    Fourth finger webbed to distal tubercule .... (*Rhacophorus*) 24

24. Rear edge of skull with bony prominences .... *R. georgi* (70)[2]
    No bony prominences on skull; adults shorter than 70 mm 25

25. Males about 27-28 mm, females about 32-35 mm; male with snout-tip truncate to rounded in profile .... *R. edentulus* (30)[2]
    Males about 34-40 mm, females about 42 mm; male with snout-tip sloping forward from nostril, projecting in profile .... *R. monticola* (40)[2]

Appendix J. Key to house geckos.[1]

1. Large, total length about 30 cm (17 cm head and body, 15 cm tail), grey above with red spots, tail with darker bands; young animals dark grey with white bands . . . . *Gekko gecko*
   Smaller 2

2. Total length about 10-12 cm 3
   Total length about 15-18 cm, tail cylindrical, slightly depressed, six longitudinal rows of long scales along tail; flaps on unwebbed toes undivided; rough skin particularly on adults, brown or grey above, spotted with darker tints, double series of dark spots along middle of bark, most obvious when young. . . . . . .
   . . . . *Gekko monarchus*

3. Tail flattened on the underside with finely-toothed edge. 4
   Tail rounded, slightly flattened, with six longitudinal rows of long scales about 2 mm apart, giving appearance of roughly-toothed edge; junction of tail and body smooth or slightly enlarged if tail has regrown greyish or pinkish brown above, either uniform or with dark markings; light-edged dark streak on side of head passing through eye, not necessarily extending to shoulder. Fingers unwebbed appearing relatively long. Call 'chi-chi-chi-chi' . . . .
   . . . . *Hemidactylus frenatus*

4. Flap of skin extending from armpit to groin, and on back of hind limb. Grey-brown above, marbled with darker grey, generally has a dark streak from the eye to shoulder. Fingers partly webbed appearing relatively short. Largely silent. . . . . *H. platurus*
   Short, broad fingers and toes, webbed at base, with flaps divided; Greyish or reddish-brown above; either uniform or variegated with darker and lighter brown spots, sometimes with longitudinal rows of white spots on back, no dark streak through eye. Call like a buzzing bee. . . . . *Gehyra mutilata*

Appendix K. Key to parrots.[1]

1. Smaller than 16 cm, feet orange 2
   Not as above 5

2. Bill black 3
   Bill not as above. . . . . *Loriculus exilis*

3. Undertail coverts green. . . . . *L. stigmatus*
   Undertail coverts red 4

4. Forehead green and red . . . . *L. catamene* (male)
   Forehead green only . . . . *L. catamene* (female) .

5. Some red on head 6
   No red on head 9

6. Bill white, grey, or white and grey. 7
   Bill not as above 8

7. Wings all green . . . . *Prioniturus flavicans* (male)
   Wings green and grey. . . . . *P. platurus* (male)

8. Wings green or green and brown. . . . . *Trichoglossus ornatus*
   Wings black and red . . . . *Eos histrio*

9. Feathers all white, or yellow and white. . . . . *Cacatua sulphurea*
   Feathers not white, or yellow and white 10

10. Bill white, grey, or white and grey. 11
    Bill not as above 13

11. Bill all white. . . . . *Tanygnathus surnatranus* (female)
    Bill grey or grey and white. 12

12. Some blue on top of head . . . . *Prioniturus flavicans* (female)
    No blue on top of head. . . . . *P. platurus* (female)

13. Some yellow on head. . . . . *Trichoglossus flavoviridis*
    No yellow on head 14

14. Head green and blue . . . . *Tanygnathus lucionensis*
    Head all green 15

15. Wings black, yellow and green . . . . *T. megalorhynchus*
    Wings all green . . . . *T. sumatranus* (male)

Appendix L. Key to bat families.[1]

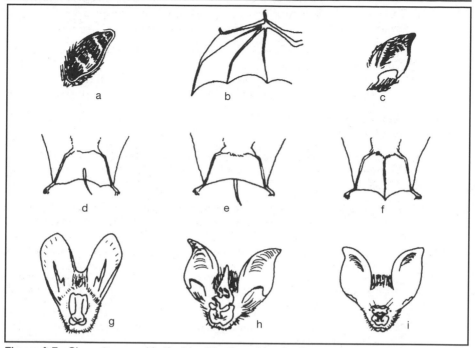

Figure A.7. Characters used in the identification of bat families.

1. Margin of ear forms a complete ring (fig. A.7a); thumb and first finger with claw (fig. A.7b) (except *Eonycteris*); relatively large, large eyes, nose normal (fruit bats) . . . . Pteropididae
   Margin of ear does not form a complete ring (fig. A.7c), only thumb with claw, relatively small, small eyes, nose sometime elaborate (insect-eating bats). . . . 2

2. Without elaborate nose 3
   Elaborate nose 5

3. Tail long and slender, completely enclosed in wide interfemoral membrane (fig. A.7d). . . . . Vespertilionidae
   Tail rather thick, projecting beyond or through interfemoral membrane . . 4

4. Tail relatively short, projecting through interfemoral membrane (fig. A.7e) .
   . . . . Emballonuridae
   Tail long and muscular, projecting beyond edge of interfemoral membrane (fig. A.7f) . . . . Molossidae

5. No tail, ears with tragus connected across top of head (fig. A.7g) . . . .
   . . . . Megadermatidae
   Ears not as above 6

6. Posterior noseleaf triangular with only one tip (fig. A.7h) . . . Rhinolophidae
   Posterior noseleaf rounded (fig. A.7i) . . . . Hipposideridae

Appendix M. Key to fruit bats.[1]

1. Upper side of muzzle with contrasting stripe or band of white fur 2
   Fur on upper side of muzzle not strongly contrasting with surrounding fur 3

2. Wing membranes from sides of back; first finger with claw; 2 lower incisors;
   forearm length 96-104 mm . . . . *Styloctenium wallacei*
   Wing membranes from near middle of back; first finger without claw; 4 lower
   incisors; forearm length about 100-110 mm . . . . *Neopteryx frosti*

3. Forearm length more than 105 mm 4
   Forearm length less than 105 mm. 10

4. First finger with claw; wing membranes from sides of back 5
   First finger without claw; wing membranes from spinal line, covering back fur
   completely 9

5. First upper molar with distinct antero-internal cusp; first and second lower
   molars with inner basal ledges (fig. A.8a); forearm length 130-141 mm . . . .
   . . . . *Acerodon celebensis*
   First upper molar without antero-internal cusp; first and second lower molars
   without inner basal ledges 6

6. Forearm length 160-175 mm . . . . *Pteropus alecto*
   Forearm length less than 155 mm 7

7. Ear relatively long and pointed, reaching back of eye when lain forward;
   forearm length 135-144 mm . . . . *Pteropus caniceps*
   Ear relatively short and not pointed, not reaching back of eye when lain forward; forearm length from 118 to 146 mm 8

8. Forearm length 128-146 mm . . . . *Pteropus hypomelanus*
   Forearm length 118-128 mm . . . . *P. griseus*

9. First lower molar with distinct antero-internal cusp or ledge (fig. A.8b) . . . .
   . . . . *Dobsonia crenulata*
   First lower molar without antero-internal cusp or ledge . . . . *D. exoleta*

10. Forearm length 38-42 mm; muzzle and tongue long and narrow . . . .
    . . . . *Macroglossus minimus*
    Forearm length 43 mm or more; if less than 60 mm, then muzzle and tongue
    short and broad. 11

11. External tail distinct. 12
    External tail rudimentary (a mere knob) or absent 19

12. First finger without claw. 13
    First finger with claw 14

13. Wing membranes from spinal line, covering back fur completely; forearm in
    single known specimen 78.5 mm; (at most) 2 upper and 2 lower incisors . .
    . . . . *Dobsonia species*
    Wing membranes from sides of back; forearm length 65-77 mm; 4 upper and 4
    lower incisors . . . . *Eonycteris spelaea*

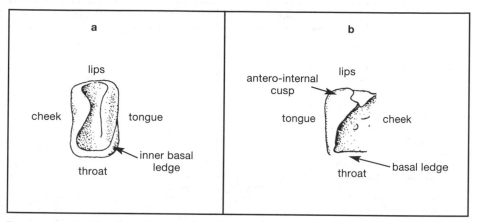

Figure A.8. Detail of the first lower molars of (a) *Acerodon celebensis* and (b) *Dobsonia crenulata*.
After W. Bergmans pers. comm.

14. Nostrils projecting as cylindrical tubes from upper surface of muzzle; ears and wing membranes spotted with yellow; no lower incisors 15
   Nostrils not projecting as cylindrical tubes; ears and wings wholly dark; 4 lower incisors 16

15. Forearm length 60-70 mm . . . . *Nyctimene cephalotes*
   Forearm length about 50 mm . . . . *N. minutus*

16. Muzzle short and broad; ear margin (and often fingers) marked with white; forearm length 59-69 mm; tail 6-11 mm; 4 upper and 5 lower cheek teeth .
   . . . . *Cynopterus brachyotis*
   Muzzle relatively long; ears and fingers wholly dark; forearm 72 mm or more; tail 12 mm or more; 5 upper and 6 lower cheek teeth 17

17. Forearm length 94-104 mm . . . . *Boneia bidens*
   Forearm length 85 mm or less 18

18. Tail membrane dorsally naked; forearm length 73-85 mm . . . .
   . . . . *Rousettus amplexicaudatus*
   Tail membrane dorsally densely furred; forearm length 72-78 mm. . . . .
   . . . . *R. celebensis*

19. Forearm length less than 50 mm; fur of breast and belly yellowish . . . .
   . . . . *Chironax melanocephalus*
   Forearm length 71 mm or more; fur of chest and belly very dark 20

20. Forearm length 71-82 mm; 4 upper and 4 lower incisors; cheek teeth relatively simple . . . . *Thoopterus nigrescens*
   Forearm length 83-93 mm; 2 upper and 2 lower incisors; cheek teeth with many cusps . . . . *Harpyionycteris celebensis*

Appendix N. Data sheet for coral reef monitoring.

	Circle 1	Circle 2	Circle 3
Locality			
Date			
Water depth			
Recorder			
General description of area			
FISH COUNTS (100 metre line)			
Predators			
Butterfly			

PERCENT COVER

Code: 0%=0   1%-5%=1   6%-30%= 2   31%-50%= 3
51%-75%= 4   76%-100%= 5

Sediment:			
mud			
sand			
rubble			
blocks			
Live hard coral			
Soft corals and sponges			
Dead standing coral			
Crustose corallines			
Marine plants			

Appendix N. Data sheet for coral reef monitoring. (Continued.)

FORMS PRESENT, DOMINANT AND SIZE (pres., dom., sz.)

Size/code: fist=1 forearm=2 armspan=3

	pres. dom. sz.	pres. dom. sz.	pres. dom. sz.
Hard corals:			
Branching			
Staghorn			
Massive			
Encrusting			
Tabulate/flat			
Erect foliose			
Cup-shaped			
Mushroom			
Soft corals and sponges:			
Massive			
Fans & whips			
Plants:			
Thick turf			
Long filaments			
Large browns			
Halimeda			
Other fleshy			
Sea grass			

COUNTS OF ANIMALS

Mushroom coral			
Giant clams			
Synaptid sea cucumbers			
Other sea cucumbers			
Acanthaster			
Other starfish			
Urchins			
Trochus			
Other (specify)			

VISIBLE POLLUTION

(Specify/count)			

OTHER NOTES

Notes

Chapter One: Physical, Biological and Human Background

1. 1 Ma = one megayear or 1 million years.

2. A geological term for the double arc of islands from Flores, through Alor and Wetar, to Banda (Inner Banda Arc), and Raijua, through Timor and the Tanimbar Islands, then north to the Kai Islands and west to Seram and Buru (Outer Banda Arc).

3. The forcing of one part of the earth's crust below another at the junction of two tectonic plates.

4. The forcing of one part of the earth's crust over another at the junction of two tectonic plates.

5. The Director-General of Geology and Mineral Resources flew over Una-una on 23 July on a reconnaissance flight and the climax of the eruption occurred only a few hours after the flight finished.

6. These ores are within the Bogani Nani Wartabone National Park and a decision has recently been taken not to exploit them (p. 640).

7. Care must be taken in these analyses that one is not measuring copper concentrations from commonly-used copper-based fungicides.

8. The Sunda Shelf is the shallow continental shelf on which Sumatra, Java, Bali and Borneo sit connected to mainland Asia. New Guinea and the Aru Islands sit on the Sahul Shelf connected to Australia.

9. A dry month is defined here as a month whose mean precipitation (mm) is less than twice the mean temperature (°C). Therefore a month with 65 mm of rain and a mean temperature of 30°C is nearly a dry month (65 is more than 30x2).

10. Species are the basic unit of the structured system of hierarchical biological classification that was developed in the eighteenth century by the Swede Carl von Linné, a name better known in its latinized form Linnaeus. Each organism is generally known by a name in two parts. For example, the flying fox is given the name *Pteropus vampyrus,* the first of which (always with a capital letter) being the 'genus' or generic name and the second the 'species' or specific epithet; both are written in italics or underlined. A genus can have any number of members (e.g., central flying fox *Pteropus alecto,* ashy-headed flying fox *Pteropus caniceps,* grey flying fox *Pteropus griseus,* etc.). In a list the genus of the second and subsequent species may be abbreviated to its initial letter (i.e., *P. caniceps*). One or more related genera (plural of genus) are grouped into a family (e.g., Pteropidae representing all the fruit bats). The family name ends in 'idae' for animals and usually in 'aceae' for plants and always starts with a capital letter. A group of related families are grouped in an order such as bats, related orders in a class such as mammals, and related classes in a phylum, such as the chordates, which includes all vertebrates. Finer divisions such as sub-families and sub-orders exist but these are used relatively rarely.

11. The assemblage of plants in a given area.

12. All plants mentioned in this book are followed by a four-letter abbreviation of the family except where this has been given in the same section or sub-section. A key to the family names is given in Appendix B.

13. A hemi-parasite is a facultative parasite which can survive in the absence of a host.

14. There is debate, as there is in all taxonomic fields, concerning the status and

names of certain animals. Thus the dwarf buffaloes are sometimes given the generic name *Anoa;* some people believe that a mountain subspecies of tarsier is a valid species *Tarsius pumilus,* and that the cuscus on Peleng Island is valid, too, as *Phalanger pelengenisis.* A list reflecting these minor differences of opinion is given elsewhere (Musser in press).

15. That is, found in Sulawesi and nowhere else.

16. The lowland anoa is relatively small, short-tailed and has smooth, round horns. The mountain anoa is relatively large, long-tailed, white-legged and has rough horns triangular in cross-section (Groves 1969).

17. An identification key is provided in Appendix I.

18. *Channa* is often referred to as *Ophicephalus* or *Ophiocephalus,* but *Channa* has priority under the rules of zoological nomenclature.

19. The arguments for species conservation are discussed on page 634.

20. A subspecies of this bird has also been recorded from the Nenusa Islands north of Talaud, but considering that almost the whole of these islands has been converted to coconut groves, the bird is probably extinct there.

21. The similarity of this word to the name of the lake might indicate the derivation of the lake's name.

22. A group of natural populations that differ constantly in their taxonomic characters from other groups of the same species, but which can interbreed successfully with those other groups.

23. Milkweed butterflies are conspicuous, medium-sized insects with bright colours separated by bold, black lines predominating in caterpillars and adults alike. The caterpillars feed on poisonous milkweeds (Ascl.) and absorb the toxins thereby making themselves unpalatable to predators. The adults retain these toxins. These butterflies are much harder to kill with a thorax pinch than are most butterflies, and many specimens, presumed dead, have 'come to life' unexpectedly.

24. The ecological role or profession of an organism in a community.

25. The floral regions of Malesia extend from a line between Kangar and Pattani in northern Peninsular Malaysia and southern Thailand respectively, across Indonesia and the Philippines, to the Bismark Islands east of New Guinea.

26. Some have maintained that mountain plants of the central Latimojong mountains have closer affinity with New Guinea than with Borneo (van Steenis 1937; van Steenis and Veldkamp 1984).

27. The Talaud Islands are inhabited by a rat *Melomys fulgens* known also from Seram. The Talaud Islands are the most westerly record of this genus which is most common in New Guinea.

28. B.P. = Before Present.

29. Taro, banana and sugarcane may have been developed earlier and independently on at least Sulawesi, and New Guinea (I. Glover pers. comm.).

30. For example, in the early 20th century many second-generation American immigrants, originally from southern Italy, were closer to the American norm in terms of measurable characteristics than to the Italian parent stock, even though they mainly married within their cultural group.

31. Chert is a hard, tough, dense and splintery rock comprising of silica, opal and quartz. It has similar mechanical properties to obsidian (pp. 74-75).

32. A midden of *Corbicula* mussel shells has been reported from Lobonto Island in Lake Lindu (Sarasin and Sarasin 1905).

33. Obsidian is a glossy, dark-coloured rock formed by the rapid cooling of lava flows. When hit, its flakes break off, leaving sharp edges. Obsidian was commonly used for prehistoric tools and such tools have been found in northern Sumatra, Central and West Java, as well as North Sulawesi.

34. Megaliths are found in several areas of Indonesia and other parts of Southeast Asia but there is no evidence that they were related or contemporary, that is, there was no single 'megalithic culture' (Glover 1979).

35. This drum can be seen in the village of Bontobangun, 3 km south of Benteng, the Salayar capital.

36. Interestingly, however, there is evidence from northern Australia that wooden-hafted, ground-edged axes were in use 23,000 years ago.

37. Sulawesi has ten species of *Dioscorea* yams, half of which are widespread in the tropics, and half of which are endemic (Burkill 1951).

38. An interesting paper on the manner and effects of this acceptance on the dynamics of Islamization in South Sulawesi is available (Pelras 1985).

39. It is of some interest that it has been remarked that tree names in the Onggak-Dumoga area of Bolaang Mongondow are quite similar to the Filipino names (Verhoef 1938).

40. The stone has been the site of five other major meetings. Some years after the first meeting it was necessary for different factions to discuss unity; in 1643 a meeting was held to determine insurrection over the Spaniards, in 1655 to determine how to respond to the threat posed by the kings of Bolaang Mongondow, in 1939 Dr. Sam Ratulangi gathered resistance forces together to plan attacks against the Dutch, and resistance leaders met there on 25th August 1945 to unite under the new national government.

41. A standard bibliography on the different groups is available (Kennedy 1974).

Chapter Two: Seashores

1. This is sometimes referred to as the littoral zone.

2. The earth and moon rotate around their common centre of gravity which, due to the much larger size of the earth, is within the mass of the earth. The rotation of the earth about this point gives rise to centrifugal force or acceleration which is the same at all points on the earth's surface. The force of the moon's gravitational pull, however, decreases in proportion to the square of its distance and so is greatest at the point of the earth closest to the moon.

3. Formerly know as *Littorina* (Reid 1986).

4. The range of this species never overlaps with that of its congener *L. littorea*.

5. The measurement of dry weight biomass of a forest requires access to drying ovens, a great deal of time, and patience. If one or more of these is not available it is possible to use allometric relationships (Ong et al. 1985). In these the above-ground dry weight (wa.g.) for *Bruguiera parviflora* trees is: wa.g. = 0.035 (girth at breast height)$^{2.424}$ and for *Rhizophora apiculata* trees is wa.g. = 0.0068 (girth at breast height)$^{2.624}$ Relationships for other species can be determined from first principles using a number of samples.

6. Perhaps the best known members of this genus are the vegetables kangkung *Ipomoea aquatica* (p. 269) and *ubi keladi* or sweet potato *I. batatas* (van Oostroom 1953).

7. Roots of *I. pes-caprae* in Tanjung Peropa Reserve were found to penetrate at least 30 cm into the sand (L. Clayton pers. comm.).

8. A plant growing on another for support rather than to gain water or nutrients.

9. The cultivated form with pale green leaves is known as 'Molucca Cabbage' which is incorrectly called *P. alba*. It belongs to the small family Nyctaginaceae, the best-known member of which is the garden plant *Bougainvillea* (Stemmerik 1964).

10. A useful guide to coastal snails is given by Roberts et al. (1982).

11. Some workers use 1.0 mm and 0.05 mm sieves.

12. In an Australian mangrove, molluscs accounted for 60% of the infauna biomass (Wells 1984).

13. A key to the mudskippers of Sulawesi coasts is provided in Appendix H.

14. The common scrub hen *Megapodius freycinet* is found in the Nicobar Islands northwest of Sumatra, Borneo and the Philippines.

15. A hen's egg is about 5 cm long and weighs about 50 g or about 3% of the adult body weight. The maleo egg weighs 16% of the adult weight, one of the highest ratios of egg-to-body weight known.

16. A curious name meaning 'white rock' since the beach is composed of black volcanic gravel.

17. That is, they may eat predators but are not eaten themselves.

18. A key to the portunid crabs (including *Scylla serrata*) of use to biologists working in Sulawesi is available (Moosa 1980).

19. Most hermit crabs found on beaches belong to the Paguridae.

20. A single specimen has also been reported from the shores of Tangkoko-Batuangus Reserve (Anon. 1980b).

21. Distance between the anterior and posterior borders of the thoracic groove.

22. Known in Indonesian as 'ikan mujair' after Mujair, a foreman of the colonial fisheries service, by whose efforts and in whose home pond this fish was first cultivated in Indonesia.

23. At the same time, however, the Directorate-General of Fisheries is encouraging the conversion of mangrove forests to ponds.

Chapter Three: Estuaries, Seagrass Meadows and Coral Reefs

1. An identification key to seagrasses is provided in Appendix D.

2. The same authors have suggested standardizing methods of seagrass analysis. These are: the collection of regular data to investigate seasonal and between-year variations, and of environmental data such as salinity, temperature, sediment type. At least six samples should be taken and each should be weighed before and after drying in an oven at 60°C for 12 hours. Epiphytic algae should be removed by gentle scraping, with the edge of a plastic ruler for example, and washing in seawater. When dry, the blades should be ground. The chemical procedures are given in their paper (Dawes and Lawrence 1983).

3. The other species are called manatees and these live in freshwater and estuarine habitats in southeast North America to northern South America, west Africa and the Amazon River.

4. Ruminant herbivores include deer, cows and goats which chew the cud and have a complex, chambered stomach.

5. Aggregations of 100 or more occur in undisturbed populations but it is unlikely whether such sights can still be seen around Sulawesi.

6. The use of light aircraft or helicopters are the most efficient means of surveying dugongs, but in the current economic climate it might be difficult to find sufficient funds.

7. Perhaps for this reason their eyes are adapted to low-intensity light (Anderson 1982).

8. Both animals died within three weeks of their capture.

9. These corals are of special interest because of their ability to shift their position.

10. Coral identification is not easy but a good field guide is available (Ditlev 1980).

11. Differences in the rate at which new species or genera are encountered over a fixed time can only be compared with results collected by a single observer. Identification, even to genus level, in the field requires considerable expertise. Such fixed-time search techniques can, however, provide an easily applied and convenient tool for coral reef survey work where quantification of community responses to factors causing disturbance are required (Harger 1984).

12. Excellent colour photographs of sea-cucumbers are available in two publications concerning New Caledonia (Anon. 1979; Guille et al. 1986), but since most of the species have wide distributions, most if not all of the species found around Sulawesi should be illustrated. The more recent publication also has hundreds of photographs of the other echinoderms.

13. A note of warning needs to be added about competitive exclusion and niche differentiation. Niche differentiation is sometimes very hard to demonstrate but it is impossible to prove its absence. If niche differentiation between two species cannot be found it can always be argued that the wrong parameters were investigated. The exclusion

principle cannot be disproved and so fails to meet the standard definition of a scientific hypothesis.

14. The next largest centre is Surabaya which in the same year exported 'only' 300 t.

15. A conventional fibreglass surveyors tape is quite suitable but it must be rinsed in fresh water and dried after use.

16. It is difficult to estimate size underwater because everything appears larger and closer than it actually is.

17. This is generally white but small plants can grow on it giving it a brown, black, pink or red tinge. This coral is in its natural position but has died recently. This should be recorded only if it is certain it is dead.

Chapter Four: Freshwater Ecosystems

1. The bottom of the lake is actually below sea level.

2. Literally 'small world' meaning a small community that can be taken as representative of a much larger ecological system.

3. A measure of the exchangeable cations that can be held by a soil expressed as milli-equivalents/100 g of soil at pH7.

4. A table for the interpretation of soil analyses is given in Appendix A.

5. The clear headwaters of the Jeneberang River examined by an EoS team had a total dissolved solid concentration of just 11mg/l.

6. Keys to the submerged, swimming and floating macrophytes of Sulawesi are given in Appendix E. A key to all aquatic plants, including emergents, of Java is provided elsewhere (Pancho et al. 1985) and this would be of some use on Sulawesi.

7. The pressure 10 m beneath the water is approximately twice that at the water surface.

8. *Chara* is of considerable use in biology teaching because the protoplasm can easily be seen streaming around inside its large cells under a low-power microscope. Village people find its main use to be as stuffing for pillows (L. Clayton pers. comm.).

9. Equivalent to 0.05 mm.

10. 'Fishes' is used in the conventional sense of meaning a number of fish species; 'fish' refers to the singular or plural of a single species of fish.

11. It should be noted that the taxonomy of molluscs is somewhat confused because of considerable variation within species in shell shape and colour. As a result some of these species, particularly among the Thiaridae and Corbiculidae, may be found to be invalid (E. Gittenberger pers. comm.; D. Dudgeon pers. comm.).

12. Adults trapped at Toraut were noticeable paler than, say, those in Australia. This may be because few Sulawesi species appear to be diurnal (A. Wells pers. comm.).

13. *Labiobarbus* is sometimes incorrectly referred to as *Dangila* (M. Kottelat pers. comm.).

14. *Channa* is often incorrectly referred to as *Ophicephalus* or *Ophiocephalus*.

15. Formerly known as *Panchax panchax* in the large family Cyprinodontidae. It is now the only representative of the Aplocheilidae in Southeast Asia.

16. A colour illustration is included with the original description (Boulenger 1897b).

17. Authors rarely if ever give details of sample depth which is clearly an important omission.

18. Saprophytes are organisms that derive their energy from dead organic matter and are the chief agents in the process of decay.

19. Recent work has shown that six to seven replicate samples of drift are required before confidence can be achieved in the quantification of drift and its comparison between sites (Allan and Russek 1985).

20. This fish is sometimes incorrectly referred to as *P. javanicus*.

21. One fisherman interviewed near Aopa Swamp by an EoS team had caught 18 large (>50 cm) eels that morning but had discarded them because they had no market value.

22. This is similar to the ornamental *Canna* hybrids. *C. edulis* is sometimes grown in gardens for its edible rhizome, and can

be recognized by the broad, brownish-purple border of its leaves and its small red flowers.

Chapter Five: Lowland Forests

1. Tropical forests are not the only ecosystems with large numbers of plant species; heathlands in South Africa have similar richness.

2. This site later used by the Royal Entomological Society of London's Project Wallace scientists is hereafter referred to simply as Toraut or the Toraut forest.

3. The equation was derived independently by Shannon and Wiener and is sometimes mistakenly called the Shannon-Weaver equation. $\mathrm{Log}_2 p_i$ can be calculated as $\mathrm{Log}_{10} p_i \div \mathrm{Log}_{10}^2$ or $\mathrm{Log}_{10} p_i \div 0.3010$.

4. In the west of Bogani Nani Wartabone National Park.

5. A fourth species *Gnetium gnemon* found on Sulawesi is a lowland forest tree whose fruit is crushed, dried and fried to make 'emping' crisps. This tree also has quite conspicuous hoops around the trunk.

6. The climbing stems of both species is used as cordage and in basket making (Heyne 1927; Burkill 1966).

7. A gap has been defined as a 'hole' in the forest extending through all levels down to an average height of 2 m above the ground. The sides of a gap are regarded as vertical, and the side at a particular point on the perimeter is located at the innermost point reached by foliage, at any level at that place. Only openings of at least 20 m^2 were considered to be gaps (Brokaw 1982a).

8. Primary succession is the sequential growth of plant species on newly formed substrates such as coastal sediments (p. 128) or volcanic ash and lava (p. 520).

9. Fine litter is leaves, thin twigs, flowers, fruits and the faeces of caterpillars and other invertebrate herbivores.

10. Coarse litter is the big twigs, branches and tree trunks.

11. Gross primary production is the total assimilation of organic matter by a plant or population of plants per unit time per unit area. Net primary production is gross primary production less that consumed by respiration processes.

12. The second category is included because, for birds that feed largely but not entirely on fruit, seeds or nectar, insects represent a protein-rich source of food essential for the energy-expansive tasks of feeding young and moulting.

13. Lipid-rich fruit, such as nutmegs, are oily to the touch and the flesh is generally drier than in carbohydrate-rich and lipid-poor fruit, such as figs. The families bearing lipid-rich fruit include Annonaceae, Burseraceae, Lauraceae, Meliaceae, Bombacaceae and Myristicaceae (Leighton 1982).

14. Snails are generalist herbivores frequently used in palatability experiments.

15. Sources of nectar, often at the base of the leaf blade, which serve to attract insects that prey on small herbivorous insects.

16. A seed has been likened to a child and its packed lunch inside a spacecraft the route and destination of which is never certain.

17. Seed predators eat seeds which are killed in the digestive tract.

18. This is now a relatively rare tree which has considerable economic value as a source of red wood used in making furniture.

19. Defined as those growing to at least 35 cm diameter at breast height.

20. *Garuga floribunda* is illustrated on page 484.

21. The persimmon fruit *Diospyros kaki* is also a member of this genus.

22. A detailed study of succession has also been conducted at a site 870-1,235 m above sea level in the south of Mindanao, the Philippine island closest to Sulawesi.

23. Interestingly, this plant is also found in southern Florida, U.S.A., but in the absence of suitable bats it is not extending its range.

24. A key to the genera of Sulawesi soldier termites is provided in Appendix G.

25. Pheromones are volatile chemicals produced by animals. The ingestion or smell of a pheromone produced by one individual of a particular species can determine or

influence the behaviour of another individual of the same species.

26. *Pitta erythrogaster* is widespread, whereas *P. sordida* is known only from the northern peninsula, and *P. moluccensis* is a rare vagrant.

27. See Appendix I for a key to the species of toads and frogs.

28. This tree is most common along the banks of certain rivers and it is probably significant that the fruits can float in water for long periods. The fruits are brown, roughly pear-shaped, 15-25cm x 7-12cm containing about 20 close-packed seeds. The fruit are a frequent food of babirusa although it is not known whether they eat just the soft pericarp or the entire fruit including seeds. This is of relevance because the seeds contain high concentrations of prussic acid or hydrocyanic acid. Boiling the seeds for an hour destroys the enzyme gynocardase which would interact with the glucoside gynocardine to produce hydrocyanic acid. The poison has been used to advantage to stupefy fish and as an antiseptic in preserving fresh fish during transport. The mature leaves are a common food in Minahasa where the shredded leaves are cooked with pig's blood and salt (Heyne 1927; Sleumer 1954; Burkill 1966).

29. A 'guild' is a group of species having similar ecological resource requirements and foraging strategies, and therefore having similar roles in the community (Lincoln et al. 1982).

30. The breeding season was found to be September to November in Tangkoko-Batuangus (Anon. 1980), but in Central Sulawesi was found to be synchonized within a population but not regular (Watling 1983).

31. When censusing hornbills it is important to remember that when they are breeding the adult females will not be seen because they are enclosed with the young in the nest hole.

32. Foraging is a term given to the activity of searching for food.

33. The generally smaller insectivorous bats are much harder to trap in conventional mist nets partly because they can detect it,

and partly because they fly slower and are therefore less likely to get entangled should they fly into a net.

34. Other sources of food are nectar, pollen and leaves, the pith of which, like fruit pulp, is spat out (Marshall 1985).

35. Stomachs of Asian and African fruit bats sometimes include small quantities of insect remains but these appear to be ingested accidentally whilst eating fruit (Thomas 1984; Marshall 1985).

36. The perception of objects using echoes of high frequency sounds emitted by some bats and swiftlets (p. 553).

37. In dry air the velocity is 330 m/s.

38. Weights were calculated from a weight/wing length curve produced from data obtained for the wood pigeon, feral pigeon, barbary dove, and diamond dove.

39. This is generally regarded as a species of lowland forest but one has been observed as 2,400 m near Lore Lindu National Park (K.D. Bishop pers. comm.).

40. This Park has three types of macaque; *Macaca nigra nigra* in the extreme east, *M. n. nigrescens* in the centre and *M. tonkeana* in the west.

41. The seriousness of soil compaction caused by bulldozers may be judged when it is considered that water was unable to infiltrate the soil of one of the main walking paths used by Project Wallace scientists in Bogani Nani Wartabone National Park, yet infiltration was immediate just 50 cm away from the path.

Chapter Six: Specialized Lowland Forest Types

1. After capture the deer are released.

2. Only a preliminary report was ever produced and since the untimely death of the principal investigator, Dr. Marius Jacobs, in 1983 the detailed manuscript that had been prepared has not been found (M. van Balgooy pers. comm.).

3. Alluvium is soil transported to flat areas by rivers.

4. Gleying is the process by which iron compounds are reduced to their ferrous

forms during inundated periods, and then partially reoxidized and precipitated during dry periods.

5. Called *Sarcocephalus* by Bloembergen (1940).

6. Pitcher plants are unusual among insectivorous plants in that there appear to be no conspicuous guidelines that show up when viewed under ultra-violet light, wavelengths that are visible to insects (Joel et al. 1985).

7. The Sopu valley *Eucalyptus deglupta* have recently been logged and exported by a forest concessionaire (N. Wirawan pers. comm.). Despite its size, *E. deglupta* is little used for its wood by village people. In Bolaang Mongondow, at least, its vernacular name refers to a fatal skin disease the appearance of which is similar to the flaking of bark of the tree (Steup 1933).

8. The average velocity between the ground and the canopy.

9. A key to toads and frogs is provided in Appendix I.

10. Alluvial soils are transported by water, colluvial soils by gravity.

11. A type of North American ground squirrel.

12. Immature soils with dark friable upper A horizon, calcareous lower A horizon over a light grey or yellow calcareous B horizon, typical of limestone areas.

13. It is said that if candlenut seeds are pounded with cotton and copra until the mix attains the consistency of stiff wax, this can be moulded around a bamboo splint to form a candle (Burkill 1966).

14. Apart from sugar and fermented juice, this palm provides very strong wood from its trunk, the leaflets stalks are made into brooms, the leaves make a temporary thatch, the black fibres at the leaf base are used as thatch, cordage, brooms, brushes, etc.

15. *Dracaena* and *Cordyline,* both large members of the lily family, are superficially similar with longish, straight leaves, leaf scars around the stem, and bifurcating stems like pandans which they resemble. *Dracaena* may

be distinguished by its orange or yellow roots which are often pleasantly scented.

16. Formerly *Papilio androcles.*

17. Sometimes referred to as *Gossampinus malabarica.* During the Hindu period in East Java, this tree used to be planted as boundary markers.

18. This is an introduced species that was cultivated near the Mediterranean four centuries B.C. There are no records of when it first arrived in Sulawesi.

19. This was probably introduced to Sulawesi centuries ago (p. 34) and its presence in many savannas is evidence of long-since failed plantations (Steup 1939).

20. The largest being these of its congener *C. umbraculifera* of southern India.

21. When crushed the red pigment cochineal adheres to the skin.

22. The insect which had had the most dramatic effect under Australian conditions was the pyralid moth *Cactoblastis cactorum* which is often quoted as one of the most impressive examples of biological control known.

23. In the last couple of years the *Leucaena farnesiana* trees of the valley have suffered considerably from insect pests.

Chapter Seven: Mountains

1. A table enabling limited interpretation of soil analysis results is provided in Appendix A.

2. Typically about 500-5,000 lux (Fletcher 1981).

3. Another scheme for Malesian mountains is 0-1,000 m Tropical Zone, 1,000-1,500 m Submountane Zone, 1,500-2,400 m Montane Zone, 2,400-4,000 m Subalpine Zone (van Steenis 1984).

4. A key to the tree ferns found, and expected to be found, in Sulawesi is provided in Appendix F.

5. This is the only species in Indonesia of the Taxaceae, a largely temperate tree family (de Laubenfels 1978).

6. The latter is the sole member in Indonesia of the Aceraceae (sycamores and

maples), a family common in the temperate regions.

7. A close relative *Macadamia ternifolia* is planted widely in Hawaii and northeast Australia for the nut which, because it contains 70% oil, is used in a growing market for confectionery. The nut of *M. hildebrandii* is probably as tasty as that of the commercial species and it might be worth examining the potential for bringing the Sulawesi species into cultivation because it fares much better in wet climates than *M. ternifolia* (Sleumer 1955).

8. Many of the pillars have been vandalized in the last few years by people who believe stories that valuable items are hidden within them.

9. *Gleichenia* and *Dicranopteris* are superficially similar ferns of the Gleicheniaceae. The former has simple or once-forked veins, the latter at least twice-forked veins. Both are found in open habitats, often on poor or leached soils, and occur on mountains. *Gleichenia* is almost exclusively found on mountains.

10. Acknowledgement must be made of the contribution of L. Coomans de Ruiter who was held by the Japanese in four civil prisoner camps on Sulawesi. The final one, in the Quarles Mts. was quite the most wretched and many people died. In the paper quoted here he states: "It was the worst ordeal in our three-and-a-half year internment–but it was also a very instructive period for the writer for the valley was inhabited by many mountain birds which are biologically almost unknown."

11. Small buckets buried in the ground such that the soil surface was level with the bucket top. They were baited with lumps of corned beef and shrimp paste.

12. This is a standard method of sampling earthworm populations. A solution of 2%-4% formalin (100% formalin = 40% formaldehyde) is generally used.

13. This secretive bird is known from less than 10 museum specimens.

14. *Zosterops chloris* is also found in lowland forest but it is more typically a bird of secondary forest and scrubby vegetation.

15. The pigeons and parrots of Sulawesi mountain and lowland forests are discussed further on pp. 428-429. A key to the parrots of Sulawesi is provided in Appendix K.

16. This rat was also caught at 3,200 m by the EoS team that climbed Mt. Rantemario. This was 1,000 m above its know altitudinal limit.

17. As mentioned earlier (p. 38) there is some doubt whether this is genuinely different from *B. depressicornis*.

18. A composite sample of dung was made from three dung heaps. The first 25 epidermal fragments passing through a 0.08 mm sieve were examined on a microscope slide. This was repeated four times, giving a total of 100 fragments.

19. This animal is sometimes referred to as the giant civet but it is in fact not unusually large (Wemmer et al. 1983).

20. Tracks of Sulawesi civet forefeet measure 47 mm x 43 mm (width x length) and are five-claw impressions whereas tracks of the introduced but also forest-living Malay Civit *Viverra tangalunga* are more oval, measure 40 mm x 47 mm and are four-claw impressions (Wemmer and Watling 1986).

21. In an ecological sense, fire may be likened to a generalist, but not entirely random, herbivore; that is, it shows little selectivity although certain plants are more susceptible than others.

Chapter Eight: Caves

1. Caves are often defined as underground chambers that can be entered by humans, but this would not include chambers or passages through which permanent cave inhabitants can pass. If chambers that can be entered by micro-invertebrates are included in the understanding then 'caves' extend to pore spaces in soil. For the purposes of this chapter caves are taken to mean both those chambers that can be entered by humans and those voids inaccessible to them but which contribute in a major way to the working of the cave ecosystem.

2. Himpunan Kegiatan Speleologi Indonesia (HIKESPI), P.O. Box 55, Bogor.

3. The very small cracks, passages and chambers which are found between the soil and the caves.

4. Hardness is a measure of the quantity of calcium, magnesium and iron compounds dissolved in water. The water flowing from a limestone area is said to be hard because of the large quantities of calcium and magnesium bicarbonates. When soap is added to such water an insoluble scum forms consisting of these salts and the fatty acids of the soap. The removal of the salts renders the water soft. Hardness can also be removed by boiling in which the soluble bicarbonates break down into insoluble carbonates, carbon dioxide and water. This carbonate is what forms the 'fur' inside kettles.

5. Association Pyrénéenne de Spéléologie, Toulouse, France.

6. Small birds of the swift family Apodidae.

7. The dark zone may be said to have been reached when a hand held in front of your face can no longer be discerned five minutes after artificial lights have been extinguished or switched off.

8. A key to bat families is provided in Appendix L.

9. Long-fingered bats *Miniopterus* can cruise at about 50 km/h.

10. In contrast, young of large fruit bats are born with hair and open eyes but it takes 9-12 weeks before they can fly and 15-20 weeks before they are weaned (Tuttle and Stevenson 1982).

11. The only species of nycteribiid fly not associated with bats was recently found in Bogani Nani Wartabone National Park during Project Wallace. This rather primitive species is new and was found inhabiting ant nests (Disney 1985).

12. Some tineid moths are known only from the fur or nests of mice, beavers and sloths, and from the nests of burrowing owls (Davis et al. 1986).

13. Tonasa I was closed in 1984 as a consequence of the recession. Tonasa II and III between them produce 1.1 million tons of cement each year which meets almost all the needs of eastern Indonesia.

Chapter Nine: Agroecosystems

1. The term 'biocide' is used in this chapter in preference to pesticide, insecticide, etc., since it serves as a reminder that these chemicals generally have some deleterious effect on all organisms.

2. This term is much used but its meaning is not consistent between authors. Some equate swidden agriculture with burned fields, others with forest farmers.

3. This means 'leaves that blow in the wind' referring to the way they generally flee from outsiders (Powell n.d.).

4. It is beyond the scope of this book to deal comprehensively with pest ecology and control. For details of the ecology of particular pest species readers should consult the agriculture literature and good general texts on pest ecology are available (e.g., Wratten and Watt 1984).

5. Unless indicated otherwise, the term 'rice' in the rest of the chapter refers to the crop grown in rain-fed and irrigated fields.

6. A list of rice field ostracods or seed shrimps from Sulawesi has been published (Victor and Fernando 1980).

7. Defined as '75%-100% of stems did not bear seed' although the utility of this definition is disputed.

8. A throaty 'kwok' (King et al. 1975).

9. This may have arisen from the frequent depiction of angels and Christ in white robes in Christian paintings.

10. It is an historical quirk that whereas oil palm, tea, sugarcane, coffee, cinchona and rubber are officially classified as plantation crops, and coconut, cotton, tobacco, nutmeg and clove are classified a industrial crops.

11. In recent years the clove price has sometimes exceeded Rp 15,000/kg, providing smallholders with extraordinarily high incomes.

12. Some insects lay eggs which hatch into juveniles looking superficially like very small adults. These juveniles grow and moult several times before becoming adults, and each of the juvenile stages is called an instar.

13. This species, identified by J. Holloway, has been named after the Coconut Research

Centre (**Ba**lai: Pene**lit**ian **K**elapa) in Manado, in recognition of its role in discovery of the pest (Hosang et al. 1986).

14. Prices on the international market have been falling over the last ten years but Indonesia desires to develop this crop for its own textile industry. At the moment however, less than 5% of the cotton used is home produced.

Chapter Ten: Urban Areas

1. Even a relatively small town such as Bitung, Minahasa, produces about 700 tons of collected waste each year (Palenewan 1983).

2. An exception to this is the effort being made by the Southeast Sulawesi Environmental Bureau to plant *Pericopsis mooniana* in Kendari.

3. This beautiful, red-flowered tree was discovered by a French botanist in Madagascar in 1824 but it is now unknown in the wild.

4. A guild is a group of species having similar roles in a community by virtue of similar ecological requirements and feeding strategies.

5. This bird list was compiled while the authors were interned in a Japanese prisoner-of-war camp (see page 677, chapter 7, note 10).

6. Keys to bat families and fruit bat species are given in Appendix L and M respectively.

7. The females of many mosquito species suck nectar rather than blood. No male mosquitoes suck blood but some may suck nectar.

8. A key to house geckos is given in Appendix J.

Chapter Eleven: Resources and the Future

1. It is possible to tread a middle path of course but even this extends beyond sustainable exploitation and for the sake of this illustration will not be considered.

2. United Nations Conference on Trade and Development.

3. This tree does best above 1,000 m and can be harvested on a 6-7 year rotation.

4. International Union for the Conservation of Nature and Natural Resources, Gland, Switzerland.

5. An additional reason is self-interest, but this only applies where individuals have control over significant areas of land. For example, early this century, the King of Bolaang Mongondow prohibited the felling of trees around the small Bunung Lake simply for the sake of the landscape which he himself enjoyed (Steup 1933).

6. This term was used in French field hospitals during World War II where severely limited medical supplies and workers had to be used to the best advantage. Patients who could probably wait and those who had little chance of survival were passed over in favour of those most likely to benefit.

Appendix A

1. Based on information from the Soil Research Centre, Bogor.

2. Bray 25%

3. Bray

4. Olsen

5. HCI 25%

Appendix C

1. Based on Wyatt-Smith, J. and Kochumen, K.M. (1979). Pocket check list of timber trees. 3rd ed. *Malay. For. Rec.* No. 17.

Appendix D

1. Based on Hartog, C. den (1957). Hydrochantaceae. *Flora Malesiana I* 5: 381-413; (1970). *Sea Grasses of the World*. North Holland, Amsterdam.

Appendix E

1. Adapted from van Bruggen (1971), den Hartog (1957a, b), Leach and Osborne (1985), van Steenis (1949a, b, 1981), de Wilde (1962).

2. The species of Najadaceae are more or less impossible to identify from sterile specimens and positive identification therefore must be based on reproductive parts; see de Wilde (1962).

3. Adapted from Backer (1951), Leach and Osborne (1985), Oostroom (1953), van der Plas (1971), Taylor (1977).

4. Adapted from den Hartog (1957b).

5. Adapted from van Steenis (1949b).

6. Adapted from Pancho et al. (1985).

7. Adapted from den Hartog (1957a).

8. Adapted from van der Plas (1971).

Appendix F

1. Provided by A.C. Jermy, British Museum (Natural History), Cromwell Road, London SW7 5BD.

Appendix G

1. Based on Collins, N.M. (1984). The termites of the Gunung Mulu National Park. *Sarawak Mus. J.* 30: 65-87; Kemner, N.A. (1934). Systematisches und biologisches Studien ueber die Termiten Javas und Celebes'. *Kungl. Svenska Vetensk. Handl.* 13: 1-241.

2. This genus has not yet been recorded from Sulawesi but it is quite likely to be present.

Appendix H

1. Supplied by A.J. Whitten using Koumans, F.P. (1953). Goboiea. X. In *The Fishes of the Indo-Australian Archipelago* by M. Weber and L.F. de Beaufort. Brill, Leiden, and Carcasson, R.H. (1977). *A Field Guide to the Coral Reef Fishes of the Indian and West Pacific Oceans.* Collins, London.

Appendix I

1. Supplied by Julian Dring.

2. Endemic species.

3. Taxonomically confused group, *zimmeri* not seen.

4. *Occidozyga* (or *Ooeidozyga*) celebensis is probably not distinct from *O. laevis.*

5. Taxonomically confused group, *heinrichi* not seen but seems to key out with *modesta;* records of additional species (*grunniens, kuhli* and *microtympanum*) probably based on misidentifications. The subgenus *Limnonectes* is not generally recognised.

6. Taxonomically confused group, Sulawesi species called *papua* but identification uncertain.

Appendix J

1. Supplied by A.J. Whitten based on specimens in the Zoology Museum Bogor,

and Rooij, N. de (1915). *Reptiles of the Indo-Australian Archipelago, Vol. 1.* Brill, Leiden.

Appendix K

1. Adapted from Nash, S.V. and Nash, A.D. (1984). *Kakatua, Pluri dan Kesturi (Psittaciformes) di Sulawesi dan Nusa Tenggara.* World Wildlife Fund/IUCN, Bogor.

Appendix L

1. Adapted from Lekagul, B. and McNeely, J.A. (1977). *The Mammals of Thailand.* Association for the Conservation of Wildlife, Bangkok.

Appendix M

1. Adapted from Bergmans, W. and Rozendaal, F.G. (in press). Notes on collections of fruit bats from Sulawesi and some off-lying islands (Mammalia, Megachiroptera). *Zool. Verh.*

Bibliography

Abdulhadi, R., Kartawinata, K. and Sukardjo, S. 1981. Effects of mechanized logging in the lowland dipterocarp forest at Lempake, East Kalimantan. *Malay. For.* 44: 407–418.

Abe, T. 1978. The role of termites in the breakdown of dead wood in the floor of Pasoh study area. *Malay. Nat. J.* 30: 391–404.

Abe, T. 1979. Food and feeding habits of termites in Pasoh Forest Reserve, Malaysia *Jap. J. Ecol.* 29: 121–136.

Abe, T. and Matsumoto, T. 1978. Distribution of termites in Pasoh Forest Reserve. *Malay. Nat. J.* 30: 325–335.

Abe, T. and Matsumoto, T. 1979. Studies on the distribution and ecological role of termites in a lowland forest in West Malesia. III. Distribution and abundance of termites at Pasoh Forest Reserve. *Jap. J. Ecol.* 29: 337–353.

Achmad, S. 1983. Studi Analisis Korelasi Curah Hujan dan Debit Sungai untuk Pendugaan Banjir pada Daerah Aliran Sungai Sa'dan. Thesis, Fakultas Pertanian, Universitas Hasanuddin, Ujung Pandang.

Achmad, S. 1986. Tambak atau hutan mangrove. Diskusi panel Dayaguna dan Lebar Jalur Hijau Hijau Mangrove, Ciloto, 27 February - 1 March 1986.

Achmad, T. and Cholik, E. 1977. Laporan penelitian beberapa aspek limnologi danau Moat dalam rangka usaha penebaran elver. Laporan No. 10, Lembaga Penelitian Perikanan Darat, Ujung Pandang.

Adisoemarto, S. 1980. *Binatang Hama.* Bogor: Lembaga Biologi Nasional.

Ahl, E. 1936. Beschreibung eines neuen Fisches der Familie Atherinidae. *Zool. Anz.* 114: 175–177.

Alabaster, J.S. and Lloyd, R.L. 1980. *Water Quality Criteria for Freshwater Fish.* London: Butterworths.

Alauddin, C. 1977. Cengkeh *Eugenia caryophyllus.* Banda Aceh.

Albrecht, G.H. 1977. The cranio-facial morphology of the Sulawesi macaques. Multivariate approaches to biological problems. *Contrib. primatol.* 151.

Aldrich-Blake, F.P.G. 1980. Long-tailed macaques. In *Malayan Forest Primates: Ten Years' Study in Tropical Rainforest,* ed. D.J. Chivers, pp. 147–167. New York: Plenum.

Alfred, E.R. 1961. Singapore fresh-water fishes. *Malay. Nat. J.* 15: 1–19.

Alfred, E.R. 1966. The freshwater fishes of Singapore. *Zool. Verh.* 78: 1–68.

Allan, J.D. and Russek, E. 1985. The quantification of stream drift. *Can. J. Fish. Aquat. Sci.* 42: 210–215.

Allaway, W.G. and Ashford, A.E. 1984. Nutrient input by seabirds to the forest on a coral island of the Great Barrier Reef. *Mar. Ecol. Progr. Ser.* 19: 297–298.

Alldredge, A.L. and King, J.M. 1977. Distribution, abundance and substrate preferences of demersal reef zooplankton at Lizard Island Lagoon, Great Barrier Reef. *Mar. Biol.* 41: 317–333.

Alrasjid, H. and Effendi, R. 1979. Pengaruh eksploitasi dengan traktor terhadap kerusakan tegakan sisa di kelompok hutan hujan tropis Pulau Pagai Selatan, Sumatera Barat. Laporan No. 293, Lembaga Penelitian Hutan, Bogor.

Alston, A.H.G. 1959. Isoetaceae. *Flora Malesiana II* 1: 63–64.

Altieri, M.A., Martin, P.B. and Lewis, W.J. 1983. A quest for ecologically based pest management systems. *Env. Mgmt.* 7: 91–100.

Altona, T. 1922. Korte aanteekeninger over Javaansche boomnamen. *Tectona.* 15: 870–875.

Amesbury, S.S. 1980. Biological studies on the coconut crab (*Birgus latro*) in the Mariana Islands. *Univ. of Guam. Techn. Rep.* No. 17.

Andaya, L.H. 1981. The heritage of Arung Palakka: A history of South Sulawesi (Celebes) in the seventeenth century. *Ver. Kon. Inst. Taal-Volk.* 91.

Anderson, J.A.R. 1965. Limestone habitat in Sarawak. In *Proc. Symp. Ecol. Res. Humit. Trop. Veg.*, pp. 49–57. Paris: UNESCO.

Anderson, J.M.and Swift, M.J. 1983. Decomposition in tropical forests. In *Tropical Rain Forests: Ecology and Management*, eds. S.L. Sutton, T.C. Whitmore and A.C. Chadwick, pp. 287–309. Oxford: Blackwell.

Anderson, N.M. and Polhemus, J.T. 1976. Water striders (Hemiptera: Gerridae, Vellidae). In *Marine Insects*, ed. L. Cheng, pp. 187–224. Amsterdam: North Holland.

Anderson, P.K. 1981. The behaviour of the dugong (*Dugong dugon*) in relation to conservation and management. *Bull. Mar. Sci.* 31: 640–647.

Anderson, P.K. 1982. Studies of dugong at Shark Bay, Western Australia. II. Surface and subsurface observations. *Aust. J. Wildl. Res.* 9: 85–99.

Anderson, P.K. and Birtles, R.A. 1978. Behaviour and ecology of the dugong *Dugong dugon* (Sirenia). *Aust. J. Wildl. Res.* 5: 1–23.

Andriesse, J.P. and Koopmans, T.Th. 1984. A monitoring study on nutrient cycles in soils used for shifting cultivation under various climatic conditions in tropical Asia. I. The influence of simulated burning on form and availability of plant nutrients. *Agric. Ecosyst. Env.* 12: 1–16.

Angermeier, P.L. and Karr, J.R. 1983. Fish communities along environmental gradients in a system of tropical streams. *Environ. Biol. Fishes* 9: 117–135.

Anggawati, S. 1985. Maros 1984: Survai potensi speleoturisme. *Warta Speleo* 1: 49–56.

Anon. 1968. *Camptostemon philippinense. Bot. Bull. Herb., Forest Dept., Sandakan* No. 10.

Anon. 1976. *Generalized soil map of Indonesia.* Bogor: Soils Research Institute.

Anon. 1976. *Pest Control in Rice.* London: Central for Overseas Pest Research.

Anon. 1977a. Survai biologi danau Tempe dan sekitarnya. BIOTROP, Bogor and FIPIA, Universitas Indonesia, Jakarta.

Anon. 1977b. Proposed Lore Kalamanta National Park Management Plan. 1978/79–1980/81. Bogor: FAO.

Anon. 1978a. *Tropical Forest Ecosystems.* Paris: UNESCO.

Anon. 1978b. Study habitat dan populasi kera endemik Sulawesi. Bogor: Direktorak Perlindungan dan Pengawetan Alam.

Anon. 1979. Jenis penyu laut di Indonesia. *Suara Alam* 5: 34–36.

Anon. 1979a. Draft final report on master plan study for the central South Sulawesi water resources development project. Tokyo: JICA.

Anon. 1979b. Ekologi Danau Tondano. Laporan Akhir, Laboratorium Biologi Wilayah, Universitas Indonesia, Jakarta.

Anon. 1979c. Data Dasar Gunung api Indonesia. Bandung: Direktorat Vulkanologi.

Anon. 1979d. *Beche-de-Mer: A Handbook for Fishermen.* Noumea: Handbook No. 18. South Pacific Commission.

Anon. 1979e. Laporan penelaahan awal di Suaka Margasatwa Lore Kalimanta, Sulawesi Tengah. Direktorat Perlindungan dan Pengawetan Alam Bogor, dan Institut Teknologi Bandung, Bandung.

Anon. 1980a. Komunitas, lingkungan regenerasi dan kemungkinan pengembangan hutan mangrove Malangke, Sulawesi Selatan Hasanuddin. Ujung Pandang.

Anon. 1980b. Cagar Alam Gn. Tangkoko-Dua Saudara, Sulawesi Utara: Management Plan 1981-1986. Bogor: World Wildlife Fund.

Anon. 1980c. *World Conservation Strategy.* Gland: International Union for the Conservation of Nature and Natural Resources.

Anon. 1980d. *Symposium on Small Mammals: Problems and Control.* BIOTROP Spec. Publ. 12, Bogor.

Anon. 1980e. *Environmental Impact Study.* Jakarta: P.T. International Nickel Indonesia.

Anon. 1981a. Pola pengendapan dan komposisi mineral berata delta di muara Sungai Rongkong. Pusat Studi Lingkungan Hidup, Universitas Hasanuddin, Ujung Pandang.

Anon. 1981b. *Penduduk Kalimantan dan Sulawesi menurut Propinsi dan Kabupaten/-Kotamadya.* Jakarta: Biro Pusat Statistik.

Anon. 1981c. Kriteria dan mutu lingkungan daerah aliran sungai di Sulawesi Selatan. Pusat Studi Sumberdaya Alam dan Lingkungan, Universitas Hasanuddin, Ujung Pandang.

Anon. 1981d. Karakteristik fisik dan kimia tanah di bawah tegakan mangrove di Malangke. Pusat Studi Lingkungan Hidup, Universitas Hasanuddin, Ujung Pandang.

Anon. 1981e. Lore Lindu National Park Management Plan. 1981-1986. Bogor: World Wildlife Fund.

Anon. 1981f. Coral reefs, associated habitats and species in the vicinity of Menado: Assessment of their conservation value. Bogor: World Wildlife Fund.

Anon. 1982a. National Conservation Plan for Indonesia Vol. V. Sulawesi, FAO: Bogor.

Anon. 1982b. National Conservation Plan for Indonesia, Vol. I., Introduction: Evaluation Methods and Overview of National Nature Richness. Bogor: FAO.

Anon. 1982c. Potensi danau, masalah dan pengembangannya. Pusat Studi Sumber Daya Alam dan Lingkungan Hidup, Universitas Hasanuddin, Ujung Pandang.

Anon. 1982d. Taka Bone Rate: assessment of marine conservation value and needs. Bogor: FAO.

Anon. 1982e. Ecological Aspects of Development in the Humid Tropics. Washington, D.C.: National Academy Press.

Anon. 1982f. Sejarah Revolusi Kemerdekaan Daerah Sulawesi Tengah. Dinas Pendidikan dan Kebudayaan, Palu.

Anon. 1982g. Hasil penelitian resettlement desa Kabupaten Banggai Propinsi Sulawesi Tengah. Direktorat Pembangunan Desa, Palu.

Anon. 1982h. National conservation plan for Indonesia. Vol. 1. Introduction. Bogor: FAO.

Anon. 1982i. Marine conservation potential of the Togian Islands, Central Sulawesi. Bogor: FAO.

Anon. 1983a. Survei pengembangan potensi perairan umum Rawa Aopa Sulawesi Tenggara. Fakultas Perikanan, Institut Pertanian Bogor, Bogor.

Anon. 1983b. Almanac of Indonesia, Statistik Indonesia. Jakarta: Biro Pusat Statistik.

Anon. 1983c. Little-known Asian Animals with a Promising Economic Future. Washington, D.C.:

National Academy of Sciences Press.

Anon. 1983d. Survai pengembangan potensi perairan umum Rawa Opa di Sulawesi Tenggara. Fakultas Perikanan IPB, Bogor.

Anon. 1983e. Kajian kemungkinan penentuan mintakat penyanggah Taman Nasional Dumoga Bone. Fakultas Pertanian, Universitas Sam Ratulangi, Manado.

Anon. 1984a. Buku Tahunan: Pertambangan Indonesia. Jakarta: Departemen Pertambangan dan Energi.

Anon. 1984b. Coral reef invertebrates in Indonesia: Their exploitation and conservation needs. Bogor: World Wildlife Fund.

Anon. 1984c. First probe of fish reaction to seismic shock waves. Arctic News Rec. Summer 1984: 23.

Anon. 1984d. Lower Jeneberang River urgent flood control works. No.1 Hydrology and geographic survey. Ujung Pandang: CTI Engineering.

Anon. 1984e. Laporan Survai Inventarisasi satwa di Rawa Opa, Sulawesi Tenggara. Direktorat Perlindungan dan Pengawetan Alam, Bogor.

Anon. 1984f. Why Conservation? Commission on Ecology Occasional Paper No. 4. International Union for Conservation and Natural Resources, Gland.

Anon. 1984g. Statistik Kehutanan Indonesia. Jakarta: Pusat Inventarisasi Hutan, Departemen Kehutanan.

Anon. 1985a. Asbuton. P.T. Sarana Karya (Persero), Jakarta.

Anon. 1985b. Bantimurung, nature reserve or nature retail. Voice of Nature 35: 5.

Anon. 1985c. Thai-Maros 1986. Rapport spéléologique et Scientifique Association Pyrénéenne de Spéléologie: Toulouse.

Anon. 1985d. Mengenal Taman Nasional Rawa Aopa-Wotumohai Sulawesi Tenggara. Kendari: PHPA.

Anon. 1985e. Laporan hasil penjajagan calon Taman Wisata Goa-goa Patunuang, Kab. Maros. Kanwil Departemen Kehutanan, Ujung Pandang.

Anon. 1985f. A review of policies affecting the sustainable development of forest lands in Indonesia. Department of Forestry, State Ministry of Population and Environment and Department of the Interior, Jakarta.

Anon. 1985g. Statistik Indonesia 1984. Biro

Pusat Statistik: Jakarta.

Anon. 1985h. Lokasi proyek transmigrasi penepatan Pelita III (1979/80–1983/84) dan Pelita IV Th. I (1984/85). Direktorat Jenderal Pengerahan dan Pembinaan Transmigrasi, Jakarta.

Anon. 1985i. Indonesia forestry project working paper No.1. FAO/World Bank Cooperative Program, Jakarta.

Anon. 1985j. Daftar proyek transmigrasi yang dibina, tahun 1985/86, keadaan Agustus 1985. Direktorat Jenderal Pengerahan dan Pembinaan Transmigrasi, Jakarta.

Anon. 1985k. FAO/World Bank Cooperative Program, Indonesia Forestry Project, Working Paper 1.

Anon. 1986a. S. Sulawesi, natural gas project postponed. *Jakarta Post* 1 Feb. 1986.

Anon. 1986b. Thai-Maros 1986. Rapport spéléologique et scientifique. Association Pyrénéenne de Spéléologie, Toulouse.

Anon. 1986c. Sebagian wilayah Sulawesi jadi kritis akibat pembabatan hutan. *Kompas* 15 July 1986.

Anon. 1986d. Empat bendungan di Gorontalo tak berfungsi. *Kompas* 9 October 1986.

Anon. 1986e. Belum jalan, program untuk pengembangan hutan industri. *Kompas* 20 June 1986.

Anon. 1986f. Penebang liar dituduh penyebab perusakan hutan di Sulteng, *Kompas* 9 October 1986.

Anon. 1986g. Terancam mati, 30.000 pohon cengkeh di Sulawesi Tengah. *Kompas* 13 September 1986.

Anon. 1986h. Kini kami makan nasi. *Kompas* 14 June 1986.

Anon. 1986i. Coral reef fishes – a case for trade controls. *Oryx* 20: 137.

Anon. 1986j. Kayu hitam senilai Rp. 55 juta disita polsus kehutanan Sulteng. *Kompas* 18 October 1986.

Anon. n.d. Proposal for a National Park Rawa Opa in Sulawesi Tenggara. Unpubl. ms.

Anwar, A., Wibowo, R. and Rachman, A.M.A. 1986. Konsepsi pembangunan wilayah pantai dalam hubungan dengan penentuan jalur hijau mangrove. Diskusi Panel Dayaguna dan Lebar Jalur Hijau Mangrove, Ciloto, 27 February – 1 March 1986.

Appanah, S. 1985. General flowering in the climax rain forests of South-east Asia. *J.*

Trop. Ecol. 1: 225–240.

Arroyo, C.A. 1979. Flora of the Philippine mangrove. In *Proceedings of the Symposium on Mangrove and Estuarine Vegetation in Southeast Asia*, ed. P.B.L. Srivastava, pp. 33–44. Bogor: BIOTROP.

Arsyad, U. 1985. Studi intersepsi curah hujan pada hutan alam di Sub DAS Malino, DAS Sa'dan. Dissertation, Universitas Hasanuddin, Ujung Pandang.

Arumugam, P.T. and Furtado, J.I. 1981. Eutrophication of a Malaysian reservoir; effects of agro-industrial effluents. *Trop. Ecol.* 22: 221–275.

Ashton, P. 1964. Ecological studies in the mixed dipterocarp forest of Brunei State. *Oxf. For. Mem.* No. 25.

Ashton, P. 1980. The biological and ecological basis for the utilization of dipterocarps. *Biol. Indonesia* 7: 43–53.

Ashton, P. 1984. Dipterocarpaceae. *Flora Malesiana I* 9: 237–552.

Askew, G.P. 1964. The montain soils of the east ridge of Mt Kinabalu. *Proc. Roy. Soc. B* 161: 65–74.

Asmar, T. 1978. The megalithic tradition. In *Dynamics of Indonesian History*, eds. Haryati Soebadio and C.A. du Marchie Sarvaas, pp. 29-40. Amsterdam: North Holland.

Atkinson, J.M. 1979. *Paths of the Spirit Familiars: A Study of Wana Shamanism*. Ph.D. thesis, Stanford University.

Atkinson, J.M. 1985. Agama dan suku Wana di Sulawesi Tengah. In *Peranan Kebudayaan Tradisional Indonesia dalam Modernisasi*, ed. M.R. Dove, pp. 3-30. Jakarta: Yayasan Obor.

Atmadja, W.S. 1977. Notes on the distribution of red algae (Rhodophyta) on the coral reef of Pari Islands, Seribu Island. *Mar. Res. Indonesia* 17: 15–27.

Audley-Charles, M.G. 1981. Geological history of the region of Wallace's Line. In *Wallace's Line and Plate Tectonics*, ed. T.C. Whitmore, pp. 24-35. Oxford: Oxford University Press.

Audley-Charles, M.G. 1983. Reconstruction of eastern Gondwanaland. *Nature* 306: 48–50.

Audley-Charles, M.G. 1987. Dispersal of Gondwanaland: Relevance to the evolution of angiosperms. In *The Biogeographic Evolution of the Malay Archipelago*, ed. T.C. Whitmore. Oxford: Clarendon.

Audley-Charles, M.G. and Hooijer, D.A. 1973. Relation of Pleistocene migrations of pygmy stegodonts to island are tectonics in eastern Indonesia. *Nature* 241: 197–198.

Augspurger, C.K. 1984. Seedling survival of tropical tree species: interactions of dispersal distance, light- gaps and pathogens. *Ecology* 65: 1705–1712.

Augspurger, C.K. and Kelly, C.K. 1983. Pathogen mortality of tropical tree seedlings: experimental studies of the effects of dispersal distance, seedling density, and light conditions. *Oecologia* 61: 211–217.

Aurich, H. 1935a. Mitteilung der Wallacea-Expedition Woltereck, XIII. Fishe I. *Zool. Anz.* 112: 97–107.

Aurich, H. 1935b. Mitteilung der Wallacea-Expedition Woltereck, XIV. Fishe II. *Zool. Anz.* 112: 161–177.

Aurich, H. 1938. Mitteilung der Wallacea-Expedition Woltereck. XXVIII. Gobiiden. *Intern. Rev. Hydrobiol.* 38: 125–183.

Ayensu, E.S., Heywood, V.H. Lucas, G.I. and Defilips, R.A. 1984. *Our Green and Living World: The Wisdom to Save it.* Cambridge: Cambridge University Press.

Backer, C.A. 1951a. Pontederiaceae. *Flora Malesiana I* 4: 255–261.

Backer, C.A. 1951b. Crassulaceae. *Flora Malesiana I* 4: 197–202.

Backer, C.A. and van Steenis, C.G.G.J. 1951. Sonneratiaceae. *Flora Malesiana I* 4: 280–289.

Baillie, I.C., and Ashton, P.S. 1983. Some aspects of the nutrient cycle in mixed dipterocarp forest in Sarawak. In *Tropical Rain Forest: Ecology and Management*, eds. S.L. Sutton, T.C. Whitmore and A.C. Chadwick, pp. 347-356. Oxford: Blackwell.

Bak, R.P.M. 1981. Survival after fragmentation of colonies of *Madracis mirabilis*, *Acropora palmata* and *A. cervicornis* (Scleractinia) and the subsequent impact of a coral disease *Proc. IVth Int. Coral Reef Symp. Manila, Vol. 2*, pp. 221-227.

Baker, A.J.M. 1981. Accumulators and excluders: strategies in the response of plants to heavy metals. *J. Pl. Nut.* 3: 643–654.

Baker, A.J.M., Brooks, P.R. and Kersten, W.J. 1985. Accumulation of nickel by *Psychotria* species from the Pacific basin. *Taxon* 34: 89–95.

Baker, H.G. 1975. Sugar concentrations in nectars from hummingbird flowers. *Biotropica* 7: 37–41.

Baker, J.M. 1982. Mangrove swamps and the oil industry. *Oil Petrochem. Poll.* 1: 5–22.

Baker, R.R. 1982. *Migration: Paths Through Time.* Hodder and Stoughton, London.

Bakhuizen van der Brink, R.C. 1936. Revisio Ebenacearum Malayensium. *Bull. Jard. Bot. Buitenzorg* 15: 1–515.

Balazs, D. 1973. Comparative morphogenetical study of karst regions in tropical and temperate areas with examples from Celebes and Hungary. *Trans. Cave Res. Grp. G.B.* 15: 1–7.

Barber, R.T. and Chavez, R.P. 1983. Biological consequences of El Nino. *Science* 222: 1203–1210.

Barnard, C.J. and Thompson, D.B.A. 1985. *Gulls and Plovers: the Ecology of Mixed Species Feeding Groups.* London: Croom Helm.

Barnes, R.D. 1968. *Invertebrate Zoology.* Saunders: Philadelphia.

Barnes, R.S.K. 1984a. *Estuarine Biology 2nd. ed.* Studies in Biology No. 49, Edward Arnold: London.

Barnes, R.S.K. 1984b. *A Synoptic Classification of Living Organisms.* Oxford: Blackwell.

Barnes, R.S.K. and Hughes, R.N. 1982. *An Introduction to Marine Ecology.* Oxford: Blackwell.

Barr, T.C. Jr. 1968. Cave ecology and the evolution of troglobites. In *Evolutionary Biology Vol. 2*, eds. T. Dobzhansky, M. Hecht and W. Streere, pp.35-102. New York: Appleton-Century-Crofts.

Barry, S.J. 1984. Small mammals in a southeast Queensland rainforest: the effects of soil fertility and past logging disturbance. *Aust. Wildl. Res.* 11: 31–39.

Bartels, H. 1952. Machaeramphus a. alcinus Westerm. Waarnemingen bij vogel, nest en ei. *Limosa* 25: 93–100.

Bartstra, G.J. 1977. Walanae formation and Walanae terraces in the stratigraphy of south Sulawesi (Celebes, Indonesia). *Quar.* 27: 21–30.

Bartstra, G.J. 1978. Notes on new data concerning the fossil vertebrates and stone tools in the Walanae valley in south Sulawesi, Celebes. *Mod. Quat. Res. S.E. Asia* 4: 71–72.

Basuni, F. 1983. Tidak ada pilihan lain: Studi tentang pandangan nelayan terhadap kerja di pulau Balang Lompo Kabupaten Daerah Tingkat II Pangkajene Kepulauan. Pusat Latihan Penelitian Ilmu-Ilmu Social, Universitas Hasanuddin, Ujung Pandang.

Bauchop, T. 1978. Digestion of leaves in vertebrate arboreal folivores. In *The Ecology of Arboreal Folivores*, ed. G.G. Montgomery, pp. 193–204. Washington, D.C.: Smithsonian Institution Press.

Beadle, D. 1985. In *Interwader Annual Report 1984*. Kuala Lumpur: Interwader.

Becker, P. and Wong, M. 1985. Seed dispersal, seed predation and juvenile mortality of *Aglaia* sp. (Meliaceae) in lowland dipterocarp rainforest. *Biotropica* 17: 230–237.

Becker, P., Leighton, M. and Payne, J.B. 1985. Why tropical squirrels carry seeds out of source crowns. *J. Trop. Ecol.* 1: 183–186.

Bedford-Russell, A. 1981. A spectacular new *Idea* from Celebes (Lepidoptera, Danaidae). *System. Entomol.* 6: 225–228.

Bedford-Russell, A. 1984. Two new skippers from Sulawesi (Celebes) (Lepidoptera: Hesperinae). *Entomol. Ber.* 44: 154.

Beisinger, S.R. and Osborne, D.R. 1982. Effects of urbanization on avian community organization. *Condor* 84: 75–83.

Bell, H.L. 1982a. A bird community of lowland rainforest in New Guinea. I. Composition and density of the avifauna. *Emu* 82: 24–41.

Bell, H.L. 1982b. A bird community of lowland rainforest in New Guinea. II. Seasonality. *Emu* 82: 65–74.

Bell, H.L. 1982c. A bird community of New Guinean lowland rainforest. III. Vertical distribution of the avifauna. *Emu* 82: 143–162.

Bell, H.L. 1982d. A bird community of lowland rainforest in New Guinea. IV. Birds of secondary vegetation. *Emu* 82: 217–224.

Bell, H.L. 1983. A bird community of lowland rainforest in New Guinea. V. Mixed-species feeding flocks. *Emu* 82 Suppl.: 256–275.

Bell, H.L. 1984. A bird community of lowland rainforest in New Guinea. VI. Foraging ecology and community structure of the avifauna. *Emu* 84: 142–158.

Bellwood, P.S. 1978. Archaeological research in Minahasa and Talaud Islands, Northeastern Indonesia. *Asian Perspect.* 19: 240–288.

Bellwood, P.S. 1980a. The peopling of the Pacific. *Scient. Amer.* 243: 174–185.

Bellwood, P.S. 1980b. Plants, climate and people: the early horticultural prehistory of Austronesia. In *Indonesia: Australia Perspectives, Vol. I Indonesia. The Making of a Culture*, ed. J.J. Fox, pp. 57-74. Canberra: Australian National University.

Bellwood, P.S. 1985. *Prehistory of the Indo-Malaysian Archipelago.* Ryde N.S.W.: Academic Press.

Bennett, C.P.A. 1986. Observations of pests and diseases of the coconut *Cocos nucifera* L., in the northern islands of Maluku. Unpubl. ms.

Bennett, C.P.A. and Godoy, R. 1986. Coconut farming systems in Indonesia. Unpubl. ms.

Bennett, C.P.A., Hunt, P. and Asman, A. 1985. Association of a xylem-limited bacterium with Sumatra disease of cloves in Indonesia. *Plant Pathol.* 34: 487–494.

Bennett, E.L. 1983. The banded langur: ecology of a colobine in West Malaysian rain-forest. Doctoral dissertation, University of Cambridge, Cambridge.

Benzie, J.A.H. 1984. The colonisation mechanisms of stream benthos in a tropical river (Menik Ganga : Sri Lanka). *Hydrobiol.* 111: 171–179.

Benzing, D.H. 1981. Mineral nutrition of epiphytes: an appraisal of adaptive features. *Selbyana* 5: 219–223.

Benzing, D.H. 1983. Vascular epiphytes: a survey with special reference to their interactions with other organisms. In *Tropical Rain Forest: Ecology and Management*, eds. S.L. Sutton, T.C. Whitmore and A.C. Chadwick, pp. 11–24. Oxford: Blackwell.

Bergmans, W. and Rozendaal, F.G. 1982. Notes on *Rhinolophus* from Sulawesi with the description of a new species. *Bijdr. Dierk.* 52: 169–174.

Bergmans, W. and Rozendaal, F.G., in press. Notes on collections of fruit bats from Sulawesi and some off-lying islands (Mammalia, Megachiroptera). *Zool. Verh.*

Bergquist, P.R. 1978. *Sponges.* London: Hutchinson.

Berhaut, R.N. 1982. *Ecology of Rocky Shores.*

Studies in Biology No. 139. Edward Arnold: London.

Berry, A.J. 1963. Faunal zonation in mangrove swamps. *Bull. natl. Mus. Singapore* 32: 90–98.

Berry, A.J. 1972. The natural history of West Malaysian mangrove faunas. *Malay. Nat. J.* 25: 135–162.

Berry, P.Y. 1965. The diet of some Singapore Anura (Amphibia). *Proc. zool. Soc. Lond.* 144: 163–174.

Berry, P.Y. and Bullock, J.A. 1962. The food of the common Malayan toad *Bufo melanostictus* Schneider *Copeia* 1962: 736–741.

Bertness, M.D. 1981a. Competitive dynamics of a tropical hermit crab assemblage. *Ecology* 62: 751–761.

Bertness, M.D. 1981b. Pattern and plasticity in tropical hermit crab growth and reproduction. *Am. nat.* 117: 754–773.

Biere, J.M. and Uetz, G.W. 1981. Web orientation in the spider *Micrathena Gracilis* (Araneae: Araneidae). *Ecology* 62: 336–344.

Birch, W.R. 1975. Some chemical and calorific properties of tropical marine angiosperms compared with those of other plants *J. Ecol.* 63: 201–212.

Birkeland, C. 1984. Sources and destinations of nutrients in coral reef. *UNESCO Rep. Mar. Sci.* 21: 3–20.

Bishop, J.E. 1973. *Limonology of a Small Malayan River, Sungai Gombak.* The Hague: Junk.

Bishop, K.D. and King, B. 1984. The Sunda serin *Serinus estherae* in Sulawesi. *Kukila* 2: 90–91.

Bismark, M. 1982a. Ekologi dan tingkahlaku *Macaca nigrescens* di Suaka Margasatwa Dumoga, Sulawesi Utara. Laporan 392. Balai Penelitian Hutan, Bogor.

Bismark, M. 1982b. Habitat dan populasi *Macaca ochreata* di Suaka Margasatwa Tanjung Peropa, Sulawesi Tengah. Laporan 408. Balai Penelitian Hutan, Bogor.

Bismark, M. 1982c. Pengaruh perladangan dan pengambilan kayu gergajian terhadap yakis (*Macaca nigrescens*) di Suaka Margasatwa Dumoga, Sulawesi Utara. *Duta Rimba* 46–48.

Bismark, M., Ginting, A.R. and Yasaf, A.K.M. 1978. Inventarisasi jenis-jenis Leguminosae yang bersifat weeds. Fakultas Peternakan, Universitas Tadulako, Palu.

Black, H.C. and Harper, K.P. 1979. The adaptive value of buttresses to tropical trees: additional hypotheses. *Biotropica* 11: 240.

Bloembergen, S. 1940. Verslag van een exploratie-tocht naar Midden Celebes *Tectona* 33: 377–418. (Partial English translation available.)

Bloembergen, S. 1941. Belklimming van den G. Ngilalaki in West Midden Celebes. *Meded. Ned- Ind. Ver. Bergsport* 17: 20–21.

Bloembergen, S. 1948. Aceraceae. *Flora Malesiana J.* 4: 366–376.

Boedhisampurno, S. 1982. Studi sisa-sisa kremasi dari Leang Pettekere, Sulawesi Selatan. Pertemuan Arkeologi ke-2, 1980.

Bosch, H.A.J. in den 1985. Snakes of Sulawesi: Checklist, key and additional biogeographical remarks. *Zool. Verh.* 217: 1–50.

Bothwell, A.M. 1981. Fragmentation, a means of asexual reproduction and dispersal in the coral genus *Acropora* (Scleractinia: Astrocorniida: Acroporidae) – a preliminary report. *Proc. IVth Int. Coral Reef Symp., Manila, Vol. 2* pp. 137–144.

Boto, K.G., Bunt, J.S. and Wellington, J.T. 1984. Variations in mangrove forest productivity in Northern Australia and Papua New Guinea. *Estuar. Coast. Shelt Sci.* 19: 321–329.

Boudowski, G. 1963. Forest succession in tropical lowlands. *Turrialba* 13: 42–44.

Boulenger, G.A. 1897a. An account of the freshwater fishes collected by Drs P. and F. Sarasin. *Proc. zool. Soc. Lond.* 13: 426–429.

Boulenger, G.A. 1897b. A catalogue of the reptiles and amphibians of Celebes with special reference to the collections made by Drs P. and F. Sarasin. *Proc. zool. Soc. Lond.* 13: 193–237.

Bouman-Houtman, A. 1926. Lente op Lompobatang. *Trop. Natuur* 15: 93–97.

Brafield, A.E. 1972. Some adaptations to low environmental oxygen in marine beaches. In *Essays in Hydrobiology*, eds. R.B. Clark and R.J. Wootton, pp. 93-106. Exeter: University of Exeter.

Brafield, A.E. 1978. *Life in Sandy Shores.* Studies in Biology No. 89. London: Edward Arnold.

Braithwaite, L.W., Turner, J. and Kelly, J. 1984. Studies on the arboreal marsupial fauna of eucalypt forests being harvested for wood pulp at Eden, New South Wales. III. Relationships between faunal densities, eucalypt occurrence and foliage nutrients and soil parent material. *Aust. Wildl. Res.* 11: 41–48.

Brakefield, P.M. and Larsen, T.B. 1984. The evolutionary significance of dry and wet season forms in some tropical butterflies. *Biol. J. Lin. Soc.* 22: 1–12.

Braley, R.D. 1984. Reproduction in the giant clams *Tridacna gigas* and *T. derasa* on the north-central Great Barrier Reef, Australia. *Coral Reefs* 3: 221–227.

Brasell, H.M. and Sinclair, D.F. 1983. Mineral elements returned to forest floor in two rainforest and three plantation plots in tropical Australia. *J. Ecol.* 71: 367–378.

Brembach, M. 1982. Drei neues *Demogenys*-Arten aus Sulawesi: *D. montanus, D. vogti* und *D. megarhamphus. Aquar. Terrar. Z* 35: 51–55.

Brockaw, N.V.L. 1982a. The definition of treefall gaps and its effect on measures of forest dynamics. *Biotropica* 14: 158–160.

Brockaw, N.V.L. 1982b. Treefalls: Frequency, timing, and consequences. In *The Ecology of a Tropical Forest: Seasonal Rhythms and Long-term Changes*, eds. E.G. Leigh, A.S. Rand and D.M. Windsor, pp. 101–108. Washington, D.C.: Smithsonian Institution Press

Brooks, J.L. 1950. Speciation in ancient lakes (concluded). *Quart. Rev. Biol.* 25: 131–176.

Brooks, R.R., Wither, E.D. and Westra, L.Y.T. 1978. Biogeochemical copper anomalies on Salajar Island Indonesia. *J. Geochem. Expl.* 10: 181–188.

Brooks, W.R. and Mariscal, R.N. 1977. The acclimation of anenome fish to sea anenomes: protection by changes in the fish's mucous coat. *J. Exp. Mar. Biol. Ecol.* 81: 277–285.

Broom, M.J. 1982. Structure and Seasonality in a Malaysian mudflat community. *Estuar. Coast. shelf Sci.* 15: 135–150.

Broome, L.S., Bishop, K.D. and Anderson, D.R. 1984. Population density and habitat use by *Megapodius frecinet eremita* in West New Britain. *Aust. Wildl. Res.* 11: 161–171.

Brouns, J.J.W.M. 1985. A preliminary study of the seagrass *Thalassodendron ciliatum* (Forrsk.) den Hartog from Eastern

Indonesia. Biological results of the Snellius II Expedition. *Aquat. Bot.* 23: 249–260.

Brown, J. 1978. Avian communal breeding systems. *Ann. Rev. Ecol. Syst.* 9: 123–155.

Brown, S. and Lugo, A.E. 1982. The storage and production of organic matter in tropical forests and their role in the global carbon cycle. *Biotropica* 14: 161–187.

Brown, S. and Lugo, A.E. 1984. Biomass of tropical forests: A new estimate based on forest volumes. *Science* 223: 1290–1293.

Bruinjnzeel, L. A. 1984. Elemental content of litterfall in a lower montane forest in Central Java, Indonesia. *Malay Nat. J.* 37: 199–208.

Buchari, U. 1981. Beberapa faktor ekologi yang memepengaruhi penyebaran dan kepadatan beberapa jenis moluska di danau Tondano, Sulawesi Utara. Tesis, Fakultas Perikanan, Unversitas Sam Ratulangi, Manado.

Buchari, U. 1984. Komunitas plankton di danau Mooat – Sulawesi Utara. Tesis Fakultas Perikanan, Universitas Sam Ratulangi, Manado.

Budiman, A. 1985. The molluscan fauna in reef-associated mangrove forests in Elpaputih and Wailale, Ceram, Indonesia. In *Coasts and Tidal Wetlands of the Australian Monsoon Region*, eds. K.N. Bardsley, J.D.S. Davie and Woodroffe, pp. 251–256. Darwin: Australian National University Mangrove Monograph No. 1.

Budiman, A. and Darnaedi, D. 1982. Struktur komunitas moluska di hutan mangrove Morowali, Sulawesi Tengah. In *Prosiding Seminar II Ekosistem Mangrove*, pp. 175–182. Jakarta: Lembaga Oseanologi Nasional.

Budiman, A. and Williams, W.T. 1981. Vegetational relationships in the mangroves of tropical Australia. *Mar. Ecol. Prog. Ser.* 4: 349–359.

Bullock, J.A. 1966. The ecology of Malaysian caves. *Malay. Nat. J.* 19: 57–63.

Bumby, M.J. 1982. A survey of aquatic macrophytes and chemicals qualities of nineteen locations in Costa Rica. *Brenesia* 19/20: 487–535.

Bunt, J.S. 1980. Degradation of mangroves. In *Marine and Coastal Processes in the Pacific: Ecological Aspects of Coastal Zone Management.* Jakarta: UNESCO.

Blunt, J.S., and Williams, W.T. 1981. Vegetational relationships in the mangroves of tropical Australia. *Mar. Ecol. Prog. Ser.* 4: 349–359.

Burbridge, P.R. and Maragos, J.E. 1985. Coastal resources managment and environmental assessment needs for aquatic resources development in Indonesia. International Institute for Environment and Development, Washington, D.C.

Burgess, P.F. 1971. The effect of logging on hill dipterocarp forests. *Malay. Nat. J.* 24: 231–237.

Burhanuddin, M. 1980. Pengamatan terhadap ikan gelodok *Periopthalmodon schlosseri* di Muara Sungai Banyuasin. In *Sumber Daya Hayati Bahari*, eds. Burhanuddin, M.K. Moosa, and H. Razak pp. 117-124. Lembaga Oseanology Nasional, Jakarta.

Burhanuddin, M. 1986. Studi penyebaran sifat tanah dan vegetasi pada gunung Rantemario Kabupaten Enrekang. Thesis, Fakultas Ilmu Pertanian, Universitas Hasanuddin, Ujung Pandang.

Burhanuddin, M. and Martosewojo, S. 1978. Pengamatan terhadap ikan gelodok. *Periophthalmus koelreuteri* (Pallas) di Pulau Pari. In *Prosiding Seminar Ekosistem Hutan Mangrove*, eds. S. Soemodihardjo, A. Nontji, A. Djamali, pp. 86–92. Lembaga Oseanologi Nasional, Jakarta.

Burkill, I.H. 1951. Dioscoreaceae. *Flora Malesiana I* 4: 293–335.

Burkill, I.H. 1966. *A Dictionary of the Economic Products of the Malay Peninsula.* Singapore: Government Printing Office.

Burkill, I.N. 1917. The flowering of the pigeon orchid, *Dendrobium crumenatum* Lindl. *Grdns. Bull. Straits Settlement* 1: 400–405.

Burleigh, R. 1981. Radiocarbon dating of freshwater shells from Leang Burung 2: Part. I. *Mod. Quat. Res. S.E. Asia* 6: 51–52.

Burnham, C.P. 1984. Soils. In *Tropical Rain Forests of the Far East*, by T.C. Whitmore, pp. 103–120. Oxford: Oxford University Press.

Buwalda, P. 1949. Umbelliferae. *Flora Malesiana I* 4: 366–376.

Calabrese, D.M. 1986. Gerridae of Sulawesi: preliminary results and exciting implications. *Antenna* Jan. 1986: 27–28.

Caldwell, G.S. and Rubinoff, R.W. 1983. Avoidance of venemous sea snakes by naive

herons and egrets. *Auk* 100: 195–198.

Cambridge, M.L. and McComb, A.J. 1984. The loss of seagrasses in Cockburn Sound, Western Australia. I. The time course and magnitude of seagrass decline in relation to industrial development. *Aquat. Bot.* 20: 229–243.

Cane, M.A. 1983. Oceanographic events during El Nino. *Science* 222: 1189–1195.

Carcasson, R.H. 1977. *A Field Guide to the Coral Reef Fishes of the Indian and West Pacific Oceans.* London: Collins.

Carney, W.P. and Sudomo, M. 1980. Schistosomiasis in Indonesia, pp. 58–63. Symposium Masalah Penyakit Parasit, Jakarta,

Carney, W.P., Peenen, P.F.D., See, R., Hagelstein, E. and Lima, B. 1977a. Parasites of man in remote areas of Central and South Sulawesi, Indonesia. S.E. Asian *J. Trop. Med. Pub. Hlth.*

Carney, W.P. Purnomo, van Peenen, P.F.D., Brown, R.J. and Sudomo, M. 1977b. *Schistosoma incognitum* from mammals of Central Sulawesi, Indonesia. *Proc. Helminth. Soc. Wash.* 44: 150–155.

Carney, W.P., Purnomo, van Peenen, P.F.D., and Sudomo, M. 1978. A mammalian reservoir of *Schistosoma japonicum* in the Napu Valley, Central Sulawesi, Indonesia. *J. Parasitol.* 64: 1138–1139.

Carney, W.P., Sudomo, M. and Purnomo 1980. Echinostomiasis: a disease that disappeared. *Trop. geogr. Med.* 32: 101–105.

Carney, W.P., Hadidjaja, P. Davies, G.M., Clarke, M.D., Djajasasmita, M. and Nalim, S. 1973. *Oncomelania hupensis* from the schistosomiasis focus in Central Sulawesi, Celebes, Indonesia. *J. Parasitol.* 59: 210–211.

Carthaus, E. 1909. Ist *Tectona grandis* ein urspruenglich im malaischen Archipel einheimischer Waldbaum? *Tectona* 2: 309–319.

Caughley, G. 1978. *Analysis of Vertebrate Populations.* Chicester: Wiley.

Chambers, M.J. 1980. The environment and geomorphology of deltaic sedimentation. (Some examples from Indonesia.) In *Tropical Ecology and Development*, ed. J. Furtado, pp. 1091–1095. University of Malaya, Kuala Lumpur.

Chan, H.T. 1985. Coastal and riverbank erosion of mangroves in Peninsular Malaysia.

In *Coasts and Tidal Wetlands of the Australian Monsoon Region,* eds. K.N. Bardsky, J.D.S. Davie and C.D. Woodroffe, pp. 43–52. Mangrove Monograph No. 1, Australian National University North Australia Research Unit, Darwin.

Chan, H.T., Ujang R. and Putz, F.E. 1982. A preliminary study on planting of *Rhizophora* species in an *Avicennia* forest at the Matang mangroves. *Proc. Seminar II Ekosistem Mangrove,* 340–345. Lembaga Oseanologi Nasional, Jakarta.

Channels, P. and Morrissey, N. 1981. Technique for analysis of seagrass genera present in dugong stomachs, including a key to north Queensland seagrasses based on cell details. In *The Dugong,* ed. H. Marsh, pp. 176–179. Townsville: James Cook University.

Chapman, E.C. 1975. Shifting agriculture in tropical forest areas of Southeast Asia. In *The Use of Ecological Guidelines for Development in Tropical Areas of Southeast Asia,* pp. 120–135. Gland: IUCN.

Chapman, P. 1984. The invertebrate fauna of the caves of Gunung Mulu National Park. *Sarawak Mus. J.* 30: 1–18.

Chapman, V.J. 1970. *Seaweeds and their Uses.* London: Methuen.

Chapman, V.J. 1977a. Introduction. In *Wet Coastal Ecosystems,* ed. V.J. Chapman, pp. 1–29. Amsterdam: Elsevier.

Chapman, V.J. 1977b. Wet coastal formations of Indo-Malesia and Papua New Guinea. In *Wet Coastal Ecosystems,* ed. V.J. Chapman, pp. 261–270. Amsterdam: Elsevier.

Chavez, F.P., Barber, R.T. and Soldi, S.H. 1984. Propagated temperature changes during onset and recovery of the 1982-83 El Nino. *Nature* 309: 47–49.

Cheng, L. 1976. Insects in marine environments. In *Marine Insects,* ed. L. Cheng, pp. 1–4. Amsterdam: North-Holland.

Cherfas, J. 1985. When is a tree more than a tree? *New Scient.* 20 June: 42–45.

Chester, R.H. 1969. Destruction of Pacific corals by the sea star *Acanthaster planci. Science* 165: 280–283.

Chiba, S. 1978. Numbers, biomass and metabolism of soil animals in Pasoh Forest Reserve. *Malay. Nat. J.* 30: 313–324.

Chin, S.C. 1977. The limestone hill flora of Malaya II. *Gards' Bull. Singapore* 30: 165–203.

Chin, S.C. 1979. The limestone hill flora of Malaya II. *Gards' Bull. Singapore* 32: 64–203.

Chin, S.C. 1983a. The limestone hill flora of Malaya III. *Gards' Bull. Singapore* 35: 137–190.

Chin, S.C. 1983b. The limestone hill flora of Malaya IV. *Gards' Bull. Singapore* 36: 31–91.

Chin, S.C. 1986. Preliminary report on a field visit to the limestone areas of South Sulawesi. Unpubl. ms.

Chou, L.M. 1975. Systematic account of the Singapore house geckos. *J. Natl. Acad. Sci. Singapore* 4: 130–138.

Chou, L.M. 1978. Some bionomic data on the house geckos of Singapore. *Malay. Nat. J.* 31: 231–235.

Christensen, B. and Wium-Anderson, S. 1977. Seasonal growth of mangrove trees in southern Thailand. I. The phenology of *Rhizophora apiculata* Bl. *Aquat. Bot.* 3: 281–286.

Christy, J.H. and Salmon, M. 1984. Ecology and evolution of mating systems of fiddler crabs (genus *Uca*). *Biol. Rev.* 59: 483–509.

Chua, N.H., Kwok, S.W., Tan, K.K., Teo, S.P. and Wong, H.A. 1972. Growths on concrete and other surfaces in Singapore. *J. S'pore Inst. Architect.* 51: 13–15.

Church, G. 1962. The reproductive cycles of the Javanese house geckos, *Cosymbotus platurus, Hemidactylus frenatus* and *Peropus mutilatus. Copeia* 2: 262–269.

Church, G. and Lim, C.S. 1961. The distribution of three species of house geckos in Bandung (Java). *Herpetologica* 17: 119–201.

Chye, Ho Sinn, and Furtado, Jose I. 1982. The limnology of lowland streams in West Malaysia. *Trop. Ecol.* 23: 84–97.

Clason, A.T. 1976. A preliminary note about the animal remains from the Leang I cave, South Sulawesi; Indonesia. *Mod. Quat. Res. S.E. Asia* 2: 53–67.

Clason, A.T. 1979. Mesolithic huntergatherers in Sulawesi. In *South Asian Archaeology 1977, Vol. I.,* ed. M. Taddei, pp. 3–6. Instituto Universitario Orientale, Naples.

Clutton-Brock, T.H. and Harvey, P.H. 1977. Primate ecology and social organization. *J. Zool., Lond.* 183: 1–39.

Cockburn, P.F. 1974. The origin of the Sook Plain, Sabah. *Malay. Forest.* 37: 61–63

Cole, B. 1981. Colonizing abilities, island size

and the number of species on archipelagoes. *Amer. Nat.* 117: 629–638.

Cole, G.A. 1983. *Textbook of Limnology*. Missouri: Masby St. Louis.

Coley, P.D. 1982. Rates of herbivory on different tropical trees. In *The Ecology of a Tropical Seasonal Forest: Rhythms and Changes*, eds. E.G. Leight, A.S. Rand, D.M. Windsor, pp. 123–132. Washington, D.C.: Smithsonian Institution Press.

Coley, P.D. 1983a. Intraspecific variations in herbivory on two tropical tree species. *Ecology* 64: 426–433.

Coley, P.D. 1983b. Herbivory and defensive characteristics of tree species in a lowland tropical forest. *Ecol. Monogr.* 53: 209–233.

Collar, N.J. 1986. Species area a measure of man's freedom: reflections after writing a Red Data Book on African birds. *Oryx* 20: 15–19.

Collins, M. 1984. *Urban Ecology: A Teacher's Resource Book*. Cambridge: Cambridge University Press.

Collins, N.M. 1980a. The distribution of soil macrofauna on the west ridge of Gunung (Mount) Mulu, Sarawak. *Oecologia* 44: 263–275.

Collins, N.M. 1980b. The habits and populations of terrestrial crabs (Brachyura: Gecarcinacoidea and Grapsoidea) in the Gunung Mulu National Park, Sarawak. *Zool. Meded.* 55: 81–85.

Collins, N.M. 1982. The importance of being a bugga-bug. *New Scient.* 94: 834–837.

Collins, N.M. 1983. Termite populations and their role in litter removal in Malaysian rain forests. In *Tropical Rain Forest: Ecology and Management*, eds. S.L. Sutton, T.C. Whitmore and A.C. Chadwick, pp. 311–325. Oxford: Blackwell.

Collins, N.M. and Morris, M.G. 1985. *Threatened Swallowtail Butterflies of the World - an IUCN Red Data Book*. Cambridge: IUCN.

Collins, R.T. Jung, R.K., Anoez, H., Sutrisno, R.H. and Putut, D. 1979. A study of the coastal malarial vectors *Anopheles sundaicus* (Rodenwaldt) and *A. subpictus crassi* in South Sulawesi. Dinas Kesehatan, Ujung Pandang.

Connell, J.H. 1972. Community interactions on marine rocky intertidal shores. *Ann. Rev. Ecol. Syst.* 3: 132–169.

Conway, G.R. ed. 1982. *Pesticide Resistance and World Food Production*. London: Imperial College, University of London.

Conway, G.R. 1985a. Agroecosystem analysis. *Agric. Admin.* 20: 31–55.

Conway, G.R. 1985b. Agricultural ecology and farming systems research. In *Agricultural Systems Research for Developing Countries*, ed. J. Remenyi, pp. 43–59. Canberra: ACIAR.

Cook, L.M. 1983. Polymorphism in a mangrove snail in Papua New Guinea. *Biol. J. Linn. Soc.* 20: 167–173.

Coomans de Ruiter, L. 1946. Over de wederontdekking van *Aramidopsis plateni* (W. Blacius) In de Minahasa (Noord-Celebes) en het voorkomen van *Gymnocrex rosenbergi* (Schlegel) aldaar. *Limosa* 19: 65–75.

Coomans de Ruiter, L. 1950. Vogels van het Quarles-Gebergte (ZW. Centraal Celebes). *Ardea* 38: 40–64. (Partial translation available.)

Coomans de Ruiter, L. 1951. Vogels van het dal de Bodjo-rivier (Zuid Celebes). *Ardea* 39: 261–318.

Coomans de Ruiter, L. 1954. Trekvogels in Sulawesi. *Trop Natuur* 34: 67–96.

Coomans de Ruiter, L. and Maurenbrecher, L.L.A. 1948. Stadsvogels van Makasar (Zuid-Celebes). *Ardea* 36: 163–198. (Partial English translation available.)

Corlett, R.T. 1986. The mangrove understorey: some additional observations. *J. Trop. Ecol.* 2: 93–94.

Corner, E.J.H. 1952. *Wayside Tress of Malaya*. *2nd ed.* (2 vols.) Singapore: Government Printer.

Corner, E.J.H. 1964. *The Life of Plants*. Chicago: University of Chicago Press.

Cossalter 1980. Location and ecological data of some provenances of *Eucalyptus deglupta* Blume in the Celebes and Ceram Islands: Characteristics of the natural stands. *For. Gen. Res. Info.* 6: 16–23.

Coster, C. 1926. Perioedische Blutterscheinungen in den Tropen. *Ann. Jard. Bot. Buitenzorg* 35: 125–162.

Cranbrook, Earl of 1981. The vertebrate faunas. In *Wallace's Line and Plate Tectonics*, ed. T.C. Whitmore, pp. 57-69. Oxford: Oxford University Press.

Crawfurd, J. 1856. *A Descriptive Dictionary of the Indian Islands and Adjacent Countries*.

London: Bradbury and Evans.

Cremer, K.W. 1977. Distance of seed dispersal in eucalyptus estimated from seed weights. *Aust. For. Res.* 7: 225–228.

Cross, J.H., Clarke, M.D., Carney, W.P., Putrali, J. Joesoef, A. and Sajidiman, H. 1975. Parasitology survey in the Palu valley, Central Sulawesi (Celebes), Indonesia. *S.E. Asian J. Trop. Med. Pub. Hlth.* 6: 366–375.

Cross, J.H., Wheeling, C.H., Stafford, E.E., Irving, G.S., Peterson, H.V. and Gindo, S. 1977. Biomedical survey of the Minahasa Peninsular of North Sulawesi, Indonesia. *S.E. Asian J. Trop. Med. Pub. Hlth.* 8: 390–399.

Crowther, J. 1981. Small-scale spatial variations in the chemistry of diffuse-flow seepages in Gua Anak Takun, West Malaysia. *Brit. Cave Res. Assoc.* 8: 168–177.

Crowther, J. 1982a. The thermal characteristics of some West Malaysian rivers. *Malay. Nat. J.* 35: 99–109.

Crowther, J. 1982b. Ecological observations in a tropical karst terrain, West Malaysia. I. Variations in topography, soils and vegetation *J. Biogeogr.* 11: 65–78.

Crowther, J. 1984. Mesotopography and soil cover in tropical karst terrain, West Malaysia. *Z. Geomorph.* 28: 219–234.

Croxall, J.P. 1976. The composition and behaviour of some mixed species bird flocks in Sarawak. *Ibis* 118: 333–346.

Dahl, A.L. 1981a. *Coral Reef Monitoring Handbook*. Noumea: South Pacific Commission.

Dahl, A.L. 1981b. Monitoring coral reefs for urban impact. *Bull. Mar. Sci.* 31: 544–551.

Dammerman, K.W. 1929. *The Agricultural Zoology of the Malay Archipelago de Bussy*. Amsterdam.

Darlington, P.J. 1957. *Zoogeography: the Geographical Distribution of Animals*. New York: Wiley.

Darnaedi, D. and Budiman, A. 1984. Analisis vegetasi hutan mangrove Morowali, Sulawesi Tengah. In *Prosiding Seminar II Ekosistem Mangrove*, pp. 162–171. Lembaga Oseanologi Nasional, Jakarta.

Darsidi, A. 1982. Pengelolaan hutan mangrove di Indonesia. In *Prosiding Seminar II Ekosistem Mangrove*, pp. 19–28. Lembaga Oseanologi Nasional, Jakarta.

Darsidi, A. and Liang, D.H. 1986. Jalur hijau mangrove dalam kontak tata guna hutan pantai. Diskusi Panel Dayaguna dan Lebar Jalurhijau Mangrove, Ciloto, 27 February - 1 March 1986.

Davidson, J. 1977. Exploration collection, evaluation conservation and utilization of the genetic resources of tropical *Eucalyptus deglupta* Bl. Paper presented at Third World Consultation on Forest Tree Breeding, Canberra.

Davie, J.D.S. 1984. Structural variation, litter production and nutrient status of mangrove vegetation in Moreton Bay. In *Focus on Stradbroke*, eds. R. Coleman, J. Covacevich, and P. Davie, pp. 208–223. Brisbane: Royal Soc. Queensland.

Davie, J.D.S. and Mustafa, M. 1984. Socioeconomic factors as environmental constraints to sustainable fisheries management on inhabited coral cays in South Sulawesi, Indonesia, pp. 149-152. Proc. MAB/COMAR Seminar, Tokyo.

Davis, C.C. 1955. *The Marine and Freshwater Plankton*. Michigan: Michigan University Press.

Davis, D.R., Clayton, D.H., Janzen, D.H. and Brooke, A.P. 1986. Neotropical Tiniedae II: Biological notes and descriptions of two new moths phoretic on spiny pocket mice in Costa Rica (Lepidoptera: Tineoidea). *Proc. Entomol. Soc. Wash.* 88: 98–109.

Davis, G. 1976. Parigi: A Social History of the Balinese Movement to Central Sulawesi, 1907-1974. Ph.D. thesis, Stanford University.

Davis, T.A., Sudasrip, H. and Darwis, S.N. 1985. *Coconut Research Institute, Manado, Indonesia*. Manado: Coconut Research Institute.

Dawes, C.J. and Lawrence, J.M. 1983. Proximate composition and calorific content of seagrasses. *Mar. Tech. Soc. J.* 17: 53–58.

Dawson, T.J. and Degabrielle, R. 1973. The cuscus (*Phalanger maculatus*) - A marsupial sloth? *J. Comp. Physiol.* 83: 41–50.

Dawson, T.J., Sudasrip, H. and Darwis, S.N. 1985. *Coconut Research Institute, Manado, Indonesia: An Overview of Research Activities*. Manado: CRI.

Day, R.W. 1983. Effects of benthic algae on sessile animals: observational evidence from coral reef habitats. *Bull. Mar. Sci.* 33: 53–58.

de Klerk, L.G. 1983. Zeespiegels, riffen en

kustvlakten in Zuidwest Sulawesi, Indonesia; eenmorphogenetisch-bodemkundige studie. Thesis, University of Utrecht.

de Korte, K. 1984. Status and conservation of seabird colonies in Indonesia. *ICBP Techn. Publ.* 2: 527–545.

de Laubenfels, D.J. 1969. A revision of the Malesian and Pacific rainforest conifers, I. Podocarpaceae, in part. *J. Arn. Arb.* 50: 274–369.

de Laubenfels, D.J. 1978. The taxonomy of Philippine Coniferae and Taxaceae. *Philipp J. Biol.* 7: 117–152.

de Leeuw, H. 1931. *Crossroads of the Java Sea.* New York: Garden City.

de Nève, G.A. 1982. Development and origin of the Sangkarang reef archipelago (South Sulawesi, Indonesia). Proc. PIT X Ikatan Ahli Geologi Indonesia, 8-10 December, 1981, pp. 102-III, Bandung.

de Rooij, N. 1915. *The Reptiles of the Indo-Australian Archipelago Vol. I. Laceritilia, Chelonia and Emydosauria.* Leiden: Brill.

de Rooij, N. 1917. *The Reptiles of the Indo-Australian Archipelago Vol. II.* Leiden: Brill.

de Vent, C. 1948. Het Paloedal in Midden Celebes: Dood land. *Tijds. Kon. Ned. Aadr. Gen.* 65: 842–843.

de Vogel, E.E. 1986. The vegetation of the area west of Soroako. In van Balgooy and Tantra (1986).

de Wilde, W.J.J.O 1962. Najadaceae. *Flora Malesiana I* 6: 155–171.

de Wilde, W.J.J.O. 1981. Supplementary data on Malesian *Knema* (Myristicaceae), including three new taxa. *Blumea* 27: 223–234.

Dean, J.M. and Smith, A.P. 1978. Behavioral and morphological adaptations of a tropical plant to high rainfall. *Biotropica* 10: 152–154.

den Hartog, C. 1957a. Alismataceae. *Flora Malesiana I* 5: 317–334.

den Hartog, C. 1957b. Hydrocharitaceae. *Flora Malesiana I* 5: 381–413.

den Hartog, C. 1970. *Sea Grasses of the World.* Amsterdam: North Holland.

Denton, L.G., Marsh, H., Heinsohn, G.E. and Burdon-Jones, C. 1980. The unusual metal status of the dugong *Dugong dugon. Mar. Biol.* 57: 201–219.

Diamond, J.M. 1973. Distributional ecology of New Guinea birds. *Science* 179: 759–765.

Dickerson, R.E. 1928. *Distribution of Life in the Philippines.* Manila: Bureau of Printing.

Dieterlen, F. 1982. Fruiting seasons in the rain forest of Eastern Zaire and their effect upon reproduction. Poster presentation at Tropical Rain Forest Symposium, 1982. Leeds University, Leeds.

Ding Hou 1957. Centrolepidaceae. *Flora Malesiana I* 5: 421–428.

Ding Hou 1958. Rhizophoraceae. *Flora Malesiana I* 5: 429–493. Introductory section on ecology (pp. 431–441) by C.G.G.J. van Steenis.

Ding Hou 1978. Anacardiaceae. *Flora Malesiana I* 8: 395–465.

Ding Hou 1984. Aristolochiaceae. *Flora Malesiana I* 10: 53–108.

Disney, H. 1985. Watching scuttle flies on Project Wallace. *Antenna* 9: 139–141.

Disney, H. 1986a. A new genus and two new species of Phoridae (Diptera) from nests of ants (Hymenoptera: Formicidae) in Sulawesi. *J. Nat. Hist.* 20: 777–787.

Disney, H. 1986b. Two remarkable new species of scuttle fly (Diptera: Phoridae) that parasitise termites (Isoptera) in Sulawesi. *System. Zool.* 11: 413–422.

Disney, H. 1986c. Morphological and other observations and phylogenetic implications for the Cyclorrhapha (Diptera.). *J. Zool., Lond.* A 210: 77–87.

Disney, H., in prep. Biology and taxonomy of Old-World Puliciphora (Diptera: Phoridae) with description of nine new species.

Disney, R.H.L. 1986. A new genus and three new species of Phoridae (Diptera) parasitizing ants (Hymenoptera) in Sulawesi. *J. Nat. Hist.* 20: 777–787.

Ditlev, H. 1980. *A Field Guide to the Reef-Building Corals of the Indo-Pacific.* Backhuys: Rotterdam.

Djajasasmita, M. 1972. A list of the freshwater molluscs of Sulawesi (=Celebes). Unpubl. ms.

Djajasasmita, M. 1975. On the species of the genus *Corbicula* from Celebes, Indonesia (Mollusca, Corbiculidae). *Bull. Zool. Mus.* 4: 83–88.

Djajasasmita, M. 1982. The occurrence of *Anodonta woodiana* Lea 1837 in Indonesia

(Pelecypoda: Unionidae). *Veliger* 25: 175.

Docters van Leeuwen, W.M. 1911. Ueber die Ursache der wiederhotten Verzweigung der Stuetzwurzeln van *Rhizophora. Ber. Deut. Bot. Ges.* 29: 476–478.

Doctors van Leeuwen, W.M. 1937. Botanical results of a trip to the Salajar Islands. *Blumea* 2: 239–277.

Doda, J. 1980. Studi kelimpahan dan keragaman jenis serangan di daerah pertanian desa transmigrasi Mopuya, Kabupaten Bolaang Mongondow. Dissertation, Institut Pertanian Bogor, Bogor.

Doty, M.S. Soeriaatmadja, R.E. and Soegiarto, A., 1963. Penelitian laut di Indonesia. *Mar. Res. Indonesia* 5: 1–25.

Dove, M.R. 1985a. *Swidden Agriculture in Indonesia: The Subsistence Strategies of the Kalimantan Kantu.* Berlin: Mouten.

Dove, M.R. 1985b. The agroecological mythology of the Javanese and the political economy of Indonesia. *Indonesia* 39: 1–36.

Dove, M.R. 1985c. *Swidden Agriculture in Indonesia: The Subsistence Strategies of the Kalimantan Kantu.* Berlin: Mouton.

Doyle, M.E. 1969. Factors affecting the distribution of fauna in Gua Pondok. *Malay. Nat. J.* 23: 21–26.

Dransfield, J. 1976. A note on the habitat of *Pigafatta filaris* in North Celebes. *Principes* 20: 48.

Dransfield, J. 1981. Palms and Wallace's Line. In *Wallace's Line and Plate Tectonics,* ed. T.C. Whitmore, pp. 43-56. Oxford: Clarendon.

Dransfield, J. 1987. Bicentric distribution in Malesia as exemplified by palms. In *The Biogeographic Evolution of the Malay Archipelago,* ed. T.C. Whitmore. Oxford: Clarendon.

Dransfield, S. 1980. Three new Malesian species of Graminae. *Reinwardtia* 9: 385–392.

Dransfield, S. 1983. The genus *Racemobambos* (Graminae : Bambusoideae). *Kew Bull.* 37: 661–679.

Dudgeon, D. 1982a. An investigation of physical and biological processing of two species of leaf litter in Tai Po Forest Stream New Territories, Hong Kong. *Arch Hydrobiol.* 96: 1–32.

Dudgeon, D. 1982b. The life history of *Brotia hainanensis* (Brot. 1872) (Gastropoda:

Prosobranchia: Thiaridae) in a tropical forest stream. *Zool. J. Linn. Soc.* 76: 141–154.

Dudgeon, D. 1983a. Preliminary measurements of primary production and community respiration in a forest stream in Hong Kong. *Arch Hydrobiol.* 98: 287–298.

Dudgeon, D. 1983b. An investigation of the drift of aquatic insects in Tai Po Kan Forest Stream New Territories, Hong Kong. *Arch. Hydrobiol.* 96: 434–447.

Dudgeon, D. 1983c. The importance of streams in tropical rain-forest systems. In *Tropical Rain Forest: Ecology and Management,* eds. S.L. Sutton, A.C. Chadwick and T.C. Whitmore, pp. 71-82. Oxford: Blackwell.

Dudgeon, D. 1986a. The life cycle, population dynamics and productivity of *Melanoides tuberculata* (Muller, 1774) (Gastropoda: Prosobranchia: Thiaridae) in Hong Kong. *J. Zool.* 208: 37–53.

Dudgeon, D. 1986b. Research on Sulawesi stream insects. *Antenna* January 1986: 24–26.

Dudgeon, D. and Yipp, M.W. 1983a. A report on the gastopod fauna of aquarium fish farms in Hong Kong, with special reference to an introduced human schistosome host species *Biomphalaria straminea* (Pulmonata: Planorbidae). *Malacol. Rev.* 16: 93–94.

Dudgeon, D. and Yipp, M.W. 1983b. The diets of Hong Kong freshwater gastropods. In *Proceedings of the Second International Workshop on the Malaco-fauna of Hong Kong and Southern China,* eds. B. Morton and D. Dudgeon, pp. 491–509. Hong Kong: Hong Kong University Press.

Duffels, J.P. 1983. Distribution patterns of Oriental *Cicadoidea* (Homoptera) east of Wallace's Line and plate tectonics. *Geo J.* 7: 491–498.

Duke, N.C., Bunt, J.S. and Williams, W.T. 1981. Mangrove litter fall in North-eastern Australia. I. Annual totals and components in selected species *Aust. J. Bot.* 29: 547–553.

Duke, N.C., Bunt, J.S. and Williams, W.T 1984. Observations on the floral and vegetative phenologies of north-eastern Australian mangroves. *Aust. J. Bot.* 32: 87–99.

Dunne, T. and Leopold, L.B. 1983. *Water in Environmental Planning.* San Francisco: Freeman.

Durden, L.A. 1985. Rats and ectoparasites on Project Wallace. *Antenna* 10: 29–30.

Durden, L.A. 1986. Ectoparasites and other

arthropod associates of tropical rain forest mammals in Sulawesi Utara, Indonesia. *Nat. Geog. Res.* 2: 322–331.

Durden, L.A., in press. The reinfestation of forest rats (*Maxomys musschenbroeckii*) by epifaunistic arthropods in Sulawesi, Indonesia. *J. Trop. Ecol.*

Eccles, D.H., Jensen, R.A.C. and Jensen, M.K. 1969. Feeding behaviour of the bat hawk. *Ostrich* 40: 26–27.

Eck, S. 1976. Die Voegel der Banggai Inseln, inbesonderes pelengs. *Zool. Abh.* 34: 53–98.

Edwards, P.J. 1977. Studies of mineral cycling in a montane rain forest in New Guinea. II. The production and disappearance of litter. *J. Ecol.* 65: 971–992.

Edwards, P.J. 1982a. Studies of mineral cycling in a montane forest. IV. Soil characteristics and the division of mineral elements between the vegetation and the soil. *J. Ecol.* 70: 649–666.

Edwards, P.J. 1982b. Studies of mineral cycling in a montane rain forest in New Guinea. V. Rates of cycling in throughfall and litter fall. *J. Ecol.* 70: 507–528.

Edwards, P.J. and Wratten, S.D. 1980. *Ecology of Insect-Plant Interactions*. London: Edward Arnold.

Edwards, W.E. 1967. The late pleistocene extinction and diminution in size of many mamalian species. In *Pleistoacene Extinctions: A Search for a Cause*, eds. P.S. Martin and H.E. Wright, pp. 141–154. New Haven: Yale University Press.

Ehrenfeld, D. 1986. Thirty million cheers for diversity. *New Scient.* 12 June: 43–88.

Ehrlich, P.R. and Ehrlich, A.H. 1982. *Extinction: The Causes and Consequences of the Disappearance of Species*. New York: Random House.

Eisner. T. and Nowick, S. 1983. Spider web protection through visual advertisement: role of the stabilimentum. *Science* 219: 185–186.

Erkert, H.G. 1982. Ecological aspects of bat activity rhythms. In *The Ecology of Bats*, ed. T.H. Kunz, pp. 201–242. New York: Plenum.

Erwin, T.L. 1982. Tropical forests: their richness in Coleoptera and other arthropod species. *Coleopt. Bull.* 36: 74–75.

Erwin, T.L. and Scott, J.C. 1980. Seasonal and size patterns, trophic structure, and richness of Coleoptera in the tropical arboreal ecosystem: the fauna of the tree *Luehea seemanii* Triana and Planch in the Canal Zone of Panama. *Coleopt. Bull.* 34: 305–322.

Escott, C.J. and Holmes, D.A. 1980. The avifauna of Sulawesi, Indonesia: faunistic notes and additions. *Bull. Brit. Orn. Cl.* 100: 189–194.

Evans, G.O., Sheds, J.G. and Macfarlane, D. 1961. *The Terrestrial Acari of the British Isles. Vol. I.* London: British Museum (Natural History).

Ewel, J. 1980. Tropical succession: mani-fold routes to maturity *Biotropica* (Tropical Succession Suppl.) pp. 2–7.

Ewel, J. 1983. Succession. In *Tropical Rain Forest Ecosystems. A. Structure and Function*, ed. F.B. Golley, pp. 217–223. Amsterdam: Elsevier.

Ewel, J., Berish, C., Brown, B., Price, N., and Reach, J. 1981. Slash and burn impacts on a Costa Rican wet forest site. *Ecology* 62: 816–829.

Fain, A. and Hyland, K.E. 1974. The listrophorid mites in North America. II. The family Listrophoridae. *Bull. Inst. R. Sci. Nat. Belg.* 50: 1–69.

Fasola, M. 1982. Feeding dispersion in the night heron *Nycticorax* and little egret *Egretta garzetta* and the information centre hypothesis. *Boll. Zool.* 49: 177–186.

Fenton, M.B. 1983. *Just Bats*. Toronto: University of Toronto Press.

Fenton, M.B., and Fullard, J.H. 1981. Moth hearing and the feeding strategies of bats. *Amer. Scient.* 266–275.

Fenton, M.B., Boyle, N.G.H., Harrison, T.M. and Oxley, D.J. 1979. Activity patterns, habitat use, and prey selection by some African insectivorous bats. *Biotropica* 9: 73–85.

Ferguson, M.W.J. and Joanen, T. 1982. Temperature of egg incubation determines sex in *Alligator mississippiensis*. *Nature* 296: 850–853.

Fernando, C.H. 1963. Notes on aquatic insects colonizing an isolated pond in Mawai, Johore. *Bull. Nat. Mus. Singapore* 32: 80–89.

Fernando, C.H. 1977. Investigations on the aquatic fauna of ricefield with special reference to Southeast Asia. *Geo-Eco-Trop.* 3: 169–188.

Fernando, C.H. 1980. Ricefield ecosystems: a synthesis. In *Tropical Ecology and Development*, ed. J.I. Furtado, pp. 939–942. Kuala Lumpur: Universiti Malaya Press.

Fernando, C.H. 1984. Reservoirs and lakes of the Southeast Asia (Oriental Region). In *Lakes and Reservoirs*, ed. F.B. Taub, pp. 411–446. Amsterdam: Elsevier.

Fernando, C.H., Furtado, J.I. and Lim, R.P. 1980. The ecology of ricefield with special reference to the aquatic fauna. In *Tropical Ecology and Development*, ed. J.I. Furtado, pp. 943–951. Kuala Lumpur: Universiti Malaya Press.

Finlayson, M.C., Farrell, T.P. and Griffths, D.J. 1984. Studies of the hydrobiology of a tropical lake in Northwestern Queensland III. Gouth, Chemical composition and potential for harvesting of the aquatic vegetation. *Aust. J. Inter. Freshw. Res.* 31: 584–596.

Fishelson, L., Montgomery, W.L. and Myrberg, A.A. 1985. A unique symbiosis in the gut of tropical herbivorous surgeonfish (Acanthuridae: Teleostei) from the Red Sea. *Science* 229: 49–51.

Fisher, J.B. 1976. Adaptive value of rotten tree cores. *Biotropica* 8: 261.

Fitting, H., Schumacher, H.W., Harder, R. and Firbas, F. 1954. Lehrbuch der Botanik für Hochsschule. Gustav Fischer: Stuttgart.

Fleming, T.H. 1981. Fecundity, fruiting pattern, and seed dispersal in *Piper amalago* (Piperaceae), a bat-dispersed tropical shrub. *Oecologia* 51: 42–46.

Fleming, T.H. 1986. Opportunism versus specialization: the evolution of feeding strategies in frugivorous bats. In *Frugivores and Seed Dispersal*, eds. A. Estrada and T.N. Fleming, pp. 105-118. Dordrecht: Junk.

Fleming, T.H. and Heithaus, E.R. 1981. Frugivorous bats, seed shadows and the structure of tropical forests. *Biotropica* (Reprod. Bot. Suppl.), pp. 45–53.

Flenley, J.R. 1980. *The Equatorial Rain Forest: A Geological History*. London: Butterworth.

Flenley, J.R. 1980. The Quaternary history of the tropical rain forest and other vegetation of tropical mountains. *IV Int. Palyn. Conf. Lucknow* (1976-77) 3: 21–27.

Fletcher, N. 1981. Leaf size and leaf temperature in tropical vines. *Am. Nat.* 11: 1011–1014.

Fogden, M.P.L. 1972. The seasonality and population dynamics of equatorial forest birds in Sarawak. *Ibis* 114: 307–343.

Fogden, S.C.L. and Proctor, J. 1985. Notes on the feeding of land leeches (*Haemadipsa zeylandica* Moore and *H. picta* Moore) in Gunung Mulu National Park. *Biotropica* 17: 172–174.

Fontanel, J. and Chantefort, A. 1978. *Bioclimates of the Indonesian Archipelago*. Institut Francais de Pondicherry: Pondicherry.

Fooden, J. 1969. Taxonomy and evolution of the monkeys of Celebes. *Bibl. primatol.* 10: 1–148.

Forshaw, J.M. and Cooper, W.T. 1978. *Parrots of the World*. David and Charles, Newton Abbot.

Francis, P. 1978. *Volcanoes*. London: Penguin.

Frank, R. 1981. Sediments from Leang Burung 2. *Mod. Quat. Res. S.E. Asia* 6: 39–44.

Frankel, O.H. and Soulé, M.E. 1981. *Conservation and Evolution*. Cambridge: Cambridge University Press.

Franken, N.A.P. and Roos, M.C. 1981. Studies in lowland equatorial forest in Jambi Province. Unpublished report, BIOTROP.

Frankie, G.W. and Daly, H.V. 1983. *Xylocopa gualanensis* (Xicote, Avispa Carpentera, Carpenter Bee). In *Costa Rican Natural History*, ed. D.H. Janzen, pp. 777–779. Chicago: University of Chicago Press.

Freeland, W.J. and Janzen, D.H. 1974. Strategies in herbivory by mammals: the role of plant secondary compounds. *Am. Nat.* 108: 269–289.

Freeland, W.J. and Winter, J.W. 1975. Evolutionary consequences of eating: *Trichosurus vulpecula* (Marsupalia) and the genus *Eucalyptus*. *J. Chem. Ecol.* 4: 439–455.

Freeze, R.A. and Cherry, J.A. 1979. *Groundwater*. New Jersey: Prentice-Hall.

Frith, D.W., Tantanasiriwong, R. and Bhatia, O. 1976. Zonation of macrofauna on a mangrove shore, Phuket Island. *Phuket Mar. Biological Center, Res. Bull. No. 10*.

Gaisler, J. 1979. The ecology of bats. In *Ecology of Small Mammals*, ed. D.R. Stoddart, pp. 281–342. London: Chapman and Hall.

Ganning, B., Reish, D.J. and Straughan, D. 1984. Recovery and restoration of rocky shores, sandy beaches, tidal flats, and shallow subtidal bottoms impacted by oil spills. In *Restoration of Habitats Impacted by Oil Spills,* eds. J. Cairns and A.L. Buikema, pp. 7–36. Ann Arbor, Michigan.

Gaskell, B.H. 1984. Flying fruit-bat faunas of the upper canopy in two palaeotropical rain forests. In *Tropical Forest: The Leeds Symposium,* eds. A.C. Chadwick and S.L. Sutton, pp. 303. Leeds: Leeds Philosopical and Literary Society.

Geertz, C. 1963. Agricultural Involution: *The Processes of Ecological Change in Indonesia.* Berkeley: University of California Press.

Getter, C.D., Cintron, G., Dicks, B. and Lewis, R.R. 1984. The recovery and restortion of salt marshes and mangroves following an oil spill. In *Restoration of Habitats Impacted by Oil Spills,* eds. J. Cairns and A.L. Buikema, pp. 65–114. Stoneham: Butterworth.

Gilbert, O. 1983. The wildlife of Britain's wasteland. *New Scient.* 97: 824–829.

Gill and Rasmusson 1983. *Nature* 306: 229–234.

Gittins, S.P. 1983. Road casualties solve toad mysteries. *New Science.* 97: 530–531.

Glover, E. 1981. Leang Burung 2: Shell analysis. *Mod. Quat. Res. S.E. Asia* 6: 45–50.

Glover, I. C. 1976. Ulu Leang cave, Maros: a preliminary sequence of post-Pleistocene cultural development in South Sulawesi. *Archipel.* 11: 113–154.

Glover, I.C. 1977. The late stone age in eastern Indonesia. *World Archaeol.* 9: 42–61.

Glover, I.C. 1978a. Survey and excavation in the Maros District, South Sulawesi, Indonesia: The 1975 field season. Indo-Pac. Prehist. *Ass. Bull.* 1: 60–102.

Glover, I.C. 1978b. Leang Burung 2: An upper Palaeolithic rock shelter in South Sulawesi, Indonesia. *Mod. Quat. Res. S.E. Asia* 6: 1–38.

Glover, I.C. 1979a. The late prehistoric period in Indonesia. In *Early South East Asia,* eds. R.B. Smith and W. Watson, pp. 167–184. Oxford: Oxford University Press.

Glover, I.C. 1979b. Prehistoric plant remains from southeast Asia, with special reference to rice. In *South Asian Archaeology, Vol. I.,* ed. M. Taddei, pp. 7–37. Naples: Instituto Universitario Orientales.

Glover, I.C. 1981. Leang Burung 2: An upper palaeolithic rock shelter in South Sulawesi, Indonesia. *Mod. Quat. Res. S.E. Asia* 6: 1–38.

Glover, I.C. 1985. Some problems relating to the domestication of rice in Asia. In *Recent Advances in Indo-Pacific Prehistory,* eds. V.N. Misra and P. Bellwood, pp. 265–274. New Delhi: Oxford and IBH.

Glover, I.C. and Presland, G. 1985. Microliths in Indonesian flaked stone industries. In *Recent Advances in Indo-Pacific Prehistory,* eds. V.N. Misra and P. Bellwood, pp. 185–195. New Delhi: Oxford and IBH.

Goff, M.L., Durden, L.A. and Whitaker, J.O. 1986. A new species of *Schoengastia* (Acari: Trombiculidae) from mammals collected in Sulawesi, Indonesia. *Int. J. Acarol.* 12: 91–93.

Goldsmith, E. and Hildyard, N. 1984. The Social and Environmental Effects of Large Dams. Wadebridge Ecological Centre, Wadebridge, Cornwall.

Gong, W.K. and Ong, J.E. 1983. Litter production and decomposition in a coastal hill dipterocarp. forest. In *Tropical Rain Forest: Ecology and Management,* eds. S.L. Sutton, T.C. Whitmore and A.C. Chadwick, pp. 275–285. Oxford, Blackwell.

Goodland, R.J.A. 1985. Wildlands management in economic development. IX World Forestry Congress and the Wildlife Society of Mexico, 14 May 1983. Mexico City.

Goodland, R.J.A., Watson, C. and Ledec, G. 1984. *Environmental Management in Tropical Agriculture.* Boulder: Westview Press.

Goodwin, D. 1970. *Pigeons and Doves of the World.* London: British Museum (Natural History).

Goodwin, D. 1976. *Crows of the World.* Ithaca: Cornell University Press.

Goodwin, R.E. 1979. The bats of Timor: Systematics and ecology. *Bull. Amer. Mus. Nat. Hist.* 163: 1–122.

Gorman, M. 1979. *Island Ecology.* Chapman and Hall: London.

Gould, E. 1978. Foraging behaviour of Malaysian nectar-feeding bats. *Biotropica* 10: 184–193.

Gould, E. 1978. Opportunistic feeding by tropical bats. *Biotropica* 10: 75–76.

Grant, J. 1984. Sediment microtopography and shorebird foraging. *Mar. Ecol. Progr. Ser.* 19: 293–296.

Green, J., Corbet, S.A., Watts, E. and Oey, B.L. 1976. Ecological studies on Indonesian lakes: Overturn and restratification of

Ranu Lamongan. *J. Zool., Lond.* 180: 315–354.

Greenslade, P. and Deharveng, L. 1986. *Psammisotoma,* a new genus of Isotomidae (Collembola) from marine littoral habitats. *Proc. Royal. Soc. Queensland* 97: 89–95.

Greer, A.E. 1971. Crocodilian nesting habits and evolution. *Fauna* 3: 20–28.

Greer, A.E. 1974. On the maximum total length of the salt-water crocodile (*Crocodylus porosus*). *J. Herpetol.* 8: 381–384.

Gremmen, W.H.E. 1990. Palynological investigations in the Danau Tempe depression, Southwest Sulawesi (Celebes), Indonesia. *Mod. Quat. Res. S.E. Asia* 11: 123–134.

Gressitt, J.L. 1961. Problems in zoogeography of Pacific and Antarctic insects. *Pac. Ins. Monogr.* 2: 1–94.

Grigg, G.C. 1981. Plasma homeostasis and cloacal urine composition in *Crocodylus porosus* caught along a salinity gradient. *J. comp. Physiol. B.* 144: 261–270.

Groombridge, B. 1982. *The IUCN Amphibia-Reptilia Red Data Book. Part I.* Testudines Crocodylia, *and* Rhynchocephalia. Conservation Monitoring Centre, IUCN, Cambridge.

Groves, C.P. 1969. Systematics of the anoa. *Beaufortia* 17: 1–11.

Groves, C.P. 1976. The origin of the mammalian fauna of Sulawesi (Celebes). *Z. Saugetierk.* 41: 201–216.

Groves, C.P. 1980a. Notes on the systematics of *Babyrousa* (Artiodactyla, Suidae). *Zool. Med.* 55: 29–46.

Groves, C.P. 1980b. Speciation in *Macaca*: the view from Sulawesi. In *The Macaques: Studies in Ecology, Behaviour and Evolution,* ed. D.G. Lindburg, pp. 84–124. New York: Van Nostrand Reinhold.

Groves, C.P. 1981. Ancestors for the pigs. Tech. Bull. No. 3, Dept. Prehistory, Australian National University, Canberra.

Groves, C.P. 1985. The Sulawesi 'specials': Archaeic, strange, endemic. *Aust. Nat. Hist.* 21: 442–444.

Grubauer, A. 1923. *Celebes: Ethnologische Streifzuege in Suedost und Zentral Celebes.* Darmstadt: Hagen.

Grubb, P.J. 1971. Interpretation of the 'Massenerhebung' effect on tropical mountains. *Nature* 229: 44–46.

Grubb, P.J. 1974. Factors controlling the distribution of forest types on tropical mountains: new factors and a new perspective. In *Altitudinal Zonation in Malesia,* ed. J. R. Flenley, pp. 13–46. Hull: University of Hull.

Grubb, P.J. 1977. Control of forest growth and distribution on wet tropical mountains with special reference to mineral nutrition. *Ann. Rev. Ecol. Syst.* 8: 83–107.

Guille, A., Laboute, P. and Menou, J.L. 1986. *Guide des Etoiles de Mer, Oursins et Autres Echinoderms du Lagon de Nouvelle-Calédonie.* Paris: ORSTOM.

Guillemard, F.H.H. 1889. *The Cruise of the Marchesa to Kamschatka and New Guinea with notices of Formosa, Liu-Liu, and Various Islands of the Malay Archipelago.* London: Murray.

Guppy, N. 1983. The case for an Organization of Timber Exporting Countries (OTEC). In *Tropical Rain Forest: Ecology and Management,* eds. S.L. Sutton, T.C. Whitmore and A.C. Chadwick, pp. 459–463.

Guppy, N. 1984. Tropical deforestation a global view. *For. Affairs* Spring 1984.

Hadi, S., Hanson, A.J., Koesoebiono, Mahlan, M., Purba, M. and Rahardjo, S. 1977. *Mar. Res. Indonesia,* 19: 109–135.

Hadi, T.R. and Tenorio, J. M. 1982. A new species of *Laelaps* (Acari: Laelapidae) from Indonesia, with notes on the *juxtapositus* species-group. *J. Med. Entomol.* 19: 728–733.

Hadidjaja, P. 1982. Beberapa penelitian mengenai aspek biologik dan klinik schistosomiasis di Sulawesi Tengah, Indonesia. Thesis, Fakultas Kedokteran, Universitas Indonesia, Jakarta.

Hadikoesworo, H. 1977. The exploitation of the estuaries in the Kendal area (northern coast of Java) by artisinal fishermen. *Mar. Res. Indonesia* 19: 95–100.

Hadimuljono and Muttalib 1979. *Sejarah Kuno Sulawesi Selatan.* Kantor Suaka Peninggalan Sejarah dan Purbakala, Ujung Pandang.

Hadisumarno, S. 1977. The geomorphology of Palu area, Sulawesi from LANDSAT I. *Indon. J. Geog.* 7: 45–59.

Hadisumarno, S. 1978. Landsat imagery for land use/land cover mapping at the island of Sulawesi (Indonesia) and the problem of data transfer into the existing topographic base. *Indon. J. Geog.* 8: 41–49.

Hadiwijaya, S. 1981. Status perikanan

perairan umum di Sulawesi Utara. Prosiding Seminar Perikanan Perairan Umum, Jakarta, 19-20 August, 1981.

Haffer, J. 1982. General aspects of the Refuge Theory. In *Biological Diversification in the Tropics*, ed. G. Prance, pp. 6–24. New York: Columbia University Press.

Haile, N.S. 1978. Reconnaissance palaeomagnetic results from Sulawesi, Indonesia, and their bearing on palaegeographic reconstructions *Tectonophysics* 46: 743–771.

Hails, A.J. and Yaziz, S. 1982. Abundance, breeding and growth of the ocypodid crab *Dotilla myctiroides* (Milne Edwards) on a West Malaysian beach. *Estuar. Coast. Shelf Sci.* 15: 229–239.

Hails, C.J. 1984. The breeding biology of the Pacific swallow *Hirundo tahitica* in Malaysia. *Ibis* 126: 198–211.

Hails, C.J. and Amiruddin, A. 1981. Food samples and selectivity of white-bellied swiftlets *Collocalia esculenta*. *Ibis* 123: 328–333.

Hails, C.J. and Turner, A.K. 1984. The breeding biology of the Asian palm swift *Cypsiurus balasiensis*. *Ibis* 126: 74–81.

Hamilton, L.S. and King, P.N. 1983. *Tropical Forested Watersheds: Hydrologic and Soils Response to Major Uses or Conversions*. Boulder: Westview Press.

Hamilton, W. 1979. *Tectonics of the Indonesian Region*. Washington, D.C.: United States Goverment Printing Office.

Hammond, R. and King, M. 1984. *Nature by Design: A Teachers' Guide to Practical Nature Conservation*. Birmingham: Urban Wildlife Group.

Hamzah, Z. 1978. Some observations on the effects of mechanical logging on regenaration, soil and hydrological conditions in East Kalimantan. *BIOTROP Spec. Publ.* 3: 73–78.

Hanski, I. 1983. Distributional ecology and abundance of dung and carrion feeding beetles (Scarabaeidae) in tropical rain forest in Sarawak, Borneo. *Acta Zool. Fenn.* 167: 1–45.

Hanski, I. 1985. What does a shrew do in an energy crisis? In *Behavioural Ecology Ecological Consequences of Adaptive Behaviour*, eds. R.M. Sibly and R.H. Smith, pp. 247–252. Oxford: Blackwell.

Hanson, A.J. and Koesoebiono 1977. Settling

coastal swamplands in Sumatra: a case study for integrated resource management. PSPSP Research Report 004, Institut Pertanian, Bogor.

Harcombe, P.A. 1980. Soil nutrient loss as a factor in early tropical secondary succession. *Biotropica* 12 (*Tropical Succession Suppl.*), 8–15.

Hardjasoemantri, K. 1985. *Environmental Legislation in Indonesia*. Yogyakarta: Gadjah Mada University Press.

Harger, J.R.E. 1984. Rapid survey techniques to determine distribution and structure of coral communties. UNESCO *Rep. Mar. Sci.* 21: 83–91.

Harries, H.C. 1983. The coconut palm, the robber crab and Charles Darwin: April Fool or a curious case of instinct? *Principles* 27: 131–137.

Harris, L.D. 1984. *The Fragmented Forest*. Chicago: Chicago University Press.

Harrison, J.L. 1955. Data on the reproduction of some Malayan mammals. *Proc. zool. Soc. London.* 125: 445–460.

Harrison, J.L. 1956. Survival rates of Malayan rats. *Bull. Raffles Mus.* 27: 5–26.

Harrison, J.L. 1958. Range of movement in some Malayan rats. *J. Mamm.* 39: 190–206.

Harrison, J.L. 1968. The effect of forest clearance on small mammals. In *Conservation in Tropical Southeast Asia*. Gland: IUCN.

Harrison, P.L., Babcock, R.C., Bull, G. D., Oliver, J.K., Wallace, C.C. and Wiler, B.L. 1984. Mass spawning in tropical reef corals. *Science* 223: 1186–1189.

Hartert, E. 1896. Ornithological collections made by A. Everett in Celebes. *Novit. zool.* 3: 148–183.

Harun, W.K. and Tantra, I.G.M. 1983. Flora dan analisa vegetasi Cagar Alam Karaenta, Sulawesi Selatan. Laporan No. 423, Pusat Penelitian dan Pengembangan Hutan, Bogor.

Hasan, M. 1975. Percobaan penanaman rumput laut *Eucheuma spinosum* (Rhodophyta: Gigartinales) di Pulau Samaringa, Kecamatan Menui Kepulauan, Sulawesi Tengah. *L.P.P.L.* 1975: 78–101.

Hatcher, B.G. 1981. The interaction between grazing organisms and the epilithic algal community of a coral reef; a quantitative assessment. *Proc. 4th. Int. Coral Reef Symp., Manila.* 2: 515–524.

Hatcher, B.G. 1983. Grazing in coral reef ecosystems. In *Perspectives on Coral Reefs*, ed. D.J. Barnes, pp. 164–179. Manuka: Cloustan.

Hatcher, B.G. and Larkum, A.W.D. 1983. An experimental analysis of factors controlling the standing crop of the epilithic algal community on a coral reef. *J. Exp. Mar. Biol. Ecol.* 69: 61–84.

Haugum, J., Ebner, J. and Racheli, T. 1980. The Papillionidae of Celebes (Sulawesi). Lepidoptera Group of 1968, Supplement 9.

Hayes, A.H. 1983. A striking new species of *Gnathothlitus* (Lepidoptera: Sphingidae) (Macroglossinae) from Sulawesi. *Entomol. Rec.* 95: 19–20.

Heckman, C.W. 1979. *Rice Field Ecology in Northeastern Thailand. The Effect of Wet and Dry Seasons on a Cultivated Aquatic Ecosystem.* The Hague: Junk.

Heinrich, G. 1939. Mitteilungen zur Oekologie der Voegel von Celebes. In Die Voegel von Celebes (by E. Streseman), *J. Orn.* 87: 299–425.

Heinsohn, G.E. 1981a. Aerial survey techniques for dugongs. In *The Dugongs*, ed. H. Marsh. pp. 125–129. Townsville: James Cook University.

Heinsohn, G.E. 1981b. The dugong in the seagrass ecosystem. *Aquaculture* 12: 235–248.

Helfman, G.S. 1977. Agonistic behaviour of the coconut crab *Birgus latro* (L). *Z. Tierpsychol.* 43: 425–438.

Helfman, G.S. 1982. Coconut crabs and cannibalism. *Nat. Hist.* xx: 76-83.

Hendrix, S.D., and Marquis, R.J. 1983. Herbivore damage to three tropical ferns. *Biotropica* 15: 108–111.

Hendrokusumo, S., Sumitro, D. and Tas'an 1981. The distribution of the dugong in Indonesian waters. In *The Dugong*, ed. H. Marsh, pp. 5–10. Townsville: James Cook University.

Henrey, L. 1982. *Coral Reefs of Malaysia and Singapore.* Kuala Lumpur: Longman.

Henwood, K. 1973. A structural model of forces in buttressed tropical rain forest trees. *Biotropica* 5: 83–93.

Herbst, L.H. 1986. The role of nitrogen from fruit pulp in the nutrition of the frugivorous bat *Carollia perspicillata*. *Biotropica* 18: 39–44.

Heringa, P.K. 1921. Rapport over de begroeing van de Sangi-en Talaud-Eilanden. *Tectona* 14: 733–746.

Herrera, C.M. 1982. Defense of ripe fruit from pests: its significance in relation to plant-disperser interactions. *Am. Nat.* 12: 218–241.

Heyne, K. 1927. Nuttige Planten van Nederlandsch Indie.

Hickling, C.F. 1957. Tropical Inland Fisheries. Longmans, London.

Hickson 1889. *A Naturalist in Celebes.* London, Murray.

Hidayan, Z., n.d. The Wana people of Sulawesi. Report to Operation Drake.

Highsmith, R.C. 1982. Reproduction by fragmentation in corals. *Mar. Ecol. Prog. Ser.* 7: 207–226.

Hii, J., Meck, S. and Vun, Y.S. 1985. Mosquito surveys in the Dumoga-Bone National Park, North Sulawesi, Indonesia. Unpubl. report.

Hill, J.E. 1983. Bats (Mammalia: Chiroptera) from Indo-Australia. *Bull. Brit. Nat. Hist. Mus.* 45: 103–208.

Hill, J.E. and Smith, J.D. 1984. *Bats: A Natural History.* London: British Museum (Natural History).

Holdway, P., n.d. A biological survey of the western reef of the Vesuvius group. Unpubl. ms.

Holloway, J.D. 1987. Lepidoptera pattern involving Sulawesi : What do they indicate of past geography? In *Biogeographic Evolution of the Malay Archipelago*, ed. T.C. Whitmore. Oxford: Clarendon.

Holmes, P.R. and Wood, H.J. 1979. The Report of the Ornithological Expedition to Sulawesi, 1979. Unpubl. ms.

Holmes P.R. and Holmes, H.J. 1985. Notes on *Zosterops* spp. from the Lake Matano area of Southeast Sulawesi, Indonesia. *Bull. Brit. Orn. Cl.* 105: 136–140.

Holthuis, L.B. 1979. Caverniculous and terrestrial decapod crustaceans from northern Sarawak, Borneo, *Zool. Verh. Leiden.* 171: 1–47.

Holthuis, L.B. 1984. Freshwater prawns (Crustacea Decapoda: Natantia) from subterranean waters of the Gunung Sewu area, Central Java, Indonesia. *Zool. Meded.* 58: 141–148.

Holttum, R.E. 1954. *Plant Life in Malaya.* Kuala Lumpur: Longman.

Holttum, R.E. 1959. Gleicheniaceae. *Flora Malesiana II* 1: 1–36.

Holttum, R.E. 1963. Cyathaceae. *Flora Malesiana II* 1: 65–176.

Hoogmoed, M.S. and Crumly, C.R. 1984. Land tortoise types in the Rijksmuseum van Natuurlijke Historie with comments on nomenclature and systematics (Reptilia: Testudines: Testudinidae). *Zool. Med.* 58: 241–259.

Hoogstraal, H. and Wassef, H.J. 1977. *Haemaphysalis* (*Ornithophycalis*) *kadarsani* sp. n. (Ixodoidea: Ixodidae), a rodent parasite of virgin lowland forests in Sulawesi (Celebes). *J. Parasitol.* 63: 1103–1109.

Hoogstraal, H., Trapido, H. and Kohls, G.M. 1963. Studies on Southeast Asian *Haemaphysalis* ticks (Ixodidae). The identity, distribution and hosts of *H.* (*Kaiseriana*) *hystricis* Supino. *J. Parasitol.* 51: 467–480.

Hooijer, D.A. 1948a. Pleistocene vertebrates from Celebes. 3. *Anoa depressicornis* (Smith) subsp. and *Babyrousa babirusa beruensis* nov. subsp. *Proc. Kon. Ned. Akad. Wet.* 51: 1322–1330.

Hooijer, D.A. 1948b. Pleistocene vertebrates from Celebes. 2. *Testudo margae* n. sp. Proc. *Kon. Ned. Akad. Wet.* 51: 1169–1182.

Hooijer, D.A. 1949. Pleistocene vertebrates from Celebes. 3. *Archidiskodon celebensis* nov. spec. *Zool. Meded.* 30: 205–226.

Hooijer, D.A. 1950. Man and other mammals from Toalian sites in south western Celebes. *Verh. Kon. Ned. Akad. Wet.* 46: 1–165.

Hooijer, D.A. 1954. Crocodilian remains from the Pleistocene in Celebes. *Copeia* 4: 263–266.

Hooijer, D.A. 1958. The Pleistocene vertebrate fauna of Celebes. *Asian Persp.* 2: 71–76.

Hooijer, D.A. 1964. Pleistocene vertebrates from Celebes. 12. Notes on pygmy stegodonts. *Zool. Meded.* 40: 37–44.

Hooijer, D.A. 1967. Indo-Australian insular elephants. *Genetica* 38: 143–162.

Hooijer, D.A. 1969. Pleistocene vertebrates from Celebes. 13. *Sus celebensis* Muller and Schlegel, 1845. *Beaufortia* 16: 215–218.

Hooijer, D.A. 1970. Pleistocene south-east pygmy stegodonts. *Nature* 225: 474–475.

Hooijer, D.A. 1972. Pleistocene vertebrates from Celebes. 14. Additions to the *Archidiskodon-Celebocherus* fauna. *Zool. Meded.* 46: 1–16.

Hooijer, D.A. 1974. *Elephas celebensis* (Hooijer) from the Pleistocene of Java. *Zool. Meded.* 48: 86–93.

Hooijer, D.A. 1982. The extinct giant land tortoise and the pygmy stegodont of Indonesia. *Mod. Quat. Res. S.E. Asia* 7: 171–176.

Hope, G. 1986. Vegetation change in Sulawesi provinces. Unpubl. ms.

Hopkins, M.S. and Graham, A.W. 1983. The species composition of soil seed banks beneath lowland tropical rainforests in North Queensland, Australia. *Biotropica* 15: 90–99.

Hosang, M.L.A., Bennett, C.P.A., Holloway, J.D. 1986. *Parasa balitkae*, suatu species baru dari hama *Parasa* yang menyerang daun kelapa di Sulawesi Utara. Unpubl. ms.

Howarth, F.G. 1983. Ecology of cave arthropods. *Ann. Rev. Entamol.* 28: 365–389.

Howe, H.F. 1980. Monkey dispersal and waste of a neotropical fruit. *Ecology* 61: 944–959.

Howe, H.F. 1981. Dispersal of a neotropical nutmeg (*Virola sebifera*) by birds. *Auk* 98: 88–98.

Howe, H.F. 1984. Implications of seed dispersal by animals for tropical reserve management. *Biol. Conserv.* 30: 261–281.

Howe, H.F. and Smallwood, J. 1982. Ecology of seed dispersal. *Ann. Rev. Evol. Syst.* 13: 201–228.

Howe, H.F. and van de Kerckhove, G.A. 1980. Nutmeg dispersal by tropical birds. *Science* 210: 925–927.

Hubbell, S.P. and Johnson, L.K. 1977. Competition and nest spacing in a tropical stingless bee community. *Ecology* 58: 949–963.

Huc, R. 1981. Preliminary studies on pioneer trees in the dipterocarp forest of Sumatra. Unpublished report, BIOTROP, Bogor.

Huc, R. and Rosalina, U. 1981. Chablis and primary dynamics in Sumatra. Unpubl. BIOTROP report, Bogor.

Hutchinson, G.E. 1959. Homage to Santa Rosalia, or why are there so many kinds of animals? *Amer. Nat.* 93: 145–149.

Hutomo, M. and Djamali, A. 1980. Komunitas ikan pada padang 'seagrass' di pantai selatan pulau tengah, Pulau pulau Seribu. In *Sumber Daya hayati Bahari*, eds. Burhanuddin, M.K. Moosa, and H. Razak, pp. 97–108. Lembaga Oseanologi Nasional, Jakarta.

Hutomo, M. and Martosewojo, S. 1977. *Mar. Res. Indonesia* 77: 147–172.

Hutomo, M. and Naamin, N. 1982. Pengamatan pendahuluan tentang ikan gelodok, *Boleophtalmus boddaerti* Pallas dan catatan singkat tentang *Periophtalmus koelruteri* (Pallas). *Proc. Seminar II* Ekosistem Mangrove, pp. 243–249. Lembaga Oseanologi Nasional, Jakarta.

Iskandar, D. 1979. A second specimen of the Matanna water snake, *Enhydris matannesnsis* (Blgr.), from Raha, Muna Island, Indonesia. *Brit. J. Herpetol.* 5: 849–850.

Jacob, T. 1967. Some problems pertaining to the racial history of the Indonesian region. Netherlands Technical Assistance Bureau, Utrecht.

Jacobs, M. 1977. Report on northern extension of Lore Kalimanta proposed National Park, Celebes. Bogor: World Wildlife Fund.

Jacobs, M. 1978. Preliminary report on a first botanical exploration of the Opa Swamp and surrounding forests in S.E. Sulawesi. Unpubl. ms.

Jacobs, M. 1982. The study of minor forest products. *Flora Malesiana Bull.* 35: 3768–3782.

Jaffre, T. 1976. Composition chemique et conditions de l'alimentation minerale des plantes sur roches ultrabasiques (Nouvelle Caledonie). *C. ORSTOM ser. Biol.* 11: 53–63.

Janos, D.P. 1980. Mycorrhizae influence tropical succession. *Biotropica* (Trop. Succession Suppl.), pp. 56–64.

Janos, D.P. 1983. Tropical mycorrhizas, nutrient cycles and plant growth. In *Tropical Rain Forest: Ecology and Management,* eds. S.L. Sutton, T.C. Whitmore and A.C. Chadwick, pp. 327–345. Oxford: Blackwell.

Janzen, D.H. 1970. Herbivores and the number of tree species in tropical forests. *Am. Nat.* 104: 501–528.

Janzen, D.H. 1973a. Dissolution of mutualism between *Cecropia* and its *Azteca* ants. *Biotropica* 5: 15–28.

Janzen, D.H. 1973b. Rate of regeneration after a tropical high-elevation fire. *Biotropica* 5: 117–122.

Janzen, D.H. 1974a. Tropical blackwater rivers, animals and mast fruiting by the Dipterocarpaceae. *Biotropica* 6: 69–103.

Janzen, D.H. 1974b. Epiphytic myrmecophytes in Sarawak: mutualism through the feeding of plants by ants. *Biotropica* 6: 237–259.

Janzen, D.H. 1975. *Ecology of Plants in the Tropics.* London: Edward Arnold.

Janzen, D.H. 1976. Why tropical trees have rotten cores. *Biotropica* 8: 110.

Janzen, D.H. 1977. Why fruit rots, seeds mold and meat spoils. *Am. Nat.* 111: 691–713.

Janzen, D.H. 1978a. Complications in interpreting the chemical defenses of trees against tropical arboreal plant-eating vertebrates. In *The Ecology of Arboreal Folivores,* ed. G.G. Montgomery, pp. 73–84. Washington, D.C.: Smithsonian Institution Press.

Janzen, D.H. 1978b. A bat-generated fig seed shadow in rain forest. *Biotropica* 10–121.

Janzen, D.H. 1979. How many babies do figs pay for babies? *Biotropica* 11: 48–50.

Janzen, D.H. 1981. The defences of legumes against herbivores. In *Advances in Legume Systematics,* eds. R.M. Polhill and P.H. Raven, pp. 951–977. Kew: Royal Botanic Gardens.

Janzen, D.H. 1982. Simulation of *Andira* fruit pulp removal by *Cleogonus* weevils. *Brenesia* 19/20: 165–170.

Janzen, D.H. 1983. Insects - Introduction. In *Costa Rican Natural History,* ed. D.H. Janzen, pp. 619–645. Chicago: Chicago University Press.

Janzen, D.H. 1983a. Food webs: who eats what, why, how and with what effects in a tropical forest? In *Tropical Rain Forest Ecosystems,* ed. F.B. Colley, pp. 167–182. Amsterdam: Elsevier.

Janzen, D.H. 1983b. Seed and pollen dispersal by animals: convergence in the ecology of contamination and sloppy harvest. *Biol. J. Linn. Soc.* 20: 103–113.

Janzen, D.H. 1983c. Physiological ecology of fruits and their seeds. In *Physiological Plant Ecology III,* eds. O.L. Lange, P.S. Nobel, C.B. Osmond and H. Ziegler, pp. 625–655. Berlin: Springer-Verlag.

Janzen, D.H. 1984a. Dispersal of small seeds by big herbivores: foliage is the fruit. *Am. Nat.* 123: 338–353.

Janzen, D.H. 1984b. Weather related color polymorphism of *Rothschildia lebeau* (Saturniidae) *Bull. Entomol. Soc. Am.* 30: 16–20.

Janzen, D.H. 1985a. Coevolution as a process: What parasites of animals and plants do not have in common. In *Coevolution of Par-*

asitic Arthropods and Mammals, ed. Ke Chung Kim, pp. 83–100. New York: Wiley.

Janzen, D.H. 1985b. A host plant is more than its chemistry. III. *Nat. Hist. Surv. Bull.* 33: 141–174.

Janzen, D.H. 1985c. Mangroves: where's the understory? *J. Trop. Ecol.* 1: 89–92.

Janzen, D.H. 1986. Guanacaste National Park: Tropical ecological and cultural restoration. Report to National Park of Costa Rica.

Janzen, D.H. and Waterman, P.G. 1984. A seasonal census of phenolics, fibre and alkaloids in foliage of forest trees in Costa Rica: Some factors influencing their distribution and relation to host selection by Sphingidae and Saturniidae. *Bot. J. Linn. Soc.* 21: 439–454.

Janzen, D.H., Miller, G.A., Hackforth-Jones, J., Pond, C.M., Hooper, K., and Janos, D.P. 1976. Two Costa Rican bat-generated seed shadows of *Andira inermis* (Leguminosae). *Ecology* 57: 1068–1075.

Jenkins, P. and Hill, J.E. 1981. The status of *Hipposideros galeritus* Cantor, 1946 and *Hipposideros cervinus* (Gould, 1854) (Chiroptera, Hipposideridae). *Bull. Br. Mus. nat. Hist.* (Zool.) 41: 279–294.

Jennings, J.N. 1971. *Karst.* Cambridge, Mass.: MIT Press.

Jennings, J.N. 1972. The character of tropical humid karst. *Z. Geomorph.* 16: 336–341.

Jermy, C. and Walker, T.G. 1975. *Lecanopteris spinosa* a new ant-fern from Indonesia. *Fern Gaz.* 11: 165–176.

Jimanez, J.A., Lugo, A.E. and Cintron, G. 1985. Tree mortality in mangrove forest. *Biotropica* 17: 177–185.

Johannes, R.E. and Rimmer, D.W. 1984. Some distinguishing characteristics of nesting beaches of the green turtle *Chelonia mydas. Mar. Biol.* 83: 149–154.

Johansen, H.W. 1981. *Coralline Algae: A First Synthesis.* Florida: CRC.

Johns, A.D. 1985. Selective logging and wildlife conservation in tropical rain-forest: problems and recommendations *Biol. Conserv.* 31: 355–375.

Johns, A.D., Pine, R.H. and Wilson, D.E. 1985. Rain forest bats: an uncertain future. *Bat News* 5: 4–5.

Johnson, A. 1979. The algae of Singapore mangrove. In *Proceedings of the Symposium on Mangrove and Estuarine Vegetation in southeast Asia,* ed. P.B.L. Srivastava, pp. 45–49. Bogor: BIOTROP.

Johnson, B. 1985. Chimera or opportunity? An environmental appraisal of the recently concluded international timber agreement. *Ambio* 14: 42–44.

Johnson, D.L. 1980. Problems in the land vertebrate zoogeography of certain islands and the swimming powers of elephants. *J. Biogeog.* 7: 383–398.

Johnson, D.S. 1961. An instance of large-scale mortality of fish in natural habitat in S. Malaya. *Malay. Nat. J.* 15: 160–162.

Johnson, D.S. 1973. Equatorial forest and the inland aquatic fauna of Sundania. In *Nature and Conservation in the Pacific,* eds. A.B. Costin and R.H. Groves, pp. 111–116. Gland: IUCN.

Johnson, D.S., Soon, M.H.H. and Wee, E.I. 1979. Freshwater streams and swamps in the tree country of southern and eastern Malaya with special reference to aquarium fish. In *Natural Resources of Malaysia and Singapore,* ed. B. Stone. Kuala Lumpur: University of Malaya.

Johnson, L.K. 1983. *Trigona fulviventris.* In *Costa Rican Natural History,* ed. D.H. Janzen, pp. 770–772. Chicago: University of Chicago Press.

Johnson, L.K. and Hubbell, S.P. 1975. Constrasting foraging strategies and coexistence of two bee species on a single resource. *Ecology* 56: 1398–1406.

Johnson, L.K. and Hubbell, S.P. 1984. Nest tree selectivity and density of stingless bee colonies in a Panamanian forest. In *Tropical Rain Forest: Ecology and Management,* eds. S.L. Sutton, T.C. Whitmore and A.C. Chadwick, pp. 147–154. Oxford: Blackwell.

Johnstone, I.M. 1981. Consumption of leaves by herbivores in mixed mangrove stands. *Biotropica* 13: 252–259.

Johnstone, I.M. and Hudson, B.E.T. 1981. The dugong diet: mouth sample analysis. *Bull. Mar. Sci.* 31: 681–690.

Johnson, M.P., Keith, A.D. and Ehrlich, P.R. 1968. The population biology of the butterfly *Euphydryas editha* VII. *Evolution* 22: 422–423.

Jones, G.W. 1977. The Population of North Sulawesi. Yogyakarta: Gajah Mada University Press.

Jordan, C.F. 1985. *Nutrient Cycling in Tropical Forest Ecosystems: Principles and their Application in Management and Conservation.* Wiley: Chichester.

Jordano, P. 1983. Fig-seed predation and dispersal by birds. *Biotropica* 15: 138–41.

Kadir, H. 1980. Aspek megalitik de Toraja Sulawesi Selatan. Pertenunan Ilmiah Arkeologi, 1977.

Kalshoven, L.G.E. 1981. *Pests of Crops in Indonesia,* Rev. ed., Ichtiar Baru - van Hoeve, Jakarta.

Kartawinata, K. 1980a. Classification and utilization of Indonesian forest. *Bio Indonesia* 7: 95–106.

Kartawinata, K. 1980b. The environmental consequences of tree removal from the forest in Indonesia. In *Where Have All the Flowers Gone? Deforestation in the Third World,* pp. 191-214. Williamsburg: College of William and Mary.

Kartawinata, K., Rochadi, A. and Turkirin, P. 1981. Composition and structure of a lowland dipterocarp forest at Wanariset, East Kalimantan. *Malay. For.* 44: 397–406.

Kartawinata, K., Adisoemarto, S., Soemodihardjo, S., and Tantra, I.G.M. 1979. Status pengetahuan hutan bakau di Indonesia. In *Prosiding Seminar Ekosistem Hutan Mangrove,* eds. S. Soemodihardjo, A. Nontji and A. Djamali, pp. 21–39. Lembaga Oseanologi Nasional, Jakarta.

Katili, J.A. 1975. Volcanism and plate tectonics in the Indonesian island arcs. *Tectonophysics* 26: 165–188.

Katili, J.A. 1978. Past and present geotectonic position of Sulawesi, Indonesia. *Tectonophysics* 45: 289–322.

Katili, J.A. and Sudrajat 1984. The devastating 1983 eruption of Colo volcano, Una-Una Island, Central Sulawesi Indonesia. *Geol. Jahrb. A.* 75: 27–47.

Kato, R., Tadaki, V. and Ogawa, H. 1978. Plant biomass and growth increment studies in Pasoh forest. *Malay. Nat. J.* 39: 211–224.

Kaudern, W. 1938. Megalithic finds in Central Celebes. *Etnogr. Stud. in Celebes. No. 5,* Gotenburg.

Kay, R.F. and Hylander, W.L. 1978. The dental structure of mammalian folivores with special reference to primates and phalangeroids (Marsupialia). In *The Ecology of Arboreal folivores,* ed. G.G. Montgomery, pp. 173–191. Washington, D.C.: Smithsonian Institution Press.

Kellman, M.C. 1970. Secondary plant succession in tropical montane Mindanao. Research School of Pacific Studies, Australian National University, Canberra.

Keng, H. 1972. Coniferae. In *Tree Flora of Malaya. Vol. I.,* ed. T.C. Whitmore, pp. 39–53. Longman, Kuala Lumpur.

Keng, H. 1978. The genus *Phyllocladus* (Phyllocladaceae). *J. Arn. Arb.* 59: 249–273.

Kennedy, R. 1974. *Bibliography of Indonesian Peoples and Culture.* Yale: Southeast Asia Studies, Yale University.

Kevan, P.G. and Gaskell, B.H. 1986. The awkward seeds of *Gonystylus macrophyllus* (Thymelaceae) and their dispersal by the bat *Rousettus celebensis* in Sulawesi, Indonesia. *Biotropica* 18: 76–78.

Khlebovich, V.V. 1968. Some peculiar features of the hydrochemical regime and the fauna of meso-haline waters. *Mar. Biol.* 2: 47–49.

Kikkawa, J. and Williams, W.T. 1971. Altitudinal distribution of land bird in New Guinea. *Search* 2: 64–65.

King, B., Woodcock, M. and Dickinson, E.C. 1975. *A Field Guide to the Birds of Southeast Asia.* London: Collins.

King, W.B. 1979. *Red Data Book. Vol. 2. Aves.* Morges: IUCN.

Kira, T. 1978. Primary productivity of Pasoh Forest. *Malay. Nat. J.* 30: 291–297.

Kirk, M. 1982. *That Greater Freedom.* Singapore: OMF.

Kistner, D.H. 1984. A revision of the termitophilous genus *Termitodiscus* with an analysis of the relationships of the species and a review of their behaviour and relationship to their termite hosts (Coleoptera: Staphylinidae; Isoptera, Termitdae). *Sociobiol.* 8: 225–285.

Kitching, R. and Schofield, C. 1986. Every pitcher tells a story. *New Scient.* 109: 48–50.

Kitching, R.L. 1986. A dendrolimnetic dragonfly (Anisoptera: Libellulidae) from Sulawesi. *Odontologica* 15: 203–209.

Kitting, C.L., Fry, B. and Morgan, M.D. 1984. Detection of inconspicuous epiphytic algae supporting food webs in seagrass meadows. *Oecologia* 62: 145–149.

Klapste, J. 1982a. Notes on the Celebes bee-eater. *Aust. Bird Watcher* 9: 252–259.

Klapste, J. 1982b. The bat-hawk in Sulawesi. *Dutch Birding* 4: 29–30.

Ko, R.T.V. 1983. Gua. *Kompas* Nov. 1985.

Ko, R.K.V. 1986. Conservation and environmental management of subterranean biota. Paper presented at BIOTROP Symposium on the Conservation and Management of Endangered Plants and Animals. June 18-20, 1986, Bogor.

Ko, R.T.V., ed. 1985. *Simposium Nasional Lingkungan Karst.* Himpunan Kegiatan Speleologi Indonesia, Bogor.

Koesoebiono, Collier, W.L. and Burbridge, P.R. 1982. Indonesia: Resources' use and management in the coastal zone. In *Man. Land and Sea,* eds. C. Soysa, L.S. Chia and W.L. Collier, pp. 115–133. Agricultural Development Council, Bangkok.

Kohn, A.J. 1959. The ecology of *Conus* in Hawaii. *Ecol. Monogr.* 29: 47–90.

Kohn, A.J. 1979. Ecological shift and release in an isolated population: *Conus miliaris* at Easter Island. *Ecol. Monogr.* 48: 323–336.

Kooders, S.H. 1895. Iets over een vindplaats van fossiele planten en dieren bij Sonder in de Minahassa (Celebes). *Tijds. Kon. Ned. Aard. Gen.* 12: 395–398.

Kopstein, F. 1927. Die Reptilienfauna der Sula-Inseln. *Treubia* 9: 437–446.

Kosasih, S.A. 1983. Tradisi berburu pada lukisan gua di pulau Muna, Sulawesi Tenggara. Rapat Evaluasi hasil Penelitian Arkeologi Nasional, Jakarta.

Kosasih, S.A. 1984. Hasil penelitian lukisan-lukisan pada beberapa gua dan cerak di Pulau Muna (Sulawesi Tenggara). Rapat Evaluasi Hasil Penelitian Arkeologi II, Cisarua, 5-10 March 1983. Pusat Penelitian Arkeologi Nasional, Jakarta.

Koumans, F.P. 1953. X. Goboidea. In *The Fishes of the Indo-Australian Archipelago,* eds. M. Weber and L.F. de Beaufort. Leiden: Brill.

Koyama, H. 1978. Photosynthesis studies in Pasoh Forest. *Malay. Nat. J.* 30: 253–258.

Kramer, D.L. and Mehegan, J.P. 1981. Aquatic surface respiration, an adaptive response to hypoxia in the guppy *Poecilia reticulata* (Pisces, Poeciliidae). *Env. Biol. Fish.* 6: 299–313.

Kramer, K.V. 1971. Lindsaea-group. *Flora Malesiana II* 1: 177–254.

Krishna, A. and Dominic, C.J. 1983. Growth of young and sexual maturity in three species of Indian bats. *J. Anim. Morphol. Physiol.* 30: 162–168.

Kruyt, A.C. 1929. The influence of western civilization on the inhabitats of Poso (Central Celebes). In *The Effect of Western Influence on Native Civilizations in the Malay Archipelago,* ed. B. Schrieke, pp. 1–9. Batavia: Kolff.

Kullberg, R.G. 1982. Algal succession in a hot spring community. *Amer. Midl. Nat.* 108: 224–244.

Kvalvågnaes, K. 1980. The ornamental fish trade in Indonesia. UNDP/FAO National Parks Development Project, Bogor.

La Caro, F. and Rudd, R.L. 1985. Leaf litter disappearance rates in Puerto Rican montane rain forest. *Biotropica* 17: 269–276.

Lack, A., n.d. Botanical studies in Sulawesi. Unpubl. ms.

Lack, A.J. 1984. Occurrence and distribution of *Syzygium* sp. In *Tropical Rain Forest: The Leeds Symposium,* eds. A.C. Chadwick and S.L. Sutton, p. 309. Leeds: Leeds Philosophical and Literary Society.

Lack, A.J. and Kevan, P.G. 1984. On the reproductive biology of a canopy tree, *Syzygium syzygioides* (Myrtaceae) in a rain forest in Sulawesi, Indonesia. *Biotropica* 16: 31-36. See also Erratum *Biotropica* 17: 14.

Lack, D. 1971. *Ecological Isolation in Birds.* Blackwell: Oxford.

Ladiges, W. 1972. Zwei neue Hemirhamphiden von Celebes und Cebu. *Mitt. Hamburg. Zool. Mus. Inst.* 68: 207–212.

Lalamentik, L. 1985. Karang batu didaerah rataan pantai timur pulau Bunaken: identifikasi, kepadatan, pola penyebaran dan keragaman. Thesis, Faculty of Fisheries, Sam Ratulangi University, Manado.

Lam, H.J. 1945. Notes on the historical phytogeography of Celebes. *Blumea* 5: 600–640.

Lam, T.M. 1983. Reproduction in the rice field rat *Rattus argentiventer. Malay. Nat. J.* 36: 249–282.

Law, A.T. and Mohsin, M.A.K. 1980. Environmental studies of Kelang River I. Chemical, physical and microbiological parameters. *Malay. Nat. J.* 33: 175–187.

Lawton, R.O. 1982. Wind stress and elfin stature in a montane rain forest tree: an adaptive explanation. *Amer. J. Bot.* 69:

1224–1230.

Lawton, R.O. 1984. Ecological constraints on wood density in a tropical montane rain forest. *Amer. J. Bot.* 71: 261–267.

Leach, G.J. and Osborne, P.L. 1985. *Freshwater Plants of Papua New Guinea.* Port Moresby: The University Press of Papua New Guinea.

Lee, D.W. and Lowry, J.B. 1980. Solar ultraviolet radiation on tropical mountains: can it affect plant specification? *Am. Nat.* 115: 880–883.

Lee, J., Brooks, R.R., Reeves, R.D., Boswell, C.R. and Jaffre, T. 1977. Plant-soil relationships in a New Caledonian serpentine flora. *Plant and Soil* 46 : 675–650.

Lee, K.B. and Wee, Y.C. 1982. Algae growing on walls in Singapore. *Malay. Nat. J.* 35: 125–132.

Leefmans, S. 1927. Herinneringen aan het natuur monument Bantimoeroeng bij Makassar. *Trop. Natuur* 16: 92–101.

Leefmans, S. 1930. Een bezoek aan de Mampoegrotten bij Pompanoea (Zuid Celebes). *Trop. Natuur* 19: 33–40.

Leenhouts, P.W. 1956. Burseraceae *Flora Malesiana I* 5: 209–296.

Leenhouts, P.W. 1957. Goodeniaceae. *Flora Malesiana I* 5: 335–344.

Leigh, E.G. 1975. Structure and climate in tropical rain forest. *Ann. Rev. Ecol. Syst.* 6: 67–86.

Leightbody, J.P. 1985. Distribution of leaf shapes of *Piper* sp. in a tropical cloud forest: evidence for the role of drip tips. *Biotropica* 17: 339–342.

Leighton, M. 1982. Fruit resources and patterns of feeding: spacing and grouping among sympatric Bornean hornbills. Doctoral dissertation, University of California, Davis.

Leighton, M., and Leighton, D.R. 1983. Vertebrate responses to fruit seasonality within a Bornean rain forest. In *Tropical Rain Forest Ecology and Management,* eds. S.L. Sutton, T.C. Whitmore and A.C. Chadwick, pp. 181–196.

Lenton, G.M. 1983. Wise owls flourish among the oil palms. *New Scient.* 97: 436–437.

Lenton, G.M. 1984. The feeding and breeding ecology of barn owls *Tyto alba* in Peninsular Malaysia. *Ibis* 126: 551–575.

Levington, J.S. 1982. *Marine Ecology.* Printice-Hall, New Jersey.

Lewin, R. 1984. Parks: how big is big enough? *Science* 225: 611.

Lewis, W. M. Jr. 1973. The thermal regine of Lake Lanao (Philippines) and its theoretical implications for tropical lakes. *Limnol. Oceanogr.* 18: 200–217.

Lieberman, D., Lieberman, M., Hartshorn, G. and Peralta, R. 1985. Growth rates and age-size relationships of tropical wet forest trees in Costa Rica. *J. Trop. Ecol.* 1: 97–109.

Lim, B.L. 1974. Small mammals associated with rice fields. *MARDI Res. Bull.* 3: 25–33.

Lim, M.T. 1978. Litterfall and mineral-nutrient content of litter in Pasoh Forest Reserve. *Malay. Nat. J.* 30: 375–380.

Lincoln, R.J., Boxshall, G.A. and Clark, P.F. 1982. *A Dictionary of Ecology, Evolution and Systematics.* Cambridge: Cambridge University Press.

Lock, M. 1980. The layers of the jungle. In *Jungles,* ed. E.S. Ayensu, pp. 30–31. Jonathan Cape: London.

Louw, G.N. and Nicholson, S.W. 1983. Thermal, energetic and nutritional considerations in the foraging and reproduction of the carpenter bee *Xylocopa capitata. J. ent. Soc. Sth. Afr.* 46: 227–240.

Lovejoy, T. 1976. We must decide which species will go forever. *Smithson. Mag.* July 1976: 52–58.

Lowder, G.G. and Dow, J.A.S. 1978. Geology and explorations of porphyry copper deposits in North Sulawesi, Indonesia. *Econ. Geol.* 73: 628–644.

Lugo, A.E. and Snedaker, S.C. 1974. The ecology of mangroves. *Ann. Rev. Syst. Ecol.* 5: 39–64.

Lugo, A.E., Snedaker, S.C, Cintron, G. and Geonaga, C. 1978. Mangrove ecosystems under stress. In *Stress Effects on Natural Ecosystems,* eds. G.W. Barrett and R. Rosenberg, pp. 1–32.

Lugo, A.E., Snedaker, S.C., Evink, G., Brinson, M.M., Broce, A. and Snedaker, S.C. 1975. Diurnal rates of photosynthesis, respiration and transpiration in mangrove forests of south Florida. In *Tropical Ecological Systems,* eds. F.B. Golley and E. Medina, pp. 335–350. New York: Springer-Verslag.

Lumingas, L. 1983. Beberapa aspek biologi ikan lele putih, *Claris batrachus* Linneaus di danau Tondano, Sulawesi Utara. Thèsis, Fakultas Perikanan, Universitas Sam Ratulangi, Manado.

Mac Arthur, R.H., Diamond, J.M. and Karr, J.R. 1972. Density compensation in island faunas. *Ecology* 53: 330–342.

Mac Nae, W. 1968. A general account of the fauna and flora of mangrove swamps and forests in the Indo-Pacific region. *Adv. Mar. Biol.* 6: 73–270.

MacKinnon, J.R. 1978. Sulawesi megapodes. *World Pheasant Assoc. J.* 3: 96–103.

MacKinnon, J.R. 1981a. Methods for the conservation of maleo birds, *Macrocephalon maleo* on the island of Sulawesi, Indonesia, Biol. *Conserv.* 20: 183–193.

MacKinnon, J.R. 1981b. On the structure and function of the tusks of babirusa. *Mammal. Rev.* 11: 37–40.

MacKinnon, J.R. and MacKinnon, K.S. 1980a. Niche differentiation in a primate community. In *Malayan Forest Primates: Ten Years' Study in Tropical Rain Forest*, ed. D.J. Chivers, pp. 167–190. New York: Plenum.

MacKinnon, J.R. and MacKinnon, K.S. 1980b. The behaviour of wild spectral tasiers. *Int. J. Primatol.* 1: 361–379.

MacKinnon, J.R. and MacKinnon, K.S. 1984. Territoriality, monogamy and song in gibbons and tarsiers. In *The Lesser Apes: Evolutionary and Behavioural Biology*, eds. H. Preuschoft, D.J. Chivers, W.R. Brockelman and N. Creel, pp. 291–297. Edinburgh: Edinburgh Univ. Press.

MacKinnon, K.S. 1983. Report of a World Health Organization (WHO) consultancy to Indonesia to determine population estimates of the Cynomolgus or Long-tailed macaque *Macaca fascicularis* (and other primates) and the feasibility of semi-wild breeding projects of this species. WHO Primate Resources Programme Feasibility Study: Phase II. Paris: WHO.

MacKinnon, K.S. 1986. The conservation status of Indonesian primates. *Primate Eye* 29: 30–35.

Maiorana, V.C. 1979. Non-toxic toxins: the energetics of coevolution. *Biol. J. Linn. Soc.*

Maiorana, V,C. 1981. Herbivory in sun and shade. *Biol. J. Linn. Soc.* 15: 151–156.

Makaliwe, W., Hafid, A., Saleh, A.K. and Sallatang, M.A. 1985. An introduction to the socio-economic aspects of a developing coastal area in South Sulawesi, Indonesia. In *The Traditional Knowledge and Management of Coastal Systems in Asia and the Pacific,* eds. K. Ruddle and R.E. Johannes, pp. 265–278. Jakarta: UNESCO.

Malley, D.F. 1977. Adaptations of decapod crustaceans to life in mangrove swamps. *Mar. Res. Indonesia.* 18: 63–72.

Manggabarani, H. 1981. Status perikanan perairan umum di Sulawesi Selatan. Prosiding Seminar Perikanan Perairan Umum, Jakarta, 19-20 August 1981.

Mani, M.S, 1980. The animal life of highlands. In *Ecology of Highlands*, eds. M.S. Mani and L.E. Giddings, pp. 149–159. The Hague: Junk.

Mann, K.H. 1982. *Ecology of Coastal Waters: A Systems Approach.* Oxford: Blackwell.

Manokaran, N. 1980. The nutrient content of precipitation, throughfall and stemflow in a lowland tropical rainforest in Peninsular Malaysia. *Malay. Forester* 43: 266–289.

Marn, H.M. and Jonkers, W. 1982. Logging damage in tropical high forest. In *Tropical Forests: Source of Energy Through Optimization and Diversification*, pp. 27–38. Kuala Lumpur: Universiti Pertanian Press.

Marsh, C.W. and Wilson, W.L. 1981. A Survey of Primates in Peninsular Malaysian Forests. Universiti Kebangsaan Malaysia, Bangi.

Marsh, H. 1981. The food of the dugong. *Aust. J. Wildl. Res.* 9: 55–68.

Marsh, H. 1986. Dugong ecology as a basis for management. Paper presented at the Symposium on Ecology of Australia's Wet Tropics, 25–27 August, 1986, Brisbane.

Marsh, H., Heinsohn, G.E. and Marsh, L.M. 1984. Breeding cycle, Life history and population dynamics of the dugong, *Dugong dugon* (Sirenia: Dugongidae). *Aust. J. Zool.* 32: 767–788.

Marshall, A.G. 1971. The ecology of *Basilla hispida* (Diptera: Nycteribiidae) in Malaysia. *J. Anim. Ecol.* 40: 141–154.

Marshall, A.G. 1980. The comparative ecology of insects ectoparasticit upon bats in West Malaysia. In *Proceedings of the 5th International Bat Research Conference*, eds. D.E. Wilson and A.C. Gardner, pp. 135–142. Lubbock, Texas: Texas Technical Press.

Marshall, A.G. 1985. Old World phytophagous bats bats (Megachiroptera) and their food plants: a survey. *Zool. J. Linn. Soc.* 83: 351–369.

Marshall, J.T. 1978. Systematic of smaller Asian night birds based on voice. *Ornith. Monogr.* 25.

Martin, T.E. 1985. Selection of second-growth woodlands by frugivorous migrating birds in Panama: an effect of fruit size and plant density? *J. Trop. Ecol.* 1: 157–170.

Martosewojo, S., Burhanuddin and Sutomo, A.B. 1982. Makanan ikan gelodok *Boleophthalmus boddaerti* dari muara sungai Banyuasin dan Sungai Jeneberang. In *Prosiding Seminar II Ekosistem Mangrove*, pp. 259–269. Lembaga Oseanologi Nasional, Jakarta.

Mathias, J.A. 1977. The effect of oil on seedlings of the pioneer *Avicennia intermedia* in Malaysia, *Mar. Res. Indonesia* 18: 17.

Matsumoto, T. 1976. The role of termites in an equatorial rain forest ecosystem of West Malaysia I. Population density, biomass, carbon, nitrogen and calorific conten and respiration rate. *Oecologia* 22: 153–178.

Matsumoto, T. 1978. The role of termites in the decomposition of leaf litter on the forest floor of Pasoh study area. *Malay. Nat. J.* 30: 405–413.

Mattulada 1978. Pre-Islamic South Sulawesi. In *Dynamics of Indonesian History*, eds. Haryati Soebadio and C.A. du Marchie Sarvaas, pp. 123–140. Amsterdam: North Holland.

Mattulada 1979. South Sulawesi: its ethnicity and culture. *Southeast Asian Studies* 20: 4–22.

Mattulada 1985. *Latoa: Satu Lukisan Analitis terhadap Politik Orang Bugis*. Yogyakarta: Gadjah Mada University Press.

May, V. 1981. Long-term variation in alga intertidal floras. *Aust. J. Ecol.* 6: 329–343.

May, V., Collins, A.J. and Collett, L.C. 1978. A comparative study of epiphytic algal communities on two common genera of seagrasses in Eastern Australia. *Aust. J. Ecol.* 3: 91–104.

Mayr, E. 1944. Wallace's line in the light of recent zoogeographic studies. *Q. Rev. Biol.* 19: 1–14.

McCaffrey, R. and Sutarjo, R. 1982. Reconnaisance microearthquake survey of Sulawesi Indonesia. *Geophys. Res. Let.* 9: 793–796.

McCaffrey, R., Silver, E.A. and Raitt, R.W. 1981. Seismic refraction studies in the east arm, Sulawesi - Banggai Islands region of eastern Indonesia. *Geol. Res. Dev. Centre, Spec. Publ.* 2: 321–325.

McClure, H.E. 1961. Garden birds in Kuala Lumpur, Malaya. *Malay. Nat. J.* 15: 111–135.

McClure, H.E. 1966. Flowering, fruiting and animals in the canopy of a tropical rain forest. *Malay. Forester* 29: 182–203.

McClure, H.E., Lim, B.L., and Winn, S.E. 1967. Fauna of the Dark Cave, Batu Caves, Kuala Lumpur, Malaysia. *Pac. Insects* 9: 399–428.

McDonald, R.C. 1976. Limestone morphology in South Sulawesi, Indonesia. *Z. Geomorph.* (Suppl). 26: 79–91.

McIntosh, D.J. 1979. Predation of fiddler crabs (*Uca* spp.) in estuarine mangroves. In *Mangrove and Estuarine Vegetation in Southeast Asia*, eds. P.B.L. Srovastrava, A.M. Ahmad, G. Dhanarajan and I. Hamjah. BIOTROP Spec. Publ. 10: 101–110.

McIntyre, A.D. 1968. The microfauna and macrofauna of some tropical beaches. *J. Zool., Lond.* 156: 377–392.

McKenzie, N.L. and Rolfe, J.K. 1986. Structure of bat guilds in the Kimberley mangroves, Australia. *J. Anim. Ecol.* 55: 401–420.

McKey, D. 1978. Soils, vegetation and seed eating by black colobus monkeys. In *The Ecology of Arboreal Folivores*, ed. G.G. Montgomery, pp. 423–437. Washington, D.C.: Smithsonian Institution Press.

McKnight, C.C. 1983. The rise of agriculture. In South Sulawesi before 1600. *Rev. Indon. Mal. Aff.* 17: 92–116.

McKnight, C.C. and Bulbeck, F.D. 1985. Brief report of an archaeological visit to Jakarta and South Sulawesi, June and July, 1985. Unpubl. ms.

McLay, C.L. 1970. A theory concerning the distance travelled by animals entering the drift of a stream. *J. Fish. Res. Board. Can.* 27: 359–370.

McWhirter, N.D., ed. 1985. *Guinness book of Records 1985*. Enfield, Essex: Guinness Books.

Mears, L.A. 1981. *The New Rice Economy of Indonesia*. Yogyakarta: Gajah Mada University Press.

Medway, Lord 1962. The swiftlets (Collocalia) of Niah Cave, Sarawak. *Ibis* 104: 228–245.

Medway, Lord 1966. Field characters as a

guide to the specific relations of swiftlets. *Proc. zool. Soc. Lond.* 177: 151–172.

Medway, Lord 1967. A bat-eating bat, *Megaderma lyra* Geoffrey. *Malay. Nat. J.* 20: 107–110.

Medway, Lord 1968. Field characters as a guide to the specific relations of swiftlets. *Proc. Linn. soc. Lond.* 177: 151–172.

Medway, Lord 1969. Studies on the biology of the edible nest swiftlest of South-east Asia. *Malay. Nat. J.* 22: 57–63.

Medway, Lord 1971. Observations of social and reproductive biology of the bent-winged bat *Miniopterus australis* in northern Borneo. *J. Zool., Lond.* 165: 261–275.

Medway, Lord 1972. Phenology of a tropical rain forest in Malaya. *Biol. J. Linn. Soc.* 4: 146–177.

Meijer, W. 1970. Regeneration of tropical lowland forest in Sabah, Malaysia, forty years after logging. *Malay. For.* 33: 204–229.

Meijer, W. 1974. *Field Guide to the Trees of West Malesia.* Kentucky: University of Kentucky.

Meijer, W. 1982. Plant refuges in the Indo-Malesian region. In *Biological Diversification in the Tropics,* ed. G. Prance, pp. 576–584. New York: Columbia University Press.

Meijer, W. 1983. Botanical explorations in Celebes and Bali. *Nat. Geog. Soc. Res. Rep. 1976 Projects* 583–605.

Meijer, W. 1984. Botanical explorations in Celebes and Bali. *Nat. Geog. Soc. Rep.* 1976: 588–605.

Meise, W. 1939. *Eutrichomyias* novum genus Muscicapidarum. *Ornith. Monatsb.* 47: 134–136.

Menge, B.A., Ashkens, L.R., Matson, A. 1983. Use of artificial holes in studying community development in cryptic marine habitats in a tropical rocky intertidal region. *Mar. Biol.* 77: 129–142.

Mercer, D.E. and Hamilton, L.S. 1984. Mangrove ecosystems: some economic and natural benefits. *Ambio* ?? 14–19.

Metzner, J. 1981. Palu (Sulawesi): Problems of land utilization in a climatic dry valley on the equator. *Appl. Geog. Dev.* 18: 45–62.

Meyer, A.B. 1878. Description of two species of birds from the Malay Archipelago. *Ornith. Misc.* 3: 163–164.

Meyer, A.B. and Wigglesworth, L.W. 1898. *The Birds of Celebes and the Neighbouring Islands.* Berlin: Friedlander.

Miles, M.Y. and Semadi, I.G.M. 1981. Kegiatan resetelmen penduduk di Kabupaten Dati II Gorontalo, Sulawesi Utara dalam rangka mengulangi perluasan perladangan berpindah-pindah. Laporan No. 381, Balai Penelitian Hutan, Bogor.

Miller, R.R. 1977. *Red Data Book. Vol. 4. Pisces.* Morges: IUCN.

Minshall, G.W., Petersen, R.C. Jr., and Nime, F. 1985. Species richness in streams of different size from the same drainage basin. *Amer. Nat.* 125: 16–38.

Mitchell, A.W. 1982. *Reaching the Rainforest Roof.* Leeds: Leeds Philosophical and Literary Society.

Mitchell, D.S. 1985. Distribution of aquatic vegetation in the tropics and sub-tropics. Workshop on the Ecology and Management of Aquatic Weeds, Jakarta, Indonesia, March 26-29, 1985.

Miyazaki, N., Itano, K. Fukushima, M., Kawai, S., Handa, K. 1979. Metals and organochlorine compounds in the muscle of dugong from Sulawesi Island. *Sci. Rep. Whales Res. Inst.* 31: 125–128.

Moeliono, B. and Tuyn, P. 1960. Campanulaceae. *Flora Malesiana I* 6: 107–141.

Mogea, J. 1986. Notes on succession of opened leaves in the tree palm *Pigafetta filaris* and *Livistona rotundifolia.* Unpubl. ms.

Mogea, J.P. and Suhardjono 1982. Komposisi flora pohon di Gunung Malemo, Sulawesi Tengah. Kongres Nasional Biologi V, 26-28 Juni, 1982.

Mohsin, M.A.K. and Ambak, A.K. 1983. *Freshwater Fishes of Peninsular Malaysia.* Kuala Lumpur: Universiti Pertanian Press.

Mohsin, M.A.K. and Law, A.T. 1980. Environmental studies of Kelang River. II. Effects on Fish. *Malay. Nat. J.* 33: 189–199.

Moll, H. 1983. Zonation and diversity of scleractinian on reefs off southwest Sulawesi. Thesis, Leiden University.

Moll, H. 1985. Snellius II, the line-transect and coral systematics. *Proc. 5th Int. Coral Reef Symposium, Tahiti.*

Moll, H. and Borel-Best, M. 1984. New scleractinian corals (Anthozoa: Scleratinia) from the Spermonde Archipelago, South Sulawesi, Indonesia. *Zool. Med.* 58: 47–58.

Mook, W.G. 1981. Radio-carbon dating of freshwater shells from Leang Burung Cave

2: part 2. *Mod. Quat. Res. S.E. Asia* 6: 53–54.

Moore, P. 1986. Why are rain forests so special? *New Scient.* 21 August: 38–40.

Moosa, M.K. 1980. Beberapa catatan mengenai rajungan dari Teluk Jakarta dan pulau-pulau Seribu. In *Sumber Daya Hayati Bahari,* eds. Burhanuddin, Moosa, M.K. and Razak, H., pp. 57–79. Lembaga Oseanologi Nasional, Jakarta.

Morley, R.J. and Flenley, J.R. 1987. Late Cenozoic vegetational and environmental changes in the Malay Archipelago. In *The Biogeographic Evolution of the Malay Archipelago,* ed. T.C. Whitmore. Oxford: Clarendon.

Morton, E.S. 1978. Avian arboreal folivores: Why not? In *The Ecology of Arboreal Folivores,* ed. G.G. Montgomery, pp. 123–130. Washington, D.C.: Smithsonian Institution Press.

Moss, B. 1980. *Ecology of Freshwaters.* Oxford: Blackwell.

Moss, M.R. and Raimadoya, M. 1985. Land resource bibliography: Provinces of South and South-east Sulawesi with emphasis on the Sanrego and Gumas pilot areas. Sulawesi Regional Development Project, Jakarta.

Moyle, P.B. and Senanayake, F.R. 1984. Resource partitioning among the fishes of rainforest streams in Sri Lanka. *J. Zool.* 202: 195–223.

Mrosovsky, N., Hopkins-Murphy, S.R. and Eichardson, J. I. 1984a. Sex ratio of sea turtles: seasonal changes. *Science* 225: 739–741.

Mrosovsky, N., Hopkins-Murphy, S.R., Eichardson, J. I., Dutton, P.H. and Whitmore, C.P. 1984b. Sex ratios of two species of sea turtle nesting in Suriname. *Can. J. Zool.* 62: 2227–2239.

Muchmore, W.B. 1972. A remarkable pseudoscorpion from the hair of a rat (Pseudoscorpinida, Chernetidae). *Proc. Biol. Soc. Wash.* 85: 427–432.

Mulcahy, M.J. and Humphries, A.W. 1967. Soil classifications, soil surveys and land use. *Soils Fertil.* 30: 1–8.

Mulvaney, D.J. and Soejono, R.P. 1970. The Australian-Indonesian archaeological expedition to Sulawesi. *Asian Perspect.* 13: 163–177.

Mundy, G.R. 1848. *Narrative of Events in Borneo and Celebes down to the Occupation of Labuan from the Journals of James Brooke, Esq.*

London: Murray.

Murray, A.M., Marsh, H., Heinsohn, G.E. and Spain, A.V. 1977. The role of the mid-gut caecum in the digestion of seagrasses by the dugong. *Comp Biochem. Physical.*

Musser, G.G. 1971a. The taxonomic association of *Mus Faberi* Jentink with *Rattus xanthurus* (Gray), a species known only from Celebes (Rodentia: Muridae). *Zool. Med.* 45: 107–118.

Musser, G.G. 1971b. Results of the Archbold Expeditions, No. 94. Taxonomic status of *Rattus tatei* and *Rattus frosti,* two taxa of murid rodents known from middle Celebes. *Amer. Mus. Novit.* 2454: 1–19.

Musser, G.G. 1973. Zoogeographical significance of the rice field rat. *Rattus argentiventer* on Celebes and New Guinea and the identity of *Rattus pesticulus. Amer. Mus. Novit.* 2511: 1–30.

Musser, G.G. 1977. *Epimys benguentesis* a composite and one zoogeographic view of rat and mouse faunas in the Philippines and Celebes. *Amer. Mus. Novit.* 2624: 1–15.

Musser, G.G. 1981a. Result of the Archbold Expeditions No. 105. Notes on systematics of Indo-Malayan murid rodents and descriptions of new genera and species from Sri Lanka, Sulawesi and the Philippines. *Bull. Amer. Mus. nat. Hist.* 168: 229–334.

Musser, G.G. 1981b. Notes on systematics of Indo-Malayan murid rodents and descriptions of new genera and species from Ceylon, Sulawesi and the Philippines. *Bull. Amer. Mus. nat. Hist.* 168: 229–334.

Musser, G.G. 1982a. Result of the Archbold Expeditions No. 110. *Crunomys* and the small-bodied shrew rats native to the Philippine Islands and Sulawesi (Celebes) *Bull. Amer. Mus. nat. Hist.* 174: 1–95.

Musser, G.G. 1982b. *Crunomys* and the small bodied shrew-rats native to the Philippine Island and Sulawesi (Celebes). *Bull. Amer. Mus. nat. Hist.* 174: 1–95.

Musser, G.G. 1984. Identities of subfossil rats from caves in southwestern Sulawesi. *Mod. Quat. Res. S.E. Asia* 8: 61–64.

Musser, G.G. 1987. The mammals of Sulawesi. In *The Biogeographic Evolution of the Malay Archipelago,* ed. T.C. Whitmore. Oxford: Clarendon.

Musser, G.G., Koopman, K.F. and Califa, D.

1982. The Sulawesi *Pteropus arquatus* and *P. argentatus* are *Acerodon celebensis*: the Philippine *P. Leucotis* is an *Acerodon. J. Mammal.* 62: 319–328.

Mustafa, M., Zubair, H. and Gusli, S. 1981. Masalah-masalah sumberdaya alam dan lingkungan hidup di Sulawesi Selatan. Pusat Studi Sumberdaya Alam dan Lingkungan Hidup, Universitas Hasanuddin, Ujung Pandang.

Musters, C.J.M. 1983. Taxonomy of the genus *Draco* L. (Agamidae, Lacertilia, Reptilia). *Zool. Verh.* 199: 1–120.

Muul, L. and Lim, B.L. 1978. Comparative morphology, food habits and ecology of some Malaysian arboreal rodents. In *The Ecology of Arboreal Folivores*, ed. G.G. Montgomery, pp. 361–368. Washington, D.C.: Smithsonian Institution.

Myers, G.S. 1949. Salt tolerance of freshwater fish groups in relation to zoogeographical problems. *Bijd. Dierk.* 28: 315–322.

Myers, J.P., Williams, S.L. and Pitelka, F.A. 1980. An experimental analysis of prey availability for sanderlings (Aves: Scolopacidae) feeding on sandy beach crustaceans. *Can. J. Zool.* 58: 1564–1574.

Myers, N. 1983. *A Wealth of Wild Species: Storehouse for Human Welfare*. Westview, Boulder.

Myers, N. 1984a. *The Primary Source: Tropical Forest and Our Future*. New York: Norton.

Myers, N. 1984b. Plants–an embarassment of choices. *IUCN Bull.* 15: 16–18.

Myers, N. 1985a. The end of the lines. *Nat. Hist.* 94: 2,6,12.

Myers, N. 1985b. *The Gaia Atlas of Planet Management*. London: Pan.

Nadkarni, N.M. 1981. Canopy roots: convergent evolution in rain forest nutrient cycles. *Science* 214: 1023–1024.

Naess, A. 1986. Intrinsic value: will the defenders of nature please rise? In *Conservation Biology*, ed. M.E. Soulé, pp. 504–515. Sunderland: Sinauer.

Nance, J. 1975. *The Gentle Tasaday*. New York: Gollancz.

Napompeth, B. 1985. Biological methods of aquatic vegetation management in Tropical Asia. Workshop on The Ecology and Management of Aquatic Weeds. Jakarta, March 26-29, 1985.

Nelson, B. 1980. *Seabirds: their Biology and Ecology*. London: Hamlyn.

Nelson, W.A. 1984. Effects of nutrition of animals on their ectoparasites. *J. Med. Entomol.* 21: 621–635.

Nessa, M.N. 1985. Pengaruh faktor pengelolaan dan lingkungan terhadap daya hasil tambak (Kasus Kabupaten Pangkep Sulawesi Selatan). Doctorate Thesis, Institut Pertanian Institut Bogor, Bogor.

Ng, F.S.P. 1983. Ecological principles of tropical lowland rain forest conservation. In *The Tropical Forest: Ecology and Management,* eds. S.L. Sutton, T.C. Whitmore and A.C. Chadwick, pp. 359–375. Oxford: Blackwell.

Nisbet, I.C.T. 1968. The utilization of mangrove by Malayan birds. *Ibis* 110: 345–352.

Nishiwaki, M., Kasuyu, T., Miyasaki, N., Toboyama, T. and Kataoko, T. 1979. The distribution of dugong in the world. *Sci. Rep. Whales Res. Inst.* 31: 133–141.

Nooteboom, H.P. 1962. Simaroubaceae. *Flora Malesiana I* 6: 193–226.

Nuitja, I.N.S. and Uchida, I. 1983. Studies in the sea turtles. II. The nesting site characteristics of the hawksbill and green turtles. *Treubia* 29: 63–79.

Nurkin, B 1979. Beberapa catatan tentang aspek pengusahaan hutan mangrove di Sulawesi Selatan. In *Prosiding Seminar Ekosistem hutan Mangrove,* eds. S. Soemodihardjo, A. Nontji, and A. Djamali, pp. 120–125. Lembaga Oseanologi Nasional, Jakarta.

Nursall, J.R. 1981. Behaviour and habitat affecting the distribution of five species of sympatric mudskippers in Queensland. *Bull. mar. Sci.* 31: 730–735.

O'Connor, C.T. and Sopa, T. 1981. A checklist of the mosquitoes of Indonesia. Special Publ. No. 2, U.S. Naval Medical Research Unit, Jakarta.

Oates, J.F., Swain, T. and Zantovska, J. 1977. Secondary compounds and food selection by colobus monkeys. *Biochem. Syst. Ecol.* 5: 317–321.

Oates, J.F., Waterman, P.G. and Choo, G.M. 1980. Food selection by the South Indian leaf-monkey. *Presbytis johnii,* in relation to leaf chemistry. *Oecologia* 45: 45–46.

Ogawa, H. 1978. Litter production and carbon cycling in Pasoh Forest. *Malay. Nat. J.* 30: 367–373.

Ohler, J.G. 1984. Coconut, tree of life. *FAO. Pl. Prod. Prot. Paper 57.*

Oldemann, L.R. and Darmiyati, S. 1977. An agroclimatic map of Sulawesi. *Contr. Centr. Res. Inst. Agric. Bogor* 60: 1–32.

Oldemann, L.R., Irsal, L. and Muladi 1980. The agroclimatic maps of Kalimantan, Maluku, Irian Jaya and Bali, West and East Nusa Tenggara. *Contr. Centr. Res. Inst. Agric. Bogor* 60: 1–32.

Omar, S.A. 1985. Komposisi jenis dan jumlah plankton di perairan tambak desa Tasiwalie, Kecamatan Suppa, Kabupaten Pinrang. Thesis, Hasanuddin University, Ujung Pandang.

Ong, J.E., Gong, W.K. and Wong, C.H. 1980a. *Ecological Survey of the Sungei Merbok Estuarine Mangrove Ecosystem.* Universiti Sains Malaysia, Penang.

Ong, J.E., Gong, W.K. and Wong, C.H. 1980b. Contribution of aquatic productivity in a managed mangrove ecosystem in Malaysia. Paper presented at the UNESCO Symposium on *'Mangrove Environment: Research and Management'.* Kuala Lumpur: Universiti Malaya Press.

Ong, J.E., Gong, W.K. and Wong, C.H. 1985. Seven years of productivity studies in a Malaysian Managed mangrove forest. Then what? In *Coasts and Tidal Wetlands of the Australian Monsoon Region,* ed. K.N. Bardsley, J.D.S. Davie and C.D. Woodroffe, pp. 213–223. Mangrove Monograph No. 1, Australian National University North Australia Research Unit, Darwin.

Opler, P.A., Frankie, G.W. and Baker, H.G. 1980. Comparative phenological studies of treelet and shrub species in tropical wet and dry forests in the lowlands of Costa Rica. *J. Ecol.*

Ostrofsky, M.L. and Zettler, E.R. 1986. Chemical defences in aquatic plants. *J. Ecol.* 74: 279–289.

Otofuji, Y., Sasajima, S., Nishimura, S., Dharma, S. and Hehuwat, F. 1981. Palaeomagnetic evidence for clockwise rotation of the northern arm of Sulawesi, Indonesia. *Earth Planet. Sci. Let.* 54: 272–280.

Paine, R.T. 1969. A note on trophic complexity and community stability. *Amer. Nat.* 103: 91–93.

Palenewan, J. 1983. Pengelolaan sampah di kota Bitung Daerah Tingkat I Sulawesi Utara. Unpubl. ms.

Palenewan, J.L. 1984. Daya pulih alamiah air sungai Tondano terhadap pencemaran oleh air limbah pabrik minyak kelapa. Universitas Sam Ratulangi, Manado.

Palmieri, M.D., Palmieri, J.R. and Sullivan, J.T. 1980. A chemical analysis of the habitat of nine commonly-occurring Malaysian freshwater snails. *Malay. Nat. J.* 34: 39–45.

Pancho, J.V., Tjitrosoepomo, G. and Megia, R. 1985. Some major aquatic plants in the tropics. Workshop on the Ecology and Management of Aquatic Weeds, Jakarta, Indonesia, March 26-29, 1985.

Parkhurst, D.F. and Loucks, O.L. 1972. Optimal leaft size in relation to environment. *J. Ecol.* 60: 505–537.

Parrish, J.D. 1980. Effects of exploitation upon reef and lagoon communities. In *Marine and Coastal Processes in the Pacific: Ecological Aspects of Coastal Zone Management* UNESCO, Jakarta.

Parry, P.E., n.d. Ultramafic soils in the humid tropics with particular reference to Indonesia. Hunting Technical Services Ltd., Jakarta.

Payne, J., Francis, C.M. and Philips, K. 1985. *A Field Guide to the Mammals of Borneo.* The Sabah Society, Kota Kinabalu, and World Wildlife Fund Malaysia, Kuala Lumpur.

Payne, J.B. 1980. Competitors. In *Malayan Forest Primates: Ten Years' Study in Tropical Rain Forest,* ed. D.J. Chivers, pp. 261–277. New York: Plenum.

Peck, S.B. 1981. A new cave-inhabiting *Ptomaphaginus* beetle from Sarawak (Leiodidae: Cholevinae). *Syst. Entomol.* 6: 221–224.

Peckarsky, B.L. 1979. Biological interactions as determinations of distribution of benthic invertebrates within the stony substrate of steams. *Limnol. Oceanogr.* 24: 59–68.

Pelras, C. 1981. Celèbes-Sud avant l'Islam, selon les premiers temoignages etrangers. *Archipel* 21: 163.

Pelras, C. 1985. Religion, tradition and dynamics of Islamization in South Sulawesi. *Archipel* 29: 107–135.

Perry, D.R. 1978. A method of access into the crowns of emergent and canopy trees. *Biotropica* 10: 155–157.

Perry, D.R. and Williams, J. 1981. The tropical rainforest canopy: a method of providing total access. *Biotropica* 13: 283–285.

Pethick, J. 1984. *An Introduction to Coastal Geomorphology*. London: Edward Arnold.

Pettigrew, J. 1986a. Flying primates? Megabats have the advanced pathway from eye to mid brain. *Science* 231: 1304–1306.

Pettigrew, J. 1986b. Are flying foxes really primates? *Bats* 3: 5–6.

Phillipson, J. 1966. *Ecological Energetics*. Studies in Biology No. 1. London: Edward Arnold.

Pickett, S.T.A. 1983. Differential adaptation of tropical tree species to canopy gaps and its rule in community dynamics. *Trop. Ecol.* 24: 68–84.

Pienkowski, M.W. 1983. Surface activity of some intertidal invertebrates in relation to temperature and the foraging behaviour of their shorebird predators. *Mar. Ecol. Prog. Ser.* 11: 141–150.

Pierson, E.D. 1984. Can Australia's flying foxes survive? *Bats.* Sept. 1984: 1–3.

Pigafetta, A. 1906. *Magellan's Voyage around the World*. English translation by J.A. Robertson, Arthur H. Clark. Cleveland.

Piggot, A. 1979. *Common Epiphytic Ferns of Malaysia and Singapore*. Singapore: Heinemann.

Pires, A. 1944. *The Suma Oriental*. English translation by Armando Cortesao. London: Hakluyt Society.

Pirzan, A.M.N. and Wardoyo, S.E. 1979. Penelitian sumber benih sidat (*Anguilla* spp.) di sungai Poso Sulawesi Tengah. Lembaga Penelitian Perikanan Darat Cabang Ujung Pandang.

Pócs, T. 1982. Tropical forest bryophytes. In *Bryophyte Ecology*, ed. A.J.F. Smith, pp. 59–104. London: Chapman and Hall.

Polhemus, D. and Polhemus, J.T. 1986. The zoogeography and phylogeny of the genus *Ptilomera* Amyet and Serville (Hemiptera: Gerridae). *Abstrac. Bull. N. Amer. Benth. Soc.* 3: 67.

Polunin, N.V.C. 1983. The marine resources of Indonesia. *Oceanogr. Mar. Biol. Ann. Rev.* 21: 455–531.

Polunin, N.V.C. 1984. Do traditional marine reserve conserve? A view of Indonesian and New Guinean evidence. *Senri Ethmol. Stud.* 17: 13–15.

Poore, M.E.D. 1968. Studies in Malaysian rain forest. I. The forest on Triassic sediments in Jengka Forest Reserve. *J. Ecol.* 56: 143–196.

Popta, C.M.L. 1905. *Haplochilus sarasinorum*, n. sp. *Notes Leyden Mus.* 25: 239–247.

Porter, J.W. Porter, K.G. and Batac-Catalan, Z. 1977. Quantitative sampling of Indo-Pacific demersal reef plankton. *Proc. Third Intnl. Coral Reef Symp.* I : 105–112.

Potts, D.C. 1983. Evolutionary disequilibrium among Indo-Pacific corals. *Bull. Mar. Sci.* 33: 619–632.

Potts, D.C. 1984. Generation times and the quaternary evolution of reef-building corals. *Paleobiol.* 10: 48–58.

Poulson, T.I. and White, W.B. 1969. The cave environment. *Science* 165: 971–981.

Powell, R., n.d. Journey to Uewaja. Report to Operation Drake.

Power, M.E. 1983. Grazing responses of tropical freshwater fishes to different scale of variation in their food. *Env. Biol. Fishes* 9: 103–115.

Pranowo, H.A. 1985. *Manusia dan Hutan. Proses Perubahan Ekologi di Lereng Gunung Merapi*. Yogyakarta: Gadjah Mada University Press.

Pratt, T.K. and Stiles, E.W. 1983. How long fruit-eating birds stay in the plants where they feed: Implications for seed dispersal. *Am. Nat.* 122: 797–805.

Prendergast, H.D.V. 1982. Pollination of *Hibiscus rosa-sinensis*. *Biotropica* 14: 287.

Prescott-Allen, R. and Prescott-Alen, P.A.C. 1982. *What's Wildlife Worth? Economic Contribution of Wild Animals to Devloping Countries*. Earthscan/International Institut for Environment and Development, London and Washington, D.C.

Preston-Martin, R. and Preston-Martin, K. 1984. *Spiders of the World*. Poole: Blandford Press.

Primack, R.B. and Tomlinson, P.B. 1978. Sugar secretions from the buds of *Rhizophora*. *Biotropica* 10: 74–75.

Proctor, J. 1983. Tropical forest litterfall. I. Problems of data comparison. In *Tropical Rain Forest: Ecology and Management*, eds. S.L. Sutton, T.C. Whitmore and A.C. Chadwick, pp. 267–273. Oxford: Blackwell.

Proctor, J. and Whitten, K. 1971. A population of the valley pocket gopher (*Thomomys bottae*) on a serpentine soil. *Am. Midl. Nat.* 78: 176–179.

Proctor, J. and Woodell, S.R.J. 1975. The ecology of serpentine soils. *Adv. ecol. Res.* 9: 255–366.

Proctor, J., Anderson, J.M., Chai, P. and Vallack, H.W. 1983. Ecological studies in four contrasting lowland rain forests in Gunung Mulu National Park, Sarawak. I. Forest environment, structure and floristics. *J. Ecol.* 71: 237–260.

Proctor, J., Anderson, J.M., Fogden, S.C.L. and Vallack, H.W. 1983. Ecological studies in four contrasting lowland rain forests in Gunung Mulu National Park, Sarawak. II. Litterfall, litter standing crop and preliminary observations on herbivory. *J. Ecol.* 71: 261–283.

Prowse, G.A. 1968. Pollution in Malayan waters. *Malay. Nat. J.* 21: 149–158.

Pryor, D. 1976. *Biology of Eucalypts.* Studies in Biology No. 61, London: Edward Arnold.

Ptroli, J., Dajo, B.C., Hardjawidjaja, L., Sudomo, M. and Barodji, A. 1980. A schistosomiasis pilot control project in Lindu valley, Central Sulawesi, Indonesia. *S.E. Asian J. Trop. Med. Pub. Hlth.* 11: 480–486.

Pudjiharta, A.G. and Achmad, A. 1978. Kondisi tata air DAS Jenekelara. Laporan No. 273, Lembaga Penelitian Hutan, Bogor.

Pudjiharta, A.G. and Mile, Y.M. 1981. Karakteristik daerah aliran Sungai Jeneberang, Sulawesi Selatan. Laporan No. 360, Lembaga Penelitian Hutan Bagor.

Purchon, R.D. and Enoch, I. 1954. Zonation of the marine fauna and flora on a rocky shore near Singapore. *Bull. Raffles Mus. S'pore,* No. 25: 47–65.

Purseglove, J.W. 1968. *Tropical Crops: Dicotyledons.* London: Longman.

Putrali, J., Carney, W.P., Stafford, E.E. and Tubo, S. 1977. Intestinal and blood parasites in the Banggai Kabupaten Central Sulawesi, Indonesia. *S.E. Asian J. Trop. Med. Pub. Hlth.* 8.

Putrali, J., Dajo, B.C., Hardjawidjaja, L., Sudomo, M. and Barodji, A. 1980. A schistosomiasis plot control project in Lindu valley, Central Sulawesi, Indonesia. *S.E. Asian J. Trop. Med. Pub. Hlth.* 11: 480–486.

Putz, F.E. 1984a. How trees avoid and shed lianas. *Biotropica* 16: 19–25.

Putz, F.E. 1984b. The natural history of lianas on Barro Colorado Island, Panama. *Ecology.* 45: 1713–1724.

Putz, F.E. and Parker, G.G. 1984. Mechanical abrasion and intercrown spacing. *Amer. Mid. Nat.* 112: 24–28.

Putz, F.E., Coley, P.D., Lu, K., Mantalvo, A. and Aiello, A. 1983. Uprooting and snapping of trees: structural determinants and ecological consequences. *Can. J. For. Res.* 13: 1011–1020.

Quammen, M.L. 1982. Influence of subtle substrate differences on feeding by shorebirds on intertidal mudflats. *Mar. Biol.* 71: 339–343.

Rabinowitz, D. 1978. Early growth of mangrove seedlings in Panama and an hypothesis concerning the relationship of dispersal and zonation. *J. Biogeogr.* 5: 113-133.

Raemaekers, J.J., Aldrich-Blake, F.P.G. and Payne, J.B. 1980. The forest. In *Malayan Forest Primates: Ten Years' Study in Tropical Rain Forest,* ed. D.J. Chivers, pp. 29–61. New York: Plenum.

Rambo, A.T. 1979. Primitive man's impact on genetic resources of the Malaysian tropical rain forest. *Mal. Appl. Biol.* 8: 59–65.

Rand, A.S. 1978. Reptilian arboreal folivores. In *The Ecology of Arboreal Folivores,* ed. S.G. Montgomery, pp. 115–122. Washington, D.C.: Smithsonian Institution Press.

Rasmussen, R.A. and Khalil, M.A.K. 1983. Global production of methane by termites. *Nature* 301: 704–705.

Rasmusson, E.M. and Wallace, J.M. 1983. Meteorological aspacts of the El Nino/-Southern oscillation. *Science* 222: 1195–1202.

Ratag, V.F.R.C. 1981. Suatu penelitian tentang hubungan antara kepadatan dan komposisi phytoplankton dengan zooplankton di Danau Tondano, Kabupaten Minahasa, Propinsi Sulawesi Utara. Thesis, Fakultas Perikanan, Universitas Sam Ratulangi, Manado.

Rauf, D.A. 1982. Kapasitas tampung dan komposisi botanis tanah-tanah kritis yang digunakan sebagai lapangan penggembalaan ternak di Kelurahan Kawatuna. Kecamatan Palu Timur. Fakultas Peternakan, Universitas Tadulako, Palu.

Raunkier, C. 1934. *The Life Forms of Plants and Statistical Plant Geography.* Oxford: University Press.

Raven, P.H. 1977. Onagraceae. *Flora Male-*

siana I 8: 98–113.

Rees, C.J.C. 1983. Microclimate and the flying Hemiptera fauna of a primary lowland rain forest in Sulawesi. In *Tropical Rain Forest: Ecology and Management*, eds. S.L. Sutton, T.C. Whitmore and A.C. Chadwick, pp. 121–136. Oxford: Blackwell.

Rees, C.J.C., n.d. Caves in the Morowali region. Unpubl. ms.

Reese, E.S. 1981. The ecology of the coconut crab, *Birgus latro* (L). *Bull. Ecol. Soc. Amer.* 46: 191–192.

Regenass, U. and Kramer, E. 1981. Zur Systematik der gruenen Grubenotter der Gattung *Trimeresurus* (Serpentes, Crotalidae). *Rev. Suisse Zool.* 88: 163–205.

Reid, A. 1983. The rise of Makassar. *Rev. Indon. Mal. Aff.* 17: 117–160.

Rerung, R. 1983. Struktur populasi tumbuhan berkayu di areal bekas perladangan di kawasan hutan Amaro, Kabupaten Barru, Sulawesi Selatan. Dissertation, Universitas Hasanuddin, Ujung Pandang.

Reyne, A. 1938. On the distribution of *Birgus latro* L. in the Dutch East Indies. *Arch. Neerl. Zool.* (Suppl.) 38: 239–247.

Reyne, A. 1939. On the food habits of the coconut crab *Birgus latro* L. with notes on its distribution. *Arch. Neerl. Zool.* 39: 283–247.

Rhyther, J.H. and Dunstan, W.M. 1971. Nitrogen and phosphorus and eutrophication in the coastal marine environment. *Science* 171: 1008–1013.

Ribi, G. 1981. Does the wood boring isopod *Sphaeroma terbrans* benefit red mangroves (*Rhizophora mangle*)? *Bull. Mar. Sci.* 31: 925–928.

Ribi, G. 1982. Differential colonization of roots of *Rhizophora mangle* by the wood boring isopod *Sphaeroma terebrans* as a mechanism to increase root density. *Mar. Ecol.* 3: 13–19.

Richards, P.W. 1983. The three-dimensional structure of tropical rain forest. In *Tropical Rain Forest: Ecology and Management*, eds. S.L. Sutton, T.C. Whitmore, and A.C. Chadwick, pp. 287–309. Oxford: Blackwell.

Ricklefs, R.E. 1977. Environmental heterogeneity and plant species diversity: a hypothesis. *Am. Nat.* 111: 376–381.

Rijksen, H.D. 1978. *A Field Study on Sumatran Orang-utans (*Pongo pygmaeus abelli *Lesson 1827): Ecology, Behaviour and Conservation.* Wageningen: Veenman and Zonen.

Riswan, S., Kenworthy, J.B. and Kartawinata, K. 1985. The estimation of temporal processes in tropical rain forest: a study of primary mixed dipterocarp forest in Indonesia. *J. Trop. Ecol.* 1: 171–182.

Robbins, R.G. and Wyatt-Smith, J. 1964. Dry land forest formations and forest types in the Malay Peninsula. *Malay. For.* 27: 188–217.

Robinson, G.S. 1980. Cave-dwelling tineid moths: A taxanomic review of the world species (Lepidoptera: Tineidae). *Brit. Cave Res. Assoc.* 7: 83–120.

Robinson, K. 1983. Living in the hutani jungle village life under the Darul Islam. *Rev. Indon. Mal. Aff.* 17: 208–279.

Robson, T.O. 1985. Chemical control of aquatic weeds in the tropics. Workshop on the Ecology and Management of Aquatic Weeds, Jakarta, March 26-29, 1985.

Rodelli, M.R., Gearing, J.N., Gearing, P.J., Marshall, N. and Sasekumar, A. 1984. Stable isotope ratio as a tracer of mangrove carbon in Malaysian ecosystems. *Oecologia* 61: 326–333.

Rodenburg, W. and Palete, R. 1981. Proposed National Park Dumoga-Bone Management. Plan 1981/82–1982/83. Bogor: World Wildlife Fund.

Rombe, P. 1982. Spektrum 'life-form' pada berbagai tingkat perkembangan vegetasi di bekas-bekas ladang, sekitar Taman Nasional Lore Lindu Sulawesi Tengah. Dissertation, Universitas Hasanuddin, Ujung Pandang.

Rondo, M. 1977. Beberapa aspek biologi dan keadaan lingkungan ikan mujair *Tilapia mossambica* Peters di danau Tondano, Sulawesi Utara. Thesis, Fakultas Perikanan, Universitas Sam Ratulangi, Manado.

Rondo, M. and Sondakh, L.W. 1984. Identifikasi ketergantungan berbagai tindakan pemanfaatan karang sebagai sumberdaya alam di sekitar pulau Bunaken. Fakultas Perikanan, Universitas Sam Ratulangi, Manado.

Rosen, B.R. 1981. The tropical high diversity enigma – the corals-eye view. In *Change and Challenge: The Evolving Biosphere*, ed. P.L. Forey, pp. 103–129. London: British Museum (Natural History).

716

Rosen, D.E. and Parenti, L.R. 1981. Relationships of *Oryzias* and the groups of Atherinomorph fishes. *Am. Mus. Novit.* 2719: 1–25.

Ross, M.S. and Donovan, D.G. 1985. An overview of major issues and constraints related to land clearing in the conversion of tropical moist forests. Paper presented at the International Inaugural Workshop on Landclearing and Development for Increased Agricultural Production, August 27, September 3, 1986, Jakarta.

Roth, L.M. 1980. Cave-dwelling cockroaches from Sarawak, with one new species. *Syst. Entomol.* 5: 97–104.

Roubik, D.W. and Aluja, N. 1983a. Flight ranges of *Melipona* and *Trigona* in tropical forest. *J. Kansas Entamol. Soc.* 56: 217–222.

Roubik, D.W. and Aluja, N. 1983b. Nest and colony characteristics of stingless bees from Panama (Hymenoptera: Apidae). *J. Kansas Entamol.* 56: 327–355.

Ruttner, F. 1931. Hydrographische und hydrochemische Beobachtungen auf Java, Sumatra und Bali. *Arch. Hydrobiol. Suppl.* 5: 197–454.

Sabar, F., Djajasasmita, M. and Budiman, A. 1979. Susunan dan penyebaran moluska dan krustasea pada beberapa hutan rawa payau: suatu studi pendahuluan. In *Prosiding Seminar Ekosistem Hutan Mangrove,* eds. S. Soemodihardjo, A. Nontji and A. Djamali, pp. 120–125. Lembaga Biologi Nasional, Jakarta.

Saenger, P., Hegerl, E.J. and Davie, J.D.S. 1981. First Report of the Global Status of Mangrove Ecosystems. Commission on Ecology. Gland, IUCN.

Salick, J. 1983. Natural history of crop-related wild species: Uses in pest habitat management. *Env. Mgmt.* 7: 85–90.

Salm, R.V. and Halim, M. 1984. Marine Conservation Atlas: Planning for the Survival of Indonesia's Seas and Coasts. Bogor: World Wildlife Fund.

Sarasin, P. and Sarasin, F. 1905. *Reisen in Celebes ausgefahrt in den Jahren 1893-1896 und 1902-1903.* Wiesbaden: Kriedel.

Sarnita, A. 1973. Laporan survey perikanan Danau Lindu dan Poso. Lembaga Penelitian Perikanan Darat, Bogor.

Sarnita, A. 1974. Beberapa aspek Limnologis

tentang Danau Towuti, Matano dan Danau Mahalona (Sulawesi Selatan). Lembaga Penelitian Perikanan Darat, Bogor.

Sartono, S. 1979. The age of the vertebrate fossils and artefacts from Cabenge in South Sulawesi, Indonesia. *Mod. Quat. Res. S.E. Asia* 5: 45–81.

Sartono, S. 1982. Genesa Danau Tempe, Sulawesi Selatan. Pertemuan Arkeologi ke-2, 1980. pp. 555–560.

Sasajima, S., Nishimura, S., Hirooka, K., Otofuji, Y., van Leeuwen, T., and Hehuwat, F. 1980. Palaeomagnetic studies combined with fission-track datings on the western arc of Sulawesi, east Indonesia. *Tectonophysics* 64: 163–172.

Sasekumar, A. and Loi, J.J. 1983. Litter production in three mangrove forest zones in the Malay Peninsula. *Aquat. Bot.* 17: 283–290.

Sastrapradja, S. and Kartawinata, K. 1975. Leafy vegetables in the Sundanese diet. In *South-east Asian Plant Genetic Resources,* ed. J.T. Williams, pp. 166–170. Bogor: Lembaga Biologi Nasional.

Sastrapradja, S., Soenartono, A., Kartawinata, K. and Tarumingkeng, R.C. 1980. The conservation of forest animal and plant genetic resources. *BioIndonesia.* 7: 1–42.

Sastroutomo, S.S. 1985. The role of aquatic vegetation in the environment. Paper presented at the Workshop on the Ecology and Management of Aquatic Weeds, Jakarta, March 26-29, 1985.

Satjapradja, O. and Mas'ud, F. 1978. Kegiatan resetelment penduduk dataran Palolo sebagai salah usaha untuk pengamanan hutan di daerah Sulawesi, Tengah. Laporan No. 286, Lembaga Penelitian Hutan, Bogor.

Sato, K. 1984. Studies on the protective functions of the mangrove forest against erosion and destruction (IV). The protective functions of the mangrove forest and a preliminary hydraulic model experiment on the root system of mangrove. *Sci. Bull. Coll. Agric. Univ. Ryukyus* 31: 189–200.

Schaeffer, F.A. 1970. *Pollution and the Death of Man: The Christian View of Ecology.* Wheaton: Tyndale House.

Scheffer, M., Achterberg, A.A. and Beltman, B. 1984. Distribution of macroinvertebrates in a ditch in relation to the vegetation. *Freshw. Bol.* 14: 367–370.

Schemske, D.W. and Brokaw, N. 1981. Treefalls and the distribution of understory birds in a tropical forest. *Ecology* 62: 938–945.

Schmidt, F.H. and Ferguson, J.H.A. 1951. Rainfall types based on wet and dry period ratios for Indonesia and Western New Guinea. *Verh. Djawatan Met. dan Geofisik, Jakarta* 42.

Schroeder, R.E. 1980. *Philippine Shore Fishes of the Western Sulu Sea.* NMPC Books, Manila.

Schuchmann, K.L. and Wolters, H.E. 1982. A new subspecies of *Serinus estherae* (Carductidae) from Sulawesi. *Bull. Brit. Orn. Cl.* 102: 12–14.

Schuster, W.H. 1950. Pemeliharaan ikan dalam perempangan di Djawa. Pengumuman No. 2, Urusan Perikanan Darat, Kementrian Pertanian, Jakarta.

Schuster, W.H. 1950. Comments on the importation and transplantation of different species of fish into Indonesia. *Contr. gen. agr. Res. Stn. Bogor* 111: 1–31.

Schuster, W.H. 1952. *Local Common Names of Indonesian Fishes.* W. van Hoeve, Bandung.

Selmier, V.J. 1983. Berstandsgroesse und Verhalten des Hirschebers (*Babyrousa babyrussa*) auf den Togian-Inseln. *Bongo* 7: 51–64. (Translation available.)

Sidiyasa, K. and Tantra, I.G.M. 1984. An analysis of vegetation in the Poboya Nature Reserve. Laporan No. 451, Lembaga Penelitian Hutan, Bogor.

Sidiyasa, K. and Tantra, I.G.M. 1984. Analisis vegetasi Cagar Alam Poboya, Sulawesi Tengah. Laporan No. 451, Pusat Penelitian dan Pengembangan Hutan, Bogor.

Silvius, M., Verheught, W. and Iskandar, J. 1985. Coastal surveys in south-east Sumatra. In *Interwader Annual Report 1984*, eds. D. Parish and D.R. Wells, pp. 133–142. Kuala Lumpur: Interwader.

Simberloff, D. and Boecklen, W. 1981. Santa Rosalia reconsidered: size ratios and competition. *Evolution* 35: 1206–1228.

Simpson, G.G. 1977. Too many lines: the limits of the Oriental and Australian zoogeographic regions. *Proc. Amer. Phil. Soc.* 121: 102–120.

Sinha, P. and Glover, I.C. 1984. Changes in stone tool use in Southeast Asia 10,000 years ago: a microwear analysis of flakes with use gloss from Leang Burung 2 and Ulu Leang 1 caves, Sulawesi, Indonesia. *Mod. Quat. Res. S.E. Asia* 8: 137-164.

Sleumer, H. 1954. Flacourtiaceae. *Flora Malesiana I* 5: 1–106.

Sleumer, H. 1955. Proteaceae. *Flora Malesiana I* 5: 147–206.

Sleumer, H. 1966-1967. Ericaceae. *Flora Malesiana I* 6: 469–914.

Smith, A.P. 1972. Buttressing of tropical trees: a descriptive model and new hypotheses. *Am. Nat.* 106: 32–46.

Smith, D. 1984. *Urban Ecology.* George Allen and Unwin, London.

Smith, J.M.B. 1970. Herbaceous plant communities in the summit zone of Mount Kinabalu. *Malay. Nat. J.* 24: 16–29.

Smith, J.M.B. 1981. Vegetation and flora of Gunung Rantemario, Latimojong Range, Sulawesi (Celebes). Unpubl. ms.

Smith, J.M.B. and Guyer, I.J. 1983. Rain forest-eucalypt forest interactions and the relevance of the biological nomad theory. *Aust. J. Ecol.* 8: 55–60.

Smith, T.J. 1986. The influence of seed predators on the structure of tropical tidal forests. Paper presented at Symposium on The Ecology of Australia's Wet Tropics, 25-27 August, 1986, Brisbane.

Snedaker, S. 1982a. A perspective on Asian mangroves. In *Man, Land and Sea*, eds. C. Soysa, L.S. Chic, and W.L. Collier, pp. 65–74. Agricultural Development council, Bangkok.

Snedaker, S. 1982b. Mangrove species zonation: why? In *Tasks for Vegetation Science, Vol. 2*, eds. D.N. Sen and K.S. Rajpurohit. The Hague: Junk.

Sneed, M.W. 1981. *Pigafetta* and other palms in Sulawesi (Celebes). *Principles* 25: 106–119.

Soegiarto, A. 1985. The mangrove ecosystem in Indonesia: its problems and management. In *Coasts and Tidal Wetlands of the Australian Monsoon Region*, eds. K.N. Bardsley, J.D.S. Davie and C.D. Woodroffe, pp. 313–326. Mangrove Monograph No. 1, Australian National University North Australia Research Unit, Darwin.

Soegiarto, A. and Polunin, N.V.C. 1980. *The Marine Environment of Indonesia.* Bogor: World Wildlife Fund.

Soejono 1978. Prehistoric Indonesia. In *Dynamics of Indonesian History,* eds. H. Soebadio and C.A. du Marchie Sarvaas, pp. 1–28. Amsterdam: North-Holland.

Soepadmo, E. 1972. Fagaceae *Flora Malesiana I* 7: 265–403.

Soeriaatmadja, A. 1977. Laporan penelaahan awal di Suaka Margasatwa Lore Kalamanta Sulawesi Tengah. Unpubl. ms.

Soerianegara, I. 1970. Soil investigation in Mount Hondje Forest Reserve, West Java. *Rimba Indonesia.* 15: 1–16.

Soerjani, M. 1978. Pengelolaan danau di Indonesia. Seminar Pengelolaan Waduk dan danau, 7-9 Nopember, 1978.

Soerjani, M. 1985. Principles of aquatic vegetation management. Workshop on The Ecology and Management of Aquatic Weeds, Jakarta, March 26-29. 1985.

Soewanda A.P.R., n.d.a. Daftar nama pohon-pohonan Sulawesi Selatan, Tenggara dan Sekitarnya. Lembaga Penelitian Hutan, Bogor.

Soewanda A.P.R., n.d.b. Daftar nama pohon-pohonan Menado. Lembaga Penelitian Hutan, Bogor.

Soewanda A.P.R. and Tantra, I.G.M., n.d.a. Daftar nama-nama pohon-pohonan. Menado (Sulawesi Utara). Lembaga Penelitian Hutan, Bogor.

Soewanda A.P.R. and Tantra, I.G.M., n.d.b. Daftar nama pohon-pohonan. Sulawesi Selatan, Tenggara dan Sekitarnya. Lembaga Penelitian Hutan Bogor.

Sohma, K. 1973. *Florshuetzia:* a fossil sonneratoid pollen genus from Sulawesi, Indonesia. *Sci. Rep. Tohuku Univ. 4th. ser.* 36: 261–264.

Sopher, D.E. 1978. *The Sea Nomads.* Singapore: National Museum.

Spain, A.V. 1984. Litterfall and the standing crop of litter in three tropical Australian rainforests. *J. Ecol.* 72: 947–961.

Spears, J. 1982. Preserving watershed environments. *Unasylva* 34: 10–14.

Spellerberg, I.F. 1972. Temperature tolerances of southeast Australian reptiles examined in relation to thermoregulatory behaviour and distribution. *Oecologia* 9: 23–46.

Stafford, E.E., Dennis, D.T., Masri, S. and Sudomo, M. 1980. Intestinal and blood parasites in the Torro valley, Central Sulawesi, Indonesia. *S.E. Asian J. Trop. Med. Pub. Hlth.*

11: 468–472.

Stager, K.E. and Hall, L.S. 1983. A cave-roosting colony of the black flying fox (*Pterpus alecto*) in Queensland, Australia. *J. Mammal.* 64: 523–525.

Start, A.N. 1974. More cave earwigs transported by a fruit bat. *Malay. Nat. J.* 27: 170.

Start, A.N. and Marshall, A.G. 1975. Nectarivorous bats as pollinators of trees in West Malaysia. In *Tropical Trees: Variations, Breeding and Conservation,* eds. J. Burley and B.T. Styles, pp. 141–150. London: Academic Press.

Stebbins, R.C. and Kalk, M. 1961. Observations on the natural history of the mudskipper *Periophthalmus sobrinus. Copeia* 1961: 18–27.

Stemmerik, J.F. 1964. Nyctaginaceae. *Flora Malesiana I* 6: 450–468.

Stephenson, A.G. 1981. Flower and fruit abortion: proximate causes and ultimate functions. *Ann. Rev. Ecol. Syst.* 12: 253–279.

Stern, V.M., Smith, R.F., van den Bosch, R., and Hagen, K.S. 1959. The Integrated control concept. *Hilgardia* 29: 81–101.

Steup, F.K.M. 1929. Plantengeografische schets van het Paloedal. *Tectona* 22: 576–596. (English translation available.)

Steup, F.K.M. 1930-33. Bijdragen tot de kennis der bosschen van Noord-en Midden-Celebes. *Tectona* 31: 7–29.

Steup, F.K.M. 1931. Bijdragen tot de kennis der bosschen van Noord-en Midden Celebes. II. Een verkenningstocht door Midden Celebes. *Tectona* 24: 1121–1135.

Steup, F.K.M. 1932. Bijdragen tot de kennis der bosschen van Noord-en Midden-Celebes. III. Het zoogenaamde tjempaka-hoetan complex in de Minahasa. *Tectona* 25: 119-147.

Steup, F.K.M. 1933a. Bijdragen tot de kennis der bosschen van Noord-en Midden-Celebes. IV. Over de boschgesteldheid in de onderafdeeling Bolaang Mongondow. *Tectona* 26: 26–49. (Partial English translation available.)

Steup, F.K.M. 1933b. Bijdragen tot de kennis der bosschen van Noord-en Midden-Celebes. IV. Over de boschgesteldheid in de onderafdeeling Bolaang Mongondow. *Tectona* 31: 7-29. (Partial English translation available.)

Steup, F.K.M. 1935. De lasi-bosschen van de

onderafdeeling Boalemo. *Tectona* 28: 95–107. (Partial English translation available.)

Steup, F.K.M. 1939a. Vegetatieschetsen uit Zuid-Celebes. *Trop. Natuur* 27: 140–146. (Translation available.)

Steup, F.K.M. 1939b. Over vegetatie typen op Celebes *Natuurk. Tijds. Ned. Ind.* 98: 283–293. (English translation available.)

Steup, F.K.M. 1941. Kustaanwas en mangrove. *Natuurk. Tijds. Ned. Ind.* 101: 353–355.

Stocker, G.C. 1981. Regeneration of a North Queensland rain forest following felling and burning. *Biotropica* 13: 86–92.

Stocker, G.C., Unwin, G.L. and West, P.W. 1985. Measures of richness, evenness and diversity in tropical rainforest. *Aust. J. Bot.* 33: 131–137.

Stone, B.C. 1983. Studies in Malesian Pandanaceae 19. New species of *Freycinetia* and *Pandanus* from Malesia, Southeast Asia. *J. Arn. Arb.* 64: 309–324.

Strahler, A.N. 1957. Quantitative analysis of watershed geomorphology. *Trans. Amer. Geophys. Union* 38: 413-920.

Stresemann, E. 1939-41. Die Vogel von Celebes 1-3. *J. Ornithol.* 87: 299–425; 88: 1–135; 89: 1–102.

Sugondo, H. 1978. Studi pendahuluan mengenai ekosistem hutan mangrove daerah aliran sungai Rongkong di Kabupaten Luwu Sulawesi Selatan. Universitas Hasanuddin, Ujung Pandang.

Sukamto, R. 1975a. Perkembangan tektonik di Sulawesi dan daerah sekitarnya: suatu sitnesis berdasarkan tektonik lempeng. *Geol. Indon.* 2: 1–13.

Sukamto, R. 1975b. Peta geologi Indonesia, Lembar Ujung Pandang: Geologi map of Indonesia, Ujung Pandang sheet. Direktoral Geologi, Bandung.

Sukendar, H. 1976. Obyek kepurbakalaan di Palu Sulawesi Tengah. *Kalpataru* 3: 61–104.

Sukendar, H. 1980a. Mencari peninggalan nenek moyang, pendukung tradisi megatitik di Tanah Bada (Sulteng). *Kalpataru* 5: 1–63.

Sukendar, H. 1980b. Tinjauan tentang peninggalan tradisi megalitik di daerah Sulawesi Tengah. Pertemuan Ilmiah Arkeologi, 1977.

Sunartadindja, M.A. and Lehmann, H. 1960.

Der tropische Karst van Maros und Nord-Bone in SW Celebes (Sulawesi). *Z. Geomorph. (Suppl.)* 2: 49–65.

Sungkawa, W. 1975. Pengamatan habitat anoang dataran rendah (*Anoa depressicornis* H. Smith) di daerah Tanjung Amolengo, Sulawesi Tenggara. Laporan No. 205. Lembaga Penelitin Hutan, Bogor.

Susanto, M. 1984a. Rawa Aopa-Watumohae punya cerita. *Suara Alam* 27: 34–35.

Susanto, M. 1984b. Laporan inventarisasi fauna di Taman Nasional Rawa Aopa-Watumohai. PPA, Kendari.

Susanto, M. 1985. Laporan inventarisasi flora di Taman Nasional Rawa Aopa-Watumohai. Taman Nasional Rawa Aopa-Watumohai, Kendari.

Sutisna, U. and Soeyatman, H.C. 1984. Komposisi jenis pohon hutan bekas tebangan di Malili, Sulawesi Selatan: deskripsi dan analisa. Laporan No. 430, Pusat Penelitian dan Pengembangan Hutan, Bogor.

Sutopo, M. and Suayeb, D. 1980. Pelestarian lingkungan hidup dalam pandangan Islam. Badan Koordinasi Kemehasiswaan, IKIP Ujung Pandang Cabang Palu.

Sutton, D.L. 1985. Aquatic vegetation for fish production. Workshop on The Ecology and Management of Aquatic Weeds, Jakarta, March 26-29, 1985.

Sutton, S.L. 1983. The spatial distribution of flying insect in tropical rain forests. In *Tropical Rain Forest: Ecology and Management*, eds. S.L. Sutton, T.C. Whitmore, and A.C. Chadwick, pp.77–91. Oxford: Blackwell.

Suwardjo, H., Sudjadi, M. and Ross, M.S. 1985. Potentials and constraints and development strategies for agricultural land development in Indonesia. Paper presented at the International Inaugural Workshop on Land Clearing and Development for Sustained Agricultural Production, Oct. 1985, Jakarta.

Suwignyo, P. 1978. Kasus perencanaan Danau Tempe ditinjau dari aspek biologi/ekologi perairannya. Bogor: BIOTROP.

Suwignyo, P. 1979. Lake Tempe the fish bowl of Indonesia. *BIOTROP Newsl.* 17: 19.

Swarbrick, J.T., Finlayson, C.M. and Cauldwell, A.J. 1982. *The Biology and Control of Hydrilla verticillata (L.f.) Roule.* BIOTROP

Spec. Publ. 16, Bogor.

Sweeting, M. 1972. *Karst Landforms*. London: MacMillan.

Swennen, C. and Marteijn, E. 1985. Feeding ecology studies in the Malay Peninsula. In *Interwader East Asia/Pacific Shorebird Study Programme Annual Report 1984*, eds. D. Parish and D. Wells, pp. 13–26. Kuala Lumpur: Interwader.

Symington, C.F. 1933. The study of secondary growth in rain forest sites in Malaya. *Malayan Forest.* 2: 107–117.

Takeda, H., Prachaiyo, B. and Tsutsumi, T. 1984. Comparison of decomposition rates of several tree leaf litter in a tropical forest in the north-east Thailand. *Jap. J. Ecol.* 34: 311–319.

Takenaka, O. 1982. Kyoto University overseas research report of studies on Asian non-human primates. Kyoto University Primate Research Institute, Kyoto.

Takenaka, O. and Brotoisworo, E. 1982. Preliminary report on Sulawesi macaques - their distribution and interspecific difference. *Stud. Asian Non-Human Primates* 2: 11–22.

Takenaka, O., Watanabe, T., Watanabe, K., Kawamoto, Y., Hamada, Y., Ishikawa, K., Brotoisworo, E. and Suryobroto, B. 1985. Tentative final report (Sulawesi Survey). Unpubl. ms., Kyoto University.

Tanner, E.V.J., 1983. Leaf demography and growth of the tree fern *Cyathea pubsecens* in Jamaica.

Tanner, E.V.J. and Kapos, V. 1982. Leaf structure of a Jamaican upper montane rainforest tree. *Biotropica* 14: 16–24.

Taufik, A.G., Yacob, J., Abas, R. and Wardoyo, S.E. 1980. Penelitian pendahuluan pengaruh waya massapi terhadap ruaya sidat dari dan ke Danau Poso. Lembaga Penelitian Perikanan Darat Ujung Pandang.

Taulu, H.M. 1981. *Sejarah dan Anthropologi Budaya Minahasa*. Tunas Harapan, Manado.

Tay, S.W. and Khoo, H.W. 1984. The distribution of coral reef fishes at Pulau Salu, Singapore. *BIOTROP Spec. Publ.* 22: 27–40.

Taylor, P. 1977. Lentibulariaceae. *Flora Malesiana I* 8: 275–300.

Tee, A.C.G. 1982. Some aspects of the ecology of the mangrove forest at Sungai

Buloh, Selangor II. Distribution pattern and population dynamics of tree-dwelling fauna. *Malay Nat. J.* 35: 267–278 (12.9.47).

Teo, L.W. and Wee, Y.C. 1983. *Seaweeds of Singapore*. Singapore: Singapore University Press.

Thaithong, O. 1984. Bryophytes of the mangrove forest. *J. Hattori Bot. Lab.* 56: 85–87.

Thana, D. and Wardoyo, S.E. 1980. Penelitian perikanan Danau Tondano Sulawesi Utara. Lembaga Penelitian Perikanan Darat Ujung Pandang.

Thana, D., Abbas, R. and Wardoyo, S.E. 1981. Penelitian perikanan Rawa Aopa dalam rangka usaha peningkatan daya gunaannya. Lembaga Penelitian Perikanan Darat, Bogor.

Thomas, D.W. 1984. Fruit intake and energy budgets of frugivorous bats. *Physiol. Zool.* 57: 457–467.

Thorhaug, A. 1983. Habitat restoration after pipeline construction in a tropical estuary: seagrasses. *Mar. Poll. Bull.* 14: 422–425.

Thorhaug, A 1985. Largescale seagrass restoration in a damaged estuary. *Mar. Poll. Bull.* 16: 55–62.

Thorhaug, A., Miller, B., Jupp, B. and Booker, F. 1985. Effects of a variety of impacts on seagrass restoration in Jamaica. *Mar. Poll. Bull.* 16: 355–360.

Thornback, L.J. 1978. *Red Data Book. Vol. 1. Mammalia*, Revision of 1972 edition. Morges: IUCN.

Thornback, L.J. 1983. *Wild Cattle, Bison and Buffaloes: Their Status and Potential Value.* Cambridge: IUCN.

Tilaar, F.F. 1982. Suatu penelaahan kebiasaan makanan ikan-ikan karang dominan diperairan sekitar Lapango-Mahumu, Kabupaten Sangihe-Talaud. Thesis, Fakultas Perikanan, Universitas Sam Ratulangi, Manado.

Tinal, U.K. and Palinewan, J.C. 1978. Mechanical logging damage after selective cutting in the lowland dipterocarp forest at Beloro, East Kalimantan. *BIOTROP Spec. Publ.* 3: 91–96.

Tjia, H.D. 1980. The Sunda Shelf, Southeast Asia. *Z. Geomorph.* 24: 405–427.

Tjia, H.D., Sujitno, S., Suklija, Y., Harsono, R.A.F., Rachmat, A., Hainim, J. and Djunaedi 1984. Holocene shorelines in the Indonesian tin islands. *Mod. Quat. Res. S.E.*

Asia 8: 103–117.

Tomlinson, P.B., Primack, R.B. and Bunt, J.S. 1979. Preliminary observations on floral biology in mangrove Rhizophoraceae. *Biotropica* 11: 256–277.

Townsend 1980. *The Ecology of Streams and Rivers.* London: Edward Arnold.

Troll, D. and Dragendorf, O. 1931. Uber die Luft wurzeln von *Sonneratia* Linn. f. und Ihre biologische Bedeutung. *Planta* 13: 311–473.

Trueman, E.R. 1975. *The Locomotion of Soft-bodied Animals.* Edward Arnold: London.

Turkay, M. 1974. Die Gecarcinidae Asiens und Ozeaniens. *Senckenbergiana Biol.* 55: 223–259.

Tuttle, M.D. and Ryan, M.J. 1981. Bat predation and the evolution of frog vocalizations in the Neotropics. *Science.* 214: 677–678.

Tuttle, M.D. and Stevenson, D. 1982. Growth and survival of bats. In *Ecology of Bats,* ed. T.H. Kunz, pp. 105–150. New York: Plenum.

Tweedie, M.W.F. 1983. *The Snakes of Malaya.* Singapore: Singapore National Printers.

Uhl, C. and Jordan, C.F. 1984. Succession and nutrient dynamics following forest burning and cutting in Amazonia. *Ecology* 65: 1476–1490.

Uhl, C., Clark, H., Clark, K. and Murphy, P. 1981. Early plant succession after cutting and burning in the Upper Rio Negro of the Amazon basin. *J. Ecol.* 69: 631–649.

Uhlig, H. 1980. Man and tropical karst in South East Asia: Geo-ecological differentiation, land use and rural development potentials in Indonesia and other regions. *Geo. J.* 4: 31-44.

Uktolseya, H.L. 1977. Some water characteristics of estuaries in Indonesia. *Mar. Res. Indonesia* 20: 39–50.

Umbgrove, J.H.F. 1930. De Koraalriffen van den Spermonde Archipel, Z. Celebes. *Leidsche Geol. Meded.* 3: 227–247.

Umbgrove, J.H.F. 1939. De atollen en barriere-riffen der Togian-Eilanden. *Leidsche Geol. Meded.* 11: 132–187.

Ungemach, H. 1969. Chemical rain studies in the Amazon region. In *Simposio y Foro de Biologia Tropical Amazonica,* pp. 354–358. Assaciation pro Biologia Tropical.

Uno, A. 1949. Het Natuurmonument Panoea (N. Celebes) en het maleohen (*Macrocephalon maleo* Sal. Muller) in het bijzonder. *Tectona* 39: 151–165.

Uttley, J. 1986. Survey of Sulawesi Selatan to assess the status of wetlands and to identify key sites for breeding and migratory waterbirds. Preliminary report, Kuala Lumpur.

Vaas, K.F. 1956. Laporan pemeriksaan Rawa Opa (Sulawesi Tenggara). Laporan No. 8, Balai Penyelidikan Perikanan Darat, Bogor.

Valentijn, F. 1924-26. *Oud en Nieuw Oost-Indien.* Amsterdam: Dordrecht.

van Balgooy, M.M.J., in press. A plant geographical analysis of Sulawesi. In *The Biogeographic Evolution of the Malay Archipelago,* ed. T.C. Whitmore. Oxford: Clarendon.

van Balgooy, M.M.J., in press. The floristic position of Sulawesi. In *The Biogeographic Evolution of the Malay Archipelago,* ed. T.C. Whitmore. Oxford: Clarendon.

van Balgooy, M.M.J. and Tantra, I.G.M. 1986. The vegetation in two areas in Sulawesi, Indonesia. *Bull. Balai Pen. Hutan* 1986: 1–61.

van Baren, F.A. 1975. The soil as an ecological factor in the development of tropical forest areas. In *The Use of Ecological Guidelines for Development in Tropical Forest Areas of South East Asia,* pp. 88–98. Gland: IUCN.

van Beers, W.F.J. 1962. *Acid Sulphate Soils.* Amsterdam: Veenan and Zonen.

van Bemmelen, R.W. 1970. *The Geology of Indonesia.* Government Printing Office, The Hague.

van Bruggen, H.W.E. 1971. Aponogetonaceae. *Flora Malesiana I* 7: 213–218.

van der Goot, P. 1940. De Biologische bestrijding van de cactus plaag in het Paloe-dal. *Meded. Alg. Proefst.* 43: 1–17.

van der Koppel, C. 1982. De rotan van Celebes. *Tectona* 21: 61–94.

van der Plas, F. 1971. Lemnaceae. *Flora Malesiana I* 7: 213–218.

van der Vecht, J. 1953. The carpenter bees (*Xylocopa* Latr.) of Celebes. *Idea* 9:57-59.

van der Vlies, A. P. 1940. De Agathisbosschen in de Afdeeling Poso. *Tectona* 33: 616–640. (Translation available.)

van Emden, H.F. 1974. *Pest Control and Its Ecology.* London: Edward Arnold.

van Halteren, P. 1979. The insect pest complex and related problems of lowland rice cultivation in South Sulawesi, Indonesia.

Meded. Landbouwhogeschool Wageningen 79: 1–112.

van Heekeren, H.R. 1958a. The Tjabenge flake industry from south Celebes. *Asian Persp.* 2: 77–81.

van Heekeren, H.R. 1958b. The bronze-iron age of Indonesia. *Ver. Kon. Inst. Taal, Land Volk.* 22: 1–90.

van Heekeren, H.R. 1972. The stone age of Indonesia. *Ver. Kon. Inst. Taal-, Land-Volk* 61: 1–230.

van Kampen, P.N. 1923. *The Amphibia of the Indo-Australian Archipelago.* Leiden: Brill.

van Meeuwen, M.S., Nooteboom, H.P. and van Steenis, C.G.G.J. 1961. Preliminary revisions of some genera of Malaysian Papillionaceae. *Reinwardtia* 5: 419–456.

van Ooststroom, S.J. 1953. Convolvulaceae. *Flora Malesiana I* 4: 388–512.

van Schaik, C.P. and Mirmanto, E. 1985. Spatial variation in the structure and litterfall of a Sumatran rainforest. *Biotropica* 17: 196–205.

van Schouwenburg, J.C. 1915. Boschverkenning van de Tijger-Eilanden, 965 ha, Juni 1915. Unpubl. ms.

van Schouwenburg, J.C. 1916a. Boschverkenning van de glarangschappen Tamboelongan en Kajoeadi. Unpubl. ms.

van Schouwenburg, J.C. 1916b. Boschverkenning van de glarangschappen Kalao-Toea. Unpubl. ms.

van Schouwenberg, J.C. 1916c. Boschverkenning van de Postilijon-en Paternoster-Eilanden. Unpubl. ms.

van Shaik, C.P. 1983. Why are diurnal primates living in groups? *Behaviour* 87: 120–144.

van Steenis, C.G.G.J. 1936. On the origin of the Malaysian mountain flora. 3. Analysis of floristical relationships (first installment). *Bull. Jard. Bot. Buitenzorg ser.* 3 14: 56–72.

van Steenis, C.G.G.J. 1937. Naar den hoogsten top van Celebes. *Actueel Wereldnieuws* 14: 4 pp. (Translation available.)

van Steenis, C.G.G.J. 1949a. Ceratophyllaceae. *Flora Malesiana I* 4: 41–42.

van Steenis, C.G.G.J. 1949b. Podostemaceae. *Flora Malesiana I* 4: 65–68.

van Steenis, C.G.G.J. 1950. The delimitation of Malaysia and its main plant geographical divisions. *Flora Malesiana I* 1: 70–75.

van Steenis, C.G.G.J. 1951. Dipsacaceae.

Flora Malesiana I 4: 290–292.

van Steenis, C.G.G.J. 1953. Droseraceae. *Flora Malesiana I* 4: 377–381.

van Steenis, C.G.G.J. 1957. Outline of vegetation types in Indonesia and some adjacent regions. *Proc. Pac. Sci. Cong.* 8: 61–97.

van Steenis, C.G.G.J. 1972. *The Mountain Flora of Java.* Brill: Leiden.

van Steenis, C.G.G.J. 1981a. *Rheophytes of the World.* Sijthoff and Noordhoff, Alphen aan den Rijn.

van Steenis, C.G.G.J. 1981b. *Flora untuk Sekolah.* Jakarta: Pradnya Paramita.

van Steenis, C.G.G.J. 1984. Floristic altitudinal zones in Malesia. *Bot. J. Linn. Soc.* 89: 289–292.

van Steenis, C.G.G.J. and Schippers Lammertse, A.F. 1965. Concise plant geography of Java. In *Flora of Java, Vol. 2*, eds. C.A. Backer and R.C. van der Brink Bakhuizen. Groningen: Noordhoff.

van Steenis, C.G.G.J. and Veldkamp, J. 1984. Miscellaneous botanical notes XXVII. *Blumea* 29: 399–408.

van Stein Callenfels, P.V. 1951. Prehistoric sites on the Karama River. *J. E. Asiatic Soc.* 1: 82–97.

van Strien, N.J. 1986. *The Sumatran Rhinoceros Dicerorhinus sumatrensis (Fischer 1814) in the Gunung Leuser National Park, Sumatra, Indonesia; Its Distribution, Ecology and Conservation.* Berlin: Parey.

van Zijll de Jong, J. 1934. Een tocht naar den Lompobatang (Piek von Bonthain, 2971 m.b.z.) vanuit het Zuidwestern. *Meded. Ned. Ind, Ver. Bergsport* 8: 11–15.

Verhoef, L. 1938. Bijdragen tot de kennis der bosschen van Noord-en Midden-Celebes. V. *Tectona* 13: 7–29.

Vermuelen, J.W.C. 1985. Ornithological reflections in Indonesia: Watch your herons. *Voice of Nature* 34: 9–10.

Verstappen, H.T. 1957a. Some observations on karst development in the Malay Archipelago. *J. Trop. Geogr.* 14: 1–10.

Verstappen, H.T. 1957b. Een en ander over het rifpantser van het eiland Muna. *Tijds. Kon. Ned. Aardr. Gen.* 74: 441–449.

Verstappen, H.T. 1969. *Tropical Rainforests of the Far East.* Oxford: Oxford University Press.

Verstappen, H.T. 1980. Quaternary climatic changes and natural environment in *SE*

Asia. Geo. J. 4. 45–54.

Victor, R. and Fernando, C.H. 1980. Freshwater Ostracoda from the ricefields of Southeast Asia. In *Tropical Ecology and Development*, ed. J. Furtado, pp. 957–970. Universiti Malaya, Kuala Lumpur.

Villaluz, D.K., Villaluz, A., Ladrera, B., Sheik, M. and Gonzaga, A. 1977. Production, larval development, and cultivation of sugpo *Penaeus monodon* Fabricus. In *Readings in Aquaculture Practice*, pp. 1–15. Southeast Asian Fisheries Development Centre, Tigbauan.

Visser, G. 1984. Husbandry and reproduction of the sail-tailed lizard, *Hydrosaurus amboinensis* (Schlosser, 1768) (Reptilia: Sauria: Agamidae), at Rotterdam Zoo. In *Maintenance and Reproduction of Reptiles in Captivity, Vol. 1*, eds. V.L. Bels and Sande, P. van de, pp. 129–148.

Vita-Finzi, C. 1981. X-ray diffraction and S.E.M. analysis of freshwater shells from Leang Burung 2. Mod. *Quat. Res. S.E. Asia* 6: 55–56.

Vitt, L.J. 1983. Tail loss in lizards: the significance of foraging and predator escape modes. *Herpetol.* 39: 151–162.

Vitt, L.J., Congdon, J.D. and Dickson, N.A. 1977. Adaptive strategies and energetics of tail autonomy in lizards. *Ecology.* 58: 326-337.

Vogt, R.C. and Bull, J.J. 1982. Temperature controlled sex determination in turtles. *Herpetologica* 38: 154–164.

Voris, H.K. and Voris, H.H. 1983. Feeding strategies in marine snakes: an analysis of evolutionary, morphological, behavioural and ecological relationships. *Amer. Zool.* 23: 411–425.

Walang, A. 1984. Studi tentang daerah aliran sungai (DAS) Toraut ditinjau dari segi pengawetan tanah. Fakultas Pertanian, Universitas Sam Ratulangi, Manado.

Walker, D. 1982. Speculations on the origin and evolution of Sunda Sahul rain forests. In *Biological Diversification in the Tropics*, ed. G. Prance, pp. 554–575. New York: Columbia University Press.

Walker, J.R.L. 1975. *The Biology of Plant Phenolics*. London: Edward Arnold.

Wallace, A.R. 1859. Letter from Mr. Wallace concerning the geographical distribution of birds. *Ibis* 1: 449–454.

Wallace, A.R. 1860. The ornithology of northern Celebes. *Ibis* 1860: 140–147.

Wallace, A.R. 1862. List of birds from the Sula Islands (east of Celebes), with descriptions of the new species. *Proc. zool. Soc. Lond.* 1862: 333–346.

Wallace, A.R. 1863. On the physical geography of the Malay Archipelago. *J. Royal Geog. Soc.* 33: 217–234.

Wallace, A.R. 1869. *The Malay Archipelago: The Land of the Orangutan and the Bird of Paradise. A Narrative of Travel with Studies of Man and Nature*. London: MacMillan. Republished 1962 by Dover, New York.

Wallace, A.R. 1910. *The World of Life*. London: Chapman and Hall.

Walsh, G.E. 1974. Mangroves: A review. In *Ecology of Halophytes*, eds. R.J. Reinmold and W.M. Queen, pp. 51–174. New York: Academic.

Walton, O.E. 1978. Substrate attachment by drifting aquatic insect larvae. *Ecology* 59: 1023–1030.

Ward, P. 1968. Origin of the avifauna of urban and suburban Singapore. *Ibis* 110: 239–255.

Ward, P. 1969. The annual cycle of the yellow-vented bulbul *Pycnonotus goiaver* in a humid equatorial environment. *J. Zool.* 157: 25–45.

Ward, P. 1970. Seasonal and diurnal changes in the fat content of an equatorial bird. *Physiol. Zool.* 85–95.

Ward, P. and Poh. G.E. 1968. Seasonal breeding in an equatorial population of the tree sparrow *Passer montanus. Ibis* 110: 359–363.

Wardoyo, S.E. 1978. Penelitian perikanan Danau Matano dan Towuti Sulawesi Selatan. Laporan No. 17, Lembaga Penelitian Perikanan Darat Cabang Ujung Pandang.

Wardoyo, S.E. and Thana, D. 1978. Penelitian Perikanan Danau Limboto Sulawesi Utara. Laporan No. 20. Lembaga Penelitian Perikanan Darat. Ujung Pandang.

Wasscher, J. 1941. The genus *Podocarpus* in the Netherlands Indies *Blumea* 4: 359–542.

Watanabe, K. and Brotoisworo, E. 1982. Field observation of Sulawesi macaques. In *Kyoto University Overseas Research Report on Studies on Non-human Primates.* 2: 3–9. Kyoto: Kyoto

University.

Waterman, P.G. 1983. Distribution of secondary metabolites in rain forest plants: toward an understanding of cause and effect. In *Tropical Rain Forest: Ecology and Management*, eds. S.L. Sutton, T.C. Whitmore and A.C. Chadwick, pp. 167–179. Oxford: Blackwell.

Waterman, P.G. and Choo, G.M. 1981. The effects of digestibility-reducing compounds in leaves on food selection by some Colobinae. *Malays. appl. Biol.* 10: 147-162.

Watling, R.J. 1983a. Ornithological notes from Sulawesi. *Emu* 83: 247–261.

Watling, R.J. 1983b. Sandbox incubator. *Anim. Kingdom* June/July 28-35.

Watling, R.J., in press. Mineral composition of the water and the social significance of springs used by anoa *Anoa* sp. in the Lore Lindu Reserve, Sulawesi. *Biotropica*.

Watson, J.G. 1928. Mangrove forests of the Malay Peninsula. *Malay. For. Rec.* No. 6.

Waworoentoe, W.J. 1984. Urban Minahasa- a non-involutionary alternative. Unpubl. ms.

Webb, G.J.W., Sack, G.C., Buckworth. R. and Manolis, S.C. 1983. An examination of *Crocodylus porosus* nests in two northern Australian freshwater swamps, with an analysis of embryo mortality. *Aust. Wildl. Res.* 10: 571–605.

Webb, L.J. 1959. A physiognomic classification of Australian rain forests. *J. Ecol.* 47: 551–570.

Weber, M. 1904. *Die Saeugetiere. Einfuenrung in die Anatomie und Systematik der Recenten und Fossiflen Mammalia*. Jena: Fischer.

Weber, M. and de Beaufort, C.F. 1922. IV. Heteromi, Solenichthyes, Synentognathi, Percesoces, Labyrinthici, Microcyprini. In *The Fishes of the Indo-Australian Archipelago*, eds. M. Weber and L.F. de Beaufort. Leiden: Brill.

Wee, Y.C. 1982. Airborne algae around Singapore. *Intern. Biodeter. Bull.* 18: 1–5.

Wegman, B. 1983. Present and proposed land use of the planned bufferzone in the Dumoga Valley. College of Forestry and Land and Water Management, Velp.

Wells, D.R. 1974. Resident birds. In *Birds of the Malay Peninsula* (by Lord Medway and D.R. Wells). London: Witherby.

Wells, F.E. 1984. Comparative distribution of macromolluscs and macrocrustaceans in a North-western Australian mangrove system. *Aust. J. Mar. Freshw. Res.* 35: 591–596.

Wells, S.M., Pyle, R.M. and Collins, N.M. 1983. *The IUCN Invertebrate Red. Data Book*. Conservation Monitoring Centre, IUCN, Cambridge.

Wemmer, C. and Watling, D. 1986. Ecology and status of the Sulawesi palm civet *Macrogalidia muschenbroekii* Schlegel. *Biol. Cons.* 35: 1–17.

Wemmer, C. and Watling, R. 1982. Eye colour polymorphism in the babirusa pig. *Malay. nat. J.* 36: 135–136.

Wemmer, C., West, J., Watling, D., Collins, L. and Lang, K. 1983. External characters of the Sulawesi palm civet, *Macrogalidia musschenbeoeckii* Schlegel, 1879. *J. Mammal.* 64: 133–136.

Wheelwright, N.T. and Orians, G.H. 1982. Seed dispersal by animals: contrasts with pollen dispersal, problems of terminology and contraints on co-evolution. *Am. Nat.* 119: 402–423.

Whitacre, D.F. 1981. Additional techniques and safety hints for climbing tall trees and some equipment and information sources. *Biotropica* 13: 286–291.

White, A. 1985. *Nat. Res.* 11: 13–20.

White, C.W.M. 1974. Three water birds of Wallacea. *Bull. Brit. Orn. Cl.* 94: 9–11

White, C.W.M. 1975. Migration of Palaearctic waders in Wallacea. *Emu* 75: 37–39.

White, C.W.M. 1976. Migration of Palaearctic non-passerine birds in Wallacea. *Emu* 6: 79–82.

White, C.W.M. 1977. Migration of Palaearctic passerine birds in Wallacea. *Emu* 7: 37–38.

White, C.W.M. and Bruce, M. 1986. *The Birds of Wallacea*. Checklist No. 7. British Ornithologsts' Union.

Whitmore, T.C. 1972a. Euphorbiaceae. In *Tree Flora of Malaya Vol. 2*, ed. T.C. Whitmore, pp. 34–136. Kuala Lumpur: Longman.

Whitmore, T.C. 1972b. Aceraceae. In *Tree Flora of Malaya, Vol. 2*, ed. T.C. Whitmore, pp. 1–2. Longman, Kuala Lumpur.

Whitmore, T.C. 1977a. *Palms of Malaya*. Revised ed., Oxford: Oxford University Press.

Whitmore, T.C. 1977b. A first look at *Agathis*. *Trop. For. Papers 11*.

Whitmore, T.C. 1980. Potentially economic

species of Southeast Asian forests. *BioIndonesia* 7: 65–74.

Whitmore, T.C. 1981. Palaeoclimate and vegetation history. In *Wallace's Line and Plate Tectonics,* ed. T.C. Whitmore, pp. 36–42. Oxford: Oxford University Press.

Whitmore, T.C. 1982. On pattern and process in forests. In *The Plant Community as a Working Mechanism,* ed. E.I. Newman, pp. 45–49. Oxford: Blackwell.

Whitmore, T.C. 1984a. *Tropical Rain Forests of the Far East, 2nd ed.* Oxford: Clarendon.

Whitmore, T.C. 1984b. A vegetation map of Malesia at scale 1:5 million. *J. Biogeog.* 11: 461–471.

Whitmore, T.C. and Page, C.N. 1980. Evolutionary implications of the distribution and ecology of the tropical conifer *Agathis. New Phytol.* 84: 407–416.

Whitmore, T.C. and Sidiyasa, K. 1986. Report on the forests at Toraut, Dumoga-Bone proposed National Park. *Kew Bull.* 41: 747–756.

Whitmore, T.C. and Tantra, I.G.M., in press. *Annotated checklist to the timber trees of Sulawesi.* Bogor: Penelitian dan Pengembangan Hutan.

Whitmore, T.C., Peratta, R. and Brown, K. 1986. Total species count in a Costa Rica tropical rain Forest. *J. Trop. Ecol.* 1: 375–376.

Whitten, A. J. 1980a. *Arenga* fruit as a food for gibbons. *Principies* 26: 143–146.

Whitten, A.J. 1980b. The Kloss Gibbon in Siberut Rain Forest. Ph.D. dissertation, University of Cambridge.

Whitten, A.J. 1981. Notes on the ecology of *Myrmecodia tuberosa* Jack on Siberut Island. *Ann. Bot.* 47: 525-526.

Whitten, A.J. 1986. Ecological impacts of the Indonesian transmigration program. Paper given at 4th International Congress of Ecology 11-15 August 1986, Syracuse, New York.

Whitten, A.J. and Whitten, J.E.J. 1982. Preliminary observations of the Mentawai macaque on Siberut Island, Indonesia. *Int. J. Primatol.* 3: 445–459.

Whitten A.J., Bishop, K.D., Nash, S. and Clayton, L., in press. One or more extinctions from Sulawesi, Indonesia? *Conserv. Biol.* 1

Whitten, A.J., Damanik S.J., Anwar, J. and

Hisyam, N. 1998. *The Ecology of Sumatra, 3rd ed.* Singapore: Periplus Editions Ltd.

Whitten, J.E.J. and Whitten, A.J., in press. Analysis of bark eating in a tropical squirrel. *Biotropica.*

Wiersum, K.F. 1979. *Introduction to principles of forest hydrology and erosion with special reference to Indonesia.* Bandung: Lembaga Ekologi, Universitas Padjadjaran.

Wilcox, B.A. 1986. Extinction models and conservation. *Trends Ecol. Evol.* 1: 46–48.

Williams, A.M. and Holland, L. 1967. Investigations into the 'wall-fungus' found in caves. *Trans. Cave Res. Grp. G.B.* 9: 3 pp.

Williams, D.D. 1981. Migrations and distributions of stream benthos. In *Perspectives in Running Water Ecology,* eds. M.A. Lock and D.D. Williams), pp. 155–207. New York: Plenum.

Wilson, W.L. and Johns, A.D. 1982. Diversity and abundance of selected animal species in undisturbed forest, selectively logged forest and plantations in East Kalimantan, Indonesia. *Biol. Conserv.* 24: 205–218.

Wind, J. 1977. Rawa Aopa or Rawa Opa. Unpubl. ms.

Wind, J. 1984. *Management Plan 1984-1989: Dumoga-Bone National Park.* Bogor: World Wildlife Fund.

Wind, J. and Amir, M. 1978. Some notes on anoa (*Bubalus* spp.) in Lore Kalamanta. Unpubl. ms.

Wint, G.R.W. 1983. Leaf damage in tropical rain forest canopies. In *Tropical Rain Forest: Ecology and Management,* eds. S.L. Sutton, T.C. Whitmore, and A.C. Chadwick, pp. 229–239. Oxford: Blackwell.

Wirawan, N. 1981. *Ecological survey of the proposed Lore Lindu National Park, Central Sulawesi.* Ujung Pandang: Universitas Hasanuddin.

Wiriosoepartho, A.S. 1979. Pengamatan habitat dan tingkah laku bertelur maleo (*Macrocephalon maleo*) di kompleks hutan Dumoga, Sulawesi Utara. Laporan No. 315. Bogor: Lembaga Penelitian Hutan.

Wiriosoepartho, A.S. 1980. Penggunaan habitat dalam berbagai macam aktivitas oleh *Macrocephalon maleo* Sal. Muller di Cagar Alam Panua, Sulawesi Utara. Laporan No. 356. Lembaga Penelitian Hutan, Bogor.

Witkamp, H. 1940. Langs de Lariang-River.

Tijds. Kon. Ned. Aardr. Gen. 57: 581–600.

Wium-Anderson, S. 1981. Seasonal growth of mangrove trees in southern Thailand. III. Phenology of *Rhizophora mucronata* Lamk. and *Scyphiphora hydrophyllacea. Gaertn. Aquat. Bot.* 10: 371–376.

Wium-Anderson, S. and Christensen, B. 1978. Seasonal growth of mangrove trees in southern Thailand. II. Phenology of *Bruguiera cylindrica, Ceriops tagal, Lumnitzera littorea* and *Avicennia marina. Aquat. Bot.* 5: 383-390.

Wolf, E.C. 1985. Conserving biological diversity. In *State of the World 1985,* ed. L.R. Brown, pp. 124–146. New York: Norton.

Wolverton, B.C. 1985. The use of aquatic plants for treating wastewater: a review workshop on The Ecology and Management of Aquatic Weeds, Jakarta, March 26-29, 1985.

Wong, M. 1983. Understory phenology of the virgin and regenerating habitats in Pasoh Forest Reserve, Negeri Sembilan, Malaysia. *Malay. For.* 46: 197–223.

Wong, M. 1985. Understorey birds as indicators of regeneration in a patch of selectively logged West Malaysian rainforest. *ICBP Tech. Publ.* 4: 249–263.

Wong, M., in press a. Understory foliage arthropods in the virgin and regenerating habitats of Pasoh Forest Reserve, West Malaysia. *Malay. For.*

Wong, M., in press b. Trophic organization of understorey birds in a Malaysian dipterocarp forest. *Auk.*

Wong, P.P. 1978. The herbaceous formation and its geomorphic role, East Coast, Peninsular Malaysia. *Malay. Nat. J.* 32: 129–141.

Wood, T.G. 1978. The termite (Isoptera) fauna of Malesian and other tropical rainforests. In *The Abundance of Animals in Malesian Rain Forest,* ed. A.G. Marshall, pp. 113–132. Hull: University of Hull.

Woodard, D. 1805. *The narrative of Captain David Woodard and Four Seamen who lost their Ship While in a Boat at Sea, and Surrendered Themselves up to the Malays in the Island of Celebes.* London: Johnson. Reprinted 1969, Dawsons, London.

Woodroffe, C.D. 1985. Variability in detrital production and tidal flushing in mangrove swamps. In *Coasts and Tidal Wetlands of the Australian Monsoon Region,* eds. K.N. Bard-

sley, J.D.S. Davie and C.D. Woodroffe, pp. 201-212. Mangrove Monograph No. 1, Australian National University North Australia Research Unit, Darwin.

Wratten, S.D. and Watt, A.D. 1984. *Ecological Basis of Pest Control.* London: George Allen and Unwin.

Wright, D.G. 1982. A discussion paper on the effects of explosives on fish and marine mammals in the water of the Northwest Territories. *Can. Techn. Rep. Fish. Aqua. Sci.* No. 1052.

Wyatt-Smith, J. 1963. Manual of Malayan silviculture for inland forests. *Malay. For. Rec.* No. 23.

Wyrtki, K. 1961. *Physical Oceanography of the Southeast Asian Waters.* La Jolla: Scripps Institute for Oceanography.

Yalden, B.W. and Morris, P.A. 1975. *The Lives of Bats.* London: David and Charles.

Yamada, I. 1976. Forest ecological studies of the montane forest of Mt Pangrango, West Java. III. Litterfall of the tropical montane forest near Cibodas. *South East Asian Studies* 14: 193–229.

Yap, S.P. 1976. The feeding biology of some padi field anurans. B.Sc. thesis, University Malaya, Kuala Lumpur.

Yoda, K. 1978a. Organic carbon, nitrogen and mineral nutrients stock in the soil of Pasoh Forest. *Malay. Nat. J.* 30: 229–251.

Yoda, K. 1978b. Respiration studies in Pasoh forest. *Malay. Nat. J.* 30: 259–279.

Yoneda, T., Yoda, K. and Kira, T. 1978. Accumulation and decomposition of wood litter in Pasoh Forest. *Malay. Nat. J.* 30: 381–389.

Yorke, C.D. 1984. Avian community structure in two modified Malaysian habitats. *Biol. Conserv.* 29: 345–362.

Young, A. 1976. *Tropical Soils and Soil Survey.* Cambridge: Cambridge University Press.

Yunus, A. and Lim, G.S. 1971. A problem in the use of insecticides in paddy fields in West Malaysia. A case study. *Malay. Agric. J.* 48: 168–178.

Zimmerman, P.R. and Greenberg, J.P. 1983. Termites and methane. *Nature* 302: 354–355.

Zucca, C.P. 1982. Effects of road construction on a mangrove ecosystem. *Trop. Ecol.* 23: 105–124.

Index